Fluid-Mineral Interactions: A Tribute to H. P. Eugster

Special Publications Series Editor: D. G. Brookins
                                        Department of Geology
                                        University of New Mexico
                                        Albuquerque, New Mexico 87131 U.S.A.

Fluid-Mineral Interactions: A Tribute to H. P. Eugster

R. J. Spencer and I-Ming Chou

Editors

Department of Geology and Geophysics, The University of Calgary,

Calgary, Alberta T2N 1N4 Canada

and

U.S. Geological Survey, Reston, Virginia 22092 U.S.A.

Special Publication No. 2

THE GEOCHEMICAL SOCIETY

Library of Congress Catalogue Number 89-82754

ISBN 0–941809–01-3

Printed in The United States of America
by Lancaster Press, Inc.

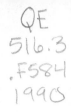

Fluid-Mineral Interactions: A Tribute to H. P. Eugster
© The Geochemical Society, Special Publication No. 2, 1990
Editors: R. J. Spencer and I-Ming Chou

Preface

Fluid-Mineral Interactions: A Tribute to H. P. Eugster was organized to honor Hans' many contributions to geology. The volume contains 23 papers by many former students and colleagues of Hans Eugster, which cover a spectrum of topics from experimental studies to igneous, metamorphic and sedimentary petrology. This spectrum was deemed appropriate for this volume because of the wide range of topics that Hans was interested in and published on.

Hans worked on metamorphic rocks in the Swiss Alps for his Ph.D. under the supervision of Paul Niggli. After receiving his Ph.D. in 1951, and following a brief stay at MIT, Hans went to The Geophysical Laboratory, where he developed the oxygen buffer technique for studies of redox equilibria at high temperatures and pressures. Hans became an Associate Professor in experimental petrology at the Johns Hopkins University in 1958, the same year he became interested in the petrology of evaporite deposits of the Green River Formation. His interest in the Green River Formation led him to Wyoming, where he also developed an interest in ore deposits. Until his death in December of 1987, Hans and his students extended the high-temperature high-pressure buffer techniques to more and more chemical systems and continued to study non-marine evaporite petrology and sedimentology, marine evaporite petrogenesis, and the origin of ore deposits.

A central theme in Hans' work was the interaction of minerals with aqueous fluids. Hans gained insight for experimental studies from field observations and used the results of his experimental studies to better understand the processes involved in the formation of rocks. He demonstrated that the underlying principles of experimental petrology and theoretical thermodynamics can be applied to sediments as well as to igneous and metamorphic rocks.

We have attempted to show some of the diversity of Hans' interests in this volume. It begins with four papers on experimental techniques and buffers, a field in which Hans probably made his most significant contributions. The topics covered include both high-temperature high-pressure buffer techniques and low-temperature solubility measurements. The second part of the volume contains papers on igneous and metamorphic systems. Hans' early interest in understanding metamorphic rocks led to many of the experimental studies he was involved with. He encouraged all of his students to work both in the field and in the laboratory. The third part of the volume covers hydrothermal and ore systems. It includes papers on speciation in chloride-rich hydrothermal fluids, a topic Hans was interested in and working on at the time of his death.

The fourth part of the volume covers diagenetic systems, an area where fluid-mineral interactions, buffering by solid and organic compounds, and fluid flow are important. The final section is devoted to surface environments. Hans much enjoyed his work on modern saline lake systems and was instrumental in changing evaporite petrology from a predominantly descriptive science to one that uses experimental and theoretical chemistry to increase our understanding of evaporite petrogenesis.

We thank The Geochemical Society for making available its resources allowing the publication of this tribute. The support and advice of Dr. A. A. Levinson on editorial matters is greatly appreciated. As editors we also thank the following people who served as reviewers for the papers contained within this volume: Hugh J. Abercrombie, Robert A. Ayuso, Mary Jo Baedecker, Lukas Baumgartner, John Bloch, Elizabeth Burton, Edward Callender, Philip A. Candela, C. Page Chamberlain, Peter Davies, Walter Dean, Jeremy B. Fein, George W. Fisher, Michael E. Fleet, Grant Garven, Mark S. Ghiorso, Pierre D. Glynn, Hugh J. Greenwood, William D. Gunter, Jane M. Hammarstrom, H. Tren Haselton, Julian J. Hemley, J. Stephen Huebner, Blair F. Jones, Rama K. Kotra, Charles W. Kreitler, Tim K. Lowenstein, Hans G. Machel, Margaret T. Mangan, C. M. Molenaar, H. Wayne Nesbitt, Roger L. Nielsen, Gordon L. Nord, Jr., William H. Orem, L. Neil Plummer, J. Donald Rimstidt, Malcolm J. Rutherford, Richard O. Sack, Richard Sanford, Everett Shock, George B. Skippen, Sorena Sorensen, Glenn A. Wilson, Scott A. Wood, Warren W. Wood and E-an Zen.

RONALD J. SPENCER
I-MING CHOU
February, 1990

Hans Peter Eugster, 1925–1987.

Fluid-Mineral Interactions: A Tribute to H. P. Eugster
© The Geochemical Society, Special Publication No. 2, 1990
Editors: R. J. Spencer and I-Ming Chou

HANS PETER EUGSTER
Teacher and Colleague

To UNDERSTAND the academic career of Hans Eugster it is necessary to know something of the circumstances that prevailed at the time Hans arrived at The Johns Hopkins University.

Ernst Cloos became Chairman of the Department of Geology in 1951 and, in consultation with Aaron Waters, made plans for upgrading the Hopkins program in geology. The Department underwent a renaissance. In 1952 Aaron Waters, already on board, was given a permanent appointment. Also appointed were James Gilluly and Francis Pettijohn. Pettijohn accepted; Gilluly did not. In 1956, when Cloos was on leave, Aaron Waters, who was Acting Chairman, invited Eugster, then a visiting investigator at the Geophysical Laboratory in Washington, to come to Hopkins to teach a class in geochemistry. In 1958, Eugster left the Laboratory and came to Hopkins where he spent the rest of his life. The arrival of Eugster completed the Cloos program for the enhancement of geology at Hopkins.

With the arrival of Eugster, the Department consisted of Hans Eugster, Ernst Cloos, Aaron Waters, Francis Pettijohn, J. D. H. Donnay, and Harold Vokes—six professors—and Tom Amsden, Assistant Professor. By national standards this was a small Department. It remained so but it was a distinguished group; four members (Cloos, Waters, Pettijohn and Eugster) were elected to the National Academy of Sciences.

The program in geochemistry and Eugster's appointment would not have happened except for Aaron Waters. More than anyone he was the catalyst that brought Hopkins into the main stream of geology. In the fall of 1956, when he was Acting Chairman, he sent a memorandum to the Provost, P. Stewart Macauley, outlining a plan for geochemistry that included renovation of the third floor of Latrobe Hall (where geology was housed) and conversion to geochemical laboratories. It was after this plan was approved that Eugster came to Hopkins.

The appointment of Eugster had the most profound and lasting effect on the Department for it ushered in the "geochemical revolution" that had altered the nature and course of geology in America. The coming of Eugster added a new dimension to the Hopkins program. It had a two-fold effect, first on teaching and second on research. The first led to an upgrading of the curriculum and student proficiencies in chemistry, especially greater emphasis

on physical chemistry and thermodynamics. The second, the research of Eugster and his students, made Hopkins one of the leading centers of experimental petrology in American universities. As Eugster himself expressed it, he devoted his career to "building a bridge from chemistry to geology." The central theme was the interaction of minerals with aqueous fluids, from surface waters to geochemical brines, to metamorphic fluids to igneous gases. He was eminently successful. This is all the more remarkable because Eugster was largely self-taught. His training at Zurich was largely in classical and analytical chemistry—not thermodynamics.

Eugster had a knack for recognizing new opportunities. He did not hesitate to switch fields of interest—to leave a field where further work yielded diminishing returns and to move into a new area—to "fish in other waters" as he once expressed it. He first worked on mineral systems at elevated temperatures and pressures and later became interested in brines and nonmarine evaporites. He left his furnaces to wade around in the muds of Lake Magadi in the African Rift Valley, then to Lake Chad in northern Africa and to salt lakes in Bolivia. And later he turned to ore deposits in China and elsewhere. In all these fields of science he left his mark.

Eugster attracted a brilliant group of students. Included among these were I-Ming Chou, J. D. Frantz, Bevan French, W. O. Gunter, Steve Huebner, Frank Kujawa, James Munoz, M. Rieder, M. S. Rutherford, H. W. Nesbitt, George Skippen, R. J. Williams, Glenn Wilson, Allan Turnock, Joseph Domagalski and Sigurdur Gíslason. Among his students also was Gary Ernst, member of the National Academy of Sciences and one-time president of the Geological Society of America. Among the students who worked on brines and evaporite deposits were L. A. Hardie, Blair Jones, Ronald Spencer and James Wood. In addition, Eugster had, from time to time, various post-doctoral investigators working in his laboratory. Among these were Anthony Beswick and John Nolan from England, Hans Fuchtbauer from Germany, and Milan Rieder from Czechoslovakia.

Eugster not only left an imprint on Hopkins, but was Adjunct Professor at the University of Wyoming. He also taught at the Swiss Federal Institute of Technology at Zurich—the place where he had received his doctoral degree in 1951.

Upon his promotion to Full Professor in 1960, Eugster had an increasing influence on the departmental policies and, most especially, on faculty appointments. In 1983 Hans Eugster became Chairman of what was then, and still is, an enlarged Department of Earth and Planetary Sciences. It is a vastly different Department from that which Eugster joined in 1958. Eugster played a significant role—perhaps the major role—in the selection of the faculty of the Department as it now exists. Eugster had the highest standards. He had a remarkable ability to recognize original and significant work—even in fields somewhat removed from his own. He left his imprint on the Hopkins Department as well as on the science to which he devoted his life. He came to Hopkins to teach a course in geochemistry; he left a changed Department.

In addition to his on-campus career as teacher, investigator and administrator, he had a full and interesting life outside the walls of academia. He was a talented musician—a violinist, an accomplished painter, a potter, and an avid theater and concert goer. In all of these, as in his science, he was always searching out for the new, the novel, the cutting edge of each area of interest—the avant garde. He loved to ski—a holdover, perhaps, from his field studies of snow and avalanches in Switzerland. He enjoyed country living—an enjoyment fulfilled by an old farm in rural Maryland where, despite a long commute to his office, he and his wife lived.

But it was in science that he made his contribution. The Department he left behind and his students are a monument to his academic career.

January, 1990   F. J. Pettijohn

Hans Peter Eugster, self portrait, 1962

Fluid-Mineral Interactions: A Tribute to H. P. Eugster
© The Geochemical Society, Special Publication No. 2, 1990
Editors: R. J. Spencer and I-Ming Chou

HANS PETER EUGSTER
Chemical Geologist*

HANS PETER EUGSTER was born in Landquart, Switzerland on November 19, 1925, the third of five children. He died suddenly of a ruptured aorta at the Johns Hopkins Hospital in Baltimore on December 17, 1987. He is survived by his wife, Elaine H. Koppelman, the James Beall professor of Mathematics and Computer Science at Goucher College, his three daughters by a previous marriage, Rachel, Erica and Sandra, and two brothers, Conrad and Carl in Switzerland.

He was raised in a middle class family, with emphasis on frugality. Intensive learning was considered an invaluable investment for the future, playing music its most suitable and compatiable compliment. Hans hardly ever went out for competitive sports; instead enjoying bicycle tours, mountaineering and skiing, downhill and on seals-skins. Endurance was considered important for making headway and building of character.

Hans interest in geology was evident at an early age. On a family outing to climb to the high Alp near Davos, a brisk pace was set. As a result Hans was repeatedly lagging behind. The rest of the family grumbled but adjusted their pace down. However, the situation did not improve, Hans was still not coping. His father, who was a methodical engineer, inquired into the deeper reasons for the delay. Hans diligently had been collecting stones on the way and he had filled his pockets to the point where he could no longer walk. The Grisons, a paradise of petrography for an eight year old.

Hans was drafted near the end of the second world war although he was never militaristic in attitude. Thanks to his patience and his expert knowledge with instruments, he distinguished himself as a gun-layer and crew leader in the artillery as well as being an excellent instructor.

Hans lived with his brothers in Zurich, while studying. When not engaged in his studies, the brothers formed a threesome to hike or climb in the mountains. Cony made most of the decisions, Carl organized the details and took care of the supplies, while Hans was content just to follow. As a rule they walked or climbed for long hours, without ever exchanging more than the strictly necessary words or commands. Hans always valued his time.

In later life when on a visit to Switzerland to go mountaineering in the Bernina region with his two brothers, they were trapped by a snowstorm in a hut at the start of their climb of the Diavolezza. Hans was not desperate. Paper napkins were organized, so that he could begin writing a scientific article which had been lying on his chest for who knows how long.

He received a diploma in engineering geology from the Swiss Federal Institute of Technology (ETH) in 1948. Following that Hans spent two summers in field work. He assiduously cartographed and collected the hand specimens of metamorphics necessary for his PhD thesis under Paul Niggli, which during the winter time had to be analyzed chemically and viewed under the petrographic microscope. While at his field H.Q. at Alp Russein which was also the alpine dairy center for the summer season of the whole Oberland valley, he solved the "Chasteilet." The correct allocation of the total cooperative cheese output to the individual owners of the cows called for an involved calculation for which the chief of the Chalet was responsible. He felt relieved when Hans said he was willing to assist him. Thereupon, Hans became a modern sorcer. He devised a consistent algorithm and did the computing with his slide rule. Fast and precise with an impressively reliable result, the alpine hosts held him in higher regard than ever. Hans had gracefully ended his field season and, in addition to a rich harvest of rock samples meticulously catalogued, he carried on his return a rucksack full of delicious cheese.

During his PhD program he spent a year at the Institute for Snow and Avalanche Control near Davos as a crystallographer. Hans was the only theoretical researcher—with no pretence for the practical utility of his project. He spent hours and hours in a small laboratory at a temperature of $-40°C$, studying the formation of snow crystals. And there were plenty of them outside, in any condition of metamorphosis. The laboratory was at the top of a long ski run and it may have been this attraction at the end of the day that was his reward. He was usually the last one down the hill at night. Once he fell and broke his leg and had to use the two skis as a sled to make it to the bottom of the hill and first aid.

Hans received his PhD in 1951 and despite being

---

* Reprinted from The Geochemical News No. 75, Fall, 1989.

courted by several oil companies, went immediately to MIT. Hans had planned to spend 8 months at MIT, and then return to teach in Paul Niggli's institute. Earlier under Niggli's guidance, Hans had taken a course in silicate analysis. After three months of carefully following the book, and repeatedly spilling precious solutions, he produced some silicate analyses of doubtful accuracy. Deciding that was too hard, he turned to the spectrograph. At that time any insoluble geological problem was thought to be answerable by trace element analysis. So he went to MIT to study optical spectroscopy with Louis Ahrens.

At MIT, Hans found the spectrograph produced too many numbers, in sharp contrast to the few numbers produced by classical wet silicate analysis. To escape, in 1952 he accepted an invitation to study the synthesis of micas at the Geophysical Laboratory under the tutelage of Hat Yoder. While Hans was synthesizing micas; next door, Bowen and Tuttle were working on granites. Hans found it to be an exhilarating time, each day bringing new surprises. Here he was first exposed to the use of chemical thermodynamics through his collaboration with Dave Wones. He realized its potential power in application to geologic problems; a completely different approach to that of Goldschmidt who had greatly influenced his thinking while at ETH. It allowed him to develop perhaps his most significant scientific contribution, the oxygen buffers to study redox equilibria at high temperatures and pressures. Since that time Eugster and his coworkers have extended the buffer techniques to reactions with carbon, fluorine, nitrogen and sulfur species as well as acids, bases and metal chlorides in supercritical aqueous solutions. A whole new field of measuring properties of supercritical fluids was opened up when it was realized that some of the buffers could be configured slightly differently to serve as sensors. All of this led to a quantitative understanding of the role of metamorphic fluids and solutions in mineral formation within the earth's crust and mantle. Hans also indirectly influenced the directions of modern igneous petrology. His pioneering development and laboratory calibration of oxygen fugacity buffers led to the routine consideration of oxygen fugacity in volcanic rocks and the lavas from which they crystallize.

In 1958 Hans left the Geophysical Laboratory to accept an associate professorship in experimental petrology at The Johns Hopkins University in Baltimore. Shortly after Hans went to Hopkins, Phil Abelson, director of The Geophysical Laboratory, proposed a lecture series jointly sponsored by Hopkins and the Laboratory. They were published as

Researches in Geochemistry in 1959. One of the papers in this volume was by Charles Milton, on the minerals of the Eocene Green River formation of Wyoming, which Eugster was asked to review. Hans saw at once that one could go beyond cataloging and describing the unusual minerals; one could look at the mineral assemblages in term of phase equilibriia just as one would look at a mineral assemblage from a metamorphic or igneous rock. It was a natural conclusion for someone who was familiar with the pioneering work of Van't Hoff on the evaporation of sea water; and Hans gravitated back into field work. Beginning with the Green River Formation, through Saline Valley and Searles Lake in California, Lake Magadi in the rift valley of Africa, Lake Chad in the Sahara, salt lakes of the Bolivian Altiplano and the Qaidam Basin of China, and Great Salt Lake in Utah, he evaluated the hydrologic, chemical and sedimentological processes leading to evaporite formation in continental and marine basins. When necessary this work on chemical sediments was backed by experiment and modeling; finally leading to a computer solution to what Van't Hoff could not attain experimentally geologically consistent phase diagrams for the evaporation of seawater. Hence there was a choice for Han's graduate students; to be dragged in the mud pursuing the origin of chemical sediments or to be cooked at 500°C and 2000 bars calibrating metamorphic reactions.

In Lake Magadi located in the rift valley of Kenya, Africa, Hans encountered nodule-like accumulations of amorphous silica—perhaps the precursors of chert. So he became interested in the chert problem and that led to the possibility of an inorganic origin for cherty iron formations which he set in a playa-lake environment using the Green River-Magadi depositional model. At this time he also found a bridge into the origin of undersaturated volcanics. Extremely rare natrocarbonatites flows had erupted from the African volcano Oldoinyo Lengai. Alkalic volcanoes, soda-encrusted lakes and ground-water sodic brines, are all locally abundant. Hans proposed that the lavas formed by the mixing of ordinary nephelinitic silicate magma with highly sodic ground-water brines; a sharp contrast to the accepted theory of origin—as a carbonatite differentiate of nephlenitic magma. Hans' theory of origin has never been fully explored.

His interest in Green River led him to Wyoming. Here at his yearly sojourn at the University of Wyoming, Bob Houston introduced him to Indian art, particularly rugs and interested him in the origin of ore deposits. This was another turning point in his career which led to experiments in the solubility of tin minerals and the formation of hydrothermal ore deposits. He and his students measured the solubility of magnetite, cassiterite and wolframite. First he speculated on the origin of the Cornwall-type magnetite deposits and the tin deposits at Dachang in southeast China; and then on to Iceland to look at the evolution of geothermal systems in a young volcanic pile and their recharging with surface water. At the same time he became interested in the formation of the hydrothermal ore-bearing fluids from granitic silicate melts. Unfortunately he never had a chance to quantitate his ideas on the formation of hydrothermal ore-bearing solutions. Hans died unexpectly at the age of 62.

Although many of the problems Hans set about to solve were fundamental, he could probably have made major contributions to industry. In one case, a Hopkins postgraduate student wanted a more practical problem to work on. Hans suggested working with General Electric who needed substances which were perfect insulators along one axis and perfect conductors in the other. Hans thought micas could be synthesized with gold in the interstices which may have the desired properties. The student left, opting for the private sector and did not pursue the problem. Little did they know at the time they were standing on the cutting edge of the silicon chip.

Hans was an excellent teacher. Two characteristics of his teaching stood out, firstly the clarity of his highly organized lectures, many illustrated with elaborate blackboard drawings using five colors of chalk, and secondly his homework assignments that essentially asked the students to derive some of the material for his next lecture. The problems were quite demanding and although seldom completed, they still prepared one for grasping the logic and significance of the subsequent lecture. Some of his assignments included problems for which he himself had not found an answer. These were often springboards for a students PhD research. Together, Hans' clear logical lectures and his challenging homework assignments brought his students to the frontiers of his subject while providing them with the skills and background to advance these frontiers. Perhaps inadvertently, his approach also transmitted the skill and dedication to excellence in teaching that is characteristic of so many of his former students. Even students who had left Hopkins and geology altogether missed Hans' sharp wit and the mental stimulation he could bring out in others.

Hans used to hold informal seminars at his home at Queene Anne's road in Baltimore where the students would sit on the floor and sip a can of beer or two. At one such meeting the speaker was Don Lindsley who talked about his most recent experiments dealing with oxides. The experiments were performed in a piston-and-cylinder apparatus, and the main problem was to keep the conditions constant. He could maintain his pressure, temperature and oxygen fugacity for a limited time only. The key diagram Don was showing contained relatively few data points, chiefly because successful runs were not easy to obtain. At the end of Don's presentation, Hans asked in his typically quiet voice, what the run duration was. Unsuspectingly, Don pointed to individual symbols on his diagram, this run was ten minutes, the other three for fifteen minutes, and so on. Then he looked at Hans anticipating a penetrating question, but Hans merely said, "What do you do in the afternoon?"

Hans excelled in the fine arts. He was a good violinist, and one of the things he brought with him from Switzerland was his violin. A story he used to tell on himself; was that he came with the cultural snobbery that so many Europeans have for the U.S. It was quite a shock to him to find out he wasn't good enough for the MIT orchestra. In the late 60's and earlier 70's he took up the violin again and played for many years in the Goucher Hopkins Community Orchestra. The year he got the Day Medal, he attended the meetings in Washington during the day, drove back to Baltimore to attend a rehearsal of the orchestra, and returned to Washington the next day to receive the medal.

Hans was also a very talented painter, but never received the recognition that he had in the geological sciences. He was continually experimenting, and in that respect may have been ahead of his time. Donald Graf in introducing Euster for the award of the Goldschmidt medal describes one of Hans' paintings; "It was a shaped canvas portraying a young woman, and the expression on her face changed as one walked past—first Pocohontas beginning to have nagging doubts about Englishmen, then the queen of the Chamber of Commerce Apple Festival, then a foreboding creature with sunken eye and drooping lip. The person who created that canvas understood one thing very well—everything depends on where you are standing and the direction in which you are looking." This is reflected in Hans' application of thermodynamics to chemical sediments in his other career. At one point in his life he seriously considered giving up science to devote himself totally to his painting. Perhaps it was only finding a new scientific challenge—namely his work in Magadi—that stopped him from doing this. He found it hard to explain his fascination with salt lakes. They were an enormous source of fun, excitement and adventure. Hans was a rare example of a meld between the two "cultures," the arts and the sciences.

He had two exhibits of his art, but they did not attract a great response. So discouraged he turned to pottery. His primary interest was in murals and sculpture. As in his painting, faces were very important to him. He built his own kiln on his farm; the largest non-commercial kiln in Maryland. So as his interests in the sciences changed from high temperature to low temperature processes his art went in the opposite direction. The glazes used in pottery are usually mixtures of natural minerals fired to a temperature where they melt and mix. The colors of some of the prettiest glazes are controlled by oxidation-reduction reactions in which Hans was already an expert. Evidence of his adherence to the arts is seen even in his more light hearted comments. Once when looking for some rocks which had been stored for years in the heating tunnels of John Hopkins, a friend from the biology department asked him, why on earth he would be interested in a bunch of old rocks. Hans' reply was, "Rocks are like fine red wine, they just keep getting better with age."

Hans loved to entertain and cook. He always prepared too much food and the reason was that at one of the first dinners he and Elaine gave for the students at his art and pottery studio in Sparks, he ran out of food. A long narrow table was set up in his small back yard where his graduate students, postdocs and their families were sitting around. A meal was served fit for royalty accompanied by liberal quantities of red Ruffino wine. None of the guests realized that the bread and cheese served near the end of the meal were simply fillers and not a dish that was planned well in advance.

In spite of his accomplishments, Hans did not dominate arrogantly. Rather than explode in rage or fury, he developed a technique of polite yet subtle withdrawal. Hans felt that if one cannot write clearly he does not have the solution clearly organized in his head. It is true that he did not normally level with ordinary people, except for operational reasons. Yet, he possessed and expressed a well of sympathy for the competent outsider or the talented member of a minority: as long as he was decent and creative (writers, poets, artists and in later years also students).

Hans was always concerned with the historical development of science. He felt that the progress of the scientific leaders of the past was only constrained by the tools at their disposal. It was very evident in his acceptance speeches of the three medals awarded to him, that he felt that a broad-based outlook was important to progress in science.

He was awarded the Day Medal of the Geological Society of America in 1971 and in his acceptance speech he speculated on how an experimental physicist such as Day with no previous experience in Earth Sciences could have been so effective in guiding the direction of the Geophysical Laboratory. When Hans received the Goldschmidt Medal of the Geochemical Society in 1976, he traced the development of Goldschmidt, concluding that his success lay in his willingness to follow problems wherever they were taking him even to the extent of becoming a chemist. Hans received the Roebling Medal of the Mineralogical Society of America in 1983 where he noted the role of mineralogy as the bridge between such diverse geological fields as experimental petrology, geochemistry and sedimentology.

Hans frequently emphasized this bridging of artificial barriers set up by accepted boundaries of a geologic profession. "What is important is the point of view, the unifying theory and the approach rather than the rock type, the particular chemical system, the particular apparatus, or the particular quadrangle." He preferred to describe himself as a chemical geologist. He found little intrinsic difference between an igneous rock and a chemical sediment. They all form from solutions be they gases, liquids or solids. Let me quote from his acceptance speech for the Day Medal in 1971: "Unification in geology is in the air, but compartmentalization is the reality. Compartmentalization means safety. We become specialists and as such effective and respected members of a group. Compartmentalization is also repetition, and repetition is also sterility. After successfully solving a problem, how often do we find ourselves looking for a similar one. This is tempting, because it is easy, and it is encouraged by the pressures of publication and competition. We should keep reminding ourselves of a basic truth familiar to every self-respecting artist: To paint the same painting twice is certain death. By deliberating crossing traditional boundaries and searching for basic designs we can avoid falling into this trap. Fishing in many ponds is not only challenging, it is also more fun."

Hans spent 31 years at Johns Hopkins. During that time, he received the Day, Goldschmidt and Roebling Medals, he was president of the Mineralogical Society of America in 1985, chairman of the Department of Earth and Planetary Sciences at Johns Hopkins from 1983 to 1987, adjunct professor at the University of Wyoming from 1970, member of the National Academy of Sciences and the American Academy of Arts and Sciences and worked part time for the U.S. Geological Survey from 1958 to 1976. He did not consider himself a mineralogist even though he had discovered two new minerals at Magadi.

Although Hans never had a course in physical chemistry or thermodynamics, he mastered both; most of his research was the application of chemical thermodynamics to mineral systems. From the time that Hat Yoder introduced him to the world of hydrothermal synthesis, Hans' central theme remained the interaction of minerals with aqueous fluids; from metamorphic fluids to surface waters to hydrothermal brines to igneous gases. If his life had not been so short, we might have seen the cycle completed, ending at silicate melts. His excellent work in experimental petrology, metamorphism, chemical sediments and ore deposits made him one of the most broad-based and foremost contemporary petrologists in the country. With the passing of Hans Eugster, the Earth Sciences have lost an outstanding researcher and teacher who left behind an entire generation of "chemical geologists."

W. D. Gunter

TABLE OF CONTENTS

Part E. Surface Environments

Part A.
Experimental Techniques and Buffers

Fluid-Mineral Interactions: A Tribute to H. P. Eugster
© The Geochemical Society, Special Publication No. 2, 1990
Editors: R. J. Spencer and I-Ming Chou

Quantitative redox control and measurement in hydrothermal experiments

I-MING CHOU and GARY L. CYGAN

959 National Center, U.S. Geological Survey, Reston, Virginia 22092, U.S.A.

Abstract—In situ redox measurements have been made in sealed Au capsules containing the assemblage $Co-CoO-H_2O$ by using hydrogen-fugacity sensors at 2 kbar pressure with three different pressure media, Ar, CH_4 and H_2O, and between 500 and 800°C. Results show that equilibrium redox states can be achieved and maintained under Ar or CH_4 external pressure, but they cannot be achieved and maintained at temperatures above 600°C under H_2O external pressure. The conventional assumption that equilibrium redox condition is achieved at a fixed P-T condition in the presence of a redox buffer assemblage is therefore not necessarily valid, mainly due to the slow kinetics of the buffer reaction and/or the high rate of hydrogen leakage through the buffer container. The inconsistent equilibrium phase boundaries for the assemblage annite-sanidine-magnetite-fluid reported in the literature can be explained by the inadequate redox control in some of the earlier experiments. The attainment of redox equilibrium in hydrothermal experiments should be confirmed by the inclusion of a hydrogen-fugacity sensor in each experiment where the redox state is either quantitatively or semiquantitatively controlled.

INTRODUCTION

THE PIONEERING WORK OF EUGSTER (1957) on the solid oxygen buffer technique provides a convenient way to control redox states in hydrothermal experiments. Information on the technique's basic theory, the experimental setups of the original design as well as its variations, and its applications are contained in papers by HUEBNER (1971) and CHOU (1987a). The basic experimental assembly consists of two nested capsules (Fig. 1) and has often been referred to as the double-capsule technique. The mineral assemblage under study and water are sealed in the inner capsule, constructed of either Pt or one of the Ag-Pd alloys, which are highly permeable to hydrogen. This charge system, together with an oxygen buffer assemblage and water, is in turn sealed in a larger outer capsule, made of either Au or Ag, which are relatively impermeable to hydrogen. When this assembly is run at a fixed P-T condition, the unknown redox state of the charge system is compared with the known redox state of the buffer system by examining the direction of the redox reaction in the charge system after quench.

Suppose that the reversal of the redox reaction of the charge system can be demonstrated at a given pressure P within a minimum temperature bracket $\Delta T = T_2 - T_1$, where $T_2 > T_1$. Then the equilibrium of the redox reaction of the charge system is located at $T = [(T_2 + T_1)/2] \pm (\Delta T/2)$ and at the redox state of the buffer system, assuming that the equilibrium redox states of the buffer system had been achieved and maintained during the run. However, our recent redox measurements of the buffer system by using the hydrogen fugacity sensors (CHOU and EUGSTER, 1976; CHOU, 1978, 1987a,b) indicate that

this assumption is not always valid. Therefore, some of the published data on the mineral stability relations in P-T-$f O_2$ or P-T-$f H_2$ space obtained by using the oxygen-buffer technique may not be reliable, as indicated, for example, by the recent data on annite (HEWITT and WONES, 1981, 1984). Therefore, we will re-evaluate the oxygen-buffer technique based upon our $f H_2$-sensor measurements in this paper, point out some potential inherent problems, and provide remedies. Preliminary results have been presented earlier (CHOU and CYGAN, 1987, 1989; CHOU, 1988).

THE EQUILIBRIUM ASSUMPTION OF THE BUFFER SYSTEM

In the application of the oxygen-buffer technique, it is commonly assumed that the redox state of the buffer system at P and T is fixed at the equilibrium condition if the phases of the buffer assemblage (oxygen buffer plus water) are present after the experiment as demonstrated by X-ray and/or optical methods after quench. However, several factors may affect the validity of this equilibrium assumption: (1) a buffer phase may alloy with the container material, thus changing its activity (HUEBNER, 1971); (2) a buffer phase may be isolated by the armoring of the reactant by a layer of buffer reaction product; (3) hydrogen may leak through the external capsule; and (4) the buffer reactions may have slow kinetics.

Alloying of a buffer phase with the container material has been examined by the establishment and maintenance of steady-state hydrogen fluxes through the wall of the inner capsule. The experimental setup is the same as shown in Fig. 1 except that the charge system is replaced by another oxygen buffer + H_2O assemblage (CHOU, 1986). In relatively short experiments, the alloying is not a significant problem. The decay of the steady-state hydrogen fluxes observed in the same experiments with longer durations (CHOU, 1986, his Fig. 4) is at least partially attributed to the armoring of buffer phases, which prevents equilibrium redox control of the buffer assemblages on both sides of the inner membrane. The existence of the alloying and armoring problems in certain runs may be inferred from inconsistent experimental results but is generally difficult to identify unambiguously. The use of the hydrogen-fugacity sensor technique (CHOU and EUGSTER, 1976; CHOU, 1978, 1987a,b) makes it possible to identify when alloying and amoring significantly affect equilibrium redox conditions and also to identify the problems of hydrogen leakage and slow kinetics. The nature of the last two problems will be presented after a brief review of the $f\,H_2$ sensor technique.

THE HYDROGEN-FUGACITY SENSOR TECHNIQUE

A brief summary of the technique is given here. For additional information on the theory, experimental details, and applications, the reader is referred to previous publications (CHOU, 1978, 1987a,b, 1989). To prepare a typical sensor, 10–20 μL of solution, either distilled H_2O (for a type A sensor) or 3 m HCl (for a type B sensor), together with ~20 mg Ag and ~20 mg AgCl are loaded into a Pt or AgPd capsule, 1.85 mm OD, 1.54 mm ID, 19 mm long, which is then welded shut. These sensors are then exposed to a system, whose $f\,H_2$ is to be measured at P and T. It has been shown that

$$(f\,H_2)_{P,T} = (K_1)_{P,T}(m_{HCl})^2_{P,T} \qquad (1)$$

where K_1 is a constant at a fixed P-T condition and $(m_{HCl})_{P,T}$ is the molality of associated HCl at P and T calculated from the total Cl^- molarity measured after quench, $(M_{Cl^-})_{1\ atm,\ 25°C}$ (CHOU, 1987a). The attainment of H_2 osmotic equilibrium in the system is indicated by similar measured values of $(M_{Cl^-})_{1\ atm,\ 25°C}$ in sensors A and B. The $(M_{Cl^-})_{1\ atm,\ 25°C}$ values range from ~0.05 for the magnetite-hematite-H_2O assemblage at 2 kbar and 600°C to ~2.0 for Co-CoO-H_2O assemblage at 2 kbar and 800°C. For more reducing conditions, it is advantageous to use the analogous Ag-AgBr-HBr type of $f\,H_2$ sensor instead of the Ag-AgCl-HCl type, since the Br^- concentrations in the sensors are about an order of magnitude lower than those of Cl^- under the same P-T-$f\,H_2$ conditions, so that m_{Br^-} can be approximated by M_{Br^-}. To prepare the Ag-AgBr-

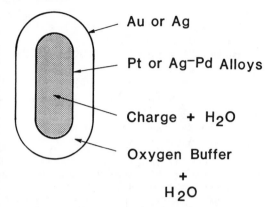

FIG. 1. Schematic diagram showing experimental arrangement of the oxygen buffer technique developed by EUGSTER (1957).

HBr type of $f\,H_2$ sensor, AgCl is replaced by AgBr and 3 m HCl is replaced by 1.5 M HBr in the procedures described above for the preparation of the Ag-AgCl-HCl type. The Ag-AgBr-HBr type of $f\,H_2$ sensor has been used to calibrate the graphite-methane buffer (CHOU, 1987b) and monitor the $f\,H_2$ in the relatively reducing assemblage WO_2-H_2O (CYGAN and CHOU, 1987) according to the relation

$$(f\,H_2)_{P,T} = (K_2)_{P,T}(m_{HBr})^2_{P,T}$$

$$\approx (K_2)_{P,T}(M_{Br^-})^2_{1\ atm,\ 25°C} \qquad (2)$$

where K_2 is a constant, $(m_{HBr})_{P,T}$ is molality of associated HBr at P and T, and $(M_{Br^-})_{1\ atm,\ 25°C}$ is molarity of Br^- in the sensor measured after quench. The above approximation relation is only valid when HBr at P and T is mostly associated; otherwise the HBr dissociation constant at P and T is needed to convert $(M_{Br^-})_{1\ atm,\ 25°C}$ to $(m_{HBr})_{P,T}$.

HYDROGEN LEAKAGE AND THE EFFECT OF THE PRESSURE MEDIUM

As mentioned earlier, Au and Ag capsules are commonly used as containers for the buffer assemblage (see Fig. 1) because they are relatively impermeable to hydrogen. However, since these materials are still permeable to hydrogen, especially at temperatures above 650°C (CHOU, 1986, his Fig. 3), it is unavoidable that the choice of the external pressure medium used in the buffer experiments will have some effect on the rate of hydrogen leakage through the wall of external capsule and hence may affect the redox state of the buffer system, especially when the buffer reaction is sluggish. To assess these effects, we performed a series of experiments using the $f\,H_2$ sensors to monitor simultaneously the redox states in the buffer system and in three different pressure media: Ar, CH_4, and H_2O. The effect of the external pressure medium on the fH_2 that can be maintained in a sealed Au capsule containing the buffer assemblage $Co\text{-}CoO\text{-}H_2O$ was examined at 2 kbar and between 500 and 800°C. The Au capsules used were 0.2 mm thick and 25.4 mm long and had an ID of either 4.0 mm or 2.7 mm. Three different capsule arrangements were used (Fig. 2). In arrangement (a) shown in Fig. 2, fH_2 sensors A and B were exposed to the $Co\text{-}CoO\text{-}H_2O$ buffer system, and the degree of attainment of osmotic equilibrium can be demonstrated in each run. In arrangement (b), only one fH_2 sensor was exposed to the buffer system in order to save more room for the buffer assemblage; the durations of the runs were longer than those required for the attainment of osmotic equilibrium established by experiments using arrangement (a). Arrangement (c) was similar to (b), except that an extra layer of Au capsule con-

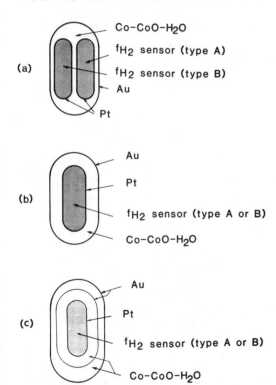

FIG. 2. Schematic diagrams showing three types of experimental arrangements used in this study to measure fH_2 in Au capsules containing the buffer assemblage $Co\text{-}CoO\text{-}H_2O$. Type A of fH_2 sensor uses H_2O as a starting solution, and type B uses acid solution (either 3 m HCl or 1.5 M HBr) as a starting solution (for detail, see text). The capsules were pressurized externally at 2 kbar by Ar, H_2O, or CH_4 and an additional fH_2 sensor (type A or B) was exposed directly to the pressure medium to measure its redox state at P and T. Results are given in Table 1 and Figs. 3 and 4.

taining the same buffer assemblage was added to minimize the hydrogen gradient and hence hydrogen transfer across the inner Au capsule wall. In all three experimental setups, an additional fH_2 sensor was placed adjacent to the capsule to monitor redox state in the pressure medium. Experiments were performed in Stellite 25* cold-seal pressure vessels using Ar, H_2O, or CH_4 as a pressure medium. The experimental setup was similar to that used to calibrate the $C\text{-}CH_4$ buffer (CHOU, 1987b, his Fig. 1), except that stainless steel instead of graphite filler rods were used when the pressure medium was Ar or H_2O. Both the Ag-AgCl-HCl type and the Ag-AgBr-HBr type of fH_2 sensor were used in this study. The experimental procedures were the same as those described previously (CHOU, 1987b). In addition, the H_2O budget in the $Co\text{-}CoO\text{-}H_2O$ assemblage was determined for some of the runs in order to monitor the rate of H_2 permeation through the Au capsule wall. According to the reaction

$$Co_{(s)} + H_2O_{(g)} = CoO_{(s)} + H_{2(g)}, \qquad (3)$$

for every mole of H_2 being added to or subtracted from the system, one mole of H_2O is created or consumed, respectively. Therefore, the average permeation rates of hydrogen through the Au capsule can be calculated from the weight change of H_2O in the capsule; the initial weight of H_2O is measured during loading and the final weight of H_2O after quench is measured by the weight-loss method (CHOU, 1986).

Experimental results are given in Table 1 and shown in Figs. 3 and 4. All of the experimental data under CH_4 external pressure shown in Fig. 3 are taken from CHOU (1987b, his Table 1) except one run (no. 28 in Table 1), in which the H_2O budget in the Au capsule was obtained to demonstrate that hydrogen diffuses into instead of out of the Au capsule. Experimental results under Ar pressure medium are also shown in Fig. 3 for comparison. The data are plotted in terms of log M_{Br^-}, which can be related to log fH_2 through Equation (2). Redox states of the pressure medium in some runs are measured by using the Ag-AgCl-HCl sensors, and the results in $(M_{Cl^-})_{1\ atm,\ 25°C}$ are converted to M_{Br^-} through the relation

$$(fH_2)_{P,T} = (K_1)_{P,T}(m_{HCl})^2_{P,T}$$

$$= (K_2)_{P,T}(M_{Br^-})^2_{1\ atm,\ 25°C} \qquad (4)$$

where $(m_{HCl})_{P,T}$ is calculated from $(M_{Cl^-})_{1\ atm,\ 25°C}$

* Registered trademark of Haynes Stellite Co. Use of trade names in this publication is for descriptive purposes only and does not constitute endorsement by the U.S. Geological Survey.

Table 1. Experimental results of the effect of pressure medium on the redox states of the buffer assemblage Co-CoO-H_2O contained in Au capsules of 0.2 mm wall thickness at 2 kbar total pressure

Run no.	Experimental setup[a]	T (°C)	Duration, t (hrs.)	Log M_{Br^-}[b]		wt. of H_2O (mg)[c]		$\Delta H_2O/t$[d] (μg/hr)
				Co-CoO-H_2O	Pressure medium	Initial	Final	
				(A) Ar pressure medium				
23	b	800	8	(−0.800)	−1.043[h]	24.95	22.47	−310.0
21	b	800	8	−0.821	n.d.[e]	24.35	21.66	−336.3
22	b	752	16	−0.852	−1.260[h]	25.28	22.03	−203.1
24A	b	701	52	−0.838	−1.040[h]	23.44	18.19	−101.1
19	b	700	68	(−0.872)	−0.942	26.18	23.05	−46.0
26	a	651	96	−0.876; (−0.911)	−1.190[h]	19.93	15.71	−44.0
27	a	601	120	−0.938; (−0.980)	−1.470[h]	25.23	24.15	−9.0
29	a	550	167	−1.093; (−1.087)	−1.520	19.80	15.91	−23.3
29A	a	550	120	−1.063; (−1.099)	n.d.	20.00	19.51	−4.1
25	a	500	288	−1.121; (−1.102)	−1.180	25.33	24.99	−1.2
				(B) H_2O pressure medium				
11	b	800	20	(−1.137)	n.d.	10.00	5.70	−215.0
12	b	800	19	−1.151	n.d.	n.d.	n.d.	n.d.
15	c	800	6	−1.204	n.d.	n.d.	n.d.	n.d.
18	c	800	10	−1.134	−1.101	16.30	10.21	−609.0
C8	b	800	25	n.l.[f]	−1.000; (−0.988)	n.a.[g]	n.a.	n.a.
2	a	750	49	−1.070; (−1.132)	−1.201	n.d.	n.d.	n.d.
7	c	750	46	−1.135	n.d.	12.45	10.08	−51.5
4	c	706	90	−1.071	n.d.	17.01	n.d.	n.d.
1	a	700	72	−1.174; (−1.180)	−1.359; (−1.348)	39.36	n.d.	n.d.
9	c	700	96	(−1.008)	n.d.	13.29	7.96	−55.5
E	b	700	48	(−1.205)	n.d.	20.95	17.07	−80.8
34	a	696	48	−1.245; (−1.327)	−1.319	20.28	18.09	−45.6
8	c	650	120	−1.117	n.d.	18.22	11.99	−51.9
5	c	600	188	−1.101	n.d.	14.77	9.38	−23.4
10	c	600	23	(−1.025)	n.d.	15.00	n.d.	n.d.
33	a	595	163	−1.148; (−1.161)	−1.581	30.12	27.70	−14.8
14	b	500	240	−1.140	n.d.	20.00	16.71	−13.7
17	b	500	240	(1.160)	−1.343	20.00	19.70	−1.25
				(C) CH_4 pressure medium				
28	a	650	120	−0.895; (−0.908)	−0.768	22.96	25.08	17.7

[a] As shown in Fig. 2.

[b] Numbers in parentheses are from sensors using 1.5M HBr as a starting solution; the rest are from sensors using H_2O as a starting solution. The relation between log M_{Br^-} and log $f H_2$ can be obtained from Equation (4). Results are shown in Figs. 3 and 4.

[c] Weight of H_2O in Au capsules containing the buffer assemblage Co-CoO-H_2O. For those using double Au capsules (setup (c) in Fig. 2), the numbers given are for the inner Au capsule; the final wt. of H_2O in the outer Au capsule was determined for run no. 18 only. The $\Delta H_2O/t$ for the outer Au capsule in this run was −1143 (μg/hr) including the H_2O loss from the inner Au capsule.

[d] Average rates of H_2O generation (positive) or consumption (negative) in the Au capsules; to obtain rates of H_2 gain or loss, see Equation (5).

[e] Not determined.

[f] Not loaded.

[g] Not applicable.

[h] Calculated from measured m_{Cl^-} using Equation (4).

and the ionization constant of HCl (for detail, see CHOU, 1987a). The data shown in Fig. 3 indicate that equilibrium redox states (shown by the open symbols) can be maintained in the Au capsule for the assemblage Co-CoO-H_2O at 2 kbar CH_4 or Ar external pressure; these redox states are represented

by the solid line, which is the least-squares fit of the data.

The reasons for choosing the Co-CoO-H_2O buffer in this study are that the buffer reaction is kinetically fast and that, at P and T, the assemblage defines redox states (open symbols in Fig. 3) that are in-

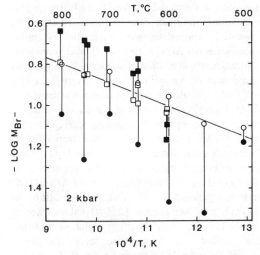

FIG. 3. Experimental results of fH_2-sensor measurements in Au capsules of 0.2 mm wall thickness containing the buffer assemblage Co-CoO-H_2O at 2 kbar Ar (open circles) and CH_4 (open squares) external pressure. The corresponding redox measurements of the pressure medium at P and T in these runs are shown by the filled symbols connected by the vertical lines. The plotted data are given in Table 1 and in CHOU (1987b, his Table 1). For runs containing both sensors A and B (type (a) in Fig. 2), average values are plotted. The solid line with a negative slope is the least-squares fit of all the data shown by the open symbols and can be represented by the equation log $M_{Br^-} = 0.1096 - 974.6/T$, K ($r^2 = 0.8928$). The relationship between fH_2 and M_{Br^-} is given in Equation (2). Equilibrium redox states for the buffer assemblage Co-CoO-H_2O were attained in these experiments in the Au capsules.

termediate between those of the pressure media, with CH_4 being more reducing (filled squares in Fig. 3 at $T \geq 650°C$) and Ar more oxidizing (filled circles in Fig. 3). Reversal of the buffer reaction (Equation 3) can be demonstrated by consideration of the H_2O budget in the system. Under Ar pressure, H_2 diffuses out of the Au capsule and the buffer reaction consumes H_2O, while under CH_4 pressure at $T \geq 650°C$, H_2 diffuses into the Au capsule and generates H_2O. The H_2O budget data are listed in Table 1. Also listed are the average rates of H_2O consumption (negative sign) or generation (positive sign), $\Delta H_2O/t$. The average permeation rates of H_2 through the Au capsules can be calculated from

$$\Delta H_2/t = (\Delta H_2O/t) \times (2.016/18.015) \quad (5)$$

where ΔH_2 and ΔH_2O are weight changes in H_2 and H_2O, respectively, in the buffer system, t is duration of the run, and 2.016 and 18.015 are molecular weights of H_2 and H_2O, respectively. A positive sign indicates influx of H_2 and a negative sign indicates outflow of H_2. The $\Delta H_2O/t$ data given in

Table 1 are obtained by ignoring the minor hydrogen exchanges between the fH_2 sensor and the Co-CoO-H_2O system. Also, these average rates of H_2O consumption or generation are not necessarily equivalent to the rates at steady-state conditions; during the run, the redox states of the pressure media and hence the H gradients in the Au membrane are not rigorously defined so the presence or absence of steady-state conditions cannot be determined. Nevertheless, the $\Delta H_2O/t$ data can provide rough estimates on the rates of H_2 transfer as indicated by the rather systematic changes in $\Delta H_2O/t$ under Ar external pressure, which range from -1.2 $\mu g/hr$ at 500°C to -310 $\mu g/hr$ at 800°C. The H_2O budget was not determined in the earlier experiments under CH_4 external pressure (CHOU, 1987b); the H_2O budget determined in run no. 28 (listed in Table 1) indicates influx of H_2 with a $\Delta H_2O/t = 1.77$ $\mu g/hr$.

Experimental results obtained by using H_2O as a pressure medium are listed in Table 1 and shown in Fig. 4. The data plotted in Fig. 4 show that at temperatures above 600°C, equilibrium redox conditions of the Co-CoO-H_2O assemblage, represented by the solid line derived from Fig. 3, cannot be achieved and maintained in the Au capsules. At these temperatures, the pressure medium exerts a

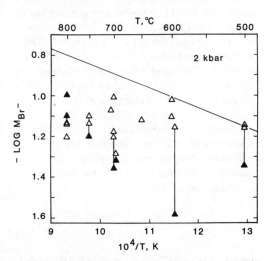

FIG. 4. Experimental results of fH_2-sensor measurements in Au capsules of 0.2 mm wall thickness containing the buffer assemblage Co-CoO-H_2O at 2 kbar H_2O external pressure (open triangles). The filled triangles represent the corresponding redox measurements of the pressure medium. The plotted data are given in Table 1. For runs containing both sensors A and B, average values are plotted. The equilibrium redox states of the Co-CoO-H_2O buffer is shown by the solid line with a negative slope, taken from Fig. 3. These equilibrium states were not attained in the Au capsules for runs at temperatures above 600°C. The relationship between fH_2 and M_{Br^-} is given in Equation (2).

strong influence on the redox states of the fluids in the Au capsules. This is evident from the closeness of the M_{Br^-} values in sensors exposed to fluids inside and outside of the Au capsules for each run; their differences as given in Table 1 are less than 0.2 log unit, even for those runs in which the Co-CoO-H_2O assemblage is protected by double layers of Au (configuration (c) in Fig. 2).

It has been shown previously (CHOU, 1986, his Equation (9)), that the rate of hydrogen transfer across a cylindrical metal membrane at steady state, dM_{H_2}/dt, can be described by

$$dM_{H_2}/dt = [2\pi\kappa l/\ln (r_e/r_i)][\,|(f^iH_2)^{1/2}$$
$$- (f^eH_2)^{1/2}\,|\,],\quad (6)$$

where κ is permeability constant of the metal membrane to hydrogen, l is the overall length of the cylinder, r_e and r_i are the outer and inner radii of the cylinder, respectively, and f^iH_2 and f^eH_2 are the internal and external hydrogen fugacities of the cylindrical container, respectively. According to Equation (6), there should have been no hydrogen transfer across the inner Au capsule in the double Au capsule experiments (type (c) in Fig. 2) because both sides of the inner membrane are exposed to the same redox buffer and therefore no driving force should exist for the diffusion and permeation of hydrogen (i.e., $(f^iH_2)^{1/2} - (f^eH_2)^{1/2} = 0$). However, this is not the case; the $\Delta H_2O/t$ data given in Table 1 indicate significant hydrogen loss from the inner Au capsule. The rapid outflow of H_2 through the outer Au capsule may generate some finite difference in fH_2 between the internal and external fluids of the inner Au capsule, such that $f^iH_2 > f^eH_2$, even though the two fluid phases were buffered by the same redox buffer assemblage. In Equation (6), equilibrium adsorption and desorption reactions are assumed at the internal or external faces of the membrane, such that

$$1/2\ H_{2(g)} = H_{(membrane)}$$

$$K_s = C_H/(f_{H_2})^{1/2}\quad (g/cm^3\ bar^{1/2})\quad (7)$$

where K_s is the equilibrium constant for the hydrogen adsorption or desorption and C_H is the hydrogen concentration in the membrane (g/cm^3). It is conceivable that for a given metal at a fixed P-T condition, the K_s value may depend upon the nature of the fluid matrix to which the metal is exposed. For example, because Ar is more inert than H_2O, the desorption constant of hydrogen at the Au/Ar interface may be greater than that at the Au/H_2O interface, such that for a given fH_2 at P and T, $(C_H)^{Au/Ar} > (C_H)^{Au/H_2O}$. Therefore, the permeability

constant, κ, derived from Au/H_2O adsorption and desorption interface experiments (CHOU, 1986) cannot be applied to cases where the desorption interface is Au/Ar; κ in the latter should be smaller. Consequently, according to Equation (6), for a given dimension of Au cylinder and a given $|(f^iH_2)^{1/2} - (f^eH_2)^{1/2}|$ value in systems having the same Au/H_2O absorption interface, dM_{H_2}/dt should be higher when the desorption interface is Au/H_2O rather than Au/Ar. In other words, Au is a more efficient hydrogen shield when Ar rather than H_2O is used as an external pressure medium. The apparent similarity in dM_{H_2}/dt values for runs under Ar and H_2O external pressures at the same P-T conditions, as indicated by their similar $|\Delta H_2/t|$ values listed in Table 1, is due to the fact that under Ar pressure, the smaller κ is compensated for by the larger $|(f^iH_2)^{1/2} - (f^eH_2)^{1/2}|$ (see Equation (6) and the fH_2 differences between two sides of Au membrane inferred from the differences in M_{Br^-} between filled and open symbols shown in Figs. 3 and 4).

It can be concluded that in hydrothermal experiments where the Co-CoO-H_2O buffer assemblage is used in the oxygen-buffer technique to control their redox states, equilibrium redox states of the buffer can be obtained in sealed Au capsules of 0.2 mm wall thickness between 500 and 800°C under 2 kbar of either Ar or CH_4 external pressure but cannot be obtained at temperatures between 600 and 800°C under 2 kbar of H_2O external pressure. For more reducing buffers, such as WO_2-WO_{2+x}-H_2O, $Fe_{1-x}O$-Fe_3O_4-H_2O, Fe-Fe_3O_4-H_2O, and Fe-$Fe_{1-x}O$-H_2O, there is no doubt that equilibrium redox states cannot be obtained in Au capsules of 0.2 mm wall thickness, as demonstrated by CYGAN and CHOU (1987) for the WO_2-WO_{2+x}-H_2O buffer under Ar and CH_4 as well as H_2O external pressure. The measured steady-state fH_2's in the Au capsules are much lower than the calculated equilibrium values, mainly due to high rates of H_2 leakage.

KINETICS OF BUFFER REACTIONS

In order to maintain the equilibrium redox states of a buffer system at P and T in the Au capsule shown in Fig. 1, the buffer reaction rate has to be fast enough that the generation or consumption of H_2 in the buffer reaction can effectively counterbalance the loss or gain of H_2 in the system due to H_2 leakage through the outer capsule. The fH_2-sensor measurements for various buffer assemblages in Au capsules under 2 kbar Ar external pressure and between 600 and 800°C made by CHOU (1978) indicate that equilibrium redox states can be maintained for the buffer assemblages MnO-Mn_3O_4-

H_2O, Ni-NiO-H_2O (NNO), and Co-CoO-H_2O. This is demonstrated by the linear relationship shown in Fig. 5 for these buffers, as follows from (1),

$$\log (m_{HCl})_{P,T} = 1/2 \log fH_2 - 1/2 \log K_1. \quad (8)$$

In Fig. 5, the chevrons indicate the reversed sensor HCl measurements for any given buffer. There are a number of chevron pairs for each buffer because different thermodynamic data from literature can be used to calculate the corresponding fH_2 value. The sloping line, on the other hand, is based upon the assumption that the NNO point is experimentally and thermodynamically sound, a case argued in CHOU (1978) that for brevity will not be repeated here, with the slope of ½ corresponding to the value given by Equation (8). The circles, then, are the implied correct fH_2 values for all other buffers, including MH (magnetite-hematite-H_2O) and FMQ (fayalite-magnetite-quartz-H_2O). It can be argued

that point 4 (HAAS and ROBIE, 1973, thermodynamic data) is essentially correct and therefore fH_2 equilibrium was attained in our sensor experiment with that buffer. However we feel, on the basis of our own experience with MH, that a somewhat lower fH_2 is probably correct and the HCl value given by the sensor was too high. From Equation (8) it is clear that higher HCl would correspond to a higher fH_2 value, and, therefore, H_2 leaked *into* the buffer assemblage tube. A similar argument applies to FMQ. In this case, however, point 10 is not based upon thermodynamic calculations; rather, it is an experimental hydrogen membrane measurement, which is apparently near the correct value. The other FMQ points show HCl values that are too low, indicating that H_2 leaked *out* of the buffer assemblage tube.

To minimize the influx of H_2 in the Au capsule containing the MH buffer and to prolong the life span of the buffer, it has been a common laboratory

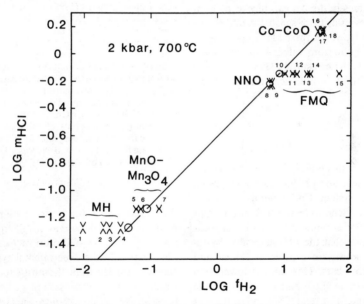

FIG. 5. The internal consistency of fH_2 values of various buffers at 2 kbar and 700°C derived from fH_2-sensor experiments (open circles) compared with values calculated from various sources of thermodynamic data (chevrons). Values of m_{HCl} are calculated for each of the redox buffers from experimental data of CHOU (1978) as described in CHOU (1987a). One set of m_{HCl} values is given for each buffer (∧ for sensor A and ∨ for sensor B). The fH_2 values are calculated from: (1) and (18) ROBIE and WALDBAUM (1968); (2) and (14) EUGSTER and WONES (1962); (3), (15), and (16) SCHWAB and KÜSTNER (1981); (4) HAAS and ROBIE (1973); (5) CHARETTE and FLENGAS (1968); (6) and (9) HAAS (in CHOU, 1987a); (7) and (8) HUEBNER and SATO (1970); (10) MYERS and EUGSTER (1983); (11) HEWITT (1978); (12) O'NEILL (1987); (13) WONES and GILBERT (1969); and (17) MYERS and GUNTER (1979). The dissociation constants for H_2O are taken from ROBIE et al. (1979) and the fH_2O data are from BURNHAM et al. (1969) in calculating fH_2 from fO_2 values. The relation between log m_{HCl} and log fH_2 given in Equation (8) is shown by the solid line assuming the fH_2 values for NNO buffer in (8) is correct; the open circles are located based on this assumption (for detail, see CHOU, 1978). Deviation of the chevrons from the solid line indicates that either the thermodynamic data for other redox buffers are not consistent with the precise NNO buffer data or the measured values for m_{HCl} are not equilibrium values (see text).

practice to use hematite instead of a magnetite-hematite mixture as the initial solid buffer material and CO_2 or Ar instead of H_2O as an external pressure medium. This practice has been quite successful because the use of CO_2 and Ar instead of H_2O lowers the fH_2 gradients across the Au membrane and also possibly lowers Au's permeability constant to hydrogen, as discussed above. However, the assumption that equilibrium redox control of the MH buffer exists in this experimental system is still questionable, as indicated in Fig. 5, even when the buffering material is still far from being exhausted.

The intrinsic fH_2's in Au capsules containing the FMQ-H_2O assemblage were measured under 2 and 4 kbar Ar external pressure and between 600 and 800°C by using the fH_2 sensors (CHOU, 1978). These data are recalculated to 1 kbar pressure in order to compare with other data in the literature, and the results are shown in Fig. 6. It should be emphasized that the fH_2-sensor measurements give the true fH_2 values in the Au capsules as demonstrated by the reversals from sensors A and B (triangles in Fig. 6). However, these fH_2 values are not necessarily the equilibrium values imposed by the buffer because (1) the measured fH_2's are built up by the decomposition of H_2O according to the reaction

$$3 \ Fe_2SiO_{4(s)} + 2 \ H_2O$$

$$= 2 \ Fe_3O_{4(s)} + 3 \ SiO_{2(s)} + 2 \ H_{2(g)}, \quad (9)$$

and the reversal of this buffer reaction has not yet been demonstrated, and (2) H_2 may have leaked out of the buffer system. Consequently, the measured fH_2's would tend to be lower than the equilibrium values. For example, the fH_2-sensor measurements indicate that the fH_2 generated by reaction (9) in a sealed Au capsule at 1 kbar and 700°C is about 5.5 bars. This value is lower than the equilibrium value of 6.8 bars derived from the data of MYERS and EUGSTER (1983), indicating that only about 80% of the equilibrium fH_2 value can be achieved. If HEWITT's (1978) equilibrium fH_2 value of 8.4 bars is adopted, then the fH_2 generated in the Au capsule is only about 65% of the equilibrium value. In other words, if the FMQ-H_2O assemblage is used as a hydrogen source in the redox experiment, H_2 generation will be sluggish and the equilibrium fH_2's will be difficult to obtain even when there is no H_2 leakage from the Au capsule. Similarly, our preliminary data indicate that the H_2 consumption reaction of the FMQ buffer (Equation (9) to the left) is even more sluggish, and the fH_2 in the system has to be considerably higher than

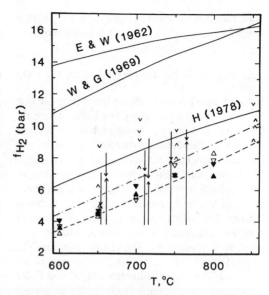

FIG. 6. Comparison of experimental reversals for the buffer assemblage Fe_2SiO_4-Fe_3O_4-SiO_2-H_2O (FMQ) at 1 kbar. Data of HEWITT (1978) are shown by chevrons and those of MYERS and EUGSTER (1983) by arrows, the two ends of arrows indicating the initial and final fH_2. Dash-dot line is a regression of MYERS and EUGSTER's data. Open and solid triangles are calculated, respectively, from 2- and 4-kbar data of CHOU (1978); right-side-up triangles are from sensor A and the inverted triangles are from sensor B. The reversals shown by the triangles indicate the attainment of osmotic equilibrium among the fH_2 sensors and the FMQ buffer system in the experiments and do not demonstrate reversals of the buffer reaction. The dashed curve is the least-squares fit of Chou's data. Solid curves are for EUGSTER and WONES (1962), WONES and GILBERT (1969), and HEWITT (1978).

the equilibrium value to drive the reaction. Erroneous interpretations may result from the invalid assumption that equilibrium redox control by the FMQ buffer in the experimental system has been achieved simply from the post-experiment observation that the buffer assemblage has not been exhausted. We should recognize the fact that at a fixed P-T condition, the assemblage FMQ-H_2O does not fix fH_2 in the experimental system at its equilibrium value. In an experimental system, the univariant curve defined by the FMQ buffer in an isobaric fO_2-T space is not a thin line but rather a wide band.

It is equally important to recognize the fact that the fH_2 level generated in a sealed Au capsule in hydrothermal experiments by the decomposition of H_2O in certain buffer assemblages depends not only on the nature of the buffer reaction but also on the ratio of H_2O to the solid buffer. For example, the fH_2-sensor measurements for the Ni-NiO-H_2O buffer at 2 kbar and 600°C are given in Table 2 for

Table 2. The effect of H_2O/Ni ratio on the fH_2 values generated in 44 hours by the assemblage $Ni-NiO-H_2O$ contained in Au capsules of 0.2 mm wall thickness at 2 kbar Ar external pressure and 600°C

Run no.	H_2O (mg)	Ni (mg)	NiO (mg)	M_{Cl^-} [a] (±0.005)	fH_2 (bar)
RR 5	17.7	196.2	197.0	0.400; (0.394)	4.2 [b]
RR 3	19.0	62.9	8.6	0.318; (0.321)	2.8 [c]

[a] Measured at 1 atm. and 25°C for the fH_2 sensors containing $Ag-AgCl-H_2O-HCl$; the numbers in parentheses are from sensors using 3m HCl as a starting solution and the rest are from sensors using H_2O as a starting solution.

[b] The equilibrium value given in Table 3.6 in CHOU (1987a) for run no. BC-76 of 4-day duration at 2 kbar and 600°C; the equilibrium M_{Cl^-} for BC-76 are 0.3899 and (0.3882), which are very close to those given above for RR 5.

[c] Calculated from Equations (3.23) and (3.26) in CHOU (1987a); S and R in Equation (3.26) represent RR 3 and BC-76, respectively.

two runs; one (RR5) has an H_2O/Ni weight ratio of 0.090, and the other (RR3) 0.302, but both have about the same mass of H_2O. After 44 hours, M_{Cl^-} values of fH_2 sensors of RR3 are smaller than the equilibrium value. Assuming the equilibrium fH_2 at 2 kbar and 600°C for the $Ni-NiO-H_2O$ assemblage is 4.2 bars (see CHOU, 1987a, his Table 3.6), then according to Equation (1) the fH_2 in RR3 is 2.8 bars, which is only 66% of the equilibrium value. It is evident from this example that when the surface area of the reactant (Ni in this case) exposed to the H_2O-rich fluid is limited, as in RR3, it will take longer to establish the equilibrium fH_2, if it can be attained at all. Again, the assumption of equilibrium redox control of a buffer system where the ratio of H_2O to solid buffer is high may result in erroneous interpretation of experimental data. The negative deviation of H_2O activity from ideality in H_2O-rich CO_2-H_2O fluids reported by CHOU and WILLIAMS (1979) is an example. We believe that equilibrium fH_2's in the H_2O-rich CO_2-H_2O fluids in that study were not established because of the high ratio of H_2O to the solid buffer in these runs.

COMMENTS ON ANNITE-SANIDINE-MAGNETITE EQUILIBRIUM

The reaction

$$KFe_3AlSi_3O_{10}(OH)_{2(s)}$$
$$\text{annite}$$

$$= KAlSi_3O_{8(s)} + Fe_3O_{4(s)} + H_{2(g)} \quad (10)$$
$$\qquad \text{sanidine} \qquad \text{magnetite}$$

was the first redox reaction investigated by EUGSTER and WONES (1962) using the oxygen buffer technique. Later, the same reaction was studied by RUTHERFORD (1969), WONES et al. (1971), and HEWITT and WONES (1981). These experimental data have been discussed by HEWITT and WONES (1984) and CHOU (1987a). The experimental difficulties described above for the oxygen buffer technique lead to additional comments.

If the solids in Equation (10) are pure,

$$\log K_{10} = \log fH_2. \quad (11)$$

This equilibrium constant as a function of T is shown in Fig. 7 by the two heavy straight lines; one from EUGSTER and WONES (1962)

$$\log fH_2 = (-9215/T, K) + 10.99, \quad (12)$$

and the other from HEWITT and WONES (1984)

$$\log fH_2 = (-8113/T, K) + 9.59$$
$$+ 0.0042(P - 1)/T, K. \quad (13)$$

It is important to note that the position of Equation (12) is entirely dependent upon the assumption of redox buffer equilibrium, whereas (13) is partially fixed by Shaw membrane experiments at approximately 750°C. The large discrepancy between these two lines, particularly at high fH_2, can be attributed to problems related to redox buffer control. For example, the reversal of reaction (10) observed by EUGSTER and WONES (1962) at 2070 bars along $Fe_{1-x}O-Fe_3O_4-H_2O$ (WM) buffer is at 825 $\pm$ 5°C and $fH_2 = 347$ bars, while the fH_2 predicted by Equation (13) at the same P-T condition is only 153 bars, a much lower value than the fH_2 buffered by the WM. Similarly, the reversal observed by EUGSTER and WONES (1962) at 1035 bars along the same buffer is at 785 $\pm$ 5°C and $fH_2 = 221$ bars, while Equation (13) predicts $fH_2 = 80$ bars. These large discrepancies can be explained if the fH_2 values actually obtained by EUGSTER and WONES (1962) in their Au capsules pressurized externally by H_2O at various P-T conditions were much less than their corresponding equilibrium values defined by the WM buffer, similar to our observations for the $Co-CoO-H_2O$ buffer (Fig. 4) and for the $WO_2-WO_{2+x}-H_2O$ assemblage (CYGAN and CHOU, 1987).

The reversal of reaction (10) reported by RUTHERFORD (1969) along the $C-CH_4$ buffer is at 830 $\pm$ 5°C and $fH_2 = 306$ bars, which is calculated from thermochemical data assuming equilibrium redox control of the buffer (EUGSTER and SKIPPEN, 1967). As shown in Fig. 7, this data point essentially agrees with the reversal point obtained along the WM buffer at 2070 bars by EUGSTER and WONES (1962).

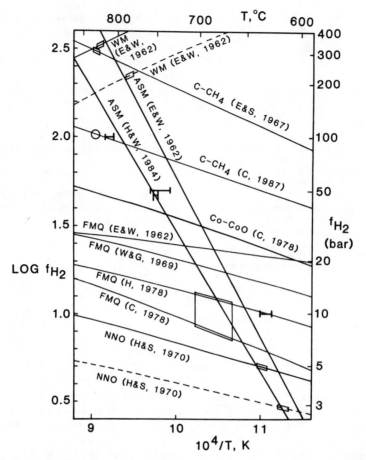

FIG. 7. Summary of experimental results for the reaction annite = sanidine + magnetite + H_2 (reaction 10). The light solid and dashed lines are buffer curves at 2 and 1 kbar, total pressure, respectively. The parallelograms are reversals for reaction (10) based on the data of EUGSTER (1959; on FMQ buffer) EUGSTER and WONES (1962; on WM, FMQ, and NNO buffers), and RUTHERFORD (1969; on C-CH_4 and FMQ buffers). Note that the parallelograms on FMQ and NNO buffers have been shifted from their original positions because of the use of more reliable buffer calibration curves. The open circle represents Rutherford's reversal on C-CH_4 buffer using the calibrated buffer curve (CHOU, 1987b) instead of the calculated one (EUGSTER and SKIPPEN, 1967). The data sources for FMQ buffer are given in Fig. 6. The Co-CoO-H_2O buffer (CHOU, 1978) is shown here for reference. The horizontal bars are reversal data for reactions (10) obtained by the Shaw bomb technique; those at 10, 50 and 100 bars $f H_2$ are from WONES et al. (1971), and the one at 47 bars is from HEWITT and WONES (1981). The two heavy lines marked ASM represent the equilibrium constant for reaction (10) given by EUGSTER and WONES (1962) (Equation 12) and HEWITT and WONES (1984) (Equation 13). For the explanation of the discrepancies between these two lines, see text.

However, the real $f H_2$ values determined by using the C-CH_4 buffer are much lower than the equilibrium values (CHOU, 1987b, also shown in Fig. 7). Therefore, the reversal point observed by Rutherford (1969) at 2 kbar and 830°C along the C-CH_4 buffer should be at $f H_2 \approx 100$ bars (open circle in Fig. 7) instead of 306 bars. The value is approximate because the C-CH_4 buffer calibration is system specific, although it may be applied in a general way to the experimental system used by RUTHERFORD (1969). This new reversal point agrees fairly well,

as shown in Fig. 7, with the reversal point observed by WONES et al. (1971) at 1 kbar and 812 ± 6°C with $f H_2 = 100$ bars. Note that the reversal point at 1 kbar, 746 ± 12°C and $f H_2 = 50$ bars observed by WONES et al. (1971) also agrees well with the point obtained by using the improved Shaw bomb technique at 1 kbar, 752 ± 3°C and $f H_2 = 47 ± 3$ bars (HEWITT and WONES, 1981; also see Fig. 7). However, the reversal point observed by WONES et al. (1971) at 1 kbar, 631 ± 6°C and $f H_2 = 10$ bars does not agree with the reversal points defined by observations along the Ni-NiO-H₂O buffer at 2070 bars and $635 \pm 5°C$ and at 1050 bars and $615 \pm 5°C$ (EUGSTER and WONES, 1962; also see Fig. 7). This is probably due to the inadequacy of the Shaw bomb technique in low $f \mathrm{H_2}$ experiments. On the other hand, at these P-T-$f \mathrm{H_2}$ conditions the use of the Ni-NiO-H₂O buffer is ideal because the hydrogen leakage and the slow kinetics of the buffer reaction problems described previously are minimal to non-existent. We agree with HEWITT and WONES (1984) that the original reversed points along the Ni-NiO-H₂O buffer (EUGSTER and WONES, 1962) should be adopted.

The relatively large reversal brackets, in terms of both T and $f \mathrm{H_2}$, for reaction (10) along the FMQ buffer, as shown in Fig. 7, reflect the inadequacy of using the FMQ-H₂O assemblage as a buffer as described previously. Similarly, the reversal point along the MH-H₂O buffer at 2 kbar and $440 \pm 5°C$, (HEWITT and WONES, 1981), not shown on Fig. 7, needs to be verified. Our initial experiments using $f \mathrm{H_2}$ sensors have been unable to verify this reversal point.

QUANTITATIVE REDOX CONTROL

Direct $f \mathrm{H_2}$-sensor measurements as well as the experimental data on annite-sanidine-magnetite equilibrium described above demonstrate that equilibrium redox conditions may not always be obtained for buffer assemblages in hydrothermal experimental systems using the oxygen-buffer technique of EUGSTER (1957). This is especially true when H₂O is used as an external pressure medium and the redox conditions of the system are more oxidizing than those defined by the MnO-Mn₃O₄-H₂O buffer or more reducing than those defined by the FMQ-H₂O buffer. The presence of the buffering phases examined after quench is a necessary condition but not sufficient proof for equilibrium redox control during the run. Therefore it is necessary to incorporate an $f \mathrm{H_2}$ sensor into the experiment design (Fig. 8) to monitor the true redox conditions. Even though continuous monitoring of redox conditions during the run is not practical, deviations from equilibrium redox conditions can be detected and subsequent erroneous assumptions of equilibrium redox control can be avoided. Furthermore, since the exact redox condition of the experimental system will be determined by the $f \mathrm{H_2}$ sensor, several semiquantitative redox control methods can be adopted such that intentional deviations from redox conditions defined by the conventional solid-oxygen buffers can be achieved. For example, the charge system and the $f \mathrm{H_2}$ sensor shown in Fig. 8 can be

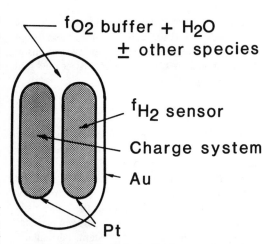

FIG. 8. Schematic diagram showing experimental arrangement for quantitative redox control and measurement. The other species commonly used to reduce $f \mathrm{H_2O}$ in the buffer system are salts or stable gases produced by decomposition of Ag₂C₂O₄ or AgN₃. Alternately, the charge system and the $f \mathrm{H_2}$ sensor can be exposed directly to the external pressure medium, the $f \mathrm{H_2}$ of which is controlled semiquantitatively (see text).

exposed directly to the pressure medium in the pressure vessel and the $f \mathrm{H_2}$ of the system can be controlled semiquantitatively by using various pressure media such as H₂O, CO₂, Ar, CH₄, and mixtures of Ar and H₂ (SHAW, 1963) or Ar and CH₄ (POPP et al., 1984). Alternatively, the $f \mathrm{H_2}$ values buffered by the external system shown in Fig. 8 can be lowered by lowering $f \mathrm{H_2O}$ through the addition of a second species in the vapor phase of the external system (WHITNEY, 1972). Silver oxalate (Ag₂C₂O₄) and silver azide (AgN₃; KEPPLER, 1989) are convenient sources for CO₂ and N₂, respectively, to be used for this purpose. Salts such as NaCl (FRANZ, 1982) and KCl can also be used. However, it should be kept in mind that unless the fluid at P and T is saturated with respect to the salt added, the $f \mathrm{H_2}$ in the experimental system is only semiquantitatively controlled instead of being defined or buffered, because the system is open to H₂O and therefore its $f \mathrm{H_2O}$ and $f \mathrm{H_2}$ are not rigorously defined. Fortunately, since the quantities of H₂O in the system before and after the run can easily be determined gravimetrically, the overall changes in $f \mathrm{H_2O}$ and $f \mathrm{H_2}$ in the run can be determined, and these changes can be reduced by minimizing the hydrogen fluxes across the inner and outer capsule walls. To determine the reaction direction in the charge system shown in Fig. 8, monitoring the gain or loss of H₂O in the charge system may be more sensitive than the conventional methods based upon analysis of the solid reaction products. This approach, together with the quantitative redox control methods described above, has been successfully applied to the studies of mineral stability relations in the skarn system Ca-Fe-Si-O-H (MOECHER and CHOU, 1989).

CONCLUSIONS

Direct $f\mathrm{H}_2$-sensor measurements in hydrothermal experimental systems using the oxygen-buffer technique of EUGSTER (1957) indicate that equilibrium redox conditions may not have been achieved and maintained at the run conditions. This is primarily due to the slow kinetics of the buffer reactions and/or the significant hydrogen leakage through the container of the system, particularly when $\mathrm{H}_2\mathrm{O}$ instead of Ar is used as an external pressure medium and when the redox states of the system are quite different from those of the pressure medium. The inconsistent experimental data reported by various investigators for the annite-sanidine-magnetite-vapor equilibria, for example, can be explained by the invalid assumption of equilibrium redox control in some of the experiments. The incorporation of an $f\mathrm{H}_2$ sensor in each redox-related experiment can ensure that proper redox control in the experiments using the conventional oxygen buffers has been achieved. It also provides a way to control and measure deviations of the redox states from those defined by these buffers.

Acknowledgements—I-Ming Chou had the privilege of working closely with Prof. Hans P. Eugster for about six years at The Johns Hopkins University, first as a graduate student and later as a postdoctoral fellow and a research associate. Without Hans' pioneering work on the solid-oxygen buffers and his constant guidance and encouragement, this contribution would not have been possible. J. S. Huebner kindly provided access to his high-pressure equipment and X-ray facilities. Reviews by H. T. Haselton, J. J. Hemley, W. D. Gunter, and G. B. Skippen were greatly appreciated.

REFERENCES

BURNHAM C. W., HOLLOWAY J. R. and DAVIS N. P. (1969) Thermodynamic properties of water to 1000°C and 10,000 bars. *Geol. Soc. Amer. Special Paper* **132**, 96pp.

CHARETTE G. G. and FLENGAS S. N. (1968) Thermodynamic properties of the oxides of Fe, Ni, Pb, Cu, and Mn by EMF measurements. *J. Electrochem. Soc.* **115**, 796–804.

CHOU I-MING (1978) Calibration of oxygen buffers at elevated P and T using the hydrogen fugacity sensor. *Amer. Mineral.* **63**, 690–703.

CHOU I-MING (1986) Permeability of precious metals to hydrogen at 2 kb total pressure and elevated temperatures. *Amer. J. Sci.* **286**, 638–658.

CHOU I-MING (1987a) Oxygen buffer and hydrogen sensor techniques at elevated pressures and temperatures. In *Hydrothermal Experimental Techniques* (eds. G. C. ULMER and H. L. BARNES), pp. 61–99. J. Wiley & Sons, New York.

CHOU I-MING (1987b) Calibration of the graphite-methane buffer using the $f\mathrm{H}_2$ sensors at 2 kilobar pressure. *Amer. Mineral.* **72**, 76–81.

CHOU I-MING (1988) Quantitative redox control in hydrothermal experiments (abstr.) *EOS* **69**, 528.

CHOU I-MING (1989) A low-temperature redox sensor (abstr.) *EOS* **70**, 507.

CHOU I-MING and CYGAN G. L. (1987) Effect of pressure medium on redox control in hydrothermal experiments (abstr.). *EOS* **68**, 451.

CHOU I-MING and CYGAN G. L. (1989) Equilibrium and steady-state redox control in hydrothermal experiments (abstr.). *28th Internat. Geol. Congr.* 1–287.

CHOU I-MING and EUGSTER H. P. (1976) A sensor for hydrogen fugacities at elevated P and T and applications (abstr.). *EOS* **57**, 340.

CHOU I-MING and WILLIAMS R. J. (1979) Activity of $\mathrm{H}_2\mathrm{O}$ in supercritical fluids: $\mathrm{H}_2\mathrm{O}$-CO_2 at 600° and 700°C at elevated pressures (abstr.). *Lunar Planet. Sci.*, **X**, 201–203.

CYGAN G. L. and CHOU I-MING (1987) Calibration of the WO_2-WO_3 buffer (abstr.). *EOS* **68**, 451.

EUGSTER H. P. (1957) Heterogeneous reactions involving oxidation and reduction at high pressures and temperature. *J. Chem. Phys.* **26**, 1760–1761.

EUGSTER H. P. and SKIPPEN G. B. (1967) Igneous and metamorphic reactions involving gas equilibria. In *Researches in Geochemistry*, Vol. 2 (ed. P. H. ABELSON), pp. 492–520. Wiley, New York.

EUGSTER H. P. and WONES D. R. (1962) Stability relations of the ferruginous biotite, annite. *J. Petrol.* **3**, 81–124.

FRANZ G. (1982) The brucite-periclase equilibrium at reduced $\mathrm{H}_2\mathrm{O}$ activities: some information about the system $\mathrm{H}_2\mathrm{O}$-NaCl. *Amer. J. Sci.* **282**, 1325–1339.

HAAS J. L., JR. and ROBIE R. A. (1973) Thermodynamic data for wüstite, $\mathrm{Fe}_{0.947}\mathrm{O}$, magnetite, $\mathrm{Fe}_3\mathrm{O}_4$, and hematite, $\mathrm{Fe}_2\mathrm{O}_3$. *EOS* **54**, 483.

HEWITT D. A. (1978) A redetermination of the fayalite-magnetite-quartz equilibrium between 650° and 850°C. *Amer. J. Sci.* **278**, 715–724.

HEWITT D. A. and WONES D. R. (1981) The annite-sanidine-magnetite equilibrium (abstr.). GAC-MAC Joint Annual Meeting, Calgary, **6**, A-66.

HEWITT D. A. and WONES D. R. (1984) Experimental phase relations of the micas. In *Mica* (ed. S. W. BAILEY), Reviews of Mineralogy **13**, pp. 201–256. Mineral. Soc. Amer.

HUEBNER J. S. (1971) Buffering techniques for hydrostatic systems at elevated pressures. In *Research Techniques for High Pressure and High Temperature* (ed. G. C. ULMER), pp. 123–177. Springer-Verlag, New York.

HUEBNER J. S. and SATO M. (1970) The oxygen fugacity-temperature relationships of manganese and nickel oxide buffers. *Amer. Mineral.* **55**, 934–952.

KEPPLER H. (1989) A new method for the generation of N_2-containing fluids in high-pressure experiments. *Eur. J. Mineral.* **1**, 135–137.

MOECHER D. P. and CHOU I-MING (1989) Experimental study of equilibria in the system Fe-Ca-Si-O, and revised estimates of $\Delta_f G^0_{298}$ for andradite and hedenbergite (abstr.). *Geol. Soc. Amer. Abstr. with Progr.* **21**, A156.

MYERS J. and EUGSTER H. P. (1983) The system Fe-Si-

O: Oxygen buffer calibrations to 1500K. *Contrib. Mineral. Petrol.* **82,** 75–90.

MYERS J. and GUNTER W. D. (1979) Measurement of the oxygen fugacity of the cobalt-cobalt oxide buffer assemblage. *Amer. Mineral.* **64,** 224–228.

O'NEILL H. ST. C. (1987) Quartz-fayalite-iron and quartz-fayalite-magnetite equilibria and the free energy of formation of fayalite (Fe$_2$SiO$_4$) and magnetite (Fe$_3$O$_4$). *Amer. Mineral.* **72,** 67–75.

POPP R. K., NAGY K. L. and HAJASH A., JR. (1984) Semi-quantitative control of hydrogen fugacity in rapid-quench hydrothermal vessels. *Amer. Mineral.* **69,** 557–562.

ROBIE R. A. and WALDBAUM D. R. (1968) Thermodynamic properties of minerals and related substances at 298.15K (25°C) and one atmosphere (1.013 bars) pressure and at higher temperatures. *U.S. Geol. Surv. Bull.* **1259,** 256pp.

ROBIE R. A., HEMINGWAY B. and FISHER J. R. (1979) Thermodynamic properties of minerals and related substances at 298.15K and 1 bar (10^5 Pascals) pressure and at higher temperatures. *U.S. Geol. Surv. Bull.* **1452** (Reprint with corrections 1979), 456pp.

RUTHERFORD M. J. (1969) An experimental determination of iron biotite-alkali-feldspar equilibria. *J. Petrol.* **10,** 381–408.

SCHWAB R. G. and KÜSTNER D. (1981) The equilibrium fugacities of important oxygen buffers in technology and petrology. *Neues Jahrb. für Mineral. Abhandl.* **140,** 111–142.

SHAW H. R. (1963) Hydrogen-water vapor mixtures: control of hydrothermal atmospheres by hydrogen osmosis. *Science* **139,** 1220–1222.

WHITNEY J. A. (1972) The effect of reduced H$_2$O fugacity on the buffering of oxygen fugacity in hydrothermal experiments. *Amer. Mineral.* **57,** 1902–1908.

WONES D. R. and GILBERT M. C. (1969) The fayalite-magnetite-quartz assemblage between 600 and 800°C. *Amer. J. Sci.* **267,** 480–488.

WONES D. R., BURNS R. G. and CARROLL B. M. (1971) Stability and properties of synthetic annite (abstr.). *EOS* **52,** 369.

Fluid-Mineral Interactions: A Tribute to H. P. Eugster
© The Geochemical Society, Special Publication No. 2, 1990
Editors: R. J. Spencer and I-Ming Chou

A new hydrothermal technique for redox sensing using buffer capsules

G. C. ULMER and D. E. GRANDSTAFF

Department of Geology, Temple University, Philadelphia, PA 19122, U.S.A.

and

J. MYERS

International Technology, 2340 Alamo St., S E Suite 30-6, Albuquerque, New Mexico 87106, U.S.A.

Abstract—A reversal of the usual double capsule technique pioneered by Eugster was used in this study: various buffer assemblages contained in small thin-walled (0.02 mm) palladium capsules were included with the more massive amounts of water and solids loaded in the gold bag of a rocking autoclave, not to control the redox potential, but to measure it. In this manner the reacting palladium capsules could be used to estimate redox conditions within the much larger gold cell of Dickson autoclaves. Experiments involving basalt-simulated groundwater mixtures were conducted at 200° and 300°C at 30 MPa for 4000 hours. The following buffers were used: $Cu-Cu_2O$, $Cu-CuO$, Fe_3O_4-Fe_2O_3, Ni-NiO, Co-CoO, Fe_2SiO_4-Fe_3O_4-SiO_2. The reacting buffers fix the $\log fO_2$ between -23 and -35 at 300°C and -40 and -43 at 200°C. These ranges are consistent with fO_2 values obtained from SO_4-H_2S and fH_2 measurements for the same experiments. This technique holds promise for redox estimates at lower temperatures than has been previously thought possible in using such buffering assemblages.

INTRODUCTION

THE GEOCHEMICAL CHARACTERIZATION of hydrothermal systems involves the determination of all pertinent equilibria. However, in experiments it is often difficult to measure or control the redox state. This paper presents results of hydrothermal experiments conducted to determine the redox conditions developed during basalt-water reactions (MOORE et al., 1985).

Historically, the double buffer capsule technique of EUGSTER (1957), and later of EUGSTER and WONES (1962), utilized a gold outer capsule containing a buffer assemblage with a platinum inner capsule containing a small amount of material to be equilibrated. In this arrangement the mass of the buffer was much larger than that of the material to be equilibrated and the diffusivity of hydrogen through platinum together with the buffer thus controlled the resulting redox state of the system.

By contrast, in this technique we have added six or seven palladium capsules, each containing small amounts of buffer materials (ca. 25 mg) and water, into the gold cell of a Dickson autoclave. This cell contained at least 15 g of basalt and 150 mL of solution. Because the relative masses involved are so different, reaction of only a small amount of the basalt could control the system redox, if allowed by reaction, permeation, or diffusion kinetics. These buffer capsules would then act to sense the redox state and not control it.

Previous investigations (reviews by HUEBNER, 1971; or by CHOU, 1987) suggest that equilibration might not take place in reasonable periods (inferred to be days to weeks) at temperatures even as low as ca. 400°C. Within our experiments which were designed for other reasons to last 3000–5000 hours at 200° and 300°C it was possible to see if equilibration would occur and, if so, to determine the redox conditions arising from basalt-water reactions under conditions simulating those which might occur in a nuclear waste repository or in a geothermal field.

EXPERIMENTAL TECHNIQUES AND MATERIALS

The hydrothermal experiments were conducted using a Dickson rocking autoclave. The design and operation of the autoclave are described in DICKSON et al. (1963), SAKAI and DICKSON (1978), SEYFRIED et al. (1979 and 1987), and MOORE et al. (1985). This apparatus consists of a flexible gold sample cell (approx. volume 250 mL) with a titanium lid connected by a gold-lined titanium capillary tube to an external sampling valve. The gold cell is contained within a pressurized steel autoclave which is heated in a furnace. Aqueous samples of 5–10 mL may be removed periodically through the external sampling valve by squeezing the gold cell with a small over-pressure within the autoclave. This can be done at any time during the experiment without seriously perturbing the pressure and temperature of the experiment. The solution aliquots were analyzed for major dissolved species including sulfate, sulfide, and pH (MOORE et al., 1985), and dissolved hydrogen using the method of KISHIMA and SAKAI (1984a and b), GRANDSTAFF et al. (1985) and ULMER et al. (1985).

The basalt used throughout the experiments was collected from the entablature of the Umtanum flow, in the Pasco Basin northwest of Hanford, WA. It is a fairly typical low-magnesium Columbia Plateau Miocene tholeiite, with 47% mesostasis (more than half of which is true glass (AL-

LEN *et al.*, 1985)). The composition and mineralogy are given in MOORE *et al.* (1985).

The solution used in experiments is a chemically prepared synthetic equivalent of groundwater found in contact with the Umtanum flow at a depth of 1 km under the Pasco Basin. This synthetic groundwater contained about 300 mg/L dissolved chloride and 30 mg/L fluoride. The need to quantitatively conserve the halide ions to satisfy the water-rock interaction aspects of this research made us cautious about introducing silver-palladium alloys which would have had higher permeability rates than pure palladium (CHOU, 1987 and 1989). The groundwater composite used is given in SMITH (1980, 1981).

The capsules were made from 3.0 mm diameter, thin-walled (0.20 mm) palladium tubing by arc welding the "ash can" shape in a technique similar to that of SNEER-INGER and WATSON (1985). This capsule shape was selected to avoid scoring of the inner wall of the gold bag during the rocking of the autoclave. The 0.20 mm wall thickness did endure the abrasion of 3000–4000 hours of rocking in the basalt powder (60–125 μm size range)-water slurry. The capsules were 12 to 18 mm long. They each contained *ca.* 25 mg of the buffer assemblages: Cu-Cu_2O, Cu-CuO, Fe_3O_4-Fe_2O_3, Fe_2SiO_4-Fe_3O_4-SiO_2, Ni-NiO and Co-CoO all prepared from reagent grade chemicals. Sufficient water (5 to 12 mg) was added to allow complete redox equilibration. Comparison of the before- and after-weights to judge reaction of the buffers was pointless due to abrasion losses of the palladium capsules.

After the experiments the capsules were visually examined for possible holes. The capsules were cut open and the contents were qualitatively examined and characterized with binocular microscope, X-ray diffraction, and electron microprobe.

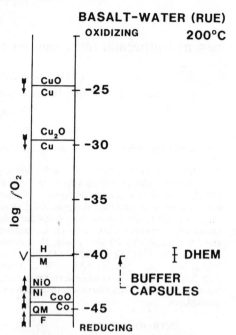

FIG. 2. Columbia River Plateau Reference Umtanum Entablature (RUE) Basalt Redox Data for 200°. Various buffer assemblages used in palladium capsules as explained in the text are summarized: arrows point in the direction of change recorded by the buffer assemblage; a recrystallization and change in the starting proportions of a buffer is recorded as a "V". Brackets to the right indicate the ranges measured by two techniques. DHEM stands for direct hydrogen evolution measurement.

RESULTS

Results of the experiments are given in Figs. 1 and 2. In each figure $\log f O_2$ (oxygen fugacity) is plotted as a vertical nomograph. The value of $f O_2$ for each buffer assemblage (as calculated from thermochemical data of HELGESON *et al.*, 1978) is plotted on this vertical axis. Buffers used varied with the experiment but included the following: the fayalite-magnetite-quartz (FMQ) buffer, the magnetite-hematite buffer (MH), the Ni-NiO buffer and the Co-CoO buffer. Oxidation of the buffer assemblage is indicated by an upward pointing arrow. Reduction is indicated by a downward pointing arrow. Buffers in which no reaction occurred are indicated by ($\times$). Also plotted on the diagrams are other redox values of pertinent buffers such as Cu-Cu_2O and/or Cu-CuO. The oxygen fugacity of the parent basalt, as extrapolated down from 900°C from our own ZrO_2 cell measurements falls approx. 1.3 log units below the (FMQ) buffer, *i.e.* −34.7 at 300°C.

Fig. 1 summarized the results of a basalt/water experiment (BR-14) conducted at 300°C. The

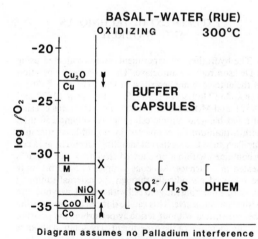

FIG. 1. Columbia River Plateau Reference Umtanum Entablature (RUE) Basalt Redox Data for 300°C. Various buffer assemblages used in palladium capsules as explained in the text are summarized: arrows point in the direction of change recorded by the buffer assemblage; no major change in the buffer assemblage is recorded as an "X". Brackets to the right show the ranges determined by the three different techniques. DHEM stands for direct hydrogen evolution measurement.

copper buffer (Cu-Cu_2O) is very likely disproportionated to the CuO-Cu buffer in that Cu_2O is reported to be unstable below $375°C$ (Hansen, 1958). Nevertheless, the inferred CuO was totally reduced while the (Co-CoO) buffer was totally oxidized (as indicated by the arrows). There was little or no reaction in the iron (Fe_3O_4-Fe_2O_3) and nickel (Ni-NiO) buffers. These results bracket the value of log fO_2 between -23 and -35. The results indicate that hydrogen diffusion across the palladium membrane was fast enough in 3000–4000 hours to have had reaction of at least some of the buffer solids.

Values of log fO_2 calculated from high-temperature sulfate-sulfide speciation (labeled SO_4^{2-}/ H_2S; MOORE et al., 1985) and from measured values of dissolved hydrogen (labeled DHEM) (GRANDSTAFF et al., 1985; ULMER et al., 1985) indicated that, assuming equilibrium, the oxygen fugacity had attained values near that defined by the magnetite-hematite (MH) equilibrium. Therefore, the results are not inconsistent over the three techniques, i.e., SO_4^{2-}/H_2S, DHEM, and the current capsule technique. The iron and nickel buffers may not have reacted because the redox state of the experiment may have been close to the equilibrium boundary and therefore, there was little thermodynamic driving force for their reaction. While KISHIMA and SAKAI (1984a) have shown that (FMQ) is quite reactive at $300°C$, results from experiments (e.g. BARTON and SKINNER, 1979) suggest that Cu-bearing phases should react faster than Ni- or Fe-bearing phases.

Results of an experiment at $200°C$ are shown in Fig. 2. Buffer reactions in that experiment constrain values of log fO_2 between -40 and -43. Most of the buffers showed complete or virtually complete reaction at $200°C$ and 5000 hours. Again, the Cu_2O may have disproportionated to CuO and Cu, but both buffers showed the disappearance of all copper oxides. Moreover, the magnetite-hematite (MH) buffer showed a dramatic octahedral recrystallization of the magnetite along with an XRD-noticeable depletion (but not annihilation) of the amount of hematite present. These observations taken together with the compelling mass of the water-rock charge, both suggest that the redox conditions were not in violation of the (MH) buffer value, but were rather close to it. Again the bracketed redox values from the redox buffers are not inconsistent with the log fO_2 value calculated from measured dissolved hydrogen concentrations (indicated by "DHEM").

DISCUSSION

Ideally, a sensor should be inert and not react with the system materials (except for hydrogen diffusion in the present experiments). This may not have been the case. The outside of one capsule from the $300°C$ experiment was examined using scanning electron microscopy with element discrimination. The palladium capsule had reacted with dissolved sulfide in solution [the synthetic groundwater has an initial concentration of 2.9 mg/L and some sulfur may be contributed from the natural basalt used] and produced palladium sulfide. Solution aliquots from this experiment contained less sulfide than had aliquots from previous experiments without the palladium capsules. This suggests that the palladium capsule reacted with and removed sulfur from solution. This may have gently perturbed the redox state of the system to slightly more oxidized values as evidenced by the production of traces of hematite as a run-by-product within the gold cell. Nevertheless, in sulfur-free systems this complication would not occur.

Additionally, the reactivity of the buffers with palladium must also be considered. With the electron microprobe, we specifically checked for any alloying of buffer materials with palladium and detected only the anticipated Cu-Pd alloy reaction as shown in the sketched version of the electron-backscatter image in Fig. 3. Note the diffusion of copper into the palladium did produce both an alpha and a beta alloy phase, qualitatively matching the anticipated alloys shown in the Cu-Pd phase diagram (HANSEN, 1958), as shown in the inset of Fig. 3. Since the activity of copper was diminished by this alloy-formation, some adjustment is necessary to the fO_2 values plotted in Figs. 2 and 3 which are those of the copper-copper oxide systems without palladium. No activity data for the Cu-Pd-O system could be found with which to estimate the necessary corrections to the fO_2 values for the Cu buffers in palladium tubes.

Despite the reactivity of palladium with some buffers and with sulfur, the conclusion is that reaction rates of the buffers in the palladium capsules were such, that by running a group of buffers in this fO_2-sensing-mode, useful redox approximations were/are still possible at temperatures as low as $200°$ and $300°C$. With our choices of specific buffers, we could only narrow down the water-rock interaction redox range to about 14 log units of oxygen fugacity. The AgPd-AgCl-Ag technique subsequently reported by CHOU (1989) allows the determination of the specific redox value in such autoclave studies. Our times for redox equilibration were defined by the experimental design to be in the order of thousands of hours, as is often the case in autoclave experiments, but shorter times are also likely to work as indicated by KISHIMA and SAKAI

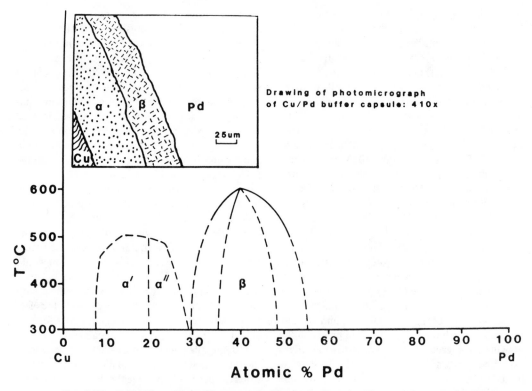

FIG. 3. The Cu-Pd Binary Phase Diagram and a Sketch of a Scanning Electron Microscope Image of the Palladium Tube Wall. The phase diagram (after HANSEN, 1958) predicts the exact phases found by the reaction of copper containing buffers with the palladium of the capsules. Further discussion is given in the text.

(1984a) and CHOU (1987, 1989). In our study the buffering capsules, the sulfate/sulfide equilibria and the hydrogen evolution methods are consistent and as has been shown in the review by ULMER et al. (1986), the fO_2 values obtained in this study do overlap those measured for basalt geothermal wells by others (ARNÓRSSON et al., 1983; and GISLASON and EUGSTER, 1984).

Acknowledgements—This research was done under a grant from DOE to Rockwell/Hanford Operations under which Temple University was a second level subcontractor. J. Myers was then employed by Rockwell/Hanford Operations. In addition to the thanks due DOE and Rockwell/Hanford for the support of this work, special thanks are due to Mike Sterner who helped with suggestions for the construction of a die for shaping of the palladium capsule caps. The year 1987–88 saw the loss to geochemistry of Marcello Carapezza, Hans Eugster, and Bob Garrels. On several occasions these three scientists synergistically influenced each other in direct and in indirect ways. As late as 1983–84, all three were together on a Carapezza-Eugster planned international field trip to Cozodizzi and to Pasquazia, Sicily, to the sulfur and potassium (sylvite) mines of the Miocene evaporites. We feel that it is appropriate to recall that all three of these men shared the global respect of the geochemical community.

REFERENCES

ALLEN C. C., JOHNSTON R. G. and STROPE M. B. (1985) Characterization of reference Umtanum and Cohassett basalt. Rockwell Hanford Operations Document SD-BWI-DP-053, 47 pp.

ARNÓRSSON S., GUNNLAUGSON E. and SVARSSON H. (1983) The chemistry of geothermal waters in Iceland. II. Mineral equilibria and independent variables controlling water compositions. *Geochim. Cosmochim. Acta* **47**, 547–556.

BARTON P. B., JR. and SKINNER B. J. (1979) Sulfide mineral stabilities. In *Geochemistry of Hydrothermal Ore Deposits* (ed. H. L. BARNES), pp. 278–403. Wiley-Interscience, New York.

CHOU I-MING (1987) Oxygen buffer and hydrogen sensor techniques at elevated pressures and temperatures. In *Hydrothermal Experimental Techniques* (eds. G. C. ULMER and H. L. BARNES), pp. 61–99. Wiley-Interscience, New York.

CHOU I-MING (1989) A low temperature redox sensor (abstr.). *EOS* **70**, 507.

DICKSON F. W., BLOUNT C. W. and TUNELL G. (1963) Use of hydrothermal solution equipment to determine the solubility of anhydrite in water from 100° to 275°C and from 1 bar to 1,000 bars pressure. *Amer. J. Sci.*, **261**, 61–78.

EUGSTER H. P. (1957) Heterogeneous reactions involving oxidation and reduction at high pressures and temperatures. *J. Chem. Phys.* **26**, 1760–1761.

EUGSTER H. P. and WONES D. R. (1962) Stability relationships of the ferruginious biotite, annite. *J. Petrology* **3**, 82–125.

GISLASON S. R. and EUGSTER H. P. (1984) Oxygen fugacities in a basalt-meteoritic water geothermal system: Krafla, N. E. Iceland (abstr.). *Geol. Soc. Amer. Abstracts with Programs* **16**, 520.

GRANDSTAFF D. E., FOSTER R. W., KORN R. A. and ULMER G. C. (1985) Measurement of hydrothermal redox using a direct hydrogen evolution method (DHEM) (abstr.). *Second International Symposium on Hydrothermal Reactions,* p. 10. The Pennsylvania State University.

HANSEN M. (1958) *Constitution of the Binary Alloys* (2nd ed). McGraw-Hill, New York 1305p.

HELGESON H. C., DELANEY J. M., NESBITT H. W. and BIRD D. K. (1978) Summary and critique of the thermodynamic properties of the rock-forming minerals. *Amer. J. Sci.* **278-A**, 1–229.

HUEBNER J. S. (1971) Buffering techniques for hydrostatic systems at elevated pressures. In *Research Techniques for High Pressure and High Temperature* (ed. G. C. ULMER), pp. 123–177. Springer Verlag, New York, 123–177.

KISHIMA N. and SAKAI H. (1984a) A simple gas analytical technique for the Dickson-type hydrothermal apparatus and its application to the calibration of MH, NNO, and FMQ oxygen buffers. *Geochem. J.* **18**, 19–29.

KISHIMA N. and SAKAI H. (1984b) Fugacity-concentration relationship of dilute hydrogen in water at elevated temperature and pressure. *Earth Planet. Sci. Lett.* **67**, 79–86.

MOORE E. L., ULMER, G. C. and GRANDSTAFF D. E. (1985) Hydrothermal interaction of Columbia Plateau basalt from the Umtanum flow (Washington, USA) with its coexisting groundwater. *Chem. Geol.* **49**, 53–71.

SAKAI H. and DICKSON F. W. (1978) Experimental determination of the rate and equilibrium fractionation factors of sulfur isotope exchange between sulfate and sulfide in slightly acid solutions at 300°C and 1000 bars. *Earth Planet. Sci. Lett.* **39**, 151–161.

SEYFRIED W. E., JR., GORDEN P. C. and DICKSON F. W. (1979) A new reaction cell for hydrothermal solution equipment. *Amer. Mineral.* **64**, 646–649.

SEYFRIED W. E., JR., JANECKY D. R. and BERNDT M. (1987) Rocking autoclaves for hydrothermal experiments: II. The flexible reaction-cell system. In *Hydrothermal Experimental Techniques* (eds., G. C. ULMER and H. L. BARNES), pp. 216–239. Wiley-Interscience, New York.

SMITH M. J. (1980) BWIP data package for reference data on groundwater chemistry. I. Reference Grande Ronde groundwater composition; II. Recipe for making synthetic Grande Ronde groundwater; III. Procedure for making synthetic groundwater. Rockwell Hanford Operations, Richland, Washington, Doc. RSD-BWI-DP-007, 13p.

SMITH M. J. (1981) BWIP data package for reference chemistry of and procedures for preparation of synthetic Grande Ronde groundwaters. Rockwell Hanford Operations, Richland, Washington, Doc. RSD-BWI-DP-007, 3p.

SNEERINGER M. A. and WATSON B. E. (1985) Milk cartons and ash cans: two unconventional welding techniques. *Amer. Mineral.* **70**, 200–201.

ULMER G. C., FOSTER R. W., KORN R. A. and GRANDSTAFF D. E. (1985) The applicability of direct hydrogen evolution measurements (DHEM) to hydrothermal studies of nuclear waste disposal (abstr.). *Second International Symposium on Hydrothermal Reactions,* p. 29. The Pennsylvania State University.

ULMER G. C., KACANDES G. H. and GRANDSTAFF D. E. (1986) Icelandic geothermal fields as an analog for nuclear waste disposal in basalt. *Adv. Ceram. Sci.* **20** *NUCLEAR WASTE MANAGEMENT II,* 755–764. Amer. Ceram. Soc.

Fluid-Mineral Interactions: A Tribute to H. P. Eugster
© The Geochemical Society, Special Publication No. 2, 1990
Editors: R. J. Spencer and I-Ming Chou

Reversed equilibrium solubility of a high-magnesium calcite

G. Michel Lafon

Exxon Production Research Company, Houston, Texas 77252, U.S.A.

Abstract—A new experimental technique has been developed to study reactions between magnesian calcite solid solutions and aqueous fluids. It results in congruent dissolution and precipitation of the mineral and yields a very accurate estimate of its solubility without relying on empirical extrapolations. Use of low fugacities of CO_2 (comparable to that of air) and varying the stirring rate provide a measure of control on reactions and permit accurate characterization of metastable states. Stirring of the solution and abrasion of the solids can be used to reach metastable supersaturation for extended periods of time. Through appropriate manipulation of experimental parameters, it is possible to reach equilibrium from both under- and super-saturation.

The reversed equilibrium solubility of a biogenic high-magnesium calcite has been experimentally determined at 25°C for the first time. A fragment of the sea urchin *Sphaerechinus granularis* containing 11.6 mol percent $MgCO_3$ was reacted with distilled water at equilibrium with a gas mixture of 377 ppm V CO_2 in N_2. Congruent dissolution, precipitation and equilibrium data are presented for this material. Its free energy of formation is: $-1,117.75$ kJ mol^{-1}.

INTRODUCTION

MAGNESIAN CALCITES CONSTITUTE an isostructural solid-solution series between end-members calcite ($CaCO_3$) and magnesite ($MgCO_3$), and represent a major fraction of recent carbonate sediments and rocks. They occur mostly as biogenic skeletal grains, but are also commonly found as pore-lining cements precipitated from sea water. They can be classified in two main groups. The low-magnesium calcites, containing less than 4 to 6 percent $MgCO_3$, are stable over periods of geologic time. High-magnesium calcites usually react with pore fluids over time spans of tens to hundreds of thousands of years to form low-magnesium calcites and possibly dolomite, and to release dissolved magnesium. This "stabilization process" plays an essential role in the early diagenesis of carbonate rocks (see, for instance BATHURST, 1976, chapters 8, 9 and 10).

The common occurrence in recent marine sediments of a variety of carbonate minerals with different structures and compositions is difficult to understand from a chemical equilibrium standpoint. The difficulty is compounded by selective reactions with pore fluids during early diagenesis. These problems have led numerous investigators to try and estimate the stability of magnesian calcite in order to answer questions such as: is sea water more supersaturated with respect to aragonite or magnesian calcite?; why are both aragonite and magnesian calcite cements precipitated during early marine diagenesis?; what is the composition of a magnesian calcite at equilibrium with sea water?; how can we explain the evolution of modern carbonate sediments toward the long-term stable assemblage calcite-dolomite-solution?

Three approaches have been used to estimate the stability of magnesian calcite: experimental precipitation from a solution of known Ca/Mg ratio to determine a partition coefficient, thermodynamic analysis of the magnesian calcite solvus, and equilibration with aqueous solution to derive a solubility. The first approach relies on a non-equilibrium process and interpreting the experimental distribution coefficients is difficult (compare WINLAND, 1969; KATZ, 1973; MUCCI and MORSE, 1983). The second approach requires long temperature extrapolations (LERMAN, 1965). Recent work has emphasized the third approach (PLUMMER and MACKENZIE, 1974; WALTER and MORSE, 1984; BISCHOFF et al., 1987; BUSENBERG and PLUMMER, 1989).

Equilibrating a high-magnesium calcite with an aqueous solution is experimentally difficult because the desired equilibrium state is generally metastable with respect to the solubility of low-magnesium calcite. Early investigators such as CHAVE et al. (1962) and CHAVE and SCHMALZ (1966) assumed that the steady-state pH they observed could be interpreted in terms of relative stability or solubility. In contrast, upon dissolving high-magnesium calcites in water at fugacities of CO_2 between 0.32 and 100 kPa, PLUMMER and MACKENZIE (1974), WALTER and MORSE (1984) and BISCHOFF et al. (1987) all initially observed congruent dissolution, soon followed by a different reaction, precipitation of low-magnesium calcite. These observations led them to conclude that previously reported steady-state pH values were misleading and could not be translated into thermochemical properties. To interpret their data in terms of equilibrium, they had to *extrapolate the dissolution kinetics* to an estimated equilibrium state. Using the subset of data corresponding to congruent dissolution, they applied an empirical linear extrapolation of solution pH against (reaction time)$^{-1/2}$.

The latter procedure was originally introduced by GARRELS et al. (1960) as a technique which, if valid, would help determine the stability of carbonate minerals that had not yet been synthesized at room temperature and pressure. The key assumption, that the asymptotic variation of pH near its equilibrium value is well represented by a time$^{-1/2}$ function, lacks theoretical justification and is not self-consistent. If we use a similar extrapolation for other solution parameters such as the Ca or Mg concentrations, we obtain markedly different results for the estimated equilibrium state (LAFON, 1978). The pH extrapolation predicts an empirical order for calcite dissolution kinetics which is not compatible with that measured by PLUMMER and WIGLEY (1976). Although it leads to the correct value for the equilibrium of calcite with one atmosphere of CO_2 (GARRELS et al., 1960; PLUMMER and MACKENZIE, 1974; BISCHOFF et al., 1987), it also predicts a clearly erroneous stability for dolomite. In summary, extrapolation of solution pH against (reaction time)$^{-1/2}$ has no fundamental justification and should be considered unreliable.

Very recently, BUSENBERG and PLUMMER (1989) have used another experimental technique to inhibit the precipitation of unwanted low-magnesium phases. They added small concentrations of orthophosphate to initial dissolution runs of high-magnesium calcites and found that they could obtain reproducible congruent dissolution leading to well-defined steady-states and ion activity products. Orthophosphate irreversibly forms a strongly chemisorbed surface complex with Ca in calcite and similar minerals and it is not yet clear that the results of BUSENBERG and PLUMMER (1989) characterize the original magnesian calcites rather than the behavior of surfaces modified by their phosphate treatment.

For a magnesian calcite with a given composition, stability estimates based on pH extrapolations obtained by different investigators often conflict (compare results of PLUMMER and MACKENZIE, 1974; WALTER and MORSE, 1984; and BISCHOFF et al., 1987). These discrepancies have been ascribed in part to differing sample treatments such as damage caused by grinding (WALTER and MORSE, 1984). Another source of uncertainty derives from the fact that biogenic materials have compositions and crystal structures which are subtly different from inorganic magnesian calcites of the same magnesium content (BISCHOFF et al., 1983; BISCH-

OFF et al., 1985; BUSENBERG and PLUMMER, 1989). Finally, the lack of consistency among extrapolations of differing chemical parameters against (reaction time)$^{-1/2}$ suggests that results from different investigators may not be directly comparable.

This report describes an experimental technique which allows an aqueous solution to reach equilibrium with magnesian calcite from both under and supersaturation, yielding a reversed solubility. This technique avoids the precipitation of low-magnesium phases and provides some measure of control on metastable supersaturation states by carefully controlling the following three experimental parameters.

1. Use of low CO_2 fugacity (comparable to that of air) leads to relatively small departures from equilibrium with low-magnesium calcite and with the high-magnesium calcite of interest.

2. A stirring effect, which inhibits precipitation from slightly supersaturated solutions, allows the persistence of metastably supersaturated states for extended periods.

3. Creation of high-energy dissolution sites by experimental abrasion of the solids allows attaining supersaturated states via congruent dissolution.

The technique has been successfully used to measure the reversed equilibrium solubilities of several magnesian calcites and is described in detail. The solubility and free energy of formation of a high-magnesium calcite containing 11.6 mol percent $MgCO_3$ are reported.

EXPERIMENTAL METHODS

Materials

The magnesian calcite studied here consisted of interambulacral plates of the recent sea urchin *Sphaerechinus granularis* from the Gulf of Corinth (Greece) kindly supplied by Dr. D. K. Richter, Geologisches Institut, Ruhr-Universität Bochum. This material had previously been found to yield very sharp X-ray powder diffraction peaks, suggesting a single phase of well-defined composition. Sizeable fragments were gently broken by hand and treated with 30% H_2O_2 to remove reactive organic matter, then rinsed in distilled water and dried. No ultrasonication or annealing was necessary as the material had not been crushed or strained (contrast WALTER and MORSE, 1984 and BISCHOFF et al., 1987). Inspection by S.E.M. confirmed that no impurities were present and that there was no significant difference between the initial material and what remained after the experiment. X-ray powder diffraction spectra with a fluorite internal standard, obtained before and after the experiment, showed no significant change due to the reactions (Table 1). A structure refinement yielded hexagonal cell parameter values of: $a = 0.4947(2)$ nm and $c = 1.6850(7)$ nm indicating a composition between 10 and 13 mol percent $MgCO_3$ if we use the data and the calibration curves of BISCHOFF et al. (1983) for synthetic magnesian calcite. A similar com-

Table 1. *d*-spacings of crystallographic planes in *Sphaere-chinus granularis*. Filtered Cu Kα radiation, fluorite internal standard

hk.l	*d* before reaction	*d* after reaction	Average
10.4	0.3006 nm	0.3004 nm	0.3005 nm
11.0	0.2473	0.2473	0.2473
11.3	0.2266	0.2267	0.22665
	0.2260	0.2262	0.2261
20.2	0.2071	n.a.	0.2071
10.8	0.1889	0.1890	0.1889

$a_0 = 0.4947(2)$ nm $c_0 = 1.6850(7)$ nm

Estimated composition: 11.5 ± 1.5 percent $MgCO_3$.

position with greater uncertainty is indicated for biogenic samples. The krypton B.E.T. surface area was determined to be 0.11 m^2 g^{-1}.

The initial aqueous solution was deionized distilled water saturated with an analyzed commercial gas mixture certified to contain 377 parts per million by volume of CO_2 (complement N_2). The gas composition was checked by precision pH measurements in standard bicarbonate solutions equilibrated with the gas mixture, and was found to be correct within analytical uncertainty.

Experimental apparatus

Experiments were conducted in a jacketed 500 cm^3 Pyrex reaction vessel maintained at $25.0 \pm 0.1°C$ by a circulating Haake water bath and pump. Temperature was monitored with a precision mercury-in-glass thermometer. The vessel was tightly covered by a Plexiglass plate with holes for sampling, pH electrodes when necessary, thermometer and gas inlet and outlet. In preliminary experiments, it became apparent that stagnant solution in a standard glass frit bubbler could easily become supersaturated with calcite and lead to precipitation of an unwanted phase. To avoid this problem, gas was introduced by means of a Pasteur pipette where precipitation did not take place. Gas was presaturated with distilled water at reaction temperature before entering the reaction vessel to avoid excessive evaporation. To obtain accurate values of the CO_2 partial pressure, room pressure was measured with a precision mercury barometer before each sampling. The overpressure required to circulate the gas was measured with a water differential barometer and found equal to 391 Pa. The solution and solids were stirred with a magnetic stirring bar at a rate previously found sufficient to maintain equilibrium between gas and solution. This procedure resulted in minor abrasion of the solids through the course of the experiment.

Analytical techniques

A Radiometer research-grade combination pH electrode and pH meter were used for precision pH measurements at $25.0°C$. Because the desired sample size was small (typically 2 to 5 cm^3) and the anticipated concentrations low, calcium and magnesium analyses were obtained by atomic absorption spectrometry using an IL 151 spectrometer. Samples and standards were prepared to contain 1.20 g l^{-1} lanthanum to suppress ionization in the calcium anal-

yses. Standard calcium and magnesium chloride solutions were prepared gravimetrically by dissolving dry reagent grade calcium carbonate and magnesium metal, respectively, in dilute HCl. These standards were run after every three unknowns to correct for instrumental drift and assess analytical precision. All samples were run in duplicate or triplicate on separate dates. Precision and accuracy are estimated to be generally better than ± 2 percent, and always better than ± 3 percent.

After equilibrium had been reached from supersaturation, and at the conclusion of the experiment, 100 cm^3 aliquots of aqueous solution were analyzed for calcium and magnesium separately, using E.D.T.A. titrations at the appropriate pH. The indicators were Calcon for calcium and Eriochrome Black T for magnesium. Alkalinity was determined by a precision Gran titration. Results of these analyses agreed with the corresponding atomic absorption results within ± 1 percent.

Experimental procedure

Distilled deionized water was equilibrated at 25.0 $\pm 0.1°C$ with the previously described CO_2/N_2 gas mixture. The gas flow rate had previously been determined sufficient to maintain the system at equilibrium while carbonate dissolution took place. Once gas-liquid equilibrium was attained, as evidenced by negligible pH drift, several plate fragments of *Sphaerechinus granularis* (a total of 180 mg) were introduced. Dissolution was monitored by sampling and analyzing aliquots of the solution. Sampled volumes were replaced by equal volumes of distilled deionized water to prevent excessive decrease of the reacting solution volume. Therefore, the dissolution and precipitation data should not be used for a kinetic analysis, which was not the purpose of this study. Samples were analyzed shortly after being obtained and permitted preliminary identification and analysis of the reactions during the run. When it became clear, based on experience gained from similar runs, that the solution had become supersaturated with the magnesian calcite, stirring was stopped and precipitation was allowed to proceed until equilibrium was reached. Several aliquots were sampled over 8 days to establish an accurate estimate of solubility reached from supersaturation. Next, 200 cm^3 of solution were removed from the reaction vessel and replaced with an equal volume of distilled deionized water, ensuring that the system was now undersaturated with the magnesian calcite. Renewed dissolution was then monitored as above, but care was taken as equilibrium was approached to minimize stirring to the extent necessary to maintain gas-liquid equilibrium. This intermittent stirring then permitted the system to reach equilibrium from undersaturation, and several aliquots of solution were obtained to accurately define this condition.

EXPERIMENTAL RESULTS AND INTERPRETATION

THORSTENSON and PLUMMER (1977) introduced the concept of "stoichiometric saturation" to describe reactions between aqueous solution and a solid-solution mineral when the latter's composition is held strictly constant. Under this constraint, the only mass transfers allowed are congruent dissolution and congruent precipitation. "Stoichiometric saturation" defines one of the two possible equilibrium states between a solid-solution mineral and aqueous fluid (THORSTENSON and PLUMMER, 1977; LAFON, 1978) and, when demonstrated in practice, provides the only unambiguous experimental evidence for equilibrium between the phases.

The concept of "stoichiometric saturation" was originally formed to rationalize the interpretation of initial dissolution data sets, the "congruent dissolution stages" observed by PLUMMER and MACKENZIE (1974). It offered a theoretical framework for the use of empirical pH extrapolations and for the meaning of the free energy estimates derived from these extrapolations. The present work reports the first experimental verification of the validity of "stoichiometric saturation", covering all the congruent reactions between aqueous fluid and a biogenic magnesian calcite: dissolution, precipitation and equilibrium. To describe how experimental conditions vary with reaction time, it is convenient to show the variation of the relative super- or undersaturation with respect to the reacting material, expressed by the logarithm of the ratio of ion activity product to equilibrium constant (Fig. 1). In a first stage, we observe congruent dissolution of the magnesian calcite. Because this fragile material is abraded during the experiment, release of Ca and Mg to the fluid persists (presumably from high-energy damage sites) even at concentrations higher than those of the final equilibrium state, as shown by ion activity products greater than the equilibrium constant. In a second stage, we observe concomitant dissolution (at high-energy sites) and precipitation (at low energy sites), which can be manipulated to

reach equilibrium from supersaturation (Fig. 1a). Finally, renewed congruent dissolution following dilution permits reaching equilibrium again, this time from undersaturation (Fig. 1b). The three stages of this experiment are discussed below. Experimental analyses are reported in Appendix 1.

Congruent dissolution

The initial reaction of CO_2-charged water with the magnesian calcite leads to release of calcium and magnesium to the aqueous solution in a constant mol ratio. Data for the first twenty solution samples (representing a reaction time of about 118 hours) are shown in Fig. 2 and accurately define the composition of the dissolving phase to be: 11.6 mol percent $MgCO_3$. The dissolution line in Fig. 2 was obtained from a least squares fit of the calcium and magnesium concentrations forced to pass through the origin. The value of 11.6 percent is in good accord with the estimate of approximately 11.5 percent derived from the X-ray diffraction data, and with semi-quantitative E.D.S. analyses. For the aqueous sample with the highest Ca and Mg concentrations, more than 27 percent of the initial solid has been dissolved. Therefore, we are observing congruent dissolution of the bulk solid, one of the requirements of "stoichiometric saturation." Congruent dissolution has also been reported in other investigations (PLUMMER and MACKENZIE, 1974; WALTER and MORSE, 1984; BISCHOFF et al., 1987) but has always been restricted to the initial stage of the reaction because precipitation of low-magnesium calcite soon caused the overall mass transfer

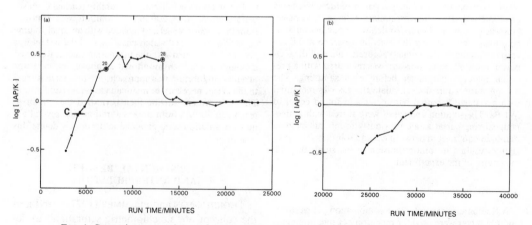

FIG. 1. Saturation state of the aqueous solution during the experiment. The saturation state is expressed by the logarithm of the ratio of ion activity product for magnesian calcite to the equilibrium constant; positive values denote supersaturation, negative values undersaturation and zero represents equilibrium. (a) Congruent dissolution and precipitation followed by equilibrium; C denotes equilibrium with respect to pure calcite; samples No. 20 and No. 28 are identified (see discussion in text). (b) Redissolution following dilution with distilled water.

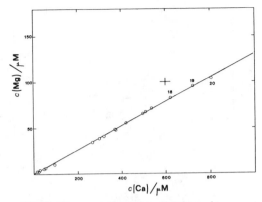

FIG. 2. Congruent dissolution of *Sphaerechinus granularis*. The dissolution line is a least squares fit to the concentration data forced to pass through the origin. Concentrations are micromolar at 22.5 ± 1°C. The error bars illustrate a ±3 percent error at typical concentrations near equilibrium.

congruent precipitation because growth on the pre-existing mineral surface can take place at low supersaturation. However, continued dissolution is still possible as a result of mechanical abrasion of the sea urchin plates. Continuous production of fractures probably creates high energy dissolution sites which permit release of calcium and magnesium to the solution even though its bulk composition is supersaturated with respect to the bulk solid.

Congruent precipitation and equilibrium solubility from supersaturation

To test the hypothesis that the aqueous solution was supersaturated with the magnesian calcite, stirring was stopped, which had two distinct consequences: first, it stopped creation of high-energy dissolution sites through abrasion and allowed precipitation to predominate; and second, it permitted precipitation to reach equilibrium because stirring is known to inhibit carbonate precipitation in slightly supersaturated Ca-Mg solutions (L. A. HARDIE, personal communication). Control of this effect has been used successfully in the past to obtain homogeneous partition coefficients between chloride brines and carbonate solid solutions (FÜCHTBAUER and HARDIE, 1976, 1980).

After withdrawing aliquot No. 28, stirring was stopped and the solution was allowed to reach equilibrium over a period of about 7 days (samples No. 29 to 38). Previous tests of this procedure had shown that, once stirring has stopped, precipitation occurs

to become incongruent. It is important to note that the calcium concentration in the present congruent dissolution process becomes significantly greater than that at equilibrium with pure calcite (condition denoted by point C on Fig. 1a) while there is no indication of the precipitation of this phase as long as the solution is sufficiently stirred. I interpret this observation as representing metastable supersaturation caused by Mg poisoning of low-magnesium calcite nuclei, and the absence of pre-existing low-energy growth sites for calcite such as could be provided by calcite seeds, together with precipitation inhibition due to a stirring effect. This interpretation is discussed below at greater length in a separate section.

Congruent dissolution and precipitation

After approximately 118 hours of reaction, the behavior of the solution chemistry changes (Fig. 1a and Fig. 3, samples No. 21 to 28). We now observe parallel increases and decreases of the calcium and magnesium concentrations which continue to obey the congruent reaction condition. It is particularly important to note that the calcium and magnesium concentrations cluster closely about the congruent dissolution line previously defined by the data in Fig. 2. This is evidence for concomitant and competing dissolution and precipitation of a composition identical to that of the bulk solid. While the variation of ion activity product *vs.* time for samples No. 21 to 28 could easily be interpreted as evidence for equilibrium, this would be a significant error as shown experimentally below. The solution is supersaturated with magnesian calcite, which permits

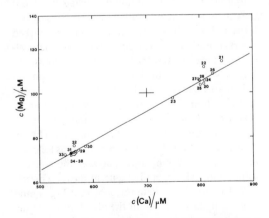

FIG. 3. Congruent dissolution and precipitation of $Mg_{0.116}Ca_{0.884}CO_3$. Note that the reaction line is taken from data in Fig. 1 and is completely independent of the data in this plot. Equilibrium from supersaturation is represented by samples No 34 through 38. Error bars represent ±3 percent. The scale is slightly different from that in Fig. 2.

over a time scale of minutes and slows to a very low rate after about 15 minutes. The additional reaction time served to ensure that the system would be near equilibrium. Carbonate precipitation releases dissolved CO_2 to the aqueous solution and upsets the previously established equilibrium between liquid and gas. In the absence of stirring, reestablishing this equilibrium requires an extended period of time because the transfer of CO_2 across the gas-liquid boundary is slow and inefficient. To ensure that equilibrium was reached, several liquid samples were obtained over a period of 171 hours. The equilibrium solubility reached from supersaturation is reported in Table 2 and represents the average of samples No. 34 through 38, all of which are very close to equilibrium (Fig. 1a). Note that the ion activity product one could derive from samples No. 21 to 28 (which could be thought to represent near equilibrium steady-state) is about three times greater (0.5 log units) than that of the actual equilibrium state. Accuracy of the equilibrium concentrations is estimated to be better than ±1 percent.

Redissolution and equilibrium from undersaturation

The preceding experimental procedure allowed attainment of equilibrium from supersaturation. To verify that this solubility was in fact correct, it was desirable to reverse the process and reach equilibrium from undersaturation. To this end, 200 cm³ of aqueous solution were removed from the reaction vessel and replaced with an equal volume of distilled deionized water. This dilution immediately reduced the calcium and magnesium concentrations by two fifths and brought the system to a state of marked undersaturation with respect to the magnesian calcite. To avoid creating fresh fractures and abrasions, stirring was then limited to the strict minimum required to maintain equilibrium between gaseous and dissolved CO_2, which resulted in markedly slower dissolution than at the beginning of the experiment (compare Fig. 1a and Fig. 1b). The renewed dissolution of magnesian calcite was followed as before by sampling and analyzing the aqueous solution for calcium and magnesium over a period of 224 hours. These data are shown in Fig. 4, and their tight clustering about the dissolution line defined at the beginning of the experiment again demonstrates congruent dissolution of a composition identical to that of the bulk solid. The reaction slowed considerably as equilibrium was approached. After successive samples no longer showed a monotonic increase in calcium and magnesium

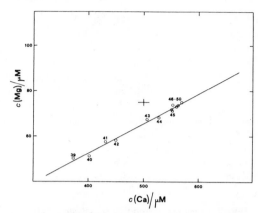

FIG. 4. Congruent redissolution of *Sphaerechinus granularis* and a precipitate of the same composition. Note that the reaction line is taken from data in Fig. 1 and is completely independent of the data in this plot. Equilibrium from undersaturation is represented by samples No. 46 through 50. Error bars represent ±3 percent. The scale is slightly different from those in Figs. 2 and 3.

concentrations, the equilibrium solubility from undersaturation was determined using the average of samples No. 46 through 50 (Table 2). This value is within analytical error of the solubility obtained from supersaturation.

Discussion of the present experimental technique and comparison with previous work

In marked contrast to previously reported dissolution experiments, this work has succeeded in experimentally maintaining "stoichiometric saturation" throughout, without modifying the magnesian calcite. Even though the aqueous solution becomes supersaturated first with pure calcite, then with the reacting magnesian calcite, all mass transfers between solution and solids remain congruent during dissolution, precipitation, and at equilibrium. Variation of the composition of the aqueous

Table 2. Solubility of *Sphaerechinus granularis*, a magnesian calcite containing 11.6 mol percent magnesium carbonate, at equilibrium with 36.81 Pa of CO_2

	[Ca]/μM	[Mg]/μM
From supersaturation, average of five samples	562	73.7
From undersaturation, average of five samples	559	73.5
Grand Average	560.5	73.6

Concentrations are micromolar at 22.5 ± 1°C, the laboratory temperature, slightly lower than the experimental temperature of 25.0°C.

solution defines three distinct experimental regimes, all consistent with congruent reaction of the bulk solid. First, the high-magnesium calcite dissolves, eventually yielding calcium and magnesium concentrations in excess of the equilibrium solubility of low-magnesium calcite and of that of the reacting solid. Second, once the aqueous solution has reached a large enough supersaturation (of the order of 3 for CO_2 contents similar to that of air), congruent precipitation begins and competes with dissolution. Third, stopping the stirring action lifts the inhibition to magnesian calcite precipitation and stops the formation of high-energy dissolution sites, which allows precipitation of magnesian calcite to proceed until equilibrium is reached. Equilibrium solubility is verified by letting the solid react with diluted solution while restricting stirring to the absolute minimum. Through appropriate manipulation of stirring and of CO_2 fugacity, equilibrium solubilities can be obtained from both supersaturation and undersaturation. Both determinations agree within better than 1 percent, which is within analytical error.

While the results presented above apply to one experiment and one biogenic magnesian calcite, I have successfully conducted over ten similar long-term experiments with a variety of materials. These results will be published elsewhere. The experimental techniques described here can be used systematically and reproducibly to control the saturation state of solutions. They yield reversed equilibrium solubilities for high-magnesium calcites without requiring modification of the carbonate surfaces and without relying on empirical extrapolations of partial dissolution kinetics. In contrast, BUSENBERG and PLUMMER (1985, 1989) had to expose their materials to relatively high concentrations of orthophosphate (100 μM) to suppress or retard the precipitation of unwanted phases, and this precluded precipitation of the magnesian calcites as well. Calcite surfaces exposed to such high concentrations of orthophosphate near equilibrium are covered by a complete monolayer of calcium phosphate and are not expected to behave identically to the original carbonate. I have discussed above the lack of foundation and the unreliability of empirical kinetic extrapolations against (time)$^{-1/2}$, used in earlier works by PLUMMER and MACKENZIE (1974), WALTER and MORSE (1984) and BISCHOFF et al. (1987). In summary, I have obtained for the first time the composition of an aqueous solution unequivocally at equilibrium with a high-magnesium calcite.

Some concern might be raised by the simultaneous occurrence of precipitation and dissolution

(R. J. SPENCER, personal communication). While the solution analytical data clearly demonstrate that the precipitated, dissolved, and original carbonate all have the same composition, it is conceivable if highly unlikely that the reversed solubility may apply to the freshly precipitated material only and not to Sphaerechinus granularis. The solubility of this material could be different from that of the sea urchin because of differing degrees of structural disorder and defect densities, as has been suggested by BISCHOFF et al. (1987). To address this issue, Fig. 5 shows S.E.M. photographs of the experimental material as received and after the end of the experiment. Under close inspection, the reacted material (Figs. 5.2 and 5.4) appears to consist solely of partially dissolved sea urchin plates, indicating that the freshly precipitated magnesian calcite was entirely redissolved in the last stage of the experiment. Therefore, the equilibrium reached from undersaturation applied strictly to the original biogenic material, while the equilibrium reached from supersaturation applied to a mixture of the biogenic material and of fresh precipitate. This indicates that there is no significant free energy difference between the two, an important observation developed further in the section on free energies of formation.

The new experimental technique introduced here has considerable potential for the thermodynamic characterization of high-magnesium calcites, and perhaps, more generally, for the study of other metastable ionic solid solutions. Other investigators should reproduce its results and verify its applicability, using a variety of single phase magnesian calcites. The three key experimental variables which permit control of the experiments are discussed below.

First, the concentration of dissolved CO_2 (equivalently, the CO_2 fugacity) plays a crucial role. Low concentrations of dissolved CO_2 (similar to that at equilibrium with air) allow solutions to remain supersaturated for long periods of time in the presence of slight surface poisoning effects, provided that the supersaturation does not exceed a threshold value. In the writer's experience, metastable supersaturation with respect to calcite and magnesian calcite can be maintained indefinitely at 25°C with air-like fugacities of CO_2 so long as the supersaturation remains below 2, corresponding to: log [IAP/K] ≤ 0.3. This observation has not been reported by other workers, which suggests that CO_2 fugacities of 0.32 kPa and greater probably accelerate the precipitation of carbonate minerals. Most earlier dissolution experiments using magnesian calcites have been conducted at fugacities of CO_2 markedly higher than that of air, probably because this choice

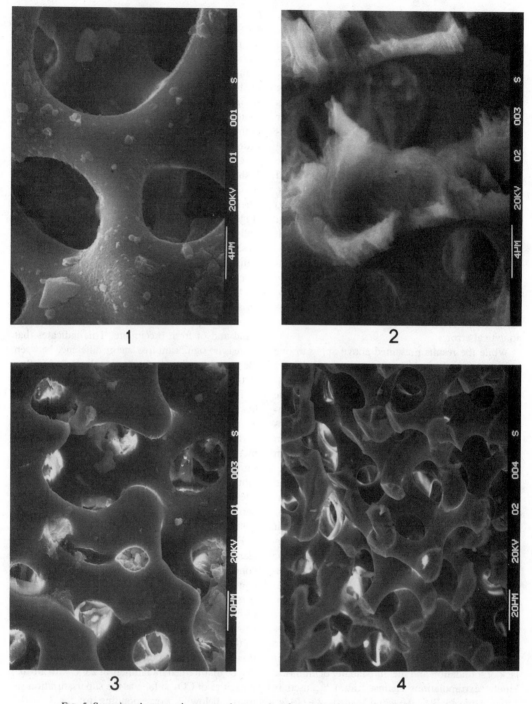

FIG. 5. Scanning electron microscope photographs of two plate fragments of *Sphaerechinus gran-ularis*. 1. unreacted material, plan view at 5,000× magnification; the small particles have the same composition as the bulk. 2. after reaction, plan view at 5,000× magnification; note the extensive exfoliation and dissolution, attributed to combined mechanical damage and normal dissolution. 3. side view of unreacted material at 2,000× magnification. 4. after reaction, side view at 1,000× mag-nification. Note that the reacted samples show no evidence of retaining freshly precipitated magnesian calcite.

also significantly accelerates dissolution and shortens experimental runs. It is likely that the common precipitation of low magnesium calcite after a short congruent dissolution phase is in part directly related to the use of high CO_2 fugacities.

Second, a proper match of substrate or seed composition to that of the potential precipitate is essential if one wants to obtain precipitation under the constraint of "stoichiometric saturation." I attempted to precipitate magnesian calcites congruently from bicarbonate solutions with a magnesium to calcium ratio of 0.10, using pure calcite as seeds. These attempts all failed, yielding only low magnesium calcite overgrowths in agreement with the observations of MUCCI and MORSE (1983). Growth of a different phase on pure calcite has to overcome an interfacial free energy barrier due to strains resulting from compositional and structural mismatch between two different lattices. Congruent precipitation seems to be greatly favored by the absence of this barrier and can presumably take place at lower supersaturation.

Third, control of stirring can be exploited to maintain the solution in a metastable state. Stirring performs three different functions in these experiments.

1. Stirring is necessary to equilibrate gas and aqueous solution faster than CO_2 is consumed or produced. This function is particularly important during the early part of the experiment, at low concentrations of bicarbonate, when relatively large amounts of CO_2 must be efficiently transferred from the gas to the liquid. Equilibration becomes much easier as saturation with the solids is approached, and can be effected with intermittent or lower energy stirring.

2. Stirring is also responsible for the mechanical abrasion of the solids. This effect has been documented for carbonate minerals by a number of investigators (R. A. BERNER, personal communication; MORSE and BERNER, 1972). Fig. 5 demonstrates the effects of abrasion on *Sphaerechinus granularis* fragments. Figs. 5.1 and 5.2 are views of the flat surface of sea urchin plate fragments before and after reaction, respectively. The flat surface which has been repeatedly exposed to shocks from the stirring bar has been the site of extensive dissolution (Fig. 5.2). By contrast, Figs. 5.3 and 5.4 are side views of the same plate fragments. While some dissolution can be observed in Fig. 5.4, it does not display the extensive exfoliation and fissuring of Fig. 5.2. These observations support the hypothesis that mechanical damage due to stirring enhances dissolution at some locations of the sample.

The amount of agitation necessary to maintain gas-liquid equilibrium readily leads to steady formation of damaged sites on crystals and, if not recognized, can lead to incorrect interpretation of experimental data. Spuriously high steady-state concentrations such as the ones observed in samples No. 21 to 28 of this work could easily have been interpreted as evidence for equilibrium. Empirical extrapolation of dissolution kinetics obtained while stirring is highly uncertain because there is no independent way to quantify abrasion damage. Discrepancies between results obtained by extrapolating pH *vs.* $(time)^{-1/2}$ may in part be attributable to these effects. On the other hand, a beneficial result of abrasion by stirring is that supersaturated states can be reached in a *dissolution* experiment, permitting reversal of reaction between solids and aqueous solutions.

3. Finally, stirring appears to inhibit homogeneous nucleation and growth in calcium-magnesium-bicarbonate-carbonate solutions. This phenomenon is what permits control of the experiment when solutions are supersaturated. FÜCHTBAUER and HARDIE (1976, 1980) exploited a similar effect to obtain reproducible homogeneous partition coefficients for magnesium between chloride solutions and magnesian calcites. Control of precipitation using stirring is crucial to obtaining reversed equilibrium solubilities of magnesian calcite and could perhaps also be used more generally in the study of other metastable ionic solid solutions. The experimental procedures described here are systematically reproducible with a variety of samples and can be widely used. However, it would be desirable to progress beyond a strictly empirical application of stirring control and to develop a more fundamental understanding of the interaction between fluid flow characteristics and precipitation kinetics. In the next section, I present a set of hypotheses which may serve as the sketch for a model of precipitation in calcium-magnesium-bicarbonate-carbonate solutions.

Toward a model for the precipitation mechanisms

A key observation is that precipitates from solutions containing calcium, magnesium, bicarbonate, carbonate and CO_2 can vary greatly according as the solution is stirred or not. This suggests that there are chemically significant mechanical interactions between ions and solute molecules or aggregates, and with water molecules. One systematic set of hypotheses explaining presently known facts follows.

1. In supersaturated solutions, aggregates or

protonuclei for potential precipitates form from cations and anions in an electrically balanced configuration, most likely as aggregates of $CaCO_3^0$ and $MgCO_3^0$ ion-pairs. They are hydrated, with two or more H_2O per cation (solvent-bridged ion-pairs) and the H_2O molecules are fairly strongly bound to Mg, less so to Ca. I assume the formation of ion-pairs and aggregates is rapid so that they are in approximate equilibrium with the monomeric ions; consequently, the Mg/Ca mol ratio of aggregates is equal to that of the bulk solution multiplied by the ratio of the association constants (about 0.5 at 25°C).

2. In many solutions of geochemical interest, such as seawater, the Mg/Ca ratio is high and the protonuclei also have a high Mg/Ca (for seawater, about 2.5). Potential precipitates with the same bulk composition are extremely unstable, much more so than commonly encountered natural solid-solutions. Precipitates that actually form will have a very much lower Mg/Ca ratio. This requires two important changes which act as kinetic barriers: loss of most of the magnesium so that the protonuclei reach a bulk composition which can be the precursor for a solid-solution, and loss of almost all the hydration water prior to growth of an anhydrous phase. Obvious modifications to the second requirement apply for precipitation of hydrous phases such as the hydrocalcites, hydromagnesite, nesquehonite, etc.

3. In the absence of strong mechanical interactions with their surroundings, free energy differences and statistical fluctuations will drive the compositions of aggregates toward much lower magnesium and H_2O contents, until the bulk composition of a nucleus is very similar to that of a potential precipitate (e.g. a magnesian calcite) and the nucleus can grow to form a macroscopic precipitate. However, this is probably a slow process (time scale on the order of milliseconds to seconds) as indicated by slow precipitation kinetics in unstirred solutions. For dolomite which requires the additional ordering of calcium and magnesium in separate planes, the time scale is of the order of days to months at higher than ambient temperatures.

4. In stirred solutions, hydrodynamic shear causes strong mechanical interactions between aggregates and water molecules. These encounters favor frequent splitting and erosion of protonuclei in addition to normal statistical fluctuations and their lifetime in stirred solutions is much shorter than in unstirred ones. While estimates of the lifetimes of aggregates are not available, I envisage that mechanical interactions in stirred solutions at low supersaturations make these lifetimes very short relative to the time necessary for cleaning the aggregates. As a result, aggregates split before they can reach a more stable composition and grow.

5. Crystal growth takes place on seeds and preexisting surfaces by adsorption of nuclei. If the seed has the same composition and structure as the potential precipitate, the activation energy is probably very small and the nucleus can consist of a small number of cations and anions. If, however, the potential precipitate has a different structure or composition, the interfacial mismatch and strain require a higher activation energy and a much larger nucleus. Inhibition of low-magnesium calcite growth in solutions containing significant magnesium ion concentrations is due to this effect.

This set of hypotheses appears to account for all the empirical observations which have served to develop the experimental technique used here. They have not yet been experimentally verified because the number and type of experiments required are beyond the scope of this research. Investigations would have to draw on modern techniques of physical chemistry and chemical physics. Some of the crucial parameters are the times required for magnesium loss and dehydration in aggregates. These times are likely to depend on the size and composition of an aggregate which are also highly time-dependent. Other crucial parameters are the size distribution of aggregates and their statistical evolution (composition and size) as function of time given a bulk supersaturation, temperature and concentration of CO_2. Experiments to quantify the properties of aggregates could measure the fluctuations and time dependence of variables such as the electrical conductivity, the viscosity and the ultrasound absorption of supersaturated stirred and unstirred solutions. In addition, the n.m.r. relaxation time of ^{13}C in the rotating frame ($T_{1\rho}$) could help directly to measure the number of C atoms in aggregates and relate it to their bulk concentration. Measurements of viscosity and conductivity as functions of shear would throw light on the relations between solution flow properties and precipitation mechanisms. Finally, it would be desirable to study precipitation kinetics in a rotating disc apparatus where flow properties are particularly well understood.

FREE ENERGY OF FORMATION OF MAGNESIAN CALCITE

The solubility obtained above can be used to calculate the standard free energy of formation for the test of *Sphaerechinus granularis*, a magnesian calcite containing 11.6 mol percent $MgCO_3$. This requires converting the measured concentrations into activities. There are two generally accepted ways of deriving free energies from solubility data. The first relies on building a chemical model where we choose *a priori* simple expressions for the activity coefficients of individual species such as ions and molecules, and we represent ion-ion interactions by assuming the presence of well-defined complex species such as ion-pairs (see, for example, PLUMMER and BUSENBERG, 1982). The other approach uses empirical models for the mean activity coefficients of major dissolved components, usually based on a leading Debye-Hückel term followed by a virial expansion (*e.g.*, ROBINSON and STOKES, 1965; PITZER and SILVESTER, 1976). In the present case, the major solutes are calcium and magnesium bicarbonate, and there are also very minor concentrations of dissolved calcium and magnesium carbonate and CO_2. While both approaches are self-consistent, they do not necessarily yield the same value for an activity product upon extrapolation to infinite dilution. For example, the activity product of calcium times carbonate for pure calcite *calculated from the same solubility data* is: 3.31×10^{-9} using the ion-pairing model (CHRIST *et al.*, 1974; PLUMMER and BUSENBERG, 1982), but 3.98×10^{-9} using the mean activity coefficient of $Ca(HCO_3)_2$ (LAFON, 1975; HARVIE, 1982; HARVIE *et al.*, 1984). The extrapolation using mean activity coefficients rests on fewer, simpler and more direct assumptions, which recommends its choice in this work. In the following, the mean activity coefficients of $Ca(HCO_3)_2$ and $Mg(HCO_3)_2$ are given by the expression:

$$\log \gamma_\pm = -2 \times 0.5094 \, \sqrt{I}/(1 + 1.5\sqrt{I})$$

$$+ 0.5536I \quad (1)$$

where I denotes the formal ionic strength of the solution on the molal scale. This expression fits the experimental solubilities of calcite in calcium bicarbonate solutions at 25°C up to an ionic strength of 0.03 molal with a standard deviation of residuals of 0.023 log units and affords a good estimate of the thermodynamic properties of dilute alkaline earth bicarbonate solutions (LAFON, 1975). For the standard state at 25°C and 100 kPa, it yields a solubility product for pure calcite of:

$$K_c = \gamma_\pm^3 [Ca][HCO_3]^2/(f_{CO_2}a_{H_2O})$$

$$= 1.27 \times 10^{-6}, \quad \text{or} \quad \log K_c = -5.894,$$

corresponding to a standard free energy of formation of $-1,129.27$ kJ mol^{-1}.

The average solubility of $Mg_{0.116}Ca_{0.884}CO_3$ at equilibrium with 36.81 Pa of CO_2 is 560.5 μM calcium and 73.6 μM magnesium (Table 2), corresponding to 561.8 and 73.8 μm on the molal scale, respectively. To obtain an activity product, we can write the reaction as:

$$Mg_{0.116}Ca_{0.884}CO_3 + CO_{2\,(gas)} + H_2O \leftrightharpoons$$

$$0.116 \, Mg^{2+} + 0.884 \, Ca^{2+} + 2 \, HCO_3^-, \quad (2)$$

with the equilibrium constant:

$$K_{mc} = \gamma_\pm^3 [HCO_3]^2 [Mg]^{0.116}[Ca]^{0.884}/(f_{CO_2}a_{H_2O}), \quad (3)$$

where brackets denote molal concentrations, f the fugacity and a the activity of the subscripted species. To obtain the concentrations of the bicarbonate components, the solubilities reported above must be corrected for the presence of trace amounts of carbonate ion and of the $CaCO_3^0$ and $MgCO_3^0$ ion-pairs. Calculations are based on thermochemical properties for CO_2, HCO_3^- and CO_3^{2-} reported by WAGMAN *et al.* (1982) and on the association constants determined by SIEBERT and HOSTETLER (1977) for $MgCO_3^0$ and by PLUMMER and BUSENBERG (1982) for $CaCO_3^0$. At the equilibrium pH of 8.35, the concentration of carbonate is 15 μm, that of $CaCO_3^0$ 8.9 μm and that of $MgCO_3^0$ 0.56 μm. The equilibrium bicarbonate concentrations are 540 μm for $Ca(HCO_3)_2$ and 71.4 μm for $Mg(HCO_3)_2$. Substituting these values into Equation (3), we obtain:

$$K_{mc} = 0.7564 \times 0.001223^2 \times 0.000540^{0.884}$$

$$\times 0.0000714^{0.116}/0.0003681$$

$$= 1.31 \times 10^{-6},$$

and $\log K_{mc} = -5.881$ for the standard state at 25°C and 100 kPa.

The corresponding value for the standard free energy of formation of $Mg_{0.116}Ca_{0.884}CO_3$ is $\Delta G_f^0 = -1\,117.75$ kJ mol^{-1}. An error analysis presented in Appendix 2 estimates the uncertainty on this value at: ± 0.10 kJ mol^{-1}, relative to the free energy of formation of pure calcite reported above ($-1,129.27$ kJ mol^{-1}).

Comparison with other free energy estimates

The free energies of formation of magnesian calcites are significantly different from that of pure calcite. Most investigators have assumed that this effect is primarily controlled by the magnesium content and that other compositional or structural factors are secondary, if at all significant. Recently, however, BISCHOFF *et al.* (1987) and BUSENBERG and PLUMMER (1985, 1989) have drawn attention to the potential roles of structural disorder and of trace concentrations of sulfate and sodium. BUSEN-BERG and PLUMMER (1985) found that trace concentrations of Na and SO_4 greatly decrease the stability of synthetic calcites. They correlated the levels of impurity concentrations in experimental precipitates with their growth rate from salt solutions containing sodium and sulfate. BISCHOFF et al. (1987) reported that magnesian calcites synthesized from $CaCO_3$ and $MgCO_3$ are more stable than biogenic materials with the same magnesium concentration. They attributed this difference to the greater structural, physical and chemical heterogeneity of biogenic samples but declined to assign the free energy differences (0.20 to 0.85 kJ mol^{-1}) to particular characteristics. BUSENBERG and PLUMMER (1989) now make a major distinction between "Group II" magnesian calcites which have significant defect densities due to trace concentrations of Na, SO_4 and vacancies and the simpler "Group I" magnesian calcites. They report two markedly different curves for the variation of apparent ion activity product versus magnesium mol fraction, with "Group II" magnesian calcites much more soluble than "Group I" materials.

Explicit consideration of components other than $CaCO_3$ and $MgCO_3$ raises new and important issues in the interpretation of experimental dissolution data and the relations between ion activity products and the stability of magnesian calcite. While a complete discussion of the thermodynamic consequences is beyond the scope of this paper, the following points are worth noting.

We can choose to consider the congruent dissolution properties of the whole solid, including sodium and sulfate in the ion activity product and using the concept of "stoichiometric saturation" to derive free energies of formation, if we assume that steady-state properties represent near-equilibrium. However, this approach has two major drawbacks. First, dissolution experiments have not led to reversible transfer of sodium and sulfate between solids and liquid because Na and SO_4 are strongly partitioned into the aqueous solution (BUSENBERG and PLUMMER, 1985 and references therein). All experimental precipitates in dissolution experiments to-date have had insignificant concentrations of sodium and sulfate. Interpretation of steady-states reported by PLUMMER and BUSENBERG (1989) as representing "stoichiometric saturation" appears unwarranted so long as such an asymmetry between dissolution and precipitation exists. Second, the free energies of formation of a complex solid-solution series with at least four end-members cannot easily be used to extract mixing properties for the $CaCO_3$ and $MgCO_3$ components, or to evaluate the roles

of the sodium and sulfate-bearing components relatively to that of $MgCO_3$. To obtain free energies for the components, we have to know the dependence of the total free energy of formation on component concentrations accurately enough that partial derivatives can be calculated at the composition of interest (DARKEN and GURRY, 1953). This requires that, for each mol fraction of $MgCO_3$, we need the variation of free energy with respect to sodium and sulfate mol fractions, separately. Such data are not available and are not likely to be collected for a considerable time. In summary, if it were necessary to incorporate sodium and sulfate in ion activity products and in free energies of formation, we would still be far from understanding the chemical properties of magnesian calcites.

We can, instead, focus our attention on the ion activity product and the free energy of formation of the idealized magnesian calcite formula with the two components $CaCO_3$ and $MgCO_3$ only. This is the approach followed by BUSENBERG and PLUMMER (1989) in their discussion of mixing properties and by all previous investigators. The reversed equilibrium demonstrated in the present work provides strong support for using the two-component magnesian calcite formula. The precipitated material must have contained negligible amounts of sodium and sulfate because the concentrations of these impurities in the liquid were very low. Taking as representative the sea urchin analyses reported by BUSENBERG and PLUMMER (1989), I estimate that the solution from which the precipitate formed contained at most about 22 μM sodium and 9 μM sulfate. Such very low concentrations of impurities could not have produced a solid carbonate with significant defects. Because solubilities reached from undersaturation and supersaturation are identical, this work experimentally demonstrates that trace amounts of sodium and sulfate in Sphaerechinus granularis did not have a significant effect on its solubility.

If we use the two-component approach to interpret experimental reaction data for magnesian calcites, we are also implicitly stating that the effects of impurities and defects can be separated from the mixing properties of the calcium and magnesium carbonate components. Instead of envisaging a complex multicomponent solid-solution, we represent magnesian calcites as groups of pseudo-binary solutions and let the properties of the end-members account for other components such as sodium, sulfate, vacancies and defects. This approach was taken by BUSENBERG and PLUMMER (1989) who fitted their ion activity products to the $CaCO_3$-$Ca_{0.5}Mg_{0.5}CO_3$ idealized solid-solution. They used

Table 3. Free energies of formation (in kJ mol^{-1}) for magnesian calcites with compositions near $Ca_{0.884}Mg_{0.116}CO_3$

Source	Material	$MgCO_3$ mol fraction	ΔG_f^0	G_{mc}
This work	*Sphaerechinus granularis*	0.116	−1117.75	−1129.34
PLUMMER and MACKENZIE	?	0.069	−1121.98	−1128.88
(1974)	?	0.104	−1119.06	−1129.46
	?	0.127	−1113.49	−1126.19
WALTER and MORSE	*Clypeaster*	0.12	−1116.9	−1128.9
	Tripneustes			
BISCHOFF *et al.* (1987)	Synthetic samples	0.08	−1121.8	−1129.8
		0.10	−1119.6	−1129.6
		0.101	−1119.6	−1129.7
		0.125	−1117.0	−1129.5
	Diadema (spine)	0.07	−1122.3	−1129.3
	Tripneustes	0.109	−1118.0	−1128.9
	Homotrema	0.124	−1116.9	−1129.3
	Diadema (plate)	0.124	−1116.1	−1128.5
	Diadema (plate)	0.128	−1115.7	−1128.5
BUSENBERG and PLUMMER	Synthetic samples	0.0753	−1121.7	−1129.2
(1989)		0.1068	−1118.2	−1128.8
		0.1467	−1113.8	−1128.5
		0.0900	−1119.9	−1128.9
	Biogenic samples	0.123	−1116.5	−1128.8
		0.108	−1118.0	−1128.8
		0.112	−1116.3	−1127.5
		0.109	−1117.6	−1128.5
		0.129	−1115.9	−1128.8
		0.127	−1118.5	−1131.2
		0.139	−1114.6	−1128.5

Notes:
1. The dissolving phases of PLUMMER and MACKENZIE (1974) had compositions markedly different from the bulk solids and probably do not represent well-defined minerals, but rather mixtures.
2. $G_{mc} = \Delta G_f^0 - 100x$, where x denotes the mol fraction of $MgCO_3$.

for the Mg-free end member either pure calcite ("Group I" materials) or a defect-rich, sodium and sulfate bearing calcite ("Group II" materials). In the following discussion of free energies of formation, I follow the binary solid-solution model and assume that the magnesium carbonate concentration is the principal controlling factor.

The standard free energy of formation of magnesian calcite has a steep upward trend with increasing magnesium carbonate mol fraction, which can mask small variations over a limited composition range. While this effect would be best compensated by using free energies of mixing, present uncertainties in the standard free energy of formation of magnesite make this approach inadvisable. A similar result can be obtained if we use an empirical correction to flatten the compositional trend. I use the quantity: $G_{mc} = \Delta G_f^0 - 100x$ (kJ mol^{-1}), where x denotes the mol fraction of magnesium carbonate in the formula $Ca_{1-x}Mg_xCO_3$, to compare estimates from other investigators to the free energy measured in this study. The mol fractions considered here range from 0.06 to 0.15, bracketing the composition of *Sphaerechinus granularis*.

The original free energy estimates of PLUMMER and MACKENZIE (1974), WALTER and MORSE (1984), BISCHOFF *et al.* (1987) and values computed from the ion activity products of BUSENBERG and PLUMMER (1989) are listed in Table 3. The values reported by PLUMMER and MACKENZIE (1974) are probably in error because the compositions interpreted as dissolving congruently were markedly different from the bulk compositions and it is likely that several distinct phases were dissolving simultaneously (LAFON, 1978), which would have made the pH extrapolations invalid. Data from the other references are plotted in Fig. 6. The results for the synthetic samples of BISCHOFF *et al.* (1987) define a consistent trend of lower free energies. The estimates of WALTER and MORSE (1984), BISCHOFF *et al.* (1987) and BUSENBERG and PLUMMER (1989) for impure and/or biogenic samples are much more widely scattered and are, on average, about 0.6 to 0.8 kJ mol^{-1} less negative. These two trends correspond to the "Group I" and "Group II" classification of BUSENBERG and PLUMMER (1989). The result of this study is intermediate between the two general trends, and somewhat closer to the estimates

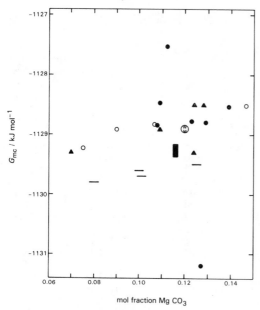

FIG. 6. Free energy of magnesian calcites with compositions near 11.6 mol percent $MgCO_3$ plotted against the mol fraction of magnesium carbonate. The quantity; $G_{mc} = \Delta G_f^0 - 100x$ is used to suppress the upward trend of the free energy of formation with increasing mol concentration of $MgCO_3$ (see text). The filled box illustrates results from this work. Other sources are: Ⓦ estimate from WALTER and MORSE (1984); (—) estimates for synthetic samples (BISCHOFF et al., 1987), and (▲) (filled triangles) estimates for biogenic samples (BISCHOFF et al., 1987); (○) (open circles) synthetic samples from BUSENBERG and PLUMMER (1989), and (●) (filled circles) biogenic samples from BUSENBERG and PLUMMER (1989).

of BISCHOFF et al. (1987) for their synthetic, more perfect samples.

The clustering of estimated free energies of magnesian calcites along two trends has been interpreted by BUSENBERG and PLUMMER (1989) as characterizing two distinct families of materials. Their "Group I" consists of the simpler magnesian calcites which approach the ideal binary $CaCO_3$-$MgCO_3$ solid-solution with pure calcite as the Mg-free end-member. "Group II" comprises the more complex, more impure materials, with significant concentrations of sodium and sulfate, structural defects and disorder. The Mg-free end-member of these samples would be an impure, defect-laden calcite with a much less negative free energy of formation. A similar but less far-reaching proposal was advanced by BISCHOFF et al. (1987) to explain the systematic difference between their estimates for synthetic and biogenic samples. Given the large uncertainties of kinetic extrapolations and the poor understanding of the effects of phosphate poisoning,

together with the considerable scatter in Fig. 6, a key question remains unanswered. Does the small but systematic difference between the average free energy trends of "Group I" and "Group II" reflect a genuine difference between the true free energies of formation of these materials? Or does it, instead, reflect how structural and compositional variations modify dissolution kinetics, leading to systematically different extrapolated values but not to reliable differences between the equilibrium free energies?

Distinguishing between these alternatives is crucial for reliable equilibrium modeling of magnesian calcites in nature. While the proposed free energy difference between "Group I" and "Group II" materials is numerically small (a fraction of one kJ mol^{-1}), it leads to large differences between the calculated activities of the $MgCO_3$ component in the solids because these activities are related to the *derivative* of total free energy with respect to the $MgCO_3$ mol fraction (THORSTENSON and PLUMMER, 1977; BUSENBERG and PLUMMER, 1989). They control important geological parameters such as the equilibrium magnesium-to-calcium ratio of fluids and the relative stabilities of different carbonate mineral phases. As was well shown by BUSENBERG and PLUMMER (1989), their "Group I" and "Group II" magnesian calcites must have markedly different stability ranges and phase relationships. If, however, the two apparent trends of free energies in Fig. 6 correspond to the effects of impurities on dissolution kinetics but not to stability differences, little can be stated at this time regarding equilibrium phase relations because the solubility reported in this work is the only one that unambiguously pertains to a documented equilibrium condition. In this case, previously reported free energy estimates for magnesian calcites would be of little value, serving at best to illustrate the effects of composition, structure and defects on dissolution kinetics.

Because the free energy differences between "Group I" and "Group II" materials are very small, we must have great confidence in extrapolations of dissolution kinetics or in the close approach of steady states to true equilibrium states before we can trust that estimates will resolve two truly distinct populations of magnesian calcites. To the contrary, a strong case can be made that the small differences between the estimated stabilities of "Group I" and "Group II" materials cannot be reliably derived from kinetic extrapolations and from reactions inhibited by phosphate. The free energy estimates obtained from these experiments are extremely sensitive to minor sample variations and differences in sample treatment. WALTER and MORSE (1984) experimentally demonstrated that cleaning and a mild temperature treatment of their biogenic materials led to marked changes in the extrapolated equilibrium pH. Estimated free energies of "Group II" materials obtained from both pH extrapolations of dissolution kinetics and steady-states with phosphate poisoning are widely scattered at a given mol fraction of $MgCO_3$ (see Fig. 6). Both experimental techniques yield data that show poor consistency within a given study and between different investigators. Several inconsistencies remain to be resolved and the following list is not exhaustive.

1. "Group II" does not correspond to a well-defined population of materials because of wide variability in sodium, sulfate and defect concentrations. Although the mol fractions of Na and SO_4 in "Group II" magnesian calcites are small, their ranges from less than 0.5 percent to nearly 3 percent are large relative to their mean values and clearly show that they cannot be taken as constant. This observation is hardly compatible with defining a pseudo-binary solid-solution and its end-member. We would have to believe that the presence of sodium and sulfate in this concentration range is enough to fix a free energy of formation for the Mg-free end member which does not depend significantly on the concentrations of impurities. In contrast, BUSENBERG and PLUMMER (1985) have reported that increasing the mol fractions of Na or SO_4 in calcite from 0.5 to 3 percent increases the apparent ion activity product by about 0.1 log unit, which is comparable to the difference between "Group I" and "Group II" materials.

2. Some of the data presented by BUSENBERG and PLUMMER (1989) in support of their interpretation can also support alternative models. For example, steady-state ion activity products for synthetic sodium-containing magnesian calcites (their Table 3) help define the stabilities of "Group II" materials. The same data converted to free energies of formation extrapolate naturally to pure calcite as the Mg-free end member. This observation once again suggests that the distinction between "Group I" and "Group II" may be more apparent than real.

3. Many of the materials used in kinetic extrapolations or phosphate-poisoned dissolution were in fact composites, not single phases. BUSENBERG and PLUMMER (1989) grew their synthetic samples from calcite seeds which constituted a non-negligible fraction of the final products. BISCHOFF et al. (1987) report that their synthetic, more perfect samples had bulk compositions slightly different from the dissolving ones. Most biogenic materials, unless carefully selected, have variable compositions on a microscopic scale (MOBERLY, 1968) and consist of a mixture of different minerals or mineral domains. We do not expect that the dissolution behavior of a composite sample can be meaningfully extrapolated to an accurate equilibrium state.

Structural disorder, impurities and compositional heterogeneities are well documented in biogenic magnesian calcites and in samples rapidly precipitated from seawater. These lattice imperfections probably modify dissolution kinetics much more markedly than equilibrium solubilities. At the present time, almost all free energy estimates for magnesian calcites are based on empirical extrapolations of dissolution kinetics or steady states obtained in strongly inhibited conditions. The clustering of these estimates and differences between groups of materials are likely to reflect kinetic variability rather than thermochemistry. Existing data are clearly not accurate enough to decide whether classification of magnesian calcites in "Group I" and "Group II" corresponds to significantly different solubilities and phase relations. We must obtain more reversed equilibrium solubilities or data of comparable quality before we can fruitfully model the mixing thermodynamics and the phase relations of magnesian calcites.

CONCLUSIONS

THORSTENSON and PLUMMER (1977) introduced the concept of "stoichiometric saturation" to describe the reactions between a solid solution and aqueous fluid under the added constraint that the composition of the solid must remain constant. Mass transfers between solid and fluid then reflect only congruent dissolution and precipitation. This work reports on a new experimental technique which allows magnesian calcite solid solutions to react with aqueous solution under the constraint of "stoichiometric saturation". Application of this technique to a biogenic magnesian calcite containing 11.6 mol percent $MgCO_3$ yielded a reversed equilibrium solubility which was used to derive an accurate free energy of formation. The free energy of formation for this composition is: $-1,117.75$ kJ mol^{-1}.

The technique used in this work exploits the long-term persistence of solution compositions which are metastable relatively to the lower-energy equilibrium states. It requires using relatively low fugacities of CO_2, comparable to that of air, and low supersaturations. In addition, it relies heavily on the fact that a stirred solution slightly supersaturated with respect to magnesian calcite will not precipitate this solid over a laboratory time scale. It also exploits the continuous creation of high energy dissolution sites by abrasion caused by mechanical stirring. Finally, it relies on the fact that precipitation on a seed with the same composition has less of a kinetic barrier than precipitation on a different substrate such as low-magnesium calcite. These observations give the experimenter a fair degree of control over dissolution and precipitation of magnesian calcites and suggest a new conceptual model to interpret precipitation observations in supersaturated calcium-magnesium-bicarbonate-carbonate solutions.

Many of the previously reported stabilities for magnesian calcites are based on empirical extrapolations of pH against reaction time during the initial dissolution stage. This procedure approximates the true reaction kinetics over a small time span only and appears to be very sensitive to the presence of minor defects in the dissolving solids. Other stability estimates recently reported by BUSENBERG and PLUMMER (1989) correspond to apparent dissolution steady-states in the presence of orthophosphate inhibition. Large discrepancies between the estimated stabilities of biogenic samples with comparable compositions probably reflect differences between dissolution kinetics rather than true differences between their free energies of formation. While estimates of free energies based on kinetic extrapolations may provide good first approximations, they are not accurate enough to permit calculation of the activities of the $CaCO_3$ and $MgCO_3$ components, to obtain reliable phase relationships, or to distinguish between the stabilities of different materials such as the "Group I" and "Group II" proposed by BUSENBERG and PLUMMER (1989).

The free energy reported here for $Ca_{0.884}Mg_{0.116}CO_3$ is the first that corresponds to a documented, reversed equilibrium state. Additional data of comparable quality are required before thermodynamic modeling of the pseudo-binary magnesian calcite solid solution can yield useful analogs to natural carbonate assemblages. Additional experiments have been conducted on single-phase biogenic magnesian calcites with a range of compositions. Free energies of formation have been derived from these equilibrium solubilities and will be reported elsewhere. These new data improve our understanding of the saturation states of sea water with respect to aragonite, magnesian calcite and low-magnesium calcite.

Acknowledgements—Professor H. Füchtbauer and Dr. D. K. Richter aroused my interest in measuring the solubilities of magnesian calcites during a stay in the Geologisches Institut, Ruhr-Universität Bochum. Dr. Richter kindly supplied the specimen used in this study. I thank V. E. Grant and A. Eaton for instruction in atomic absorption techniques, Ralph Hockett for S.E.M. photographs and A. M. Bishop for the surface area analysis. I discussed many aspects of this research with Hans Eugster over a period of years and his influence is apparent in this paper. The experimental work was carried out while I was in the Department of Earth and Planetary Sciences, The Johns Hopkins University. I thank A. H. Thompson for discussions on the physics of the proposed precipitation mechanisms. Finally, I thank Exxon Production Research Co. for permission to publish this material. Critical reviews and comments by J. J. Zullig, W. D. Gunter, E. A. Burton, D. R. Pevear and R. J. Spencer led to significant improvement of the original manuscript.

REFERENCES

BATHURST R. G. C. (1976) Carbonate Sediments and Their Diagenesis. 2nd ed., Developments in Sedimentology No. 12, xix + 658p. Elsevier.

BISCHOFF W. D., BISHOP F. C. and MACKENZIE F. T. (1983) Biogenically produced magnesian calcite: inhomogeneities in chemical and physical properties; comparison with synthetic phases. *Amer. Mineral.* **68**, 1183–1188.

BISCHOFF W. D., SHARMA S. K. and MACKENZIE F. T. (1985) Carbonate ion disorder in synthetic and biogenic magnesian calcites: a Raman spectral study. *Amer. Mineral.* **70**, 581–589.

BISCHOFF W. D., MACKENZIE F. T. and BISHOP F. C. (1987) Stabilities of synthetic magnesian calcites in aqueous solution: Comparison with biogenic materials. *Geochim. Cosmochim. Acta* **51**, 1413–1423.

BUSENBERG E. and PLUMMER L. N. (1985) Kinetic and thermodynamic factors controlling the distribution of SO_4^{2-} and Na^+ in calcites and selected aragonites. *Geochim. Cosmochim. Acta* **49**, 713–725.

BUSENBERG E. and PLUMMER L. N. (1989) Thermodynamics of magnesian calcite solid-solutions at 25°C and 1 atm total pressure. *Geochim. Cosmochim. Acta* **53**, 1189–1208.

CHAVE K. E., DEFFEYES K. S., WEYL P. K., GARRELS R. M. and THOMPSON M. E. (1962) Observations on the Solubility of Skeletal Carbonates in Aqueous Solutions. *Science* **137**, 33–34.

CHAVE K. E. and SCHMALZ R. F. (1966) Carbonate-seawater interactions. *Geochim. Cosmochim. Acta* **30**, 1037–1048.

CHRIST C. L., HOSTETLER C. B. and SIEBERT R. M. (1974) Stabilities of calcite and aragonite. *J. Res. U.S. Geol. Surv.* **2**, 175–184.

DARKEN L. S. and GURRY R. W. (1953) *Physical Chemistry of Metals.* vii + 535p. McGraw-Hill.

FÜCHTBAUER H. and HARDIE L. A. (1976) Experimentally determined homogeneous distribution coefficients for precipitated magnesian calcites: Applications to marine carbonate cements (abstr.). *Geol. Soc. Amer. Annual Meeting* p. 877.

FÜCHTBAUER H. and HARDIE L. A. (1980) Comparison of experimental and natural magnesian calcites. *Internat. Assoc. Sedimentologists,* First European Meeting, Bochum 1980, 167–169.

GARRELS R. M., THOMPSON M. E. and SIEVER R. (1960) Stability of some carbonates at 25°C and one atmosphere total pressure. *Amer. J. Sci.* **258**, 402–418.

HARVIE C. E. (1982) Theoretical Investigation in geochemistry and atom surface scattering. Unpublished Ph.D. Dissertation, University of California at San Diego, La Jolla, CA.

HARVIE C. E., MØLLER N. and WEARE J. H. (1984) The prediction of mineral solubilities in natural waters: The Na-K-Mg-Ca-H-Cl-SO$_4$-OH-CO$_3$-CO$_2$-H$_2$O system to high ionic strengths at 25°C. *Geochim. Cosmochim. Acta* **48**, 723–751.

LAFON G. M. (1975) The solubility of calcite and the calculation of chemical potentials in aqueous solutions (abstr.). *Geol. Soc. Amer. Annual Meeting*, **7**, 1157.

LAFON G. M. (1978) Discussion of: Equilibrium criteria for two component solids reacting with fixed composition in an aqueous phase—example: the magnesian calcites. *Amer. J. Sci.* **278**, 1455–1468.

LERMAN A. (1965) Paleoecology problems of Mg and Sr in biogenic calcites in light of recent thermodynamic data. *Geochim. Cosmochim. Acta* **29**, 977–1002.

MOBERLY R., JR. (1968) Comparison of magnesian calcites of algae and pelecypods by electron microprobe analysis. *Sedimentology* **11**, 61–82.

MORSE J. W. and BERNER R. A. (1972) Dissolution kinetics of calcium carbonate in sea water. II A kinetic origin for the lysocline. *Amer. J. Sci.* **272**, 840–851.

MUCCI A. and MORSE J. W. (1983) The incorporation of Mg^{2+} and Sr^{2+} into calcite overgrowths: influences of growth rate and solution composition. *Geochim. Cosmochim. Acta* **47**, 217–233.

PITZER K. S. and SILVESTER L. F. (1976) Thermodynamics of electrolytes. VI. Weak electrolytes including H$_3$PO$_4$. *J. Solution Chem.* **5**, 269–278.

PLUMMER L. N. and BUSENBERG E. (1982) The solubility of calcite, aragonite and vaterite in CO$_2$-H$_2$O solutions between 0 and 90°C, and the evaluation of the aqueous model for the system CaCO$_3$-CO$_2$-H$_2$O. *Geochim. Cosmochim. Acta* **46**, 1011–1040.

PLUMMER L. N. and MACKENZIE F. T. (1974) Predicting mineral solubility from rate data: application to the dissolution of magnesian calcite. *Amer. J. Sci.* **274**, 61–83.

PLUMMER L. N. and WIGLEY T. M. L. (1976) The dissolution of calcite in CO$_2$-saturated solutions at 25°C and 1 atmosphere total pressure. *Geochim. Cosmochim. Acta* **40**, 191–202.

ROBINSON R. A. and STOKES R. H. (1965) *Electrolyte Solutions*. 2nd ed. (Revised), Butterworths, London.

SIEBERT R. M. and HOSTETLER P. B. (1977) The stability of the magnesium carbonate ion-pair from 10 to 90°C. *Amer. J. Sci.* **277**, 716–734.

THORSTENSON D. C. and PLUMMER L. N. (1977) Equilibrium criteria for two component solids reacting with fixed composition in an aqueous phase—example: the magnesian calcites. *Amer. J. Sci.* **277**, 1203–1223.

WAGMAN D. D. and seven others (1982) The NBS Tables of chemical thermodynamic properties. *J. Phys. Chem. Ref. Data* **11**, Suppl. 2.

WALTER L. M. and MORSE J. W. (1984) Magnesian calcite stabilities: a reevaluation. *Geochim. Cosmochim. Acta* **48**, 1059–1070.

WINLAND H. D. (1969) Stability of calcium carbonate polymorphs in warm, shallow seawater. *J. Sediment. Petrol.* **39**, 1579–1587.

APPENDIX 1
SOLUTION ANALYTICAL DATA

Concentrations are micromolar at 22.5 ± 1°C (the laboratory temperature). The CO$_2$ fugacity in Pa is given by:
$$f = 377 (P + 0.391 - 3.126)/1000.$$

Sample No.	Run time minutes	[Ca]/μM	[Mg]/μM	Pressure kPa	Sample No.	Run time minutes	[Ca]/μM	[Mg]/μM	Pressure kPa
1	15	13.7	2.3		26	11,208	825.	108.5	100.10
2	30	21.8	2.6		27	12,703	797.	106.	101.76
3	60	28.2	4.2		28	13,136	799.	105.7	101.23
4	121	44.9	5.44		29	14,277	573.	74.5	99.69
5	180	51.7	6.15		30	14,737	586.	76.3	100.55
6	354	95.4	10.0		31	15,731	559.	73.4	101.24
7	546	140.	14.6		32	17,232	565.	76.7	102.00
8	1,110	244.	28.7		33	18,865	548.	72.6	101.41
9	1,315	266.	34.6		34	19,985	567.	74.0	101.20
10	1,642	298.	38.7		35	20,692	565.5	73.55	101.05
11	1,904	319.	41.1		36	21,751	568.5	74.25	99.93
12	2,591	373.	48.0		37	22,800	558.5	73.0	100.25
13	2,747	370.	48.6	99.20	38	23,397	561.	73.2	
14	3,388	420.	55.7	99.64	39	24,189	373.5	50.2	101.00
15	4,010	498.5	65.7	100.18	40	24,635	402.	51.35	100.80
16	4,277	510.5	67.7	100.04	41	25,750	431.	57.6	100.65
17	4,780	537.	71.35	100.10	42	27,311	450.	58.3	99.65
18	5,590	624.	82.9	99.87	43	28,521	506.	67.5	100.15
19	6,345	724.	95.2	100.00	44	29,356	527.5	68.3	100.67
20	7,054	807.	104.	100.42	45	30,006	552.	71.2	100.55
21	8,324	841.5	114.	99.97	46	30,788	562.5	73.6	100.03
22	8,805	808.	111.5	98.30	47	31,491	552.5	73.95	99.73
23	9,135	749.	97.6	98.25	48	32,240	559.5	73.15	99.71
24	9,750	809.	105.5	98.30	49	33,352	569.	75.0	100.82
25	10,474	801.	103.5	98.86	50	34,520	550.5	71.8	100.60

APPENDIX 2
ERROR ANALYSIS

The accuracy of the equilibrium concentrations of calcium and magnesium is estimated at better than ±1 percent, given the reproducibility of the experiment. The accuracy of the CO_2 fugacity is more difficult to evaluate because the pressure reading itself does not introduce significant error and most of the uncertainty is due to potential disequilibrium between gas and liquid. After reaching equilibrium from supersaturation, an accurate alkalinity determination on the aqueous solution yielded a computed value of the CO_2 fugacity of: 36.44 Pa vs. the measured 36.82, indicating no measurable disequilibrium for that sample within the analytical uncertainty. It is reasonable to conclude that the uncertainty of the CO_2 fugacity is given by the variation of total pressure, that is: ±0.6 percent.

Other sources of uncertainty are the composition of the solid, which is estimated accurate to better than ±0.1 mol percent $MgCO_3$, and the expression for the mean activity coefficient of calcium and magnesium bicarbonate. Estimates of uncertainty for the mean activity coefficient cannot be separated from that for the solubility product of calcite (and also its free energy of formation). When the free energy of mixing is computed, this part of the uncertainty cancels out. It is unlikely that the uncertainty of the mean activity coefficient itself exceeds 2 percent.

Combining the relative uncertainties, we obtain for the equilibrium constant:

$$\Delta K_{mc}/K_{mc} = 3 \times 0.02 + 3 \times 0.01 + 0.006 = 0.096.$$

The corresponding uncertainty for the logarithm of the solubility product is: ±0.039 and that for the free energy of formation is: ±0.10 kJ mol^{-1}, relative to a free energy of formation of pure calcite of: −1,129.27 kJ mol^{-1}.

Fluid-Mineral Interactions: A Tribute to H. P. Eugster
© The Geochemical Society, Special Publication No. 2, 1990
Editors: R. J. Spencer and I-Ming Chou

Dissolution kinetics of calcite in the H_2O-CO_2 system along the steam saturation curve to 210°C

S. J. TALMAN[1], B. WIWCHAR[2], W. D. GUNTER[2] and C. M. SCARFE[†]

[1] C. M. Scarfe Laboratory of Experimental Petrology, Department of Geology, University of Alberta, Edmonton, Alberta, T6H 2G2 Canada

[2] Alberta Research Council, Oil Sands and Hydrocarbon Recovery Department, P.O. Box 8330, Postal Station F, Edmonton, Alberta, T6H 5X2 Canada

Abstract—The dissolution kinetics of calcite is well described in aqueous solutions below 100°C, but not at higher temperatures. This work documents dissolution experiments performed using single crystals of calcite in a batch reactor at temperatures between 100 and 210°C and using various stirring rates and CO_2 partial pressures. Aqueous speciation was calculated at the temperature of each experiment based on quench measurements. The dissolution rate was found to be dependent on stirring rate, pH and p_{CO_2}. Our data are fit using the rate law proposed by PLUMMER et al. (1978) for calcite dissolution at lower temperatures, specifically,

$$\frac{dCa}{dt} = k_1[H^+] + k_2[H_2CO_3^*] + k_3[H_2O] - k_4[Ca^{2+}][HCO_3^-]$$

where k_1 is dependent on stirring rate, k_4 is a function of CO_2 pressure, the bracketed terms represent aqueous activities and $H_2CO_3^* = H_2CO_3 + CO_2(aq)$. The rate constants were determined by fitting the experimental data to an integrated form of the rate equation. The constant k_1 is poorly constrained by our experiments except at 100°C and a stirring rate of 500 RPM where it is 1.6×10^{-5} moles cm^{-2} s^{-1}. The value of k_2 changes slowly with temperature, apparently having a maximum value of about 10^{-6} moles cm^{-2} s^{-1} between 100 and 150°C. Finally, k_3 can be expressed by $\log(k_3) = -1300/T - 5.52$ in agreement with the dependence observed at lower temperatures by PLUMMER et al. (1978).

INTRODUCTION

THE ALKALINE EARTH CARBONATES, in particular calcite and dolomite, are common minerals in near surface environments. They are formed in diverse geologic settings, almost always by a reaction between aqueous CO_2 and the alkaline earths. Dissolved CO_2 is an important factor controlling pH in many near surface waters and, consequently, mineral stabilities in these waters. As a consequence, equilibrium solubilities and phase relations of carbonate minerals have been studied extensively at low temperatures (FREAR and JOHNSTON, 1929; MACKENZIE et al., 1983). The kinetics of dissolution and precipitation of calcite have also been studied extensively below 100°C (see PLUMMER et al., 1979; INSKEEP and BLOOM, 1985; MORSE, 1983, for recent reviews). These rate studies used a number of different experimental setups, each chosen to elucidate separate aspects of the dissolution reaction. In the simple Ca-CO_2-H_2O system at temperatures below 80°C, calcite dissolution kinetics has been found to be dependent on pH, p_{CO_2} and on the transport conditions in the reaction vessel (PLUMMER et al., 1978; HERMAN, 1982; RICKARD and

SJÖBERG, 1983; COMPTON and DALY, 1984). In more complex aqueous systems, the reaction kinetics can be affected by a number of other aqueous species; effects have been demonstrated with phosphate, sulphate and a host of divalent metals (MORSE, 1983). A number of rate laws have been used to interpret the kinetic data; HOUSE (1981), MORSE (1983) and COMPTON and DALY (1987) discuss the relative merits of these equations.

Carbonates are also important in higher temperature environments. They are common gangue minerals in ore deposits (HOLLAND and MALININ, 1979), they can affect porosity during diagenesis (WOOD, 1986) and they can be used as CO_2 barometers (PERKINS and GUNTER, 1989). HOLLAND and MALININ (1979) reviewed calcite solubility studies in subcritical solutions, and calcite's solubility in supercritical fluids was determined by SHARP and KENNEDY (1965) and FEIN and WALTHER (1987). The incentive for studying the high temperature kinetics of carbonates has been lacking, because most of the traditional applications assume equilibrium. However, kinetic considerations may become important when the relative reaction rates of two minerals are similar, as in the case of carbonate diagenesis, or when a solution is subjected to conditions which change more rapidly than mineral equilibria

---

[†] Deceased 20 July, 1988.

Table 1. Chemical analysis of Iceland spar (Chihuahua, Mexico)

Element	Concentration (ppm)	Element	Concentration (ppm)
Li	<1.	Mn	<1.
Na	25.	Fe	24.
K	<3.	Al	<6.
Mg	106.	Si	39.
Sr	37.	B	<1.
Ba	<1.	S	<3.
P	22.		

can be established (*e.g.* mid-ocean ridges). Kinetic data at higher temperatures are needed if computer simulations of mid-ocean ridge processes (*e.g.* BOWERS and TAYLOR, 1985) are to be improved. Industrial processes are, generally, much more rapid than geological processes, and consequently the rate data are applicable to models of these processes at most temperatures. Until recently, there were very few kinetic data on any minerals under hydrothermal conditions; however, rates are now available for a number of minerals (quartz—RIMSTIDT and BARNES, 1980; BIRD *et al.*, 1986; feldspars—LAGACHE, 1976; flourite—POSEY-DOWTY *et al.*, 1986; also HELGESON *et al.*, 1984; WOOD and WALTHER, 1983). Rate data are virtually non-existent for the carbonate system in solutions above 80°C. In this paper we report results on the dissolution kinetics of calcite at temperatures between 100 and 210°C and at steam saturation pressures.

EXPERIMENTAL DESIGN

Optical quality cleavage rhombs of Iceland spar weighing between about 1 and 13 grams from Chihuahua, Mexico (Wards Chemical of Rochester, New York) were lightly etched in dilute HCl to clean the surfaces. A sample was dissolved in nitric acid. The resulting solution was analysed by ICP and the composition of the calcite was calculated (Table 1). Magnesium, at about 100 ppm, was the principal impurity. The surface areas of the calcite crystals (between one and four cm²/g) were below the sensitivity of the BET method. The surface area was calculated from the rhomb geometry, assuming the crystal was a perfect calcite rhombohedron with angles of 75 and 105 degrees. The calculation requires that the crystal is flat; to check this assumption, calcite crystals were examined by SEM (Fig. 1). Three samples were chosen: a freshly fractured crystal, a crystal that was fractured and subsequently etched in 0.01 N HCl for five minutes, and a crystal that had been used in a dissolution experiment. The fractured surfaces show some surface roughness which disappears upon etching of the crystal. The surface of the dissolution experiment crystal appears unchanged from the etched crystal, with the exception of contamination by dust particles. A freshly cleaved surface was also examined on a Dektak II profilometer. Line profile scans of the crystal were made over

several horizontal distances to see if the surface roughness varies with resolution. A typical tracing is shown in Fig. 2. The average slope at each scan length is approximately

a

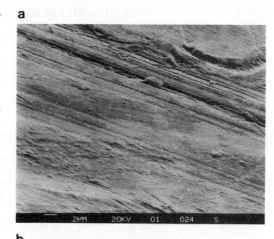

b

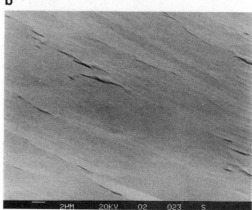

c

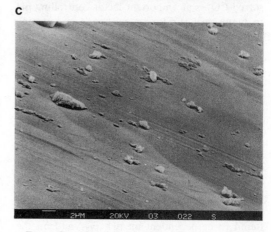

FIG. 1. Scanning electron photomicrograph of Iceland Spar used in dissolution studies a. freshly fractured surface, b. etched in 0.01 N HCl for 5 minutes, c. following a dissolution run. The scale bar in the lower left corner represents 2 μm.

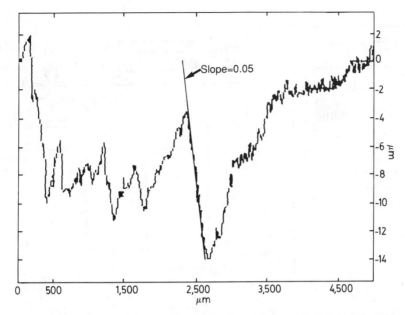

FIG. 2. Line profile scans of Iceland Spar used in kinetic experiments using a Dektak II profilometer. A typical tangent to the surface and its slope are shown. Note the difference in the horizontal and vertical scales.

0.05 from the horizontal. The greatest slope seen was 0.1. If the surface of the crystal were composed of peaks and valleys bounded by planes of slope 0.1, the surface area would only be increased by 1% compared to a perfectly flat surface; therefore, no corrections for surface roughness need be considered. The dimensions of each crystal were measured using vernier calipers.

Reactor design is important in kinetic studies (LEVENSPIEL, 1972; RIMSTIDT and DOVE, 1986; POSEY-DOWTY et al., 1986). We used a stirred batch reactor since it allows a lower surface area to solution mass ratio than does a plug flow reactor, a design which would be more appropriate to study much slower reactions. Either an open or closed stirred reactor can be used. In closed reactors the solution and the solid are simply left to react while the solution composition is monitored. Open system experiments on calcite were pioneered by MORSE (1974) who used a "pH-stat" reactor where the system is open to CO_2 and H^+ but not to calcium. PLUMMER et al. (1978) used a pH-stat reactor to study calcite dissolution kinetics far removed from equilibrium.

The rate of dissolution is measured directly in open systems (CHOU and WOLLAST, 1984). A number of difficulties arise with open reactors as the temperature of the system is increased. The pH-stat method cannot be used because of the absence of reliable high temperature pH electrodes. Other open systems are complicated by the need for precise determination of the chemical composition of quench samples. The reaction rate is determined by the difference in the composition of the input and output solutions, which makes the rate determination very sensitive to analytic error. The number of steps required to treat the quench samples leads to relatively large uncertainties in the solution composition.

In order to obtain the dissolution rate from a closed reactor, the solution data must be differentiated; errors associated with derived quantities are generally greater than those of the original data. However, if a rate law is postulated, its validity can be checked with the data from a single batch reactor run, since each experiment provides rate data for variable solution compositions. Since a number of rate laws have been proposed for calcite at lower temperatures, we can test our data against these to see if they are applicable. For these reasons, a closed batch reactor was used for all the experiments reported here.

The experiments were performed in a four litre Parr stirred stainless steel autoclave; a schematic diagram of the apparatus is shown in Fig. 3. Impellers affect stirring in both the gas and liquid phases. Two types of crystal holder were used. Initial runs used a cage holder that surrounded the crystal, and consequently restricted the circulation of the solution near the crystal. Later runs used a holder consisting of two strands of stainless steel wire which reduces the interference with the fluid flow near the sample. A screw device allows the crystal to be raised or lowered while the autoclave is at pressure. The autoclave was precharged with CO_2 and 2.0 kg of deionized water were added. Run conditions were achieved while the crystal was in the vapour. At the start of a dissolution experiment, the crystal was lowered into the aqueous phase. Samples of the solution (50 ml) in the autoclave were taken regularly through floating piston sample tubes. The floating piston ensures that there is no great pressure drop at the dip tube where the sample is removed from the autoclave, and hence no boiling. The solution removed during sampling was replaced by a solution whose CO_2 pressure was near to the starting composition of the autoclave solution. The temperature was maintained at the run temperature ± 0.2°C,

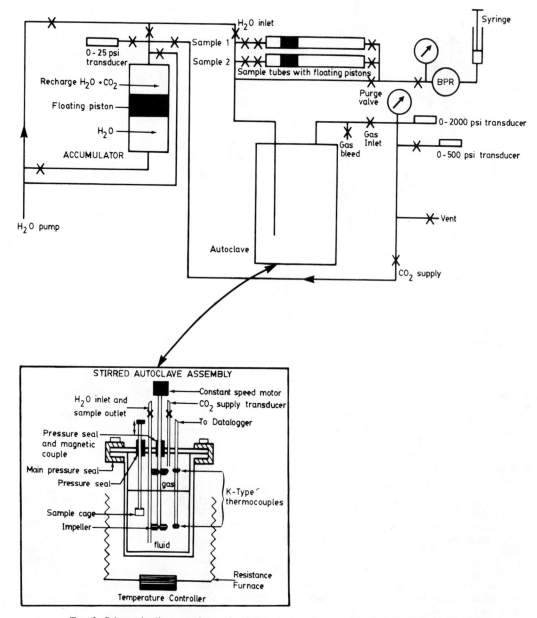

FIG. 3. Schematic diagram of the stirred autoclave system used in the experiments. The back pressure regulator is labeled BPR.

except immediately after sampling, when the temperature would drop by about 2°C.

The concentration of carbonic acid at run conditions could not be calculated from the initial charge of CO_2 since a small thermal gradient in the vapour made the partitioning of CO_2 difficult to predict. Consequently two samples were taken at each time interval. One sample was basified with NaOH to keep the CO_2 in solution and was analysed for TIC (total inorganic carbon). The other sample was split. One portion was acidified and analysed for cations by ICP and the pH and TIC were measured on the other portion after it had stabilized with respect to loss of

CO_2. The difference in TIC between the two samples represented the CO_2 lost from the unbasified quench sample.

EXPERIMENTAL RESULTS AND DATA REDUCTION

Dissolution experiments were performed at 210, 150, and 100°C and a variety of stirring rates and CO_2 pressures (Table 2 and Appendix). Plots of calcium *vs.* elapsed time for these experiments are shown in Fig. 4. The equations used to generate the

Table 2. Summary of experimental conditions*

Experiment	Temp °C	Initial TIC (mmoles/kg)	Stirring rate (rpm)	Surface area (cm²)	Crystal mass (g)	Remarks
W130587	210	31.	500	4.35	1.014	crystal 1
W150987	210	38.	750	3.99	0.895	crystal 1
W271087	210	39.	750	3.80	0.774	crystal 1
W011287	210	32.	300	3.6**	0.654	crystal 1
W150288	210	36.	500	4.61	1.518	crystal 2
W250288	210	100.	500	4.40	1.396	crystal 2
W140388	210	39.	500	2.88	0.417	crystal 1
W160588	210	5.0	500	4.09	1.208	crystal 2
W310588	210	38.	750	3.92	1.145	crystal 2
W251088	150	7.1	500	17.5	13.243	crystal 3
W141288	150	7.3	350	17.5	13.042	crystal 3
W020389	100	0.	500	17.5	12.848	crystal 3
W200389	100	16.5	500	17.4	12.790	crystal 3
W040489	100	0.	500	17.3	12.230	crystal 3 0.002 molal HCl

* Runs W130587 and W150987 used cage crystal holder, all other runs used wire holder. Solution mass 2.0 kg for all the runs.
** Estimated.

curves will be discussed later. As expected, the aqueous calcium concentration increases steadily with time, reaching a maximum value when equilibrium is approached.

The total calcium and carbon in the solution were used to calculate the distribution of species and the saturation quotient ($\Omega = [Ca^{2+}][CO_3^{2-}]/K_{sp}$) at run conditions. The calculation involves solving the mass balance equations for Ca and TIC, subject to mass action and charge balance constraints. The stability constants for the aqueous complexes and the solubility product of calcite were taken from the SOLMINEQ.88 data base (KHARAKA et al., 1988; Table 3). The stability of the complex $CaHCO_3^+$ was reduced from the value in the SOLMINEQ.88 data base. If this stability constant was not decreased, the calculated speciation was still considerably undersaturated with respect to calcite at the end of the experimental run. This suggests that this complex is not stable in dilute solutions at hydrothermal temperatures. PLUMMER and BUSENBERG (1982) present low temperature solubility data of calcium carbonates that are consistent with a lower stability constant for this complex than is in the SOLMINEQ.88 data base. FEIN and WALTHER (1987) reached a similar conclusion for supercritical H_2O-CO_2 solutions. Values of the ion activity product Q ($=K_{sp}\Omega$) at saturation for the 210°C runs varied between 2.7×10^{-12} and 3.8×10^{-12}, and were, on average only slightly higher (3.3×10^{-12}) than the value of K_{sp} in the SOLMINEQ.88 data base of 2.9×10^{-12}. Furthermore, data

from some initial runs (not discussed here) supports this value for K_{sp}. Initial runs were performed without replacing the solution lost from the system during sampling. As a consequence, the solution was depleted in CO_2 through the run, which lead to calcite precipitation late in the run. In one run dissolved calcium and TIC decreased by 7% and 25% respectively; however, Q maintained a near constant value between the previously mentioned limits. This suggests that the value of K_{sp} used here is reasonable.

An alternative method was used in some runs to check the calculated solution composition. This involved using Ca, pH and TIC from the quenched sample. The speciation was calculated with these data and then an amount of CO_2 was titrated back into the sample to make up the difference between the TIC from the basic sample and the quenched sample. The two sets of calculations were generally in agreement. However, since this calculation uses all the same data as the first calculation, as well as the pH of the quench sample and the TIC of the acidic sample we chose, in later experiments, to use only the first calculation of the solution chemistry at run conditions. A possible error in the calculation arises from the failure to consider any cations other than calcium in a charge balanced solution; however, calcium was the dominant cation in virtually all of the samples. Some of the initial samples had relatively high iron contents (see Appendix) which quickly plated out on the autoclave.

A number of empirical rate laws have been used to fit calcite dissolution data in aqueous solutions

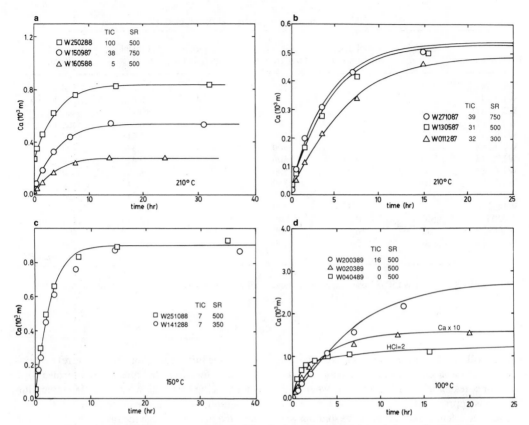

FIG. 4. Aqueous calcium concentration (millimoles/kg) versus elapsed time in hours for dissolution experiments at 100, 150 and 210°C. Curves were fit according to the rate expression of PLUMMER *et al.* (1978) using the rate constants listed in Table 4. The run conditions are shown in the figure, stirring rate is abbreviated as SR and concentration units of TIC and HCl are millimoles/kg. The run conditions are also listed in Table 2.

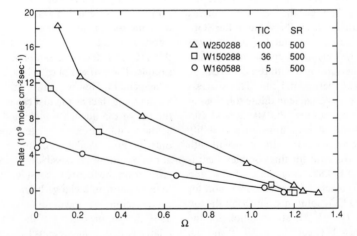

FIG. 5. Dissolution rate (in nanomoles cm^{-2} s^{-1}) plotted against Ω for three runs at various values of p_{CO_2} at a constant stirring rate (SR).

(see MORSE, 1983). Most of the rate laws are based on Ω, and are of the form:

$$\text{Rate} = k(1 - \Omega^n)^m$$

where n and m are fit parameters. To test this rate law, the reaction rate was determined by numerically differentiating the calcium concentration vs. time curve using a second degree interpolating polynomial. The rate was plotted in Fig. 5 as a function of Ω for runs at 210°C and three different p_{CO_2}'s. Errors associated with the calculated initial rate are about 10%. The uncertainty in Ω is large; as discussed earlier the average value of the solubility product calculated here is 15% greater than the value of K_{sp} used to calculate Ω. As well, since the value of $[CO_3^{2-}]$ is calculated from a charge balance equation it is sensitive to errors in $[Ca^{2+}]$ which leads to an error of $\pm 15\%$. The apparent dissolution of calcite in supersaturated solutions is a consequence of these errors. Although the errors are relatively large, it is clear that there are differences in the reaction rate. All the data should plot along the same curve if the rate is solely dependent on Ω. Clearly, this is not the case; therefore, any rate law simply based on Ω is inadequate. Fig. 5 demonstrates a dependence of the rate on p_{CO_2} which cannot be described solely in terms of Ω.

The stirring rate dependence of the rate of calcite dissolution is well documented (LUND et al., 1975; PLUMMER et al., 1978; 1979; COMPTON and DALY, 1984) and it has been demonstrated that in acidic solution it is the rate of transport of H^+ to the surface that limits the reaction. Stirring rate dependence can be seen in our data (Fig. 4b). This figure shows the calcium concentration as a function of time for three runs performed at similar p_{CO_2}'s. The upper curve was obtained using a slightly smaller crystal than the middle curve and a higher stirring rate (runs W130587 and W271087). The separation of these curves occurs early in the run, indicating that the stirring rate dependence is due to hydronium transport. The lower curve is from run W11287 at a still lower stirring rate; however, the crystal was smaller still.

To our knowledge PLUMMER et al. (1978) are the only investigators who have tried to separate the effects of both p_{CO_2} and H^+ on the rate. Their rate expression is

$$\frac{dCa}{dt} = k_1[H^+] + k_2[H_2CO_3^*]$$

$$+ k_3[H_2O] - k_4[Ca^{2+}][HCO_3^-] \quad (1)$$

where k_1–k_3 and k_4 are rate constants and the bracketed terms represent the activities of these species in the solution. This rate law is developed by assuming that calcite dissolution proceeds by three simultaneous reactions:

$$CaCO_3 + H^+ \rightarrow Ca^{2+} + HCO_3^- \quad (2)$$

$$CaCO_3 + H_2CO_3^* \rightarrow Ca^{2+} + 2HCO_3^- \quad (3)$$

and

$$CaCO_3 + H_2O \rightarrow Ca^{2+} + HCO_3^- + OH^-. \quad (4)$$

PLUMMER et al. (1978; 1979) derive Equation 1 from a mechanistic model of dissolution. The forward rate of reaction (2), $k_1[H^+]$, is limited by the rate of transport to the calcite surface, so that k_1 is dependent on stirring rate. As the stirring rate is increased, k_1 should increase until it reaches $K_{eq,1}/k_{-1}$ where $K_{eq,1}$ is the equilibrium constant for reaction 2, and k_{-1} is the rate constant for the back reaction. PLUMMER et al. (1978) grouped the rate for all three back reactions into one term involving k_4; however, in order to do this they had to introduce a p_{CO_2} dependence into k_4. Although this is not necessary since the forward and backward rate constants for elementary reactions are related to the equilibrium constant (see LASAGA, 1981), it is convenient for fitting our experimental data.

DETERMINATION OF RATE CONSTANTS

The speciation calculated by the methods described previously can be used to calculate rate constants for the Plummer rate expression. Extracting the forward rate constants for reactions 2, 3 and 4 from the batch reactor data requires either integrating the proposed rate expression or numerically differentiating the experimentally determined concentration evolution. We chose to integrate an approximation to the rate law rather than the numerical differentiation. The rate obtained by differentiation is very sensitive to errors in the solution analysis.

The complexity of the rate law requires that a simplification be introduced in order to make the expression integrable. The activity of the aqueous species in (1) (except Ca^{2+}) at any time, t, between sampling times t_1 and t_2 was estimated by a linear extrapolation of their activity, so that the various activity terms in Equation (1) can be written

$$a_i(t) = \frac{a_i(t_2) - a_i(t_1)}{t_2 - t_1}(t - t_1) + a_i(t_1), \quad (5)$$

Table 3. Values of thermodynamic constants used in the calculations. K_1 and K_2 are the first and second dissociation constants of carbonic acid, K_w is the dissociation constant of water, and $K_{CaCO_3^0}$ is the dissociation constant for $CaCO_3^0$. The stability constant for $CaHCO_3^+$ was set to -1.0 for all temperatures

T	$\log K_1$	$\log K_2$	$\log K_w$	$\log K_{CaCO_3^0}$	$\log K_{sp}$	$B^{\bullet}$
100.	-6.43	-10.16	-12.26	-4.17	-9.27	0.046
150.	-6.77	-10.39	-11.64	-4.67	-10.16	0.047
210.	-7.30	-10.87	-11.20	-5.60	-11.53	0.045

where a_i is the activity of species i and $t_1 < t < t_2$. In this case Equation (1) becomes

$$rate = \frac{M}{A}\frac{dCa}{dt} = (C_1 + D_1 t)Ca + C_2 + D_2 t \quad (6)$$

where Ca is the concentration (molal) of calcium in solution, A and M are the surface area of the crystal and the solution mass respectively and

$$C_1 = k_4\left[f_{Ca}(t_1)[HCO_3^-(t_1)]\right.$$

$$\left. - \frac{f_{Ca}(t_2)[HCO_3^-(t_2)] - f_{Ca}(t_1)[HCO_3^-(t_1)]}{t_2 - t_1}t_1\right]$$

$$D_1 = k_4\left[\frac{f_{Ca}(t_2)[HCO_3^-(t_2)] - f_{Ca}(t_1)[HCO_3^-(t_1)]}{t_2 - t_1}\right]$$

$$C_2 = k_1\left[[H^+(t_1)] - \frac{[H^+(t_2)] - [H^+(t_1)]}{t_2 - t_1}t_1\right]$$

$$+ k_2\left[[H_2CO_3^*(t_1)]\right.$$

$$\left. - \frac{[H_2CO_3^*(t_2)] - [H_2CO_3^*(t_1)]}{t_2 - t_1}t_1\right]$$

$$+ k_3\left[[H_2O(t_1)] - \frac{[H_2O(t_2)] - [H_2O(t_1)]}{t_2 - t_1}t_1\right]$$

and

$$D_2 = k_1\frac{[H^+(t_2)] - [H^+(t_1)]}{t_2 - t_1}$$

$$+ k_2\frac{[H_2CO_3^*(t_2)] - [H_2CO_3^*(t_1)]}{t_2 - t_1}$$

$$+ k_3\frac{[H_2O(t_2)] - [H_2O(t_1)]}{t_2 - t_1}$$

and f_{Ca} relates Ca to $[Ca^{2+}]$ and it is assumed to be the same for the fit and experimentally determined calcium concentration. Despite the approximations

made in integrating the rate expression, it should be more accurate than the alternative approach of differentiating the concentration data, since it uses requires one less calculated quantity (dCa/dt).

Activity coefficients, γ_i were calculated using the expression

$$\log \gamma_i = \frac{-A(T)Z_i^2\sqrt{I}}{1 + \mathring{a}_i(T)B(T)\sqrt{I}} + B^{\bullet}(T)I$$

(HELGESON, 1969), where I is the ionic strength, Z_i is the ionic charge, and $\mathring{a}$ is an ion size parameter. A and B are related to the density, ρ, and dielectric constant, ϵ of water by

$$A = \frac{1.82 \times 10^6\sqrt{\rho}}{(\epsilon T)^{3/2}}$$

and

$$B = \frac{50.3 \times 10^8\sqrt{\rho}}{\sqrt{\epsilon T}}$$

(HELGESON, 1969) and $B^{\bullet}$ is listed in Table 3. The contribution of the $B^{\bullet}$ term to the activity coefficient is small for all the samples. Based upon the linear approximation, Equation 1 can be integrated to give

$$Ca(t) = \exp\left[-\frac{D_1}{2}(t_1^2 - t^2)\right.$$

$$+ C_1(t_1 - t)\right]\left[Ca(t_1) - \frac{D_2}{D_1}\right]$$

$$+ \frac{D_2}{D_1} + \sqrt{\frac{2}{D_1}}\left(C_2 - \frac{D_2 C_1}{D_1}\right)$$

$$\times \left[Daw\left(\sqrt{\frac{D_1}{2}}t + \frac{C_1}{\sqrt{2D_1}}\right)\right.$$

$$- \exp\left[-\frac{D_1}{2}(t_1^2 - t^2) + C_1(t_1 - t)\right]$$

$$\times Daw\left(\sqrt{\frac{D_1}{2}}t_1 + \frac{C_1}{\sqrt{2D_1}}\right)\right] \quad (7)$$

where Daw refers to the Dawson integral

$$Daw(x) = \exp(-x^2)\int_0^x \exp(y^2)dy. \quad (8)$$

Equation 7 relates the concentration of calcium at time t_{i+1} to the concentration at time t_i (and, by recursion, to its initial concentration) and the activities of the other species appearing in (1). It is linear in C_2 and D_2 and non-linear in C_1 and D_1, and therefore, it is linear in the three constants k_1–k_3 and non-linear in k_4. Consequently, given a k_4, values of k_1–k_3 which minimize the expression $\sum_i (\mathrm{Ca}(t_i)' - \mathrm{Ca}(t_i))^2$, where $\mathrm{Ca}(t_i)'$ is the value from (7), can be determined. A number of values for k_4 were tried in order to obtain the one which gave lowest deviation from the data. It is also possible to fit the entire curve with an additional constant, [Ca(0)]; however, with the approximations already made, it is unlikely to provide a superior fit. Values of k_1 and k_4 were in fair agreement for different runs at the same stirring rate and p_{CO_2}; however, values of k_2 and k_3 were not consistent. This is because neither [H_2O] nor [$H_2CO_3^*$] change very much during a given run, so these parameters are very sensitive to small changes in [$H_2CO_3^*$] through the run (or random errors in the TIC data). However, the sum $(k_3 + k_2[H_2CO_3^*])$ is relatively insensitive. Therefore, the values for k_2 and k_3 were determined from pairs of experiments at different carbonic acid activities ([$H_2CO_3^*$]$_{(1)}$ and [$H_2CO_3^*$]$_{(2)}$). The net contribution to the rate by water and carbonic acid $(k_3 + k_2[H_2CO_3^*])$ should be the same with the new constants k_2 and k_3 as with the calculated values $k_{2,(1)}$, $k_{3,(1)}$, associated with [$H_2CO_3^*$]$_{(1)}$ and $k_{2,(2)}$ and $k_{3,(2)}$ associated with [$H_2CO_3^*$]$_{(2)}$, so that

$$k_3 + k_2[H_2CO_3^*]_{(1)} = k_{3,(1)} + k_{2,(1)}[H_2CO_3^*]_{(1)}$$

and

$$k_3 + k_2[H_2CO_3^*]_{(2)} = k_{3,(2)} + k_{2,(2)}[H_2CO_3^*]_{(2)}.$$

These equations are solved for k_2 and k_3. Fig. 4a demonstrates that the values calculated in this manner are consistent with rate data from a run at a third p_{CO_2}. The constant k_4 can be calculated given values of k_1, k_2, k_3 and p_{CO_2} using the relation given by PLUMMER et al. (1978) which becomes, after taking into consideration several assumptions which they discuss,

$$k_4 = \frac{K_2}{K_{sp}}\left(k_1' + \frac{1}{a_{H^+_{(s)}}}[k_2 a_{H_2CO_3^*} + k_3 a_{H_2O}]\right) \quad (9)$$

where K_2 is the second dissociation constant of carbonic acid, k_1' is the limiting value of k_1 at infinite stirring rate, and $a_{H^+_{(s)}}$ is the activity of hydronium at the crystal surface. This can be approximated by

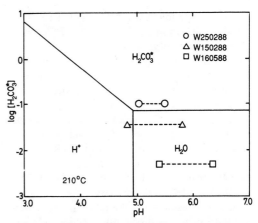

FIG. 6. The initial and final solution hydronium and $H_2CO_3^*$ activities from three runs at 210°C and different p_{CO_2}'s are shown in this figure connected by dashed lines. The solid lines divide the plot into regions where each of the three forward reactions make the dominant contribution to the net forward rate at this temperature. Reaction between the crystal and condensate is responsible for the relatively high initial pH in the experiment at high p_{CO_2}.

the equilibrium value of H^+ in the system, and its value is generally sufficiently small to ensure that k_1' is negligible compared to the second term in the parentheses.

Fig. 6 shows the solution compositions where the contributions from each of the three forward reactions are dominant at 210°C. The values of the rate constants used to develop this plot and Fig. 4 are given in Table 4. The initial and final solution compositions of three runs at different p_{CO_2}'s are also shown in Fig. 6. Each of the three rate terms predominate for at least a portion of one of the runs. The confidence in the estimates of the rate constants is roughly related to the number of data points in each area; the rate constant, k_1 is the least well defined.

Table 4. Comparison of the results of this work with values extrapolated from the Arrhenius fit given by PLUMMER et al. (1978). These equations are $\log k_1^{ext} = -2.802 - 444/T$, $\log k_2^{ext} = -0.16 - 2177/T$, and $\log k_3^{ext} = -4.10 - 1737/T$.

	Rate constants (moles cm^{-2} s^{-1})					
	This work			Extrapolated		
T	$\log k_1$	$\log k_2$	$\log k_3$	$\log k_1^{ext}$	$\log k_2^{ext}$	$\log k_3^{ext}$
100.	−4.8	−6.3	−9.0	−4.0	−6.0	−8.8
150.	−4.1	−6.2	−8.6	−3.9	−5.3	−8.2
210.	−3.3	−7.1	−8.2	−3.7	−4.7	−7.7

A single experiment was performed at 100°C in a starting solution with 2. × 10⁻³ molal HCl, which defines the value of k_1 very well at the run conditions. This experiment could not be performed at other temperatures due to the corrosive nature of the initial solutions. We are in the process of obtaining lined autoclaves in which these experiments can be performed. This will help to define k_1 with more certainty. The values of k_1 in Table 4 are based on data obtained at a stirring rate of 500 RPM. The values of log (k_1) used to generate the upper and lower curves in Fig. 4b differ from the value in Table 4 by 0.15 and −0.58 respectively.

The rate constants and the rate expression were used to simulate the dissolution reaction. The calculations require as input the initial Ca and, for those runs with an applied p_{CO_2}, the total dissolved carbonate in solution (essentially constant through

the run). The total carbonate in the runs without an applied p_{CO_2} changed appreciably through the run, so a temperature dependent parameter, β (the ratio of $H_2CO_3^*$ in solution to the mass of CO_2 in the vapour), was introduced to simulate the partitioning of CO_2 between the solution and vapour, and a mass balance on carbonate was included in the calculations. A value of β which fit the measured TIC values was determined. The data input defines the initial solution speciation and, from Equation (1), the initial rate of calcite dissolution. This rate is used to determine how much calcium will be added to the solution during a given time interval, after which the solution composition and dissolution rate are recalculated. This is repeated until the solution becomes saturated. The curves plotted on Fig. 4 are calculated in this manner. The agreement between these curves and the data is very good, with

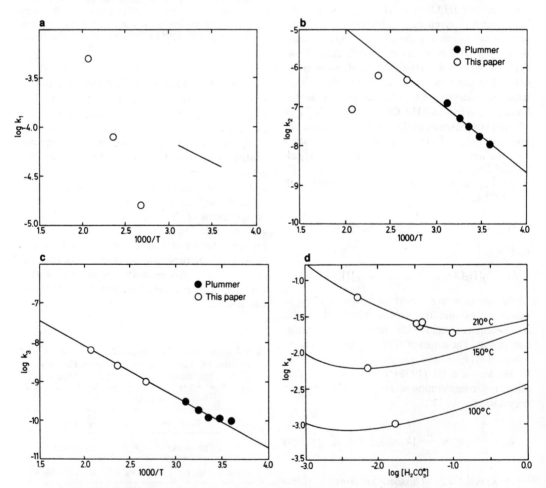

FIG. 7. Arrhenius plots of the rate constants k_1, k_2 and k_3 (in moles cm⁻² s⁻¹) are shown in figures 7a, 7b, and 7c with Plummer's rate constants from 5 to 60°C and our rate constants at 100, 150, and 210°C. Figure 7d shows the fit values of k_4 from several runs. The curves show the values of k_4 predicted by Equation 9.

the exception of some intermediate points in the two lower temperature runs; however, the fit in the region where the rate of the back reaction is small is much better. The discrepancy between the predicted and fit values may be due to an incorrect expression for the back reaction. COMPTON and DALY (1987) performed calcite dissolution experiments in a batch reactor. The fit of their data to Equation 1 showed a similar discrepancy in intermediate saturations. BUSENBERG and PLUMMER (1986) and CHOU et al. (1989) use a rate expression very similar to Equation (1) to describe the rate of dissolution of a number of single component carbonates, but with different expressions for the back reactions.

TEMPERATURE DEPENDENCE

PLUMMER et al. (1978) studied calcite dissolution kinetics as a function of solution composition for temperatures between 0 and 60°C. They fit the temperature dependence of the rate constants with an Arrhenius relationship,

$$\log k = a - \frac{b}{T} \qquad (10)$$

where T is the temperature in K. The values of a and b determined by PLUMMER et al. (1978) are given in Table 4, along with the extrapolated rate constants. The rate constants calculated here are shown on an Arrhenius plot in Fig. 7, together with the temperature dependence determined by PLUMMER et al. (1978). Errors on the 210°C points were estimated by a Monte-Carlo type calculation in which the initial calcium and TIC values were varied by a random errors of less than 5%. Two of the rate constants k_1, k_2, and k_3 were held constant while the third constant and k_4 were determined using the regression calculation described previously. After about fifty calculations the mean and standard deviation of the k values were calculated. The standard deviations determined in this way were generally within 15% of the mean.

The logarithms of our rate constants k_2 and k_3 at 100°C are about 0.25 lower than the extrapolated values; this discrepancy is consistent with errors in surface area estimates or differences in defect densities (SCHOTT et al., 1989; CHOU et al., 1989). CHOU et al. (1989) also found discrepancies between their calcite dissolution rate data and those predicted using rate constants given by PLUMMER et al. (1978) of a similar magnitude. The expression by PLUMMER et al. (1978) for k_3 above 25° extrapolates relatively well to our values; however, the behavior of k_1 and k_2 with temperature is more complex. The

value of k_2 at 100°C is essentially co-linear with the low temperature data; however, the slope of the curve decreases as the temperature increases to such an extent that the slope of the curve is negative at 210°. The values of k_1 obtained here show an irregular behaviour with temperature. For reasons described above, the value of k_1 is the least well defined by our experiments, with the exception of the 100° datum. This value is considerably lower than the value fit from the low temperature data, even if it is increased by a factor of two on the assumption that there is discrepancy due to surface area or crystal features. PLUMMER et al. (1978) observed variations in the value of k_1 by a factor of about 1.5 by changing the stirring rate. Our runs were done at a lower stirring rate, but, perhaps more importantly, on a much bigger crystal. Flow across the larger crystal surface will be less turbulent than for the smaller grains used in the low temperature studies. Until it is possible to perform the higher temperature experiments in more acidic solutions, it will be difficult to define k_1. However, the data we have (Fig. 7a) suggests that the rate increases with temperature more rapidly than was observed by PLUMMER et al. (1978). It is difficult to assign any significance to this observation, since the values are not very well defined and there are additional complications associated with the stirring rate dependence. Fig. 7d plots the values of k_4 determined here along with the values calculated from Equation 9 at the three run temperatures.

The apparent negative activation energy of k_2 at temperatures above 150°C can most easily be explained by invoking a change in reaction mechanism. A very slight decrease could be attributed to a decrease in the equilibrium constant for the hydration of $CO_2(aq)$ with temperature (PALMER and VAN ELDIK, 1983), but the observed decrease is much greater than can be attributed to this. In addition, the hydration reaction is promoted by an increase in pressure which will decrease the effect. Our data suggests a change in the mechanism of reaction (3) with increased temperature. With only the batch reactor data obtained here, it is difficult to propose, with any confidence, a new reaction mechanism for the reaction between calcite and carbonic acid at these temperatures.

Our data are consistent with the PLUMMER et al. (1978) rate equation at 210°, despite the change in mechanism of reaction (3). Their rate equation (Equation 1) will still be valid if the mechanism changes from being controlled by the rate of reaction between $H_2CO_3^*$ and the calcite surface, as postulated by PLUMMER et al. (1978), to a rate controlled by the decomposition of a surface calcite-carbonic acid complex. In this case, the rate of dissolution due to reaction with carbonic acid, r_2, will be $r_2 = k_{sc}K_{eq} [H_2CO_3^*]$ where k_{sc} is the rate of decomposition of the surface complex and K_{eq} is the equilibrium constant for the formation of the surface complex. If the complex is formed by a sufficiently exothermic reaction the apparent activation energy obtained by considering reaction (3) to be rate limiting may be negative.

CONCLUSIONS

The rate of calcite dissolution from ambient to temperatures in excess of 200°C depends on temperature, flow rate, and pressure of CO_2 and is well described by the rate law proposed by PLUMMER *et al.* (1978). The value of k_1 is poorly constrained by our experiments except at 100°C at a stirring rate of 500 RPM where a quantity of acid was added at the start of the experiment. An anomalous temperature dependence is seen in k_2 which shows very little dependence on temperature and appears to have a maximum near 150°C. This is a clear indication of a change in reaction mechanism, although the linear dependence of the rate on $[H_2CO_3^*]$ appears to remain. The rate constant, k_3, fits well with the extrapolation of PLUMMER *et al.* data above 25°C with an activation energy of about 24.7 kJ mole^{-1}. Further experiments in lined autoclaves are required to more fully understand the behaviour of k_1 with temperature.

Acknowledgements—Financial support for this work was obtained by C. Scarfe (later managed by K. Muehlenbachs) from the Alberta Oil Sands Technology and Research Authority (AOSTRA). L. Holloway, K. Montgomery, and B. Young provided assistance with the chemical analyses. Thanks are also due to Lei Chou for providing us with a preprint of her paper, J. Fein, P. Glynn, N. Plummer and D. Rimstidt for their reviews of the initial manuscript, Karen Jensen for drafting services, M. K. Allen for splicing infinitives, and finally to Jamey Hovey and Ernie Perkins for technical and logistical aid. We are honoured that this paper has been included in this volume.

REFERENCES

BIRD G., BOON J. and STONE T. (1986) Silica transport during steam injection into oil sands. I. Dissolution and precipitation kinetics of quartz: new results and review of existing data. *Chem. Geol.* **54**, 69–80.

BOWERS T. S. and TAYLOR H. P. JR. (1985) An integrated chemical and stable-isotope model of the origin of mid-ocean ridge hot spring systems. *J. Geophys. Res.* **90**, 12,583–12,606.

BUSENBERG E. and PLUMMER L. N. (1986) A comparative study of the dissolution and crystal growth kinetics of calcite and aragonite. In *Studies in Diagenesis* (ed. F. A. MUMPTON), *U.S. Geol. Surv. Bull. 1578*, pp. 139–168.

CHOU L. and WOLLAST R. (1984) Study of the weathering of albite at room temperature and pressure with a flu-

idized bed reactor. *Geochim. Cosmochim. Acta* **48**, 2205–2217.

CHOU L., GARRELS R. M. and WOLLAST R. (1989) A comparative study of the kinetics and mechanisms of dissolution of carbonate minerals. *Chem. Geol.* (submitted).

COMPTON R. G. and DALY P. J. (1984) The dissolution kinetics of Iceland spar single crystals. *J. Colloid Interface Sci.* **101**, 159–166.

COMPTON R. G. and DALY P. J. (1987) The dissolution/precipitation kinetics of calcium carbonate: An assessment of various kinetic equations using a rotating disc method. *J. Colloid Interface Sci.* **115**, 493–498.

FEIN J. B. and WALTHER J. V. (1987) Calcite solubility in supercritical CO_2-H_2O fluids. *Geochim. Cosmochim. Acta* **51**, 1665–1673.

FREAR G. L. and JOHNSTON J. (1929) Solubility of calcium carbonate (calcite) in certain aqueous solution at 25°C. *Amer. Chem. Soc. J.* **51**, 2082–2093.

HELGESON H. C. (1969) Thermodynamics of hydrothermal systems at elevated temperatures and pressures. *Amer. J. Sci.* **267**, 729–804.

HELGESON H. C., MURPHY W. M. and AAGAARD P. (1984) Thermodynamic and kinetic constraints on reaction rates among minerals and aqueous solutions. II. Rate constants, effective area, and the hydrolysis of feldspar. *Geochim. Cosmochim. Acta* **48**, 2405–2432.

HERMAN J. S. (1982) The dissolution kinetics of calcite, dolomite and dolomitic rocks in the carbon dioxide-water system. Unpublished PhD. thesis, Pennsylvania State University, 219p.

HOLLAND H. D. and MALININ S. D. (1979) The solubility and occurrence of non-ore minerals. In *Geochemistry of Hydrothermal Ore Deposits* (ed. H. L. BARNES), Chap. 9, pp. 461–508. John Wiley and Sons.

HOUSE W. A. (1981) Kinetics of crystallisation of calcite from calcium bicarbonate solutions. *J. Chem. Soc. Faraday Trans. 1* **77**, 341–359.

INSKEEP W. P. and BLOOM P. R. (1985) An evaluation of rate equations for calcite precipitation kinetics at pCO_2 less than 0.01 atm and pH greater than 8. *Geochim. Cosmochim. Acta* **49**, 2165–2180.

KHARAKA Y. K., GUNTER W. D., AGGARWAL P. K., PERKINS E. H. and DeBRAAL J. D. (1988) SOLMI-NEQ.88: A computer program code for geochemical modeling of water-rock interactions. *U.S. Geol. Surv. Water-Resources Investigations Report 88-4227.*

LAGACHE M. (1976) New data on the kinetics of the dissolution of alkali feldspar at 200°C in carbon dioxide charged water. *Geochim. Cosmochim Acta* **40**, 157–161.

LASAGA A. C. (1981) Rate laws of chemical reactions. In *Kinetics of Geochemical Processes* (eds. A. C. LASAGA and R. J. KIRKPATRICK), Chap. 1, pp. 1–68. *Reviews in Mineralogy,* **8**.

LEVENSPIEL O. (1972) *Chemical Reaction Engineering.* 2nd ed. John Wiley, 578 pp.

LUND K., FOGLER H. S. and McCUNE C. C. (1975) Acidization—II. The dissolution of calcite in hydrochloric acid. *Chem. Eng. Sci.* **30**, 825–835.

MACKENZIE F. T., BISCHOFF W. D., BISHOP F. C., LOIJENS M., SCHOONMAKER J. and WOLLAST R. (1983) Magnesian calcites: low-temperature occurrence, solubility and solid solution behavior. In *Carbonates: Mineralogy and Chemistry* (ed. R. J. REEDER), Chap. 4, pp. 97–144. *Reviews in Mineralogy,* **11**.

MORSE J. W. (1974) Dissolution kinetics of calcium carbonate in sea water. III A new method for the study of carbonate reaction kinetics. *Amer. J. Sci.* **274**, 97–107.

MORSE J. W. (1983) The kinetics of calcium carbonate dissolution and precipitation. In *Carbonates: Mineralogy and Chemistry* (ed. R. J. REEDER), Chap. 7, pp. 227–264. *Reviews in Mineralogy,* **11.**

PALMER D. A. and VAN ELDIK R. (1983) The chemistry of metal carbonato and carbon dioxide complexes. *Chem. Rev.* **83,** 651–731.

PERKINS E. H. and GUNTER W. D. (1989) Applications of SOLMINEQ.88 and SOLMINEQ.88 PC/SHELL to thermally enhanced oil recovery. *Fourth UNITAR/ UNDP International Conference on Heavy Crude and Tar Sands Proceedings* (eds. R. F. Myer and E. J. Wiggins), Vol. 3, p. 413–422.

PLUMMER L. N. and BUSENBERG E. (1982) The solubilities of calcite, aragonite and vaterite in the CO_2-H_2O solutions between 0 and 90°C, and an evaluation of the aqueous model for the system $CaCO_3$-CO_2-H_2O. *Geochim. Cosmochim. Acta* **46,** 1011–1040.

PLUMMER L. N., WIGLEY T. M. L. and PARKHURST D. L. (1978) The kinetics of calcite dissolution in CO_2-water systems at 5° to 60°C and 0.0 to 1.0 atm CO_2. *Amer. J. Sci.* **278,** 179–216.

PLUMMER L. N., PARKHURST D. L. and WIGLEY T. M. L. (1979) Critical review of the kinetics of calcite dissolution and precipitation. In *Chemical Modeling in Aqueous Systems. Speciation, Sorption, Solubility, and Kinetics* (ed. E. A. JENNE) pp. 537–573. *Amer. Chem. Soc. Symposium Series* **93.**

POSEY-DOWTY J., CRERAR D., HELLMAN R. and CHANG C. (1986) Kinetics of mineral water reactions: theory, design and application of circulating hydrothermal equipment. *Amer. Miner.* **71,** 85–94.

RICKARD D. and SJÖBERG E. L. (1983) Mixed control of calcite dissolution rates. *Amer. J. Sci.* **283,** 815–830.

RIMSTIDT J. D. and BARNES H. L. (1980) The kinetics of silca-water reactions. *Geochim. Cosmochim. Acta* **44,** 1683–1699.

RIMSTIDT J. D. and DOVE P. M. (1986) Mineral/solution reaction rates in a mixed flow reactor; wollastonite hydrolysis. *Geochim. Cosmochim. Acta* **50,** 2509–2516.

SCHOTT J., BRANTLEY S., CRERAR D., GUY C., BORCSIK M. and CHRISTIAN W. (1989) Dissolution of strained calcite. *Geochim. Cosmochim. Acta* **53,** 373–382.

SHARP W. E. and KENNEDY G. C. (1965) The system CaO-CO_2-H_2O in the two phase region calcite and aqueous solution. *J. Geol.* **73,** 391–403.

WOOD B. J. and WALTHER J. V. (1983) Rates of hydrothermal reactions. *Science* **222,** 413–415.

WOOD J. R. (1986) Thermal mass transfer in systems containing quartz and calcite. In *Roles of Organic Matter in Sediment Diagenesis* (ed., D. L. GAUTIER) pp. 169–180. SEPM Special Publication 38.

APPENDIX A

The raw data from the runs described in the text is given in the following tables. The values for the cations are all from ICP analysis, pH and C_q are the pH and the TIC of the quenched sample and C is the TIC of the basified sample, corrected for dilution with NaOH. The units of concentration are ppm. The last sample of each run is, unless otherwise noted, a standard of 26 ppm Ca. Those entries marked — were not analyzed and those listed as *nd* were below detection.

Table A.1. Raw data from experiments

Run	Sample	Time (hrs.)	Ca	C	Fe	Mg	pH	C_q
W130587	1	0.0	1.4	380.	3.4	*nd*	4.39	230.
	3	0.5	3.0	387.	1.7	*nd*	4.48	238.
	5	1.5	6.6	395.	0.9	*nd*	4.56	243.
	7	3.5	11.2	384.	0.5	0.1	4.83	213.
	9	7.5	16.7	385.	0.4	0.1	4.83	230.
	11	15.5	20.0	381.	0.4	0.2	4.76	217.
	13	31.5	21.2	391.	0.4	0.1	4.80	234.
	15	55.5	21.4	381.	0.4	0.3	4.79	224.
	17	79.5	21.0	375.	0.4	0.2	4.85	216.
	19	101.0	21.0	388.	0.6	*nd*	4.81	246.
	21	—	28.4	—	*nd*		3.1	—	—
W150987	1	0.0	1.4	475.	3.0	0.5	4.53	170.
	3	0.5	3.3	450.	0.6	0.2	4.67	118.
	5	1.5	7.5	448.	0.6	0.2	4.97	110.
	7	3.5	13.1	443.	0.5	0.3	5.28	92.
	9	6.5	17.6	448.	0.3	0.2	5.27	95.
	11	14.0	21.8	441.	0.3	0.3	5.47	99.
	13	31.0	21.6	454.	0.2	0.2	5.14	181.
	15	57.0	22.2	426.	0.3	0.2	5.22	170.
	17	81.3	21.8	450.	0.3	0.3	5.39	115.
	21	—	27.8	—	*nd*		3.1	—	—
W271087	1	0.0	0.7	464.	0.1	0.3	4.54	12.[a]
	3	0.5	3.6	458.	0.6	0.1	4.49	204.
	5	1.5	8.0	466.	0.5	0.2	4.72	191.
	7	3.5	12.4	474.	0.5	0.2	4.89	178.
	9	7.0	17.4	488.	0.5	0.2	5.19	139.
	11	15.0	20.2	471.	0.4	0.2	5.08	181.
	13	31.0	21.4	472.	0.2	0.2	5.04	205.
	15	130.4	21.4	284.	0.2	0.3	5.24	187.
	18	—	26.8	—	*nd*		3.0	—	—

S. J. Talman *et al.*

Table A.1. (Continued)

Run	Sample	Time (hrs.)	Ca	C	Fe	Mg	pH	C_q
W011287	1	0.0	0.9	390.	1.8	0.2	7.02	15.3
	3	0.5	2.1	385.	0.7	nd	7.23	2.7
	5	1.5	4.6	380.	0.4	nd	7.63	15.8
	7	3.5	8.6	386.	0.2	nd	7.71	14.6
	9	7.5	13.7	390.	0.3	nd	7.98	17.3
	11	15.0	18.5	392.	nd	nd	8.17	19.4
	13	32.0	21.8	396.	0.1	nd	8.26	23.1
	15	55.0	21.6	384.	nd	0.2	8.33	25.8
	17	79.0	21.2	385.	0.2	nd	8.23	27.0
	19	103.0	21.4	—	0.2	0.1	8.16	13.8
	21	—	28.2	—	nd	3.1	—	—
W150288	1	0.0	2.2	429.	1.8	0.1	7.39	1.7
	3	0.5	4.4	434.	0.4	nd	7.52	2.9
	5	1.5	8.8	354.[a]	0.3	nd	7.92	5.8
	7	3.5	14.1	350.[a]	0.2	nd	8.26	12.2
	9	7.5	19.3	427.	0.4	nd	8.39	16.2
	11	14.0	21.9	338.[a]	0.3	0.2	8.33	14.3
	13	32.0	22.2	414.	0.3	0.1	8.37	16.1
	15	57.0	22.1	396.	0.2	0.1	8.30	13.2
	17	128.0	22.1	385.	0.3	0.2	8.28	13.2
	21	—	28.5	—	nd	3.2	—	—
W250288	1	0.0	11.2	1140	0.8	nd	8.03	7.4
	3	0.5	14.0	1210	1.3	nd	8.12	8.9
	5	1.5	18.3	1300	0.3	nd	8.22	12.1
	7	3.5	24.8	1300	0.3	0.1	8.01	4.1
	9	7.5	30.4	1160	0.2	0.1	8.00	5.3
	11	15.0	33.2	1220	0.4	nd	8.07	5.9
	13	32.0	33.8	1230	0.4	0.2	8.01	5.4
	15	56.0	34.4	1230	0.5	nd	8.07	6.5
	17	79.5	33.6	1210	0.5	nd	8.01	5.5
	21	—	27.6	—	nd	3.1	—	—
W140388	1	0.0	5.7	469.	0.7	0.2	7.30	1.2
	3	0.5	5.9	419.	0.3	0.1	7.36	1.0
	5	1.5	8.0	403.	0.2	nd	7.35	1.3
	7	3.5	11.6	397.	0.1	0.1	7.52	2.1
	9	7.5	16.7	393.	0.2	0.1	7.69	3.0
	11	15.0	20.2	399.	0.1	0.3	7.78	3.9
	13	33.5	19.7	408.	0.3	0.1	7.62	2.2
	15	—	28.0	—	nd	3.1	—	—
	17[b]	—	nd	—	nd	nd	—	—
	19[c]	—	nd	—	nd	0.1	—	—
W160588	2	0.0	1.2	60.	0.2	0.1	6.82	0.4
	4	0.5	1.8	62.	0.2	0.3	6.92	0.7
	6	1.5	3.6	62.	0.1	nd	7.24	1.1
	10	3.5	6.6	67.	0.3	0.1	7.41	1.7
	12	7.5	9.6	69.	0.3	0.6	7.71	3.5
	14	13.8	11.2	33.[a]	0.1	0.1	7.66	3.0
	16	24.0	11.6	65.	0.2	0.4	7.68	3.1
	18	48.0	11.3	64.	nd	0.2	7.74	3.0
	20	72.0	11.7	62.	0.2	0.2	7.71	3.0
	22	—	27.8	—	0.1	3.1	—	—
W310588	2	0.0	0.7	453.	0.5	0.7	6.88	0.6
	4	0.5	3.7	470.	0.2	0.5	7.25	1.2
	6	1.5	8.8	466.	0.3	0.4	7.58	2.2
	8	3.5	14.8	460.	0.2	0.5	7.75	3.5
	10	7.5	19.4	465.	0.1	0.5	7.89	4.6
	12	13.5	21.8	467.	0.2	0.5	7.93	5.0
	14	19.8	22.4	474.	0.1	0.6	7.92	5.4
	16	30.0	22.4	474.	0.2	0.6	7.95	5.6
	18	50.0	22.8	455.	nd	0.5	7.99	6.1
	—	—	—	—	—	—	—	—

Table A.1. (Continued)

Run	Sample	Time (hrs.)	Ca	C	Fe	Mg	pH	Cl[e]
W251088	1	0.0	2.2	85.	nd	0.2	—	—
	3	0.5	6.5	88.	nd	0.1	—	—
	5	1.0	11.9	91.	nd	nd	—	—
	7	2.0	19.8	96.	nd	0.1	—	—
	9	3.5	26.4	110.	nd	nd	—	—
	11	8.0	33.4	95.	nd	0.2	—	—
	13	15.0	35.6	104.	nd	0.2	—	—
	15	35.0	37.2	96.	nd	0.3	—	—
	17	52.0	36.8	—	nd	0.3	—	—
	19	—	23.6	—	nd	0.2	—	—
W141288	2	0.0	1.2	87.	0.1	0.5	—	—
	4	0.5	6.6	77.	nd	0.3	—	—
	6	1.0	9.6	79.	nd	0.1	—	—
	8	2.0	17.9	83.	nd	0.3	—	—
	10	3.5	24.4	84.	nd	0.3	—	—
	12	7.5	30.4	87.	nd	nd	—	—
	14	14.7	34.8	91.	nd	0.3	—	—
	16	37.2	34.6	87.	0.2	0.3	—	—
	18	61.8	36.4	90.	nd	0.3	—	—
	—	—	—	—	—	—	—	—
W020389	2[d]	0.0	3.4	0.5	0.1	0.1	8.4	—
	4	0.5	1.1	0.7	0.1	nd	8.7	—
	6	1.0	2.0	1.2	0.1	nd	6.3	—
	8	2.0	2.9	1.3	0.1	0.2	7.3	—
	10	4.0	4.4	1.8	0.1	nd	7.6	—
	12	7.0	5.1	2.1	0.1	0.1	8.6	—
	14	12.0	5.9	2.4	nd	nd	8.8	—
	16	20.0	6.1	2.5	0.1	nd	9.0	—
	18	45.0	6.5	2.6	0.1	0.1	9.2	—
	20	91.0	6.7	3.5	nd	nd	9.0	—
	22	—	27.0	—	nd	3.0	—	—
W200389	2	0.0	1.8	195.	0.2	0.2	—	—
	4	0.5	8.1	200.	nd	nd	—	—
	6	1.0	14.6	194.	nd	nd	—	—
	8	2.0	24.4	200.	0.3	nd	—	—
	10	4.0	43.6	208.	nd	0.1	—	—
	12	7.0	63.0	238.	0.1	0.2	—	—
	14	12.7	86.5	203.	nd	0.1	—	—
	16	37.2	106.5	213.	nd	0.1	—	—
	18	46.0	108.0	214.	nd	nd	—	—
	20	69.0	109.0	207.	nd	0.2	—	—
	22	—	29.6	—	nd	3.4	—	—
W040489	2	0.0	4.7	0.	0.4	0.4	—	70.9
	4	0.5	19.0	1.	0.7	0.2	—	69.1
	6	1.0	27.2	1.	0.9	0.1	—	67.4
	8	1.5	32.0	2.	0.8	0.1	—	65.7
	10	2.5	35.8	7.	0.5	nd	—	64.1
	12	4.0	39.8	10.	0.2	nd	—	62.5
	14	6.5	42.0	11.	0.3	nd	—	60.9
	16	15.5	44.2	12.	0.4	0.3	—	59.4
	18	38.8	49.4	12.	0.3	0.1	—	57.9
	20	69.0	48.0	—	0.2	0.2	—	56.5
	22	—	28.6	—	nd	3.7	—	—

[a] Sample froze, CO_2 loss likely.
[b] Water from infill lines.
[c] milliQ water.
[d] Sample contaminated.
[e] From HCl, assumed to decrease by 2.5% with each sample.

Part B.
Igneous and Metamorphic Systems

Part II
Igneous and Metamorphic Systems

Fluid-Mineral Interactions: A Tribute to H. P. Eugster
© The Geochemical Society, Special Publication No. 2, 1990
Editors: R. J. Spencer and I-Ming Chou

The aluminum content of hornblende in calc-alkaline granitic rocks: A mineralogic barometer calibrated experimentally to 12 kbars*

WARREN M. THOMAS and W. G. ERNST[1]

Institute of Geophysics and Planetary Physics, University of California, Los Angeles, California 90024-1567, U.S.A.

Abstract—Partial melting experiments on a natural hornblende-bearing tonalite have been conducted in a piston-cylinder apparatus in order to determine the total aluminum content of hornblende in equilibrium with quartz + K-feldspar + plagioclase + biotite + epidote + sphene + Fe-Ti oxide + melt + fluid at 750°C as a function of pressure in the range 6–12 kbar. A hornblende of similar composition to that in the tonalite itself, but possessing a higher aluminum content, was added to the starting material. In runs of one week duration, rims formed on both types of hornblende grains, with the compositions converging from initially different values in the two hornblende populations.

Using this technique, a calibration for the total Al content of hornblende as a function of aqueous fluid pressure at a constant, near solidus temperature was derived: $P (\pm 1.0 \text{ kbar}) = -6.23 + 5.34$ Al^T. The pressures calculated from this equation are appropriate for granitoid melts crystallizing at approximately 750°C. They agree well with a previous experimental calibration, but are too low by more than a kbar when extrapolated to low pressures.

INTRODUCTION

DETERMINATION of the pressure of crystallization of calc-alkaline plutons has historically been difficult; most efforts have relied largely on constraints gleaned from appropriate mineral assemblages, if any, in the contact aureole. HAMMARSTROM and ZEN (1986) and HOLLISTER *et al.* (1987) empirically calibrated an approximately temperature-independent barometer based on the Al content of hornblende coexisting with quartz + K-feldspar + plagioclase + biotite + sphene + Fe-Ti oxide (+ melt) in calc-alkaline granitoids. In practical terms, the nearly isothermal nature of the solidus at middle and deep continental crustal levels allows Al^T in the amphibole to be employed as a mineralogic barometer. RUTTER and WYLLIE (1988) published a note describing agreement with this naturally-calibrated barometer of three experimentally altered hornblende compositions in a tonalite, but under vapor-absent conditions and with garnet in the subsolidus assemblage. JOHNSON and RUTHERFORD (1988, 1989) have recently calibrated this barometer with reversed experiments in the pressure range 2 to 8 kbars using natural starting material and an internally heated gas apparatus.

The purpose of the present study was to calibrate experimentally this hornblende barometer at higher aqueous fluid pressures using a piston-cylinder apparatus. The basic approach to the problem was to melt partially a natural hornblende-bearing tonalite to which a second, chemically similar but more aluminous hornblende had been added, and to measure convergence in rim compositions of the two hornblendes as a function of pressure.

STARTING MATERIAL

The starting material for the experiments was a tonalite from the Late Cretaceous Josephine Mountain Intrusion from the San Gabriel Mountains, California. The mineral assemblage of this sample is quartz, plagioclase, K-feldspar, hornblende, biotite, epidote, magnetite, sphene, and zircon; the rock is fresh and unaltered. Because the amount of alkali feldspar is somewhat variable from sample to sample, 10 wt% natural orthoclase of composition $Or_{95}Ab_{05}An_{00}$ was added to the crushed sample to insure its presence in each experimental charge. The natural granitic material was lightly crushed to a maximum grain size of 0.1 mm, but was not sieved to avoid eliminating minerals which might be preferentially represented in the finer-grained fractions. For most of the critical experiments, the starting material was altered by adding a second hornblende similar to that in the tonalite and crushed to the same grain size. This introduced hornblende, which has a higher aluminum content and lower total iron content than that of the indigenous hornblende, is from a high-pressure garnet hornblendite from Santa Catalina Island, California (SORENSEN, 1988; JACOBSON and SORENSEN, 1986). Although the coexisting mineral assemblage of this rock consists primarily of garnet + zoisite, it was chosen in order to obtain a hornblende with a very high aluminum content to contrast with that from the tonalite. The two hornblendes are easily distinguished in the experimental charges, both optically and by the compositions of their cores. The hornblende from the tonalite is slightly zoned, and inasmuch as spatial relationships

* Institute of Geophysics and Planetary Physics Publication No. 3241.

[1] Present address: School of Earth Sciences, Stanford University, Stanford, CA 94305-2210.

60 W. M. Thomas and W. G. Ernst

Table 1. Composition of starting material, in weight percent

Weight percent	Natural tonalite*	Added K-feldspar	Introduced hornblende	Starting material
SiO_2	57.2	64.81	42.84	56.7
TiO_2	0.78	0.02	0.58	0.69
Al_2O_3	18.0	18.15	17.59	18.0
Cr_2O_3	n.a.	0.04	0.29	0.06
FeO**	6.06	0.05	9.91	5.84
MnO	0.12	0.02	0.05	0.11
MgO	2.7	0.00	10.98	3.13
CaO	6.42	0.00	11.42	6.26
Na_2O	3.9	0.53	2.05	3.47
K_2O	1.89	16.55	0.56	3.07
P_2O_5	0.43	n.a.††	n.a.	0.36†
CO_2	0.28	n.a.	n.a.	0.23†
H_2O†	0.69	n.a.	n.a.	0.57†
Total	98.47	100.17	96.27	98.49†

* XRF analysis by A. Barth.
** All Fe reported as FeO.
† Minimum values.
†† Not analyzed (n.a.).

are partially lost through crushing, the zoning contributes to the uncertainty in original Al content of ± 0.05 (1σ) per 23-oxygen formula. The composition of this modified starting material is given in Table 1. Chemical analyses and cation proportions of hornblendes are presented in Table 2. Except for Al and Fe content, the amphiboles employed in the experiments are closely comparable. Analyses are reported with all iron treated as ferrous, although they are compatible with ferric iron contents ranging from 5–20%. Changes in the estimated amount of ferric iron generally affects the calculated Al^T by less than 2% relative. This starting material was run in sealed capsules with 10 wt% distilled H_2O, enough to saturate the melt experimentally generated by partial fusion.

EXPERIMENTAL PROCEDURE

All experiments were conducted in a piston-cylinder apparatus at UCLA using the NaCl cell furnace assembly of BOETTCHER et al. (1981); no pressure corrections were deemed necessary. The press was calibrated employing the equilibrium albite = jadeite + quartz (JOHANNES et al. 1971); nominal pressure is considered accurate to 200 bars. Temperatures were measured with $Pt-Pt_{90}Rh_{10}$ thermocouples and are thought to be known to $\pm 5°C$.

All experiments were run in Pt or Ag capsules of 11.4 mm length and 2.79 mm o.d. Because as large a charge as possible was desired, these experiments were performed

Table 2. Compositions of original and experimental hornblendes (wt.%)

	Original		6 kbar		8 kbar		10 kbar		12 kbar	
	A*	B*	A	B	A	B	A	B	A	B
SiO_2	44.12	42.84	40.66	41.86	39.44	40.33	38.68	38.71	39.00	39.40
TiO_2	1.24	0.58	0.76	0.98	1.16	1.64	0.87	1.01	0.61	0.65
Al_2O_3	9.68	17.59	13.26	12.81	15.02	14.55	16.82	14.72	20.00	19.61
Cr_2O_3	0.02	0.29	0.08	0.11	0.00	0.00	0.05	0.02	0.00	0.07
FeO†	17.72	9.91	15.44	13.78	18.11	18.47	16.74	16.66	17.06	16.30
MnO	0.54	0.05	0.40	0.30	0.36	0.24	0.33	0.27	0.20	0.24
MgO	10.48	10.98	9.28	11.18	7.90	8.15	8.57	8.48	6.34	6.66
CaO	11.76	11.42	11.88	11.96	11.45	11.50	11.34	11.31	10.86	11.04
Na_2O	1.39	2.05	1.47	1.53	1.92	1.76	1.95	1.93	2.08	2.13
K_2O	1.07	0.56	1.92	1.75	1.58	1.70	1.77	1.56	1.78	1.72
Total	98.02	96.27	95.15	96.26	96.94	98.34	97.12	94.67	97.93	97.82

Cations per 23 oxygens

Si	6.64	6.26	6.30	6.34	6.07	6.11	5.90	6.07	5.86	5.91
Al^T	1.72	3.03	2.42	2.29	2.72	2.60	3.03	2.72	3.54	3.47
Al^{IV}	1.36	1.74	1.70	1.66	1.93	1.89	2.10	1.93	2.14	2.09
Al^{VI}	0.36	1.29	0.72	0.63	0.79	0.71	0.93	0.79	1.40	1.38
Ti	0.14	0.06	0.09	0.11	0.13	0.19	0.10	0.12	0.07	0.07
Cr	0.00	0.03	0.01	0.01	0.00	0.00	0.01	0.00	0.00	0.01
Fe^{2+}	2.23	1.21	2.00	1.75	2.33	2.34	2.14	2.18	2.14	2.04
Mn	0.07	0.01	0.05	0.04	0.05	0.03	0.04	0.04	0.02	0.03
Mg	2.35	2.39	2.14	2.52	1.81	1.84	1.95	1.98	1.42	1.49
Ca	1.90	1.79	1.97	1.94	1.89	1.87	1.86	1.90	1.75	1.78
Na	0.41	0.58	0.44	0.45	0.57	0.52	0.58	0.59	0.61	0.62
K	0.21	0.10	0.38	0.34	0.31	0.33	0.34	0.31	0.34	0.33

* A = indigenous hornblende; B = introduced hornblende.
† All Fe reported as FeO.

unbuffered with regard to oxygen; f_{O_2} imposed by the furnace assembly and press is between Ni-NiO and hematite-magnetite buffers, as corroborated by running charges of those assemblages at the conditions of the experiments. Moreover, JOHNSON and RUTHERFORD (1988, 1989) stated that they found a minimal effect in their experiments by variation of oxygen fugacity between the Ni-NiO and the MnO-Mn_3O_4 buffers. No evidence was found in any of the runs of oxidation or reduction of Fe in the charges.

All experiments were of one week duration. Initially, a variety of temperatures were used, ranging from 650 to 900°C, but the final critical experiments were conducted exclusively at 750°C to increase reaction rates. Experiments were performed at pressures of 6, 8, 10, and 12 kbars, with one 2-kbar experiment conducted hydrothermally in a cold-seal pressure vessel.

At the conclusion of each experiment, the capsule was mounted on a glass slide, and ground and polished for the electron microprobe. Even small degrees of partial melting caused the sample to cohere and behave like a typical rock in the thin-sectioning process. Each experimental product was examined optically and subsequently with the electron microprobe. Amphiboles were chemically analyzed using a Cameca Camebax microprobe employing 15 kv accelerating voltage, 10 nanoamp sample current and a 1 μm beam diameter. ZAF corrections were applied and major elements are considered accurate to 1–2% relative, and minor elements to 2–10% relative.

Zoned amphiboles were analyzed—and portrayed in Fig. 1—for only those well-constrained runs in which all starting phases were present in addition to melt at the conclusion of the experiments.

RESULTS

In all experiments, melt was generated and rims of distinctive composition up to 30 μm wide were produced on hornblendes in the charge. In most cases, the change in composition appeared texturally to have been accomplished by diffusional modification of the original grain rim; in a few samples, overgrowths were observed, as evidenced by euhedral rims in contact with melt. In all cases, the rims exhibited compositions with allowed amphibole stoichiometries. The cores retained their original compositions; thus the two populations of hornblendes were readily distinguished. The total aluminum content in the indigenous hornblende was altered from an original value of 1.72(5) per 23-oxygen formula unit to a rim maximum of 3.54; for the introduced hornblende, the original value of 3.03(3) changed to a minimum of 2.29 and a maximum of 3.47. The aluminum content of the rims is a clear function of pressure, with the higher values associated with higher pressures, as seen in the empirical and earlier experimental calibrations. Note also that, although the two initial amphiboles possessed contrasting Fe as well as Al contents, Table 2 demonstrates that experimentally produced rim compositions are sensibly identical—hence chemical equilibrium is indicated.

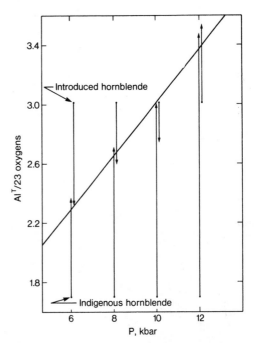

FIG. 1. Plot of the change in total Al of hornblende rims as a function of pressure in one week runs at 750°C. The base of the arrows represents the average initial Al^T of the indigenous and introduced hornblende and the heads of the arrow the *most extreme* values in the re-equilibrated rims. The line plotted represents a linear least squares fit to the midpoints between the heads of the arrows. Note that at 12 kbars, the compositions of both hornblendes have become more aluminum rich than the starting materials.

The results of the new experiments at 750°C are summarized in Fig. 1. Note that for each pressure, the base of the arrows represents the average Al content of cores for the two types of hornblendes, with the head of the arrows corresponding to the *most extreme* rim value in the runs for that pressure. In each case, excepting at 12 kbar, the aluminum compositions of the two hornblendes have converged; indeed, because the extreme values have been plotted, the compositions appear to have "passed" each other. Assuming that the one-sigma error in the formula numbers is 1%, the compositions of the runs overlap within 2σ at 6 and 8 kbar. The compositions of the hornblendes represented by the arrowheads in Fig. 1 are included in Table 2. Fitting a line to the midpoints between the arrowheads for 6, 8, 10 and 12 kbars yields a calibration of

$$P\,(\pm 1.0\ kbar) = -6.23\,(\pm 0.07)$$
$$+\ 5.34\,(\pm 0.40)(Al^T),\quad r^2 = 0.94.$$

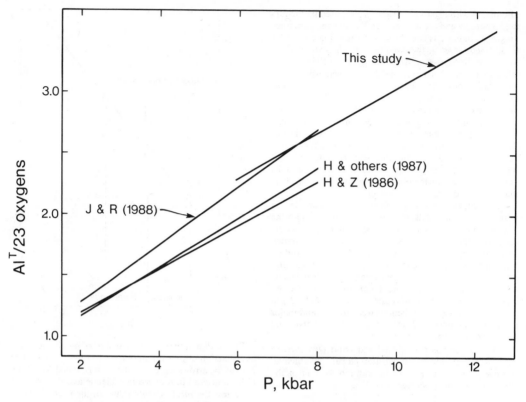

FIG. 2. Comparison of calibration from this study with those from previous studies. Sources for curves are: H & Z (1987) = HAMMARSTROM and ZEN (1987); H & others (1988) = HOLLISTER and OTHERS (1988); and J & R (1988) = JOHNSON and RUTHERFORD (1988). Curves are shown only for the pressure range from which the particular calibration was derived.

A linear relationship was assumed, and the data fitted and errors calculated using the treatment of YORK (1969). Uncertainties in the slope and intercept quoted are one sigma. The stated uncertainty in the predicted pressure is a function of the aluminum content and varies, at the 95% confidence level, from ±0.72 kbar in the midrange of the calibration to ±1.33 at the upper extreme. An unweighted mean of ±1.0 kbar has been quoted in the equation above.

DISCUSSION

Comparisons of the calibrations derived from the data of this study with the results of the three previous studies are shown in Fig. 2. The curves are shown only for the pressure ranges from which the particular calibrations were derived. It can be seen that in the range in which they overlap, that is, from 6 to 8 kbars, the present calibration and that of JOHNSON and RUTHERFORD (1988, 1989) agree closely, yielding nearly identical pressures for a given

Al content, although the slopes are different. Both experimental calibrations predict a lower pressure for a given Al^T content than the empirical curves. At Al contents below the region of calibration of this study, pressures calculated from a linear extrapolation of our results are lower by 1–2 kbars than either the empirical calibrations or the JOHNSON and RUTHERFORD (1988, 1989) experiments. Hornblendes in the 2-kbar hydrothermal run showed no adjustment of rim compositions and thus cannot be used to constrain the low-pressure end of this calibration. Comparisons of the two experimental calibrations are given in Table 3.

Characterization of the synthetic rim composition is problematic, inasmuch as the rims are gradational over several tens of micrometers. The most extreme values were chosen inasmuch as these are invariably at the edge of the grain and thus in contact with the melt. If the mean compositions of the rims from the two amphiboles comprising a bracket were actually identical, but each with a dispersion of values around the mean, the choice of the most

Table 3. Comparison of barometers

Al^T/23 oxygens	This study	J and R*
1.5	1.8 kbar	2.9 kbar
2.0	4.5	5.0
2.5	7.1	7.1
3.0	9.8	9.2
3.5	12.5	11.4

* JOHNSON and RUTHERFORD (1988, 1989).

Al-rich value for the indigenous hornblende and the least Al-rich value for the introduced hornblende would result in brackets that have apparently "passed" each other. Thus, this feature of the plot as presented may not necessarily imply disequilibrium. On the other hand, the indigenous hornblende was much closer to the composition in equilibrium with the melt than was introduced hornblende, and the latter may have initially adjusted to a more Al-poor, metastable composition. Inasmuch as all experiments in the present study were of the same duration, it cannot be determined from them if the introduced hornblende would have readjusted to more aluminous compositions with time.

Consideration of the compositions of experimentally produced amphibole rim compositions shows that both Al^{IV} and Al^{VI} increase with pressure. Fitting Al^{VI} against Al^{IV} for individual analyses in Table 2 results in the equation

$$Al^{VI} = -1.74 + 1.38Al^{IV}, \quad r^2 = 0.65,$$

suggesting that some of the variability in aluminum content is accounted for by a Tschermak type substitution, $R^{2+} + Si \rightleftarrows Al^{IV} + Al^{VI}$. Fitting the rim compositions of the indigenous hornblende and its average initial composition yield

$$Al^{VI} = -1.06 + 1.03Al^{IV}, \quad r^2 = 0.79.$$

The same treatment for rim and initial compositions of the introduced hornblende yields no significant correlation, suggesting that the introduced hornblende may be adjusting its composition by a different exchange mechanism.

In conclusion, results of the present study are consistent with the calibration of JOHNSON and RUTHERFORD (1988, 1989) and extend the exper-

imental evidence for increasing aluminum content of hornblende as a function of pressure to 12 kbar.

Acknowledgements—This experimental study was conducted at UCLA, supported by the Institute of Geophysics and Planetary Physics and by the National Science Foundation through grant EAR86-16624. Valuable reviews of the manuscript and helpful comments were provided by M. Cho, J. M. Hammarstrom, Art Montana, M. J. Rutherford, S. S. Sorensen, E. D. Young, and E-an Zen. The authors thank the above-named institutions and researchers for their support. The sample of tonalite was supplied by A. Barth and the aluminous hornblende from Catalina Island by G. Bebout.

REFERENCES

BOETTCHER A. L., WINDOM K. E., BOHLEN S. R. and LUTH R. W. (1981) Low-friction, anhydrous, low- to high-temperature furnace assembly for piston-cylinder apparatus. *Rev. Sci. Instr.* **52**, 1903–1904.
HAMMARSTROM J. M. and ZEN E-AN (1986) Aluminum in hornblende: an empirical geobarometer. *Amer. Mineral.* **71**, 1297–1313.
HOLLISTER L. S., GRISSOM G. C., PETERS E. K., STOWELL H. H. and SISSON V. B. (1987) Confirmation of the empirical correlation of aluminum in hornblende with pressure of solidification of calc-alkaline plutons. *Amer. Mineral.* **72**, 231–239.
JACOBSON C. E. and SORENSEN S. S. (1986) Amphibole compositions and metamorphic history of the Rand Schist and the greenschist unit of the Catalina Schist, southern California. *Contrib. Mineral. Petrol.* **92**, 308–315.
JOHANNES W., BELL P. M., MAO H. K., BOETTCHER A. L., CHIPMAN D. W., HAYS J. F., NEWTON R. C. and SEIFERT F. (1971) An interlaboratory comparison of piston-cylinder pressure calibration using the albite breakdown reaction. *Contr. Mineral. Petrol.* **32**, 24–38.
JOHNSON M. C. and RUTHERFORD M. J. (1988) Experimental calibration of an aluminum-in-hornblende geobarometer applicable to calc-alkaline rocks (abstr.) *EOS* **69**, 1511.
JOHNSON M. C. and RUTHERFORD M. J. (1989) Experimental calibration of the aluminum-in-hornblende geobarometer with application to Long Valley caldera (California) volcanic rocks. *Geology*, (In press).
RUTTER M. J. and WYLLIE P. J. (1988) Experimental calibration of hornblende as a proposed empirical geobarometer (abstr.). *EOS* **69**, 87.
SORENSEN S. S. (1988) Petrology of amphibolite facies mafic and ultramafic rocks from the Catalina Schist, southern California: metasomatism and migmatization in a subduction zone metamorphic setting. *J. Metamorphic Geol.* **6**, 405–435.
YORK D. (1969) Least squares fitting of a straight line with correlated errors. *Earth Planet. Sci. Lett.* **5**, 320–324.

Fluid-Mineral Interactions: A Tribute to H. P. Eugster
© The Geochemical Society, Special Publication No. 2, 1990
Editors: R. J. Spencer and I-Ming Chou

On plagioclase thermometry in island arc rocks: Experiments and theory

B. D. MARSH and JOHN FOURNELLE

Department of Earth and Planetary Sciences, The Johns Hopkins University, Baltimore, Maryland 21218, U.S.A.

JAMES D. MYERS

Department of Geology, University of Wyoming, Laramie, Wyoming 82071, U.S.A.

and

I-MING CHOU

959 National Center, U.S. Geological Survey, Reston, Virginia 22092, U.S.A.

Abstract—The abundant, anorthite-rich plagioclase phenocrysts of island arc lavas have often led to speculation on the importance of water in these magmas. A set of melting experiments at 1 atmosphere has been conducted on four Aleutian lavas (basalt to andesite) in order to calibrate plagioclase geothermometers and to obtain phase information that can be analytically extrapolated to higher pressures, temperatures and water contents. When considered on an 8-oxygen basis, both the melt and crystal compositions form a pseudo-binary that has a geometry (*i.e.,* liquidus and solidus slopes) very nearly the same as that found for the true binary by Bowen in 1913. The observed solidus slope is ~14% An (mole) per 100°C change in temperature. A number of geothermometers (Kudo-Weill, Mathez, Burnham, TRACE, SILMIN) applied to these data meet with varying success and are quantitatively evaluated. Overall the data and calculations agree in solidus slope. It is observed that the concentration of soda in the melt largely controls plagioclase composition and thus also whether or not plagioclase fractionation will either raise or lower the melt silica content. The addition of water to the system causes either a slight increase (Burnham's model) in An-content (~1% (mole) An per 10% (mole) water) or a slight decrease (SILMIN). It is clear, nevertheless, that water is not an important factor in deciding plagioclase composition in magmas of this compositional range.

INTRODUCTION

PLAGIOCLASE is exceedingly common in lavas of island arcs, generally being the liquidus phase and generally comprising 80 percent or more of the phenocrysts. Of all volcanic rocks, arc lavas contain plagioclase that can be unusually rich in anorthite. It is not uncommon to find compositions of An_{80} to An_{90} throughout the suite from basalts to andesites. Sometimes, large unzoned megacrysts can be found with compositions approaching An_{95} or more. Some compositional zoning is usually present; it is sharp but commonly not strong (*i.e.,* ±5%) in basaltic lavas and increases in amplitude with increasing silica content of the lava. The overall crystal composition and zoning records an intimate history of changes in magmatic intensive parameters (temperature, pressure, chemical potential) in the near-surface environment. This history is central to understanding the near surface behavior of the magma as well as characterizing the near-surface intensive parameters critical to understanding the ultimate origin of the magma itself.

A great deal has been said on the meaning of plagioclase composition in arc rocks. YODER (1969) has suggested that because the melting point of an-

orthite is more sensitive than that of albite to increasing water pressure, high An content reflects high magmatic water content. At the same time, however, high water content promotes normally and strongly zoned crystals whereas undersaturation with water promotes reversely and weakly zoned crystals (PRINGLE *et al.,* 1974; MARSH, 1976). The plagioclase thermometers developed by KUDO and WEILL (1970) with refinements by MATHEZ (1973) and DRAKE (1976) when applied to basaltic arc rocks sometimes give temperatures near or in excess of 1300°C, which at first seem unrealistic. To make the temperature more realistic, an arbitrary amount of water can be assumed present, but this itself begs the original question.

Much of this uncertainty in understanding plagioclase composition in basaltic arc rocks stems from the paucity of experiments explicitly designed to study plagioclase thermometry. The seminal work by EGGLER (1972; EGGLER and BURNHAM, 1973) on andesite phase equilibria has been well augmented by BAKER and EGGLER (1983; 1987), and GROVE *et al.* (1982); mainly ancillary plagioclase compositional data are available from these studies. Overall, however, there is little information

B. D. Marsh *et al.*

on plagioclase composition in terms of understanding the influence of bulk composition, water, and pressure.

We attempt here to reconcile this need by presenting data from experiments on plagioclase equilibria in four lavas (basalt to andesite) from the Aleutian Islands, Alaska. The compositions of both the plagioclase and its host liquid (glass) have been analyzed and compared with those of the lava. In each of the four compositions, the plagioclase pseudo-binary (*i.e.*, 8-oxygen mineral and melt formula basis) is found to exist in the melt with nearly the same geometry (*i.e.*, slope of liquidus and solidus) as that originally found by BOWEN (1913) for the true binary. Comparisons with the melt compositions, suggest that plagioclase composition itself is controlled largely by the soda content of the liquid; the apparent distribution coefficient is near unity. The liquidus temperature of the melt, however, is controlled largely by alumina. The application of the common thermometers to these compositions gives fair agreement, with temperature differences ranging from 10° to nearly 100°C. The slope of the solidus (or liquidus) is difficult to predict using most thermometers, but it is matched well by the silicate solution model SILMIN (*e.g.*, GHIORSO *et al.*, 1985). Using BURNHAM's (*e.g.*, 1979) and Ghiorso's solution models, the results of these 1 atmosphere experiments have been extended to higher pressures where the effects of water can be evaluated. It is found that the liquidus is of nearly constant plagioclase composition and that with increasing water content plagioclase composition becomes slightly more anorthitic at the rate of about 1% An (mole) per 10% water (mole). This small effect suggests that plagioclase composition is primarily controlled by the soda and alumina content of the original magma. Highly anorthitic crystals probably grew from melts low ($\approx$wt. 1%) in soda and high in alumina. If the soda content of the melt is beyond a critical amount, the plagioclase is more siliceous than the melt itself and plagioclase fractionation drives the magma to silica-undersaturation. Plagioclase fractionation in a single suite of lavas can produce both increasing and decreasing concentrations of silica.

MATERIALS AND EXPERIMENTAL PROCEDURES

The Aleutian lavas for this study (Table 1) were chosen as being typical and to span the range of silica exhibited by island arc lavas. These lavas are broadly similar with high alumina and lime, moderate soda and potash, and low magnesia. The two basaltic samples were chosen to accentuate the effect of increasing alumina and soda at a

Table 1. Rock compositions, norms, and activities

	Bulk composition (wt.%)			
	AT-4	AT-6	AD-61	CB-12
SiO_2	48.29	49.05	52.14	57.30
TiO_2	0.90	0.89	0.67	0.77
Al_2O_3	17.26	20.32	19.44	18.21
Fe_2O_3	6.17	3.31	4.79	2.93
FeO	4.16	6.52	4.25	4.54
MnO	0.18	0.18	0.19	0.16
MgO	7.61	3.91	3.73	3.06
CaO	11.56	10.44	9.79	8.01
Na_2O	2.42	3.36	2.92	3.04
K_2O	0.62	0.89	0.59	1.24
P_2O_5	0.19	0.27	0.19	0.19
Total	99.31	99.14	98.70	99.45

	8-Oxygen norm (mole%)			
Apatite	0.3	0.6	0.3	0.3
Ilmenite	1.2	1.1	0.9	1.0
Magnetite	5.5	2.9	4.3	2.5
Orthoclase	4.1	5.0	3.4	7.2
Albite	22.6	29.7	26.8	27.3
Anorthite	35.6	37.4	39.4	32.6
Diopside	16.3	5.4	5.8	3.3
Hedenburgite	1.1	3.5	1.5	1.5
Enstatite	12.5	3.0	7.0	6.4
Ferrosilite	0.8	1.9	1.8	3.1
Forsterite	—	3.1	—	—
Fayalite	—	2.0	—	—
Quartz	0.1	—	8.8	14.8

		Burnham activities			
a_{ab}	=	0.226	0.297	0.268	0.273
(a_{ab})*	=	0.266	0.347	0.302	0.345
a_{an}†	=	0.235	0.300	0.309	0.252

* With *ad hoc* addition of orthoclase component.
† Corrected for pyroxene interaction.

constant amount of silica. The two more andesitic compositions test the effect of increasing silica (CB-12) and the effect of low soda with high alumina and moderate silica (AD-61). A dacite (66 wt.% SiO_2) proved reluctant to crystallize regardless of run duration and history of heating.

All experimental runs were made at atmospheric pressure by the platinum loop method in a gas-mixing furnace (DONALDSON *et al.*, 1975) at the Johnson Space Center (by I-M.C.) and at Johns Hopkins (B.D.M.). The fugacity of oxygen was maintained (except as otherwise noted) at that described by the nickel to nickel oxide (NNO) synthetic buffer reaction. The temperature of the runs, their durations, and products are given by Table 2. Once the liquidus for each sample was determined (indicated as "rock liquidus" in Fig. 1), a sample was fused at a temperature slightly ($\approx$5°C) above its liquidus and then dropped to the desired run temperature. The duration of each experiment was generally about a week, which resulted in plagioclase crystals of about 0.5 × 0.1 mm. Each charge was sliced and made into a thin section, and the compositions of its crystals and glass were measured with

Table 2. Summary of experimental runs

Rock	Run no.	Temp. (°C)	Duration (days)	Average compositions (An, Ab, Or; mole%) Crystal	Liquid
AT-4	16	1233	4	79.3, 20.4, 0.3	68.9, 26.8, 4.3
	17	1223	3	78.0, 21.6, 0.4	68.2, 26.8, 4.3
	18	1213	3	75.7, 23.9, 0.4	66.0, 29.5, 4.6
	19	1204	4	75.6, 23.9, 0.5	67.8, 27.4, 4.8
	23	1177	6	72.0, 27.2, 0.7	66.2, 28.5, 5.4
	24	1165	6	72.6, 26.4, 1.0	65.2, 28.9, 5.9
	26	1192	1 hr.*	73.4, 25.9, 0.7	65.7, 29.8, 4.6
	27	1153	4	71.7, 27.3, 1.1	67.3, 27.3, 5.5
AT-6	4	1262	2	79.2, 20.4, 0.5	
	5	1262	4	79.8, 19.7, 0.5	63.2, 31.1, 5.7
	6	1226	3**	74.5, 24.7, 0.8	60.3, 32.3, 7.4
	7	1244	6	76.7, 22.8, 0.5	61.6, 31.9, 6.6
	8	1225	6	72.9, 26.3, 0.8	60.1, 32.7, 7.2
AD-61	10	1265	7	77.6, 22.1, 0.3	63.1, 32.0, 4.8
	11	1233	7	73.7, 25.9, 0.4	60.6, 33.7, 5.7
	12	1250	5	76.0, 23.6, 0.4	60.7, 34.0, 5.3
CB-12	11	1233	4	70.1, 20.3, 0.6	50.9, 30.4, 4.7
	12	1223	5	67.1, 31.1, 1.7	48.1, 41.7, 10.2

*: power failure.
**: NNO buffer for 2 days and 1 day at 2 log units higher.

the electron microprobe at the Geophysical Laboratory, Washington, D.C. (by J.D.M.). The procedures of data reduction of this instrument are described by FINGER and HADIDIACOS (1972).

Five crystals in each thin section were analyzed, each in five areas, and the glass was randomly analyzed in at least five areas. Generally no more than about 1% (mole) zoning was detected in any of the crystals; in the andesite slightly more variation occurred. The glass appears for the most part homogeneous, but losses of sodium and iron were heterogeneously distributed about the charge due to surface loss of sodium and absorption of iron by the platinum loop. Apparently, even with substantial surface loss of sodium in one run, the composition of the crystals seems unaffected.

The modal amount of crystals in each thin section was measured by point counting, and these modes were checked by mass balance calculations utilizing the known composition of the crystals, glass, and the initial (crystal-free) composition. This check generally shows fair to poor agreement between the counted and calculated modes, presumably due to the difference between surface and volumetric estimates of crystal content.

EXPERIMENTAL RESULTS

For purposes of comparison, each average glass analysis is cast into an equivalent feldspar-like mineral formula based on eight oxygens. (For convenience, iron is treated as being solely ferrous. It could just as easily be distributed according to the NNO buffer, but this makes little difference as will later be seen.) Each experiment provides a single average glass and crystal composition that are plot-

ted in Fig. 1 on a plagioclase pseudo-binary for each rock. The seven runs on AT-4 are the most complete and above 1185°C they define a straight solidus having a slope ($\approx 14\%$ An (mole)/100°C), which is similar to that determined by BOWEN (1913) in the true plagioclase binary itself. We will see that generally this same slope is a central ingredient of all plagioclase geothermometers, but its similarity to Bowen's original binary has apparently not been previously noticed. Although only three independent runs were made for two of the lavas and two for another, they are sufficiently separated in temperature to suggest solidus and liquidus slopes similar to those of AT-4. We thus assume that the following general description of crystallization holds true for each of these lavas.

Below about 1185°C the solidus and (pseudo-)-liquidus of AT-4 become vertical; the crystal composition becomes invariant with decreasing temperature. At first this might seem to reflect disequilibrium, but recalling the basic mass conservation law for crystallization in binary solid-solutions, this is to be expected. That is, equilibrium crystallization in a binary solid solution produces a solid whose final composition is the same composition as that of the initial (i.e., crystal free) liquid. At this point the liquid is exhausted of one or more of the components necessary to produce a crystal of a given stoichiometry; here the melt becomes exhausted of lime and alumina. Below this temperature further crystallization is prohibited unless the melt is replenished in the exhausted feldspar ingredients. Since in these rock compositions plagioclase crystallizes alone over a temperature interval of about 100°C, the point of exhaustion comes before the appearance of a second solid phase. When there is multiple saturation with, say, plagioclase, olivine, and/or pyroxene, the mass of melt shrinks faster and this point of exhaustion is progressively (depending on the rate of appearance of other phases) shifted to lower temperatures. In addition, the width (ΔW) of the liquidus-solidus loop also influences when this invariance is reached, and ΔW is directly proportional to the "albite" content of the melt as shown by Fig. 2.

Of the experiments themselves, no crystal is more anorthitic than about An_{80} (AT-4 at 1233°C, AT-6 at 1262°C), but if the respective solidi are extrapolated to the liquidus temperature of each rock (determined independently), the earliest plagioclase compositions can potentially become unusually anorthitic. We find for AT-4 at 1250°C, $An_{81.5}$; for AT-6 at 1300°C, An_{85}; for AD-61 at 1300°C, An_{83}; and for CB-12 at 1290°C, An_{78}. These projected

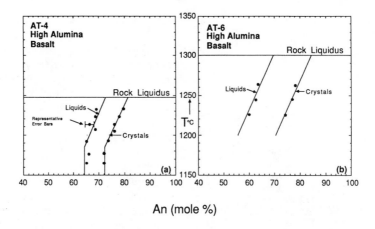

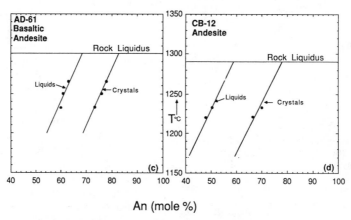

FIG. 1. Binary plots depicting the experimental crystal and liquid compositions found in the experiments. All compositions have been recalculated to an 8-oxygen basis. The liquidi and solidi have been fit by eye using the slope from the AT-4 run results with some compensation for the uncertainty of each analysis.

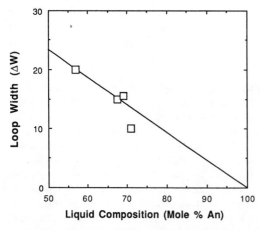

FIG. 2. The apparent liquidus-solidus loop width (ΔW) as measured from Fig. 1 plotted against the liquid composition assuming $\Delta W = 0$ at An_{100}.

anorthite-rich compositions are in part a reflection of the high (1 atm., dry) liquidus temperatures of these rocks, but it is of course both the composition of the melt and its liquidus temperature that decide the composition of the first-formed plagioclase. At a given temperature the chemical component in the melt most critical in controlling plagioclase composition is soda, whereas it is probably alumina that largely decides the liquidus temperature. This is seen by comparing the results for AT-4 and AT-6. Although they contain roughly the same amounts of silica and lime, AT-6 has substantially more alumina and soda than AT-4. At any temperature, the resulting plagioclase is found to be more An-rich in the low soda, low alumina AT-4 melt, but the liquidus temperature for AT-6 (high alumina) is significantly higher (1300°C) than that of AT-4 (1250°C). Comparing all at their respective liquidi,

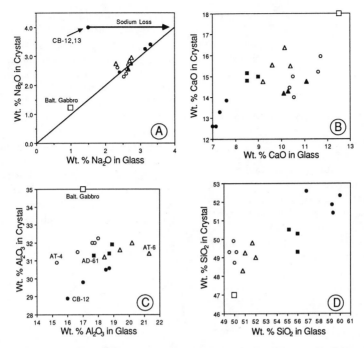

FIG. 3. The concentrations of SiO$_2$, Al$_2$O$_3$, CaO, and Na$_2$O for liquids and crystals found in these experiments. The symbol designations are given in subfigure C. The arrow in subfigure A indicates the restoration of soda due to probable surface loss.

the plagioclase of AT-6 is significantly more anorthitic. The strong effect of alumina on liquidus temperature in rocks having plagioclase as the liquidus phase has been clearly shown by FRENCH (1971), CAMERON and FRENCH (1977), and FRENCH and CAMERON (1981).

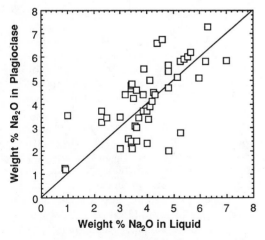

FIG. 4. Soda concentration in crystals and liquids found in Drake's experiments. A distribution coefficient (in wt.%) for soda of unity is indicated by the line.

It should be clear that if a melt were free of soda, the plagioclase would crystallize as anorthite itself. This strong effect of soda in controlling the initial plagioclase composition can be seen more quantitatively by plotting the soda content of the plagioclase against that of the accompanying melt (Fig. 3). In all except one experiment, where the analyzed glass evidently lost soda, there is nearly a one to one (in terms of wt.%, not mole fraction) partitioning of soda between melt and crystal regardless of temperature. Also shown for comparison are data from the Baltimore gabbro (SOUTHWICK, 1969), which is a rock having an unusually low soda content and An-rich plagioclase.

This apparent distribution coefficient for soda of near unity may hold best for this suite or similar suites of rocks, for a similar plot (Fig. 4) using the more varied data of DRAKE (1976) shows less regularity for soda. The lack of coherence for soda in Drake's data may reflect the large range of natural and synthetic bulk compositions used in that study; nevertheless, his data do show a broadly similar correlation. (It should also be readily obvious that this apparent distribution coefficient for soda will not be unity if plotted as mole fraction and not wt.% as in Fig. 3.) The other chemical components in plagioclase show far less coherence when plotted similarly (Fig. 3).

The steep slope of the experimental solidus ($\sim 1.4\%$ (mole)/$10°C$) shows that even rather large changes in temperature ($\pm 50°C$) will cause only modest ($\pm 7\%$) zoning in the crystal.

PLAGIOCLASE GEOTHERMOMETRY

To fully grasp the significance of these experiments in relation to the commonly used plagioclase thermometers and to further their development using these experiments, some review is necessary. The general procedure in formulating a geothermometer is first to choose an exchange reaction. Any number of reactions can be written, but since the components represented by the reaction must later be represented as activities in a thermodynamic model and these activities must be estimated from the composition of the melt, there are various advantages among the available choices. A group of exceedingly simple components can be chosen that generally lead to complicated models for representing their activities as a function of melt composition. Or a reaction can be sought that yields simple activity models, but more often than not this also calls for a complicated set of components. But if complicated components are used there may not exist the Gibbs free energy data for the standard states, thus crippling the model at its inception. Construction of a useful geothermometer must avoid these difficulties. We thus begin by considering the explicit plagioclase thermometers and then compare them and the experiments to thermometers using full magmatic phase equilibria.

Kudo-Weill model

KUDO and WEILL (1970) chose the simple exchange reaction:

(phase):   plagioclase (c)   melt (m)

$$(NaSiO_{2.5})AlSi_2O_{5.5} + (CaAlSi_{2.5})AlSi_2O_{5.5}$$

(component):   "albite" (ab)   "anorthite" (an)

plagioclase (c)   melt (m)

$$= (CaAlSi_{2.5})AlSi_2O_{5.5} + (NaSiO_{2.5})AlSi_2O_{5.5}. \quad (1)$$

"anorthite" (an)   "albite" (ab)

The exchange reaction written in this way suggests that it is the components $NaSiO_{2.5}$ and $CaAlSi_{2.5}$ that are important in deciding equilibrium. It follows that at equilibrium

$$-\frac{\Delta G^0(T, P)}{RT} = \ln K = \ln\left[\frac{a_{an}^c a_{ab}^m}{a_{an}^m a_{ab}^c}\right] \quad (2)$$

where ΔG^0 is the change in Gibbs free energy across the reaction of the components in their standard states, which are the pure components at any temperature and pressure, and the other symbols have their usual meaning.

As this equation stands there are three types of unknowns, namely, the free energy change, temperature (and pressure), and the activities. At the time Kudo and Weill formulated their thermometer free energy data did not exist for the standard state components (*e.g.*, An-glass), and little was known about the partial molar heats and entropies of mixing from which information about the activities could be obtained. They were thus forced to take a nonrigorous thermodynamic treatment and adopt "an empirical approach . . . which, *faute de mieux,* will serve . . . until additional data become available." They therefore replaced (2) with the equation

$$-\frac{\Delta G^0}{RT} = \ln\left(\frac{\lambda}{\sigma}\right) + \ln\left(\frac{\gamma_{Na}\gamma_{Si}}{\gamma_{Ca}\gamma_{Al}}\right) \quad (3)$$

which assumed that the plagioclase solid solution is ideal and where $\lambda = X_{Na}X_{Si}/X_{Ca}X_{Al}$, $\sigma = X_{ab}/X_{an}$, $\Delta G^0 = G_{an}^0 - G_{ab}^0 + G_{Na}^0 + G_{Si}^0 - G_{Ca}^0 - G_{Al}^0$, the subscripts Na, Ca, Al, and Si refer to, respectively, the oxides $NaO_{0.5}$, CaO, $AlO_{1.5}$, and SiO_2 in the melt and the subscripts an and ab refer to the "anorthite" and "albite" components in the plagioclase. The symbols X and γ, are, respectively, mole fraction and activity coefficient. Strictly speaking, (3) assumes an exchange reaction different from (1).

Because of the lack of fundamental thermodynamic data to substitute into (3), an empirical equation approximating (3) was assumed that could be fitted to data from phase equilibrium studies

$$Y(T) = \ln\left(\frac{\lambda}{\sigma}\right) + \frac{C\phi}{T} \quad (4)$$

where C is a constant and ϕ is some function of composition.

The idea to represent the activity coefficients in (3) by the last term in (4) comes from regular solution theory where $\ln \gamma \approx f(1/T)$. This formulation, however, does not assume a regular solution model and was never intended to (D. WEILL, pers. com., 1981), but is only a formulation "similar to that found for regular solutions". In fact, it can be shown that if a regular solution model itself is rigorously adopted for these multicomponent systems, only in the case where the solution is ideal can the formulation reduce to the last term in (4). The last term in (4) may, nevertheless, empirically adequately represent the data. But on the other hand, a good fit to the data is not evidence that these melts are indeed regular solutions as has sometimes been stated.

The data used by KUDO and WEILL (1970) in the regression of (4) came from experiments on natural granites and from the synthetic systems Ab-An, Ab-An-Sp, Ab-An-Di, and Ab-An-Or. In all systems the relations $X_{Na} + X_{Si} + X_{Ca} + X_{Al} = 1$ and $\phi = X_{Ca} + X_{Al} - X_{Na} - X_{Si} = 1 - 2(X_{Na} + X_{Si})$ were assumed. By working initially with different systems at the same temperature they arrived at a value for C in (4) of 1.29×10^4 (note misprint in original paper) which was subsequently used by them in all succeeding studies in the regression of (4) against temperature. In this regression the function $Y(T)$ was assumed to be of the form $Y(T) = mT + b$, where m and b are constants. Treating each set of experiments under dry conditions and water pressures of 0.5 kb, 1.0 kb, and 5.0 kb as individual sets of data, four sets of the coefficients m and b were found.

Upon application of these results to a diverse group of igneous rocks, KUDO and WEILL (1970) found generally good agreement with other estimates of temperature; the standard deviation in the original regression is 34°C. But occasionally, especially in dealing with basalts, unreasonably high temperatures (~1300°C) were found when this formulation was applied more widely. This led MATHEZ (1973) to use a wide variety of natural and experimental data from mostly basaltic systems to derive a new set of constants (m and b) for (4). (He also introduced a compositional dependence for γ that produces a double or triple-valued function over a narrow temperature interval, suggesting immiscibility in plagioclase.) What is really needed to make a tightly constrained regression over a wide compositional range is actual experimental data

over this range. DRAKE (1976) remedied this situation by providing 55 new pairs of experimental liquid (glass) and crystal compositions all at atmospheric pressure. This produced a single, tightly constrained equation with a formal uncertainty of 55°C or about 12% (mole), in solid composition.

Applying the basic Kudo-Weill formulation using these various set of constants to the Aleutian experimental results does not show (Fig. 5) particularly close agreement, except possibly for CB-12 (the andesite). Mathez's constants (assuming his $\gamma = 1$ throughout to eliminate the kinked curve) seem to fit best and Drake's constants seem to be almost an average of the other two. For AT-4 the temperature estimates are off by 50°C, or the plagioclase composition by about 10% (mole); for AT-6 the temperature is off by 100°C, or the composition by about 20% (mole); for AD-61 the temperature is off by 10–35°C, or the composition by about 5–15% (mole); and for CB-12 they are all in the same neighborhood, but each thermometer shows a slope distinctly different than that found experimentally. The agreement with Mathez's equation can be made much closer by varying γ until a close fit is found and then holding γ constant for all temperatures (Fig. 6). This method of finding a suitable value of γ and then using this fixed value is hereafter referred to as the Mathez-prime formula.

There seems to be two difficulties with the thermometers, namely, compositional dependence is not adequate and the slope of the solidus is not correct. (The slopes of the solidi found in these experiments are not unusual, for a similar one is shown by the data of BENDER et al. (1978) for a tholeiitic basalt.) The (inverse) slopes of the experimental solidi are about 14% An (mole) per 100°C, as compared with the true binary of about 18%/100°C for this region. The slopes of the thermometers are in the range of 20–27%/100°C, and thus they predict less of a temperature decrease than is needed for a corresponding change in crystal composition. The actual slope $(\partial\phi/\partial T)$ in these formulations is given by $-\sigma(C\phi/T^2 + m)$, and since $C = 1.29 \times 10^4$ and typically $T \approx 1500°K$, $|\phi| \approx 0.3$, and $m \approx 12 \times 10^{-3}$, m is typically about ten times larger than $C\phi/T^2$. Thus the slope is determined principally by m, and each succeeding formulation after Kudo and Weill has lessened the slope (increased m); to match these experiments the slope must be steepened (decrease m).

All this points to the fact that if a correct slope is initially assigned, the thermometers will be in even stronger disagreement with experiment because their dependence on melt composition is weak

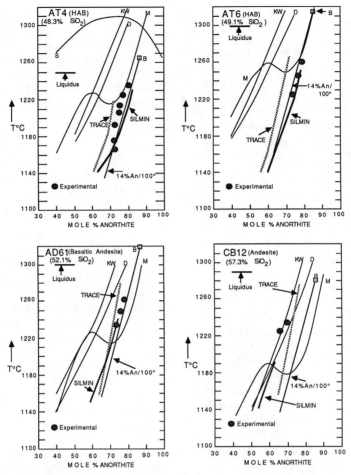

FIG. 5. Plagioclase composition from these experiments in comparison with that derived from various geothermometers. The results of the Kudo-Weill (KW), Drake (D), Mathez (M), Burnham (B), TRACE and SILMIN thermometers are indicated with symbols against each curve. The geothermometer of SMITH (1983; symbol S on AT-4) was considered for the compositions here. Its distinctive parabolic geometry and mismatch (*i.e.*, Fig. 8) led us to not consider it further.

and the thermometer solidi will never cross the now parallel experimental solidi. The slopes found by regression have absorbed some of the inadequacy of the compositional dependence of the activity model. This difficulty reflects our lack of knowledge of how to represent effectively the activities of the components $NaAlSi_3O_8$ ("albite") and $CaAl_2Si_2O_8$ ("anorthite") in the melt. The model used in these geothermometers relies almost solely on silica, alumina, lime, and soda, giving each equal importance and having little dependence on ferromagnesian components. We have, however, seen earlier that the plagioclase composition is strongly dependent on the soda content of the melt and much less so on any other components; alumina is a major influence on the liquidus. It thus seems important to

have an activity model that relies heavily on the soda content of the melt, such as in, for example, the model of BURNHAM (*e.g.*, 1979).

Burnham's model

This model rests fundamentally on two premises: (1) that near the liquidus the melt structure mimics the structure, discreteness, and stoichiometry of crystallizing phases, and (2) that properly chosen aluminosilicate components mix ideally with one another (BURNHAM, 1981). When actual magmatic compositions are recast into components based on an 8-oxygen, feldspar-like formula, BURNHAM (1975) has shown that basalts, andesites, and granites have essentially identical equimolal solubilities for water. This solubility is the same as that of water in $NaAlSi_3O_8$ melt. As the phase relations in the plagioclase binary can be adequately calculated assuming both crystals and melt to be ideal solutions (clearly understanding that both solutions are actually non-ideal), and since the solidus-liquidus relations found experimentally for plagioclase in these island arc rocks geometrically mimic those in the binary itself, it seems reasonable that these equilibria might also be calculable assuming a model like Burnham's. No attempt is made here to argue that our assumption is theoretically correct; we simply suggest that it proves to be a sensible and practical way to understand these equilibria.

Burnham's formalism is broadly similar to (1) and (2) except for the fact that separate melting reactions are written for albite and anorthite. This model insists that if the components are chosen on an 8-oxygen, feldspar-like basis, the solution is ideal and $a_i^m = X_i^m$. To calculate the mole fractions X_{ab}^m and X_{an}^m, the CIPW norm is converted to a norm of mineral components each containing 8-oxygens. That is, new (heavier) gram-formula weights based on 8-oxygens (*e.g.*, $SiO_2 = 240$) are used to change the CIPW norm (in wt.%) to moles which, when normalized to unity, give the necessary mole fractions. For the rocks of these experiments both the CIPW and 8-oxygen norms are given by Table 1. It is of interest to recall that, in computing the CIPW norm, soda alone (in common igneous rocks) completely determines the amount of albite component and hence also its activity in the melt. This is decidedly different from the earlier activity model and it is in accordance with the observed strong control of crystal composition by the soda content of the melt.

The first check on the thermometer is the agreement of calculated and measured liquidus temperatures (Fig. 5). These are all generally within about 10°C or less. Except for AT-6 which shows near perfect agreement, the calculated plagioclase compositions are generally 2–5% (mole) more anorthitic than what is observed. (The comparison with CB-12 is difficult because of the low temperature of the runs relative to the liquidus, which is where this model pertains.) Since the accuracy of the determination of plagioclase composition is probably no better than about 2 mole%, this disagreement may be somewhat less. Nevertheless, the calculated compositions are generally more anorthitic than observed. If a_{ab}^m is increased slightly, for example, by letting $a_{ab}^m = X_{ab}^m + X_{or}^m$ where the latter term is the very small orthoclase component, which might similarly influence plagioclase composition, the

disagreement lessens, but the calculated liquidus temperatures each rise 5–10°C and show less agreement.

The agreement between experiment and Burnham's method is close but not perfect. The calculated plagioclase compositions differ by about 0–5 mole%, or about 5–10°C in liquidus temperatures. More recent versions of this model may more closely fit these data. Burnham's method is sensitive enough such that high requirements are placed on the accuracy of the determination of both plagioclase and melt compositions. The simplest way to employ this method is to use the whole-rock composition to estimate the liquidus composition and then use the measured slope of the solidus ($\approx 14\%$ An(mole) per 100°C) to predict further temperatures and crystal compositions. If some other mineral phase is also crystallizing, its effect on the mass balance of soda must be monitored. For example, in tholeiitic basalts where olivine is the liquidus phase, it is important to know how much the soda concentration has been increased before plagioclase appears. The only way to handle such effects is to use a full solution model.

SILMIN and TRACE

These solution models are due, respectively, to GHIORSO (1985; GHIORSO *et al.*, 1983) and NIELSEN (1988; NIELSEN and DUNGAN, 1983). SILMIN, in its simplest description, uses a broadly regular solution model in concert with experimental phase equilibria. Given an initial liquid composition, SILMIN calculates liquid and mineral compositions as the system cools and crystallizes; either fractional or equilibrium crystallization can be followed. Equilibria at 1 bar and to moderate (~ 3 kb) pressures can be investigated with this model (with caveats regarding pyroxenes), which calculates solid and liquid compositions and the relative amounts of each.

TRACE assumes a melt consisting of two independent quasi-lattices, one network-forming and the other network-modifying, with ideal mixing within each. It also has the ability to model trace element behavior. TRACE equilibria was calibrated from 1 bar data but works on dry systems up to pressures of about 2–4 kb (NIELSEN, pers. com., 1989).

The results of using these models to predict plagioclase composition and temperature for these rocks are also shown in Fig. 5. Neither model matches perfectly the experimental data; SILMIN seems to give better agreement. Nevertheless, these two models both give better agreement than either

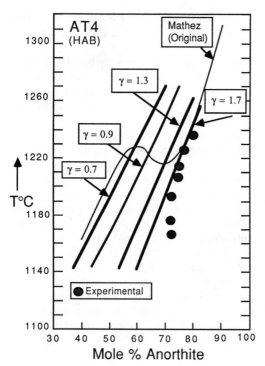

FIG. 6. The effect of varying γ in Mathez's geothermometer.

the Kudo-Weill or Drake models. The solidi slopes predicted by SILMIN and TRACE are generally close to those observed, but at lower temperatures they predict a flattening of the solidus (*e.g.,* see AT-4), which would give a stronger enrichment in the albite component, when in fact the solidus becomes vertical and plagioclase composition invariant with temperature. The liquidi predicted by these models are reasonably close except for CB-12 where SILMIN gives a strikingly low value.

Other compositions

As mentioned earlier, plagioclase-liquid composition data also exist for rocks from the Medicine Lake Highlands (GROVE *et al.,* 1982) and for rocks from the sea floor (MORB; WALKER *et al.,* 1979). Eight of these compositions, ranging from basalt to andesite are shown as Fig. 7. No single method works well for these additional compositions. The Kudo-Weill and Drake models match least well, TRACE and SILMIN overall match best, and Burnham's method gives an overall agreement intermediate to these two groups. The success of TRACE and SILMIN is partly artificial, however, because most of these data were used in calibration

of these models (unlike the case of 4 Aleutian compositions). Whereas Kudo-Weill and Drake predict too albitic solid compositions, Burnham's method most often predicts too anorthitic compositions.

Overall ratings

Figure 8 shows an overall rating of each method for each group of rocks, and the following recommendations are made for use of these geothermometers for low pressure, dry conditions.

1. For high alumina basalts, use Burnham, SILMIN, and Mathez-prime (with $\gamma = 1.7$).
2. For basaltic andesites (52–55% SiO_2) use Kudo-Weill, Drake, SILMIN, TRACE, and Mathez-prime ($\gamma = 1.3$).
3. For andesites (>56% SiO_2) use TRACE, Kudo-Weill, Drake, SILMIN and Mathez-prime ($\gamma = 0.7$).
4. For "normal" oceanic tholeiites, use those enumerated in 1 above.
5. For Fe-Ti enriched tholeiites, use SILMIN, TRACE, and Mathez-prime ($\gamma = 1.1$).

Throughout this list, however, the success of Mathez-prime is principally due to a judicious choice of the adjustable parameter (γ) and these values are so mentioned essentially as calibration factors for these compositions. The suggested approach in using these methods is to use the whole rock composition as the initial liquid composition and make a T-X_{An} plot for plagioclase composition from which the observed plagioclase compositions can be used to infer temperature. Those recommended above should, by this means, give temperatures accurate to ±30°C.

EFFECTS OF PRESSURE AND WATER ON PLAGIOCLASE COMPOSITION

The effect of water, and pressure in general, on plagioclase composition is of much concern in light of the frequent appeal to these parameters to explain the mineralogy of arc lavas. This subject is explored here analytically using Burnham's model and numerically using SILMIN.

Anhydrous melts

The liquidi for AT-4 (basalt) and CB-12 (andesite) were calculated using Burnham's method and they are shown by Fig. 9. This gives a liquidus, which may not be the true rock liquidus, of constant plagioclase composition. At lower temperatures a slope of 14% (mole) decrease in An-content per 100°C

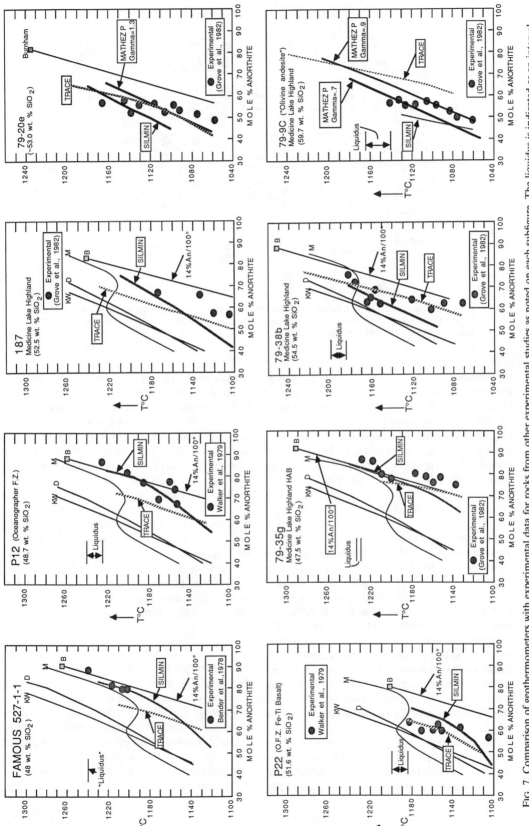

FIG. 7. Comparison of geothermometers with experimental data for rocks from other experimental studies as noted on each subfigure. The liquidus is indicated as an interval, between the lowest temperature with no reported crystals, and the temperature of first reported/analyzed plagioclase crystals.

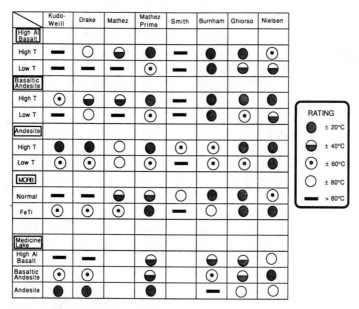

FIG. 8. Rating chart of performance of various geothermometers (across the top) when applied to experimental results for various melt compositions (left column). The various symbols indicate the closeness of fit to the experiments (*e.g.*, ±20°C).

was assumed for all pressures. The compositional contours thus necessarily assume equilibrium crystallization. SILMIN for this same calculation gives a slightly less steep liquidus slope and the liquidus plagioclase with increasing pressure decreases in An-content by about 15% (mole) per 10 kb increase in pressure. The decrease in An-content with temperature is essentially the same as given by Burnham's method, which is no surprise considering the agreement in slope with that found experimentally (*e.g.*, Fig. 5).

A comparison of experimentally determined liquidi slopes (*e.g.*, Mt. Hood andesite, EGGLER and BURNHAM, 1973) with those calculated here shows them to be close to that given by SILMIN. Adopting the SILMIN *P-T* liquidus slope throughout shows that the constant composition liquidi calculated using Burnham's method cut across the SILMIN liquidus and thus suggest a liquidus plagioclase composition that decreases in An-content by ~5% (mole) per 10 kb increase in pressure. On this account, the constant composition liquidi of Fig. 9 are probably only a close approximation to the true liquidus plagioclase composition (*i.e.*, within ~ ±2.5% An).

As seen earlier in the experiments on AT-4, however, if plagioclase crystallizes to a low enough temperature, it will exhaust the melt of one or more components critical to forming plagioclase and pla-

gioclase crystallization will cease. Below this temperature crystal composition becomes invariant with decreasing temperature. The point below the liquidus at which this happens is given by the width of the solidus-liquidus loop in the pseudo-binary for the melt composition in question. As might be expected, the loop width increases (at 1 atm) in direct proportion to the increase of ab-component in the melt. That is, as $X_{ab}^m \to 0$, the loop width $\Delta W \to 0$. This variation in ΔW with melt composition for these experiments is shown by Fig. 2, and the approximate *P-T* boundary below which the plagioclase composition becomes invariant is indicated on the *P-T* diagrams of Fig. 9. The stability fields of other phases (*e.g.* olivine and clinopyroxene in AT-4 and ortho- and clinopyroxene in CB-12) must be calculated independently.

Hydrous melts

Phase diagrams for compositions AT-4 and CB-12 containing various amounts of water are also shown by Fig. 9 as calculated using Burnham's method. It is interesting to note in these *P-T* diagrams that the plagioclase composition is constant along each solubility isopleth. Notice also that at constant pressure, as the water content is increased the composition of plagioclase on the liquidus (the temperature of which decreases with increasing X_w^m) gets *slightly* more anorthitic. The liquidus pla-

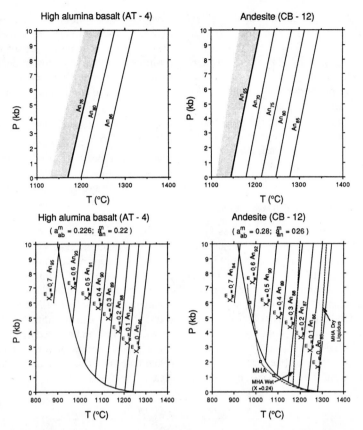

FIG. 9. The *upper* diagrams show the variation with pressure and temperature of the composition of plagioclase for rock compositions AT-4 (basalt) and andesite (CB-12) free of water as calculated using Burnham's method. The shading indicates the point at which the melt is exhausted of plagioclase component, provided that no other solid phases exist, and within which plagioclase composition is invariant with further decreases in temperature. The *lower* diagrams illustrate the same calculation involving the liquidus only and showing the effect of increasing amounts of water. The mole fraction of water in the melt and the composition of the plagioclase in equilibrium with this melt is shown against each line. The experimentally determined liquidus for the Mt. Hood andesite (MHA; EGGLER and BURNHAM, 1973) is also shown (normalized to the CB-12 liquidus). Notice that the plagioclase composition increases in anorthite component at about 1 mole% per 10% (mole) increase in melt water content.

gioclase composition becomes about 1% (mole) more anorthitic per 10% (mole) increase in water content (*i.e.*, $\partial X_{an}^c / \partial X_w^m \approx 0.1$). To increase the liquidus composition from, say, An_{80} to An_{85}, 50% (mole) water must be added to the original melt.

The Mt. Hood andesite studied by EGGLER and BURNHAM (1973) is similar overall to CB-12 and two of these liquidi at $X_w^m = 0$ and 0.24 are also plotted on the *P-T* diagram for CB-12. For the purpose of comparison, the dry liquidus at $P = 1$ atm for the Mt. Hood andesite has been set equal to that of CB-12, and the values of X_w^m for the Mt. Hood andesite have been calculated with a mean molecular weight of the melt of 279 grams for $X_w^m < 0.5$ (BURNHAM, 1975).

It is clear from this comparison that, although the measured liquidus is not of constant plagioclase composition, to pressures of 10 kb it is within about 5% An(mole) of being so. It is also clear that although a relatively small amount of water ($\approx 10\%$ (mole)) significantly lowers the liquidus temperature, the initial plagioclase composition is only slightly increased (1% An). The saturated liquidus of MHA agrees well with that calculated and, considering the uncertainty of the experimental liquidus at $X_w^m = 0.24$, so does this liquidus. Considering the difference in composition between CB-12 and the Mt. Hood andesite, this agreement is good.

Similar modeling using SILMIN shows that at any pressure adding 1.0% (wt) water to AT-4 depresses the plagioclase liquidus by 100°C and *decreases* the liquidus An-content by about 3% (mole). The possible increase in An-content by adding water is evidently almost equally offset by the strong lowering of the liquidus itself.

Although the two methods of calculating plagioclase composition disagree as to the direction of shift, there is good agreement that the influence of water on plagioclase composition is small. For the amounts of water considered reasonable in most arc magmatic systems, water should have little if any effect on plagioclase composition.

COOLING HISTORY OF THE LAVAS

At the onset of this study we mentioned as a motivating factor the need to explain the plagioclase compositions of basaltic to andesitic lavas of island arcs, such as those of the Aleutians used in this study. It is therefore of some importance to present the actual compositions of plagioclase found in these rocks and understand them in light of the foregoing presentation. Representative profiles of compositional zoning of plagioclase phenocrysts in these lavas (Fig. 10) show that all phenocrysts in each lava are not identical. The phenocrysts do show

however some common characteristics both within a single lava and among the suite itself. For example, the often strong change toward a more albitic composition near the rim most probably reflects the rapid cooling upon eruption. But overall it is the range of average compositions that requires explanation.

High An content

At their respective liquidi, either dry or with a reasonable amount of water (*i.e.,* < ~1 wt%), the experimental (and calculated) plagioclase compositions fall among those observed in the rocks. They are, however, generally less anorthitic than the most-anorthitic phenocryst composition observed. Generally this difference is about 5% (An) or less, but lava AT-4, for example, shows an experimental liquidus composition of about An_{80} whereas the observed compositions can reach about An_{90}. The experimental An-content can be increased in several ways: (1) water can be added to the melt; (2) the rock's liquidus can be raised by adding alumina; or (3) the soda content of the rock can be reduced. (Note that loss of soda in any experiment only increases the difference.)

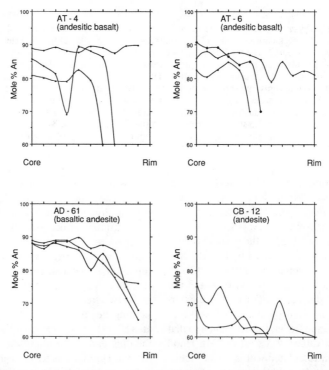

FIG. 10. Observed plagioclase phenocryst compositions found in the starting material lavas. The horizontal scale generally represents an actual distance of 1–2 mm.

(1) *Water.* To increase the crystal composition by 10% (mole) An, over 80% (mole), equivalent to ≈25 wt.%, water must be added to the melt. The addition of a more sensible amount of, say, 2 wt.% (≈8% (mole)) water increases the An-content by only 2% and this is clearly inadequate. (For conversion of wt.% to mole% water use BURNHAM's (1979) Equations (16-5; 16-6) or his Fig. 16-4.)

(2) *Alumina* and (3) *soda.* On the other hand, to raise the An-content by 10%, either about 1 wt.% soda must be removed from the melt or its liquidus temperature (*i.e.* alumina content) must be raised by about 75°C (≈4 wt.% Al_2O_3). The most rea-

sonable solution is to lower the soda content by, say 0.5 wt.% and raise the alumina content by about 2 wt.%, which can be accomplished by replenishing the magma with plagioclase lost through crystal fractionation. Introduction of a realistic amount of water (*i.e.* 1–2 wt.%) adds another 1–2% An to the liquidus phase. This strongly suggests that AT-4 has suffered a loss of plagioclase that has increased the original soda content and decreased the original alumina content. (In fact, among this suite of lavas AT-4 is unusually low in alumina.)

Upon closer inspection, this result for AT-4 may be a rather general conclusion. That is, a rock whose experimental (or calculated) plagioclase composition is *less* anorthitic than that of the actual phenocrysts has probably suffered plagioclase loss. On the other hand, if a rock represents an accumulation of plagioclase from more primitive melts (as is often suspected), the liquidus temperature will be high, the soda content low (relative to that of the closed system) and more important, the experimental (or calculated) plagioclase composition will be *more* anorthitic than that of the actual phenocrysts themselves. This assumes that the phenocryst composition records both the present and earlier melt compositions. The earlier history can only be recorded if the phenocrysts are slow to re-equilibrate with the changing composition of the melt.

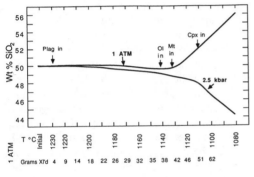

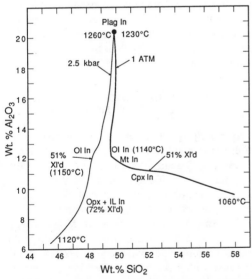

FIG. 11. *upper figure:* The change in silica content of the melt as a function of temperature and fractionation of plagioclase at pressures of 1 atmosphere (upper curve) and 2.5 kb (lower curve). *lower figure:* The variation of alumina with silica content for plagioclase fractionation. Notice that, depending on the exact plagioclase composition, the melt may become either richer or poorer in silica. All calculations were made using SILMIN (GHIORSO *et al.,* 1985).

Zoning

The ability of a zoned crystal to homogenize itself or equilibriate with surrounding melt depends critically on the rate of intra-crystalline diffusion. If charge balance is to be maintained, plagioclase homogenization is by coupled diffusion involving Na-Si and Ca-Al (GROVE *et al.,* 1984). The rate of diffusion may therefore depend on the rate of the slowest moving element, which is presumably Al. The diffusivity of Na, for example, in plagioclase is about 10^{-9} cm^2/s at 1000°C (FOLAND, 1974). If diffusion and homogenization are controlled by this element, a crystal with zones 10^{-2} mm thick will homogenize in a few days. This is probably the case for alkali feldspars (involving only Na-K exchange) that are commonly unzoned in lavas with coexisting zoned plagioclase. Even in slow cooling plutons, plagioclase often shows zoning (VANCE, 1962). This feature can be used to infer that the diffusivity (D) of Al in plagioclase must be extremely small. It is straightforward to calculate that for a pluton taking 10^6 years to cool, $D < \sim 10^{-16}$ cm^2/sec. By similar reasoning MORSE (1984) has also suggested a very small diffusivity of Al in plagioclase. It is therefore reasonable that plagioclase in lavas may record its previous history of contact with melts of a variety of compositions.

With the experimental determination of the plagioclase binary by BOWEN (1913) came many schemes to explain the zoning patterns in plagioclase. Nearly every scheme discussed today can be traced to early works by BOWEN (1928), HARLOFF (1927), HILL (1936), and PHEMISTER (1934). These have been critically discussed by VANCE (1962) and PRINGLE *et al.* (1974) and recently more quantitatively by SIBLEY *et al.* (1976) and HAASE *et al.* (1980), among others. The aim of these workers has been the explanation of oscillatory and normal zoning. We can add little new to these general schemes, but from the computed phase diagrams we can however quantitatively examine zoning of plagioclase in arc rocks as influenced by changes in (1) temperature, (2) pressure, and (3) water saturation.

(1) *Temperature.* The effect of cooling or heating at constant composition, regardless of water content, is to increase the albite content by $\approx 14\%$ (mole) per 100°C drop in temperature. If equilibrium is not maintained and the early-formed crystals do not react with the melt, the increase in albite content will be greater. Large variations in composition ($\approx 40\%$ An) as observed in some ash flows (MOROHASHI *et al.*, 1974) and non-arc lavas (*e.g.*, LARSEN *et al.*, 1938; MYERS and MARSH, 1981) can hardly be produced by temperature variations alone. The thin albitic rims of Fig. 10 might reflect a severe drop in temperature upon eruption or, perhaps more likely, the onset of vigorous vesiculation and degassing during eruption (GHIORSO and CARMICHAEL, 1987).

(2) *Pressure.* In an anhydrous melt (Fig. 9), it is clear that a magma suddenly moving nearer to the earth's surface (on its way perhaps to eruption) will cool only adiabatically (≈ 10°C/5 kb) and thus remain essentially isothermal. But if the pressure is isothermally reduced by, say, 5 kb the crystal will become about 5% (mole) more anorthitic. Under disequilibrium conditions this will produce a reversely zoned crystal, which if repeatedly compressed and decompressed (perhaps due to convection) may become oscillatory zoned. Because the effect is not large, this adiabatic pressure shift will cause only moderate ($\approx 3\%$) reversed zoning and this cannot be a general explanation of oscillatory zoning. This effect may be evident in some of the profiles of Fig. 10. Similar results are found for periodic isobaric and isochoric (constant volume)

crystallization by GHIORSO and CARMICHAEL (1987).

(3) *Water saturation.* Consider a magma initially undersaturated with water, say AT-4 of Fig. 9 where $X_w^m = 0.5$ and the liquidus plagioclase composition at 6 kb is An_{91}. As the magma moves along this liquidus to lower pressure, it will become saturated with water at 2 kb and with further decompression water will exsolve, X_w^m must decrease. With reduction of water in the melt (it must go to zero as pressure goes to 1 bar), the liquidus shifts to higher temperatures. If the temperature is maintained, the plagioclase becomes normally zoned. In fact, the heat of crystallization will raise the temperature somewhat but it will not reach the new liquidus. Once water saturation has occurred, the magma cannot again become undersaturated unless dry wall rock or drier magma is assimilated or the magma is carried deeper into the earth. Therefore, subsequent to saturation with water only normal zoning patterns can be produced. This feature has been well recognized before by VANCE (1962) among others and the point at which zoning changes from oscillatory or reversed to strongly normal may represent the onset of water-saturation in the melt. In practice, because lavas commonly contain an assortment of zoning patterns, with some crystals apparently from earlier melts, it is essential to identify those that belong to the present magma. The zoning patterns shown by these Aleutian lavas (Fig. 10) apparently represent crystallization during water-undersaturation.

SILICA VARIATION WITH PLAGIOCLASE FRACTIONATION

Because plagioclase composition is largely controlled by the soda content of the melt, crystals can form that contain more silica than their host magma. That is, if composition AT-4 contained, say, 3.0 wt.% or more soda instead of 2.42, its liquidus plagioclase would have a composition of about An_{75}, which contains about 50 wt.% SiO_2. Fractionation of this plagioclase would drive the magma to more silica-undersaturated compositions. Because with fractionation the soda content increases, the effect with time is amplified by the ever increasing difference between the silica content of the crystal and melt. A tholeiitic melt with plagioclase on the liquidus can thus be driven toward the composition of an alkali basalt. This process is in essence, akin to the "plagioclase effect" that promotes the peralkaline condition in salic lavas (*e.g.*, CARMICHAEL *et al.*, 1974). This silica condition has been apparently overlooked as it is found both in nature (MYERS and MARSH, 1981) and in the experimental products listed by WALKER et al. (1979) and BENDER et al. (1978). In each of these examples plagioclase crystallizes early and it is more siliceous than the melt. If this plagioclase is able to fractionate from the melt, the melt composition will become *less* siliceous. The condition, however, will reverse upon precipitation of olivine, which overrides the effect of plagioclase and increases the silica content of the melt. The important feature of this effect is that it is not safe to assume that calcalkaline magmas always fractionate towards increasing silica. Lavas placed in progression of ever-increasing silica (*i.e.* Harker variation diagrams) may not necessarily represent monotonically-increasing temporal samples of a fractionating magma.

This effect can be quantitatively investigated using SILMIN to follow the liquid composition during fractional crystallization of a high-alumina basalt. We have chosen a starting composition, (SH5; FOURNELLE, 1988) very similar to AT-6 where at 1-atm. the plagioclase ($\sim AN_{70}$) has a silica content ($\sim$50% wt.) almost equal to that of the rock itself. At a pressure of 2.5 kb the plagioclase is slightly less anorthitic and consequently has a silica content greater than the rock. The change in liquid composition with fractional crystallization at these pressures is shown by Fig. 11. Whereas at 1-atm. silica content increases with fractionation, at 2.5 kb the liquid composition decreases in both alumina and silica. In terms of silica alone, there is little overall change until about 50% fractionation, which reflects the initial compositional similarity. This need not generally be the case, because when the initial plagioclase is significantly more (less) silicious than the melt, strong changes in the melt composition will occur much earlier in the fractionation process.

CONCLUSIONS

Perhaps the most surprising result of these experiments is the finding that the plagioclase pseudo-binary exists in these melts in close geometrical similarity to that of Bowen's true plagioclase binary. The slope of the experimental solidus in the pseudo-binary is about 14% (mole) An per 100°C. The width of the liquidus-solidus loop varies (to a point) in direct proportion of the "albite" concentration of the melt. The position of the solidus is largely controlled by the soda (or normative albite) content of the melt, and the liquidus temperature varies directly (for the most part) with the alumina content

of the melt. The plagioclase composition is well *approximated* by an apparent distribution coefficient (wt.%) of unity. The Kudo-Weill type plagioclase thermometer, which assumes essentially an equal influence of silica, alumina, lime, and soda on "plagioclase" activity in the melt, is insensitive in these rocks to the strong control of soda on plagioclase composition.

Because of this compositional dependence on normative albite (which is of course dictated by soda content) and the geometric similarity of the pseudo-binary to the binary itself, the ideal solution model due to BURNHAM (*e.g.,* 1979) is well-posed to describe this equilibria. In this model the activity of "plagioclase" in the melt depends directly on the normative amount of albite (*i.e.* soda content) and agrees fairly well with the experiments. Additional advantages of the Burnham model are that it predicts both plagioclase composition and liquidus temperature at any water content at pressures to 10 kb. Ghiorso's SILMIN model is similarly useful and together the variation in plagioclase composition with temperature, pressure, and water content can be studied.

The liquidus of these rocks is within 5% (mole) An of being of constant composition to 10 kb. With increasing water the liquidus plagioclase becomes slightly more anorthitic at the rate of about 1% mole An per 10% (mole) increase in water. The often observed An-rich plagioclase of island arc rocks can not be explained simply by the addition of water to the melt because excessive amounts of water ($\approx$20 wt.%) are needed. Instead these anorthitic compositions probably reflect earlier equilibria in a low soda, high alumina melt. Because both the experimental and calculated plagioclase compositions for many arc lavas are less anorthitic than the observed phenocrysts themselves, these lavas must have experienced plagioclase *loss.* Their relatively high alumina and low soda contents must have been, respectively, slightly higher and lower in the more primitive parent. This strongly suggests that the high alumina character of arc rocks is a primitive feature of these melts. For if it were not, and the lavas represent plagioclase cumulates, the experimental and calculated plagioclase composition (in the basalts) should be more anorthitic than the actual phenocrysts themselves. The temperature associated with these anorthitic phenocrysts need not have been excessively high (*i.e.,* $\approx$1300°C), for addition of a sensible amount (1–2 wt.%) of water to the melt lowers the liquidus to 1200–1250°C, but hardly changes the liquidus plagioclase composition.

B. D. Marsh *et al.*

In terms of potential accuracy, because plagioclase composition is rather insensitive to temperature changes (*i.e.,* 14% mole per 100°C) it is not as useful a thermometer as might be expected.

Acknowledgements—All of us have benefited greatly from Hans Eugster in understanding experimental petrology and learning how to design useful experiments. His visits to the igneous experiments were often marked by notes with cryptic messages such as "Is this science!" and "This belongs in an art museum!". His caring cajoling consistently focused the intent of the experiments and forever endeared his spirit to us.

This work has also benefited greatly from correspondence and discussions with Daniel Weill, Michael Drake, and James Brophy and especially C. Wayne Burnham. The support of I-M.C. at the Johnson Space Center where the initial experiments were carried out is appreciated, as is the use of the electron microprobe at the Geophysical Laboratory. Useful and constructive reviews by M. Ghiorso, M. Mangan, R. Nielson, and R. Ayuso materially improved this manuscript. This work is supported by the National Science Foundation Grant EAR-8817394 to the Johns Hopkins University (B.D.M.).

REFERENCES

BAKER D. R. and EGGLER D. H. (1987) Composition of anhydrous and hydrous melts coexisting with plagioclase, augite, and olivine or low-Ca pyroxene from 1 atm to 8 kbar: application to the Aleutian volcanic center of Atka. *Amer. Mineral.* **72,** 12–28.

BAKER D. R. and EGGLER D. H. (1983) Fractionation paths of Atka (Aleutians) high-alumina basalts: constraints from phase relations. *J. Volcanol. Geotherm. Res.* **18,** 387–404.

BENDER J. F., HODGES F. N. and BENCE A. E. (1978) Petrogenesis of basalts from the Project Famous area: experimental study from 0 to 15 kbars. *Earth Planet. Sci. Lett.* **41,** 277–302.

BOWEN N. L. (1913) The melting phenomena of the plagioclase feldspars. *Amer. J. Sci.* (Fourth Ser.) **35,** 577–599.

BOWEN N. L. (1928) *The Evolution of Igneous Rocks.* 333 pp. Princeton University Press.

BURNHAM C. W. (1981) The nature of multicomponent silicate melts. In *Chemistry and geochemistry at high temperatures and pressures* (eds. D. T. RICKARAD and F. E. WICKMAN), *Phys. Chem. Earth* **13–14,** pp. 197–229.

BURNHAM C. W. (1979) The importance of volatile constituents. In *The Evolution of the Igneous Rocks: Fiftieth Anniversary Perspectives* (ed. H. S. YODER), pp. 439–482. Princeton University Press.

BURNHAM C. W. (1975) Water and magmas, a mixing model. *Geochim. Cosmochim. Acta* **39,** 1077–1084.

CAMERON E. P. and FRENCH W. J. (1977) The relationship of the order of crystallization of basalt melts to their classification and to the definition of rock series. *Mineral. Mag.* **41,** 239–251.

CARMICHAEL I. S. E., TURNER F. J. and VERHOOGEN J. (1974) *Igneous Petrology.* 739 pp. New York: McGraw-Hill.

DONALDSON C. H., USSELMAN T. M., WILLIAMS R. J. and LOFGREN G. E. (1975) Experimental modelling of the cooling history of Apollo 12 olivine basalts. *Proc. Lunar Planet. Sci. Conf. 6th,* 843–870.

DRAKE M. J. (1976) Plagioclase-melt equilibria. *Geochim. Cosmochim. Acta* **40,** 457–465.

EGGLER D. H. (1972) Water-saturated and undersaturated melting relations in a Paricutin andesite and an estimate of water content in the natural magma. *Contrib. Mineral. Petrol.* **34,** 261–271.

EGGLER D. H. and BURNHAM C. W. (1973) Crystallization and fractionation trends in the system andesite-H_2O-CO_2-O_2 at pressures to 10 kb. *Geol. Soc. Amer. Bull.* **84,** 2517–2532.

FINGER L. W. and HADIDIACOS C. G. (1972) Electron microprobe automation. *Carnegie Inst. Wash. Yb.* **71,** 269–275.

FOLAND K. A. (1974) Alkali diffusion in orthoclase. In *Geochemical Transport and Kinetics* (eds. A. W. HOFFMAN, B. J. GILETTI, H. S. YODER, JR., and R. A. YUND), pp. 77–107. Carnegie Institute of Washington.

FOURNELLE J. H. (1988) The Geology and Petrology of Shishaldin Volcano, Unimak Island, Aleutian Arc, Alaska. Ph.D. Dissertation, The Johns Hopkins University, Baltimore, Maryland, 528 pp.

FRENCH W. J. (1971) The correlation between "anhydrous" crystallization temperatures and rock composition. *Contrib. Mineral. Petrol.* **31,** 154–158.

FRENCH W. J. and CAMERON E. P. (1981) Calculation of the temperature of crystallization of silicates from basaltic melts. *Mineral. Mag.* **44,** 19–26.

GHIORSO M. S. (1985) Chemical mass transfer in magmatic processes. I. Thermodynamic relations and numerical algoithms. *Contrib. Mineral. Petrol.* **90,** 107–120.

GHIORSO M. S., CARMICHAEL I. S. E., RIVERS M. L. and SACK R. O. (1983) The Gibbs free energy of mixing of natural silicate liquids; an expanded regular solution approximation for the calculation of magmatic intensive variables. *Contrib. Mineral. Petrol.* **84,** 107–145.

GHIORSO M. S. and CARMICHAEL I. S. E. (1987) Modeling magmatic systems: petrologic applications. In *Thermodynamic Modeling of Geological Materials: Minerals, Fluids and Melts* (eds. I. S. E. CARMICHAEL and H. P. EUGSTER). *Rev. in Mineral.* **17,** pp. 467–500. Mineralogical Society of America.

GROVE T. L., GERLACH D. C. and SANDO T. W. (1982) Origin of calc-alkaline series lavas at Medicine Lake volcano by fractionation, assimilation and mixing. *Contrib. Mineral. Petrol.* **80,** 160–182.

GROVE T. L., BAKER M. B. and KINZLER R. J. (1984) Coupled CaAl-NaSi diffusion in plagioclase feldspar: experiments and applications to cooling rate speedometry. *Geochim. Cosmochim. Acta* **48,** 2113–2121.

HARLOFF C. (1927) Zonal structure in plagioclases. *Leid. Geol. Med.* ii, 99–114.

HAASE C. S., CHADAM J., FEINN D. and ORTOLEVA P. (1980) Oscillatory zoning in plagioclase feldspar. *Science* **209,** 272–274.

HILL E. S. (1936) Reverse and oscillatory zoning in plagioclase feldspars. *Geol. Mag.* **73,** 49–56.

KUDO A. M. and WEILL D. F. (1970) An igneous plagioclase thermometer. *Contrib. Mineral. Petrol.* **25,** 52–65.

LARSEN E. S., IRVING J., GONYER F. A., and LARSEN E. S. 3rd (1938) Petrologic results of a study of the minerals from the tertiary volcanic rocks of the San Juan region, Colorado. *Amer. Mineral.* **23,** 227–257.

MARSH B. D. (1976) Some Aleutian andesites: their nature and source. *J. Geol.* **84,** 27–45.

MATHEZ E. A. (1973) Refinement of the Kudo-Weill plagioclase thermometer and its application to basaltic rocks. *Contrib. Mineral. Petrol.* **41,** 61–72.

MOROHASHI T., BANNO S. and YAMASAKI M. (1974) Plagioclase zoning in the Setogawa ash-flow sheet of the Nohi Rhyolite Complex, Central Japan. *Contrib. Mineral. Petrol.* **45,** 187–196.

MORSE S. A. (1984) Cation diffusion in plagioclase feldspar. *Science* **225,** 504–505.

MYERS J. D. and MARSH B. D. (1981) Geology and Petrogenesis of the Edgecumbe Volcanic Field, SE Alaska: The interaction of basalt and sialic crust. *Contrib. Mineral. Petrol.* **77,** 272–287.

NIELSEN R. L. (1988) A model for the simulation of combined major and trace element liquid lines of descent. *Geochim. Cosmochim. Acta* **52,** 27–38.

NIELSEN R. L. and DUNGAN M. A. (1983) Low pressure mineral-melt equilibria in natural anhydrous mafic systems. *Contrib. Mineral. Petrol.* **84,** 310–326.

PHEMISTER J. (1934) Zoning in plagioclase feldspar. *Mineral. Mag.* **23,** 541–555.

PRINGLE G. J., TREMBATH L. T. and PAJARI G. J. JR. (1974) Crystallization history of a zoned plagioclase (Microprobe analysis of zoned plagioclase from the Grand Manan tholeiite sheet). *Mineral. Mag.* **39,** 867–877.

SIBLEY O. F., VOGEL T. A., WALKER B. M. and BYERLY G. (1976) The origin of oscillatory zoning in plagioclase: a diffusion and growth controlled model. *Amer. J. Sci.* **276,** 275–284.

SMITH M. P. (1983) A feldspar-liquid geothermometer. *Geophys. Res. Lett.* **10,** 193–195.

SOUTHWICK D. L. (1969) Crystalline rocks in Geology of Harford County, Maryland. *Maryland Geol. Surv.,* 1–76.

VANCE J. A. (1962) Zoning in igneous plagioclase: normal and oscillatory zoning. *Amer. J. Sci.* **260,** 746–760.

WALKER D., SHIBATA T. and DELONG S. E. (1979) Abyssal tholeiites from the Oceanographer Fracture Zone. II. Phase equilibria and mixing. *Contrib. Mineral. Petrol.* **70,** 111–125.

YODER H. S. JR. (1969) Calcalkalic andesites: experimental data bearing on the origin of their assumed characteristics. In *Andesite Conference* (ed. A. R. MCBIRNEY), *Oregon Dept. Geol., Mines and Ind. Bull.* **65,** 77–89.

Fluid-Mineral Interactions: A Tribute to H. P. Eugster
© The Geochemical Society, Special Publication No. 2, 1990
Editors: R. J. Spencer and I-Ming Chou

Crystallization history of a pyroxenite xenolith in a granulite inferred from chemical and single-crystal X-ray data

JIŘÍ FRÝDA

Geological Survey, Malostranské náměstí 19, 118 21 Praha 1, Czechoslovakia

and

MILAN RIEDER

Institute of Geological Sciences, Charles University, Albertov 6, 128 43 Praha 2, Czechoslovakia

Abstract—Porphyroclasts of orthopyroxene and clinopyroxene in the pyroxenite have bulk compositions corresponding to an equilibrium at about 1260°C and their content of Fe^{3+} indicates a relatively high oxygen fugacity during their crystallization. Later they were corroded during recrystallization by a reaction with olivine that yielded spinel grains and pyroxene neoblasts. This reaction proceeded in the subsolidus, under increasingly reducing conditions, and produced spinel of progressively smaller grain size and increasing Al content. Finally, at about 910°C, the recrystallization ceased, at approximately the same time as the porphyroclasts underwent exsolution. Orthopyroxene porphyroclasts exsolved clinopyroxene and spinel lamellae, crystallographically oriented in the host. Clinopyroxene porphyroclasts probably exsolved enstatite first and then spinel, as a result of a strong reduction probably caused by the diffusion of hydrogen. Oxygen, thus freed, combined with hydrogen, yielding about 2.5 OH per seven R^{2+} in the exsolved enstatite and converted it into an orthoamphibole-like phase that is perfectly crystallographically aligned in the host. It appears that the unmixing of spinel lamellae may be viewed as a redox reaction and that concomitantly forming hydrated phases may serve as a proof of the role of hydrogen during the evolution of mantle xenoliths.

GEOLOGICAL SETTING

THE XENOLITH of olivine websterite studied comes from the abandoned Lichtenštejn quarry located on the SSW slope of Mount Kleť, about 1.5 km from the village of Vyšný near Český Krumlov, southwestern Czechoslovakia. The xenolith is about 70 cm in diameter, embedded in a relatively homogeneous, partly recrystallized kyanite-garnet felsic granulite, which belongs to the Blanský les Mountains Massif, one of the largest granulite massifs of the Moldanubian crystalline complex. While numerous ideas on the genesis of the granulites in the area have been put forth, it is generally agreed that they formed at high pressures, near the base of the Earth's crust (VRÁNA, 1987; FIALA *et al.*, 1987). Ultramafic rocks such as dunites, lherzolites, harzburgites, and olivine websterites (most of them carrying pyrope garnet) represent about 5 vol.% of the rocks of the Blanský les Mountains Massif, and so far they have received relatively little attention. Further details about the geology of the area can be found in papers by FEDIUKOVÁ (1965, 1978), ROST (1966), KODYM (1972), KODYM *et al.* (1978), VRÁNA (1979, 1987), SLABÝ (1983), or FIALA *et al.* (1987). The papers by VAN BREEMEN *et al.* (1982) and AFTALION *et al.* (1989) deal with geochronology.

PETROGRAPHY

The studied xenolith contains orthopyroxene (OPX), clinopyroxene (CPX), olivine (OL), and spinel (SPN). The minerals vary in proportion, so the volume percentages (OPX 45–65%, CPX 15–25%, OL 15–30%, SPN < 3%) are only approximate. Nevertheless, according to Streckeisen's (1973) classification, the rock can be referred to as olivine websterite.

Orthopyroxene is present in what appears to be two generations. The first is large lamellar porphyroclasts (up to 2 cm in size, Fig. 1), the second, smaller lamellae-free neoblasts (<2 mm). The porphyroclasts are irregularly shaped because of corrosion during recrystallization, but remain elongated parallel to [001]. They are intensely bent and often kinked along deformation planes roughly perpendicular to [001], along which smaller fragments separated (Fig. 1). This deformation cannot be younger than the exsolution of lamellae; if it were, we would not observe SPN grains to follow deformation zones nor would the lamellae there be by about an order of magnitude thicker than in unaffected parts of the porphyroclasts (Fig. 2). The density of lamellae is uniform throughout the grains all the way to the border with other pyroxene grains. However, along contacts with olivine, there may be

85

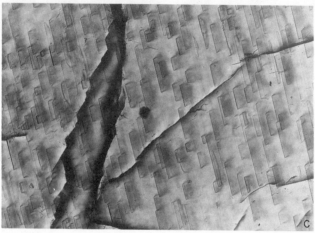

FIG. 1. a. Orthopyroxene porphyroclast, deformed and broken, surrounded by recrystallization products. Lamellae-free zones in the porphyroclast rim recrystallized olivine grains. Crossed polars, width of photograph = 4 mm. b. A plastically deformed OPX porphyroclast corroded during recrystallization. Olivine in the lower left-hand corner exhibits wavy extinction. The lamellae in OPX do not seem to disappear along $[100]_{OPX}$ as readily as along directions in the $(100)_{OPX}$ plane. Crossed polars, width of photograph = 4 mm. c. Lamellae of spinel (SPN) in OPX porphyroclast seen in a section roughly perpendicular to $[001]_{OPX}$. The SPN platelets lie parallel to $(100)_{OPX}$ and apparently contribute to the easy parting along that plane. Plane-polarized light, width of photograph = 0.5 mm.

a lamellae-free margin (0.5–1.0 mm thick). Contrary to porphyroclasts, neoblasts do not show deformation, and occur jointly with recrystallized olivine and large (vermicular) grains of spinel.

Clinopyroxene porphyroclasts reach up to 7 mm in size and contain abundant lamellae (Figs. 2 and 3) evenly distributed throughout except for lamellae-free zones, about 0.5 mm wide, that follow the border with OL grains. Clinopyroxene porphyroclasts do not show signs of plastic deformation, but many are fragmented. Small lamellae-free fragments surrounded by olivine cannot be distinguished from possible CPX neoblasts.

The mean size of OL grains is about 4 mm, larger grains are elongate and often penetrated by kink bands. The larger grains also contain zones or veils of isometric inclusions that were not studied further

(Fig. 3), but appear like CO_2- or CO-containing fluid inclusions known from mantle xenoliths (Fig. 3; KIRBY and GREEN, 1980, pl. 4-D; ANDERSON et al., 1984, Fig. 1-E, F; BERGMAN and DUBESSY, 1984, Fig. 2a, b). The zones of inclusions seem to have no crystallographic orientation and appear not to have a systematic relation to kink bands. The rock also contains small OL grains free from signs of deformation or inclusions which always accompany OPX neoblasts and vermicular spinel.

Spinel is red-brown, and its large (vermicular) grains or aggregates, which represent the vast majority of spinel in the rock, may reach 5 mm across. They do not seem to associate preferentially with any of the three primary minerals in the xenolith. Contacts of OPX and CPX porphyroclasts are punctuated with chains of SPN grains (Fig. 3) rang-

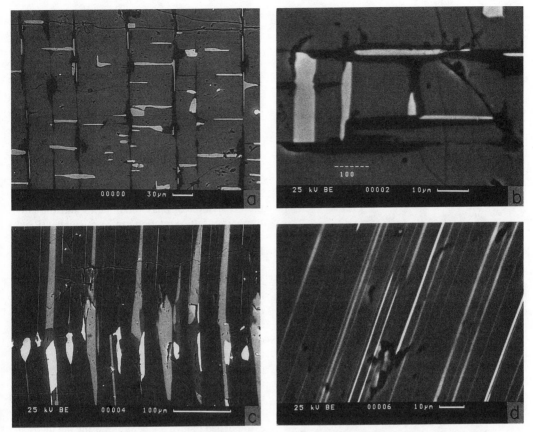

FIG. 2. Back-scattered electron images of pyroxene porphyroclasts. a, b. Section across a CPX porphyroclast, approximately perpendicular to [001]$_{CPX}$. Most spinel lamellae (white) are parallel to (010)$_{CPX}$ and lamellae of the orthoamphibole phase (OA, black) are mostly parallel to (100)$_{CPX}$, but there are some exceptions. The 'wrong' orientation of SPN lamellae seems to be associated with the OA phase. The (100)$_{CPX}$ is vertical in a. and horizontal in b. c. OPX porphyroclast, cut roughly perpendicular to [010]$_{OPX}$, showing a deformation zone with swollen CPX lamellae (light grey) and enriched in SPN grains (white). d. A section of OPX porphyroclast, approximately parallel to [001]$_{OPX}$, illustrating the distribution of lamellae of SPN (white) and CPX (light grey).

ing between 10^{-3} and 10^{-1} mm in size. Similar grains of spinel ($<10^{-2}$ mm) occur along deformation zones in OPX porphyroclasts (Fig. 2) or inside OL crystals.

In the nomenclature of MERCIER and NICOLAS (1975), the texture of the Lichtenštejn quarry xenolith is porphyroclastic. These authors cite the inclusion of spinel in olivine as evidence of a recrystallization of olivine, following an intensive deformation. However, such an effect is typical of the genetically younger equigranular texture, and thus we may refer to the present texture as transitional between porphyroclastic and equigranular.

EXPERIMENTAL

Single-crystal X-ray work was done on a Dioptra precession camera (HANIC et al., 1955). Powder data were recorded on a DRON-2,0 diffractometer (Burevestnik, Leningrad), with scanning speed 0.5° (2θ)/minute, and processed by BURNHAM's (1962) least squares program. The cell edge of spinel was based on film data extrapolated against the Nelson-Riley function (RIEDER, 1971).

An ARL-SEMQ microprobe was used to obtain chemical compositions. The accelerating voltage was 15 kV, current 50 nA, and counting time 20 seconds. Natural silicate minerals were used as standards for Ti, Cr, Mn, Si, Fe, K, Na, Mg, Ca, and Al, and synthetic Ni olivine was used for Ni. Analytical points were selected after observation in back-scattered electron mode (accelerating voltage 30–35 kV). Five to ten grains of each type of mineral (porphyroclast, neoblast, lamella) were analyzed. If homogeneous, the mineral was analyzed at three to six points, otherwise, up to 25 analyses were made. The Ca in olivine and Ni in OL, OPX, CPX, SP were determined with 100 nA current and 100 seconds counting time (used for Ca by ADAMS and BISHOP, 1986). The bulk composition of OPX and CPX porphyroclasts was analyzed either using a defocussed electron beam (35 μm) or by allowing the beam to scan over rasters, 50×50 μm, that were combined so as to pave stretches several hundred μm in diameter. Both methods yielded practically identical re-

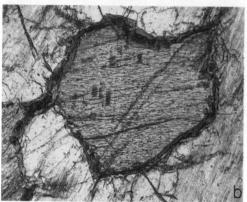

FIG. 3. a. Spinel (black) surrounds OPX neoblasts (light grey) or is enclosed in olivine neoblasts (white, lower right-hand corner). Crossed polars, width of photograph = 2 mm. b. A CPX porphyroclast with abundant lamellae, surrounded by OPX porphyroclasts. Chains of spinel grains (black) follow grain boundaries. Crossed polars, width of photograph = 2 mm. c. Veils of inclusions in a crystal of deformed olivine. Plane-polarized light, width of photograph = 0.2 mm.

sults. After corrections for drift of the instrument and for dead time of the detectors, the data were reduced according to BENCE and ALBEE (1968). When assessing the bulk chemical compositions, one must be aware that the uncertainties may exceed the errors indicated because the assumption of a homogeneous distribution of each element in the X-ray generation range is not fulfilled. Standard errors of individual determinations were computed according to KOTRBA (1989).

PHASE BOUNDARIES AND LATTICE ROTATION

Assuming coherency between lamellae and host, the orientation of phase boundaries and the corresponding lattice rotation can be calculated (ROBINSON et al., 1977; FLEET, 1981, 1982; ŠANC and RIEDER, 1983). Two models were used here and applied to the following intergrowths: CPX lamellae & OPX host, SPN lamellae & CPX host, and SPN lamellae & OPX host. No calculation involving the orthoamphibole phase (OA, see below) was performed because it appears to have formed by secondary replacement and not by a process amenable to the above numerical treatment. Relatively good precision powder cell data were used for CPX and OPX, but for SPN we had to use the less precise single-crystal cell edge because it is the only one we had for spinel in lamellae. These unit-cell data were extrapolated to high temperatures using the thermal expansion data of SKINNER (1966) and ŠANC (1982). Correction for the effect of pressure and for chemical changes after coherency had been lost were not made. The unit-cell data were then converted to equivalent subcells common to both phases in an intergrowth pair, based on near-coincidence of reflections in precession photographs, and these, in turn, were input into the calculations. The results are summarized in Table 1. Perhaps the most interesting among them is the lattice rotation of SPN in OPX host whose measured value (0.81(3)°) corresponds well to the measured 0.86(3)° reported for the same pair from Tři Studně (ŠANC and RIEDER, 1983).

ORTHOPYROXENE PORPHYROCLASTS

All six crystals examined by X-ray precession contain CPX and SPN lamellae (Fig. 2); their unit-cell data appear in Table 2. Each of the two exsolved phases is present in two orientations. The relations in this standard intergrowth are illustrated in Fig. 7 of ŠANC and RIEDER (1983). The dominant faces of all lamellae are parallel to (100) of host OPX. Clinopyroxene lamellae are less than 4 μm thick,

Table 1. Measured and calculated phase boundaries

Host Phase	OPX	OPX	CPX
Lamella	CPX	SPN	SPN
Coherent Composition Plane Parallel to	$[010]_{CPX}\|[010]_{OPX}$	$[100]_{OPX}\|[111]_{SPN}$	$[010]_{CPX}\|[\bar{1}10]_{SPN}$
Lattice Rotation	$[100]^*_{CPX} \wedge [100]^*_{OPX}$	$[010]^*_{OPX} \wedge [\bar{1}10]^*_{SPN}$	$[100]^*_{CPX} \wedge [111]^*_{SPN}$
Measurements[a]			
Lattice Rotation	0.0(3)°	0.81(3)°	0.0(3)°
Composition Plane I $\wedge$ [$[001]_{host}$	0.(5.)°	40.(20.)°	d
Composition Plane II $\wedge$ $[001]_{host}$	0.(5.)°	−40.(20.)°	d
Calculations[a,b]			
Area Robinson[c]			
Lattice Rotation I	−0.19° to −0.22°	0.52° to 0.54°	1.55° to 2.54°
Composition Plane I $\wedge [001]_{host}$	−2.95° to −6.54°	32.8° to 49.6°	49.9° to 60.3°
Lattice Rotation II		−0.52° to −0.54°	2.12° to 3.93°
Composition Plane II $\wedge [001]_{host}$		−32.8° to −49.6°	56.6° to 73.1°
Vector Robinson[c]			
Lattice Rotation I	−0.11° to −0.20°	0.0°	−0.15° to −0.19°
Composition Plane I $\wedge [001]_{host}$	−4.10° to −10.80°	0.0°	26.7° to 27.7°
Lattice Rotation II			−5.48° to −5.66°
Composition Plane II $\wedge [001]_{host}$			83.4° to 83.9°

[a] When viewed along $[010]_{host}$, angles measured in counter-clockwise direction are given as positive.
[b] Calculations are presented as intervals corresponding to 20° and 1000°C, respectively.
[c] 'Vector Robinson' is the approach of ROBINSON et al. (1977), 'area Robinson' is a modification of ŠANC and RIEDER (1983).
[d] Could not be measured.

their length along $[010]_{OPX}$ and $[001]_{OPX}$ reaches several tenths of a mm; and they are 10–30 μm apart. Spinel lamellae are thinner (≤2 μm) and distinctly longer along $[001]_{OPX}$ (up to 1 mm) than along $[010]_{OPX}$. Their termination appears to be tilted relative to $[010]_{OPX}$, towards $[010]_{OPX}$ in some lamellae, and $[0\bar{1}0]_{OPX}$ in others, under angles about 40°. The uncertainty of this angle is large, perhaps up to 20°. The observed orientations and morphologies are those predicted from the theory of optimal phase boundaries (Table 1).

The Mg/(Mg + Fe$_{tot}$) and Ca/(Ca + Mg + Fe$_{tot}$

+ Mn) ratios of OPX host are, respectively, 0.911(2) and 0.005(1) (Fig. 4). The differences between the composition of OPX host from cores and from margins of the porphyroclasts were less than the error of analysis. However, the ranges obtained for Al_2O_3 and Cr_2O_3 are wider, namely 2.95–3.85 wt.% and 0.29–0.55 wt.%, respectively, and the Cr/Al ratio varies from 0.062 to 0.102 (Fig. 5). These elements indeed are those that can be expected to vary if a lamella of SPN or CPX is accidentally present in the X-ray generation range of an analytical point in OPX host.

Table 2. Unit-cell data for phases in pyroxene porphyroclasts and spinel

	OPX porphyroclasts			CPX porphyroclasts				Spinel large grains
	OPX host	CPX lamellae	SPN lamellae	CPX host		OA lamellae	SPN lamellae	
a, nm	1.822(1)	1.829(1)	0.975(2)	0.8138(8)	0.9735(5)	0.975(1)	1.83(1)	0.815(1)	0.8156(1)
b, nm	0.8813(9)	0.882(1)	0.895(2)		0.8910(5)	0.8915(8)	1.81(1)		
c, nm	0.5195(3)	0.5188(5)	0.5252(9)		0.5251(3)	0.5258(6)	0.524(2)		
β, °	90.0	90.0	105.78(17)		105.93(2)	106.05(15)	90.0		
Method	a	b	b	b	a	b	b	b	c

[a] Powder data, DRON-2,0 diffractometer, graphite-monochromatized Cu radiation, least-squares refinement by program of Burnham (1962).
[b] Measurements of precession photographs; the data given are means weighted in proportion to the number of reciprocal rows measured in individual photographs.
[c] Cylindrical camera, 2r = 114.59 mm, extrapolation against the function of Nelson and Riley.

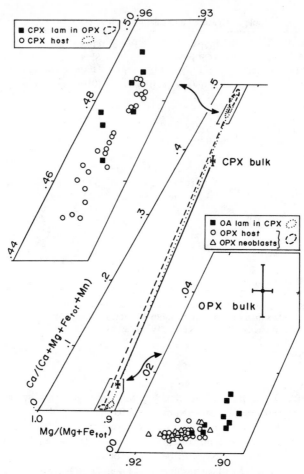

FIG. 4. Relations between phases in pyroxene porphyroclasts. The error crosses for the bulk compositions embrace ±1σ. The full tieline is that for primary equilibrium of bulk porphyroclasts, the dotted tieline is for compositions of breakdown phases in CPX porphyroclasts, the dashed tieline is for the same in OPX porphyroclasts. The points for bulk composition of the porphyroclasts fall off tielines for the breakdown phases because the bulks include spinel richer in iron.

Clinopyroxene lamellae could be analyzed only in deformation zones where their thickness increases (Fig. 2). The mean values of the $Mg/(Mg + Fe_{tot})$ and $Ca/(Ca + Mg + Fe_{tot} + Mn)$ ratios are 0.948(5) and 0.481(9), respectively (Fig. 4). The ratio Cr/Al equals 0.102–0.162 (Fig. 5).

Spinel lamellae too could be analyzed only when swollen. Their $Mg/(Mg + Fe_{tot})$ ratio falls between 0.754 and 0.777, the $Cr/(Cr + Al)$ ratio ranges from 0.144 to 0.161 (Fig. 6). The content of Fe^{3+} was calculated assuming full cation occupancy and charge neutrality (Fig. 7). Spinel grains following deformation zones in the porphyroclasts have a higher $Mg/(Mg + Fe_{tot})$ ratio (0.779–0.783), and the same applies to the $Cr/(Cr + Al)$ ratio (0.158–0.169). The contents of Mn and Fe^{3+} are identical with those of SPN lamellae.

The bulk composition of OPX porphyroclasts is well characterized by their mean values (Table 3, Fig. 4) inasmuch as individual analyses differ by less than the error of analysis. The ratio $Mg/(Mg + Fe_{tot}) = 0.908(2)$ is lower than that of the OPX host; this is due to the very low $Mg/(Mg + Fe_{tot})$ of the SPN lamellae. Mean $Ca/(Ca + Mg + Fe_{tot} + Mn) = 0.040(7)$ and $Cr/Al = 0.13(3)$. The bulk composition of cores of porphyroclasts is constant within the error of analysis, but towards contacts with olivine, the contents of Ca, Cr, and Al smoothly decrease, which is manifest as thinning and disappearance of CPX and SPN lamellae.

CLINOPYROXENE PORPHYROCLASTS

Five crystals were X-rayed to determine the mutual orientations of phases, that is the CPX host

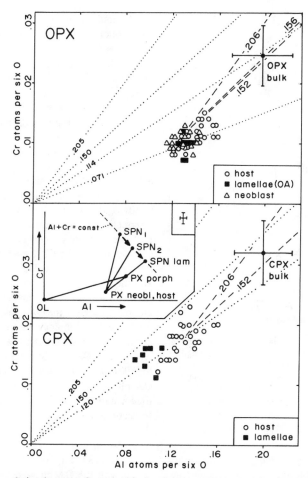

FIG. 5. The relation between Cr and Al in breakdown pyroxenes and orthoamphibole (OA); a cross of error bars 3σ in length is shown in a rectangle. The composition of bulk porphyroclasts is known with a much lower precision, the crosses indicate $\pm 1\sigma$. Dotted lines are labeled with Cr/(Cr + Al) ratios. Dashed lines emanate from points representing means for the clusters of data and are labeled with Cr/(Cr + Al) ratios of associated spinels. In the inset appears a schematic illustration of the reaction of pyroxene porphyroclasts with olivine to PX neoblasts and an older spinel (SPN$_1$), then to a younger spinel (SPN$_2$) and ending with unmixed SPN lamellae formed without the participation of olivine.

and lamellae of spinel and an orthorhombic amphibole-like phase, possibly related to anthophyllite (OA, Fig. 2). Spinel is present in two orientations whose volumes were nearly identical in OPX porphyroclasts, but clearly differ here: there is always more spinel in B orientation than in A (labeling of ŠANC and RIEDER, 1983). The intensities of reflections of SPN$_A$ correlate positively with those of OA indicating that lamellae of SPN$_A$ associate with OA lamellae. Exactly the same was observed in CPX porphyroclasts from Tři Studně (ŠANC and RIEDER, 1983; Fig. 6) except that the phase in lamellae was orthopyroxene rather than orthoamphibole.

Characterization of the orthoamphibole phase is

admittedly incomplete. First, the intensities of its reflections are much weaker than those of the host because of the relative volumes involved and, second, there is the unavoidable overlapping of reflections (Fig. 8). Nonetheless it was identified in all crystals examined. The reciprocal cell geometry and the intensity distribution agree with the Laue group *mmm*, but a complete diffraction symbol could not be assigned. There is a good agreement of the unit-cell data and a reasonably good overall agreement of intensities calculated for anthophyllite with the ones observed in the X-ray photographs, but there are some disturbing dissimilarities. For instance, the reflections 052, 072, 043, and 083, which are not

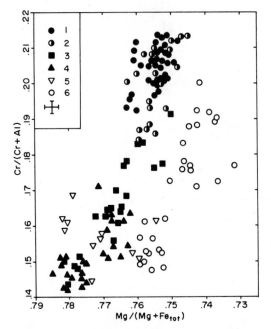

FIG. 6. The relation between chemistry and paragenetic position of spinel. The cross of error bars 3σ in length is shown in the symbol rectangle. 1 = large grains not associated with pyroxene neoblasts; 2 = rims of large grains; 3 = medium and small grains; 4 = lamellae in clinopyroxene; 5 = lamellae and grains in orthopyroxene; 6 = grains enclosed in aggregates of olivine.

permitted in space group *Pnma* (that of anthophyllite) are well visible in the photographs (Fig. 8), while several 4*k*0 reflections and the strong 161,

which should be present in patterns of anthophyllite, were not observed. Therefore, the identification remains tentative at best. Perhaps more could be learned by a high-resolution electron microscopic examination, but no such study has been contemplated yet.

The OA phase is well visible in back-scattered electron images, but it is all but invisible with petrographic microscopy. It forms lamellae with a variable thickness and irregular outline (Fig. 2); most of them are parallel to $(100)_{CPX}$. The lamellae are distributed evenly throughout the host, with a spacing of 30–80 μm. Some OA lamellae are parallel to $(010)_{CPX}$, and the largest pods of OA occur at intersections of both lamellar systems. It can be seen that there are two systems of lamellae of spinel, one associated with the CPX host, the other, with OA lamellae. The first lamellae are parallel to $(010)_{CPX}$, and their thickness is mostly between four and ten μm. The second SPN lamellae are usually thinner and often outgrow the OA phase and continue, with unchanged morphology, into CPX (Fig. 2). It is important that most SPN lamellae parallel to $(010)_{CPX}$ terminate at OA or at SPN lamellae that are parallel to $(100)_{CPX}$. This indicates that the unmixing of the OA phase (or rather its precursor's) began before the formation of SPN lamellae parallel to $(010)_{CPX}$. In sections parallel to $(010)_{CPX}$, one can see that the spinel lamellae belong in fact to two morphologies like those observed at Tři Studně (ŠANC and RIEDER, 1983, Fig. 11). The SPN lamellae associated with the OA phase belong to one morphological type, being elongated on $[001]_{CPX}$. The den-

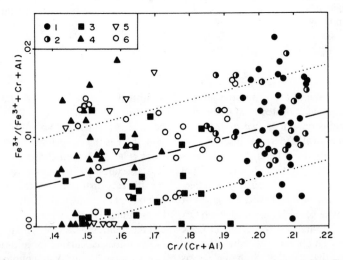

FIG. 7. Calculated trivalent iron increases in spinels richer in Cr. A few analyses whose Fe^{3+} calculated as negative were plotted at $Fe^{3+} = 0$. Correlation coefficient for the regression $r = 0.392$, dotted lines indicate $\pm 1\sigma$. Symbols as in Fig. 6.

Table 3. Chemical composition of orthopyroxene and the orthoamphibole-like phase (OA)

	OPX porphyroclast bulk mean[a]	OPX host		OA lamellae		OPX neoblast	
		Typical analysis[b]	Range	Typical analysis[b]	Range	Typical analysis[b]	Range
SiO_2	53.5(7)	55.1(4)	53.9–56.2	55.0(4)	54.7–55.5	55.5(4)	55.1–56.2
TiO_2	0.04(3)	0.04(2)	0.00–0.10	0.02(2)	0.00–0.04	0.04(2)	0.00–0.09
Al_2O_3	4.8(7)	3.40(4)	2.95–3.85	3.20(4)	3.10–3.40	3.15(4)	2.84–3.50
Cr_2O_3	0.9(2)	0.40(2)	0.29–0.55	0.37(2)	0.28–0.45	0.40(2)	0.29–0.47
FeO	5.9(1)	6.01(5)	5.65–6.24	6.38(5)	6.00–6.61	5.96(5)	5.90–6.18
MnO	0.14(2)	0.14(2)	0.06–0.18	0.14(2)	0.11–0.20	0.14(2)	0.06–0.19
MgO	32.6(5)	34.8(2)	33.5–35.7	34.1(2)	33.8–35.0	34.6(2)	34.0–35.3
CaO	2.1(4)	0.27(2)	0.19–0.41	0.46(2)	0.29–0.75	0.27(2)	0.10–0.42
NiO	[c]	0.095(5)	0.09–0.10	[c]	[c]	[c]	[c]
Na_2O	0.04(1)	0.02(1)	0.00–0.04	0.0(0)	0.0	0.0(0)	0.0
K_2O	0.0(0)	0.0(0)	0.0	0.0(0)	0.0	0.0(0)	0.0
Σ	100.(1.)	100.3(5)		99.7(5)		100.1(5)	
$Mg/(Mg + Fe_{tot})$	0.908(2)	0.9116(8)	0.907–0.916	0.9050(7)	0.901–0.910	0.9119(8)	0.906–0.919
$Ca/(Ca + Mg + Fe_{tot} + Mn)$	0.040(7)	0.0051(4)	0.0036–0.0081	0.0086(4)	0.0051–0.0147	0.0051(4)	0.0015–0.0077
Cr/Al	0.13(3)	0.079(4)	0.062–0.102	0.077(4)	0.054–0.092	0.085(4)	0.063–0.100

[a] Mean of several analyses; errors reflect variation in the set on which the mean is based.
[b] Errors given are those of individual determinations and were computed according to KOTRBA (1989).
[c] Not analyzed.

sity of both the SPN and OA lamellae throughout the porphyroclasts is uniform, but it decreases in border zones adjacent to olivine grains.

The OA lamellae exhibit a constant $Mg/(Mg + Fe_{tot})$ ratio (mean 0.905(3)), and a less constant $Ca/(Ca + Mg + Fe_{tot} + Mn)$ ratio (mean 0.009(3)), Al_2O_3 and Cr_2O_3 vary between 3.10 and 3.40 wt.% and between 0.28 and 0.45 wt.%, respectively, the mean Cr/Al ratio is 0.08(2). It should be noted that the analysis of the OA phase closely resembles that of an orthopyroxene (Fig. 4), and even the total does not indicate the presence of an undetermined water content. This is analogous to what had been observed for an Mg layer silicate in similar lamellae in CPX host from Deštná (ŠANC and RIEDER, 1983), which did have a deficient analytical total.

The composition of the CPX host does not exhibit differences in the $Mg/(Mg + Fe_{tot})$ ratio between core and margin, the mean value is 0.945(3) (Figs. 4 and 9). However, the $Ca/(Ca + Mg + Fe_{tot} + Mn)$ ratio shows considerable scatter, but there appear to be no systematic differences between core and margins. There is considerable scatter in the Cr/Al ratio, with mean 0.14(2), which may be attributable to contaminations.

Also analyzed was spinel in both systems of lamellae in OPX porphyroclasts, but lamellae of the second system (associated with the OA phase) could be analyzed at a few points only, because of insufficient thickness. No differences between the two SPN lamellar systems were found. The $Mg/(Mg$ + $Fe_{tot})$ ratio ranges from 0.762 to 0.785, $Cr/(Cr + Al)$, from 0.142 to 0.172; both correlate with each other (Fig. 6). The ratio $Fe^{3+}/(Fe^{3+} + Al + Cr)$ is below 0.02, based of course on calculated Fe^{3+}.

The bulk composition of CPX porphyroclasts (Table 4, Fig. 4) exhibits a mean $Mg/(Mg + Fe_{tot})$ = 0.926(5), which is below that of the CPX host, due to the low $Mg/(Mg + Fe_{tot})$ ratio of SPN lamellae. The mean ratios $Ca/(Ca + Mg + Fe_{tot} + Mn)$ and Cr/Al equal 0.38(1) and 0.16(3), respectively. The bulk composition of the cores is homogeneous within analytical error.

ORTHOPYROXENE NEOBLASTS

The mean ratios $Mg/(Mg + Fe_{tot})$ and $Ca/(Ca + Mg + Fe_{tot} + Mn)$ are, respectively, 0.912(3) and 0.005(2) and are practically identical with those of the host phase in OPX porphyroclasts (Fig. 9), while the mean contents of Al_2O_3 and Cr_2O_3 are generally slightly lower (Fig. 5).

SPINEL GRAINS

Spinel grains of different morphology differ also chemically (Figs. 6, 7 and 10). Large grains of vermicular SPN (several mm across) represent over 90 vol.% of all spinel in the xenolith. They show a rather narrow variability of the $Mg/(Mg + Fe_{tot})$ and $Cr/(Cr + Al)$ ratios with means 0.754(4) and 0.206(5), respectively. Individual grains appear homogeneous, and in only a few cases, asymmetric

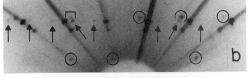

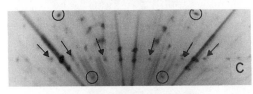

FIG. 8. Enlarged sections of precession photographs (unfiltered MoK radiation) of intergrowths involving clinopyroxene (CPX), orthoamphibole (OA), spinel (SPN), and orthopyroxene (OPX). a. Reciprocal $0kl$ net of CPX with $0k0$, $0k1$, and $0k2$ rows well visible. The hkh net of the OA phase is superposed. Reflections in the $1k1_{OA}$ row are marked by arrows; the 171 reflection is obscured by a slightly deviating pyroxene orientation. b. Reciprocal $h0l$ nets of CPX and OA; the $h02$ reciprocal rows of both phases coincide, reflections of OA are indicated by arrows. Also present is spinel in orientation B (hkk net, circles) and orientation A (small volume only) betrayed by the weak reflection within the rectangle. Reflection intensities of the $h0l_{OA}$ should obey a vertical plane of symmetry, but do not, apparently due to absorption associated with an uneven distribution of OA lamellae in the host phase. However, the symmetry plane was visible in the $h0l_{OA}$ nets of the other crystals examined. c. Reciprocal net $\bar{h}k2h$ of clinopyroxene, overlapping with the $0kl$ net of orthoamphibole and $h \cdot k \cdot -(h + k)$ net of spinel. Reflections in the $0k2$ row of OA are marked with arrows, SPN reflections appear in circles. The strongest CPX reflections are $\bar{1}32$ and $\bar{1}32$. d. Reciprocal net $0kl$ of OPX, specifically the $0k2$ and $0k3$ rows, with $\bar{h}k2h$ net of CPX superposed (reflections $\bar{1}32_{CPX}$ and $\bar{1}32_{CPX}$ are shown by arrows). Note the splitting of $h \cdot k \cdot -(h + k)$ reflections of spinel (circles) caused by lattice rotation that does not operate for SPN in CPX (preceeding photograph).

zoning was encountered. The outer zones have lower Cr/(Cr + Al) and higher Mg/(Mg + Fe$_{tot}$) ratios, but it is likely that they in fact are an accretion of a finer grained younger spinel. The large grains

have the highest ratio $Fe^{3+}/(Fe^{3+} + Al + Cr)$ of all spinel, namely 0.011(5) (Table 5).

Smaller SPN grains from contacts of pyroxene porphyroclasts exhibit more compositional variability, although individual grains are homogeneous within the error of analysis. Their Fe^{3+} calculates as lower than in large grains and the Mg/(Mg + Fe$_{tot}$) and Cr/(Cr + Al) ratios correlate well with each other (Figs. 6, 7). These spinels fit well into the relations between chemistry and grain size (Fig. 10).

Grains of spinel enclosed in olivine also display considerable chemical variability (Fig. 10), and their chemistry too varies with grain size. Moreover, individual grains are zoned, with Mg/(Mg + Fe$_{tot}$) decreasing towards contacts with olivine grains, while Cr, Al, Fe^{3+}, and Mn remain constant.

OLIVINE

Olivine occurs as large deformed grains and as smaller undeformed grains, with respective mean

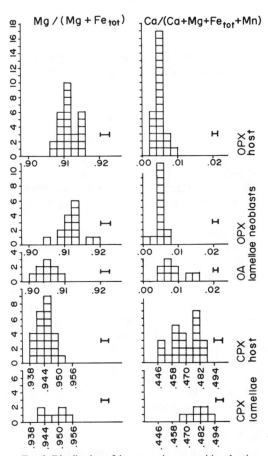

FIG. 9. Distribution of the two main compositional ratios in pyroxenes and the orthoamphibole phase (OA); one square represents one determination, the length of the error bars is 3σ.

Table 4. Chemical composition of clinopyroxene

	Porphyroclasts mean[a]	Host typical analysis[b]	Range[c]	Lamellae typical analysis[b]	Range
SiO_2	52.(1.)	53.0(4)	51.6–53.8	53.5(4)	53.0–54.4
TiO_2	0.11(4)	0.15(2)	0.10–0.21	0.11(2)	0.10–0.14
Al_2O_3	4.6(6)	3.09(4)	2.60–3.74	2.35(3)	2.12–2.65
Cr_2O_3	1.1(2)	0.63(2)	0.42–0.79	0.49(2)	0.39–0.56
FeO	2.9(2)	1.78(3)	1.64–2.04	1.68(3)	1.56–1.85
MnO	0.10(2)	0.10(2)	0.07–0.14	0.07(2)	0.04–0.12
MgO	20.2(7)	17.7(1)	17.3–18.5	17.2(1)	16.6–17.8
CaO	19.(1.)	23.1(1)	22.3–24.0	23.6(1)	22.7–24.6
NiO	[d]	0.044(5)	0.038–0.050	[d]	[d]
Na_2O	0.28(3)	0.43(2)	0.32–0.45	0.23(2)	0.19–0.27
K_2O	0.0(0)	0.0(0)	0.0	0.0(0)	0.0
Σ	100.(2.)	100.1(4)		99.2(4)	
$Mg/(Mg + Fe_{tot})$	0.926(5)	0.9466(9)	0.938–0.952	0.9480(9)	0.942–0.956
$Ca/(Ca + Mg + Fe_{tot} + Mn)$	0.38(1)	0.470(2)	0.451–0.485	0.483(2)	0.465–0.492
Cr/Al	0.16(3)	0.136(5)	0.109–0.171	0.139(6)	0.102–0.162

[a] Mean of several analyses; errors reflect variation in the set on which the mean is based.

[b] Errors given are those of individual determinations and were computed according to KOTRBA (1989).

[c] The range includes also analyses of lamellae-free fragments of porphyroclasts.

[d] Not analyzed.

values of $Mg/(Mg + Fe_{tot})$ equal to 0.907(1) and 0.9102(5). CaO is slightly higher in deformed grains, while the mean MnO and NiO contents of both types are identical (Table 6, Fig. 11). Individual olivine grains are chemically homogeneous, the only exception being zones in large deformed grains, about 0.15 mm thick, adjacent to spinel. In these zones, the $Mg/(Mg + Fe_{tot})$ increases outwards, but MnO and CaO remain constant (we did not examine the behavior of NiO).

PETROGENESIS

Textural and chemical features of the rock make it clear that at least two distinctly different processes took place.

The first process of which there is evidence involved an Al-rich orthopyroxene, an Al-rich clinopyroxene, and olivine. Despite some difficulties in obtaining the present bulk compositions of porphyroclasts and the possibility of chemical changes during recrystallization processes, there is a reasonable certainty about the temperature of primary equilibrium: the calculated temperatures based on OPX and CPX porphyroclasts fall between 1220 and 1280°C (Table 7). Such an equilibrium was not necessarily magmatic, but may have been the result of an older metamorphic-recrystallization process.

The oldest spinel (large homogeneous grains) probably did not participate in the above phase assemblage. It may have formed later by either of two processes. First, it may have been one of the products of incongruent melting of aluminous pyroxenes during partial melting, together with olivine and melt (DICKEY et al., 1971; DICK, 1977a,b; MYSEN and KUSHIRO, 1977; JAQUES and GREEN, 1980). Second, it could have formed by a subsolidus metamorphic recrystallization with Al-rich pyroxene and olivine as reactants and spinel and Al-poor pyroxene as products (GREEN, 1964; OBATA, 1980; KUO and ESSENE, 1986). A variant of the second path is an unmixing process from Al-rich pyroxenes (BASU and MACGREGOR, 1975; MERCIER and NICOLAS, 1975). Spinel formed in the subsolidus would have to be in close spatial association with pyroxene neoblasts, but we observe the latter to associate with finer grained younger spinels and not with large grains (Fig. 3). This lends indirect support to the first alternative. Also the remarkable homogeneity of large SPN grains and their low $Cr/(Cr + Al)$ ratio suggest that these spinel grains formed during a weak partial melting (DICK and BULLEN, 1984; OZAWA, 1988). Inasmuch as there was a later reequilibration between olivine and spinel, we can hardly assign meaningful temperatures to the event.

The second process which affected the rock extensively was an intensive deformation and a subsequent recrystallization. It took place at a time when the aluminous porphyroclasts became unstable and underwent the reaction,

Al-PX porphyroclast + olivine →

PX neoblast + spinel

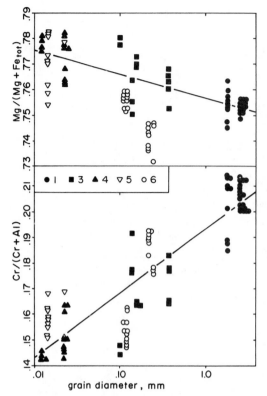

FIG. 10. The relation of the composition of spinel to grain diameter. Grain diameter is plotted logarithmically and was calculated as $(d_{min} \cdot d_{max})^{1/2}$. Inasmuch as there apparently was a cation-exchange reaction between spinel and olivine that involved Mg and Fe, points for spinel enclosed by olivine (symbol 6) were not included when the best-fit line was calculated for the top plot. No reaction with olivine involved Cr and Al, so the best-fit line in the bottom plot is based on all data points. Symbols as in Fig. 6.

(see KUO and ESSENE, 1986). The product spinel is spatially closely associated with pyroxene neoblasts (Fig. 3). The newly formed spinel exhibits mutually correlated variation of $Mg/(Mg + Fe_{tot})$ and $Cr/(Cr + Al)$ (Fig. 6), which correlates also with grain size. The chemical development in time of spinel in the Lichtenštejn quarry xenolith can be referred to as an Al trend, because the smaller (and thus younger) grains exhibit progressively lower $Cr/(Cr + Al)$ ratios. Inasmuch as one commonly finds large SPN grains with $Cr/(Cr + Al) = 0.21$ side-by-side with SPN lamellae whose $Cr/(Cr + Al) = 0.15$, the correlation of composition with grain size of spinel cannot be the result of an exchange of Al and Cr between SPN and PX similar to what had been postulated by OZAWA (1983) for Mg and Fe between SPN and OL. Also eliminated appears to be the possibility that $Cr/(Cr + Al)$ decreased by a

reaction of spinel with a trapped basaltic melt (IRVINE, 1967) or with an aluminum-rich silicate (HENDERSON, 1975). An interesting point is that the chemical composition of the smallest spinel grains is identical with that of SPN lamellae in pyroxene porphyroclasts (Fig. 6).

The set of spinel analyses from the xenolith under study falls into the fields of spinel composition from alpine-type peridotites and abyssal peridotites (DICK and BULLEN, 1984). The negative correlation between the $Cr/(Cr + Al)$ and $Mg/(Mg + Fe^{2+})$ ratios in SPN from these rocks has been interpreted as a consequence of partial melting under mantle conditions (GREEN and RINGWOOD, 1967; DICK, 1977a,b,; DICK and BULLEN, 1984; OZAWA, 1988). The $Cr/(Cr + Al)$ of spinel is thought to be an indicator of depletion of the mantle material (DICK and BULLEN, 1984). Contrary to this, at subsolidus temperatures, the distribution of Al and Cr between pyroxene and spinel appears to be independent of temperature (EALES and MARSH, 1983; OZAWA, 1988). Also the dependence of the $Cr/(Cr + Al)$ ratio in spinel on pressure is uncertain. BASU and MACGREGOR (1975) studied spinels from ultramafic xenoliths and concluded that the $Cr/(Cr + Al + Fe^{3+})$ ratio increases with the depth (and thus pressure) of origin of the xenoliths, but they admit that the ratio may be a function of the bulk composition of the rocks. Experimental data of DICKEY and YODER (1972) indicate that the Cr/Al ratio of the rock dictates the distribution of Cr and Al between subsolidus SPN and CPX, and that the latter does not change with temperature or pressure.

If we accept the idea that spinel formed, together with Al-poor pyroxene, from an Al-rich pyroxene plus olivine, we must accept its consequences. Pyroxenes have two cations per three oxygens, but there are three cations per four oxygens in both olivine and spinel. Hence the quantity of product spinel must equal the quantity of reactant olivine (Fig. 5). If, however, the reaction proceeds under reducing conditions, the equation must be rewritten so that less olivine reacts and more spinel is formed. If the composition of pyroxene neoblasts remains the same (actually, it is identical with the composition of PX host) and if there is a constant sum of Cr + Al in the spinel, the younger spinels must be progressively richer in Al and poorer in Cr. This is a clear consequence of the gradually shifting intersection point of tielines OL-PX$_{porph}$ and SPN$_{lam}$-PX$_{neobl}$ (Fig. 5, inset). Finally, the last spinel to form is the one that exsolved from porphyroclasts in whose breakdown to PX host and SPN lamellae no olivine participates. Inasmuch as the reconstructed original composition of pyroxene porphyroclasts is

Table 5. Chemical composition of spinel

	Cores of large grains		Lamellae in CPX		Lamellae in OPX		Medium and small grains		Grains enclosed in olivine	
	Typical analysis[a]	Range	Typical analysis[a]	Range	Typical analysis[a]	Range	Typical analysis[a]	Range	Typical analysis[a]	Range
SiO_2	0.0(0)	0.0	0.0(0)	0.0	0.0(0)	0.0	0.0(0)	0.0	0.0(0)	0.0
TiO_2	0.06(3)	0.00–0.10	0.03(3)	0.00–0.09	0.0(0)	0.00–0.05	0.0(0)	0.00–0.06	0.0(0)	0.0
Al_2O_3	49.5(2)	49.1–51.5	54.4(2)	53.1–55.7	54.0(2)	53.4–55.2	53.2(2)	52.0–55.7	54.3(2)	50.4–54.7
Cr_2O_3	19.38(9)	18.4–20.0	14.48(8)	13.7–16.4	14.94(8)	13.8–15.3	15.57(8)	14.0–18.4	14.71(8)	14.1–18.8
Fe_2O_3[b]	1.33	0.08–1.87	0.57	0.00–1.60	1.09	0.00–1.20	0.91	0.00–1.07	0.54	0.00–1.50
FeO	10.10(6)	9.8–10.8	9.68(6)	9.4–10.4	9.59(6)	9.5–10.7	9.91(6)	9.9–10.8	10.72(6)	10.3–11.2
MnO	0.13(2)	0.08–0.19	0.10(2)	0.06–0.16	0.10(2)	0.07–0.16	0.15(2)	0.09–0.17	0.10(2)	0.09–0.20
MgO	19.2(1)	18.9–19.6	19.9(1)	19.5–20.4	20.0(1)	19.2–20.2	19.6(1)	18.6–20.0	19.3(1)	18.8–19.7
NiO	0.341(6)	0.33–0.36	0.35[c]		0.35[c]		0.348(6)	0.34–0.37	0.35[c]	
Σ	100.1(2)		99.4(2)		100.1(2)		99.7(2)		100.0(2)	
$\dfrac{Mg}{Mg+Fe_{tot}}$	0.752(1)	0.748–0.763	0.776(1)	0.762–0.785	0.771(1)	0.754–0.777	0.765(1)	0.750–0.780	0.754(1)	0.732–0.759
$Cr/(Cr+Al)$	0.208(1)	0.193–0.214	0.1516(9)	0.142–0.172	0.1565(9)	0.144–0.161	0.1640(9)	0.144–0.191	0.1536(9)	0.148–0.200
$\dfrac{Fe^{3+}}{Fe^{3+}+Cr+Al}$	0.0134	0.0008–0.0212	0.0056	0.0000–0.0186	0.0108	0.0000–0.0172	0.0090	0.0000–0.0130	0.0053	0.0000–0.0162

[a] Errors given are those of individual determinations and were computed according to KOTRBA (1989).
[b] Fe_2O_3 was calculated; accordingly, ratios involving Fe^{3+} do not have errors.
[c] A mean of several determinations most of which come from other analytical points.

J. Frýda and M. Rieder

known with large errors, there are only mild constraints on the composition of spinel that participated in the reaction, but the above scheme seems to explain the observations.

An interesting insight into the crystallization history of the rock in the xenolith comes from an analysis of spinel exsolution in CPX porphyroclasts. First, the bulk composition recalculated to six oxygens assuming full cation occupancy yields an acceptable pyroxene formula only if all iron is trivalent (Fig. 12, origin of arrow A). Second, spinel in lamellae is very low in (calculated) Fe^{3+} (Table 5, Fig. 7), and the OA phase, when recalculated as orthopyroxene, appears to contain divalent iron only. Third, as can be seen in Fig. 12, if we try to reconstruct the composition of clinopyroxene after unmixing of OPX (OA) lamellae but before unmixing of spinel, we arrive at a point (origin of arrow B) located on a tieline CPX host-SPN lamellae, which lies off the tieline DI-COR. In other words, such a point does not represent a stoichiometric pyroxene formula because it has more than two cations per three oxygens. Thus what has been said indicates that the conditions must have been more oxidizing during crystallization of the porphyroclast (and during the partial melting event?) than during unmixing of lamellae. A reduction during the process would explain all observations, and it may even be argued that the exsolution of spinel is a redox reaction. But what was the reductant?

Let us recall that the composition of clinopyroxene assigns the olivine websterite in our xenolith to mantle xenoliths type A (FREY and PRINZ, 1978). Admittedly, the prevalence of CPX and OPX is not typical of depleted type A mantle xenoliths, but such rocks have been reported (e.g., FUJII and SCARFE, 1982). ARCULUS et al. (1984) generalized that oxygen fugacity in type A xenoliths is very low, near

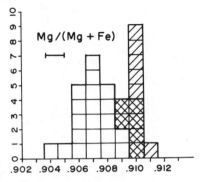
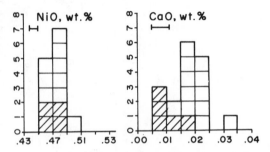

FIG. 11. Distribution of compositional characteristics of olivine; one square represents one determination, the error bar is 2σ in length. Open squares = deformed grains (i.e., cores of grains or outer zones adjacent to pyroxene or another olivine grain); hatched squares = non-deformed grains, 'neoblasts'; cross-hatched squares = outer zones along contact with grains of spinel.

that of the iron-wüstite buffer, while $f(O_2)$ in type B should correspond to the quartz-fayalite-magnetite buffer. Diffusion of hydrogen from a volatile-rich magmatic diapir is thought to be responsible for the low oxygen fugacity in type A xenoliths.

If hydrogen indeed was the all-penetrating reducing agent, we can write an equation for the pro-

Table 6. Chemical composition of olivine

	Deformed grains, cores		Deformed grains, outer zones in contact with SPN		Neoblasts	
	Typical analysis[a]	Range	Typical analysis[a]	Range	Typical analysis[a]	Range
SiO_2	40.8(3)	40.3–41.3	40.9(3)	40.4–41.2	40.7(3)	40.6–41.1
FeO	9.11(6)	8.90–9.20	8.86(6)	8.78–9.05	8.79(6)	8.78–8.89
MnO	0.12(2)	0.06–0.16	0.12(2)	0.06–0.18	0.13(2)	0.12–0.19
MgO	50.0(2)	49.4–50.3	49.9(2)	49.5–50.4	50.0(2)	50.0–50.5
CaO	0.020(3)	0.015–0.032	0.019(3)	0.014–0.020	0.011(3)	0.008–0.015
NiO	0.475	0.458–0.500	0.470(6)	0.467–0.490	0.472(6)	0.455–0.482
Σ	100.5(4)		100.3(4)		100.1(4)	
Mg/(Mg + Fe)	0.9074(6)	0.904–0.909	0.9093(6)	0.909–0.910	0.9102(6)	0.910–0.911

[a] Errors given are those of individual determinations and were computed according to KOTRBA (1989).

Table 7. Calculated temperatures and pressures of equilibration

Phase assemblage	OPX bulk & CPX bulk	OPX host & CPX lamellae	CPX host & OA lamellae	OPX host & CPX host	OPX neoblast
T, °C	1260(100)[a]	880(140)[a]	940(150)[a]	920(140)[a]	900(70)[b]
p, GPa	2.7(7)[c]	1.3(6)[c]	1.7(3)[c]	1.4(6)[c]	1.1(3)[b]

[a] Mean of temperatures according to WELLS (1977), BERTRAND and MERCIER (1986), and MERCIER (1980). As input served mean compositions and their errors. The error for the Bertrand and Mercier thermometer reflects also the error of pressure estimate. The errors given here reflect both the error of the mean and the mean errors for individual algorithms.
[b] Calculated according to MERCIER (1980).
[c] Mean of pressures calculated according to MERCIER (1980) for CPX and OPX.

cess, starting with CPX 'bulk' (obtained from bulk CPX by subtracting appropriate quantity of OA as OPX) and using real compositions of the other phases,

100 CPX 'bulk' + 2.5 H_2 →

96.4 CPX host + 4.8 SPN lam + 2.5 H_2O.

The feasibility of this reaction may be limited by the amount of iron available in the system, but in the present case, this is not a problem. If all iron is reduced to divalent, the projection point of such a composition shifts to the endpoint of the arrow in Fig. 12, beyond the intersection with the CPX host-SPN lamellae tieline, so the above reaction can proceed without restriction. Corresponding to 100 CPX 'bulk' is 7.02 OPX lamellae (now OA), which means that the reaction produces 2.5 OH per seven R^{2+} in OPX lamellae. The ideal stoichiometry of anthophyllite is 2.0 OH per seven Mg. If, moreover, the Fe in OPX lamellae too was reduced from Fe^{3+}, which appears inevitable, the process would free an additional 0.4 OH, making a total of about 2.9 OH available per seven R^{2+}. The lingering uncertainties concerning the identity of the orthoamphibole-like phase notwithstanding, the process offers a convenient sink for some of the hydrogen migrating through the rock. The synthesis of water at the location where it is used to form a hydrated phase makes the case appealing. If the water needed for the conversion of the thin OPX lamellae to OA in the present xenolith or to a sheet silicate at Deštná (ŠANC and RIEDER, 1983) were supplied to the large porphyroclasts from outside, one would expect the process to freeze at some distance from grain boundaries, at least in some grains, and one would not obtain a product in the form of single crystals perfectly crystallographically aligned, but rather a polycrystalline aggregate.

We do not see an analogous process in OPX porphyroclasts, and the reasons are easy to understand.

Clinopyroxene in lamellae (like elsewhere) is resistant to hydration and remains intact, and the quantity of water formed by a reaction analogous to the one above is apparently insufficient to cause changes in the OPX host which are detectable by X-ray diffraction.

Interestingly, the same chemical trends that were observed in spinel can be observed in spinel enclosed in olivine, which means that volatiles must have had access to that spinel, perhaps during an event like recrystallization of olivine. It should be mentioned that deformed olivine grains contain veils of bubbles (Fig. 3), apparently filled with volatiles, but not so olivine neoblasts. If the volatiles were produced in situ by a drop in pressure, then the neoblasts either contained no volatiles or formed at lower pressures (see KIRBY and GREEN, 1980).

The exsolution in porphyroclasts and the recrystallization process took place at temperatures we attempted to calculate using three geothermometers (Table 7) applied to pairs, OPX host & CPX lamellae, CPX host & OA lamellae (recalculated as OPX), OPX host & CPX host, and OPX neoblast alone. The resulting temperatures are quite close to one another, with a mean 910°C. It is interesting that the temperature involving the OA phase (recalculated as OPX) fits well into the range obtained. Thus the alteration of the original orthopyroxene must have proceeded with little or no cation exchange. The same was true of the layer silicate in CPX from Deštná (ŠANC and RIEDER, 1983).

As an independent check of these temperatures, we considered the pair olivine & spinel. A number of thermometers have been put forth (EVANS and FROST, 1975; FUJII, 1978; FABRIÈS, 1979; ROEDER et al., 1979) and reviewed (OZAWA, 1983), but the results must be interpreted carefully because of the relatively easy re-equilibration along contacts of OL and SPN grains. In order to arrive at meaningful compositions, we had to have an idea about the chemical variations in both phases. The cores of

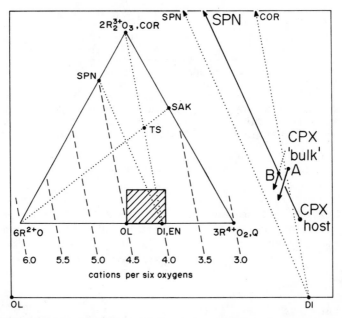

FIG. 12. This figure is an enlargement of the hatched rectangle in the inset and shows that during the unmixing of spinel from clinopyroxene, oxygen is removed from the system (see dashed isolines). DI = diopside; EN = enstatite; Q = quartz; SAK = Al_2SiO_5 phases; COR = corundum; TS = Tschermak's molecules. The CPX host and SPN lamellae are joined with the full line, and point B represents the composition of their intergrowth. Point A was obtained by subtracting orthoamphibole lamellae (as OPX) from the analytical bulk composition of CPX porphyroclasts. The endpoints of arrows represent compositions with all iron recalculated as divalent. The two arrows would coincide if not for errors associated (mainly) with planimetering the decomposed porphyroclasts. Points A and CPX host fall off the DI-COR line because no correction was made for the univalent cations present in small quantities.

deformed olivine grains (about 90 vol.% of all olivine) appear homogeneous, but in the border zone, there is an increase of forsterite component towards contacts with spinel (Fig. 13). The composition of these zones is like that of small undeformed olivine grains, and both apparently formed at the same time; neither was usable for our purposes. Spinel grains exhibit a small but systematic decrease of the

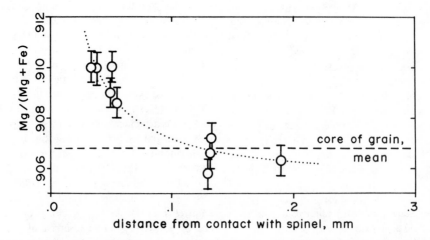

FIG. 13. The composition of outer zones of olivine grains changes towards their contacts with grains of spinel. The error bar represents ±1σ.

Mg/(Mg + Fe$_{tot}$) ratio towards neighboring olivine grains and they always show a lower Mg/(Mg + Fe$_{tot}$) than spinel grains of the same size surrounded by pyroxenes only (Fig. 10). These too had to be avoided. In the end, we used the algorithm of FABRIÈS (1979) and applied it to the composition of cores of deformed olivine grains and the composition of large spinel grains surrounded by pyroxene. The resulting temperatures have a smaller scatter about 830(±25)°C, and the similarity of this temperature to that based on the other thermometers suggest that all these data pertain to the last detectable thermal event in the history of the xenolith.

The chemical changes observed, particularly those along contacts of grains of different minerals, are probably due either to reactions during cooling (OZAWA, 1983) or during a short-term low-temperature event such as metamorphism of the granulite. Indeed, temperatures based on the compositions of border zones of olivine and spinel are lower than for recrystallization, between 680 and 800°C, but our data do not confirm the relation of temperature to the size of spinel grains (OZAWA, 1983).

There is no dependable geobarometer for the association OPX + CPX + OL + SPN, and thus the processes described above are difficult to characterize by reliable estimates of pressure. The absence of plagioclase and the very low contents of Ca in olivine nevertheless indicate that the pressures must have been those typical of the stability field for spinel peridotite. An estimation of pressure from the Ca content of olivine according to ADAMS and BISHOP (1986) might be incorrect since the barometer is based on experimental data for high temperatures. True, there are pressure estimates based on the one-pyroxene barometer of MERCIER (1980), but the results must be viewed with caution, particularly for the reconstructed bulk compositions of the porphyroclasts. Let it be mentioned though that pressures obtained for subsolidus phases in porphyroclasts (Table 7) are close to those reported by PIN and VIELZEUF (1988) for granulites from Zrcadlová huť (L2) and Blanský les (L9). Their pressures are based on the garnet-Al$_2$SiO$_5$-quartz geobarometer, and the mean corresponding to 800°C is 1.57 GPa. While the exsolution in pyroxenes in the xenolith may or may not have taken place at the time of crystallization of the enclosing granulite, it is likely that both events proceeded at similar depths.

Acknowledgements—Stimulating discussions on some aspects of analytical work were held with Ing. Z. Kotrba and Ing. R. Rybka. The manuscript was critically read by Dr. S. Vrána and Dr. K. Vokurka (all Geological Survey, Prague). The reviews by Dr. M. E. Fleet and Dr. G. L. Nord helped clarify presentation and resulted in a general improvement. This paper is an extention of the senior author's RNDr. Thesis defended at Charles University in 1985.

REFERENCES

ADAMS G. E. and BISHOP F. C. (1986) The olivine-clinopyroxene geobarometer: experimental results in the CaO-FeO-MgO-SiO$_2$ system. *Contrib. Mineral. Petrol.* **94,** 230–237.

AFTALION M., BOWES D. R. and VRÁNA S. (1989) Early Carboniferous U-Pb zircon age for garnetiferous, perpotassic granulites, Blanský les massif, Czechoslovakia. *N. Jb. Miner. Mh.* **1989,** 145–152.

ANDERSEN T., O'REILLY S. Y. and GRIFFIN W. L. (1984) The trapped fluid phase in upper mantle xenoliths from Victoria, Australia: implications for mantle metasomatism. *Contrib. Mineral. Petrol.* **88,** 72–85.

ARCULUS R. J., DAWSON J. B., MITCHELL R. H., GUST D. A. and HOLMES R. D. (1984) Oxidation states of the upper mantle recorded by megacryst ilmenite in kimberlite and type A and B spinel lherzolite. *Contrib. Mineral. Petrol.* **85,** 85–94.

BASU A. R. and MACGREGOR I. D. (1975) Chromite spinels from ultramafic xenoliths. *Geochim. Cosmochim. Acta* **39,** 937–945.

BENCE A. E. and ALBEE A. L. (1968) Empirical correction factors for the electron microanalysis of silicates and oxides. *J. Geol.* **76,** 382–403.

BERGMAN S. C. and DUBESSY J. (1984) CO$_2$-CO fluid inclusions in a composite peridotite xenolith: implications for upper mantle oxygen fugacity. *Contrib. Mineral. Petrol.* **85,** 1–13.

BERTRAND P. and MERCIER J.-C. C. (1986) The mutual solubility of coexisting ortho- and clinopyroxenes: toward an absolute geothermometer for the natural systems? *Earth Planet. Sci. Lett.* **76,** 109–122.

BURNHAM C. W. (1962) Lattice constant refinement. *Carnegie Inst. Wash. Yearb.* **61,** 132–135.

DICK H. J. B. (1977a) Evidence for partial melting in the Josephine Peridotite. *Oregon Dept. Geol. Min. Ind. Bull.* **96,** 59–63.

DICK H. J. B. (1977b) Partial melting in the Josephine Peridotite I, the effect on mineral composition and its consequence for geobarometry and geothermometry. *Amer. J. Sci.* **277,** 801–832.

DICK H. J. B. and BULLEN T. (1984) Chromian spinel as a petrogenetic indicator in abyssal and alpine-type peridotites and spatially associated lavas. *Contrib. Mineral. Petrol.* **86,** 54–76.

DICKEY J. S., JR. and YODER H. S., JR. (1972) Partitioning of chromium and aluminum between clinopyroxene and spinel. *Carnegie Inst. Wash. Yearb.* **71,** 384–392.

DICKEY J. S., JR., YODER H. S., JR. and SCHAIRER J. F. (1971) Chromium in silicate-oxide systems. *Carnegie Inst. Wash. Yearb.* **70,** 118–122.

EALES H. V. and MARSH J. S. (1983) Al/Cr ratios of coexisting pyroxenes and spinellids in some ultramafic rocks. *Chem. Geol.* **38,** 57–94.

EVANS B. W. and FROST B. R. (1975) Chrome-spinel in progressive metamorphism—a preliminary analysis. *Geochim. Cosmochim. Acta* **39,** 959–972.

FABRIÈS J. (1979) Spinel-olivine geothermometry in peridotites from ultramafic complexes. *Contrib. Mineral. Petrol.* **69,** 329–336.

FEDIUKOVÁ E. (1965) Ultrabasic xenoliths in the granulite

at Mt. Kleť near Český Krumlov. *Acta Univ. Carol., Geol.* **1965**, 189–202.

FEDIUKOVÁ E. (1978) Mafic minerals from granulites of the borehole Holubov (South Bohemian Moldanubicum). *Sbor. geol. věd, lož. geol., mineral.* **19**, 169–198.

FIALA J., MATĚJOVSKÁ O. and VAŇKOVÁ V. (1987) Moldanubian granulites and related rocks: petrology, geochemistry and radioactivity. *Rozpravy Českosl. akad. věd, řada mat. přír. věd* **97**, 1–102.

FLEET M. E. (1981) The intermediate plagioclase structure: an explanation from interface theory. *Phys. Chem. Minerals* **7**, 64–70.

FLEET M. E. (1982) Orientation of phase and domain boundaries in crystalline solids. *Amer. Mineral.* **67**, 926–936.

FREY F. A. and PRINZ M. (1978) Ultramafic inclusions from San Carlos, Arizona: Petrologic and geochemical data bearing on their petrogenesis. *Earth Planet. Sci. Lett.* **38**, 129–176.

FUJII T. (1978) Fe-Mg partitioning between olivine and spinel. *Carnegie Inst. Wash. Yearb.* **76**, 563–569.

FUJII T. and SCARFE C. M. (1982) Petrology of ultramafic nodules from West Kettle River, near Kelowna, southern British Columbia. *Contrib. Mineral. Petrol.* **80**, 297–306.

GREEN D. H. (1964) The petrogenesis of the high-temperature peridotite intrusion in the Lizard area, Cornwall. *J. Petrol.* **5**, 134–188.

GREEN D. H. and RINGWOOD A. E. (1967) The genesis of basaltic magmas. *Contrib. Mineral. Petrol.* **15**, 103–190.

HANIC F., MAĎAR J. and KISS V. (1955) Precession X-ray diffraction camera. *Czechoslovak Patent 89363* (in Slovak).

HENDERSON P. (1975) Reaction trends shown by chrome-spinels of the Rhum layered intrusion. *Geochim. Cosmochim. Acta* **39**, 1035–1044.

IRVINE T. N. (1967) Chromium spinel as a petrogenetic indicator. Part 2. Petrologic applications. *Can. J. Earth Sci.* **4**, 71–103.

JAQUES A. L. and GREEN D. H. (1980) Anhydrous melting of peridotite at 0–15 kb pressure and the genesis of tholeiitic basalts. *Contrib. Mineral. Petrol.* **73**, 287–310.

KIRBY H. S. and GREEN H. W. (1980) Dunite xenoliths from Hualalai volcano: evidence for mantle diapiric flow beneath the Island of Hawaii. *Amer. J. Sci.* **280-A**, 550–575.

KODYM O. (1972) Multiphase deformation in the Blanský les granulite massif (South Bohemia). *Krystalinikum* **9**, 91–105.

KODYM O., JAKEŠ P. and SCHOVÁNEK P. (1978) Granulite und ultramafische Gesteine aus der Strukturbohrung Holubov. *Sbor. geol. věd, Geol.* **32**, 7–41.

KOTRBA Z. (1989) Recent advances in X-ray microanalysis. *Rudy* **37**, 15–19 (in Czech).

KUO L.-C. and ESSENE E. J. (1986) Petrology of spinel harzburgite xenoliths from the Kishb Plateau, Saudi Arabia. *Contrib. Mineral. Petrol.* **93**, 335–346.

MERCIER J.-C. C. (1980) Single-pyroxene thermobarometry. *Tectonophysics* **70**, 1–37.

MERCIER J.-C. C. and NICOLAS A. (1975) Textures and fabrics of upper-mantle peridotites as illustrated by xenoliths from basalts. *J. Petrol.* **16**, 454–487.

MYSEN B. O. and KUSHIRO I. (1977) Compositional variations of coexisting phases with degree of melting of peridotite in the upper mantle. *Amer. Mineral.* **62**, 843–865.

OBATA M. (1980) The Ronda peridotite: garnet-, spinel-, and plagioclase-lherzolite facies and the *P-T* trajectories of a high-temperature mantle intrusion. *J. Petrol.* **21**, 533–572.

OZAWA K. (1983) Evaluation of olivine-spinel geothermometry as an indicator of thermal history for peridotites. *Contrib. Mineral. Petrol.* **82**, 52–65.

OZAWA K. (1988) Ultramafic tectonite of the Miyamori ophiolitic complex in the Kitakami Mountains, Northeast Japan: hydrous upper mantle in an island arc. *Contrib. Mineral. Petrol.* **99**, 159–175.

PIN C. and VIELZEUF D. (1988) Les granulites de haute-pression d'Europe moyenne témoins d'une subduction éo-hercynienne. Implications sur l'origine des groupes leptyno-amphiboliques. *Bull. Soc. géol. France (8)* **IV**, 13–20.

RIEDER M. (1971) A FORTRAN IV computer program for calculating high-precision cubic cell edges by extrapolation. *Věst. Ústřed. ústavu geol.* **46**, 65–76.

ROBINSON P., ROSS M., NORD G. L., JR., SMYTH J. R. and JAFFE H. W. (1977) Exsolution lamellae in augite and pigeonite: fossil indicators of lattice parameters at high temperature and pressure. *Amer. Mineral.* **62**, 857–873.

ROEDER P. L., CAMPBELL I. H. and JAMIESON H. E. (1979) A re-evaluation of the olivine-spinel geothermometer. *Contrib. Mineral. Petrol.* **68**, 325–334.

ROST F. (1966) Über ultrabasische Einschlüsse in metamorphen Gesteinen des südlichen Moldanubikums. *Krystalinikum* **4**, 127–162.

ŠANC I. (1982) Mean thermal expansion coefficients of diopside and enstatite cell parameters and their petrologic significance. *Acta Univ. Carol., Geol.* **1981**, 427–435.

ŠANC I. and RIEDER M. (1983) Lamellar pyroxenes and their petrogenetic significance: three examples from the Czech Massif. *Contrib. Mineral. Petrol.* **84**, 73–83.

SKINNER B. J. (1966) Thermal expansion. In *Handbook of Physical Constants* (ed. S. P. CLARK, JR.), Sec. 6, pp. 75–96. *Geol. Soc. Amer. Mem.* **97**.

SLABÝ J. (1983) Modal composition and petrochemistry of granulites of the Lišov and Blanský les Mts. massifs, southern Bohemia. *Čas. mineral. geol.* **28**, 41–60 (in Czech).

STRECKEISEN A. (1973) Classification and nomenclature of plutonic rocks: recommendations. *N. Jb. Mineral. Mh.* **1973**, 149–164.

VAN BREEMEN O., AFTALION M., BOWES D. R., DUDEK A., MÍSAŘ Z., POVONDRA P. and VRÁNA S. (1982) Geochronological studies of the Bohemian massif, Czechoslovakia, and their significance in the evolution of Central Europe. *Trans. Roy. Soc. Edinburgh, Earth Sci.* **73**, 89–108.

VRÁNA S. (1979) Polyphase shear folding and thrusting in the Moldanubicum of southern Bohemia. *Věst. Ústřed. ústavu geol.* **54**, 75–86.

VRÁNA S. (1987) Garnet-fassaitic pyroxene skarn from the granulite complex of southern Bohemia. *Věst. Ústřed. ústavu geol.* **62**, 193–206.

WELLS P. R. A. (1977) Pyroxene thermometry in simple and complex systems. *Contrib. Mineral. Petrol.* **62**, 129–139.

Fluid-Mineral Interactions: A Tribute to H. P. Eugster
© The Geochemical Society, Special Publication No. 2, 1990
Editors: R. J. Spencer and I-Ming Chou

Fe-Ti oxide-silicate equilibria: Assemblages with orthopyroxene

DONALD H. LINDSLEY

Department of Earth and Space Sciences, State University of New York, Stony Brook, New York 11794, U.S.A.

B. RONALD FROST

Department of Geology and Geophysics, University of Wyoming, Laramie, Wyoming 82071, U.S.A.

DAVID J. ANDERSEN

P.O. Box 121, Van Buren, Maine 04785, U.S.A.

and

PAULA M. DAVIDSON

Department of Geological Sciences, University of Illinois at Chicago, Chicago, Illinois 60680, U.S.A.

Abstract—In this paper equilibria are calibrated for assemblages containing orthopyroxene and two Fe-Ti oxides in the system Fe-Mg-Ti-Si-O. The isobarically univariant equilibria in the system Fe-Ti-Si-O that contain orthopyroxene (opx) are:

$$2\ SiO_2 + 2\ Fe_2TiO_4 = 2\ FeTiO_3 + Fe_2Si_2O_6 \quad (QUIlOp)$$
$$\text{quartz} \quad \text{ulvöspinel} \quad \text{ilmenite} \quad \text{opx}$$

$$Fe_2Si_2O_6 + 2\ Fe_2TiO_4 = 2\ FeTiO_3 + 2\ Fe_2SiO_4 \quad (OpUIlO)$$
$$\text{opx} \quad \text{ulvöspinel} \quad \text{ilmenite} \quad \text{fayalite}$$

The important isobaric divariant equilibria in this system include: titanomagnetite-ilmenite + orthopyroxene or olivine, titanomagnetite-orthopyroxene + quartz or olivine, ilmenite-orthopyroxene + quartz or olivine.

In the QUIlOp and OpUIlO assemblages oxygen fugacity is over-determined. If the oxides retain their original composition (as in many volcanic rocks) then these assemblages can be used to obtain oxygen fugacity, temperature, pressure, and silica activity. If the oxides have re-equilibrated these equilibria can still be used to obtain oxygen fugacity, temperature, and silica activity, provided the pressure can be estimated. In a similar manner the isobaric divariant assemblages can be used to obtain two of the intensive variables: temperature, pressure, oxygen fugacity, or silica activity (if the other two variables can be estimated).

Although strictly applicable only to Ca-free orthopyroxenes, these equilibria have important applications to silicic volcanic rocks that contain low-Ca orthopyroxenes. A porphyritic obsidian from Little Glass Mountain, California, contains two oxides and orthopyroxene. Silica activity and pressure co-vary; a_{SiO2} would be 0.83 at 1 kbar and 0.63 at 6 kbar. A rhyolite from Taupo is over-determined; it contains two oxides, two pyroxenes, olivine, and quartz. All phases except augite appear to have equilibrated at 870°C and 1.6 kbar. A porphyritic rhyolitic obsidian from Inyo Craters that contains quartz, Opx, and two oxides appears to have equilibrated at 8.7 kbars, a scarcely credible pressure—but one that should be investigated further. Orthopyroxenes from the Bishop Tuff are not in Fe-Mg exchange equilibrium with the oxides. Apparent pressures range from 0.75 to *minus* 1.7 kbar. If we adjust the orthopyroxene compositions to be in Fe-Mg exchange equilibrium with the Ti-magnetite, the corresponding pressures range from 1650 to 2575 bars; values that are less scattered and considerably more plausible!

INTRODUCTION

IN A PREVIOUS PAPER (FROST *et al.*, 1988) we noted that oxygen fugacity in rocks is reflected both in the composition of the iron-titanium oxides and in the composition of the coexisting ferromagnesian silicates. We also showed how, in iron-rich systems, the equilibrium

$$SiO_2 + 2\ Fe_2TiO_4 = 2\ FeTiO_3 + Fe_2SiO_4 \quad (1)$$
$$\text{quartz} \quad \text{ulvöspinel} \quad \text{ilmenite} \quad \text{fayalite}$$

which we abbreviate as QUIlF, relates the composition of the Fe-Ti oxides to that of olivine. This equilibrium is the algebraic sum of the Fe-Ti oxide

103

geothermometer with the displaced FMQ equilibrium (or with the displaced metastable FHQ equilibrium). The importance of this equilibrium lies in the fact that it reduces the uncertainty inherent in the Fe-Ti oxide thermometer by an order of magnitude and it allows the petrologist to "see through" much of the re-equilibration that naturally affects Fe-Ti oxides in plutonic rocks. It is well known that fayalite + quartz are topologically equivalent to orthopyroxene, which becomes stable with increasing pressure or Mg content. In this paper we calculate oxide-silicate equilibria involving orthopyroxene, either with or without olivine or quartz, and discuss the petrologic implications of the resulting equilibria. We use the mineral abbreviations listed in Table 1. In these abbreviations we distinguish between Ca-free or Ca-poor orthopyroxene (Op, Opx) and Ca-bearing pyroxenes (P). Although the activity of $Fe_2Si_2O_6$ is defined in pigeonite and augite as well as in orthopyroxene, we have yet to wed the complex expressions that do this (DAVIDSON and LINDSLEY, 1989) with the QUIlF expressions. Thus in this paper we consider only orthopyroxene. We discuss in depth the phase equilibria of systems with orthopyroxenes and two oxides, with or without olivine or quartz. The presence of Ca in natural rocks alters the location of the pyroxene-bearing equilibria in $\log fO_2 - a_{SiO_2} - \mu MgFe_{-1}$ space but does not change the topologic relations between these surfaces. Our companion papers on the Ca-bearing system will calculate the Ca-bearing equilibria and will deal more extensively with application of these equilibria to natural systems. In this paper we use the term pyroxene QUIlF in a broad sense to include equilibria involving orthopyroxene and two or more of the remaining phases Ti-magnetite, ilmenite, olivine, and quartz.

TOPOLOGIC RELATIONS

To illustrate how the topology of the Mg-bearing systems containing orthopyroxene relates to the topology of QUIlF in the iron-rich system, we first address the system Fe-Mg-Si-O (SPEIDEL and NAFZIGER, 1968), after which it will be easier to discuss the topology of the titanium-bearing system.

Table 1. Mineral abbreviations used in this paper

F = fayalite ($X_{Fa} > 0.9$)	P = pyroxene (Ca-bearing)
O = olivine ($X_{Fa} < 0.9$)	Op, Opx = orthopyroxene
H = hematite	(little or no Ca)
M = Ti-free magnetite	Q = quartz
Il = ilmenite$_{ss}$	U = ulvospinel$_{ss}$ (Ti-bearing Mt)

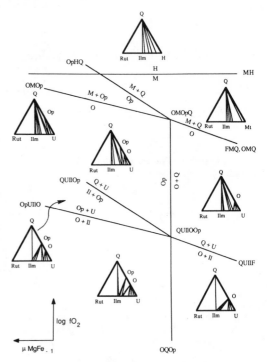

FIG. 1. Schematic isobaric, isothermal, $\log fO_2 - \mu MgFe_{-1}$ diagram showing topologic relations in the system Fe-Mg-Ti-Si-O. Abbreviations as in Tables 1, 2 and 3. The upper part of the diagram refers to the Ti-free portion of the system, and the lower part to the Ti-bearing portion. The vertical OQOp curve is degenerate; none of those phases contains appreciable Ti. Thus the M-absent curve from the upper invariant point coincides with the Il-absent curve from the lower point. Compositions of phases are projected from O and Mg onto the Fe-Si-Ti plane.

System Fe-Mg-Si-O

The best way to envision this system is to consider isobaric-isothermal $\log fO_2$-X_{Fe} sections. Because we are interested in the composition of orthopyroxene, which is not present throughout the section, we plot $\mu MgFe_{-1}$ instead of X_{Fe} as the compositional variable (Fig. 1). The numerical value of the exchange operator $\mu MgFe_{-1}$ is simply the difference $\mu_{Mg_2SiO_4} - \mu_{Fe_2SiO_4}$ (or $\mu_{MgSiO_3} - \mu_{FeSiO_3}$, since these differences must be equal under conditions of exchange equilibrium) (THOMPSON *et al.*, 1982). The equilibria involved in this system are listed in Table 2. The substitution of Mg into olivine displaces the FMQ equilibrium (which becomes OMQ) to higher oxygen fugacities. At some critical value of $\mu MgFe_{-1}$, olivine + quartz is no longer stable and orthopyroxene appears instead (the OQOp equilibrium). The intersection of this reaction with the OMQ surface generates two other equilibria:

Table 2. Buffers and equilibria in the system Fe-Mg-Si-O

$2 Fe_3O_4 + 3 SiO_2 = 3 Fe_2SiO_4 + O_2$	(FMQ; OMQ)
$2 Fe_2O_3 + 2 SiO_2 = 2 Fe_2SiO_4 + O_2$	(FHQ; OHQ)
$2 Fe_3O_4 + 6 SiO_2 = 3 Fe_2Si_2O_6 + O_2$	(OpMQ)
$2 Fe_3O_4 + 3 Fe_2Si_2O_6 = 6 Fe_2SiO_4$	
$\quad + O_2$	(FMOp; OMOp)
$2 Fe_2O_3 + 4 SiO_2 = 2 Fe_2Si_2O_6 + O_2$	(OpHQ)
$2 Fe_2O_3 + 3 Fe_2Si_2O_6 = 6 Fe_2SiO_4$	
$\quad + O_2$	(FHOp; OHOp)
$6 Fe_2O_3 = 4 Fe_3O_4 + O_2$	(MH)

Plus exchange equilibria between any two pairs of Op, O, and oxides.

$$2 Fe_3O_4 + 6 SiO_2$$
magnetite   quartz

$$= 3 Fe_2Si_2O_6 + O_2 \quad (OpMQ)$$
orthopyroxene

$$2 Fe_3O_4 + 3 Fe_2SiO_2O_6$$
magnetite   orthopyroxene

$$= 6 Fe_2SiO_4 + O_2 \quad (OMOp).$$
olivine

With increasing $\mu MgFe_{-1}$ both of these equilibria move to higher oxygen fugacities, the OpMQ equilibrium plotting above the OMOp equilibrium. Although it is not important to the present discussion, at very high $\mu MgFe_{-1}$ both the OpMQ and the OMOp surfaces must intersect the HM buffer. The assemblage orthopyroxene-hematite-magnetite-quartz is known from metamorphosed iron-formations (BUTLER, 1969), but the authors are not aware of any stable association of olivine with hematite, indicating that the intersection of OMO with HM lies at values of $\mu MgFe_{-1}$ that are not commonly attained in geologic environments.

System Fe-Mg-Ti-Si-O

In the Ti-bearing system the OMQ, OpMQ, and OMOp curves, which are isobarically, isothermally univariant in the Fe-Mg-Si-O system, become divariant surfaces that curve toward lower oxygen fugacities with increasing value of the $TiFe^{2+}Fe_{-2}^{3+}$ exchange vector. This occurs because Ti preferentially substitutes into the oxides relative to the silicates and into hematite relative to magnetite, lowering the ferric/ferrous ratio in the oxides. At sufficiently high values of $\mu TiFe^{2+}Fe_{-2}^{3+}$, the displaced HM equilibrium intersects the OMQ, OpMQ, and OMOp surfaces, generating the family of orthopyroxene-QUI1F equilibria: QUI1F, OpUI1O, and QUI1Op (Table 3). When discussing these pyrox-

ene-bearing equilibria in general we refer to them as pyroxene QUI1F to emphasize the relation between them and the initial QUI1F equilibrium described by FROST *et al.* (1988).

Topologically the relation between QUI1Op and OpUI1O is similar to that between OMQ and OMOp (Fig. 1). Both equilibria move to higher oxygen fugacities with increasing $\mu MgFe_{-1}$, with QUI1Op lying above OpUILO. However, unlike the OMOp and OpMQ surfaces, the pyroxene-QUI1F surfaces cannot attain oxygen fugacities above those of the HM buffer, since they are defined by the intersection of the *displaced* HM equilibrium with the OMOp or OpMQ surfaces.

The relationship between the isobarically, isothermally univariant equilibria and the isobarically, isothermally divariant surfaces that lie between them is shown in Fig. 1 by use of chemographic projections from oxygen and magnesium onto the Si-Fe-Ti plane. In these chemographies the redox assemblages listed in Table 2 are shown as three-phase assemblages. It is important to note that each specific three-phase field shown in the chemographic triangles on Fig. 1 is valid only for a given oxygen fugacity (as well as fixed T and P); the positions of the tie lines and tie triangles shift to higher Ti in ilmenite and magnetite with decreasing oxygen fugacity. With adequate thermodynamic data, one could use any of these assemblages as a monitor of oxygen fugacity, provided the assemblage happens to be stable. The assemblage I1-U (±Q, O, or Opx) is, of course the ilmenite-magnetite oxygen barometer. The manner in which the compositions of the two Fe-Ti oxides vary as a function of T and fO_2 (at constant pressure) has been calculated by AN-DERSEN and LINDSLEY (1988). How the other redox surfaces behave in isobaric, isothermal $\mu MgFe_{-1}$ $- \mu TiFe^{2+}Fe_{-2}^{3+}$ space is presented in this paper.

In addition to acting as monitors of oxygen fugacity, some of the equilibria described above are also important indicators of silica activity. Silica activity is fixed at unity along all quartz-bearing curves (FMQ, OMQ, QUI1F, and QUI1Op). It is also buffered in any assemblage containing olivine and orthopyroxene (displaced OMOp, OpUI1O, or displaced OHOp), with a_{SiO_2} defined by the O-Op equilibria decreasing as $\mu MgFe_{-1}$ increases. Finally, silica activity and oxygen fugacity are co-variants

Table 3. QUI1F-related equilibria

$SiO_2 + 2 Fe_2TiO_4 = 2 FeTiO_3 + Fe_2SiO_4$	(QUI1F)
$2 SiO_2 + 2 Fe_2TiO_4 = 2 FeTiO_3 + Fe_2Si_2O_6$	(QUI1Op)
$Fe_2Si_2O_6 + 2 Fe_2TiO_4 = 2 FeTiO_3$	
$\quad + 2 Fe_2SiO_4$	(OpUI1O)

in assemblages that contain two oxides and only one ferromagnesian silicate, such as UI1O (QUI1F equilibrium displaced to $aSiO_2 < 1.0$) and UI1Op (QUI1Op equilibrium displaced to $aSiO_2 < 1.0$).

The OQOp assemblage shifts to lower values of $\mu MgFe_{-1}$ with increasing pressure and thus serves as a barometer (SPEIDEL and NAFZIGER, 1968; DAVIDSON and LINDSLEY, 1989). This pressure dependence carries over to other orthopyroxene-bearing assemblages, which can therefore provide barometric information to varying extents.

CALCULATION OF THE PYROXENE-QUI1F EQUILIBRIA

At first glance, calculation of the pyroxene-bearing QUI1F equilibria seems to be a straightforward process; it appears that all one need do is to obtain expressions for the variation of olivine and pyroxene compositions as a function of $\mu MgFe_{-1}$ and incorporate them into the QUI1F expression (FROST *et al.*, 1988). In actuality the calculations are far more complex, for at moderate to high $\mu MgFe_{-1}$, considerable Mg substitutes into the oxides as well. Furthermore, in natural systems one must also consider the presence of Ca in the pyroxenes (and even in the olivine!). As noted above, in this paper we have opted to present the Ca-free system. Ca contents up to approximately 2 mole% Wo in orthopyroxene have negligible effects, so the results presented here can also be applied to assemblages containing low-Ca orthopyroxene. (We have used the model of DAVIDSON and LINDSLEY (1989) to verify that the effect of ignoring small amounts of Ca is small. For the natural samples discussed later in the paper, we eliminated any possible errors by using activities calculated for Ca-bearing orthopyroxenes.) Orthopyroxenes having higher Wo contents may also be used, provided that the appropriate activities of Fs and En components are calculated. The sources of the solution parameters used to derive the equilibria used in this paper are shown

Table 4. Sources of solution parameters used to calculate orthopyroxene-QUI1F

Solution	Reference
Fe-Mg in olivine	DAVIDSON and LINDSLEY, 1989
Fe-Mg in orthopyroxene	DAVIDSON and LINDSLEY, 1989
Fe-Mg in spinels	ANDERSEN, 1988
Fe-Mg in ilmenite	ANDERSEN, 1988
Fe-Ti in ilmenite and spinel	ANDERSEN and LINDSLEY, 1988
olivine-orthopyroxene-quartz	DAVIDSON and LINDSLEY, 1989

in Table 4. Because the Fe-Mg-Ti oxide model is currently available only in thesis format (ANDERSEN, 1988), we summarize it in Appendix 1. We chose the models listed in Table 4 for reasons of consistency. For example, ANDERSEN (1988) and DAVIDSON and LINDSLEY (1989) used the same olivine model and values of parameters specifically to allow pyroxene QUI1F equilibria to be calculated. Presumably other models could be substituted, provided that standard states are made consistent; we have not done so, and thus cannot comment on possible results.

SOURCES OF ERROR; UNCERTAINTIES IN THE RESULTS

In a system as complicated as pyroxene QUI1F, it is difficult to assess the uncertainties in the values of the calculated parameters, but they stem from three main sources: errors in the thermodynamic solution models, errors in chemical analysis of the phases, and problems with reducing analyses of complex natural minerals to the more simplified systems treated by the models. A further source of possible error is the assumption that a collection of phases in a rock represents an *equilibrium assemblage;* this is mainly the job of the petrographer. Where the pyroxene QUI1F assemblage is overdetermined, the redundant information can be used as a test of equilibrium, as illustrated in several examples in a later section.

Uncertainties introduced through the solution models cannot be treated simply, especially in the case for the Fe-Mg-Ti oxides where the parameters were extracted by the method of linear programming. That technique does not directly yield estimates of uncertainty; instead it solves for sets of parameters that are consistent with all the data *and* an arbitrarily chosen objective function. By choosing a wide range of objective functions, ANDERSEN and LINDSLEY (1988) and ANDERSEN (1988) were able to obtain the possible ranges of the model parameters, but these are *not* uncertainties in the usual sense, for the maximum value of one parameter might be possible only with, say, the minimum value of another. Fig. 3 of ANDERSEN and LINDSLEY (1988, p. 720) illustrates the ranges of temperature and oxygen fugacity allowed by the model for the Fe-Ti oxides. Within the range of experimental calibration (600–1200°C and from approximately 2 log units above to 3 units below the fO_2 of the FMQ buffer), the errors from the model itself are very small—generally less than ±10°C and ±0.1–0.2 log units for fO_2. The addition of Mg to the system (ANDERSEN, 1988) should increase these values but slightly.

As mentioned above, the models of ANDERSEN (1988) and of DAVIDSON and LINDSLEY (1989) use the same expressions for olivine, which is the common phase in the two studies. Thus errors in precision are minimized; errors in accuracy contribute ±0.1 log units in fO_2 and ±100–200 bars in calculated pressures. Errors associated with the pyroxene models are similar. Overall uncertainties resulting from the models alone are thus approximately $\pm10°C$, ±0.2–0.3 log units in fO_2, and ±400 bars in pressure.

Typical errors associated with routine but careful microprobe analyses of oxide minerals lead to uncertainties of approximately ±0.01 in the mole fractions of the oxide components; with extreme care one can reduce the uncertainties to half that value. The larger value leads to uncertainties in temperature of ±30–$100°C$ and in fO_2 of ±0.4–0.5 log units. Most of that range is associated with errors in the ilmenite compositions. FROST et al. (1988) showed how use of the QUIIF equilibrium can greatly reduce those uncertainties; the same holds

for pyroxene QUIIF. It is important to note that errors become ever more important for ilmenites that contain small amounts of ferric iron (0.01 mole fraction or less), because the hematite component reflects the difference between Ti and Fe and thus is subject to the combined errors in those values. Typical errors of ±0.01 in mole fractions of pyroxene and olivine components introduce uncertainties of ±300–400 bars in pressure and ±0.1–0.2 log units in fO_2.

As to uncertainties introduced by projecting complex natural compositions to simpler compositional space, these are obviously small if the amounts of other components are small. The treatment of oxides outlined in Appendix 2 is effective for phases containing up to approximately 10% Mn components, commonly the most abundant "others". The treatment of Al is less soundly based, and one should have increasingly less confidence in the results when Al_2O_3 contents exceed 2 wt.% in Ti-magnetites. We have obtained good results for pyroxenes when the projection (LINDSLEY and AN-

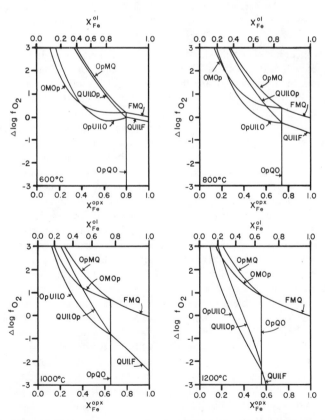

FIG. 2. $\Delta \log fO_2 - \mu MgFe_{-1}$ diagrams showing the stability of isobarically, isothermally univariant reactions in the system Fe-Mg-Ti-Si-O at 3 kilobars and 600, 800, 1000, and 1200°C. $MgFe_{-1}$ has been orthogonalized to X_{Fe}^{Opx}; $\Delta \log fO_2$ is the deviation from the FMQ buffer ($= \log fO_2 - \log fO_2$ [FMQ buffer]).

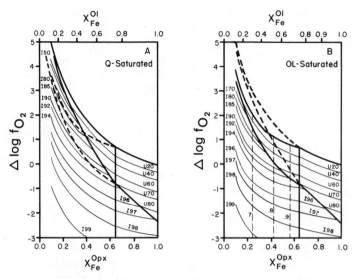

FIG. 3. $\Delta \log fO_2 - \mu MgFe_{-1}$ diagrams showing oxide compositions in some isobarically, isothermally divariant assemblages from the system Fe-Mg-Ti-Si-O at 3 kilobars and 1000°C. A = Quartz-saturated assemblages. B = Assemblages with olivine + orthopyroxene or olivine + quartz. See Fig. 1 for phases stable along univariant curves (heavy lines) and in divariant regions. Light dashed lines in B are contours for silica activity in assemblages containing olivine and orthopyroxene. Heavy dashed lines in A show metastable equilibria with olivine (see Fig. 2). Heavy dashed lines in B show metastable equilibria with quartz.

DERSEN, 1983) yields 5% or less of non-quadrilateral components. Analyses with 5–10% "others" should be viewed with caution, and those with greater than 10% should probably be used only as qualitative constraints.

APPLYING PYROXENE QUILF EQUILIBRIA

In theory, one could manipulate the expressions of the solution models to solve directly for T, P, fO_2, and a_{SiO_2}, as appropriate, for various assemblages (Table 2) and compositions. In practice, this is difficult because in the solution models, the site distributions in orthopyroxene and olivine are temperature-dependent, and thus the equations are non-linear in temperature. [By using the Akimoto-type expression rather than the site-mixing model (AN-DERSEN, 1988), we avoid that problem for the Fe-Mg-Ti spinels.] It is therefore necessary to start with an initial estimate and iterate to get the final temperature. To generate the QUIIF-related equilibria and isopleths shown in Figs. 2 through 6, we let our computer programs cycle through ranges of P, T, and compositions, as deemed appropriate from natural occurrences and experimental results. The intersections of isopleths for three-phase assemblages define points on the isobarically and isothermally univariant four-phase curves. For quartz-free orthopyroxene-olivine assemblages, we used the

model of DAVIDSON and LINDSLEY (1989) to calculate values of a_{SiO_2} that were then substituted into the QUIIF equilibria to solve for fO_2.

Our pyroxene-QUIIF program is not yet user-friendly and thus not ready for distribution. It requires as input values of temperature and of pressure for all calculations, and of silica activity for all equilibria that include SiO_2. Also needed are the compositions of the oxide minerals, olivine, and orthopyroxene, as required. Then one selects the appropriate equilibria from Tables 2 and 3 and iterates through values of pressure and temperature until the various equilibria yield closely similar values for oxygen fugacity, etc. An example of the end stage of this process is given in Appendix 3.

For the reasons given by FROST *et al.* (1988), in this paper we use the variable $\Delta \log fO_2$ [defined as $\log fO_2$ (assemblage) $- \log fO_2$ (FMQ buffer)] to express oxygen fugacity. Readers wishing to obtain "absolute" values of fO_2 should add back the values of FMQ as calculated from Table 1 of FROST *et al.* (1988) inasmuch as these are the terms used in calculating the oxide solution models. Note one important difference: most diagrams in FROST *et al.* (1988) use a standard state of *1 bar* and the temperature of interest. For this paper we adopt both *pressure* and temperature of interest as our standard state, a change mandated by the fact that pyroxene QUIIF reactions are more pressure-sensitive than

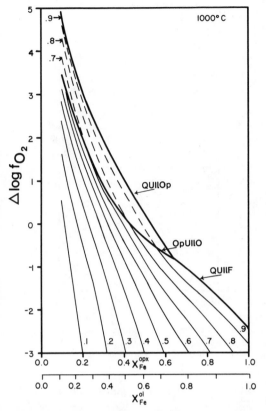

FIG. 4. $\Delta \log fO_2 - \mu MgFe_{-1}$ diagram contouring silica activity in assemblages with two Fe-Ti oxides at 3 kilobars and 1000°C. Dashed contours give silica activity where orthopyroxene coexists with Ti-magnetite and ilmenite; solid contours give silica activity in the assemblage olivine-Ti-magnetite-ilmenite.

could have contoured both X_{FeTiO_3} and X_{MgTiO_3}, but chose to keep the diagrams as uncluttered as possible. The use of X_{Ti} is especially appropriate for the Ti-magnetites, inasmuch as the Fe-Mg-Ti spinels are reciprocal, having only three compositional variables but four possible end members (Fe_3O_4, $FeMg_2O_4$, Fe_2TiO_4, and Mg_2TiO_4). For the spinels, we define X_{Ti} as the number of Ti cations in a stoichiometric spinel having 3 cations per 4 oxygens. (We are well aware that spinels can be cation-deficient, especially at high temperatures and very low pressures. We do not consider such spinels here because the model of ANDERSEN (1988) excludes them. He did not consider nonstoichiometry, partly because pressure in many geological environments will restrict cation deficiency, but mostly because of the difficulty in analyzing spinels for vacancies. Commonly, spinels in thin-section are analyzed for Fe, Mg, Ti, Al, . . . ; then a formula based on 3 cations per 4 oxygens is calculated. The model of ANDERSEN (1988) works moderately well for experimental spinels known to be nonstoichiometric (*e.g.*, TAYLOR, 1964), when recast as 3 cations per 4 oxygens. Nevertheless, readers should be cautious in applying our methods to spinels known to have *equilibrated*—not merely to have been emplaced—at high temperatures near the Earth's surface.) Under this definition, X_{Ti} [$= Ti/(Fe^{2+} + Fe^{3+} + Mg + Ti)$] ranges from 0 to 1. By plotting X_{Ti}, we avoid possible ambiguities in assigning Ti to ulvospinel and qandelite (Mg_2TiO_4) components.

RESULTS

The calculated shapes of the pyroxene-QUI1F surfaces in $\Delta \log fO_2 - \mu MgFe_{-1}$ diagrams (equivalent to the topology in Fig. 1) are shown for a series of temperatures at 3 kilobars (Fig. 2). For the sake of easy visualization, the $\mu MgFe_{-1}$ axis of this figure has been orthogonalized with respect to a linear scale for X_{Fe}^{Opx}. This is not a traditional way of plotting X_{Fe}^{Opx}, for when one equates X_{Fe}^{Opx} to $\mu MgFe_{-1}$, X_{Fe}^{Opx} has a meaning even at Fe values where orthopyroxene is intrinsically unstable. A similar scale for X_{Fe}^{Ol} is also shown; this scale is nonlinear because the K_D between olivine and orthopyroxene is not unity. From these figures it is evident that the width of the orthopyroxene-magnetite-ilmenite field (bounded by the QUI1Op and OpUI1O curves) decreases markedly with increasing temperature (at constant $\Delta \log fO_2$) while the intersection at which the assemblage QUI1OOp is stable migrates toward higher $\mu MgFe_{-1}$ (lower X_{Fe}^{Opx}) and lower oxygen fugacity. Decreasing pressure also tends to shrink the stability field for opx-mag-ilm, although the effect

those in the Mg-free system. Results from the two papers can be reconciled by adding the pressure effect on the FMQ buffer [essentially $C*(P(bars) - 1)/T(K)$, where $C = 0.092$ for low quartz and 0.110 for high quartz] to the $\Delta \log fO_2$ scales (FROST et al., 1988).

There are so many variables in the pyroxene QUI1F expressions that it is impractical to show diagrams appropriate for all petrologically interesting conditions. Accordingly, we have chosen to illustrate some of the implications of pyroxene QUI1F using isobaric, isothermal X_{Fe} (or $\mu FeMg_{-1}$) vs $\Delta \log fO_2$ diagrams (Figs. 2, 3 and 4) and isobaric $T - \Delta \log fO_2$ plots (Figs. 5 and 6). Detailed discussions of the topologies of these figures are given in the ensuing pages.

Readers should especially note that the isopleths for oxides in Fig. 3 are for X_{Ti} in spinel and ilmenite. Thus for ilmenite, the isopleths show X_{FeTiO_3} *plus* X_{MgTiO_3}, which is equivalent to $(1 - X_{Fe_2O_3})$. We

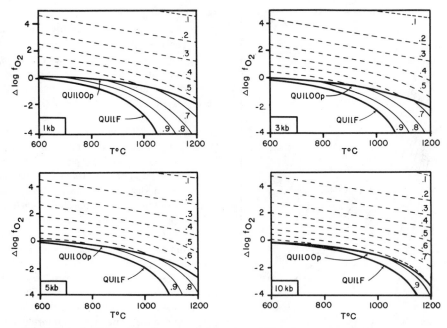

FIG. 5. Isobaric T-Δ log fO_2 diagrams showing the stability of the quartz-saturated assemblages QUI1F and QUI1Op at 1, 3, 5, and 10 kilobars. Solid contours give X_{Fe} of olivine in the QUI1F assemblage and dashed contours give X_{Fe} of orthopyroxene in the QUI1Op assemblage.

is not nearly as strong as is the effect of decreasing temperature. The major effect of decreasing pressure is to increase the stability field for olivine + quartz by driving the curve for the reaction OpQO to lower X_{Fe}.

Compositional changes of the oxides in some of the isobarically divariant assemblages can be shown by contours in the log fO_2 − μMgFe$_{-1}$ diagrams at 1000°C, with Q-saturated and OpO-saturated surfaces on separate diagrams for clarity (Figs. 3A and B). The contours mark the positions of the FMQ, OpMQ, FHQ, OpHQ, OMOp, and OHOp equilibria when the silicates are in equilibrium with the oxides whose composition are shown by the contours. The divariant assemblages OpUI1 and OUI1 are not shown in these figures because the oxide compositions in these assemblages are merely variants of the Fe-Ti oxide thermometer (ANDERSEN and LINDSLEY, 1988).

Also shown on Fig. 3B are contours for silica activity in the assemblages with olivine and orthopyroxene but lacking quartz. Since the subassemblage olivine-orthopyroxene defines silica activity, the contours for this assemblage are independent of oxygen fugacity. At a given oxygen fugacity, the assemblages orthopyroxene + two oxides and olivine + two oxides also define silica activity. The way silica activity varies in these assemblages as a function of oxygen fugacity is shown on a

Δ log fO_2 − μMgFe$_{-1}$ diagram at 3 kb and 1000°C (Fig. 4).

The manner in which changes in μMgFe$_{-1}$ affect the QUI1F, QUI1Op and OpUI1O equilibria can be shown on isobaric, $T - \Delta$ log fO_2 diagrams by contouring these surfaces for the composition of olivine or orthopyroxene (Figs. 5 and 6). On these diagrams there are only two univariant curves. One is the curve for QUI1F with pure fayalite and the other is the curve for the assemblage olivine-quartz-orthopyroxene-ilmenite-magnetite (QUI1OOp). The lighter curves are contours of the QUI1F or pyroxene-QUI1F surfaces projected onto the $T - \Delta$ log fO_2 surface from $T - \Delta$ log $fO_2 - \mu$MgFe$_{-1}$ space at constant X_{Fe} in olivine or orthopyroxene. As with relations shown in Fig. 2, increasing μMgFe$_{-1}$ drives the QUI1F surface to higher fO_2 until it intersects the OQOp equilibrium. This produces the univariant assemblage QUI1OOp. With further increases in μMgFe$_{-1}$ two different QUI1F surfaces are generated: QUI1Op (Fig. 5) and OpUI1O (Fig. 6). Increasing pressure pushes the QUI1F and pyroxene-QUI1F surfaces to higher Δ log fO_2 (compare with Fig. 2). More importantly, increasing pressure also drives the QUI1OOp surface to lower oxygen fugacity such that it truncates more of the QUI1F surface. Indeed, by approximately 11–16 kbar (depending on temperature), the QUI1F surface is completely eliminated and both

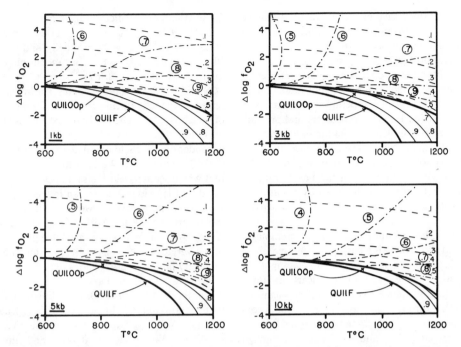

FIG. 6. Isobaric T-$\Delta \log fO_2$ diagrams showing the stability of the olivine-saturated assemblages QUI1F and OpUI1O at 1, 3, 5, and 10 kilobars. Solid contours give X_{Fe} of olivine in the QUI1F assemblage and dashed contours give X_{Fe} of orthopyroxene in the OpUI1O assemblage. Dash-dot lines are contours for silica activity in the OpUI1O assemblage. Circles values refer to isopleths for silica activity; uncircled values to isopleths of Opx or olivine composition.

the QUI1Op and the OpUI1O equilibria become stable with pure ferrosilite.

Also shown on Fig. 6 are contours for a_{SiO_2} in the assemblage OpUI1O. Of course, by definition, a_{SiO_2} for the assemblages QUI1F and QUI1OOp is unity. Accordingly, there is a strong convergence of the contours for a_{SiO_2} at low temperatures, because the slope of the OpUI1O surface in $T - \Delta \log fO_2 - \mu MgFe_{-1}$ space flattens with decreasing temperature. This flattening is so extreme that at low temperatures there are considerable portions of the OpUI1O surface, with its attendant contours of silica activity, that cannot be usefully depicted on a $T - \Delta \log fO_2$ diagram.

APPLICATIONS

The amount of information that one can obtain from assemblages (Table 5) with orthopyroxene and one or more Fe-Ti oxides is, of course, dependent on the variance of the assemblage. The assemblage olivine-orthopyroxene-quartz-ilmenite-magnetite has a phase-rule variance of two. However, the large number of equilibria among the five phases provides redundant information on the intensive variables: pressure from the silicate assemblage, and also two

means of determining temperature (Fe-Ti oxide thermometer and Fe-Mg distribution in the silicates) and oxygen fugacity (Fe-Ti oxides and the position of the QUI1O or QUI1Op surface). As a result one can obtain the oxygen fugacity and temperature for this assemblage even if one or both of the oxides have re-equilibrated. Conversely, as illustrated be-

Table 5. Orthopyroxene-oxide assemblages and information they provide

Assemblage	Information
I. Assemblages in which information is over-determined (conditions can be estimated even if one phase has undergone extensive re-equilibration).	
Opx-O-Q-U-Il	P, T, fO_2
Op-Q-U-Il	P, T, fO_2
Op-O-U-Il	P, T, fO_2, a_{SiO_2}
II. Assemblages that are not over-determined	
Op-Il-U	T, fO_2, a_{SiO_2} (need P estimate)
Op-O-U	fO_2, a_{SiO_2} (need P, T estimates)
Op-O-Il	fO_2, a_{SiO_2} (need P, T estimates)
Op-Q-U	fO_2 (need P, T estimates)
Op-Q-Il	fO_2 (need P, T estimates)

low, if the oxides have not equilibrated, it may be possible to demonstrate that the entire assemblage was not simultaneously at equilibrium.

The assemblages olivine-orthopyroxene-magnetite-ilmenite and orthopyroxene-quartz-magnetite-ilmenite are of higher variance, but are still very useful, for they yield temperature, pressure, and oxygen fugacity. If both temperature and pressure are known independently, then the Fe/Mg ratio of the silicates can provide the oxygen fugacity of formation, without any chemical data from the oxides (compare Fig. 2). In addition the Fe/Mg ratio of the silicates from the assemblage olivine-orthopyroxene-magnetite-ilmenite gives the silica activity (Fig. 3).

Along with the four- and five-phase assemblages described above there is a large array of three-phase assemblages (Fig. 1 and Table 5). Those assemblages that contain two oxides can give nearly as much data as the four-phase assemblages, although one cannot "see through" any reequilibration of the oxides. The other assemblages can give oxygen fugacity and silica activity if P and T are known (although the silica activity of the quartz-bearing assemblages is not a particularly valuable piece of information!). If only P is known, then these assemblages can give one a range of oxygen fugacities and silica activities.

In the absence of comprehensive solution models for all components of the phases in the pyroxene QUI1F equilibria, it is necessary to project analyses of the phases to appropriate values of the components. For olivines, we simply normalize mole fractions of Fe_2SiO_4, Mg_2SiO_4, and Ca_2SiO_4 to 1.0. This procedure ignores Mn, which may be important in some samples. We are uncertain as to its exact effects, but we strongly advise against combining Mn with Fe as some schemes do, because to do so would inappropriately increase the calculated activity of fayalite component. We use the pyroxene projection of LINDSLEY and ANDERSEN (1983) to obtain the appropriate Wo, En, and Fs endmembers. For the oxides we form end members following LINDSLEY and SPENCER (1982), but we treat the components in a different manner. In ilmenites, we normalize $FeTiO_3$, $MgTiO_3$, and Fe_2O_3 mole fractions to 1.0. Spinels are a bit trickier because there are four (rather than three) Fe-Ti-Mg end members. We discard Mn_2TiO_4 and any $FeAl_2O_4$; we then combine Fe_2TiO_4 with Mg_2TiO_4, and twice Mg_2TiO_4 with $MgFe_2O_4$ to obtain the X_{Ti} and X_{Mg} components (X_2 and X_3 of Andersen, 1988). Details of the oxide calculations are given in Appendix 2.

To illustrate the use of pyroxene QUI1F, we apply it to several volcanic assemblages as examples. Sample Cam 49 is a porphyritic obsidian from Little

Glass Mountain, Medicine Lake, California; it contains Opx ($Wo_{0.022}En_{0.553}Fs_{0.425}$), Ti-magnetite ($X_{Ti}$ = X_2 = 0.413; X_{Mg} = X_3 = 0.048), and ilmenite ($Ilm_{0.785}Hem_{0.115}Gk_{0.100}$) but no modal olivine or quartz (CARMICHAEL, 1967). Using the BUDDINGTON and LINDSLEY (1964) curves, Carmichael inferred an oxide temperature of 880°C and log fO_2 of -12.2. Applying the model of ANDERSEN (1988), we obtain 860°C. If we knew either a_{SiO_2} or P independently, we could calculate the other from the OpMQ equilibrium, the appropriate value being that which would yield the same fO_2 as that indicated by the oxides. Lacking an independent estimate of either variable, we can calculate only the covariation of a_{SiO_2} with P (Fig. 7). Values of Δ log fO_2 are shown for the calculated points; most of the variation results from the pressure effect on the FMQ buffer. Log fO_2 varies only from -12.61 (1 kbar) to -12.56 (6 kbar).

Another example is A-10, a rhyolite from Taupo that contains quartz, olivine ($Fo_{0.135}La_{0.003}$), augite, Opx ($Wo_{0.040}En_{0.256}Fs_{0.704}$), Ti-magnetite ($X_{Ti}$ = X_2 = 0.594; X_{Ti} = X_3 = 0.021), and ilmenite ($Ilm_{0.916}Hem_{0.056}Gk_{0.028}$) reported by EWART *et al.* (1975). This is a particularly intriguing assemblage, for it should be overdetermined. EWART *et al.* inferred 900°C and log fO_2 = -12.95 from the oxides; they also inferred a pressure of 4640–4955 bars from the Opx-Ol-Q assemblage. FROST *et al.* (1988) applied the QUI1F equilibrium to this sample and

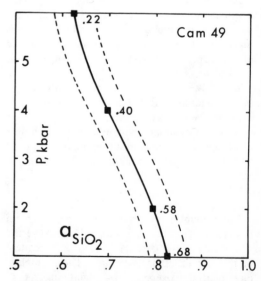

FIG. 7. Isothermal a_{SiO_2} − P plot for sample Cam 49 (CARMICHAEL, 1967). Calculated points are labelled with corresponding values of Δ log fO_2. Dashed lines show variation in silica activity corresponding to uncertainties of ±1% in the compositions of the oxides.

found 870°C, with $\Delta \log fO_2 = -0.3$, and a maximum pressure of 2500 bars. This temperature cannot apply to *both* pyroxenes, for pigeonite would form from augite plus Opx at temperatures slightly above 800C (DAVIDSON and LINDSLEY, 1989, Fig. 10). Although we are not concerned here with two-pyroxene equilibria, we note in passing that the graphical thermometer of LINDSLEY (1983) gives 805 and 850°C for the augite and opx, respectively. Thus the orthopyroxene may well have been in equilibrium with the olivine, quartz, and oxides; evidently the magma did not become saturated with augite until the temperature was just below the stability of pigeonite. We obtain close agreement between oxygen fugacities indicated by the oxides ($\Delta \log fO_2 = -0.59$; $\log fO_2 = -13.62$) and by the OMQ equilibrium ($\Delta \log fO_2 = -0.58$; $\log fO_2 = -13.62$) at 870°C, the temperature indicated by the oxides, with comfortably close correspondence between the observed and calculated compositions (Table 6).

We used the model of DAVIDSON and LINDSLEY (1989) to calculate the activities of pyroxene components for comparison with the other phases in A-10, and obtained fair agreement among the compositions of olivine, augite, and Opx (with quartz) at 850°C and 1.6–1.8 kbar (Table 6). As noted above, this temperature is still too high for the augite and Opx to have coexisted without forming pigeonite. EWART et al. (1975, p. 10) suggested that the pyroxenes and olivine might not represent equilibrium compositions. One explanation might be admixture of a slightly different magma during eruption. However, a simpler interpretation would be that the oxides, olivine, quartz, and orthopyroxene were all in equilibrium at 850–870°C, whereas augite only joined the assemblage as the magma cooled towards 800°C.

To test our expressions for the QUI1Op equilibrium, *without* olivine, we also calculated conditions for A-10 omitting all equilibria involving olivine. The temperature and oxygen fugacity—dominated by the oxides—remain unchanged; the inferred pressure is 1.3 kbar—slightly lower but within error of that inferred using the olivine. We conclude that,

for this case, the QUI1Op assemblage gives reasonable pressures, an observation that is important for the next two examples.

Cam 86, a porphyritic rhyolitic obsidian from Inyo Craters, California, contains quartz, Opx ($Wo_{0.030}En_{0.317}Fs_{0.653}$), Ti-magnetite ($X_{Ti} = X_2 = 0.420$; $X_{Mg} = X_3 = 0.074$), and ilmenite ($Ilm_{0.803}Hem_{0.150}Gk_{0.047}$) (CARMICHAEL, 1967). Carmichael reported analyses of two distinct magnetites for this sample; he inferred temperatures of 920 and 960°C with corresponding $\log fO_2$ values of -11.2 and -10.6. For simplicity we compare here only with the lower-temperature (lower-X_{Ti}) sample. Because the presence of quartz fixes silica activity, we can solve for the pressure at which the oxides, quartz, and Opx would have been in equilibrium. We plot several pyroxene QUI1F equilibria in Fig. 8. The curves intersect closely at 926 ± 10°C and 8.7 ± 0.5 kbar, the latter result being totally unexpected and perhaps implausible. (Combined uncertainties from the solution models yield a total uncertainty of ± 1 kbar for this calculation. Use of the other magnetite composition changes the calculated pressure by less that 200 bars.) We note that Carmichael reported 1% each of plagioclase and alkali feldspar for this sample but gave no analyses for them. It would be most interesting to obtain analyses and to apply the two-oxide, two-feldspar barometer of STORMER and WHITNEY (1984) to compare with this surprisingly high pressure. The oxides yield $\Delta \log fO_2 = 0.29$ ($\log fO_2 = -10.93$), while for the same temperature and 8.7 kbar, the displaced OpMQ equilibrium gives $\Delta \log fO_2 = 0.28$ ($\log fO_2 = -10.92$). In our calculations, we did not fix the magnesium contents of the oxides, but rather allowed them to vary so as to remain in exchange equilibrium with the Opx. The calculated $X_{Mg} (=X_3)$ of the spinel is 0.037, comfortably within the range of values reported by Carmichael, and the calculated X_{Gk} is 0.038, probably within error of the reported value. This is permissive evidence that the phases were all in equilibrium and that the high pressure may be real; we remain skeptical of it.

Our last example is for the Bishop Tuff, for which analyses of numerous opx-Q-two oxide assemblages

Table 6. Reported and calculated* compositions for sample A10 (EWART et al., 1975)

	Olivine		Opx		Ti-Mt		Il		
	Fo	Fa	En	Fs	X_{Ti}	X_{Mg}	Il	Hem	Gk
Reported	0.135	0.862	0.256	0.704	0.594	0.021	0.916	0.056	0.028
Calculated*	0.144	0.849	0.246	0.719	0.594	0.029	0.916	0.059	0.025

* Compositions calculated for 870°C, 1.6 kbar using the solution models; X_{Ti} and Il were fixed in the calculations.

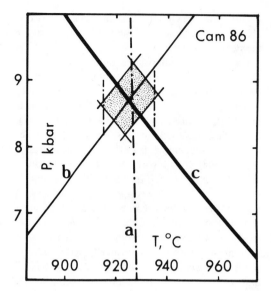

FIG. 8. *P-T* plot for sample Cam 86 (CARMICHAEL, 1967). Line **a** is for the two-oxide exchange equilibrium, which is virtually independent of pressure. Line **c** is for the displaced OpMQ equilibrium. Line **b** shows temperatures that yield the best composition for ilmenite when it is not fixed but is calculated from combined two-oxide, displaced OpMQ, and Fe-Mg exchange equilibria. The shaded area shows the effect of varying the compositions of the oxides by 1%, yielding a range in calculated temperature of 916–936°C and pressure of 8.2 to 9.3 kbar.

are available (HILDRETH, 1977). Of the 27 samples for which analyses are given, we rejected 13 because the oxides did not pass the Mg-Mn exchange test of BACON and HIRSCHMANN (1988). Of the remaining 14, we chose the seven that had the highest Mg contents reported for the oxides, because preliminary tests suggested that the oxides and Opx might not be in Fe-Mg exchange equilibrium. We first solved for the temperature and fO_2 using the oxides, and then searched for the pressure at which the OpUQ equilibrium would yield the same fO_2. The pressures ranged from 0.75 to *minus* 1.7 kbar (Fig. 9)! In all cases, the predicted Mg contents of the oxides as calculated by Fe-Mg exchange with the Opx were higher than the measured values—discrepancies that are larger than the error of the analyses. Thus we would suspect the calculated pressures even if most were not negative. As an experiment, we adjusted the compositions of the Opx until they were in Fe-Mg exchange equilibrium with the Ti-magnetites and calculated the pressures again. (Most required a decrease in X_{En} of 0.041 to 0.06; one required 0.098. These adjustments are well in excess of the probable analytical uncertainty of ±0.01.) The revised pressures (Fig. 9) range from 1.65 to 2.58 kbar, with a precision of ±0.6 kbar and

an overall uncertainty of ±1.1 kbar. Thus the revised pressures show less scatter than the original values and are in good agreement with the 2–3 kbar inferred by GRUNDER *et al.* (1987) using an updated form of the two-oxide two-feldspar barometer of STORMER and WHITNEY (1985).

Unfortunately, these revised pressures are meaningless unless we assume that the Opx changed in composition by $0.041–0.098X_{En}$ *after* equilibration with the oxides and quartz. One possibility would be Fe-Mg exchange with the liquid after eruption, but the very low MgO contents of the glass (≤0.33 wt.%) would argue against that possibility. It would be interesting to analyse the glass surrounding Opx to search for gradients in MgO and FeO; such gradients would surely exist if post-eruption exchange had occurred. Any other explanation—such as incorporation of Opx "xenocrysts" upon eruption—would render the revised pressures meaningless. All we can conclude for certain at this stage is that the reported compositions of the oxides and Opx are not equilibrium values.

Thus the QUIlOp assemblage yields very reasonable pressures for the one example (Taupo A-10) for which we have an independent estimate of pressure, but bizarre results for the other two. Inyo Craters (Cam 86) appears to have been an equilibrium assemblage, but the calculated pressure is implausibly high. The Bishop Tuff examples, on the other hand, show evidence of non-equilibrium and give implausibly low pressures—but give reasonable pressures if the Opx compositions are adjusted so as to be in Fe-Mg exchange equilibrium with the Ti-magnetite. Clearly it will be necessary to test more examples before we can assess the utility of this assemblage.

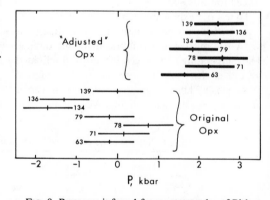

FIG. 9. Pressures inferred for seven samples of Bishop Tuff from Op-Q-U-Il equilibria using the original compositions for Op (lower left) and compositions of Op revised so as to be in exchange equilibrium with the Ti-magnetite (upper right). Numerals are sample numbers.

CONCLUSIONS

Although this paper specifically omits Ca, therefore limiting the applicability to natural systems of the diagrams presented here, a number of useful petrologic observations can be made. One is to confirm the observation of CARMICHAEL (1967) that opx + magnetite is a relatively oxidizing assemblage, particularly if the opx is magnesian or coexists with quartz. However, the assemblage orthopyroxene-quartz-ilmenite-Ti-magnetite is not restricted to lie on a curve in log fO_2-T space, as implied by CARMICHAEL (1967); rather it occupies a wide band. The exact fO_2 at which a pyroxene-bearing assemblage plots in that band is dependent on the Fe/Mg ratio of the orthopyroxene (Fig. 5). Orthopyroxene QUILF can be useful in inferring conditions of crystallization of a number of rocks, and can be especially helpful in estimating silica activities of quartz-free rocks. We plan a more complete discussion of the implications of pyroxene QUIlF in our next paper that includes the effects of Ca.

Acknowledgements—Both DHL and BRF wish to acknowledge the profound influence Hans Eugster had upon our careers, not only because Hans introduced the concept of oxygen fugacity and buffers to petrology, but also because of the personal friendships we were privileged to share with him. We have tried always to heed his exhortation: "It's all very simple: all it takes is the First and Second Law—and a little grey matter!" Research for this paper arose as part of the Laramie Anorthosite project and was funded on the following NSF grants: EAR-8416254, EAR-8618480, and EAR-8816040 (to Lindsley) and EAR-8409663, EAR-8617812, and EAR-8816604 (to Frost). Much of the solution modelling was supported by NSF grant EAR 8720185 and its predecessors. We are grateful for all this support. We also thank A. Ewart who provided us with unpublished analyses for use in this paper. We are grateful to George Fisher, Mark Ghiorso, Steve Huebner, and Richard Sack for constructive reviews that materially improved the paper.

REFERENCES

ANDERSEN D. J. (1988) Internally Consistent Solution Models for Fe-Mg-Mn-Ti Oxides. Ph.D. Dissertation, State University of New York at Stony Brook, xii + 202p.

ANDERSEN D. J. and LINDSLEY D. H. (1988) Internally consistent solution models for Fe-Mg-Mn-Ti oxides: Fe-Ti oxides. *Amer. Mineral.* **73**, 714–726.

BACON C. R. and HIRSCHMANN M. M. (1988) Mg/Mn partitioning as a test for equilibrium between coexisting Fe-Ti oxides. *Amer. Mineral.* **73**, 57–61.

BUDDINGTON A. F. and LINDSLEY D. H. (1964) Iron-titanium oxide minerals and synthetic equivalents. *J. Petrol.* **5**, 310–357.

BUTLER P. JR. (1969) Mineral compositions and equilibrium in the metamorphosed iron formation the Gagnon Region, Quebec, Canada. *J. Petrol.* **10**, 56–101.

CARMICHAEL I. S. E. (1967) The iron-titanium oxides of salic volcanic rocks and their associated ferromagnesian silicates. *Contrib. Mineral. Petrol.* **14**, 36–64.

DAVIDSON P. M. and LINDSLEY D. H. (1989) Thermodynamic analysis of pyroxene-olivine-quartz equilibria in the system CaO-MgO-FeO-SiO$_2$. *Amer. Mineral.* **74**, 18–30.

EWART A., HILDRETH W. and CARMICHAEL I. S. E. (1975) Quaternary acid magmas in New Zealand. *Contrib. Mineral. Petrol.* **51**, 1–27.

FROST B. R., LINDSLEY D. H. and ANDERSEN D. J. (1988) Fe-Ti oxide-silicate equilibria: Assemblages with fayalitic olivine. *Amer. Mineral.* **73**, 727–740.

GRUNDER A. L., HADJIGEORGIOU C. H., LINDSLEY D. H. and ANDERSEN D. J. (1987) Two-oxide, two-feldspar barometry revisited: I. Effect of revised calibrations (abstr.) *Geol. Soc. Amer. Abstr. Prog.* **19**, 686.

HILDRETH W. (1977) The magma chamber of the Bishop Tuff: Gradients in pressure, temperature, and composition. Ph.D. Dissertation, University of California, 328p.

LINDSLEY D. H. (1983) Pyroxene thermometry. *Amer. Mineral.* **68**, 477–493.

LINDSLEY D. H. and ANDERSEN D. J. (1983) A two-pyroxene thermometer. *J. Geophys. Res. Suppl.* **88**, A887–A906.

LINDSLEY D. H. and SPENCER K. J. (1982) Fe-Ti oxide geothermometry: Reducing analyses of coexisting Ti-magnetite (Mt) and ilmenite (Ilm). *EOS* **63**, 471.

SACK R. O. (1982) Spinels as petrogenetic indicators: activity-composition relations at low pressures. *Contrib. Mineral. Petrol.* **79**, 169–186.

SPEIDEL D. H. and NAFZIGER R. H. (1968) P-T-fO_2 relations in the system Fe-O-MgO-SiO$_2$. *Amer. J. Sci.* **266**, 361–379.

STORMER J. C. and WHITNEY J. A. (1984) Two-feldspar and iron-titanium oxide equilibria in silicic magmas and the depth of origin of large volume ash-flow tuffs. *Amer. Mineral.* **70**, 52–64.

TAYLOR R. W. (1964) Phase equilibria in the system FeO-Fe$_2$O$_3$-TiO$_2$ at 1300°C. *Amer. Mineral.* **49**, 1016–1030.

THOMPSON J. B. JR., LAIRD J. and THOMPSON A. B. (1982) Reactions in amphibolite, greenschist and blueschist. *J. Petrol.* **23**, 1–27.

APPENDIX 1: SUMMARY OF MODEL FOR Fe-Mg-Ti OXIDES

We briefly outline here the thermodynamic treatment of Fe-Mg-Ti oxides (ANDERSEN, 1988), which builds on that of SACK (1982) and ANDERSEN and LINDSLEY (1988). G_{excess} is modelled as an asymmetric Margules solution

$$G_{excess} = \sum_i \sum_{j,j\neq i} W_{ij}X_iX_j\left(X_j + \frac{1}{2}\sum_{k,k\neq i,j} X_k\right)$$
$$+ \sum_i \sum_{j,j\neq i} \sum_{k,k\neq i,j} W_{ijk}X_iX_jX_k. \quad \text{(A1-1)}$$

Expressions for the activity coefficients derived from Equation (A1-1) are then

$$\alpha RT \ln (\gamma_n) = \sum_i \sum_{j, j \neq i} W_{ij}(X_i X_j (X_j - X_i + 1)$$

$$- \sum_{m, m \neq n} X_m (Q_j (2X_j - X_i + 1)$$

$$+ Q_i (X_j - 2X_i + 1)))$$

$$+ \sum_i \sum_{j, j \neq i} \sum_{k, k \neq i, j} W_{ijk}(X_i X_j X_k$$

$$- \sum_{m, m \neq n} X_m (Q_i X_j X_k + Q_j X_i X_k + Q_k X_i X_j))) \quad \text{(A1-2)}$$

where Q_i is a term related to $\partial X_i / \partial X_m$ and $Q_i = 1$ ($m = i$), -1 ($i = n$), 0 ($m \neq i, i \neq n$). The temperature and pressure dependencies of the W terms are defined as

$$W = W_H - TW_S + (P - 1)W_V.$$

For ilmenite, the configurational entropy is expressed as

$$S_{\text{conf}} = -R(X_{\text{il}} \ln (X_{\text{il}}(X_{\text{il}} + X_{\text{gk}}))$$

$$+ X_{\text{gk}} \ln (X_{\text{gk}}(X_{\text{il}} + X_{\text{gk}})) + 2X_{\text{hem}} \ln (X_{\text{hem}})).$$

Activity expressions for il$_{\text{ss}}$ are then

$$RT \ln (a_{\text{il}}) = RT \ln (X_{\text{il}}(X_{\text{il}} + X_{\text{gk}})) + RT \ln (\gamma_{\text{il}})$$

$$RT \ln (a_{\text{gk}}) = RT \ln (X_{\text{gk}}(X_{\text{il}} + X_{\text{gk}})) + RT \ln (\gamma_{\text{gk}})$$

$$RT \ln (a_{\text{hem}}) = RT \ln (X_{\text{hem}}^2) + RT \ln (\gamma_{\text{hem}})$$

where $RT \ln (\gamma_n)$ is derived from Equation (A1-2).

ANDERSEN (1988) used two different expressions for spinel thermodynamics; we use the simpler modified Akimoto-type model.

Because the composition of a spinel in the system Fe^{2+}-Fe^{3+}-Mg-Ti can be defined in terms of only two variables (or three components), the choice of the compositional terms is somewhat arbitrary. If N_{Ti} and N_{Mg} are taken as the independent variables, then the compositional variables can be defined as

$$X_2 = N_{\text{Ti}}$$

$$X_3 = N_{\text{Mg}}.$$

The remaining compositional variables are then

$$N_{Fe^{2+}} = 1 + N_{\text{Ti}} - N_{\text{Mg}}$$

and

$$N_{Fe^{3+}} = 2 - 2N_{\text{Ti}},$$

where the sum of the cations is given by

$$N_{Fe^{2+}} + N_{Fe^{3+}} + N_{\text{Mg}} + N_{\text{Ti}} = 3.$$

Configurational entropy for spinel is expressed as

$$S_{\text{conf}} = -R(X_2(1 + X_2 - X_3)/(1 + X_2) \ln (X_2(1 + X_2$$

$$- X_3)/(1 + X_2)) + (1 - X_2) \ln (1 - X_2)$$

$$+ X_2 X_3/(1 + X_2) \ln (X_2 X_3/(1 + X_2))$$

$$+ (1 + X_2 - X_3)/(1 + X_2) \ln ((1 + X_2$$

$$- X_3)/(1 + X_2)) + (1 - X_2) \ln ((1 - X_2))$$

$$+ X_3/(1 + X_2) \ln (X_3/(1 + X_2))$$

$$+ X_2 \ln (X_2) - 2 \ln (2))$$

and the nonconfigurational energy G^* as

$$G^* = G^*_{Fe_3O_4}(1 - X_2 - X_3) + G^*_{Fe_2TiO_4}X_2 + G^*_{MgFe_2O_4}X_3$$

$$- \Delta \mu^*_{23} X_2 X_3 - \frac{1}{2} \Delta \mu^*_{2q} X_2 X_3 (1 + X_2 - X_3)$$

$$+ W_{12} X_2 (1 - X_2)(X_2 - X_3) + W_{21} X_2 (1 - X_2)$$

$$\times (1 - X_2 + X_3) + W_{13} X_3 (1 + X_2 - X_3)$$

$$\times (X_3 - X_2) + W_{31} X_3 (1 + X_2 - X_3)(1 - X_3)$$

$$+ \Delta W_{q3} X_2 X_3 (X_2 - 1).$$

This yields three end-member energies ($G^*_{Fe_3O_4}$, $G^*_{Fe_2TiO_4}$, and $G^*_{MgFe_2O_4}$), and one reciprocal term, $\Delta \mu^*_{23}$, which is the energy difference for the reciprocal exchange

$$Fe_3O_4 + \frac{1}{2} Mg_2TiO_4 = MgFe_2O_4 + \frac{1}{2} Fe_2TiO_4$$

or

$$\Delta \mu^*_{23} = \frac{1}{2} (G^*_{Fe_2TiO_4} - G^*_{Mg_2TiO_4}) + G^*_{MgFe_2O_4} - G^*_{Fe_3O_4}.$$

The term, $\Delta \mu^*_{2q}$, is a Bragg-Williams type ordering term for mixing of (Mg, Fe)$_2$TiO$_4$ spinels,

$$(Fe^{2+})[Fe^{2+}Ti]O_4 + (Mg)[MgTi]O_4$$

$$= (Fe^{2+})[MgTi]O_4 + (Mg)[Fe^{2+}Ti]O_4$$

or

$$\Delta \mu^*_{2q} = (G^*_{Fe_2TiO_4} + G^*_{Mg_2TiO_4}) - (G^*_{(Fe^{2+})[MgTi]O_4}$$

$$+ G^*_{(Mg)[Fe^{2+}Ti]O_4}).$$

The four possible binaries would yield eight asymmetric binary coefficients, of which only five are independent since there are only ten terms in the expansion of G^*. We have chosen to define G^* in terms of the binaries Fe$_3$O$_4$-Fe$_2$TiO$_4$ (W_{12} and W_{21}), Fe$_3$O$_4$-MgFe$_2$O$_4$ (W_{13} and W_{31}) and MgFe$_2$O$_4$-Mg$_2$TiO$_4$ (ΔW_{q3}) where

$$\Delta W_{q3} = W_{q3} - W_{3q}$$

and then

$$W_{3q} = \Delta \mu^*_{23} - 2\Delta W_{q3} + 2W_{21} - W_{12}$$

$$W_{q3} = \Delta \mu^*_{23} - \Delta W_{q3} + 2W_{21} - W_{12}.$$

The subscripts for the W terms are: 1 = mt, 2 = usp, 3 = magnesioferrite, and q = Mg$_2$TiO$_4$ (qandelite). This leads to the activity expressions:

$$RT \ln (a_{Fe_3O_4}) = \ln ((1 + X_2 - X_3)(1 - X_2)^2/(1 + X_2)) + \frac{1}{2} \Delta \mu^*_{2q} X_2 X_3 (1 + 2X_2 - 2X_3)$$

$$+ \Delta \mu^*_{23} X_2 X_3 + W_{12} X_2 X_3 - X_2)(1 - 2X_2) + W_{21} X_2 (2X_2 (1 - X_2 + X_3 (2X_2 - 1))$$

$$+ W_{13} X_3 (2X_3 - 2X_2 - 1)(X_3 - X_2) + W_{31} X_3 (2X_3 (1 + X_2 - X_3) - X_2) + \Delta W_{q3} X_2 X_3 (1 - 2X_2)$$

$$RT \ln (a_{\text{Fe}_2\text{TiO}_4}) = 2 \ln (X_2(1 + X_2 - X_3)/(1 + X_2)) + \frac{1}{2} \Delta\mu_{2q}^* X_3((X_3 - X_2)(1 - 2X_2) - 1)$$

$$+ \Delta\mu_{23}^* X_3(X_2 - 1) + W_{12}(1 - X_2)(2X_2(1 - X_2 + X_3) - X_3) + W_{21}(1 - 2X_2)(1 - X_2)$$

$$\times (1 - X_2 + X_3) + W_{13}X_3((1 + 2(X_3 - X_2))(X_3 - X_2) - 1$$

$$+ W_{31}X_3(1 + (X_3 - X_2)(1 - 2X_3)) + \Delta W_{q3}X_3(1 - 2X_2)(X_2 - 1)$$

$$RT \ln (a_{\text{MgFe}_2\text{O}_4}) = \ln (X_3(1 - X_2)^2/(1 + X_2)) + \frac{1}{2} \Delta\mu_{2q}^* X_2(1 + X_2 - X_3)(2X_3 - 1) + \Delta\mu_{23}^* X_2(X_3 - 1)$$

$$+ W_{12}X_2((X_2 - X_3)(2X_2 - 1) + X_2 - 1) + W_{21}X_2((1 + 2X_2)(1 - X_2)$$

$$+ X_3(2X_2 - 1)) + W_{13}(1 + X_2 - X_3)(2X_3(1 + X_2 - X_3) - X_2)$$

$$+ W_{31}(1 + X_2 - X_3)(2X_3 - 1)(X_3 - 1) + \Delta W_{q3}X_2(X_3(1 - 2X_2) + X_2 - 1)$$

$$RT \ln (a_{\text{Mg}_2\text{TiO}_4}) = 2 \ln (X_2X_3/(1 + X_2)) + \Delta\mu_{2q}^*(1 + X_2 - X_3)(X_2(X_3 - 1) - \frac{1}{2}X_3)$$

$$+ \Delta\mu_{23}^*(X_2 - 1)(X_3 - 2) + W_{12}(X_2 - 1)(2X_2(X_2 - X_3) + X_3)$$

$$+ W_{21}(1 - X_2)(1 + (X_3 - X_2)(1 - 2X_2)) + W_{13}(1 + X_2 - X_3)(2(1 - X_3)(X_3 - X_2)$$

$$+ X_3) + W_{31}(1 + X_2 - X_3)(X_3 - 2)(2X_3 - 1) + \Delta W_{q3}(1 - X_2)(2X_2(X_3 - 1) - X_3).$$

Table A1-1 gives the values of the solution parameters we use.

Table A1-1. Model parameters for Fe-Mg-Ti oxides

Parameter	Preferred value	Minimum value	Maximum value
Exchange equilibria			
ΔH_{OLIL}	−1.4661527E+04	−1.4723021E+04	−1.4483356E+04
ΔS_{OLIL}	1.5187286E+01	1.5147428E+01	1.5309038E+01
ΔV_{OLIL}	−9.7300000E−02		
ΔH_{FETI}	2.9435301E+04		
ΔS_{FETI}	4.5123501E+00		
ΔH_{MGFE}	−2.8368547E+04	−2.9919742E+04	−2.7179059E+04
ΔS_{MGFE}	−1.3222971E+01	−1.4203763E+01	−1.2489809E+01
Ilmenite			
$W_{H,\text{ig}}$	8.4055215E+03	8.3486436E+03	8.4621592E+03
$W_{S,\text{ig}}$	3.0423203E+00	2.9888186E+00	3.0766625E+00
$W_{V,\text{ig}}$	1.0800000E−02		
$W_{H,\text{gi}}$	7.3635693E+03	7.3621216E+03	7.6529805E+03
$W_{S,\text{gi}}$	3.4959583E+00	3.4948959E+00	3.6954560E+00
$W_{V,\text{gi}}$	1.0800000E−02		
W_{gh}	2.6651402E+04	2.6608336E+04	2.7090381E+04
W_{hg}	2.6651402E+04	2.6608336E+04	2.7090381E+04
$W_{H,\text{ih}}$	4.4204801E+04		
$W_{S,\text{ih}}$	1.2274390E+01		
$W_{H,\text{hi}}$	1.2634250E+05		
$W_{S,\text{hi}}$	1.0060010E+02		
Spinel (Akimoto distribution)			
$\Delta\mu_{H,23}^*$	2.2323242E+04	1.6138391E+04	2.8282242E+04
$\Delta\mu_{S,23}^*$	1.3994102E+01	9.7778091E+00	1.7738436E+01
$\Delta\mu_{2q}^*$	0.0000000E+00		
W_{12}	1.5748030E+04		
$W_{H,21}$	4.6175480E+04		
$W_{S,21}$	2.3076500E+01		
W_{13}	0.0000000E+00		
W_{31}	0.0000000E+00		
$\Delta W_{H,\text{q3}}$	3.9471707E+04	3.9471707E+04	4.8350406E+04
$\Delta W_{S,\text{q3}}$	2.3178127E+01	2.3178127E+01	3.1921598E+01

```
use feti, femgilsp
tk=1043; p=2200; x2=0.2573; x3=0.034; xhem=0.105?; xgk=0.044?;
tk=1048; p=2200; x2=0.2573; x3=0.034; xhem=0.105?; xgk=0.044?;
tk=1053; p=2200; x2=0.2573; x3=0.034; xhem=0.105?; xgk=0.044?;

use domq, femgilsp, femgopxil
tk=1043; p=2200; x2=0.2573; x3=0.034?; xhem=0.105?; xgk=0.044?; xopx=0.446; yopx=0.021; aqtz=1.0;
tk=1048; p=2200; x2=0.2573; x3=0.034?; xhem=0.105?; xgk=0.044?; xopx=0.446; yopx=0.021; aqtz=1.0;
tk=1053; p=2200; x2=0.2573; x3=0.034?; xhem=0.105?; xgk=0.044?; xopx=0.446; yopx=0.021; aqtz=1.0;

use feti, femgilsp, femgopxil
tk=1043; p=2200; x2=0.2573; x3=0.034?; xhem=0.105?; xgk=0.044?; xopx=0.446; yopx=0.021; aqtz=1.0;
tk=1048; p=2200; x2=0.2573; x3=0.034?; xhem=0.105?; xgk=0.044?; xopx=0.446; yopx=0.021; aqtz=1.0;
tk=1053; p=2200; x2=0.2573; x3=0.034?; xhem=0.105?; xgk=0.044?; xopx=0.446; yopx=0.021; aqtz=1.0;

use feti, domq, femgilsp, femgopxil
tk=1043; p=2200; x2=0.2573; x3=0.034?; xhem=0.105?; xgk=0.044?; xopx=0.446?; yopx=0.021; aqtz=1.0;
tk=1048; p=2200; x2=0.2573; x3=0.034?; xhem=0.105?; xgk=0.044?; xopx=0.446?; yopx=0.021; aqtz=1.0;
tk=1053; p=2200; x2=0.2573; x3=0.034?; xhem=0.105?; xgk=0.044?; xopx=0.446?; yopx=0.021; aqtz=1.0;
```

tk	p	fo2	dfmq	ti	mg	xmg	xil	xgk	xhem	xmg			
1043.0	2200.0	-14.128	0.922	0.2573	0.0340	0.0270	0.8395	0.0571	0.1034	0.0637			
1048.0	2200.0	-13.960	0.977	0.2573	0.0340	0.0270	0.8360	0.0567	0.1073	0.0635			
1053.0	2200.0	-13.794	1.030	0.2573	0.0340	0.0270	0.8323	0.0563	0.1114	0.0633			

tk	p	fo2	dfmq	ti	mg	xmg	xil	xgk	xhem	xmg			
1043.0	2200.0	-14.128	0.922	0.2573	0.0340	0.0270	0.8395	0.0571	0.1034	0.0637			
1048.0	2200.0	-13.960	0.977	0.2573	0.0340	0.0270	0.8360	0.0567	0.1073	0.0635			
1053.0	2200.0	-13.794	1.030	0.2573	0.0340	0.0270	0.8323	0.0563	0.1114	0.0633			

tk	p	fo2	dfmq	ti	mg	xmg	xil	xgk	xhem	xmg	xopx	yopx	aqtz
1043.0	2200.0	-14.073	0.977	0.2573	0.0334	0.0265	0.8388	0.0560	0.1052	0.0626	0.4460	0.0210	1.0000
1048.0	2200.0	-13.966	0.970	0.2573	0.0339	0.0270	0.8364	0.0565	0.1071	0.0633	0.4460	0.0210	1.0000
1053.0	2200.0	-13.861	0.963	0.2573	0.0345	0.0274	0.8339	0.0571	0.1090	0.0641	0.4460	0.0210	1.0000

tk	p	fo2	dfmq	ti	mg	xmg	xil	xgk	xhem	xmg	xopx	yopx	aqtz
1043.0	2200.0	-14.129	0.921	0.2573	0.0334	0.0265	0.8407	0.0560	0.1033	0.0624	0.4460	0.0210	1.0000
1048.0	2200.0	-13.960	0.976	0.2573	0.0339	0.0270	0.8362	0.0565	0.1073	0.0633	0.4460	0.0210	1.0000
1053.0	2200.0	-13.793	1.031	0.2573	0.0345	0.0274	0.8314	0.0571	0.1115	0.0642	0.4460	0.0210	1.0000

tk	p	fo2	dfmq	ti	mg	xmg	xil	xgk	xhem	xmg	xopx	yopx	aqtz
1043.0	2200.0	-14.132	0.918	0.2573	0.0318	0.0253	0.8438	0.0532	0.1030	0.0593	0.4354	0.0210	1.0000
1048.0	2200.0	-13.960	0.977	0.2573	0.0341	0.0271	0.8358	0.0568	0.1073	0.0637	0.4471	0.0210	1.0000
1053.0	2200.0	-13.790	1.034	0.2573	0.0365	0.0290	0.8276	0.0605	0.1119	0.0682	0.4585	0.0210	1.0000

FIG. A3-1. Example of a pyroxene QUIIF calculation. The upper portion is input to the program; the lower portion is the results. The temperature 1048 K provides the best calculated composition of the ilmenite; the guess of 2200 bars provides a moderately good agreement between fO_2 from the oxides and that from the OpUQ equilibrium. 2250 bars (not shown) provided the best agreement, and was adopted is the pressure. The abbreviations are defined in the text. "?" in the input means that the preceeding value is allowed to vary during the calculations.

APPENDIX 2: CALCULATION OF OXIDE COMPONENTS

In the recalculation schemes described below, we follow the procedure of LINDSLEY and SPENCER (1982) until the last steps. We assume that the analyses have been made by electron microprobe, with no independent determination of Fe^{3+}. Even in those cases where Fe^{3+} has been determined independently, it is preferable to combine ferrous and ferric iron at the beginning. The procedure:

1. Convert analyses to atomic proportions. Convert sufficient Fe to Fe^{3+} to yield 3 cations per 4 oxygens (spinel) or 2 cations per 3 oxygens (ilmenite).
2. Form and discard Mn_2TiO_4 (spinel) or $MnTiO_3$ (ilmenite). Ignore Si if it is likely that it represents contamination from silicates. Otherwise form and discard Fe_2SiO_4 or $FeSiO_3$ if you believe the Si is actually in the oxide minerals.
3. For spinels, form $MgAl_2O_4$ and either Mg_2TiO_4 or $FeAl_2O_3$ to use up all Mg and Al. For ilmenites, assign Al to Al_2O_3 and discard; form $MgTiO_3$.
4. For spinels, form Fe_2TiO_4. The remaining Fe should be in the ratio of 1 Fe^{2+} to 2 Fe^{3+}; combine to form Fe_3O_4. For ilmenites, form $FeTiO_3$ and Fe_2O_3.
5. For spinels, normallize $MgAl_2O_4$, Mg_2TiO_4, Fe_2TiO_4, and Fe_3O_4 to 1.0. Use the normallized values to obtain X_1 = Fe_3O_4; $X_{Ti} = X_2$ = $Mg_2TiO_4 + Fe_2TiO_4$; $X_{Mg} = X_3 = 2$ $Mg_2TiO_4 + MgAl_2O_4$. For ilmenites, normalize $MgTiO_3$, $FeTiO_3$, and Fe_2O_3 to 1.0, and use the normallized components.

APPENDIX 3: AN EXAMPLE OF PYROXENE-QUIlF CALCULATIONS

We illustrate our calculation methodology with an example; the data are for Bishop Tuff sample 136 (HILDRETH, 1977), with the Opx composition adjusted as discussed in the text. The upper part of Fig. A3-1 shows the input to the program. Each set of data is headed by a "use" statement of the equilibria that are to be calculated. "feti" refers to the two-oxide equilibria: the FeTi-Fe$_{-2}^{3+}$ exchange and the oxidation reaction $4\ Fe_3O_4 + O_2 = 6\ Fe_2O_3$; "domq" is the OpMQ equilibrium (Table 2) displaced by ulvospinel component in the spinel; and "femgilsp" and "femgopxil" are FeMg exchange between ilmenite and spinel and Opx and ilmenite respectively. "tk" and "p" are the trial values of temperature in kelvins and pressure in bars; "x2" and "x3" are X_{Ti} and X_{Mg} in the spinel; "xhem" and "xgk" are Fe_2O_3 and $MgTiO_3$ components of ilmenite; "xopx" and "yopx" are X_{En} and X_{Wo} in Opx. All Fe components are calculated by difference. The silica activity ("aqtz") is 1.0 because quartz is present. If there were olivine rather than quartz, the values of "aqtz" would be silica activities calculated from the programs of DAVIDSON and LINDSLEY (1989). A "?" following a number means that the value is allowed to vary in the calculation; there must be one "?" for each equilibrium that is "use"d. Thus in the first set (following "use feti, femgilsp"), the spinel is fixed and the ilmenite composition is allowed to vary so as to satisfy all three equilibria. The first three lines of numbers in the bottom part of the table are the calculated results for this set. The Fe_2O_3 content of the ilmenite is best matched at 1048 K; the calculated geikielite content is marginally high but probably within analytical error. Thus we accept 1048 K and $\log fO_2 = -13.96$ as the best values based on the oxides.

The remaining three sets of input "use" different subsets of pyroxene QUIlF equilibria; the best answer is found when all three give similar values for $\log fO_2$ *and* calculate the correct compositions. Note, for example, that x3 ($=X_{Mg}$ in spinel) is allowed to vary, but for 1048 K the calculated values (0.0339; 0.0341) are very close to the measured value (0.0340). This example is the penultimate of a number of attempts using various pressures; pressures lower than 2200 bars yield values of $\log fO_2$ for "domq" that are lower than those for "feti"; pressures greater than 2250 have the opposite effect, while the final trial value 2250 bars produced the best agreement and is accepted as the best answer. The method is tedious but effective.

Fluid-Mineral Interactions: A Tribute to H. P. Eugster
© The Geochemical Society, Special Publication No. 2, 1990
Editors: R. J. Spencer and I-Ming Chou

Control of material transport and reaction mechanisms by metastable mineral assemblages: An example involving kyanite, sillimanite, muscovite and quartz

C. T. FOSTER, JR.

Geology Department, University of Iowa, Iowa City, Iowa 52242, U.S.A.

Abstract—Metastable mineral assemblages strongly influence reaction mechanisms and material transport when a new mineral grows in a metamorphic rock. The effects exerted by the metastable assemblages on the reactions that take place when sillimanite grows in a kyanite-bearing rock are examined using metastable elements of activity diagrams and irreversible thermodynamic principles. The results show that a commonly inferred reaction mechanism, where muscovite assists in the growth of sillimanite at the expense of kyanite, is a consequence of material transport constraints imposed by a metastable mineral assemblage in the matrix that separates growing sillimanite from dissolving kyanite.

INTRODUCTION

ONE OF THE PRIMARY controls on mineral textures that develop during metamorphism is the distribution of minerals in a rock at the time when a new mineral nucleates. The distribution of the new mineral is strongly influenced by the abundance and location of other minerals with favorable nucleation sites for it. Once the new mineral has nucleated and begun to grow, the material transport and reaction mechanisms that develop are controlled by local equilibrium with metastable mineral assemblages along the transport path.

For example, consider the reaction mechanisms that develop in a rock containing the phases kyanite + muscovite + quartz + water when the temperature changes along the isobaric heating path shown on the *P-T* diagram in Fig. 1a. This path crosses the sillimanite isograd by passing from the kyanite stability field into the sillimanite stability field. At temperatures in the kyanite field, before the sillimanite isograd is crossed (point I, Fig. 1a), the stable mineral assemblage will be kyanite + muscovite + quartz + water. When the temperature rises to the point where the univariant curve representing equilibrium between kyanite and sillimanite is reached (the theoretical sillimanite isograd: point II, Fig. 1a), sillimanite becomes stable with kyanite. However, no sillimanite forms at the theoretical isograd because the reaction kyanite → sillimanite must be overstepped by a finite amount to form sillimanite nuclei (FISHER, 1977; RIDLEY and THOMPSON, 1986). The absence of sillimanite nuclei allows the metastable kyanite-bearing assemblage to persist into the sillimanite stability field without reacting (point III, Fig. 1a). A schematic illustration of the textures present in the rock at the

pressures and temperatures represented by points I, II, and III (Fig. 1a) is shown in Fig. 1b.

Sillimanite nuclei first form in sites in the rock with the lowest activation energy for nucleation of sillimanite. Under many metamorphic conditions, the low energy sites for sillimanite nucleation in pelites appear to be in micas, because this is where sillimanite is commonly first observed with increasing metamorphic grade (CHINNER, 1961; CARMICHAEL, 1969; YARDLEY, 1989). Once sillimanite nuclei form in the rock, either of two things happen. If the nucleation rate is low, and only a few nuclei form before appreciable reaction takes place in the rock, reaction cycles develop that involve material transport between the few sites where sillimanite is growing and the sites where kyanite is dissolving (Fig. 1c). This type of reaction mechanism, which uses micas as catalysts in the reaction, was first recognized by CARMICHAEL (1969) and it probably indicates that the sillimanite forming reaction is never overstepped by a large amount (point IV, Fig. 1a). If the nucleation rate is high, and little or no reaction takes place until nuclei have formed in most sites in the rock, nuclei of sillimanite will be present in the immediate vicinity of kyanite. Then the reaction mechanism will involve direct replacement without much material transport or the use of mineral catalysts (Fig. 1d).

REACTION MECHANISMS, MATERIAL TRANSPORT, AND METASTABLE ASSEMBLAGES

When sillimanite nucleates and begins to grow, many of the local mineral assemblages present in the rock are metastable with respect to the assemblage sillimanite + muscovite + quartz + water. These metastable assemblages exert a strong control

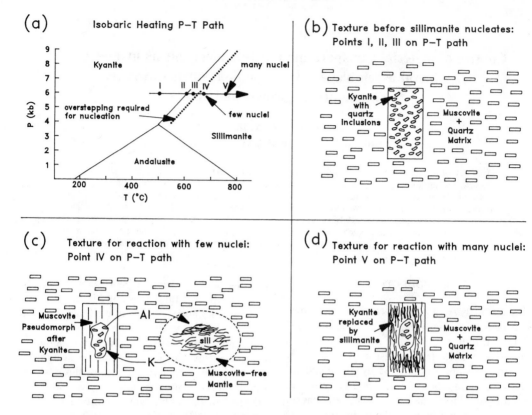

FIG. 1. (a) Pressure-temperature path followed by a kyanite-bearing rock that passes into the sillimanite field. Points I, II, III, IV, and V represent P, T conditions along the path that are discussed in the text. The dotted line schematically illustrates that some overstepping is needed to form sillimanite nuclei. The exact amount of overstepping required to form sillimanite nuclei is presently unknown. (b) Sketch of the texture existing prior to the nucleation and growth of sillimanite (points I, II, and III on Fig. 1a). The large vertical rectangle represents a kyanite poikiloblast with quartz inclusions. The many small horizontal rectangles represent muscovite crystals surrounded by quartz. (c) Sketch of the texture that develops if reaction takes place when nuclei have only formed in the sites most favorable for nucleation (point IV on Fig. 1a). The vertical line pattern inside the large vertical rectangle represents a muscovite pseudomorph after kyanite. The irregular lines inside the ellipse represent fibrolitic sillimanite. The blank area within the ellipse is a muscovite-free quartz mantle that surrounds the fibrolite. Kyanite poikiloblast and matrix muscovite patterns are the same as in Fig. 1b. Arrows labelled Al and K represent material transport of aluminum and potassium between reaction sites. (d) Sketch of the texture that develops if reaction takes place when nuclei have formed at many sites in the rock (point V on Fig. 1a). The irregular lines inside the vertical rectangle represent fibrolitic sillimanite that has replaced kyanite. Matrix muscovite and kyanite poikiloblast patterns are the same as in Fig. 1b.

over the type of textures that develop during the conversion of kyanite to sillimanite, particularly when the reaction mechanisms involve material transport under local equilibrium conditions. The effects of the metastable assemblages can be illustrated by considering the activity diagrams shown in Fig. 2. These diagrams show the distribution of stable mineral phases as a function of the activity of potassium and aluminum when temperature changes from conditions where kyanite is stable to conditions where sillimanite is stable. The diagrams in Fig. 2 were constructed using Schreinemakers'

methods (ZEN, 1966; YARDLEY, 1989) and the thermodynamic data base of HELGESON et al. (1978). The reactions represented by univariant lines in these diagrams have been written to conserve silica, conforming to a silica-fixed reference frame (BRADY, 1975), which is equivalent to an inert marker reference frame in quartz-bearing rocks under a hydrostatic stress (FOSTER, 1981). This reference frame was chosen because many geologists intuitively think of material transport with respect to an inert marker reference frame. Water is not labeled as a phase on Fig. 2, but the diagrams are

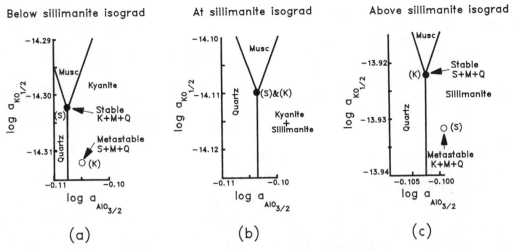

FIG. 2. Diagrams showing stable univariant lines, stable divariant fields, the sillimanite-absent (S) invariant point and the kyanite-absent (K) invariant point as a function of the activities of aluminum and potassium. S + M + Q stands for the co-existence of sillimanite + muscovite + quartz. K + M + Q stands for the co-existence of kyanite + muscovite + quartz. Water is assumed to be in equilibrium with all assemblages. See text for explanation of choice of components and standard states. Note that the tick marks on the axes have different values on each diagram. (a) Conditions 10°C below the sillimanite isograd: 607°C, 6 kb. The kyanite-absent invariant point is metastable. (b) Conditions at the theoretical sillimanite isograd: 617°C and 6 kb. The sillimanite-absent and kyanite-absent invariant points are coincident; both are stable. (c) Conditions 10°C above the sillimanite isograd: 627°C, 6 kb. The sillimanite-absent invariant point is metastable.

calculated for the situation where the solid phases were in equilibrium with pure water at the T and P of interest. The activities of potassium and aluminum have been expressed in terms of oxide components with the standard state chosen so that $KO_{1/2}$ has unit activity when the system is saturated with respect to potassium oxide and $AlO_{3/2}$ has unit activity when the system is saturated with respect to corundum. This convention has been chosen for convenience to avoid complications involving the uncertainties of speciation, particularly of aluminum, at metamorphic temperatures and pressures (WALTHER, 1986; WOODLAND and WALTHER, 1987; EUGSTER and BAUMGARTNER, 1987). If one wishes to specify a dominant species at the pressure and temperature of interest, the diagrams can be easily converted to aqueous species by calculating equilibrium constants for reactions such as $KO_{1/2}$ + HCl → KCl + 0.5 H_2O and $AlO_{3/2}$ + 1.5 H_2O → $Al(OH)_3$. The activities can then be recast in terms of the new species.

At temperatures in the kyanite stability field (point I, Fig. 1a), local equilibrium will keep the activities of aluminum and potassium in the rock at the stable invariant point where the minerals kyanite + muscovite + quartz + water coexist. This invariant point is labelled (S) on Fig. 2a because it represents the conditions where the sillimanite-ab-

sent assemblage is in equilibrium. All sillimanite-bearing assemblages are metastable at temperatures and pressures where kyanite is stable. For example, on Fig. 2a, the activities where the assemblage sillimanite + muscovite + quartz + water is in equilibrium are given by the position of the kyanite-absent (K) invariant point. It is a metastable invariant point located in the kyanite stability field because sillimanite + muscovite + quartz + water is less stable than kyanite + water at 607°C and 6 kb. As the temperature rises and conditions in the rock approach the sillimanite stability field, the differences in chemical potentials between the kyanite-absent (K) invariant point and the sillimanite-absent (S) invariant point decrease. When the theoretical sillimanite isograd is reached (point II, Fig. 1a), the two invariant points (K) and (S) are coincident, indicating that both kyanite-bearing and sillimanite-bearing assemblages are stable (Fig. 2b). However, no sillimanite forms at this time because the reaction has not been overstepped and no sillimanite nuclei are present in the rock. As the temperature continues to rise, sillimanite becomes more stable than kyanite (point III, Fig. 1a); the sillimanite-absent (S) invariant point becomes metastable while the kyanite-absent (K) invariant point becomes stable (Fig. 2c).

If sillimanite has not nucleated in the rock, the

activities of potassium and aluminum are constrained to lie at the metastable sillimanite-absent invariant point (S) on Fig. 2c because the most stable assemblage present in the rock is kyanite + muscovite + quartz + water. Eventually, sillimanite will form in micas, the site most favorable for nucleation, and begin to grow (point IV, Fig. 1a). When sillimanite grows from a mica, the reaction consumes aluminum and produces potassium, changing the conditions in the vicinity of the sillimanite. This will establish local equilibrium with the mineral assemblage muscovite + quartz + water + sillimanite. When this happens, the conditions near sillimanite will lie at the invariant point (K) on Fig. 2c, while the conditions near kyanite will lie at the invariant point (S) on Fig. 2c. The buffering of the activities of potassium and aluminum by local reactions creates chemical potential gradients between the kyanite and the sillimanite assemblages, resulting in the diffusion of aluminum from (S) to (K) and the diffusion of potassium in the opposite direction. The addition of aluminum and removal of potassium by transport through the matrix around the sillimanite results in the precipitation of sillimanite and dissolution of micas at the (K) invariant point. Similarly, the addition of potassium and removal of aluminum by transport through the matrix around kyanite causes kyanite to dissolve and micas to precipitate as the local mineral assemblage buffers the activities of aluminum and potassium at the (S) invariant point.

If the reaction and transport rates are high relative to the rate at which heat is being supplied to the rock, the local reactions will be able to consume heat at the same rate it enters the rock (FISHER, 1978) and keep the temperature close to the theoretical isograd (point III or IV, Fig. 1a). Under these conditions, nuclei will only form in a few of the most favorable sites, (YARDLEY, 1977; RIDLEY, 1985), local equilibrium will be maintained and a reaction mechanism will develop that uses material transport and mica catalysts to convert kyanite to sillimanite (Fig. 1c). If, on the other hand, the reactions are not able to buffer the temperature and it continues to rise (point V, Fig. 1a), sillimanite nuclei will form at many sites in the rock. Eventually, the reaction will be sufficiently overstepped to allow sillimanite to nucleate directly on kyanite (Fig. 1d). When this happens, local equilibrium will not be maintained because kyanite is in equilibrium with species having activities at (S) while the adjacent sillimanite will be in equilibrium with species having activities at (K) (Fig. 2c). The reaction will proceed by a non-equilibrium path, probably at conditions that lie in-between (K) and (S), until all

the kyanite in contact with sillimanite has been consumed. After kyanite is consumed, the conditions will migrate to (K) and local equilibrium will be re-established.

The factors controlling reaction mechanisms that take place in a rock when local equilibrium is maintained (e.g. Fig. 1c) can be examined in detail by constructing activity diagrams that show relationships among metastable phases in addition to those of the stable phases. An example of this type of diagram is shown in Fig. 3, which is the same activity diagram as Fig. 2c except that the metastable univariant lines and divariant fields are shown along with the stable ones. As in Fig. 2, the system is saturated with water at the T and P of interest and the standard states for the activities of potassium and aluminum are potassium oxide and corundum, respectively.

A total of twelve divariant fields are shown on Fig. 3. Each field represents a region in activity space where the mineral assemblages muscovite + water, quartz + water, sillimanite + water, and kyanite + water have a specific stability sequence. Following the notation convention of KUJAWA and EUGSTER (1966), the stability sequence of the four solid phases kyanite, sillimanite, muscovite, and quartz coexisting with water are shown by listing the first letters of the mineral names in a column in each of the twelve divariant fields. The letter representing the most stable mineral is at the bottom of the column and the letter representing the least stable mineral is at the top.

The twelve fields are separated by five univariant lines that represent equilibria among water and the mineral pairs muscovite + sillimanite [M + S], sillimanite + quartz [S + Q], muscovite + quartz [M + Q], kyanite + muscovite [K + M] and kyanite + quartz [K + Q]. When one of these lines is crossed, the stability sequence between the two minerals in equilibrium along the line reverses. For example, the fields separated by the line S + Q have quartz more stable than sillimanite on the side of the line where the aluminum activity is low and sillimanite more stable than quartz on the side of the line where aluminum activity is high. The segments of the univariant lines can be grouped into three types. Stable or first-order line segments are those which involve a switch in relative stability between the phases occupying the first and second stability levels in the divariant fields adjacent to the line. First-order line segments are shown as solid lines on Fig. 3. Second-order lines are those metastable segments of univariant lines that involve a switch in stability sequence between phases occupying the second and third stability levels in fields

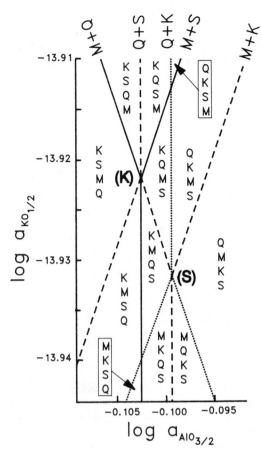

FIG. 3. Activity diagram at 627°C, 6 kb showing all stable and metastable invariant points, univariant lines and divariant fields involving the phases kyanite, sillimanite, muscovite and quartz. Water is present in all assemblages. K, S, M, and Q stand for kyanite, sillimanite, muscovite and quartz, respectively. The relative stability sequence of the phases in each divariant field is given by the columns of letters in each field. The mineral at the bottom of the list is the most stable, the one at the top is least stable. Each univariant line is labelled at the top of the diagram with the letters of the two minerals that are in equilibrium along it (*e.g.* M + Q). Stable (first-order) segments of univariant lines are shown as solid lines, second-order (metastable) segments of univariant lines are shown as dashed lines, and third-order (least stable) segments of univariant lines are shown as dotted lines. The kyanite-absent (K) invariant point is stable, the sillimanite-absent (S) invariant point is metastable.

adjacent to the line. Second-order line segments are shown as dashed lines on Fig. 3. Third-order lines are those that involve a switch in relative stability between phases occupying the third and fourth stability levels in divariant regions adjacent to the line. Third-order line segments are shown as dotted lines in Fig. 3.

Intersections of any two univariant lines involv-

ing a common phase produce invariant points representing three solid phases (plus water) in equilibrium. Invariant points always involve univariant line segments of two adjacent orders. There are two invariant points on Fig. 3: a stable one (K) formed by the intersection of first-order and second-order line segments and a metastable one (S) formed by the intersection of second-order and third-order line segments. Crossing of univariant lines that do not involve a common phase, such as M + K and S + Q, do not produce an invariant point because their energy levels differ by more than one order, so they do not intersect in free-energy space (KUJAWA and EUGSTER, 1966).

GROWTH OF SILLIMANITE SEGREGATIONS

The constraints on material transport provided by local equilibrium with metastable mineral assemblages shown on Fig. 3 can be used to explain the reaction mechanisms that form sillimanite segregations in a rock containing kyanite porphyroblasts (Fig. 1c). At temperatures in the sillimanite stability field, prior to the nucleation of sillimanite (point III, Fig. 1a), the mineral assemblage kyanite + muscovite + quartz + water buffers the chemical potentials of aluminum and potassium, keeping conditions in the rock at the metastable sillimanite-absent invariant point (S). This invariant point slowly moves to higher activities of aluminum and potassium as temperature increases. Little reaction among the solid phases is required to maintain equilibrium because the fluid phase along the grain boundaries is not a large source or sink for components, due to its relatively small volume. When sillimanite nucleates in the rock (point IV, Fig. 1a), reactions take place that drive local conditions toward equilibrium with sillimanite. Although the amount of overstepping required for sillimanite nuclei to form is presently unknown, many workers (*e.g.* CHINNER, 1961) have made the observation that it is common for sillimanite at the lowest metamorphic grades to be only present in micas, suggesting micas have sites with activation energies that are most favorable for the formation of sillimanite nuclei. The reaction mechanisms and transport paths described below depend primarily on the types of minerals present at the site of nucleation rather than on the absolute amount of overstepping required to form the nuclei. Therefore, to facilitate discussion, an overstepping of 10°C was arbitrarily assumed to be required for sillimanite nuclei to form in muscovite.

Using this assumption, the first sillimanite to form in a rock containing kyanite + muscovite

+ quartz + water that is being metamorphosed along the isobaric path shown in Fig. 1a nucleates on muscovite at 627°C. The sillimanite-forming reaction will establish local equilibrium by driving the activities from those at (S) to those at (K). Due to the configuration of the metastable elements at 627°C and 6 kb (Fig. 3), changing from conditions at (S) to conditions at (K) must involve a decrease in the log of aluminum activity from a value of −0.0994 to a value of −0.1026 and an increase in the log of the potassium activity from a value of −13.931 to a value of −13.922. The only reaction between sillimanite, muscovite, quartz and water that can produce these changes is a reaction that precipitates sillimanite and dissolves muscovite. The precise stoichiometry of the reaction depends upon the actual concentration changes of aluminum and potassium species in the grain boundary fluid that are required to produce the requisite changes in the activities of the components $AlO_{3/2}$ and $KO_{1/2}$.

Once conditions corresponding to those at the invariant point (K) are established in the vicinity of sillimanite, the chemical potentials in the muscovite + quartz matrix surrounding sillimanite are forced by local equilibrium to lie along the metastable second-order portion of the univariant line M + Q that lies between (S) and (K) on Fig. 3. As long as cross-term diffusion coefficients (ANDERSON, 1981) are not important, this constraint on the chemical potentials in the matrix causes transport of aluminum toward the sillimanite and the transport of potassium away from it. The addition of aluminum and removal of potassium results in a reaction that forms sillimanite at the expense of muscovite.

This reaction keeps conditions at invariant point (K) by consuming the newly arrived aluminum and producing potassium to replace the amount that recently departed. The precise stoichiometry of this local reaction must be a linear combination of muscovite, sillimanite, quartz, and water plus the fluxes of material in and out of the volume of rock where the reaction takes place (FISHER, 1975). In the simplest case, where cross terms in the diffusion coefficient matrix are not important, the flux of a component i (J_i) is related to the chemical potential gradient of the component i ($d\mu_i/dx$) by a single thermodynamic diffusion coefficient (L_i):

$$J_i = -L_i(d\mu_i/dx).$$

Because the ratio of chemical potential gradients transporting constituents between the invariant points (K) and (S) is fixed by the metastable second-order segment of the univariant line M + Q, the stoichiometry of the reaction at (S) is unique if the ratio of the diffusion coefficients is specified (FISHER, 1975). For example, if the ratios of diffusion coefficients given by FOSTER (1981) are used, the ratio of the fluxes of aluminum and potassium in the M + Q matrix surrounding sillimanite are given by:

$$\frac{J_{AlO_{3/2}}}{J_{KO_{1/2}}} = \frac{-L_{AlO_{3/2}} \text{ grad } \mu_{AlO_{3/2}}}{-L_{KO_{1/2}} \text{ grad } \mu_{KO_{1/2}}}$$

$$= (6/1)(-1/3) = -2.0. \qquad (1)$$

Equation (1) means that two aluminum atoms are being supplied for every potassium atom removed from the region where sillimanite + muscovite + quartz is buffering the chemical potentials at (K). The only reaction between muscovite, quartz and sillimanite that can balance the material transport required by this flux ratio is:

$$0.4 \text{ KAl}_3\text{Si}_3\text{O}_{10}(\text{OH})_2 + 0.8 \text{ AlO}_{3/2} \rightarrow$$

muscovite

$$1.0 \text{ Al}_2\text{SiO}_5 + 0.2 \text{ SiO}_2 + 0.4 \text{ KO}_{1/2} + 0.4 \text{ H}_2\text{O}.$$

sillimanite   quartz

$$(A)$$

This reaction, the overall reaction in the sillimanite segregation, consumes slightly more than 1.1 cc of muscovite and produces slightly less than 0.1 cc of quartz for every 1.0 cc of sillimanite produced.

Figure 4 shows how variation of the diffusion coefficient ratio of aluminum to potassium affects the stoichiometric coefficients for the overall reaction in the sillimanite segregation. The main effect of increasing the diffusion coefficient ratio of aluminum to potassium is to decrease the amount of

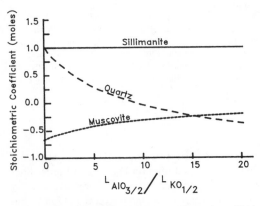

FIG. 4. The stoichiometry of the overall reaction within the sillimanite segregation as a function of the ratio of the aluminum and potassium diffusion coefficients. The stoichiometry given for this reaction in the text (reaction (A)) was calculated using $(L_{AlO_{3/2}})/(L_{KO_{1/2}}) = 6$.

quartz being produced on the sillimanite side of the reaction. Eventually, at an L ratio between 9 and 10, quartz shifts to the opposite side of the reaction and is consumed as sillimanite forms. Muscovite always remains on the opposite side of the reaction from sillimanite because the constraints forced by the metastable M + Q assemblage do not allow chemical potential gradients of potassium and aluminum to be in the same direction. This precludes transport of aluminum and potassium in the same direction unless cross terms are important in the diffusion coefficient matrix or unless the system is flow dominated.

Another effect of changing the diffusion coefficient ratios is to produce variations in the volume gain or loss among the solid phases taking part in the reaction. If the diffusion coefficient ratio of aluminum to potassium is 0.01, the production of sillimanite is accompanied by a volume loss of 33% among the solid phases. If the diffusion coefficient ratio is 20, the production of sillimanite is accompanied by a volume gain of 32%. Many workers (e.g. CARMICHAEL, 1987) have observed that most metamorphic reactions seem to take place without large volume losses or gains. A value of seven for $L_{AlO_{3/2}}/L_{KO_{1/2}}$ gives stoichiometric coefficients for the overall reaction in the sillimanite segregation that conserves volume among the solid phases.

The overall reaction in the sillimanite segregation seldom consumes or produces phases in the same proportions that they are present in matrix surrounding the reaction site. Typically, one phase will be consumed before another in the region around the reaction site, producing a mantle that is free of one matrix phase (FISHER, 1975, 1977). For example, muscovite is the only phase consumed in reaction (A). If the matrix around the growing sillimanite is composed of equal volumes of quartz and muscovite, muscovite from 2.2 cc of matrix around a growing sillimanite is required to provide the 1.1 cc of muscovite consumed to grow 1.0 cc of sillimanite. This produces a muscovite-free, quartz mantle around a sillimanite core as shown in Fig. 5. When this happens, the activities of components needed for the sillimanite-forming reaction inside of the muscovite-free mantle are no longer constrained by the muscovite Gibbs-Duhem relation. The only reaction possible between quartz and sillimanite at the quartz mantle/sillimanite interface is:

$$2\ AlO_{3/2} + 1.0\ SiO_2 \rightarrow 1.0\ Al_2SiO_5. \quad (B)$$
$$\text{quartz} \qquad \text{sillimanite}$$

The aluminum in (B) comes from diffusion of

$AlO_{3/2}$ through the quartz mantle and the SiO_2 comes from the dissolution of quartz at the sillimanite core/quartz mantle interface. Reaction (B) has been written for one mole of sillimanite to correspond to the amount of sillimanite produced in reaction (A). The material balance and transport constraints in the matrix require that reaction (A) be the net reaction taking place within the entire sillimanite segregation because this is the only reaction possible between sillimanite, quartz and muscovite that satisfies the flux ratio given by equation (1). The reaction at the muscovite + quartz matrix/quartz mantle interface is calculated by subtracting reaction (B) from (A):

$$0.4\ \text{muscovite} \rightarrow 1.2\ \text{quartz} + 0.4\ KO_{1/2}$$
$$+ 1.2\ AlO_{3/2} + 0.4\ H_2O. \quad (C)$$

The potassium produced by reaction (C) diffuses away from the reaction interface, down the potassium chemical potential gradient through the matrix, toward dissolving kyanite. The aluminum produced by reaction (C) diffuses away from the reaction interface in the opposite direction, down the chemical potential gradient through the quartz mantle, toward the growing sillimanite. The total flux of aluminum through the quartz mantle is 2.0 $AlO_{3/2}$ per mole of sillimanite that grows in the core of the segregation; the 1.2 moles of aluminum produced by reaction (C) are added to the 0.8 moles of aluminum which diffused to the segregation through the matrix. Thus, 60% of the aluminum for the growth of sillimanite comes from the local breakdown of muscovite at the margin of the sillimanite segregation while 40% is supplied by transport through the muscovite + quartz matrix. The segregation morphology, local reactions and material transport between reaction sites in the sillimanite segregation are shown on the right portion of Fig. 5.

DEVELOPMENT OF MICA PSEUDOMORPHS OF KYANITE

The Gibbs-Duhem constraints of the metastable matrix assemblage M + Q around the dissolving kyanite require that the overall reaction that forms a mica pseudomorph after kyanite be the reverse of (A):

$$1.0\ Al_2SiO_5 + 0.2\ SiO_2$$
$$\text{kyanite} \qquad \text{quartz}$$
$$+ 0.4\ KO_{1/2} + 0.4\ H_2O \rightarrow$$
$$0.4\ KAl_3Si_3O_{10}(OH)_2 + 0.8\ AlO_{3/2}. \quad (D)$$
$$\text{muscovite}$$

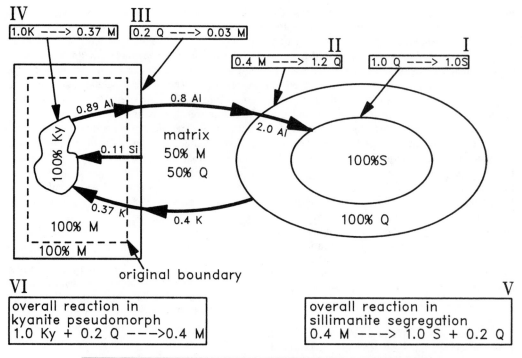

FIG. 5. Quantitative reaction model showing the sillimanite segregation and the kyanite pseudomorph produced when one mole of sillimanite grows at the expense of one mole of kyanite. The original boundary of the kyanite porphyroblast is given by a dashed line. Reaction interfaces in the segregation and pseudomorph are shown as thin lines and curves. The minerals involved in the reaction at each interface are given in the numbered boxes above the segregation and pseudomorph. These boxes, numbered I, II, III, and IV represent the local reactions (B), (C), (F) and (E), respectively. Material transport paths between reaction interfaces are shown by heavy lines with arrows giving the direction of transport for each component. Numbers and letters next to transport paths give moles of each component that diffused through each region of the rock when one mole of sillimanite is produced. Mineral modes for each part of the segregation and pseudomorph are given in volume %. The minerals involved in the overall reactions for each segregation are given in boxes labelled V and VI, representing reactions (A) and (D), respectively. Ky, S, M, and Q stand for kyanite, sillimanite, muscovite and quartz, respectively. Al, Si and K stand for the transported components $AlO_{3/2}$, SiO_2 and $KO_{1/2}$, respectively. Diffusion coefficient ratios of $(L_{AlO_{3/2}})/(L_{KO_{1/2}}) = 6$ and $(L_{SiO_2})/(L_{KO_{1/2}}) = 5$ were used for the calculation of this diagram.

If the rock is closed to material transport beyond the hand-specimen scale, this reaction produces 1.1 cc of muscovite and consumes 0.9 cc of kyanite and 0.1 cc of quartz for every 1.0 cc of sillimanite that grows in the rock at another location. In this case, the amount of muscovite consumed by sillimanite growth exactly balances the amount of muscovite produced by kyanite dissolution. Similarly, the amount of quartz locally produced by sillimanite growth will be exactly balanced by the quartz consumed locally by kyanite dissolution. Thus, there is no net gain or loss of muscovite or quartz in the rock, they both serve as mineral catalysts which help the reaction kyanite → sillimanite proceed.

The quartz required for reaction (D) is supplied either from quartz inclusions within the kyanite porphyroblast or from quartz in the matrix adjacent to the kyanite porphyroblast. If the kyanite porphyroblast contains more than 10% quartz inclusions, all of the quartz required for reaction (D) is available within the kyanite poikiloblast and reaction (D) will form a quartz + muscovite pseudomorph after the kyanite poikiloblast. The amount of quartz in the pseudomorph depends upon the volume percent of quartz inclusions in the kyanite poikiloblast that is replaced. For example, if the quartz inclusions make up 10% of the kyanite poikiloblast, then there will be 0.1 cc of quartz available locally for every 0.9 cc of kyanite that is consumed by reaction (D). Since reaction (D) consumes 0.1 cc of quartz for every 0.9 cc of kyanite, the amount of quartz consumed by reaction (D) exactly equals the amount available and the pseudomorph that replaces kyanite is composed entirely of muscovite. However, if the modal amount of quartz in the poikiloblast is 25%, then there is 0.3 cc of quartz present locally for every 0.9 cc of kyanite. Reaction (D) consumes 0.1 cc of this quartz while consuming 0.9 cc of kyanite, leaving 0.2 cc of quartz mixed with the 1.1 cc of muscovite produced by reaction (D). A pseudomorph containing 15% quartz and 85% muscovite would result. In this circumstance, where there is quartz in the poikiloblast in excess of the amount required for reaction (D), the mineral assemblage within the pseudomorph and in the matrix is the same: muscovite + quartz. Coexisting muscovite and quartz cause the ratio of chemical potential gradients outside of the pseudomorph and within the pseudomorph to be identical because they both lie on the second order segment of the univariant line M + Q (Fig. 3). This constraint requires that no reaction takes place at the matrix/pseudomorph boundary. The original boundary of the kyanite poikiloblast is marked by the change in modes of muscovite and quartz between the amounts present in the matrix and the amounts produced in the pseudomorph by reaction (D). This boundary serves as an inert marker in the system, representing the bulk compositional difference between the matrix and the kyanite poikiloblast. In some instances, the matrix and pseudomorph modes will be nearly the same, making the pseudomorph difficult to recognize unless the pseudomorph micas are different in size or orientation from the matrix micas. Generally, the amount of muscovite in the pseudomorph is much higher than the amount in the matrix so the original boundary of the kyanite poikiloblast is easily recognized.

If there is no quartz in the kyanite porphyroblast, the silica for reaction (D) comes from the quartz in the matrix around the porphyroblast. In this circumstance, when kyanite first begins to dissolve, reaction (D) consumes quartz and kyanite at the porphyroblast boundary and replaces it with muscovite. A quartz-free muscovite rim will form between the dissolving kyanite and the matrix. The material transport to and from the kyanite/muscovite interface will be constrained by the Gibbs-Duhem relation of muscovite and the requirement that the reaction must be a linear combination of muscovite and kyanite. These constraints force the reaction to always produce muscovite when kyanite dissolves. The precise stoichiometry of the reaction depends upon the relative size of the diffusion coefficients of silica, aluminum and potassium (Fig. 6). Values of $L_{AlO_{3/2}}/L_{KO_{1/2}} = 6$ have been found to be compatible with a variety of textures in amphibolite facies pelites (FOSTER 1981, 1982, 1983, 1986). Figure 6 shows that if this value is used for $L_{AlO_{3/2}}/L_{KO_{1/2}}$, the reaction replacing kyanite with muscovite in the absence of quartz is relatively insensitive to the value of the diffusion coefficient of silica. For example, if $L_{AlO_{3/2}}/L_{KO_{1/2}} = 6$, the amount of muscovite produced per mole of kyanite consumed is 0.33 moles when $L_{SiO_2}/L_{KO_{1/2}} = 0.001$ while it is 0.39 moles when $L_{SiO_2}/L_{KO_{1/2}} = 50$.

For a given set of diffusion coefficients, the reaction at the muscovite rim/matrix boundary can be calculated by subtracting the reaction at the kyanite/muscovite rim boundary from reaction (D), the net reaction for the entire pseudomorph. The exact stoichiometry of the reaction at the mica rim/matrix boundary depends upon the diffusion coefficient ratios. Figure 7 shows the effect that varying $L_{SiO_2}/L_{KO_{1/2}}$ has on the stoichiometry of the reaction at the mica rim/matrix boundary of the kyanite pseudomorph. If the silica diffusion coefficient is low, the local reactions essentially conserve silica. If the silica diffusion coefficient is high, the local reactions produce silica at the mica rim/matrix boundary of the kyanite pseudomorph and transport the silica to a major sink at the kyanite/mica rim boundary in the interior of the pseudomorph. The texture produced in either case is very similar:

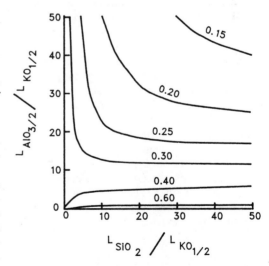

FIG. 6. Contours showing the moles of muscovite produced per mole of kyanite dissolved at the inner boundary of the quartz-free mica rim in the kyanite pseudomorph as a function of the ratio of the aluminum, potassium and silica diffusion coefficients. The stoichiometry of this reaction given in the text (reaction (E)) was calculated using $(L_{AlO_{3/2}})/(L_{KO_{1/2}}) = 6$ and $(L_{SiO_2})/(L_{KO_{1/2}}) = 5$.

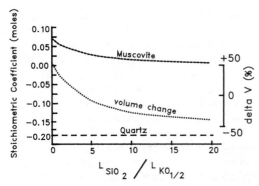

FIG. 7. The stoichiometry and volume change of the reaction at the mica rim/matrix boundary of the quartz-free mica pseudomorph after kyanite as a function of the ratio of the silica and potassium diffusion coefficients. The diffusion coefficient ratio of aluminum to potassium was held constant at a value of 6 to construct this diagram. The stoichiometry of this reaction given in the text (reaction (F)) was calculated using $(L_{AlO_{3/2}})/(L_{KO_{1/2}}) = 6$ and $(L_{SiO_2})/(L_{KO_{1/2}}) = 5$.

kyanite is replaced by muscovite and a small amount of quartz in the matrix is converted to muscovite, forming a mica rim between the kyanite and matrix. The position of the original boundary between the kyanite and matrix is within the mica rim. The exact location of the original boundary depends upon the amount of quartz in the matrix and the stoichiometry of the reaction at the mica rim/matrix boundary.

The diffusion coefficient ratios also affect the delta V of reaction at the mica rim/matrix boundary. Low values of the silica diffusion coefficient result in reactions that produce substantially more muscovite than the amount of quartz that dissolves, resulting in a volume increase when matrix is converted to muscovite rim. High values of the silica diffusion coefficient result in reactions that produce almost no muscovite while dissolving substantial amounts of quartz, causing a volume loss when matrix is converted to muscovite rim. As shown in Fig. 7, a value of 5 for $L_{SiO_2}/L_{KO_{1/2}}$ produces a matrix/pseudomorph reaction that essentially conserves volume. Using a value of 6 for the aluminum to potassium diffusion coefficient ratio and a value of 5 for the silica to potassium diffusion coefficient ratio gives:

$$1.0 \text{ kyanite} + 0.37 \text{ KO}_{1/2} + 0.11 \text{ SiO}_2$$

$$+ 0.37 \text{ H}_2\text{O} \rightarrow 0.37 \text{ muscovite} + 0.89 \text{ AlO}_{3/2} \quad \text{(E)}$$

for the reaction at the kyanite/mica rim interface and:

$$0.03 \text{ KO}_{1/2} + 0.09 \text{ AlO}_{3/2} + 0.03 \text{ H}_2\text{O}$$

$$+ 0.2 \text{ quartz} \rightarrow 0.03 \text{ muscovite} + 0.11 \text{ SiO}_2 \quad \text{(F)}$$

for the reaction at the mica rim/matrix boundary. The SiO$_2$ in reactions (E) and (F) is the silica that is transported through the mica rim by diffusion. This silica is liberated by reaction (F) and consumed by reaction (E).

Reaction (F) consumes 0.1 cc of quartz and produces 0.1 cc of muscovite for every 1.0 cc of kyanite replaced by reaction (E). In a rock containing equal parts of muscovite and quartz, 0.2 cc of matrix will be required to supply the quartz for reaction (F). After the matrix has been converted to muscovite rim by reaction (F) it will contain 0.2 cc of muscovite, half of which was made by reaction (F). The other half is the 0.1 cc of muscovite originally in the matrix. If the muscovite produced by reaction (F) had a different grain size or orientation than the matrix muscovite, one would be able to identify the two different types. If the early and late muscovite were similar in size and orientation, then the old and new muscovites could not be distinguished in the mica rim that replaced matrix. The pseudomorph morphology, local reactions and material transport produced by reactions (E) and (F) are shown on the left side of Fig. 5.

CHEMICAL POTENTIAL PROFILES

The chemical potential gradients that drive material transport around the growing sillimanite and dissolving kyanite are primarily governed by the Gibbs-Duhem constraints provided by the matrix mineral assemblage muscovite + quartz + water. In the example shown in Fig. 5, the local reactions producing sillimanite and consuming kyanite form a muscovite-free zone around the sillimanite and a quartz-free zone around the kyanite. The removal of a matrix phase allows the chemical potentials to leave the M + Q metastable line between the (K) and (S) invariant points and pass through divariant fields, as shown in Fig. 8.

The chemical potentials within the muscovite-free mantle around the sillimanite pass through the divariant field where quartz is the second-most stable phase. The inner boundary (point I, Fig. 8) of the muscovite-free mantle is constrained to lie on the sillimanite + quartz line while the outer boundary of the muscovite-free mantle (point II, Fig. 8) is constrained to lie on the muscovite + quartz line. The chemical potential gradient of potassium within the muscovite-free mantle is zero because no potassium-bearing phases are present within the sillimanite segregation. The chemical potential gra-

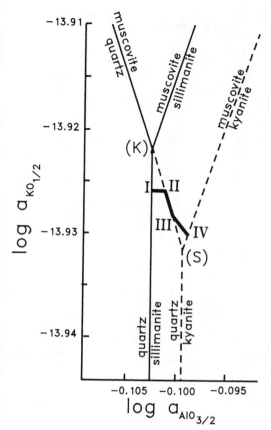

FIG. 8. Activity diagram showing conditions in the model illustrated in Fig. 5. Only selected portions of stable univariant lines (solid) and second-order univariant lines (dashed) are shown. Refer to Fig. 3 for complete diagram. Points I, II, III and IV correspond to reaction interfaces with the same numbers given in Fig. 5. The kyanite-absent and sillimanite-absent invariant points are labelled (K) and (S), respectively.

dient of aluminum within the muscovite-free mantle is governed by the width of the mantle and the difference in activities represented by the length of the line between points I and II on Fig. 8. As the sillimanite segregation grows and the width of the muscovite-free mantle increases, points I and II on Fig. 8 will gradually slide down the S + Q and M + Q univariant lines. This increases the chemical potential differences across the muscovite-free mantle so that the chemical potential gradients remain properly balanced with the gradients in the matrix to supply the constituents needed for reactions (B) and (C).

The chemical potentials within the quartz-free mica rim of the kyanite pseudomorph are represented by a line passing through the divariant field where muscovite is the second-most stable phase. Chemical potentials at the inner boundary of the

mica rim (point IV, Fig. 8) are constrained by the muscovite + kyanite assemblage, and the outer boundary (point III, Fig. 8) is constrained by the muscovite + quartz assemblage. The slope of the chemical potential profile between points III and IV is determined by the diffusion coefficient ratios and will not change as the pseudomorph grows. The magnitudes of the chemical potential gradients in the mica rim are determined by the width of the mica rim and the difference in chemical potentials corresponding to the length of the line between points III and IV. As the pseudomorph grows and the width of the mica rim increases, points III and IV will slide up the M + Q and M + K univariant lines to keep the chemical potential gradients in the mica rim balanced with the gradients in the matrix to supply the components needed for reactions (E) and (F). Eventually, when the last of the kyanite has been consumed, the chemical potentials in all domains of the rock will migrate back to values represented by invariant point (K) and reactions will cease.

CONCLUSIONS

Most metamorphic reactions involve material transport on a local scale because the product minerals are only rarely the same composition as the substrate they grow on. The reaction mechanisms and transport paths that allow the overall reaction to proceed are usually strongly influenced by the constraints on material transport provided by mineral assemblages that are metastable with respect to the growing minerals. Construction of activity diagrams that include the geometric elements involving the metastable mineral assemblages can provide useful insight into the reaction mechanisms and the processes which control them. This approach can be coupled with the irreversible thermodynamic method devised by FISHER (1975, 1977) to develop quantitative models of reaction mechanisms that can be tested against textures observed in natural samples. Eventually, it should be possible to use these techniques to extract detailed pressure-temperature-time histories from textural features in metamorphic rocks.

Acknowledgements—I am indebted to Hans Eugster for teaching me about metastable elements of phase diagrams. Perceptive reviews by Bill Briggs, I-Ming Chou, George Fisher and Rick Sanford improved early versions of this manuscript. Calculations for this work were performed using computer equipment purchased with funds provided by a grant from the Amoco Foundation.

REFERENCES

ANDERSON D. E. (1981) Diffusion in electrolyte mixtures. In *Kinetics of Geochemical Processes* (eds. A. C. LASAGA and R. J. KIRKPATRICK), Reviews in Mineral. 8, Ch. 6, pp. 211–260. Mineral. Soc. Amer.

BRADY J. B. (1975) Reference frames and diffusion coefficients. Amer. J. Sci. 275, 945–983.

CARMICHAEL D. M. (1969) On the mechanism of prograde reactions in quartz-bearing pelitic rocks. Contrib. Mineral. Petrol. 20, 244–267.

CARMICHAEL D. M. (1987) Induced stress and secondary mass transfer: thermodynamic basis for the tendency toward constant-volume constraint in diffusion metasomatism. In Chemical Transport in Metasomatic Processes (ed. H. C. HELGESON), pp. 239–264. Reidel.

CHINNER G. A. (1961) The origin of sillimanite in Glen Cova, Angus. J. Petrol. 2, 312–323.

EUGSTER H. P. and BAUMGARTNER L. (1987) Mineral solubilities and speciation in supercritical metamorphic fluids. In Thermodynamic Modeling of Geological Materials: Minerals, Fluids, and Melts (eds. I. S. E. CARMICHAEL and H. P. EUGSTER), Reviews in Mineral. 17, Ch. 10, pp. 367–404. Mineral. Soc. Amer.

FISHER G. W. (1975) The thermodynamics of diffusion-controlled metamorphic processes. In Mass Transport Phenomena in Ceramics (eds. A. R. COOPER and A. H. HEUER), pp. 111–122. Plenum.

FISHER G. W. (1977) Nonequilibrium thermodynamics in metamorphism. In Thermodynamics in Geology (ed. D. G. FRASER), pp. 381–403. Reidel.

FISHER G. W. (1978) Rate laws in metamorphism. Geochim. Cosmochim. Acta 42, 1035–1050.

FOSTER C. T. (1981) A thermodynamic model of mineral segregations in the lower sillimanite zone near Rangeley, Maine. Amer. Mineral. 66, 260–277.

FOSTER C. T. (1982) Textural variation of sillimanite segregations. Can. Mineral. 20, 379–392.

FOSTER C. T. (1983) Thermodynamic models of biotite pseudomorphs after staurolite. Amer. Mineral. 68, 389–397.

FOSTER C. T. (1986) Thermodynamic models of reactions involving garnet in a sillimanite/staurolite schist. Mineral. Mag. 50, 427–439.

HELGESON H. C., DELANY J. M., NESBITT H. W. and BIRD D. K. (1978) Summary and critique of the thermodynamic properties of rock-forming minerals. Amer. J. Sci. 278A, 1–229.

KUJAWA F. B. and EUGSTER H. P. (1966) Stability sequences and stability levels in unary systems. Amer. J. Sci. 264, 620–642.

RIDLEY J. (1985) The effect of reaction enthalpy on the progress of a metamorphic reaction. In Metamorphic Reactions: Kinetics, Textures and Deformation (eds. A. B. THOMPSON and D. C. RUBIE), pp. 80–97. Springer-Verlag.

RIDLEY J. and THOMPSON A. B. (1986) The role of mineral kinetics in the development of metamorphic microtextures. In Fluid-Rock Interactions during Metamorphism (eds. J. V. WALTHER and B. J. WOOD), pp. 154–193. Springer-Verlag.

WALTHER J. V. (1986) Mineral solubilities in supercritical H_2O solutions. Pure and Appl. Chem. 58, 1585–1598.

WOODLAND A. B. and WALTHER J. V. (1987) Experimental determination of the solubility of the assemblage paragonite, albite, and quartz in supercritical H_2O. Geochim. Cosmochim. Acta 51, 365–372.

YARDLEY B. W. D. (1977) The nature and significance of the mechanism of sillimanite growth in the Connemara Schists, Ireland. Contrib. Mineral. Petrol. 65, 53–58.

YARDLEY B. W. D. (1989) An Introduction to Metamorphic Petrology. 248 pp. Longman.

ZEN E-AN (1966) Construction of pressure-temperature diagrams for multi-component systems after the method of Schreinemakers—a geometric approach. U.S. Geol. Surv. Bull. 1125.

Fluid-Mineral Interactions: A Tribute to H. P. Eugster
© The Geochemical Society, Special Publication No. 2, 1990
Editors: R. J. Spencer and I-Ming Chou

The exploration of reaction space

GEORGE W. FISHER

Department of Earth and Planetary Sciences, The Johns Hopkins University, Baltimore, Maryland 21218, U.S.A.

Abstract—An efficient method of determining whether or not two multicomponent mineral assemblages can have equilibrated at the same metamorphic grade is to explore the reaction space defined by a matrix representing the compositions of the phases of both assemblages. If reactions having the minerals of one assemblage as products and the other as reactants (termed incompatibility reactions) can be found, the assemblage cannot have equilibrated under the same conditions.

Examination of the relationship between composition space and reaction space of the system SiO_2-MgO-FeO shows that assemblages which intersect or overlap in composition space are characterized by a reaction space containing incompatibility reactions. The regions of reaction space characterized by incompatibility reactions are bounded by univariant equilibria which are themselves incompatibility reactions. If mineral compositions change so as to reduce the area of overlap in composition space, the area of reaction space containing incompatibilities also diminishes, and finally disappears when the compositional overlaps disappear. At that point, the bounding univariant reactions change from incompatibility reactions to compatibility reactions. Consequently, the nature of reactions throughout reaction space can be determined simply by investigating the character of the univariant reactions defined by a composite assemblage.

INTRODUCTION

GREENWOOD (1967) SHOWED that we can determine whether or not a pair of mineral assemblages could have equilibrated at the same metamorphic grade by analyzing the form of reactions between phases of the two assemblages. Two cases must be distinguished.

1. If a reaction having the minerals of one assemblage as reactants and those of the other as products can be written, the tie lines defining the assemblages intersect in composition space (*e.g.* Fig. 1A). The bulk compositions at the intersections can then be represented by either assemblage. But one assemblage must be more stable than the other under arbitrary values of *P*, *T* and activities of externally fixed components (KORZHINSKII, 1959, p. 62), so both cannot have equilibrated under the same arbitrary conditions. In this paper, such assemblages are referred to as *incompatible,* and reactions which have minerals of one assemblage as reactants and the other as products are designated *incompatibility reactions.*

2. If no reactions reflecting an incompatibility between two assemblages can be found, the assemblages do not intersect in composition space (*e.g.* Fig. 1B); they simply have different compositions, and could have equilibrated under the same conditions. Such assemblages are here termed *compatible,* and reactions involving minerals of two assemblages on at least one side are termed *compatibility reactions.*

Singular value decomposition (see PRESS *et al.,*

1986, p. 52–64 for a lucid discussion of this matrix technique) provides a simple and efficient method both for modeling mineral assemblages in multicomponent space and for determining whether or not two assemblages are compatible. FISHER (1989) showed that taking the singular value decomposition (SVD) of a composite matrix containing the phases of two assemblages can be used to: (1) construct a composite model fitting both assemblages within any desired level of analytical uncertainty; (2) determine a set of basis vectors for the composition space needed to represent the model; and (3) determine a set of basis vectors for the reaction space needed to characterize reactions between the minerals of the model assemblages. The reaction space of this paper differs somewhat from that used by THOMPSON (1982) because the reactions considered here express relations between phases of metamorphic assemblages, while those discussed by Thompson express relations among components of metamorphic phases.

In order to determine whether or not the model assemblages intersect in composition space, it is necessary to explore the reaction space of the composite to see whether it contains a reaction indicating an incompatibility. FISHER (1989) stated but did not demonstrate that this exploration can be carried out simply by writing all possible univariant reactions involving the phases of the model composite assemblage. Given the importance of this point, it seems worthwhile to provide further justification for this statement.

This paper analyzes three hypothetical assemblages in the system SiO_2-MgO-FeO to illustrate

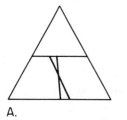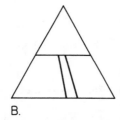

A.   B.

FIG. 1. Mineral assemblages in a hypothetical three-component system. A shows two two-phase assemblages, which intersect, and so cannot have equilibrated under the same external conditions; B shows two two-phase assemblages which do not intersect, and so can have equilibrated under the same conditions.

how the set of possible univariant reactions provides a map to all parts of reaction space. The generalization to more complex systems is straightforward.

Hans Eugster devoted much of his life to exploring P-T-activity space, often using the principles worked out by SCHREINEMAKERS (1915–1925), as an aid to navigation. This paper shows that the same rules which govern P-T-activity space can be used to explore reaction space.

A SIMPLE EXAMPLE

Consider four hypothetical assemblages (M1, M2, M3, and M4) from the system SiO_2-MgO-FeO, with mineral compositions listed in Table 1. A plot of these assemblages in composition space (Fig. 2), shows that: (1) M1 and M3 do not intersect, and can have formed under the same external conditions; (2) M1 intersects a part of M2, and so cannot have formed under the same conditions; (3) M4 completely encloses M2, and cannot have formed under the same conditions. Our task is to find a general method for distinguishing these three cases in multicomponent systems using SVD.

Reactions between M1 and M2

To determine a set of basis vectors for the reaction space defined by M1 and M2, we first write a matrix representing the composite M1 + M2:

$$\mathbf{M12} = \begin{matrix} 1.00 & 1.00 & 1.00 & 1.00 & 1.00 \\ 0.00 & 0.60 & 0.75 & 0.40 & 0.70 \\ 0.00 & 0.40 & 1.25 & 0.60 & 1.30, \end{matrix}$$

in which rows represent the components SiO_2, MgO and FeO, the first three columns represent the minerals of M1 (qz_1, op_1 and ol_1 in that order), and the last two the minerals of M2 (op_2 and ol_2). The abbreviations used in this paper are qz = quartz, ol

= olivine, op = orthopyroxene and ox = oxide; subscripts indicate the number of the mineral assemblage. Taking the SVD of **M12** gives three matrices which are useful for interpreting M1 and M2 (FISHER, 1989):

$$\mathbf{W} = \begin{matrix} 3.072 & 0.000 & 0.000 & 0.000 & 0.000 \\ 0.000 & 0.912 & 0.000 & 0.000 & 0.000 \\ 0.000 & 0.000 & 0.276 & 0.000 & 0.000 \\ 0.000 & 0.000 & 0.000 & 0.000 & 0.000 \\ 0.000 & 0.000 & 0.000 & 0.000 & 0.000, \end{matrix}$$

$$\mathbf{U} = \begin{matrix} -0.697 & 0.706 & -0.126 & 0.000 & 0.000 \\ -0.394 & -0.231 & 0.890 & 0.000 & 0.000 \\ -0.599 & -0.670 & -0.439 & 0.000 & 0.000 \\ 0.000 & 0.000 & 0.000 & 0.000 & 1.000 \\ 0.000 & 0.000 & 0.000 & 1.000 & 0.000, \end{matrix}$$

$$\mathbf{V} = \begin{matrix} -0.227 & 0.774 & -0.456 & 0.000 & -0.376 \\ -0.382 & 0.329 & 0.839 & 0.172 & -0.111 \\ -0.567 & -0.333 & -0.028 & -0.686 & -0.310 \\ -0.395 & 0.232 & -0.122 & -0.172 & 0.863 \\ -0.570 & -0.358 & -0.268 & 0.686 & -0.066. \end{matrix}$$

The diagonal elements of **W** are the singular values of **M12**; the fact that three are non-zero shows that the matrix is of rank three (written $R(\mathbf{M12}) = 3$), thereby confirming that M1 and M2 can be represented by a three component system, and indicating that the composite assemblage of five phases involves a reaction space of $5 - R(\mathbf{M12}) = 2$ dimensions. The first three columns of U (those corresponding to the non-zero elements of **W** constitute an orthonormal basis for composition space; the last two columns of V (corresponding to the zero elements of **W**) give the coefficients of two

Table 1. Compositions of minerals in hypothetical SiO_2-MgO-FeO assemblages

		qz	op	ol	ox
M1	SiO_2	1.000	1.000	1.000	
	MgO	0.000	0.600	0.750	
	FeO	0.000	0.400	1.250	
M2	SiO_2		1.000	1.000	
	MgO		0.400	0.700	
	FeO		0.600	1.300	
M3	SiO_2		1.000	1.000	
	MgO		0.300	0.500	
	FeO		0.700	1.500	
M4	SiO_2	1.000		1.000	0.000
	MgO	0.000		1.000	0.325
	FeO	0.000		1.000	0.675

ol = olivine, op = orthopyroxene, ox = oxide, qz = quartz; compositions expressed in oxide formula units (*e.g.* $op_1 = Mg_{0.6}Fe_{0.4}SiO_3$).

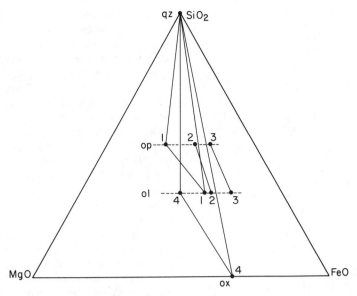

FIG. 2. The assemblages of Table 1 plotted in SiO_2-MgO-FeO space. Abbreviations are ol = olivine, op = orthopyroxene, ox = oxide, and qz = quartz; numerals show assemblage to which minerals belong (*e.g.* op_1 belongs to M1).

reactions which provide an orthonormal basis for reaction space:

V4: $0.686 \, ol_1 + 0.172 \, op_2$

$$= 0.172 \, op_1 + 0.686 \, ol_2,$$

V5: $0.376 \, qz_1 + 0.111 \, op_1 + 0.310 \, ol_1$

$$+ 0.066 \, ol_2 = 0.863 \, op_2,$$

where negative coefficients in the basis vectors are taken to represent reactants, and positive coefficients products.

V4 has phases of both M1 and M2 as both reactants and products, and V5 has phases from both assemblages as reactants, so neither represents an incompatibility between the two assemblages. But we need to determine whether reactions reflecting an incompatibility exist anywhere in the space spanned by V4 and V5.

The reaction coefficients of each mineral are obviously linear functions of the basis vectors V4 and V5, and so can be represented by contours in V4–V5 space (Fig. 3). The five zero contours in Fig. 3 have special significance. They represent lines in V4–V5 space where each of the five phases of the matrix **M12** have zero coefficients, and do not participate in the reaction. They therefore involve only four phases in the system SiO_2-MgO-FeO and correspond to the five univariant reactions defined by

the five sets of four phases in **M12**. These zero contours can be labeled by enclosing the phases not participating in each reaction in brackets, as is commonly done with univariant reactions, and the reaction coefficients can be calculated from **M12** using the method of KORZHINSKII (1959, p. 103ff) or computer programs such as MULTI of FISHER (1989):

$[qz_1]$ $4.00 \, ol_1 + 1.00 \, op_2 = 1.00 \, op_1 + 4.00 \, ol_2$;

$[op_1]$ $1.00 \, qz_1 + 2.00 \, ol_1 = 2.00 \, op_2 + 1.00 \, ol_2$;

$[ol_1]$ $2.50 \, op_2 = 1.00 \, qz_1 + 0.50 \, op_1 + 1.00 \, ol_2$;

$[op_2]$ $2.00 \, op_1 + 9.00 \, ol_2 = 1.00 \, qz_1 + 10.00 \, ol_1$;

$[ol_2]$ $1.00 \, qz_1 + 0.25 \, op_1 + 1.00 \, ol_1 = 2.25 \, op_2$.

Each zero contour or reaction line in Fig. 3 has a positive and a negative direction, characterized by reaction coefficients of opposite sign. We are free to designate either direction as positive. However, it will be convenient to define the positive direction of one reaction arbitrarily, then let the positive directions of the other reactions be established by placing labels designating products and reactants on opposite sides of each reaction element, and applying the rules used to distinguish stable and metastable portions of univariant reactions in *P-T*-activity diagrams (SCHREINEMAKERS, 1915–1925;

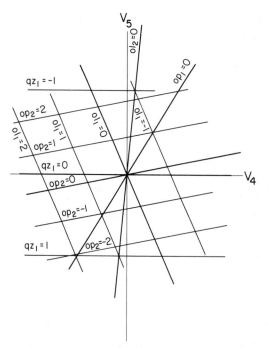

FIG. 3. V4–V5 space, contoured to show values of reaction coefficients. Zero contours (heavy lines) shown for all phases; non-zero contours (light lines) are omitted for op_1 and ol_2 for clarity.

cording to the convention adopted above—must be a product in all reactions below $[qz_1]$. Consequently, reactions at points in reaction space just below $[qz_1]$ must have the form $ol_1 + op_2 = qz_1 + op_1 + ol_2$. Similar reasoning shows that reactions immediately to the right of $[ol_1]$ have the same form, so reactions of this form characterize the entire space between $[ol_1]$ and $[qz_1]$.

Of the five reactions in Fig. 4, two ($[op_1]$ and $[ol_2]$) have the minerals of M1 and M2 on opposite sides. They are therefore incompatibility reactions, and confirm algebraically that M1 and M2 do intersect in composition space. Reactions in the sector between these two reactions are also incompatibilities, as can be seen by inspection of the coefficients of $[ol_2]$ and $[op_1]$. Ol_2 is a product of reaction $[op_1]$, and so must be a product in all parts of reaction space to the right of $[ol_2]$; consequently reactions immediately right of $[ol_2]$ must have the form of the incompatibility reaction $qz_1 + op_1 + ol_1 = op_2 + ol_2$. Likewise, op_1 is a reactant of $[ol_2]$, and so must be a reactant in reactions to the left of $[op_1]$. Consequently, reactions immediately to the left of $[op_1]$ represent the same incompatibility reaction.

Similar reasoning shows incompatibility reactions occur *only* between $[op_1]$ and $[ol_2]$. The sectors bounded by $[op_1]$ or $[ol_2]$ and one compatibility

ZEN, 1966).[1] Because $[qz_1]$ and V4 are identical, it is sensible to choose the positive direction of $[qz_1]$ to coincide with the positive direction of V4. With this convention, Schreinemakers' rules are obeyed if and only if reactants are written on the counterclockwise side of each reaction, and products on the clockwise side (Fig. 4).

The reactions plotted in Fig. 4 divide the plane representing V4–V5 space into 10 sectors, each bounded by the positive or negative ends of two reactions. The character of the reactions within each sector can be inferred by inspection of the bounding reactions. For example, consider reactions in the space between $[qz_1]$ and $[ol_1]$. The line $[qz_1]$ represents the zero contour for the qz_1 coefficient. Reactions plotting just below $[qz_1]$ have a form similar to $[qz_1]$, but involve a small amount of qz_1. Because qz_1 lies on the clockwise side of reactions $[ol_1]$ and $[op_2]$, it is a product in those reactions, and—ac-

[1] There is no requirement that the positive ends of reactions in reaction space be defined by the Schreinemakers rule, because they are not constrained by any condition comparable to that of minimizing Gibbs free energy (cf. ZEN, 1966, p. 7–9). However, there is no reason not to adopt the convention used here, and doing so leads to an internally consistent way of writing reactions which is useful in analyzing the geometry of reaction space.

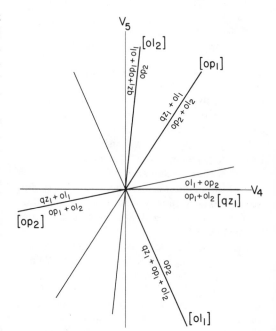

FIG. 4. Configuration of zero contours of reaction coefficients for **M1M2** composite assemblage in V4–V5 space, drawn using conventions adopted in text. Positive ends of zero contours shown by heavy lines, negative ends by light lines.

reaction contain only compatibility reactions, as can be seen by considering the space just to the right of [op$_1$]. Reactions in this area have coefficients similar to [op$_1$], but with small op$_1$ coefficients. Because op$_1$ is a product in [qz$_1$], it must be a product everywhere to the right of [op$_1$], and reactions just to the right of that reaction must have the form qz$_1$ + ol$_1$ = op$_1$ + op$_2$ + ol$_2$, a compatibility reaction. Comparable arguments show that compatibility reactions characterize reaction space immediately to the left of [ol$_2$].

The form of the reactions in each sector remains constant so long as no zero coefficient contours (either positive or negative) are crossed. But because one coefficient must change sign as each zero contour is crossed, the character of reactions must change whenever one is crossed. For example, the large arc between [ol$_1$] and [op$_2$] is made up of three sectors, two outer ones in which the reactions are compatibilities, and a central sector in which the reactions are incompatibilities. This central sector is simply the negative portion of the sector between [ol$_2$] and [op$_1$], analyzed above.

These arguments demonstrate that the reaction space defined by assemblages M1 and M2 contains incompatibility reactions in the sectors between both positive and negative portions of [op$_1$] and [ol$_2$], and only in those sectors; all other portions of reaction space are occupied by compatibility reactions.

Reactions between M1 and M3

To search for incompatibilities between M1 and M3, we form a composite matrix giving the composition of the minerals of both M1 and M3 and take its SVD, which gives a reaction space with two basis vectors,

V4′:   0.472 qz$_1$ + 0.449 ol$_1$ + 0.023 ol$_3$

$$= 0.217 \, op_1 + 0.727 \, op_3;$$

V5′:   0.543 ol$_1$ + 0.453 op$_3$

$$= 0.453 \, op_1 + 0.543 \, ol_3.$$

Using the conventions adopted above, we divide this reaction space into sectors bounded by four-phase reactions (Fig. 5). All the reactions of Fig. 5 involve phases of both assemblages as reactants or products, and so represent compatibilities. Because compatibility reactions cannot change to incompatibilities when another phase is added, all sectors of reaction space in Fig. 5 must contain compatibility reactions. This observation confirms algebraically the message conveyed by Fig. 2; M1 and M3 do not intersect in composition space.

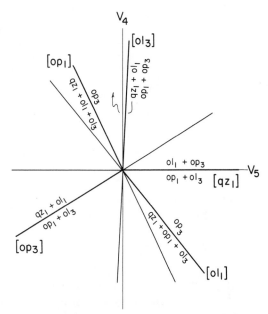

FIG. 5. Configuration of zero contours of reactions coefficients for **M1M3** composite assemblage in V4–V5 space. Positive ends of zero contours shown by heavy lines, negative ends by light lines. Note that the positions of V4 and V5 are reversed from that of Fig. 4, so as to maintain [qz$_1$] in the same orientation.

The reasons for the different character of reactions in **M1M2** and **M1M3** can be seen by comparing the chemography of M1, M2 and M3 (Fig. 2) with the form of reaction space (Figs. 4 and 5). As op$_2$ approaches the qz$_1$-ol$_1$ tie line in composition space, reactions [op$_1$] and [ol$_2$] of Fig. 4 approach one another in reaction space, narrowing the sectors characterized by incompatibilities. When the composition of op$_2$ becomes collinear with qz$_1$ and ol$_1$, reactions [op$_1$] and [ol$_2$] become degenerate and merge, eliminating the sector occupied by incompatibilities altogether. Once op$_2$ moves outside the qz$_1$-op$_1$-ol$_1$ triangle, reaction [op$_1$] moves to the left of [ol$_2$], giving the reaction space geometry of Fig. 5, which contains compatibility reactions only. The key point here is that the univariant reactions change from incompatibilities to compatibilities at precisely the point where the sector of reaction space characterized by incompatibility reactions disappears.

Reactions between M4 and M2

To search for incompatibilities between M4 and M2, we form a composite matrix giving the composition of the minerals of both M4 and M2 and take its SVD, which gives a reaction space with two basis vectors,

V4″: $0.609\ op_2 + 0.104\ ol_2$

$$= 0.567\ qz_4 + 0.145\ ol_4 + 0.526\ ox_4;$$

V5″: $0.593\ ox_4 + 0.593\ op_2$

$$= 0.049\ ol_4 + 0.543\ ol_2.$$

Using the conventions adopted above, we divide this reaction space into sectors bounded by four-phase reactions (Fig. 6). Two of these reactions ([op₂] and [ol₂]) are incompatibility reactions, reflecting the fact that M4 encloses M2 in composition space (Fig. 2). Inspection of reaction coefficients shows that two sectors of reaction space are characterized by incompatibility reactions: that between the positive end of [op₂] and the negative end of [ol₂], and that between the positive end of [ol₂] and the negative end of [op₂].

The differences between the reaction space geometries of Figs. 4 and 6 can be understood by imagining the changes that would occur were the compositions of op_1 and ol_1 to approach the compositions of ol_4 and ox_4, respectively, so that ol_2 crosses into the qz_1-op_1-ol_1 triangle across the qz_1-ol_1 tie line (Fig. 2). Were this to happen, reaction [op₂] of Fig. 4 would rotate counterclockwise and merge with the negative end of [op₁] when ol_2 becomes collinear with qz_1-ol_1 (to avoid a further degeneracy, we assume that ol_1 becomes deficient in SiO_2). The reaction rotates further when ol_2 enters the M1 phase triangle, and the reactions assume the configuration of Fig. 6. At the degeneracy the form of both reactions changes from A + B = C + D to A + B + C = D, and [op₁] becomes a compatibility, while [op₂] becomes an incompatibility. Consequently the sector of reaction space containing incompatibilities becomes bounded by the reactions [op₂] and [ol₂] rather than by [op₁] and [ol₂].

This example shows how the differences in chemography of M1 + M2 and M4 + M2 are reflected in the form of the incompatibility reactions. In M4 + M2, both M2 minerals are enclosed in the M4 triangle, so both incompatibilities are terminal reactions. In M1 + M2, only op_2 is enclosed in the M1 triangle, and only op_2 participates in a terminal reaction; the form of the incompatibility involving ol_2 is a direct reflection of the tie line intersections of Fig. 2.

DISCUSSION AND CONCLUSIONS

These examples demonstrate four points governing relations between the univariant reactions defined by a composite mineral assemblage and the vector space representing reactions within that composite.

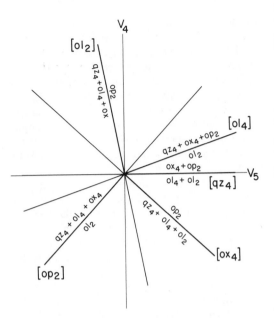

FIG. 6. Configuration of zero contours of reaction coefficients for **M4M2** composite assemblage in V4–V5 space. Positive ends of zero contours shown by heavy lines, negative ends by light lines. Note that the positions of V4 and V5 are reversed from that of Fig. 4, so as to maintain [qz₁] in the same orientation.

1. The univariant reactions represent lines in reaction space along which one or more reaction coefficients change sign, and reactions change character; they divide reaction space into sectors characterized by reactions of uniform character.

2. If the assemblages of the composite do not overlap in composition space, none of the univariant reactions represent incompatibilities, and no sectors of reaction space contain incompatibility reactions.

3. If the assemblages do overlap in reaction space, the univariant reactions include incompatibility reactions, and those reactions bound sectors of reaction space characterized by incompatibilities.

4. As changes in mineral composition narrow the overlap of assemblages in composition space, the sectors of reaction space characterized by incompatibilities also narrow. At the point where mineral compositions become collinear and univariant reactions become degenerate, the part of reaction space characterized by incompatibilities collapses into that one degenerate reaction. Any further compositional changes eliminate the collinearity and the degeneracy (thereby eliminating the last vestiges of reaction space characterized by incompatibilities) and simultaneously change the univariant reactions to compatibilities.

In addition, the nature of any overlap in composition space is reflected by the form of incompatibility reactions: minerals of one assemblage enclosed in the polyhedron defined by the minerals of the other assemblage participate in terminal reactions, while those outside the polyhedron are involved in reactions whose form reflects intersections of tie lines or tie surfaces.

These conclusions emerged from consideration of a three-component composite assemblage defining a two-dimensional reaction space, but they are easily extended to include multicomponent assemblages and multidimensional reaction space.

In a reaction space of three dimensions, the zero contour lines of Figs. 4, 5 and 6 become zero contour surfaces in that reaction space. Those surfaces intersect in lines along which two minerals have zero reaction coefficients; those lines correspond to univariant reactions, and within each zero contour surface, they have properties exactly like those of Figs. 4, 5 and 6, dividing each contour surface into sectors characterized by compatibility reactions or incompatibility reactions. These sectors then govern the character of reactions in the volumes between the surfaces just as the univariant lines of Figs. 4, 5 and 6 govern the character of reactions in the intervening sectors. And just as in the examples discussed above, changes in the chemography of composition space produce changes in reaction character as compositional collinearities lead to reaction degeneracies, reflected by the collapse of three-dimensional elements of reaction space into surfaces and lines.

Relations in reaction spaces of more than three dimensions are still more complex. But despite these complexities and the difficulty of visualizing either compositional relations in multicomponent assemblage diagrams or univariant reactions in multidimensional reaction space, the mathematical relations between mineral composition and reaction stoichiometry which give rise to the points summarized above still hold true. Univariant reactions defined by a composite assemblage are lines in reaction space along which the reaction coefficients of one or more minerals equal zero. They bound multi-dimensional sectors of reaction space, and the character of reactions in those sectors ultimately reflects the character of the bounding univariant reactions just as in the simple examples analyzed above. Consequently, even the most complex reaction space can be explored by systematically investigating the form of the univariant reactions defined by a composite mineral assemblage.

Acknowledgements—I am grateful to Jack Rice for having raised the question that lead to the writing of this paper, and to Hugh Greenwood and Page Chamberlain for careful reviews of a preliminary version of this manuscript. The research reported here was supported by NSF grant EAR-8903954.

REFERENCES

FISHER G. W. (1989) Matrix analysis of metamorphic mineral assemblages and reactions. *Contrib. Mineral. Petrol.* **102,** 69–77.

GREENWOOD H. J. (1967) The *N*-dimensional tie line problem. *Geochim. Cosmochim. Acta* **31,** 467–490.

KORZHINSKII D. S. (1959) *The Physico-Chemical Basis of the Analysis of the Paragenesis of Minerals.* 142 pp. Consultants Bureau, New York.

PRESS W. H., FLANNERY B. P., TEUKOLSKY S. A. and VETTERLING W. T. (1986) *Numerical Recipes.* 919 pp. Cambridge University Press, London.

SCHREINEMAKERS F. A. H. (1915–1925) In-, mono-, and divariant equilibria. *Koninkl. Akad. Wetenschappen te Amsterdam Proc.*, English ed., **18–28** (a series of 29 articles).

THOMPSON J. B. JR. (1982) Reaction space: an algebraic and geometric approach. In *Characterization of Metamorphism Through Mineral Equilibria*, (ed. J. M. FERRY), Reviews in Mineral. 10, Ch. 2, pp. 33–52. Mineral. Soc. Amer.

ZEN E-AN (1966) Construction of pressure-temperature diagrams for multicomponent systems after the method of Schreinemakers—a geometric approach. *U.S. Geol. Surv. Bull.* 1225.

Part C.
Hydrothermal and Ore Systems

Fluid-Mineral Interactions: A Tribute to H. P. Eugster
© The Geochemical Society, Special Publication No. 2, 1990
Editors: R. J. Spencer and I-Ming Chou

Ironstones of mixed sedimentary and hydrothermal origin in the Archean greenstone belt at Bird Lake, Manitoba

ALLAN C. TURNOCK and DAVID L. TRUEMAN*

Department of Geological Sciences, University of Manitoba, Winnipeg R3T 2N2, Canada

Abstract—This paper describes the forms and associations of aluminous ironstones in volcaniclastic comglomerates in a zone of proximal felsic volcanism, and from 14 bulk rock analyses and element correlations we assign Fe, Mn, Mg, Ca, to a chemical precipitate-exhalative origin, Al, Zr, K, Rb, Si, to a clastic felsite origin, alkali losses to hydrothermal leaching, and variable Ti, Cu, Zn, Mo, Co, V, to unexplained diagenesis.

Iron formations of three facies, chert banded silicate, sulfide ironstone, and aluminous ironstones, are found in an area 1 × 2 km of "Algoma-type" association, with clastic sedimentary rocks and felsic volcanics.

The aluminous ironstones contain iron (as FeO) 16 to 47 wt%. They are garnet + cummingtonite + biotite + hornblende as staurolite-grade metamorphic minerals. They occur as (1) beds and lenses 2 to 60 cm thick, 1 to 30 m long, interbedded in conglomerates; (2) matrix in bimodal conglomerates, *i.e.* mafic matrix to felsite fragments. The mafic matrix has a patchy distribution in conglomerates which have felsic fragments and felsic matrix; (3) filling fractures in a dome of QFP (quartz-felspar-porphyry), that has intruded explosively into the floor of the basin, and, (4) veins (rare) that cut across psammitic beds in the area at the flank of the QFP dome. Types 1 and 2 are mixtures of Fe and the debris of the felsic volcanic rocks. Types 3 and 4 are deposition in fractures from hydrothermal solutions; their presence strengthens the theory of hydrothermal origin.

INTRODUCTION

IRON FORMATIONS are widely distributed in the supracrustal greenstone belts of Archean age of the Canadian Shield (GROSS, 1965; GOODWIN, 1973). The chert-banded-oxide-facies is most widespread, but the other facies (sulfide, silicate, carbonate) are found, especially in the proximal volcanic association (SHEGELSKI, 1975; GROSS, 1980). Seeking to characterize the basins of deposition, GROSS (1980) defined "Algoma type" iron formations as those associated with volcanic and detrital rocks, and in the classification of KIMBERLEY (1978) they are type SVOP-IF (shallow-volcanic-platform iron formation). The source of the iron is attributed to the exhalative output of hydrothermal systems that go with volcanic processes (RIDLER, 1971; GOODWIN, 1973; SHEGELSKI, 1975; GROSS, 1980). All iron formations, including the sulfide facies, are exhalative components in the syngenetic theory of origin of stratiform sulfide deposits (RICHARDS, 1960; HUTCHINSON, RIDLER and SUFFEL, 1971; STANTON, 1976; STANTON and VAUGHAN, 1979).

The intent of this paper is to describe an "Algoman" association of iron formations, felsic volcanics, and detrital sediments, and, for the aluminous ironstones, to try to identify the origin of the elements as belonging to an exhalative component

and a detrital component. This is done by examining mixing relationships and correlations in the chemical analyses of 14 ironstones and 4 ferruginous shales from an area 0.5 × 1 km of an Archean greenstone belt, at a felsic volcanic center.

Following the generalized nomenclature of KIMBERLEY (1978), "ironstone" is a sedimentary rock which contains Fe > 15 wt.% (=FeO* > 19.3); and "iron formation" is a mappable rock unit composed mostly of ironstone. An ironstone is here called "aluminous" if Al_2O_3 > 4 wt.%, an arbitrary limit based on inspection. (Note: an asterisk (*) following Fe* or FeO* indicates total iron as either element or oxide).

GENERAL GEOLOGY

The Bird River greenstone belt is one of the east–west trending belts of the Archean Superior Province (Fig. 1A). It is infolded into the English River subprovince between the Manigotagan-Ear Falls paragneiss belt and the Winnipeg River batholithic belt (BEAKHOUSE, 1977). BEAKHOUSE (1985) noted only slight differences between the Bird River belt and other greenstone belts of the province, in greater abundance of felsic volcanics and clastic sedimentary rocks in the Bird River belt.

TRUEMAN (1980) has mapped the Bird River belt and discussed the origin of the major rock types and their tectonic history. The supracrustal formations (Fig. 1B) include the Lamprey Falls me-

---

* *Present address:* 5433 Collingwood St., Vancouver V6N 1S9, Canada.

144 A. C. Turnock and D. L. Trueman

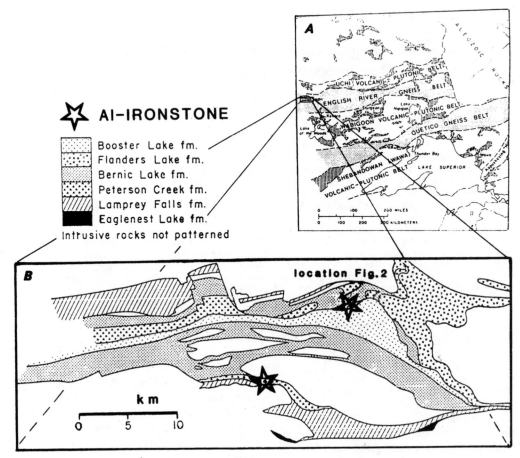

FIG. 1. A. Location of the Bird River Subprovince in the Superior Province of the Precambrian Shield of Canada. B. Map of the lithostratigraphic formations of the Bird River greenstone belt.

tabasalts and the Petersen Creek metarhyolites, a bimodal volcanic suite, and their disposition defines a major east–west synclinorium. The overlying Bernic Lake Formation includes conglomerates, psammites, iron formations, and minor intermediate volcanics. All these formations were deposited subaqueously.

These rocks have been complexly folded and recrystallized at medium grade of metamorphism, and intruded by pre-, syn-, and late-tectonic granitoid stocks and batholiths (TRUEMAN, 1980). There is a dominant east–west S2 foliation. Refolded F1 folds are found in the north-east portion of Fig. 1B, and they are a complication in the study area. It is here assumed that the primary compositions of cation ratios were not changed by metamorphism.

THE BIRD LAKE IRON FORMATIONS

Two localities of abundant iron formations have been found, both are in areas of mostly felsic vol-

canic rocks and thus mapped as Petersen Creek Fm. (Fig. 1B).

At the Bird Lake locality, three types of iron formations are found interbedded in volcaniclastic conglomerates and psammites, and felsic volcanic rocks (Figs. 2 and 3). They are: (1) Garnet ironstone, found mostly in the lower part of the section (Fig. 3) near the QFP dome, and described below; (2) Sulfide ironstone, traced by magnetometer for 1.5 km as the "SI ANOMALY" across the center of the section, in a bed 7 to 15 m thick (3 cored drill holes, courtesy Tanco Corp.). It is pyrrhotite, with irregular inclusions of garnet-rich aluminous ironstone. (3) CBSIF = chert banded silicate iron formation, found mostly at the top of the section in a bed 50 to 100 m estimated thickness. The outcrop section is thicker, but the individual bands, thickness 1 to 18 cm, are complexly folded in the style called "soft-sediment slump", and this is structural thickening. The silicate bands are mostly por-

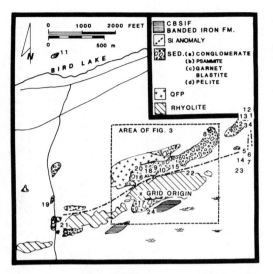

FIG. 2. Area south of Bird Lake, with numbered sample sites. GRID ORIGIN is for sample locations, which are given in Table 1 in feet from origin. CBSIF = chert banded silicate iron formation. SI = sulfide ironstone (anomaly location from magnetometer survey). QFP = quartz feldspar porphyry.

phyroblastic grunerite plus rare actinolite and sulfide, or, hornblende plus cummingtonite. A chemical analysis of a grunerite band is given (Table 1B, #24).

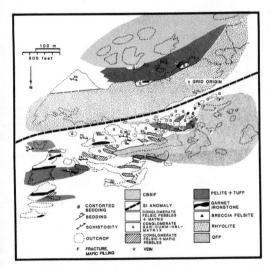

FIG. 3. Detail of the area marked near the center of Fig. 2. CBSIF = chert banded silicate iron formation. SI = sulfide ironstone (anomaly location from magnetometer survey). QFP = quartz feldspar porphyry. GRID ORIGIN is reference 0E, 0N, for locations, distances in feet in text. The map is oriented with north down, section facing up, based on intrusive structures of QFP, and rare graded beds.

Table 1A. List of samples. Analyses in Table 1B

Sample	No.	Location†	Type and reference
1	1–77	3200E, 2280N	ironstone bed
2	1–127	3200E, 2250N	ironstone bed
3	2–201	3540E, 2000N	ironstone bed
4	2–505	3751E, 2000N	ironstone bed
5	3–237	3200E, 1134N	ironstone bed
6	3–267	3200E, 1113N	ironstone bed
7	3–276	3200E, 1107N	ironstone bed
8	2854b	1980E, 1700N	ironstone bed
9	2653d	480E, 600N	ironstone bed
10	2875a	480E, 600N	ironstone bed
11	2844	2700W, 4100N	ironstone bed
12	1–34	3200E, 2325N	ironstone in conglomerate
13	1–45	3200E, 2300N	ironstone in conglomerate
14	3–384	3200E, 1031N	ironstone in conglomerate
15	2877	1100E, 600N	ironstone in conglomerate
16	2845	0E, 500N	ironstone in conglomerate
17	2872a	0E, 245S	ironstone in conglomerate
18	2886a	45W, 610N	ironstone in conglomerate
19	2649b	2720W, 60S	ferruginous pelite
20	2846	0E, 760N	felsite
21	2652c	2400W, 650S	felsite
22	2876	1600E, 600N	felsitic cobble in conglomerate
23	2866	2800E, 700N	schist, graphite + sulfide
24	2874a	190E, 400S	grunerite layer in BIF
25	FeR-1	Bathurst N.B.	BIF, magnetite, Algoma type*
26	FeR-2	Bruce L., Ont.	BIF, magnetite, Algoma type*
27	FeR-3	Temagami, Ont.	BIF, magnetite, Algoma type*
28	FeR-4	Temagami, Ont.	BIF, magnetite, Algoma type*
29	DG-8	Wallace L., Man.#	chloritic layer, magnetite BIF
30	1530b	Snow L., Man.	chloritic altered rock

* Ref. ABBEY et al. 1983.
† Shown in Fig. 2, in feet from grid origin.
# From GABOURY 1984.

By association of the iron formations with volcanics and clastic sediments, they are the "Algoma-type" of GROSS (1980). The felsite dome with breccia top, the tabular, stratiform rhyolite, and angular felsite-block conglomerates (Fig. 3), define a "proximal" and explosive volcanic environment. The occurrence of multifacies of iron formations is also a characteristic of proximity to volcanic centers (GROSS, 1980).

The sedimentary rocks are mostly conglomerates, with clast-supported frameworks and angular fragments. Sorting is absent, and bedding can be

Table 1B. Chemical composition of ironstones

	1	2	3	4	5	6	7	8	9	10	11	12	13	14	15
wt.%															
SiO_2	50.67	36.41	38.95	46.52	65.59	45.98	40.96	27.09	49.94	56.62	45.65	38.80	46.16	45.72	54.58
Al_2O_3	9.18	6.85	13.03	8.07	6.60	2.33	0.55	4.84	13.51	9.85	15.20	8.54	8.48	0.72	14.25
Fe_2O_3	5.44	29.69	0.73	9.25	1.82	23.09	0.64	30.19	4.06	8.59	3.86	14.55	4.33	1.83	2.64
FeO	21.88	14.48	34.00	21.92	15.14	13.80	39.62	19.68	22.72	19.24	26.42	25.28	26.28	35.66	16.84
MgO	2.86	2.77	2.46	3.72	2.84	4.46	3.93	5.79	3.01	2.80	2.43	2.86	2.86	4.37	2.78
CaO	2.58	1.00	3.76	4.65	2.34	4.68	5.78	5.83	1.91	1.75	1.99	1.61	4.51	4.09	2.49
Na_2O	0.08	0.15	0.03	0.26	0.02	0.04	0.01	0.16	0.08	0.01	0.29	0.04	0.38	0.00	1.24
K_2O	1.82	0.92	0.12	1.38	2.36	0.63	0.03	0.01	2.36	0.34	0.64	1.68	0.90	0.05	1.14
H_2O^+	2.37	1.78	1.40	1.82	1.44	0.92	1.73	1.30	1.53	2.78	0.91	1.66	1.36	1.94	2.20
CO_2	0.44	0.01	1.41	0.50	0.12	1.00	0.66	0.54	0.54	0.34	0.29	0.18	0.25	0.02	0.35
TiO_2	0.10	0.16	0.61	0.21	0.12	0.17	0.01	0.20	0.41	0.22	0.74	0.37	0.19	0.01	0.55
P_2O_5	0.03	0.36	0.21	0.05	0.03	0.03	0.00	0.02	0.07	0.02	0.02	0.08	0.07	0.01	0.06
MnO	2.35	2.81	2.00	1.80	1.04	2.44	2.17	2.87	1.01	1.54	0.75	2.37	2.21	2.60	0.47
S	0.35	3.49	1.25	0.47	1.03	0.64	6.56	0.04	0.007	0.05	0.124	3.62	2.74	4.72	0.205
ppm															
Zr	82	76	74	70	100	32	9	32	114	94	57	135	140	11	155
Li	24	13	19	19	35	15	7	2	47	6	39	19	12	5	19
Ba	158	169	157	65	—	122	60	119	184	99	160	235	124	64	240
Be	—	—	1	—	<1	2	—	<1	1	1	1	<1	<1	2	1
Co	3	6	42	42	4	3	4	7	41	6	47	25	14	7	17
Cu	18	21	76	73	95	46	31	34	17	101	120	38	37	41	182
Mo	5	6	11	3	3	12	—	3	7	9	7	11	12	<1	—
Ni	3	26	71	75	12	51	30	18	82	13	88	43	29	18	41
Pb	—	—	—	—	—	—	—	—	—	—	—	—	—	—	—
Rb	73	34	10	10	121	36	1	2	98	22	70	32	28	4	41
V	—	29	109	—	107	1	1	24	35	6	183	60	2	—	64
Zn	17	33	55	93	39	53	41	64	124	84	60	93	37	34	118

	16	17	18	19	20	21	22	23	24	25	26	27	28	29	30
wt.%															
SiO_2	61.44	50.70	48.13	58.86	71.90	73.78	75.28	86.29	51.50	16.99	49.03	53.51	50.41	59.93	27.26
Al_2O_3	11.16	11.25	8.78	12.31	13.87	14.30	13.58	4.43	1.04	0.55	5.20	0.03	1.75	15.78	19.76
Fe_2O_3	4.06	4.71	3.07	3.65	1.12	0.85	1.51	2.76	3.82	49.93	22.53	28.65	23.23	2.52	4.82
FeO	12.70	19.32	26.78	12.88	2.74	1.16	0.84	0.62	31.64	23.20	14.85	14.10	15.10	10.96	14.76
MgO	2.77	3.19	3.07	3.33	0.73	0.54	0.51	0.38	3.54	0.24	2.02	0.96	1.39	1.43	18.26
CaO	0.31	4.92	5.02	5.49	2.46	0.46	0.29	0.94	3.17	3.37	2.18	0.89	2.29	3.83	1.51
Na_2O	0.03	1.55	0.13	0.58	4.01	4.96	3.76	0.66	0.02	0.05	0.48	0.03	0.04	1.43	0.11
K_2O	4.31	0.74	0.23	0.78	1.48	2.43	2.88	0.34	0.04	0.01	1.29	0.03	0.26	0.49	0.28
H_2O^+	1.87	1.47	1.69	1.10	0.78	0.81	1.00	0.64	2.55	0.32	0.93	0.25	0.83	3.31	10.36
CO_2	0.25	0.25	0.28	0.16	0.27	0.09	0.11	0.28	1.38	1.38	0.04	1.21	4.88	0.10	0.94
TiO_2	0.24	0.41	0.25	0.52	0.42	0.16	0.14	0.18	0.02	0.03	0.21	0.01	0.07	0.37	0.68
P_2O_5	0.01	0.11	0.06	0.14	0.08	0.02	0.02	0.04	0.02	2.37	0.32	0.07	0.12	0.12	0.03
MnO	0.86	1.16	1.90	0.45	0.18	0.01	0.03	0.23	1.99	0.26	0.12	0.08	0.20	0.20	0.47
S	0.039	0.002	0.015	0.014	0.006	0.004	0.003	0.324	0.06	0.26	0.18	0.02	0.09	0.006	0.481
ppm															
Zr	131	92	74	152	294	213	182	161	3	10	35	—	51	223	171
Li	54	24	5	12	18	21	22	4	<1						46
Ba	136	57	35	31	121	80	90	240	87	1030	229			206	117
Be	<1	1	1	<1	1	<1	1	<1	1						1
Co	17	11	9	27	6	6	8	5	1					4	6
Cu	27	15	18	133	13	15	13	17	12					11	53
Mo	1	4	5	2	6	—	<1	19	3					3	13
Ni	18	39	26	89	10	1	7	3	7					18	2
Pb	—	—	—	—	—	9	6	4	—						34
Rb	231	23	14	19	62	109	111	6	2					15	7
V	11	40	—	56	—	—	—	63	—	95	34	8	10	28	40
Zn	114	61	57	108	32	34	29	10	21					34	2040

Table 1C. Correlation of elements† $r > 0.55$

	Positive		Negative	
	#1–11	#1–30	#1–11	#1–30
Si	Rb, K, Zr, Li	Zr, Na	Fe^*, Fe^3, Mn	Fe^*, Fe^3
Al	Co, V, Zr, Li, Ba	Zr, Li	Mg, S, Ca	
Fe^3	Mg	P, Ba, Fe^*	Fe^2, Si	Si
Fe^2		Mn, Fe^*	Fe^3	Zr, Si, Na
Fe^*	Mn	Fe^2, Fe^3	Si	Si, Zr, Na
H_2O		Zn, Mg	Ti	
CO_2	Mo			
Ti	Mo, Ni		Mn, H_2O	
P		Ba, Fe^3		
Mn	Fe^*	Fe^2	Ti, Si	
Zr	K, Al, Rb, Si, Li, Ba	Al, Na, Si	Ca, Mg, S	Fe^2, Fe^*
Rb	K, Li, Si, Ba, Zr	K, Li	Ca, Mg	
K	Rb, Li, Zr, Si	Rb, Li		
Na	Co, V	Zr, Si		Fe^2, Fe^*
Li	Rb, K, Al, Ba, V, Zr, Si, Co	Rb, Al, K	Mg, Mn	
Mg	Ca, Fe^3	Zn, H_2O	Zr, Al, Li, Rb	
Ca	Mg		Zr, Ba, Al, Rb	
Ba	Rb, Al, Li, Zr	P, Fe^3, Mo	Ca	
S			Al, Zr	
Co	Ni, Al, V, Zn, Li	Ni		
Cu	V			
Mo	Ti, CO_2	Ba		
Ni	Co, Ti, Zn	Co		
V	Cu, Co, Al, Li			
Zn	Ni, Co	H_2O, Mg		

† SAS 1983 proc corr.
Fe^* = total iron.

mapped only by the presence of clasts of mafic volcanic rock in uncommon beds (Fig. 3). These beds are lensoid, mostly 5×50 m, rarely 10×100. Uncommon psammitic and pelitic beds are lenses less than 100 m in length (Fig. 3). Such sediments are high-energy types, suggesting active faulting. The clasts are volcanic, there is no evidence here for a hinterland of gneissic or granitic rocks. The clasts are mostly felsic volcanic rock, plus mafic in some beds (Fig. 3), plus rare cobbles of ironstone and sandstone.

The environment of deposition can therefore be interpreted to be a lake, fault-bounded, with internal and external felsic volcanoes. The most extensive beds are the sulfide ironstone and the chert banded silicate iron formation. These chemical precipitates must form in the basin in a short time between episodes of debris flows and volcanism, which requires an input of fluid rich in Fe, S and Si. The fluid is theoretically hydrothermal exhalations, from a major system centered on the submarine volcanoes.

GARNET IRONSTONES

The garnet-rich rocks of the Bird Lake area occur in 4 structures, viz: (1) lensoid beds; (2) matrix of conglomerates; (3) filling fractures in QFP; (4) veins. Sample locations are shown in Fig. 2; Fig. 3 is the portion mapped at $1 = 6000$.

Type 1, lensoid beds

Lensoid beds of ironstone make up ½–1% of a wedge of clastic sediments beside an intrusive dome (QFP) and below a bed of sulfide ironstone (Fig. 3). The longest and most regular is 0.3 m thick and 20 m long, stratiform in shape and in strike parallel to other sedimentary beds. Several smaller ones, which may be disrupted beds, are lenses or irregular pods, with sizes varying from 1×2 m, to 0.2×0.2 m. There are also a few small pods in that part of the section above the rhyolite (top of Fig. 3). These ironstones are aluminous, with 10 to 80% garnet, and Al_2O_3 5 to 15% (Table 1, #1–11), except for #6 and 7; they are cummingtonite ironstone (#6) and a cummingtonite-sulfide ironstone (#7) with Al_2O_3 < 4 wt%.

A lens of aluminous ironstone 1×5 m, surrounded by cm-scale bedded psammites and conglomerates (Fig. 4A), has 60% garnet and has contents of FeO^* (26–27 wt%) and Al_2O_3 (10–14%) (analyses #9 and 10, Table 1). The range in composition of 3 crystals of garnet (#9 in Table 2) are almandine 83 to 90 mol%, spessartine 0 to 7% and zoned, grossularite 1 to 8% with one crystal showing zoning, pyrope 1 to 8% also zoned in one crystal.

Figures 4B + 4C show garnet-rich beds 1 to 4 cm thick, interbedded with psammites. They are <2 m long, and appear to have been contorted and disrupted in a soft-sediment style. The location is on the flank of the QFP dome (the area marked "B" in Fig. 3), and these contortions and differences in strike of the bedding (Fig. 3) is evidence of forcible intrusion of the QFP dome.

Fig. 4D shows a pod of garnetiferous-mafic ironstone 2×3 m, with an unusual lobate contact with QFP (the contact not shown is with conglomerate) plus included fragments of QFP. The lobate edge is the flank of the QFP dome (close to "B" in Fig. 3, and the features shown in Figs. 4B and 4C), and is interpreted as evidence of liquid (the QFP magma) being squeezed into ferruginous mud.

One of the smaller pods of garnet ironstone, 0.2×0.5 m, is shown in Fig. 4E. The irregular shape could be explained by the disruption of a bed by the debris-flow of the surrounding conglomerates, or slumping after their first deposition on a slope. There is chert in this pod, but this is not typical.

FIG. 4. Ironstones in metasedimentary rocks. Locations are given in feet from "GRID ORIGIN" in Figs. 2 and 3. A: Lens of garnet ironstone in conglomerates and sandstones, 465E, 600N. B: Beds 1 to 5 cm thick of garnet ironstone, pelite and metaconglomerate, cut by veins of garnet type 4, 5E, 560N. C: Beds of garnet ironstone, disrupted; 20W, 580N. D: Pod of garnet ironstone with lobate felsite edge, 35W, 585N. E: Pod of garnet ironstone 0.2 × 0.5 m, with chert edge, 825E, 625N.

Elongation of the felsite clasts, and the ironstone pod, is metamorphic S2.

Smaller pods of garnet ironstone, <0.2 m, are found in the conglomerates, where they may be matrix, pieces of disrupted beds, or pebbles.

Type 2, matrix in conglomerates

The garnetiferous, mafic matrix of some conglomerates is aluminous ironstone (Table 1, #12–18). It is therefore part of a mixed chemical and clastic sedimentary rock, with a bimodal composition (mafic matrix vs. felsic clasts).

These bimodal conglomerates are found in patches of irregular size and shape, in conglomerates and usually close to type 1 lensoid beds of garnetiferous ironstone, at locations marked "G" in Fig. 3. They are not large enough to be mappable, but they make up approximately 1% of the sedimentary pile. The clasts are mostly QFP or rhyolites, the lithologies of the surrounding volcanics. There are

Table 2. Garnet analyses, microprobe*

No.[#]	Sample	Location	Type	Alm	Spe	Gro	Pyr
1	1–77	3200E 2247N	bed	77	17	2	4
				62	21	15	1
				61	21	16	3
				58	24	17	1
				55	28	18	0
				64	20	15	1
9	2653d	480E 600N	bed	87	3	8	2
				90	1	8	2
				84	6	8	2
				83	6	8	2
				86	4	8	2
				85	5	8	2
				87	3	8	2
9	2653c	480E 600N	bed	84	5	8	3
				88	2	8	3
				89	1	8	3
9	74–79	480E 600N	bed	90	1	2	8
				89	2	5	4
				85	7	7	1
				90	3	6	2
				90	0	3	6
				90	1	1	7
11	2844	2700W 4800N	bed	90	2	2	5
				90	2	2	6
				90	2	2	6
				90	2	2	5
				86	6	5	3
12	1–34	3200E 2276N	in conglom- erate	67	18	12	2
				44	41	14	0
				45	42	13	0
				60	26	14	0
				77	13	7	3
19	2649c	2520W	pelite	83	6	5	6
				82	7	5	6
				81	8	5	6
				84	5	5	5
—	2654c	100E 660N	pelite	77	18	2	3
				78	16	2	4
				78	16	2	4
				77	17	2	4

* University of Manitoba MAC-5. Beam accelerating voltage 15 kV, at 10 nA. EDS spectra were collected by a Kevex 7000, and corrected by Colby-MagicV program. Analyzed standards: pyrope, grossularite, spessartine, diopside.

[#] The No. refers to bulk chemical analyses in Table 1.

also rare clasts of other types, *i.e.* felsic tuff or sandstone, garnet ironstone, and sulfide, these are intraformational clasts. The primary conglomerates were clast-supported, clast sizes 1 to 300 mm, angular to subrounded; metamorphic flattening has produced minor to severe elongation (Fig. 5A *vs.* Fig. 4A).

The conglomerates which surround them are not bimodal. They are similar in type and shape of clasts, but the matrix of fine-grained quartz and feldspar is felsic and similar in composition to the clasts. These conglomerates also have a clast-supported framework. They are without internal structure, and beds are only rarely outlined by a lens of ironstone or psammite. They are mappable in the lower part of the section (Fig. 3) by the content of clasts of mafic-volcanic rock. These beds are mostly 5 × 20 m, the largest is 20 × 100.

The largest patch of bimodal conglomerate was 5 × 8 m. The smallest was impossible to define, because the host conglomerates include scattered small patches of mafic matrix, sizes in cm and mm, many of which were rounded and are interpreted as intraformational clasts of ironstone. There were no vein structures of garnetiferous material, or continuity of the patches.

The matrix is 5 to 50% garnet, + hornblende, cummingtonite, quartz, biotite. The composition of a garnet (#12 in Table 2) is richer in Mn than the type 1 garnets. The crystal is well-zoned, with contents of spessartine 42 mol% in the core, 13% rim.

Fig. 5A shows a bimodal conglomerate with clast/matrix about 60/40. The clasts are felsite, with continuous size variation from 1 mm to 25 cm. The surface shows mild elongation of the smaller pebbles in metamorphic S2. Fig. 5B shows a similar rock, mafic matrix 35%, in which the clasts are tectonically flattened to 2:1 elongation in S2 (a foliation, strike 90–100°, dip 70–80°S, Fig. 3).

Fig. 5C shows a transition from bimodal conglomerate to brecciated QFP with garnetiferous material filling interstices or fractures. The breccia is at the top of the QFP dome (Fig. 3), and the elongation of the fragments is not tectonic, but due to a primary fracture pattern in the QFP. Fig. 5D is an area adjacent to Fig. 5C, it shows the brecciated QFP with a prominent 15 cm vein of garnet ironstone which contains 25% felsite fragments.

Chemical analyses of mafic matrix were done after cutting or picking out felsite clasts, but clasts smaller than 4 mm were not separated. These compositions (Table 1, analyses #12–18) are similar to those of the lensoid beds (type 1), *i.e.* aluminous ironstones, with two qualifications. Sample #14 is a silicate-sulfide ironstone of cummingtonite and pyrrhotite; this rock type is found in cm-scale lenses in the mafic matrix intermixed with the garnetiferous types. #16 has a content of FeO* (16.3 wt.%) slightly less than the definitive amount for ironstone, and should be called ferruginous pelite.

An hypothesis of origin for these bimodal conglomerates is by the mechanical mixing of felsite

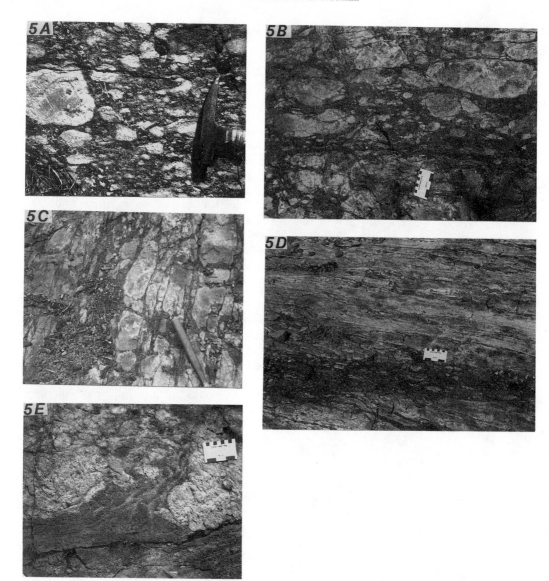

FIG. 5. Ironstone in metaconglomerates and breccias. Locations are given in feet from "GRID ORIGIN" in Figs. 2 and 3. A: Conglomerate, felsic fragments in garnet-mafic matrix, 200E, 450N. B: Same, 100W, 300N. C: brecciated QFP and related conglomerate, both with garnet-mafic matrix, 100W, 300N. D: same, 150W, 300N. E: Flame-like structure of ferruginous pelite bed into felsic conglomerate, 2150W, 815S.

fragments with ferruginous mud. A flame-like structure with swirly mixing lines (Fig. 5E) illustrates the incorporation of mafic mud into a conglomerate. Such a process should be favored by the alternate deposition of a squishy ironstone mud and a conglomerate debris flow. It explains the patchy distribution of the bimodal conglomerate and its association with ironstone lenses and beds.

The origin of bimodal breccias at the top of the QFP dome (Figs. 5C, 5D, 6C) is not known, and may be complex. Ferruginous mud may move down

fractures during brecciation, and/or hydrothermal precipitation and alteration may be active.

An origin of the mafic matrix of conglomerates by hydrothermal alteration of the chloritization type is not favored, because the contents of Fe and Ca are too high, and there is no apparent alteration of the fragments. For example, chloritic pipes at Noranda have "major additions of Fe and Mg, and depletions of Ca, Na and Si" (LYDON, 1984). The chlorite alteration pipe at the Anderson Mine has Fe/Mg = 1 (Table 1, #30), versus ratios of 5 to 10

FIG. 6. Discontinuous fractures and brecciation in QFP, filled with aluminous ironstone. A: 400W, 490N. B: 100W, 600N. C: 100W, 350N. Locations are given in feet from "GRID ORIGIN" in Fig. 3.

for samples 12–18. In contrast to the "siderite enrichment zone" in the alteration pipe at the Mattabi mine (FRANKLIN *et al.,* 1975), the bimodal conglomerates at Bird Lake contain twice as much Fe, and a third as much Ti. Also, in "pseudoconglomerates" the clasts are relics of incomplete alteration. For example, the pseudo-clasts at Sunny Corner are described by SECCOMBE *et al.* (1984): "rounded

blocks of altered acid volcanic rock, within a limonitic matrix . . . are pervasively silicified and rimmed by chalcedony and limonite." At Bird Lake, there is no appearance in the clasts of veins or rims of silicification or clay alteration. Also, a chemical analysis of a felsic clast (Table 1, #22) is similar in all elements to the felsites of the QFP dome (#20) and the rhyolite (#21); it shows no leaching of alkalis or additions of the ironstone hydrothermal elements Fe, Mn, Mg and Ca. Also, there are no obvious channels of hydrothermal flow in the conglomerates.

Bimodal conglomerates may form by mixing of felsic lapilli and basaltic sand in a source area, and be transported as lahars (FRANKLIN, 1976). The ironstone matrix of Bird Lake conglomerates is too low in Ti for this origin.

Type 3, fracture filling

Type 3 structures of garnetiferous rocks are fractures in QFP that are filled with mafic material (garnet, hornblende, cummingtonite, biotite, quartz). The fractures are irregular and discontinuous (Fig. 6A, 6B), in the dome of QFP (Fig. 3), or more orthogonal patterns (Fig. 6C, 5C, 5D) found in the breccia felsite on top of the dome. Both include small patches of ironstone-breccia in wider parts of the fractures (Fig. 5D, 6C). These internal breccias have angular fragments, and are not clast-supported framework types.

The origin of the fractures appears to be from the explosive effects of water getting into the QFP magma as the dome intruded seafloor sediments. The infilling of the fractures by iron-rich material is considered to be contemporaneous with the formation of the fractures, because of suspension of fragments in the breccia patches. These veins are the paths of hydrothermal solutions, but the direction of movement is not known.

Type 4, veins in sedimentary rocks

Type 4 structures of garnetiferous rocks are in rare small veins which cut the bedding of metased-

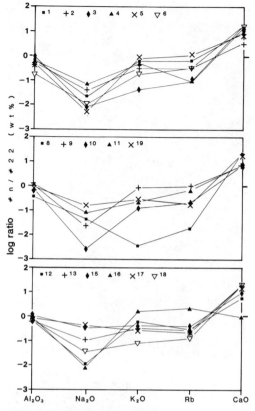

FIG. 8. Comparative contents of alumina, alkalis and lime. Data in wt% (Table 1) normalized against the values for felsite #22. Aluminous ironstones type 1 (#1–4, 8–11) and type 2 (#12–13, 15–18). Pelite (#5). Cummingtonite ironstone (#6).

imentary rocks. They are found in the flank of the QFP dome (the area around "V" in Fig. 3). They are 1 to 5 cm wide, 0.5 to 2 m long, and not continuous or connected (Fig. 4B). These veins strike 130°–160°; in direction they fan out from the QFP dome. They are older than the regional metamorphism on the basis that they have M2 minerals and S2 structures. They are interpreted to have formed during the intrusion of the QFP dome, and to indicate that the period of hydrothermal iron metasomatism was active before and after the intrusion.

ALUMINOUS IRONSTONES: ORIGIN AND CHEMICAL VARIATION

JAMES (1966) noted the association of aluminous ironstones with fine-grained clastic rocks, and concluded that they formed by interaction between iron-rich water and clastic particles. EWERS and MORRIS (1981) proposed the settling of felsitic pyroclastic dust into precipitating iron oxides or carbonates to explain the aluminous ironstone layers

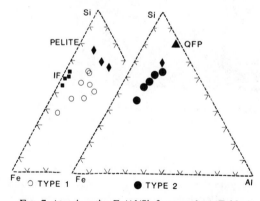

FIG. 7. Atomic ratios Fe/Al/Si, from analyses Table 1. Open circles = aluminous ironstones type 1 (anal. #1–4, 8–11). Closed circles = aluminous ironstones type 2 (anal. #12, 13, 15, 17, 18). Diamonds = PELITE, (anal. #5, 16, 19, 29). Solid triangle = QFP and felsites, 3 analyses (#20, 21, 22). Solid squares = IF iron formation (anal. #6, 7, 14, 24).

in the Brockman IF. These theories may be applied to the rocks at Bird Lake, where the supply of iron-rich water is manifest in the presence of the stratiform iron formations, and where volcaniclastic felsic detritus was abundant.

Compositions in Fe-Al-Si are shown in Fig. 7. There are 8 points (circles) for the lensoid beds (type 1) of aluminous ironstone (Table 1, #'s 1–4 and 8–11), they have a large variation in Fe/Si but only a small variation in Si/Al. For the other analyses in the Type 1 group, #5 had only 16.8% FeO*, so it is plotted as pelite (Fig. 7, diamond symbol); #6 and 7 are cummingtonite-hornblende-pyrrhotite ironstones (Fig. 7, IF squares). The field of type 1 aluminous ironstones is distinct from the low-alumina iron formations (Fig. 7, IF squares, #'s 6, 7, 14, 24), but this may be due to the small number of samples. They also differ from reference oxide iron formations (analyses 25–28, Table 1) in having more Al and Mn, but less P. The type 1 aluminous ironstones and the pelites (#5, 19, 29) have a trend that is interpreted as mixing of exhalative Fe with the Si + Al of the clastic debris of the felsic volcanic rocks.

This mixing trend is similarly defined by the points for the ironstone matrix of conglomerates (Fig. 7, solid dots, analyses #12, 13, 15–18, + pelite diamond #16). The trend is from Fe towards the point QFP (Fig. 7, triangle), which includes 3 analyses of felsites (#'s 20–22). The felsites have Si/Al = 4, and this ratio controls the trend with minor scatter. This trend is not the same as the mixing of Fe and kaolinite (Si/Al = 1) to form chamosite (MAYNARD, 1983 p. 44).

Also, the Fe end-member of the trend, the hydrothermal-exhalative component, is not the non-aluminous ironstones (Fig. 7), even though three of these samples (#6, 7, 14) are patches rich in cummingtonite in garnet-ironstone, and the fourth is a silicate band in CBSIF. They have Si/Fe ratios 6/4, which reflects the dominant cummingtonite now present, and which suggests that the original mineral was greenalite (or its precursar, as discussed by EWERS, 1983). The definition of the Fe end-member may be approached in sample #8, which is high in FeO*, Fe^{+3}, Mg, Ca, Mn, and low in Si, Na, K, Rb, Zr, Li, Co, compared to the other aluminous ironstones. Such an element distribution suggests that it was a precipitate of carbonates and iron hydroxides. Carbonates are common in the type section at Algoma (GOODWIN, 1973), where they are preserved in low-grade metamorphism. The exhalative elements Fe, Mn, Mg, Ca, are inter-correlated (Table 1C), but Ca fails to correlate with Fe. The theory of EWERS and MORRIS (1981) applies, that the addition of pyroclastic dust changes the chemistry of precipitation. The reaction is the hydrolysis of feldspar, which decreases the acidity and promotes the precipitation of carbonates. In the Bird Lake aluminous ironstones the reaction is localized, in view of inclusions of cummingtonite (samples #6, 7, 14) and unaltered clasts in the conglomerates, and therefore more of an "in situ" or diagenetic style, than the non-localized, basin-wide change in chemistry proposed by EWERS and MORRIS (1981) for the Brockman IF.

Correlation of Al, Zr and Ti was used by EWERS and MORRIS (1981) to define the pyroclastic component in the Brockman IF. For the Bird Lake aluminous ironstones, there is a correlation of Al with Zr, also a negative correlation of Al with the exhalative elements Mg, S, Ca (but not with Fe). Ti is a clastic component element also at Mt. Isa (FINLOW-BATES, 1979), and in the aluminous ironstones at Bathurst (TROOP and SCOTT, 1985). Ti correlates with Al in the "Fe shales" in the BIF of greenstone belts in the Yilgarn Block (GOLE, 1981); and in the sulfide ores and iron formations at Broken Hill (STANTON, 1976). Ti is part of the tuff component in the "tuffaceous exhalites" at Redstone (ROBINSON, 1984). At Bird Lake, however, TiO_2 (0.10–0.74 wt.%) has a random variation vs Al in samples 1–11, and cannot be assigned to the detrital component. Ti has a positive correlation with Mo and Ni, and a negative correlation with Mn (Table 1C). This could be explained by movements of Ti, Mo and Ni in a diagenetic process after deposition, or a hydrothermal alteration process in felsite before or during deposition.

Some transition metals in the metalliferous sediments of the area of the East Pacific Rise have been assigned with Fe and Mn to an origin from hydrothermal exhalations. They are Cu, Cr, Ni, Pb (BOSTROM and PETERSON, 1966), Cu, Zn (DYMOND, 1981), and Cu, Zn, Ni, Co, Pb, (BARRETT et al. 1987). In the Bird Lake ironstones, however, no correlation is found between the transition elements and the exhalative group Fe, Mn, Mg, Ca. The transition elements correlate with each other, and, for Co and V, with Al, and, for Mo and Ni, with Ti. Their behaviour, like Ti, is complex, and related to hydrothermal alteration of the felsite.

The variability of alkali contents in the Bird Lake ironstones is shown by ratio comparison, normalizing with compositions of the felsic volcanic #22 in Table 1 (Fig. 8). The ratios for Al are slightly less than 1, this is dilution of felsite by the hydrothermal component on the assumption that Al is an inert (stable) element. The ratios for Na are consistently lower, therefore Na is interpreted to have been leached from the felsite detritus. The ratios for K are similar to those of Al, or lower, but not as low as for Na. Therefore K is interpreted to be variably leached, but resistant to the leaching relative to Na. The behaviour of Rb is similar to that of K (Fig. 8) and they correlate (Table 1C). Leaching of Na and variable leaching of K is typical of alteration zones of hydrothermal systems (MEYER and HEMLEY, 1967), but whether this occurred during precipitation or diagenesis, or whether the felsite was attacked before the detrital stage, is not known. For Ca contents, the ratios to felsite are positive (Fig. 8); it is interpreted as a precipitated element.

CONCLUSIONS

Iron formations of the chert banded silicate, sulfide, and aluminous silicate types formed in proximity to centers of felsite volcanism, interbedded with volcaniclastic tuffs, conglomerates, and breccias. Vein fillings of ironstone in a felsite dome intrusion indicate that hydrothermal systems were active in transporting iron.

Aluminous ironstone formed by the mixture of felsite tuff with exhalative Fe-rich solution and their Si-Fe precipitates on the sea floor, and with Fe-Mn-Mg-Ca precipitates formed on the sea floor and in diagenesis. Mixing was accompanied by hydrothermal leaching of Na, K and Rb from the felsite. Transition metals were not precipitated with Fe.

The aluminous ironstone formed beds which are smaller than 0.3×20 m; possibly they were larger but they were subject to disruption. Rip-up of these beds by volcaniclastic debris flows created conglomerates with felsic cobbles in a matrix of mafic ironstone.

The Bird Lake aluminous ironstones are similar, except for less Cu + Ni, than the "Fe shales" of the Archean Yilgarn Block in Australia (GOLE, 1981). Possibly, higher Cu + Ni contents in the Yilgarn may result from a source area rich in ultramafic rocks.

The average contents of the Bird Lake aluminous ironstones are similar in Al, Si, Fe, Mg, Na, Ti, Mn, S, Zr, Ba, Cu, and V to, greater in Ca and K than, and lower in P, Co, Ni, and Zn than, the average Algoma silicate IF of the Superior Province of the Canadian Shield (as compiled by GROSS and McLEOD, 1980). In each of the "similar", "greater", and "lower" groups, there are elements that are assigned to the exhalative component (Fe, Mn, Mg, Ca, P?, S?), the detrital felsite component (Al, Zr, Si, K, Rb, Ba?), and the leached or fixed in felsite elements (Na, K, Rb, Co, V, Ti, Mo, Ni, Cu?, Zn?). Considering the known variability of composition of hydrothermal systems and their deposits, and the complexity of the processes interpreted for the origin of these aluminous ironstones, the clustering

of the "similar" elements is remarkable. It indicates that there are constraints on the composition of hydrothermal solutions and their deposits. There is some uniformity of host rocks, which are the igneous types of the Archean mafic-felsic cycle, and their clastic derivatives. Felsic volcanism may be a dominant factor in several ways, such as source of heat and pyroclastic debris, and source of hydrothermal solutions rich in Fe and HCl (WHITNEY et al. 1985).

The environment of deposition is interpreted to have been an aqueous basin, with explosive felsic volcanoes in and around it, and active hydrothermal systems and fault scarps. Beds of chert banded silicate iron formation, and sulfide ironstone, precipitated during quiet periods, and aluminous ironstones during periods of volcanic activity. A similar model of origin was proposed for the Helen Iron Formation (GOODWIN et al., 1985). The hydrothermal systems, to carry abundant iron, would be acid and reducing (EWERS, 1983; WHITNEY et al. 1985). Aluminous ironstones would form diagenetically and by precipitation, where the exhaled hydrothermal solutions cool, mix, and react with felsite detritus.

The Bird Lake occurrences strengthen the genetic linkage between iron formations and igneous activity, but the source of iron is not obvious. STANTON (1985) has suggested that it should be sought in the igneous processes that produce felsic magma, and a mechanism for the concentration and transportation of iron in felsic rocks and magma is given by WHITNEY et al. (1985).

Acknowledgements—Allan Turnock acknowledges the debt owed to Hans Eugster for his teaching of geochemistry in 1957. The Geological Survey of Canada provided a grant for chemical analyses and some of the field work. We thank Drs. G. A. Gross and C. R. McLeod for encouragement and the supply of rock standards for chemical analysis, and G. A. Gross, N. M. Halden, J. J. Hemley, I-Ming Chou, and Wayne Nesbitt for reviews.

REFERENCES

ABBEY S., McLEOD C. R. and LIANG-GUO (1983) Four Canadian iron-formation samples prepared for use as reference materials. Geol. Survey Can. Paper 83-19.

BARRETT T. J., TAYLOR P. N. and LUGOWSKI J. (1987) Metalliferous sediments from DSPD leg 92: the East Pacific Rise transect. *Geochim. Cosmochim. Acta* **51**, 2241–2253.

BEAKHOUSE G. P. (1977) A subdivision of the western English River Subprovince. *Can. J. Earth Sci.* **14**, 1481–1489.

BEAKHOUSE G. P. (1985) The relationship of supracrustal sequences to a basement complex in the western English River Subprovince. In *Evolution of Archean Supracrustal Sequences,* (eds., L. D. AYRES, P. C. THURSTON, K. D. CARD and W. WEBER), Special Paper 22, pp. 169–178. Geol. Assoc. Can.

BOSTROM K. and PETERSON M. N. A. (1966) Precipitates from hydrothermal exhalations on the East Pacific Rise. *Econ. Geol.* **61**, 1258–1265.

CLIFFORD P. M. (1986) Petrological and structural evolution of the rocks in the vicinity of Killarney, Ontario. Geol. Surv. Can. Paper 86 1B, 147–155.

DYMOND J. (1981) Geochemistry of Nazca plate surface sediments: An evaluation of hydrothermal, biogenic, detrital, and hydrogenous sources. *Geol. Soc. Amer. Memoir 154*, 133–154.

EDWARDS G. R. (1983) Geology of the Bethune Lake area. Ontario Geol. Surv. Report 201.

EWERS W. E. (1983) Chemical factors in the deposition and diagenesis of banded iron formation. In: *Iron Formation Facts and Problems.* (eds., A. F. TRENDALL and R. C. MORRIS) pp. 491–512 Elsevier.

EWERS W. E. and MORRIS R. C. (1981) Studies of the Dales George member of the Brockman Iron Formation, Western Australia. *Econ. Geol.* **76**, 1929–1953.

FINLOW-BATES T. (1979) Chemical mobilities in a submarine exhalative hydrothermal system. *Chem. Geol.* **27**, 65–83.

FRANKLIN J. M. (1976) Role of laharic breccias in genesis of volcanogenic massive sulfide deposits. Geol. Surv. Can. Paper 76-1A, 293–300.

FRANKLIN J. M., KASARDA J. and POULSEN K. H. (1975) Petrology and chemistry of the alteration zone of the Mattabi massive sulfide deposit. *Econ. Geol.* **70**, 63–79.

GABOURY D. (1984) A petrologic study of an Archean banded iron formation, Wallace Lake, Manitoba. Unpublished B.Sc. thesis, Geological Sci., University of Manitoba.

GOLE M. J. (1981) Archean banded iron-formations, Yilgarn Block, Western Australia. *Econ. Geol.* **56**, 897–915.

GOODWIN A. M. (1961) Some aspects of Archean structures and mineralization. *Econ. Geol.* **56**, 897–915.

GOODWIN A. M. (1973) Archean iron-formations and tectonic basins of the Canadian Shield. *Econ. Geol.* **68**, 915–933.

GOODWIN A. M., THODE H. G., CHOU C.-L. and KARKHANSIS S. N. (1985) Chemostratigraphy and origin of the late Archean siderite-pyrite rich Helen Iron Formation, Micipicoten belt, Canada. *Can. J. Earth Sci.* **22**, 72–84.

GROSS G. A. (1965) Geology of iron deposits in Canada. I. General geology and evaluation of iron deposits. Geol. Surv. Can. Econ. Geol. Rept. 22.

GROSS G. A. (1980) A classification of iron deposits based on depositional environments. *Can. Mineral.* **18**, 223–229.

GROSS G. A. and McLEOD C. R. (1980) A preliminary assessment of the chemical composition of iron formations in Canada. *Can. Mineral.* **18**, 223–230.

HUTCHINSON R. W., RIDLER R. H. and SUFFEL G. G. (1971) Metallogenic relationships in the Abitibi belt, Canada: a model for Archean metallogeny. *Bull. Can. Inst. Min. Metall.* **64**(No. 708), 48–57.

JAMES H. L. (1966) Chemistry of the iron-rich sedimentary rocks. U. S. Geol. Surv. Prof. Paper 440-W.

KIMBERELY M. M. (1978) Paleoenvironmental classification of iron formations. *Econ. Geol.* **73**, 1796–1817.

LYDON J. W. (1984) Ore deposit models 8. Volcanogenic massive sulfide deposits Part 1: A descriptive model. *Geosci. Can.* **11**, 195–202.

MAYNARD J. B. (1983) *Geochemistry of Sedimentary Ore Deposits.* Springer-Verlag.

MEYER C. and HEMLEY J. J. (1967) Wall rock alteration. In *Geochemistry of Hydrothermal Ore Deposits.* (ed., H. L. BARNES) pp. 166–235. Holt, Rinehart & Winston.

RICHARDS S. M. (1960) The banded iron formations at Broken Hill, Australia, and their relationship to the lead-zinc orebodies. *Econ. Geol.* **61**, 72–96 and 257–294.

RIDLER R. H. (1971) Analysis of Archean Volcanic Basins in the Canadian Shield using the exhalite concept. *Bull. Can. Inst. Min. Metall.* **64**(No. 714), 20.

ROBINSON D. J. (1984) Silicate facies iron-formation and strata-bound alteration: tuffaceous exhalites derived by mixing: evidence from Mn-garnet stilpnomelane rocks at Redstone, Timmins, Ontario. *Econ. Geol.* **79**, 1796–1815.

SECCOMBE P. K., LAU J. L., LEA J. F. and OFFLER R. (1984) Geology and ore genesis of Ag-Pb-Zn-Cu sulfide deposits, Sunny Corner, N.S.W. *Proc. Australian Inst. Min. Metall.* **289**, 51–57.

SHEGELSKI R. J. (1975) Geology and geochemistry of iron formations and their host rocks in the Savant Lake - Sturgeon Lake greenstone belts. In *Geotraverse Workshop 1975.* (ed., A. M. GOODWIN) Paper 34. University of Toronto.

SHEGELSKI R. J. (1976) The geology and geochemistry of Archean iron formations and their relations to reconstructed terrains in the Sturgeon Lake and Savant Lake greenstone belts. In *Geotraverse Conference 1976.* (ed., A. M. GOODWIN) pp. 29-1 to 29-22 University of Toronto.

STANTON R. L. (1960) General features of the conformable "pyritic" orebodies. *Trans. Can. Inst. Min. Metall.* **63**, 22–36.

STANTON R. L. (1976) Petrochemical studies of the ore environment at Broken Hill N.S.W. 3-banded iron formations and sulfide orebodies: contitutional and genetic ties. *Trans. Inst. Mining Metall. (Sect. B)* **85**, 132–142.

STANTON R. L. (1985) Stratiform ores and geological processes. *Proc. Roy. Soc. New South Wales* **118**, 77–100.

STANTON R. L. and VAUGHAN J. P. (1976) Facies of ore formation: a preliminary account of the Pegmont deposit as an example of potential relations between small iron formations and stratiform sulfide ores. *Proc. Australasian Inst. Min. Metall.* **270**, 25–37.

TROOP D. G. and SCOTT S. D. (1985) Banded iron formation and sea floor hydrothermal sediments: an Ordivician analog from New Brunswick, Canada. (abstr.). *Geol. Assoc. Can./Mineral. Assoc. Can., Program with Abstracts* **10**, A64.

TRUEMAN D. L. (1980) Structure, stratigraphy and metamorphic petrology of the Archean greenstone belt at Bird River, Manitoba. Unpubl. Ph.D. thesis, University of Manitoba.

WHITNEY J. A., HEMLEY J. J. and SIMON F. O. (1985) The concentration of iron in chloride solutions equilibrated with granitic compositions: the sulfur-free system. *Econ. Geol.* **80**, 444–460.

Fluid-Mineral Interactions: A Tribute to H. P. Eugster
© The Geochemical Society, Special Publication No. 2, 1990
Editors: R. J. Spencer and I-Ming Chou

Partitioning of base metals between silicates, oxides, and a chloride-rich hydrothermal fluid. Part I. Evaluation of data derived from experimental and natural assemblages

EUGENE S. ILTON

Department of Mineral Sciences, American Museum of Natural History,
Central Park West at 79th St., New York, New York 10024, U.S.A.

and

HANS P. EUGSTER

Department of Earth and Planetary Sciences, The Johns Hopkins University, Baltimore, Maryland 21218 U.S.A.

Abstract—Exchange reactions for base metals between common Fe-Mg silicates, oxides and a chloride-rich hydrothermal/metamorphic fluid were calibrated by combining experimentally determined apparent equilibrium constants (Kd(magnetite/fluid) and Kd(biotite/fluid)) with apparent mineral/magnetite-biotite equilibrium constants derived from natural assemblages and experiments.

The resulting distribution coefficients suggest that the rock-forming silicates and oxides will preferentially partition (Cu > Cd) ≫ Zn ≫ Mn ≫ Fe ≫ Mg into the fluid phase. The results for copper and cadmium are bracketed to indicate that they are statistically inconclusive, but suggestive of potentially strong fractionation.

INTRODUCTION

A MAJOR PROBLEM concerning the formation of hydrothermal ore deposits involves the manner and timing of metal enrichment. Enrichment can occur at any stage, from the formation of the source rock or magma to the depositional event. It has become apparent, however, that hydrothermal ore deposits need not be associated with source rocks or magmas that have anomalously high concentrations of metals. SKINNER (1979) states: ". . . any rock can serve as a source of geochemically scarce metals provided a hydrothermal solution undergoes the reactions that will extract them."

This contribution attempts to elucidate the role of exchange reactions in the enrichment of manganese, zinc, copper, and cadmium during the interaction of hydrothermal/metamorphic fluids with common rock-forming silicates and oxides. Iron serves as the standard of comparison because the base metals of interest may be expected to form solid solutions with iron; the oxidation states and ionic radii of Mn^{2+}, Zn^{2+}, and Cu^{2+} are similar to those of Fe^{2+}, and HAACK et al. (1984) and HEINRICHS et al. (1980) have suggested that cadmium is associated with the Fe^{2+}-bearing silicates. Since iron is considerably more abundant than zinc, manganese, cadmium, and copper, the manner in which these minor-trace metals become enriched relative to iron poses a significant problem. The relevant partition coefficients are derived here in Part I, and applied in Part II (this volume).

There are a variety of geochemical approaches to the study of base metal enrichment: partitioning experiments (mineral/fluid—ILTON and EUGSTER, 1989; melt/fluid—HOLLAND, 1972; CANDELA and HOLLAND, 1984), solubility experiments (see reviews by BARNES, 1979; EUGSTER, 1986), whole-rock/fluid interaction experiments (e.g., ELLIS, 1968; BISCHOFF and DICKSON, 1975; HAJASH, 1975; MOTTL et al., 1979; SEYFRIED and BISCHOFF, 1981; SEYFRIED and MOTTL, 1982; SEYFRIED and JANECKY, 1985), theoretical predictions of solid/fluid metal partitioning (SVERJENSKY, 1984, 1985), studies of fluid inclusions (review by ROEDDER, 1984), and observations of modern geothermal and hydrothermal fluids (review by WEISSBERG et al., 1979; HENLEY and ELLIS, 1983).

Experimental results for Zn-Fe, Mn-Fe, Cu-Fe and Cd-Fe partitioning between magnetite and a supercritical (H_2O-KCl) solution (ILTON and EUGSTER, 1989) imply very strong partitioning of zinc and manganese into the fluid phase relative to iron and even stronger partitioning of copper and cadmium into the fluid relative to zinc and manganese. Preliminary experiments indicate that biotite strongly partitions zinc into the fluid relative to iron (ILTON, 1987). The results are applicable in a variety of geological environments since magnetite occurs in ultramafic to granitic igneous rocks and in different grades and compositions of metamorphic rocks.

This paper generalizes the results of ILTON and

EUGSTER (1989), and ILTON (1987) by deriving $Kd(\text{fl/min}) = (M/\text{Fe})^{\text{fl}}/(M/\text{Fe})^{\text{min}}$, where M represents the metal of interest, and the superscripts fl and min represent the fluid and mineral of interest, respectively, by combining $Kd(\text{fl/mt})$ and $Kd(\text{fl/biot})$ with magnetite-biotite/mineral distribution coefficients from natural rocks and mineral/mineral exchange experiments. The method has been described by EUGSTER and ILTON (1983).

This simple empirical approach provides information concerning the influence of the rock-forming silicates and oxides on the relative mobility of base metals for a variety of rock environments. Derived $Kds(\text{fl/min})$ should be considered preliminary order-of-magnitude calculations. Uncertainties in temperature estimates and compositional differences between experimental and natural assemblages increase the uncertainty of the calculations. Uncertainties due to pressure estimates and extrapolations, however, are considered relatively minor. Moreover, derived $Kds(\text{fl/min})$ may be appropriate only for fluids that are similar to the fluid in the fl/ mt-biot experiments (*i.e.,* $\text{Cl}_T = 2$ molal, $P = 2$ kb, $f\text{O}_2 = \text{NNO}$), and for minerals with compositions near those used in the derivations. Because the partition coefficients for Mn-Fe and Zn-Fe involve the *ratios* of similar metals, the partition coefficients will be valid over the range of fluid compositions where metal speciation is similar and $\gamma_M/\gamma_{\text{Fe}}$ is constant or near unity. This condition may occur in the lower pressure, higher temperature portion of the supercritical region where metal chloride speciation tends towards neutral complexes (*i.e.,* the experimental conditions, see discussion in IL-TON and EUGSTER, 1989).

HOST MINERALS FOR Mn, Zn, Cu AND Cd

The importance of a mineral as a host for an element depends on the minerals abundance and its affinity for a particular element. The dominant host for a given element changes from rock to rock as a function of mineral modes, composition, pressure and temperature. In many rock types, the bulk of the whole-rock manganese and zinc is contained in the rock-forming ferromagnesian silicate fraction. Olivine and pyroxene are the dominant hosts for manganese in ultramafic to mafic rocks, whereas hornblende and biotite are the dominant hosts for manganese *and* zinc in more felsic rocks. Unlike manganese, the dominant host for zinc in basaltic rocks is magnetite (WEDEPOHL, 1972). Hydrothermal alteration introduces other important hosts for manganese and zinc such as chlorite and serpentine. In metamorphic rocks, garnet and staurolite become important hosts for manganese and zinc respectively, as do amphibole, biotite and chlorite.

The low concentrations of cadmium in average crustal rock pose a significant problem for identifying its dominant hosts. DISSANAYAKE and VINCENT (1972), and HEINRICHS *et al.* (1980) give the fraction of whole-rock cadmium held

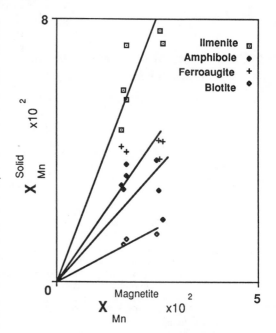

FIG. 1. Mole fraction diagram illustrating Mn-Fe^{2+} partitioning, at $\sim$700°C, between magnetite and the minerals ilmenite, ferroaugite, amphibole, and biotite. The solid lines mark mean Kd for each mineral. The compositions of the minerals are reported as the mole fraction X_{Mn} = Mn/(Mn + Fe^{2+}). Mineral compositions are from FERRY (1985b).

in specific silicates and oxides in their tables VI and 3, respectively. Such work suggests that the silicate-oxide fraction is the dominant host for cadmium. Alternatively, it is possible that cadmium might occur mainly in the sulfide fraction (MAROWSKY and WEDEPOHL, 1971).

It is possible that copper is mainly present as sulfides in geological environments. HASLAM (1968) presents mineral modes and whole rock and mineral analyses for copper from a suite of granodiorites to granites. Among the ferromagnesium-silicates, biotite carries most of the whole rock copper. Most of the samples, however, indicate that silicates only account for $1/3$ or less of the whole rock copper. Calculations using data from DE ALBUQUERQUE (1971, 1973) indicate that biotite, for the most part, accounts for $1/3$ or less of the whole rock copper. Data from GRAYBEAL (1973) indicate that chlorite + biotite + hornblende account for roughly 50% of the whole rock copper in granodiorites associated with porphyry copper deposits.

Considering that the cleanest silicate separate may contain inclusions of sulfide minerals and alteration products (*e.g.,* BANKS, 1974), the above calculations are consistent with the commonly-held belief that sulfides are indeed the dominant copper hosts. Further complications are discussed later, and in ILTON and VEBLEN (1988a and 1988b). Regardless, numerous workers have suggested that silicates such as biotite are sources of copper in porphyry copper systems (*e.g.,* GRAYBEAL, 1973; HENDRY *et al.,* 1981). Models of copper transport that involve the oxide-silicate rock fraction as copper hosts may be useful to test the degree of sulfide control on the mobility and fractionation of copper in metasomatic systems.

Table 1. List of values for $Kd = (Mn/Fe^{2+})^{min}/(Mn/Fe^{2+})^{mt}$

	800°C	725°C	700°C			500°C
	B	Hd	F	B	P	F'
calcite						1.8 **
garn/ilm					5.05 ± 0.54 (±0.45)	
ilmenite		2.04 ± 0.18 (±0.07)	3.47 ± 0.24 (±0.20)			
cpx	2.02 1.5–2.6		1.97 ± 0.49 (±0.61)	4.49 3.3–5.8		
amphibole			1.55 ± 0.49 (±0.45)			11.8 ± 1.8 *
talc						6.98 ± 5.45 *
chlorite						6.63 ± 4.94 *
biotite		0.527 ± 0.125 (±0.080)	0.652 ± 0.085 (±0.212)			2.20 **

Errors are reported as one standard deviation. The 95% confidence intervals, calculated from the students t distribution, are bracketed. Errors for cpx data from Burton given as range. Mineral compositions from B: BURTON et al. (1982); P: POWNCEBY et al. (1987); F: FERRY (1985b); F': FERRY (1985a); Hd: HILDRETH (1977).
* Small sample sizes, 95% confidence interval exeeds 100% of partition coefficient.
** Only one sample.

The preceeding discussion highlights the need for a general understanding of mineral/fluid exchange reactions. Given numerous Kds(mineral/fluid) it should be possible to model the exchange of base metals between hydrothermal fluids and a variety of rock types.

ERROR ANALYSIS

Standard deviations and 95% confidence intervals are given with each intermineral partition coefficient. Uncertainties associated with Kd(fl/min) are calculated by propagating the errors associated with Kd(fl/mt-biot) and Kd(mt-biot/min). Small sample sizes required the use of the students t distribution for calculating 95% confidence intervals.

Uncertainties associated with temperature estimates, the degree of equilibrium attained in natural and experimental assemblages, and compositional differences between experimental assemblages and natural assemblages are noted and qualitatively assessed.

The appendix contains more detailed petrographic information and assessments of equilibrium.

Mn-Fe PARTITIONING

In geological environments at higher P and T, manganese is commonly restricted to the divalent state (WEDEPOHL, 1978). One should expect Mn^{2+} to substitute preferably for Fe^{2+} since the majority of the crystal chemical properties of Mn^{2+} are closer to those of Fe^{2+} than Mg^{2+} or Ca^{2+}. Accordingly, when sufficient data has been provided, reactions are formulated as $Mn\text{-}Fe^{2+}$ exchanges, where $Kd = (Mn/Fe^{2+})^A/(Mn/Fe^{2+})^B$. Intermineral partition

coefficients are listed in Tables 1 and 2. More detailed petrographic information for each assemblage is located in the appendix.

FERRY (1985b) provides mineral analyses of co-existing magnetites, ilmenites, biotites, amphiboles, and ferroaugites in granites from the Isle of Skye. Calculated distribution coefficients, Kd(min/mt), are illustrated in Fig. 1 and listed in Table 1. Whereas the data indicate that the partitioning behavior of these minerals vary, the range of compositions are too narrow for speculations concerning attainment of equilibrium. The minerals, in the samples used, have compositions that record temperatures in the range 670–720°C.

POWNCEBY et al. (1987) experimentally calibrated Mn-Fe partitioning between ilmenite and garnet (alm-spess solid solutions) from 900–600°C, 2–5 kb, and over a wide range of Mn/(Mn + Fe) bulk compositions. Kd(garn/ilm) values were neither significantly dependent on pressure nor, for the purposes of this paper, on composition. The experiments closely bracketed equilibrium at all temperatures. Kd(garn/ilm), at 700°C, is given in Table 1.

Combining, at 700°C, Kds(min/mt) from FERRY (1985b) and Kd(garn/ilm) from POWNCEBY et al. (1987), with Kd(mt/fluid) given by ILTON and EUGSTER (1989) yields a host of mineral-fluid exchange reactions. The results are listed in Table 3. Given the precision of the data, we conclude that

Table 2. List of values for $Kd = (M/\text{Fe})^{\text{min A}}/(M/\text{Fe})^{\text{min B}}$ at $\sim 500°C$, where M is the moles of the metal of interest

	Mn-Fe		Zn-Fe		Cu-Fe		Cd-Fe
			1	2	1	2	
biot/mt	25.3 ± 17.1 (±13.2)		109 ± 227 (±162)	23 ± 19 (±16)	22 ± 29 (±22)	9.1 ± 8.7 (±7.3)	10.1 ± 9.2 (±6.3)
	AC	H					
amph/mt	51.5 ± 23.9 (±25.1)	22.5 ± 4.9 (±12.2)	17.3 ± 15.9 (±12.2)		6.8 ± 5.8 (±4.5)		13.5 ± 12.9 (±10.7)
biot/amph			1.72 ± 1.06 (±0.89)		1.30 ± 1.43 (±1.20)		0.39 ± 0.23 (±0.19)

All values derived from mineral compositions in ANNERSTEN and EKSTROM (1971). Errors are reported as one standard deviation. The 95% confidence intervals, calculated from the students t distribution, are bracketed. AC—actinolite; H—hornblende. See text for a discussion of values 1 and 2.

Mn-Fe partitioning between minerals and fluid varies greatly. This implies that given the same whole rock Mn/Fe ratios, mineralogically different rock types can impart vastly different Mn/Fe ratios to the fluid. Note that garnet might partition iron into the fluid relative to manganese!

These numbers should be considered order-of-magnitude calculations due to imperfect temperature estimates, lack of criteria that prove equilibrium, and possible compositional controls. The compositions of natural magnetites are perhaps critically important. The major component other than Fe_3O_4 in the magnetites is Fe_2TiO_4 (FERRY, 1985b), with the concentration of TiO_2 varying from 8–15 wt%. Charge balance effects were accounted for by formulating Mn-Fe^{2+} exchange reactions.

HILDRETH (1977, 1979) reports mineral compositions from the high silica rhyolites of the Bishop Tuff over a T range 720–790°C. Magnetite, ilmenite, and biotite compositions, for $T = 720$–730°C, are similar to those from FERRY (1985b), and are plotted in Fig. 2. Figure 2 reveals a strong correlation between X_{Mn} in ilmenite and X_{Mn} in magnetite,

consistent with exchange equilibrium. Biotite/magnetite Mn-Fe^{2+} partitioning displays a similar correlation, also consistent with exchange equilibrium (see Fig. 2). Kd(ilm-biot/mt) values are given in Table 1. Values for Kd(fl/ilm) and Kd(fl/biot), listed in Table 3, were derived from the low temperature data (720–730°C) using Kd(fl/mt) at 725°C.

Results derived from a plutonic environment (FERRY, 1985b) are consistent with results from an extrusive environment (HILDRETH, 1977; 1979) where exchange equilibrium has been better demonstrated.

ANNERSTEN and EKSTRÖM (1971) report compositions of coexisting magnetites, Ca-amphiboles (hornblende and actinolite), and biotites from a metamorphosed iron formation. Magnetites are nearly pure Fe_3O_4. Temperature and pressure conditions of metamorphism were homogeneous, whereas fO_2 and whole-rock composition are variable throughout the formation. Temperature is estimated at $\sim 500°C$ (ANNERSTEN, pers. comm.). Mn-Fe partitioning for all pairs mineral/magnetite

Table 3. List of values for $Kd = (\text{Mn}/\text{Fe})^{\text{fl}}/(\text{Mn}/\text{Fe}^{2+})^{\text{min}}$

	800°C	725°C	700°C		
	B	Hd	F	B	P-F
garnet					0.61 ± 0.07
ilmenite		4.4 ± 0.2	3.1 ± 0.2		
cpx	3.0		5.5 ± 1.7	2.4	
	2.3–3.9			1.9–3.3	
amphibole			6.9 ± 2.0		
biotite		17.1 ± 2.7	16.5 ± 5.4		

Errors are reported as 95% confidence intervals. Range of data given for B. Kd'(fl/mt) from ILTON and EUGSTER (1989) combined with Kd(min/mt) from B: BURTON et al. (1982); P-F: Kd(garn/ilm) from POWNCEBY et al. (1987) combined with Kd(ilm/mt) from F; F: FERRY (1985b); Hd: HILDRETH (1977).

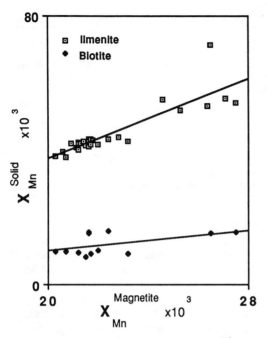

FIG. 2. Mole fraction diagram illustrating Mn-Fe^{2+} partitioning between magnetite, and the minerals ilmenite and biotite. The compositions of the minerals are reported as the mole fraction $X_{Mn} = Mn/(Mn + Fe^{2+})$. Solid lines mark the partitioning curves for the mean Kd associated with each mineral. Mineral compositions are from HILDRETH (1977) for $T \approx 720-730°C$.

roughly 450–550°C. Combination of Kd(mt/fl), at 500°C, with Kd(mt/min) yields Kd(Mn/Fe)^{fl}/(Mn/Fe^{2+})^{min}. The results, listed in Table 4, suggest that lower temperature hydrothermal alteration assemblages likewise favorably partition manganese into the fluid relative to iron. The derivations, although not strictly comparable, are consistent with bulk Mn-Fe partitioning described in basalt-seawater experiments (see introduction for reference list), and with the enrichment of manganese relative to iron in fluids associated with mid ocean ridge hydrothermal systems (SKINNER, 1979). Although these derivations are not complicated by impurities in magnetite, there is insufficient data to suggest the establishment of exchange equilibrium. Furthermore, the results are statistically inconclusive because of small sample sizes. Much more data is needed on such assemblages.

Comparison of the low with the high temperature derivations is complicated by a significant Fe_2TiO_4 component in the high temperature magnetites. Since charge balance effects have been screened by calculating Mn-Fe^{2+} exchange reactions where necessary, only possible structural effects remain. Indirect evidence comes from the compositions of magnetite-hematite assemblages in a regionally metamorphosed quartzite. The assemblages re-

is illustrated in Fig. 3. If one excludes the points in brackets, which represent magnetites contaminated by sphene, then the scatter is greatly reduced and the plot suggests that Mn-Fe exchange equilibria have been approached. Kd(min/mt) values are listed in Table 2. ANNERSTEN and EKSTRÖM conclude, based on a broader data set, that the rocks approached chemical equilibrium. Excluding the contaminated pairs, and combining average Kds(min/mt) with Kd(mt/fl) yields, at ~500°C, values for Kd(fl/min) listed in Table 4. Kd(fl/hbl) and Kd(fl/act) are significantly different (see appendix). The presence of fO_2 gradients and variable Fe/Mg in the silicates neither noticably effect Mn-Fe partitioning between magnetite and the silicates nor among the silicates.

FERRY (1985a) reports compositions of coexisting magnetite (nearly pure Fe_3O_4), biotite, chlorite, talc and calcite from hydrothermally altered gabbro. Kd(min/mt) values are given in Table 1. The associated 95% confidence intervals are greater than 100% of the partition coefficients, primarily because of very small sample sizes (i.e., 2–3). Kd(calcite/mt) and Kd(biotite/mt) are derived from only one sample. The temperature of alteration is estimated at

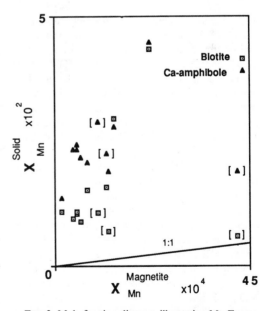

FIG. 3. Mole fraction diagram illustrating Mn-Fe partitioning between magnetite, and the minerals Ca-amphibole and biotite, at ~500°C, where $X_{Mn} = Mn/(Mn + Fe)$. Points in brackets indicate that magnetite is contaminated with sphene. The solid line marks equal partitioning. Mineral compositions are from ANNERSTEN and EKSTRÖM (1971).

Table 4. List of values for $Kd = (M/Fe)^{fl}/(M/Fe)^{min}$ at $\sim 500°C$, where M represents the moles of the metal of interest

	Mn-Fe			Zn-Fe		Cd-Fe	Cu-Fe
	F			1	2		
amphibole	4.3*	3.0A ± 1.6	7.0H ± 4.0	70 ± 38	20 ± 15	1160*	11300*
biotite	23*		6.2 ± 3.4		40.3 ± 7.6	1550*	3550*
chlorite	7.7*						
talc	7.3*						
calcite	2.0*						

Errors are 95% confidence intervals. F: $Kd(fl/mt)$ from ILTON and EUGSTER (1989) combined with $Kd(min/mt)$ from (FERRY, 1985a). $Kd(biot/fl)$ for Zn-Fe partitioning from experiments (ILTON, 1987). All other values derived by combining partition coefficients from Table 2 (data from ANNERSTEN and EKSTRÖM, 1971) with $Kd(fl/mt$-annite). Explanation of values 1 and 2 for $Kd(fl/amph)$ is given in text.
    * Values are statistically inconclusive, see text.

corded fO_2 gradients in the rock, whereas temperature and pressure were homogeneous (RUMBLE 1973; RUMBLE and DICKENSON, 1986). RUMBLE (1973) notes that the systematic partitioning of elements between oxide phases suggests that chemical equilibrium was attained. Magnetite compositions are nearly pure Fe_3O_4 and constant, whereas wt% TiO_2 in the coexisting hematites is variable. $Kd = (Mn/Fe^{2+})^{mt}/(Mn/Fe^{2+})^{hem}$ varies from 1.2 to 46 as wt% TiO_2 in hematite decreases from a high of 16.1 to 3.89. Calculation of $Kd(hem/mt)$ with total iron increases the variance fourfold. The change in hematite/fluid partitioning is attributed to variable X_{Ti}^{hem} because magnetite compositions are fairly constant. This suggests that manganese follows titanium, in hematite, for structural reasons as well as for charge balance. DASGUPTA (1967) offers crystal chemical reasons for the correlation of titanium and manganese in magnetite. It is possible that higher concentrations of titanium in magnetite may decrease $(Mn/Fe)^{fl}/(Mn/Fe^{2+})^{mt}$ and that derived $(Mn/Fe)^{fl}/(Mn/Fe^{2+})^{min}$ values, at 700°C, are perhaps too high.

Experimental work, by BURTON et al. (1982), on the reaction clinopyroxene = andradite + magnetite + quartz provides further constraints on high temperature mineral-fluid partitioning. BURTON et al. (1982) suggest that manganese distribution closely approached equilibrium at 700° and 800°C, but not at 600°C. Product clinopyroxene compositions range from $hd_{87}jo_{13}$ to $hd_{89}jo_{11}$, whereas magnetite is nearly pure Fe_3O_4 with minor manganese. $Kd(cpx/mt)$ values are given in Table 1. Combination of $Kd(mt/cpx)$ with $Kd(fl/mt)$, at 700 and 800°C, yields $Kd(fl/cpx)$ values listed in Table 3. Ln $Kd(fl/cpx)$, along with the variance in the data, is plotted $v.s.$ $1/T$ in Fig. 4. Neglecting the point at 600°C (non equilibrium), and assuming constant ΔH of reaction, yields ln $Kd(fl/cpx) = -2073/T(K)$

+ 3.04. Given the uncertainties, this temperature dependence should be viewed cautiously. Such a temperature dependence, however, implies that fluids of identical compositions can impart higher $(Mn/Fe)^{cpx}$ ratios at lower temperatures. This may have interesting consequences for interpreting the mineralogy of some metasomatic skarns (e.g., MEINERT, 1987).

$Kd(fl/cpx)$, at 700°C, is lower than the value obtained from FERRY (1985b). This is consistent with the hypothesis that $Kds(fl/min)$ derived from natural magnetites containing significant titanium, are too high. Although Kds derived from BURTON et al. (1982) are better contrained than those derived from natural assemblages, a strict comparison of the data sets is compromised by differences in clinopyroxene compositions.

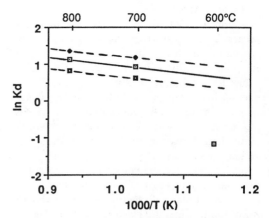

FIG. 4. Ln Kd $v.s.$ $1/T$ plot for Mn-Fe partitioning between clinopyroxene and a chloride-rich fluid. The solid line is calibrated to the points at 700°C and 800°C, whereas the point at 600°C has not achieved equilibrium. The dashed lines mark the range of the data. See text for details. Data for cpx-mt taken from BURTON et al. (1982).

Summary

The data suggest that Mn-Fe exchange reactions between minerals can approach equilibrium in nature (see ILTON, 1987, for an extensive review of mineral/mineral partitioning). It is, therefore, reasonable to assume that equilibrium Mn-Fe partitioning between minerals and hydrothermal fluids is a viable natural process. More specifically:

(1) Mn-Fe partitioning between minerals varies by over an order of magnitude. This implies that mineral/fluid partitioning should also vary widely. Combination of Kd(min/mt) with Kd(fl/min) yields a host of mineral/fluid exchange reactions that are distinct and that do vary considerably. This in turn implies that the Mn/Fe compositions of hydrothermal/metamorphic fluids can be strongly influenced by silicate-oxide mineralogy. Tables 3 and 4 are summaries of Kd(fl/min) values.

(2) Solubility studies by BOCTOR (1985) suggest that Mn-endmembers are generally more soluble than Fe^{2+}-endmembers under similar conditions. This is consistent with the mt/fl experiments in ILTON and EUGSTER (1989) and with the majority of Kd(fl/min) values derived in this paper. Within the statistical error of these calculations, however, almandine/spessartine-garnet may partition manganese into the *solid* phase relative to iron.

(3) The accuracy of derived mineral/fluid exchange reactions may be compromised by uncertain temperature estimates, compositional effects, and lack of definitive proof of exchange equilibrium. Of particular concern are the compositions of natural magnetites which can contain considerable TiO_2. At 700°C, significant titanium in magnetite yielded Kds(fl/min) that are possibly too high. Although the 700°C derivations, using natural assemblages (*i.e.*, FERRY, 1985b; HILDRETH, 1977 and 1979), are semi-quantitative, the *relative* behavior of the minerals are consistent. Note that direct comparisons of minerals in the laboratory with those in nature will always be compromised by compositional differences. At lower temperatures, where magnetite compositions are nearly pure Fe_3O_4, the derivations assume greater accuracy given fair temperature estimates and equilibrium.

Zn-Fe PARTITIONING

The existence of zincian biotite (FRONDEL and ITO, 1966), zincian serpentine (BADELOW, 1958), and zincian amphiboles (KLEIN and ITO, 1968) suggests that trace and minor concentrations of zinc will occupy regular octahedral sites in the hydrous silicates. Table 30-D-1 from WEDEPOHL (1972) indicates that the majority of crystal chemical properties of Zn^{2+} are much closer to those of Fe^{2+} than Mg^{2+}. This suggests that Zn^{2+} should substitute for Fe^{2+} preferentially. Alternatively, TAUSON and KRAVECHENKO (1956) claim that a portion of the zinc associated with silicates such as biotite is extra-structural. If true, this "extra-structural" zinc may have "exolved" at very low temperatures, and may not be representative of high temperature hydrothermal processes. Since the available data does not distinguish between Fe^{2+} and Fe^{3+}, distribution coefficients are formulated as Zn-Fe exchange reactions. This is not considered a significant problem since magnetite compositions in the ensuing calculations are nearly pure Fe_3O_4 (*i.e.*, Kd(fl/mt) from ILTON and EUGSTER, 1989, is employed, instead of Kd'(fl/mt)).

There is far less data on the zinc contents of coexisting minerals than for manganese. One may also assume less accuracy for zinc measurements compared to manganese since zinc is usually a trace element in average rock-forming minerals. Values for $Kd = (Zn/Fe)^{min}/(Zn/Fe)^{mt}$ and $Kd = (Zn/Fe)^{biot}/(Zn/Fe)^{amph}$ derived from mineral compositions in ANNERSTEN and EKSTRÖM (1971), are plotted in Figs. 5 and 6, respectively, and listed in Table 2 (see appendix and previous section for petrological details). Kd(silicate/mt) values for zinc are considerably more variable than those for manganese. Zn/Fe tends to be higher in biotite and amphibole relative to magnetite. Kd(biot/amph) values exhibit much better precision than Kd(silicate/mt) values (see Table 2).

Two values for Kd(biot/mt) are given in Table 2. Value 1 encorporates all the data, whereas value 2 excludes two data points that are associated with very low Fe/Mg ratios in biotite. Value 2 is considerably more precise than value 1. There is insufficient data, however, to verify a correlation between Kd(biot/mt) and biotite composition. Kd(biot/mt) is excluded from further consideration. Instead, an experimentally determined value for Kd(biot/fl), from ILTON (1987), is given in Table 4 and used in the ensuing derivations. Kd(biot/fl) values are plotted in Fig. 7. The data are preliminary and unreversed (perhaps maximum values). Furthermore, biotite in the experimental assemblage is annite, whereas biotite in the natural assemblage is a Fe-Mg solid solution.

On these cautionary notes, Kd(amph/mt) and Kd(amph/biot) are combined with Kd(fl/mt) and Kd(fl/annite), respectively, to yield Kd(fl/amph) values listed in Table 4. Values 1 and 2 are derived via Kd(amph/biot) and Kd(amph/mt), respectively.

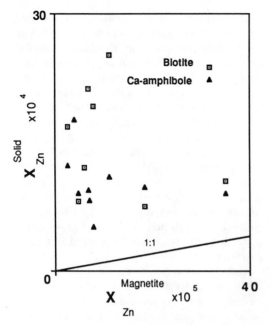

FIG. 5. Mole fraction diagram illustrating Zn-Fe partitioning between magnetite, and the minerals biotite and Ca-amphibole, at ~500°C, where $X_{Zn} = Zn/(Zn + Fe)$. The solid line marks equal partitioning. Mineral compositions from ANNERSTEN and EKSTRÖM (1971).

The results are similar, given their associated uncertainties.

Comparison of the natural data to the magnetite/fluid and annite/fluid experiments suggests that biotite and amphibole should strongly partition zinc into the fluid relative to iron. One might expect minerals with similar octahedral sites, such as chlorite and pyroxene, to exhibit not too disimilar partitioning behavior.

Cd-Fe PARTITIONING

The geochemistry of cadmium is not well understood. The low concentrations of cadmium, in all rock types, accentuates analytical problems. Whereas zinc is usually reported in the ppm range, cadmium is reported at the ppb concentration level (see HEINRICHS et al., 1980, Table 6). According to HEINRICHS et al. (1980, Table 5) cadmium is most closely correlated with zinc (ultramafic rocks), magnesium (mafic rocks), zinc (granitic and gneissic rocks), and iron (sedimentary rocks). The similarity in ionic radius for both Cd^{2+} and Ca^{2+} has suggested to some the possibility of a Cd-Ca substitution. VINCENT and BILEFIELD (1960), however, state that "Cadmium appears in the ordinary minerals of the gabbros to follow iron rather than calcium . . . neither does cadmium in the Skaergaard example show any chalcophile tendency." On the other hand, for acidic rocks, DOSTAL et al. (1979) claim that "feldspars (especially plagioclase) have the highest content of cadmium." Whereas HEINRICHS et al. (1980) state that "Cadmium has a distinct affinity for the six-coordinated Fe^{2+}-Mg sites in biotite, chlorite, and pyroxene . . . cadmium does not replace calcium preferentially in its silicates and in apatite."

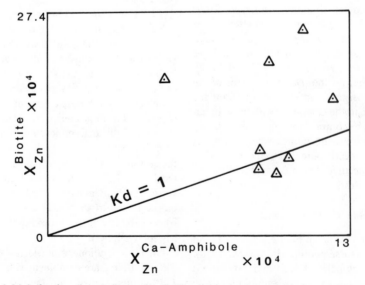

FIG. 6. Mole fraction diagram illustrating Zn-Fe partitioning between biotite and Ca-amphibole, at ~500°C, where $X_{Zn} = Zn/(Zn + Fe)$. The solid line marks equal partitioning. Mineral compositions from ANNERSTEN and EKSTRÖM (1971).

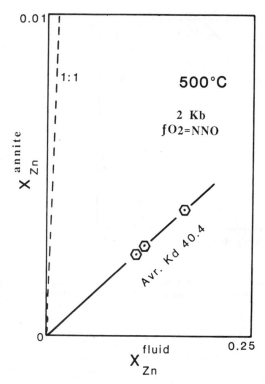

FIG. 7. Mole fraction diagram illustrating experimental Zn-Fe partitioning between annite and a 2 mol chloride solution at 500°C, 2 kb, and fO_2 = NNO. The hexagons represent the experimental distribution coefficients. Average Kd(exp) = 40, where $Kd = (Zn/Fe)^{fl}/(Zn/Fe)^{biot}$. Data from ILTON (1987).

HAACK et al. (1984) conclude that the Cd-Ca controversy is still unresolved.

ANNERSTEN and EKSTRÖM (1971) report the concentration of cadmium in coexisting magnetite, biotite and Ca-amphibole (see appendix and Mn-Fe section for petrological details). Kd(min/mt) values are listed in table 2 and illustrated in Fig. 8. Analytical difficulties and variable silicate compositions may have increased the variance of the data. There is insufficient data, however, to test the latter possibility. Despite the scatter, the data describe the tendency for cadmium to partition preferentially, with respect to iron, into biotite and Ca-amphibole relative to magnetite. Cd-Fe partitioning is considerably more systematic between biotite and amphibole (see Fig. 9 and Table 2). Since cadmium is extremely partitioned into the fluid relative to magnetite at 700°C (ILTON and EUGSTER, 1989), we assume that cadmium is partitioned at least as strongly at 500°C (although the experiments are unreversed). This is a fair assumption since the primary reason for such behavior is the relatively large size of the cadmium ion. Given this assumption,

and assuming a Cd-Fe exchange, combination of Kd(fl/mt) with average Kd(mt/min) yields Kd(fl/min) values listed in Table 4. Although the partition coefficients indicate that biotite and amphibole should very strongly partition cadmium into the fluid relative to iron, large uncertainties associated with the experiments and natural assemblages render the results statistically inconclusive.

Cu-Fe PARTITIONING

One might expect Cu^{2+} to preferentially substitute for Fe^{2+} in minerals such as biotite (Cu^{1+} will be discussed later). ILTON (1987) detected weak correlations between iron and copper in biotites and amphiboles, whereas GRAYBEAL (1973) and HAACK et al. (1984) found no correlation between copper and any of the major elements. Copper's preference for distorted octahedral sites (the Jahn Teller effect), and the possible presence of submicroscopic sulfide inclusions in silicates complicates its geochemistry.

Despite the uncertainties, mineral analyses from ANNERSTEN and EKSTRÖM (1971) were used to estimate the partitioning of copper and iron between magnetite, and Ca-amphibole and biotite (see appendix and Mn-Fe section for petrological details).

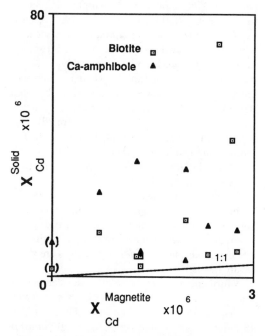

FIG. 8. Mole fraction diagram illustrating Cd-Fe partitioning between magnetite, and the minerals biotite and Ca-amphibole, at ~500°C, where X_{Cd} = Cd/(Cd + Fe). Points in parentheses were judged anomalous. The solid line marks equal partitioning. Mineral compositions from ANNERSTEN and EKSTRÖM (1971).

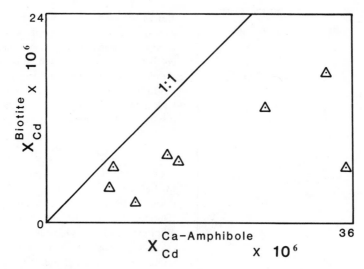

FIG. 9. Mole fraction diagram showing Cd-Fe partitioning between biotite and Ca-amphibole, at ~500°C, where $X_{Cd} = Cd/(Cd + Fe)$. The solid line marks equal partitioning. Mineral compositions from ANNERSTEN and EKSTRÖM (1971).

$Kd(min/mt)$ values are illustrated in Fig. 10 and listed in Table 2. As expected the data shows considerable scatter. The data suggest that copper, relative to iron, has a tendency to be preferentially partitioned into biotite and Ca-amphibole relative to magnetite. $Kd(biot/mt)$ exhibits the greatest variance. If two values for $Kd(biot/mt)$, associated with very low Fe/Mg ratios, are excluded from the mean, then the precision is considerably enhanced (value 1 *v.s.* value 2 in Table 2). There is insufficient data, however, to test for a correlation between $Kd(biot/mt)$ and biotite composition. Assuming a Cu-Fe exchange and combining $Kd(mt/fl)$ with $Kd(mt/min)$, given that $Kd(mt/fl)$ is insensitive to temperature within experimental error (ILTON and EUGSTER, 1989) yields $Kds(fl/min)$ values listed in Table 4.

The derivations suggest that biotite and Ca-amphibole, as well as magnetite extremely partition copper into the fluid relative to iron. Propagation of errors associated with $Kd(fl/mt)$ and $Kd(min/mt)$, however, indicate that $Kd(fl/min)$ values are statistically inconclusive.

A further complication involves the significance and nature of the copper species in these silicates. ILTON and VEBLEN (1988a,b) used TEM methods to show that *anomalous* copper ($\geq$ ~500 ppm) in biotites from rocks associated with porphyry copper deposits could be accounted for by submicroscopic inclusions of native copper along the interlayers. Textural evidence suggested that copper may have substituted for K^+ in the interlayer sites, *perhaps*

during hydrothermal activity or incipient weathering, prior to reduction and precipitation to native copper. Although the biotites used in this study contain only trace concentrations of copper (6–18

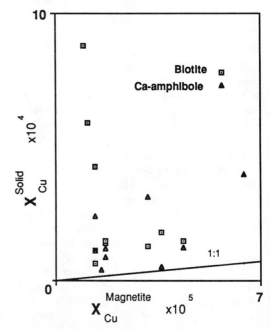

FIG. 10. Mole fraction diagram illustrating Cu-Fe partitioning between magnetite, and the minerals biotite and Ca-amphibole, at ~500°C, where $X_{Cu} = Cu/(Cu + Fe)$. The solid line marks equal partitioning. Mineral compositions from ANNERSTEN and EKSTRÖM (1971).

ppm) and probably have not experienced the same history as the biotites studied by ILTON and VEBLEN (1988a,b), the TEM evidence questions the assumption that copper substitutes for octahedrally coordinated iron in porphyry copper environments. Such evidence is consistent with observations that Cu^{1+}, not Cu^{2+}, is the dominant oxidation state of copper in the hypogene environment (CANDELA and HOLLAND, 1984; WEDEPOHL, 1974).

CONCLUSION

The partition coefficients derived in this contribution and in EUGSTER and ILTON (1983) for Mg-Fe, indicate that the common rock-forming silicates and oxides tend to partition (Cu > Cd) ≫ Zn ≫ Mn ≫ Fe ≫ Mg into chloride-rich hydrothermal/metamorphic fluids. A possible exception for manganese is almandine-garnet. Copper and cadmium are bracketed because the 95% confidence intervals exceed 100% of their associated partition coefficients. The results for copper and cadmium are intriguing but, we must conclude, statistically inconclusive.

The partition coefficients contain information on the relative mobility of these base metals. Note that, in general, the less abundant the base metal the greater its partitioning into the fluid. In Part II (ILTON, 1990), the partition coefficients are used in a simple fluid flow model to demonstrate that, under certain conditions, congruent exchange reactions can strongly fractionate the base metals, and enrich the geochemically scarce base metals relative to iron.

The partition coefficients are preliminary values, subject to revision by further experimental work and greater information concerning base metals in naturally occurring silicates and oxides. Our knowledge is still extremely limited on both accounts.

Acknowledgements—We thank our colleagues Siggi Gislason and Glenn Wilson for many fruitful discussions. Hans Annersten and John Ferry provided constructive reviews of an early version of the manuscript. ESI thanks Phil Candela, and Lukas Baumgartner for constructive reviews and comments. This research was funded, in part, by NSF grants EAR-8206177 and EAR-8411050 awarded to Hans P. Eugster.

REFERENCES

ANNERSTEN H. and EKSTRÖM T. (1971) Distribution of major and minor elements in coexisting minerals from a metamorphosed iron formation. *Lithos* **4**, 185–204.

BADELOW S. T. (1958) Pyrochroit, zinkhaltiger serpentin und allophan aus der Lagerstätte Almalyk (Usbekistan). *Sapiski Wsesojusn. Miner. Obsch.* **87**, 698.

BANKS N. G. (1974) Distribution of copper in biotite and biotite alteration products in intrusive rocks near two Arizona porphyry copper deposits. *J. Res. U.S. Geol. Surv.* **2**, 195–211.

BARNES H. L. (1979) Solubilities of ore minerals. In *Geochemistry of Hydrothermal Ore Deposits* (ed. H. L. BARNES), pp. 404–454. John Wiley & Sons.

BISCHOFF J. L. and DICKSON F. W. (1975) Seawater-basalt interaction at 200°C and 500 bars: implications for origin of seafloor heavy metal deposits and regulation of seawater chemistry. *Earth Planet. Sci. Lett.* **25**, 385–397.

BOCTOR N. Z. (1985) Rhodonite solubility and thermodynamic properties of aqueous $MnCl_2^0$ in the system MnO-SiO_2-HCl-H_2O. *Geochim. Cosmochim. Acta* **49**, 565–575.

BURTON J. C., TAYLOR L. A. and CHOU I-M. (1982) The fO_2-T and fS_2-T stability relations of hedenbergite and hedenbergite-johannsenite solid solutions. *Econ. Geol.* **77**, 764–783.

CANDELA P. A. and HOLLAND H. D. (1984) The partitioning of copper and molybdenum between silicate melts and aqueous fluids. *Geochim. Cosmochim. Acta* **48**, 373–380.

DASGUPTA H. C. (1967) Intracrystalline element correlation in magnetite. *Econ. Geol.* **62**, 487–493.

DE ALBUQUERQUE C. A. R. (1971) Petrochemistry of a series of granitic rocks from northern Portugal. *Bull. Geol. Soc. Amer.* **82**, 2738–2798.

DE ALBUQUERQUE C. A. R. (1973) Geochemistry of biotites from granitic rocks, northern Portugal. *Geochim. Cosmochim. Acta* **37**, 1799–1802.

DISSANAYAKE C. B. and VINCENT E. A. (1972) Zinc in rocks and minerals from the Skaergaard intrusion, east Greenland. *Chem. Geol.* **9**, 285–297.

ELLIS A. J. (1968) Natural hydrothermal solutions and experimental hot-water/rock interaction: Reactions with NaCl solutions and trace metal extraction. *Geochim. Cosmochim. Acta* **32**, 1356–1363.

EUGSTER H. P. (1986) Minerals in hot water. *Amer. Mineral.* **71**, 655–673.

EUGSTER H. P. and ILTON E. S. (1983) Mg-Fe fractionation in metamorphic environments. In *Kinetics and Equilibrium in Mineral Reactions* (ed. S. K. SAXENA), pp. 115–140. Springer-Verlag.

FERRY J. M. (1985a) Hydrothermal alteration of Tertiary igneous rocks from the Isle of Skye, northwest Scotland I. Gabbros. *Contrib. Mineral. Petrol.* **91**, 264–282.

FERRY J. M. (1985b). Hydrothermal alteration of Tertiary igneous rocks from the Isle of Skye, northwest Scotland II. Granites. *Contrib. Mineral. Petrol.* **91**, 283–304.

FRONDEL C. and ITO J. (1966) Hendricksite, a new species of mica. *Amer. Mineral.* **51**, 1107.

GRAYBEAL F. T. (1973) Copper, manganese, and zinc in coexisting mafic minerals from Laramide intrusive rocks in Arizona. *Econ. Geol.* **68**, 785–798.

HAACK U., HEINRICHS H., BONEB M. and SCHNEIDER A. (1984) Loss of metals from pelites during regional metamorphism. *Contrib. Mineral. Petrol.* **85**, 116–132.

HAJASH A. (1975) Hydrothermal processes along mid-ocean ridges: an experimental investigation. *Contrib. Mineral. Petrol.* **53**, 205–226.

HASLAM H. W. (1968) The crystallization of intermediate and acid magmas at Ben Nevis, Scotland. *J. Petrol.* **9**, 84–104.

HEINRICHS H., SCHULZ-DOBRICK B. and WEDELPOHL K. H. (1980) Terrestrial geochemistry of Cd, Bi, Tl, Pb, Zn, and Rb. *Geochim. Cosmochim. Acta.* **44**, 1519–1533.

168	E. S. Ilton and H. P. Eugster

HENDRY D. A. F., CHIVAS A. R., REED S. J. B. and LONG J. V. P. (1981) Geochemical evidence for magmatic fluids in porphyry copper mineralization. Part II. Ion probe analysis of Cu contents of mafic minerals, Koloula igneous complex. *Contrib. Mineral. Petrol.* **78**, 404–412.

HENLEY R. W. and ELLIS A. J. (1983) Geothermal systems ancient and modern: a geochemical review. *Earth-Science Rev.* **19**, 1–50.

HILDRETH E. W. (1977) The magma chamber of the Bishop Tuff: Gradients in temperature, pressure and composition. Ph.D. thesis. University of California, Berkeley, California. 328p.

HILDRETH E. W. (1979) The Bishop Tuff: Evidence for the origin of compositional zonation in silicic magma chambers. *Geol. Soc. Amer. Spec. Paper* 180, 43–75.

HOLLAND H. D. (1972) Granites, solutions and base metal deposits. *Econ. Geol.* **67**, 281–301.

ILTON E. S. (1987) Base metal exchange between rock-forming silicates, oxides, and hydrothermal/metamorphic fluids. Ph.D. Dissertation. Johns Hopkins University, Baltimore, Maryland. 230p.

ILTON E. S. (1990) Partitioning of base metals between silicates, oxides, and a chloride-rich hydrothermal fluid. Part II. Some aspects of base metal fractionation during isothermal metasomatism. In *Fluid-Mineral Interactions: A Tribute to H. P. Eugster* (eds. R. J. SPENCER and I-MING CHOU), Spec. Publ. 2, pp. 171–177. The Geochemical Society.

ILTON E. S. and EUGSTER H. P. (1989) Base metal exchange between magnetite and a chloride-rich hydrothermal fluid. *Geochim. Cosmochim. Acta* **53**, 291–301.

ILTON E. S. and VEBLEN D. R. (1988a) Cu in sheet silicates from rocks associated with porphyry copper deposits: A transmission electron microscopy study. (abstr.). *The V. M. Goldschmidt Conference-Symposiom: Ore-Forming Processes.* The Geochemical Society. Abstracts and Program, p. 49.

ILTON E. S. and VEBLEN D. R. (1988b) Cu inclusions in sheet silicates from porphyry copper deposits. *Nature* **334**, 516–518.

KLEIN C. and ITO J. (1968) Zincian and manganoan amphiboles from Franklin, New Jersey. *Amer. Mineral.* **53**, p. 1264.

MAROWSKY G. and WEDEPOHL K. H. (1971) General trends in the behavior of Cd, Hg, Tl and Bi in some major rock forming processes. *Geochim. Cosmochim. Acta* **35**, 1255–1267.

MEINERT L. D. (1987) Skarn zonation and fluid evolution in the Groundhog Mine, Central Mining District, New Mexico. *Econ. Geol.* **82**, 523–545.

MOTTL M. J., HOLLAND H. D. and CORR R. F. (1979) Chemical exchange during hydrothermal alteration of basalt by seawater–II Experiments for Fe, Mn and sulfur species. *Geochim. Cosmochim. Acta* **43**, 869–884.

POWNCEBY M. I., WALL V. J. and O'NEILL H. ST. C. (1987) Fe-Mn partitioning between garnet and ilmenite: experimental calibration and applications. *Contrib. Mineral. Petrol.* **97**, 116–126.

ROEDDER E. (1984) *Fluid Inclusions.* Reviews in Mineralogy 12. 643pp. Mineralogical Society of America.

RUMBLE D. (1973) Fe-Ti oxide minerals from regionally metamorphosed quartzites of western New Hampshire. *Contrib. Mineral. Petrol.* **42**, 181–195.

RUMBLE D. and DICKENSON M. P. (1986) Field trip guide to black mountain Wildwood road cut and Beaver Brook, Mount Moosilauke area, New Hampshire. *Field Trip Guide Book: Regional Metamorphism and Metamorphic Relations in N. Western and Central New England.*

SEYFRIED W. E. and BISCHOFF, J. L. (1981) Experimental seawater-basalt interaction at 300°C, 500 bars, chemical exchange, secondary mineral formation and implications for the transport of heavy metals. *Geochim. Cosmochim. Acta* **46**, 985–1002.

SEYFRIED W. E. and JANECKY D. R. (1985). Heavy Metal and sulfur transport during subcritical and supercritical hydrothermal alteration of basalt; influence of fluid pressure and basalt composition and crystallinity. *Geochim. Cosmochim. Acta* **49**, 2545–2560.

SEYFRIED W. E. and MOTTL M. J. (1982) Hydrothermal alteration of basalt by seawater under seawater-dominated conditions. *Geochim. Cosmochim. Acta* **46**, 985–1002.

SKINNER B. J. (1979) The many origins of hydrothermal mineral deposits. In *Geochemistry of Hydrothermal Ore Deposits,* (ed. H. L. BARNES), pp. 1–21. J. Wiley & Sons.

SVERJENSKY D. A. (1984) Prediction of Gibbs free energies of calcite-type carbonates and the equilibrium distribution of trace elements between carbonates and aqueous solutions. *Geochim. Cosmochim. Acta* **48**, 1127–1134.

SVERJENSKY D. A. (1985) The distribution of divalent trace elements between sulfides, oxides, silicates and hydrothermal solutions: I. Thermodynamic basis. *Geochim. Cosmochim. Acta* **49**, 853–864.

TAUSON L. V. and KRAVCHENKO L. A. (1956) Characteristics of lead and zinc distribution in minerals of the Caledonian granitoids of the Susamyr batholith in central Tian-Shan. *Geochemistry* **1**, 78–88.

VINCENT E. A. and BILEFIELD L. I. (1960) Cadmium in rocks and minerals from the Skaergaard intrusion, East Greenland. *Geochim. Cosmochim. Acta* **19**, 63–69.

WEDEPOHL K. H. (1972) Zinc. In *Handbook of Geochemistry* (ed. K. H. WEDEPOHL), p. 30-D-3. Springer-Verlag.

WEDEPOHL K. H. (1974) Copper. In *Handbook of Geochemistry* (ed. K. H. WEDEPOHL), p. 29-D-14. Springer-Verlag.

WEDEPOHL K. H. (1978) Manganese. In *Handbook of Geochemistry* (ed. K. H. WEDEPOHL), p. 25-D-3. Springer-Verlag.

WEISSBERG B. G., BROWNE P. R. L. and SEWARD T. M. (1979) Ore metals in active geothermal systems. In *Geochemistry of Hydrothermal Ore Deposits* (ed. H. L. BARNES), pp. 738–780. John Wiley & Sons.

APPENDIX
MINERAL COMPOSITIONS AND ZONING

FERRY (1985b)

Biotite compositions used in the derivations are iron rich, with Fe/Mg ratios = 2.25–1.14, and Al(VI) poor. Si/Al ratios = 2.65–2.3 and (Si + Al)IV totals are slightly less than 4. Compositional zoning is *not* reported for biotite. Amphiboles are either ferro-hornblende, ferro-edenite, or ferro-edenitic hornblende (nomenclature of Leake, 1978). One sample contains a Na rich amphibole. Fe²⁺/(Mg + Fe²⁺) = 0.56–0.92. Pyroxenes are ferroaugites with 0.78–0.86 Ca/60 and Fe²⁺/(Mg + Fe²⁺) = 0.65–0.88.

Chemical zoning was minor for pyroxene and amphibole in most samples. Some biotite, amphibole, and pyroxene have been partially altered to chlorite. The compositions of unaltered and partially altered silicates are identical.

Magnetites are near magnetite-ulvospinel solid solutions, where wt.% TiO_2 = 8.36–15.8. Most magnetites show signs of incipient oxidation-exsolution to submicroscopic intergrowths of ilmenite and magnetite. The process is considered to be isochemical with respect to cations, and averaged compositions are reported. Ilmenites are homogenous (zoning is not reported) and unaltered, with minor Mn and Fe^{3+}. Ferry concludes that the compositions of these minerals were unaffected by later hydrothermal activity and probably record conditions near crystallization temperatures (650–750°C).

HILDRETH (1977)

Mineral compositions are broadly similar to those described in Ferry (1985b). Microprobe analyses indicate that magnetite, ilmenite, and biotite are *un*zoned with respect to major elements and Mn.

The majority of individual biotite crystals indicate some degree of inhomogeneity; interlayer cations can vary by 20%, FeO and MgO by ±5–10%, whereas MnO is fairly homogenous. Lower T biotites exhibit varying degrees of oxidation. Biotites in devitrified samples are severely oxidised and iron oxides are present along cleavage planes. Hildreth, however, confined his study to vitric samples. Furthermore, biotite octahedral site compositions are strongly correlated to T and Kd(biot/mt) values indicate that Fe and Mn were systematically partitioned between biotite and magnetite (see Table 1 and Fig. 2, this paper). Hildreth concludes that whereas interlayer and hydroxyl sites record post magmatic compositions, octahedral site compositions generally maintain their magmatic signatures (although he warns for a—"cautious approach to the data").

Magnetite and ilmenite compositions are strongly correlated to T. Kd(ilm/mt) values indicate that Mn and Fe were systematically partitioned between ilmenite and magnetite. Furthermore, Kd(ilm/mt) exhibits a significant T dependence, where Kd(720–730°C) = 2.04 and Kd(765–790°C) = 1.55. A t-test indicates that the difference between the two Kds is highly significant, where $P \ll 0.001$.

The precision and systematic nature of the data suggests that Mn-Fe partitioning between magnetite, and ilmenite and biotite approached equilibrium. The evidence is stronger for mt-ilm than mt-biot.

ANNERSTEN and EKSTRÖM (1971)

Magnetites are nearly pure Fe_3O_4. Amphiboles are either hornblende or actinolite with Fe/Mg = 3.6–0.0017. Fe/Mg = 2.0–0.0084 for biotite. The Fe/Mg ratios reflect fO_2 gradients.

Microprobe analyses indicate that biotite and amphibole are chemically homogenous with respect to major elements and Mn. Chemical zoning is *not* reported. Trace element analyses for magnetite (Mn, Zn, Cu, Cd) and both silicates (Zn, Cu, Cd) were obtained from mineral separates and AAS. A few magnetite grains were contaminated by sphene. Mn seemed to be most effected by sphene contamination.

Analysed pairs of biotite and amphibole were in close contact, whereas the textural relationship of mt-silicate pairs was not reported.

Kd(mt/silicate) for Mn-Fe partitioning is largely independent of Fe/Mg, whereas Cd, Zn, and Cu partitioning indicates some correlation to Fe/Mg ratio. Excluding the oxidised samples increases precision, but more data is required to provide an adequate test of this hypothesis. There is a significant difference between Kd(hbl/mt) and Kd(act/mt) for Mn-Fe partitioning, where a t-test yields $0.05 \gg P > 0.02$.

The large variance in Kd(silicate/mt) for Cd, Zn, and Cu indicates either analytical problems, hidden correlation factors, lack of M-Fe exchange equilibrium, or some combination of the above.

FERRY (1985a)

The pertinent alteration reactions are olivine to talc + magnetite, and talc and magnetite to chlorite or montmorillinite. Magnetite grains are intimately associated with talc and chlorite (Ferry—"magnetite . . . (is) . . . set in a matrix of talc and chlorite). Biotite usually occurs along grain boundaries in close contact with chlorite and amphibole. Amphibole occurs in a variety of textural relationships including as crystals within chlorite. Calcite occurs within plagioclase and along grain boundaries. The exact textural relationships, however, are not given for any one sample. One might assume that magnetites had a better chance of equilibrating with chlorite and talc.

$Fe^{2+}/(Mg + Fe^{2+})$ = 0.03–0.07, 0.23–0.49, 0.20–0.53, and 0.13–0.44 for talc, chlorite, amphibole, and biotite. Amphiboles are primarily actinolites, biotites have low interlayer site occupencies and low Al(VI), chlorites have 1.2–2.5 Al/140, and calcites are nearly pure $CaCo_3$. Zoning and inhomogeneity within single crystals are not reported.

RUMBLE (1973)

Oxide phases are in intimate contact. Zoning is *not* reported. Hematite often contains submicroscopic-microscopic lamellae of ilmenite. Rumble used a broad electron beam to get a bulk analysis. He reasoned that bulk compositions were representative of homogenous ilmeno-hematite solid solutions at recrystallization conditions.

BURTON et al. (1982)

Andradite and magnetite exhibited minor compositional zoning. Clinopyroxene grains were to small for accurate microprobe analysis. Burton *et al.* (1982) calculated run-product clinopyroxene compositions from mass balance.

Fluid-Mineral Interactions: A Tribute to H. P. Eugster
© The Geochemical Society, Special Publication No. 2, 1990
Editors: R. J. Spencer and I-Ming Chou

Partitioning of base metals between silicates, oxides, and a chloride-rich hydrothermal fluid. Part II. Some aspects of base metal fractionation during isothermal metasomatism

EUGENE S. ILTON

Department of Mineral Sciences, The American Museum of Natural History,
Central Park West at 79th St., New York, New York 10024, USA

Abstract—The distribution coefficients described in ILTON and EUGSTER (1990) are used in a simple isothermal fluid flow model to demonstrate that congruent exchange reactions involving the silicate rock fraction can enrich the geochemically scarce base metals relative to iron in more distal portions of chloride-rich metasomatic fluids.

INTRODUCTION

THE FORMATION of economically viable manganese, zinc, copper, cadmium and copper hydrothermal ore deposits depends on their enrichment relative to iron, as well as relative to other common rock-forming elements. Fractionation and enrichment can occur during magmatic vapor phase separation, during removal and acquisition by the fluid of the metals from rocks, and during the depositional event due to differing ore-mineral solubilities. Ore-grade metallization often requires the focusing of fluids that have interacted with large volumes of rock or magma, with subsequent precipitation of metals over compressed distances from high fluid fluxes with relatively low metal concentrations, or from lower fluid fluxes with relatively higher metal concentrations. In any case, hydrothermal ore deposits are the end products of a complex series of enrichment events that extend over wide ranges of temperature and pressure.

The present paper has modest goals and cannot address the complexity of ore formation. Instead, this contribution develops a simple isothermal model for the sulfur free or sulfide undersaturated system, that gives some physical meaning to the partition coefficients derived in ILTON and EUGSTER (1990). The model illustrates that congruent base metal exchange reactions between silicates and a flowing chloride-bearing hydrothermal fluid can strongly fractionate manganese, zinc, cadmium, copper and iron, either during metal acquisition by the fluid or rock. Given the proper conditions, strong fractionation can lead to the enrichment of manganese, zinc, copper, cadmium and copper *relative* to iron. The model may yield insights concerning the formation of distal zinc skarns, and manganese enrichment in distal skarns as well as in distal portions of skarns. Alternatively, fraction-

ation patterns that differ from those predicted would indicate, in part, either the influence of more complex physical conditions and reactions, perhaps involving the non-silicate rock-fraction, or that the partition coefficients need to be revised. It is stressed that the model does *not* address the question of *absolute* enrichment. Furthermore, the data and therefore the model is limited, and errors associated with the partition coefficients can be large.

DEVELOPMENT OF THE MODEL

The model illustrates the endmember case of isothermal, one dimensional fluid flow with trace base metal concentrations in the fluid and solid phases. The Mg-endmember serves as the reference component. Phases are conserved, and mineral/fluid interactions are restricted to congruent ion exchange reactions. (It is assumed that the fluid and aquifer are equilibrated with respect to major components.) Eventual inclusion of alteration and precipitation-dissolution reactions will require further experimental data. The model employs a form of chromatographic theory (*e.g.*, WILSON, 1940). The fluid and solid form an interconnected continuum, and the condition of instantaneous local equilibrium is imposed (THOMPSON, 1959). A pressure gradient forces the fluid from a reservoir of constant composition into an aquifer. The distribution coefficients are assumed to be invariant with respect to the pressure gradient required for fluid flow, and infiltration is the only transport mechanism (*i.e.* dispersion and diffusion are neglected). Our hypothetical system is undersaturated with respect to sulfides, the fluid is chloride-rich and the reactive minerals are phlogopite, tremolite, pyrope, and diopside. The partition coefficients are strictly associated with a 2 M KCL solution, $P = 2$ Kb, and $fO_2 \approx NNO$. However, ILTON and EUGSTER (1989, 1990) argue they might be valid, for manganese and zinc, over some range of chloride concentrations, pH, P, and fO_2.

Regarding the geological literature, the reader is referred to KORZHINSKI (1970), HOFFMAN (1972), FLETCHER and HOFFMAN (1974), and BAUMGARTNER and RUMBLE (1988) for a detailed development of the relevant transport equations. An abbreviated derivation, closely following BAUMGARTNER and RUMBLE (1988) but with some variations on the theme, is presented here.

Mass balance requires that the rate of change for the concentration of base metal B, $\partial B/\partial t$, within a given volume be equal to the integral of the flux of B across the surface enclosing the volume. This is expressed by the continuity equation,

$$\partial B/\partial t = -\text{div } \vec{J} \tag{1}$$

for an infinitesimally small volume element dV. The total concentration of a base metal (B) in a given volume composed of minerals and fluid is given by

$$B = (1 - \beta) \sum_{n=1}^{x} m_n b^n + \beta m_{fl} b^{fl} \tag{2}$$

where β is the porosity completely filled by fluid, m_n and m_{fl} are the number of moles of mineral n and fluid, respectively, per unit volume of pure rock and fluid, and b^n and b^{fl} are the number of moles of the base metal in one mole of mineral n and fluid, respectively. Substituting equation 2 into equation 1 yields

$$\partial/\partial t[(1 - \beta) \sum_{n=1}^{x} m_n b^n + \beta m_{fl} b^{fl}] = -\text{div } \vec{J}. \tag{3}$$

Given the condition of instantaneous equilibrium, the composition of the fluid and a mineral are related by the distribution coefficient Kd. The base metals of interest are treated as trace elements, whereas magnesium is the major element in both the fluid and solid, such that the distribution coefficient $Kd = b^{fl} Mg^n/b^n Mg^{fl}$. Since the base metals are assumed to be trace elements, the distribution coefficients are independent of composition, $Mg^n/Mg^{fl} = M$ is considered a constant and $b^n = b^{fl}M/Kd$. Given unidirectional fluid flow, equation 3 can be modified to

$$\partial b^{fl}/\partial t[(1 - \beta) \sum_{n=1}^{x} m_n M/Kd_n + \beta m_{fl}]$$

$$= -\partial \beta \vec{v} b^{fl}/\partial Z \tag{4}$$

where the constant $\vec{v}$ is the fluid velocity and $\beta \vec{v} b^{fl} = \vec{J}$. Here, the porosity is completely accessible to infiltration. Solving equation 4 for the distance

traveled by the compositional front (Z) for the base metal of interest yields

$$Z = \beta \vec{v} t/[(1 - \beta) \sum_{n=1}^{x} m_n M/Kd_n + \beta m_{fl}]. \tag{5}$$

For small porosities equation 5 reduces to

$$Z = \beta \vec{v} t/(M \sum_{n=1}^{x} m_n/Kd_n), \quad \text{and} \tag{6}$$

$$Z = \beta \vec{v} t/(mM/Kd) \tag{7}$$

for monomineralic aquifers. The result is substantively identical to equation 38 in BAUMGARTNER and RUMBLE (1988), where the rare isotope is treated as a trace element. Given the linear relationship between metal contents in the fluid and solid, pure infiltration forms compositional plateaus along the length of the aquifer, separated by sharp reaction fronts, whereas combined infiltration, dispersion, and diffusion develop compositional plateaus separated by diffuse fronts (FLETCHER and HOFFMAN, 1974). Consequently, for monomineralic aquifers, the relative distances (Z) traveled by the base metal compositional fronts are given by the ratios of their distribution coefficients:

$$Z(b)/Z(Fe) = \frac{\beta \vec{v} t/(mM/Kd[b\text{-Mg}])}{\beta \vec{v} t/(mM/Kd[Fe\text{-Mg}])} \tag{8}$$

$$= Kd[b\text{-Mg}]/Kd[Fe\text{-Mg}], \tag{9}$$

and because magnesium is common

$$Z(b)/Z(Fe) = Kd[b\text{-Fe}]. \tag{10}$$

Therefore, the relative positions of the fronts are given by the partition coefficients in ILTON and EUGSTER (1990).

Uncertainties associated with the model

The 95% confidence intervals for the positions of the fronts are identical to those associated with the distribution coefficients in Tables 3 and 4 from ILTON and EUGSTER (1990). The reader is reminded that the confidence intervals only reflect the variance of the data. Uncertainties regarding temperature estimates, compositional differences between minerals in experimental and natural assemblages, and lack of definitive proof of exchange equilibrium in the natural assemblages are not accounted for. Moreover, the minerals are modeled as near Mg-endmembers, whereas the Kds are derived from more complex solid solutions. Copper and cadmium are modeled despite the large statistical uncertainties associated with their Kds. Although the positions of the copper and cadmium fronts are statistically inconclusive, the results are considered sufficiently informative to present.

MONOMINERALIC AQUIFERS

Silicates as sinks for base metals

At $t = 0$, assume that the aquifer, composed of either phlogopite, tremolite, diopside *or* pyrope is undersaturated with respect to the concentrations of base metals in the reservoir fluid. Further, for clarity and simplicity, we assume the special case of base metal concentration $= 0$ in the aquifer at $t = 0$. As infiltration proceeds, the metals are stripped from the *fluid* with varying degrees of efficiency; iron is removed from the fluid most efficiently, followed by manganese, zinc, cadmium, and copper. At $500°C$, the relative distances (Z) traveled by the fronts are—$\sim Z(Fe):6Z(Mn):40Z(Zn):1600Z(Cd):3600Z(Cu)$ for a phlogopite aquifer, and $Z(Fe):7Z(Mn):70Z(Zn):1200Z(Cd):11000Z(Cu)$ for a tremolite aquifer, using the distribution coefficients in Tables 1 and 2 from ILTON and EUGSTER (1990). The fractionation pattern is illustrated in Fig. 1.

The copper and iron fronts travel the fastest and slowest, respectively. If the iron front has progressed 0.1 km, then the manganese, zinc, cadmium, and copper fronts traveled 0.6, 4.0, 160 and 360 km, and 0.7, 7.0, 120, and 1100 km for phlogopite and tremolite aquifers, respectively. Since the fronts travel at different velocities, the distance *between* the fronts increases with time. Therefore, compared to the reservoir, the leading fluid packets are enriched in copper, cadmium, zinc, and manganese relative to iron.

If such a fluid reaches a distal reactive rock such as a carbonate, which removes the metals without further fractionation, then the enrichment sequence would be copper followed by cadmium → zinc → manganese → iron. Iron starts to precipitate *only* when the iron front crosses the carbonate unit. A very high fluid flux would overwhelm the exchange capacity of the aquifer, and would simply impose the base metal composition of the reservoir on the carbonate unit. Alternatively, if the fluid supply is exhausted shortly after the zinc and manganese fronts cross the carbonate boundary, then zinc, manganese, cadmium, and copper will be enriched relative to iron at that site, compared to the reservoir.

Longer fluid path lengths, prior to deposition, will increase the separation of the fronts, and

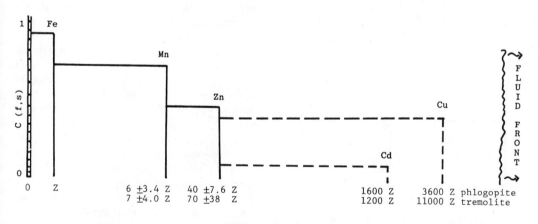

KILOMETERS

FIG. 1. Schematic diagram illustrating the development of base metal compositional fronts, at $500°C$, during infiltration of a chloride-rich fluid into an aquifer containing phlogopite or tremolite. The reservoir fluid is Mg-rich, but contains trace concentrations of Fe, Mn, Zn, Cd, and Cu, whereas initially there are no base metals in the aquifer. The base metals are stripped from the *fluid* as infiltration proceeds. The vertical axis represents the concentrations of the metals in both the fluid and solid (the fluid and solid scales differ). The horizontal scale marks distance from the fluid source. As discussed in the text, the relative positions of the fronts are given by the distribution coefficients. The 95% confidence intervals are given for the positions of the fronts. The Cu and Cd fronts are dashed as a reminder of the large uncertainties associated with their positions (see ILTON and EUGSTER, 1990, for details).

thereby increase the likelihood and degree of minor base metal enrichment relative to iron, in distal portions of the metasomatic system. Greater porosities or channelized fluid flow would require longer fluid path lengths to achieve the same degree of fractionation.

Other silicates, such as chlorite and talc, may cause similar fractionation (see Tables 1 and 2 in ILTON and EUGSTER, 1990).

The fact that copper is usually proximal, *for those parts of metasomatic systems usually sampled,* is consistent with copper mobility controlled by more complex reactions involving sulfides. Alternatively, given the large uncertainty associated with $Kd(Cu/Fe)$, the relative position of the copper front is *not* well defined.

At 700°C, partition coefficients (see Table 1 in ILTON and EUGSTER, 1990) yield metasomatic fronts for manganese and iron in aquifers composed of phlogopite, tremolite, diopside, or pyrope, as shown in Fig. 2. The manganese front leads the

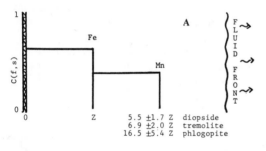

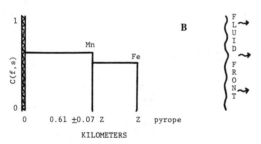

KILOMETERS

FIG. 2. Schematic diagrams illustrating the development of base metal compositional fronts, at 700°C, during infiltration of a chloride-rich fluid into an aquifer containing (A) phlogopite, tremolite, or diopside and (B) pyrope. The reservoir fluid is Mg-rich, but contains trace concentrations of Fe, Mn, Zn, Cd, and Cu, whereas initially there are no base metals in the aquifer. The base metals are stripped from the *fluid* as infiltration proceeds. The vertical axis represents the concentrations of the metals in both the fluid and solid (the fluid and solid scales differ). The horizontal scale marks distance from the fluid source. As discussed in the text, the relative positions of the fronts are given by the distribution coefficients. The 95% confidence intervals are given for the positions of the fronts.

iron front for phlogopite-, tremolite- and diopside-rich rocks. The sense of fractionation, however, is reversed by an aquifer composed of pyrope, where distal portions of the fluid are enriched in manganese relative to iron. Pyrope/almandine-rich garnets may be the only common rock-forming silicate to produce such fractionation. Although the model is limited, it may yield insights concerning the fractionation of manganese and iron in reducing skarn environments, such as deeper level tungsten skarns associated with almandine/pyrope-rich garnets, compared to higher level more oxidizing skarn environments associated with andradite-rich garnets.

Skarns distal to possible but unknown plutonic fluid sources (distal skarns), such as the Paymaster skarn (MEINERT, 1980), characteristically are associated with zinc deposits, and silicates unusually enriched in manganese (EINAUDI *et al.,* 1981). BURT (1977), MEINERT (1980, 1984, 1987), and YUN and EINAUDI (1982) have suggested that extended fluid interaction with non-carbonate rock prior to contact with a carbonate unit may be responsible for the high degree of zinc and manganese enrichment in distal skarns and more distal portions of skarns. The model presented here, although broadly consistent with their interpretations, assuming gentle temperature gradients, is obviously too limited and simplistic for direct application. It does suggest, however, that simple exchange reactions between flowing chloride-rich hydrothermal fluids and silicates may enhance, under certain conditions, such fractionation and enrichment. Obviously, proximal precipitation of chalcopyrite could produce similar enrichment of manganese and zinc, relative to iron, in distal skarns. Complex reactions involving ore minerals such as sphalerite, pyrrhotite, and pyrite may also strongly influence base metal fractionation and enrichment patterns in distal skarns.

MEINERT (1984) has documented distal clinopyroxenes (cpx) from the Iron Hill skarn, British Columbia that are unusually enriched in manganese. Furthermore, the cpx rims have higher Fe/Mn ratios than the cores. Apparently, the fluids had to flow through a considerable thickness of volcanic rock before encountering a limestone unit and precipitating Mn-enriched cpx. Given the data, MEINERT (1984) suggested that distal cpx formed from fluids that evolved from higher Mn/Fe ratios to lower Mn/Fe ratios. This is broadly consistent with the fractionation predicted by the model, where the Fe-rich front follows the Mn-rich front. Thus, it is possible that ion exchange reactions could be partially responsible for such patterns of manganese enrichment. Alternatively, more complex reactions involving precipitation of iron-rich minerals such as andradite in relatively proximal locations (YUN and EINAUDI, 1982), decreasing fO_2, and decreasing temperature (MEINERT, 1987; ILTON and EUGSTER, 1990) may be the primary cause of extreme manganese enrichment in distal portions of the skarn.

In the foregoing analysis, the base metals have been modeled as trace elements, and magnesium served as a reference frame. This approach yielded simple analytical solutions to the transport equations. Some metasomatic fluids, however, may contain high concentrations of metals. Further, silicates in metasomatic systems are often zoned. Consequently, magnesium eventually needs to be included in a more generalized transport model, that treats non-trace concentrations of base metals, and that allows for non-equilibrium partitioning. Treating the base metals as "major" elements would effect the model quantitatively, but *not* necessarily qualitatively. Since magnesium is preferentially partitioned into silicates relative to iron (EUGSTER and ILTON, 1983), one might expect magnesium to be depleted relative to iron in distal fluids. Once again this is in qualitative agreement with the skarn literature (*e.g.,* BURT, 1977; DICK and HODGESON, 1982). Enlarging the data base may eventually allow one to couple the partition coefficients with alteration and precipitation-dissolution reactions occurring along various temperature and pressure paths.

Silicates as sources of base metals

This model is identical to the model of the previous section, except the initial conditions are reversed, such that the reservoir fluid contains initial base metal concentration = 0, whereas the aquifer contains trace concentrations of base metals. As infiltration proceeds, the base metals are stripped from the *aquifer* with varying degrees of efficiency. Given an arbitrary amount of fluid flux, copper is stripped from the greatest rock volume, followed in decending order by cadmium, zinc, manganese, and iron (Fig. 3). Enrichment patterns in the fluid are spatially zoned. For example, there is no zinc in the fluid and rock *up to* the zinc front, whereas, compared to the original rock, zinc is enriched in the fluid relative to manganese and iron *between* the zinc front and fluid front. In this scenario, a distal carbonate unit would record an abrupt decrease or cessation of base metal precipitation as its associated stripping front encountered the reactive rock.

Despite the manifest differences between silicates acting as sources *vs* sinks of base metals, the stripping fronts travel with velocities directly propor-

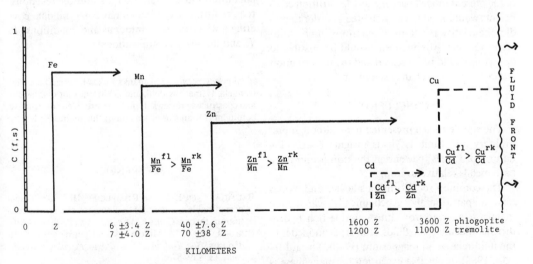

FIG. 3. Schematic diagram illustrating the development of base metal compositional fronts, at 500°C, during infiltration of a chloride-rich fluid into an aquifer composed of phlogopite or tremolite. The reservoir fluid is Mg-rich, but contains no base metals, whereas initially the aquifer contains trace concentrations of Fe, Mn, Zn, Cd, and Cu. The base metals are stripped from the *aquifer* as infiltration proceeds. The vertical axis represents the concentrations of metals in both the fluid and solid (the fluid and solid scales differ). The horizontal scale marks distance from the fluid source. As discussed in the text, the relative positions of the fronts are given by the distribution coefficients. The 95% confidence intervals are given for the positions of the fronts. The Cu and Cd fronts are dashed as a reminder of the large uncertainties associated with their positions (see ILTON and EUGSTER, 1990, for details).

176   E. S. Ilton

tional to the distribution coefficients. Consequently, the implications for enrichment are similar to those described in the previous model, where in this case too much fluid flow strips the entire aquifer of base metals, and simply imposes the base metal composition of the aquifer on the the carbonate unit. In which case no relative enrichment occurs.

POLYMINERALIC AQUIFERS

Equation 6 gives the position of the compositional front for a polymineralic aquifer. The relative position of two fronts is given by

$$Z(b)/Z(Fe) = (\sum_{n=1}^{x} X_n/Kd_n^{Fe-Mg})/(\sum_{n=1}^{x} X_n/Kd_n^{b-Mg}).$$

(11)

Therefore, the relative positions of the fronts are a simple function of the distribution coefficients and mole fractions (X) of each mineral in the aquifer. Since the silicates exhibit similar partitioning behavior, the combination of biotite, amphibole, pyroxene, and perhaps chlorite, and talc, would effect the model quantitatively, but *not* necessarily qualitatively (see Tables 1 and 2 in ILTON and EUGSTER, 1990). The addition of significant almandine-pyrope garnet, however, could reverse the fractionation behaviors of manganese and iron. Within the context of the present model, base metal-Mg partition coefficients would need to be calculated in order to predict the relative positions of the fronts in polymineralic aquifers. Although this would be possible to do, errors would be increased and the fractionation *pattern* would not change significantly.

CONCLUSION

The simple model presented here, although preliminary and limited, yields insights concerning natural processes that enrich the minor and trace base metals relative to iron.

The common rock-forming silicates and oxides tend to partition the geochemically scarce base metals, relative to iron, into chloride-rich hydrothermal-metamorphic fluids. Iron is partitioned into the fluid relative to magnesium (EUGSTER and ILTON, 1983). A possible exception for manganese is almandine-garnet. It is interesting to note that, in general, the less abundant the base metal the greater its partitioning into the fluid.

Consequently, exchange reactions between flowing Cl-rich hydrothermal fluids, and rock-forming silicates tend to enrich (Cu > Cd) ≫ Zn ≫ Mn ≫ Fe ≫ Mg in more distal portions of the fluid. Copper

and cadmium are bracketed because of the large uncertainties associated with their partition coefficients (see ILTON and EUGSTER, 1990). Longer fluid path lengths and smaller porosities increase the degree of fractionation and possibilities for enrichment of the less abundant base metals relative to iron.

The degree to which this is not the case could reflect a violation of one or more of the conditions and simplifying assumptions built into the model, including differing ore mineral solubilities during deposition, and the percentage of base metals in the whole-rock *not* contained in the Fe-Mg oxide-silicate fraction.

The application of any such model requires, among other things, rigorous identification of the host phases and site occupancies for the base metal of interest; not necessarily a trivial matter when dealing with trace-minor concentrations and hydrothermally altered assemblages. As discussed in ILTON and EUGSTER (1990) the transmission electron microscope (TEM) is a powerful tool for addressing this problem. In fact TEM observations cast doubt on the validity of such modeling for copper (ILTON and VEBLEN, 1988).

Although the results have indicated some tendencies with respect to congruent exchange reactions, more rigorous isothermal models, as well as more general models involving *T-P* gradients, and alteration and precipitation-dissolution reactions, require further experimental and natural data covering a wider range of minerals and conditions (*P, T,* and fluid-solid compositions).

Acknowledgements—ESI thanks Lukas Baumgartner, Phil Candela, Demetrius Pohl, and Ed Mathez for discussions and constructive reviews. John Ferry provided constructive comments for an earlier version of this manuscript.

REFERENCES

BAUMGARTNER L. P. and RUMBLE D., III (1988) Transport of stable isotopes: I: Development of a kinetic continuum theory. *Contrib. Mineral. Petrol.* **98,** 417–430.

BURT D. M. (1977) Mineralogy and petrology of skarn deposits. *Soc. Italiana Mineralogia Petrologia Rendiconti* **33,** 859–873.

DICK L. A. and HODGSON C. J. (1982) The Mactung W-Cu (-Zn) contact metasomatic and related deposits of the northeastern Canadian Cordillera. *Econ. Geol.* **77,** 845–867.

EINAUDI M. T., MEINERT L. D. and NEWBERRY R. J. (1981) Skarn deposits. *Economic Geology, 75th Anniversary Volume,* 317–391.

EUGSTER H. P. and ILTON E. S. (1983) Mg-Fe fractionation in metamorphic environments. In *Kinetics and Equi-*

librium in Mineral Reactions (ed. S. K. SAXENA), pp. 115–140. Springer-Verlag.

FLETCHER R. C. and HOFFMAN A. W. (1974) Simple models of diffusion and combined diffusion-infiltration metasomatism. In *Geochemical Transport and Kinetics* (eds. A. W. HOFFMAN, B. J. GILETTI, H. S. YODER JR. and R. A. YUND), pp. 243–260. Carnegie Institution of Washington.

HOFFMAN A. W. (1972) Chromatographic theory of infiltration metasomatism and its application to feldspars. *Amer. J. Sci.* **272,** 69–90.

ILTON E. S. and EUGSTER H. P. (1989) Base metal exchange between magnetite and chloride-rich hydrothermal fluid. *Geochim. Cosmochim. Acta* **53,** 291–301.

ILTON E. S. and EUGSTER H. P. (1990) Partitioning of base metals between silicates, oxides, and a hydrothermal fluid. Part I. Evaluation of data derived from natural and experimental assemblages. In *Fluid-Mineral Interactions: A Tribute to H. P. Eugster* (eds. R. J. SPENCER and I-MING CHOU), Spec. Publ. 2, pp. 157–169. The Geochemical Society.

ILTON E. S. and VEBLEN D. R. (1988) Cu inclusions in sheet silicates from porphyry copper deposits. *Nature* **334,** 516–518.

KORZHINSKI D. S. (1970) *Theory of Metasomatic Zoning* (translated by Jean Agrell). 162 pp. Oxford, Clarendon Press.

MEINERT L. D. (1980) Evolution of metasomatic fluids by transport over large distances: An example from the Paymaster zinc skarn, Esmeralda Co., Nevada (abstr.). *Geol. Soc. Amer. Abst. Progr.* **12,** 482.

MEINERT L. D. (1984) Mineralogy and petrology of iron skarns in western British Columbia, Canada. *Econ. Geol.* **79,** 869–882.

MEINERT L. D. (1987) Skarn zonation and fluid evolution in the Groundhog Mine, Central Mining District, New Mexico. *Econ. Geol.* **82,** 523–545.

THOMPSON J. B. (1959) Local equilibrium in metasomatic processes. In *Researches in Geochemistry* (ed. P. H. ABLESON), pp. 427–457. J. Wiley and Sons.

WILSON J. N. (1940) A theory of chromatography. *Amer. Chem. Soc. J.* **62,** 1583–1591.

YUN S. and EINAUDI M. T. (1982) Zinc-lead skarns of the Yeonhwa-Ulchin district, South Korea. *Econ. Geol.* **77,** 1073–1032.

Fluid-Mineral Interactions: A Tribute to H. P. Eugster
© The Geochemical Society, Special Publication No. 2, 1990
Editors: R. J. Spencer and I-Ming Chou

Cassiterite solubility and tin speciation in supercritical chloride solutions

GLENN A. WILSON* and HANS P. EUGSTER†

Department of Earth and Planetary Sciences, The Johns Hopkins University, Baltimore, Maryland 21218, U.S.A.

Abstract—In order to model the evolution of hydrothermal cassiterite deposits, a quantitative understanding of the chemistry of cassiterite-fluid reactions under supercritical conditions is needed. To obtain this information, the solubility of cassiterite in HCl solutions was measured in closed-system experiments from 400 to 700°C, 1.5 kb, with the oxygen fugacity controlled by nickel-nickel oxide (NNO) and hematite-magnetite (HM). From the measured total chloride and tin concentrations, and the conditions of electrical neutrality at P and T, a set of equations was established for each run. By solving the equations simultaneously and finding the best-fit solution for all runs at the same set of P and T conditions, data were obtained on the identity of the dominant aqueous tin-chloride species and the equilibrium constants for the cassiterite dissolution reaction,

$$SnO_2 + XH^+ + nCl^- \rightleftarrows SnCl_n^{X-n} + \frac{X}{2} H_2O + \frac{4-X}{4} O_2,$$

where X is the oxidation state of aqueous tin and n is the ligation number of the tin-chloride species.
With the f_{O_2} of NNO at 400 to 600°C, it was found that $SnCl^+$ and $SnCl_2^0$ are dominant in solution, and at 700°C only $SnCl_2^0$ was detected. With the f_{O_2} of HM at 500 and 600°C, $SnCl_3^+$ was found to be dominant. The equilibrium constants for the cassiterite dissolution reaction are given by

$$\log K = -3.217 + \frac{2.640 \times 10^4}{T} - \frac{2.143 \times 10^7}{T^2},$$

for X = 2 and n = 1 (T in Kelvin);

$$\log K = 70.080 - \frac{8.246 \times 10^4}{T} + \frac{1.891 \times 10^7}{T^2},$$

for X = 2 and n = 2; and

$$\log K = 59.064 - \frac{3.511 \times 10^4}{T},$$

for X = 4 and n = 3.
The bivalent species dominate under natural ore-forming conditions. The solubility of cassiterite is a strong function of temperature, f_{O_2}, and pH. In order to obtain significant concentrations of tin in solution it is necessary to have temperatures above about 400°C, f_{O_2}'s at or below those defined by NNO, and pH's at or below those defined by the K-feldspar + muscovite + quartz assemblage. Appropriate changes in any of these conditions can cause cassiterite precipitation. The experimental results are in general agreement with studies of natural systems where the high temperature, low pH, and low f_{O_2} environment is indicated by phase relations and fluid inclusion data.

INTRODUCTION

CASSITERITE, SnO_2, is the principle ore mineral of tin, and its primary deposits are usually found to be associated with granitic intrusions. The ore may be found within altered portions of the granite, the so-called greisens, or in hydrothermal veins within the granite or the country rock. Some of the best-known deposits are those of Cornwall (southeast England), Erzgebirge (Czechoslovakia), the Iberian Peninsula, Bolivia, Malaysia, Korea, and southeast

China. The extensive literature on tin deposits is summarized in books by TAYLOR (1979), ISHIHARA and TAKENOUCHI (1980), EVANS (1982), HALLS (1985), and KWAK (1987). Many individual deposits have been studied in detail with contributions by KELLY and TURNEAURE (1970, Bolivia), BAUMANN et al. (1974, Czechoslovakia), SILLITOE et al. (1975, Bolivia), KELLY and RYE (1979, Portugal), COLLINS (1981, Tasmania), PATTERSON et al. (1981, Tasmania), ZHANG and LI (1981, China), JACKSON et al. (1982, England), CLARK et al. (1983, Peru), EADINGTON (1983, Australia), BRAY and SPOONER (1983, England), KWAK (1983, Tasmania), VON GRUENEWALDT and STRYDOM (1985, S. Africa), JACKSON and HELGE-

* Present address: Shell Development Company, Bellaire Research Center, P.O. Box 481, Houston, Texas 77001.
† Deceased.

SON (1985b, Malaysia), PETERSON (1986, Canada), PUCHNER (1986, Alaska), SUN and EADINGTON (1987, Australia), and many others. Although an extensive set of data is available now on mineral associations, fluid inclusions, and stable isotopes, little is known with respect to the hydrothermal transport of tin and the factors which initiate and sustain the precipitation of cassiterite and the associated sulfides. This information, which is a crucial ingredient of any genetic model, may be obtained from solubility measurements carried out under carefully controlled conditions. We report here the results of such experiments between 400°C and 700°C at 1.5 kb pressure in the presence of chloride-bearing fluids.

Previous cassiterite solubility determinations at temperatures up to 400°C were carried out in solutions of sodium hydroxide (KLINTSOVA and BARSUKOV, 1973), silicic chloride (NEKRASOV and LADZE, 1973), acidic and alkaline fluoride (KLINTSOVA et al., 1975), and chloride and nitrate (DADZE et al., 1982). Unfortunately, oxygen fugacity was not controlled in these studies, and hence the nature of the tin solutes cannot be established. KOVALENKO et al. (1986) measured cassiterite solubility in HCl solutions at 500°C, 1.0 kb with the NNO oxygen fugacity buffer. They interpreted their results in terms of the species, $SnCl_2^0$, for solutions of 0.01–0.50 m HCl. Recently, PABALAN (1986) has measured cassiterite solubility in subcritical solutions.

Cassiterite solubility has been predicted from theoretical considerations. PATTERSON et al. (1981) concluded that $SnCl_3^-$ is the most abundant species between 30 and 350°C with minor amounts of $SnCl^+$ and $SnCl_2^0$. EADINGTON (1982) suggested that between 250 and 400°C under acid conditions $SnCl_2^0$ and SnF^+ are dominant, whereas hydroxide species dominate under basic conditions. JACKSON and HELGESON (1985a) found that between 250 and 350°C under slightly acid to alkaline conditions hydroxide complexes of tin predominate, whereas under very acid conditions, $SnCl_2^0$ and $SnCl_3^-$ are abundant. Obviously, these predictions must be checked by direct experiment before they can be used with any degree of confidence.

EXPERIMENTAL PROCEDURES AND RESULTS

General procedures used in solubility and speciation studies have been discussed by WILSON (1986), WILSON and EUGSTER (1984), and EUGSTER et al. (1987). To permit speciation calculations, the total metal concentration must be measured as

a function of free ligand concentration at pressure and temperature. Multiple regression analysis is used to account for the metal concentration in terms of several metal-ligand species. Alternatively, if only one species is present, a graphical analysis suffices. In order to define solubility with respect to several species, the free ligand concentration must be varied as widely as possible between measurements. Applications of these procedures in room temperature to subcritical hydrothermal solutions have been discussed by ROSSOTTI and ROSSOTTI (1961), JOHANSSON (1970), SEWARD (1973, 1976), CRERAR et al. (1978), HARTLEY et al. (1980), WOOD and CRERAR (1985), and others.

Experiments carried out under supercritical conditions require special care. Small-volume, sealed noble metal capsules are used, providing usually no more than 80 μl solution. The control of the chemical conditions, such as pH, is difficult, and in situ conditions cannot be monitored directly. The only measurements that can be made are those done after the system is quenched to room temperature. The in situ conditions must then be back-calculated from those measurements. In addition to pressure and temperature, the oxygen fugacity, f_{O_2}, must be controlled, using the appropriate oxygen buffer assemblage. This is particularly important for multivalent metals. In order to minimize back reactions during quench, rapid-quench techniques are used (EUGSTER et al., 1987).

Chloride solutions were chosen for the present study due to their known dominance in ore-forming environments (HAAPALA and KINNUNEN, 1982; COLLINS, 1981; and many others). To keep the solution composition as simple as possible, the experiments were performed in HCl solutions. From these experiments, the dominant tin-chloride speciation and the equilibrium constants for the cassiterite dissolution reactions can be determined. Since tin exists in either the bivalent or quadruvalent state, the reactions that govern the solubility of cassiterite in HCl solutions are

$$SnO_2 + 2H^+ + nCl^- \rightleftarrows SnCl_n^{2-n} + H_2O + \frac{1}{2}O_2$$
$$(1,n)$$

and

$$SnO_2 + 4H^+ + nCl^- \rightleftarrows SnCl_n^{4-n} + 2H_2O, \quad (2,n)$$

where n is the ligand number of an individual mononuclear tin-chloride species and is a small integer. For different ligand concentrations, different species (hence, different values of n) can exist. Also, since there are likely to be more than one species in solution at a given set of conditions, a number of reactions of this form will represent the system. The equilibrium constants for these reactions are expressed as

$$K_{1,n} = \frac{(SnCl_n^{2-n})f_{O_2}^{1/2}}{(H^+)^2(Cl^-)^n} \qquad (3,n)$$

and

$$K_{2,n} = \frac{(SnCl_n^{4-n})}{(H^+)^4(Cl^-)^n}, \qquad (4,n)$$

where brackets indicate molality at pressure and temperature, and H_2O and SnO_2 are in their standard states of pure liquid and solid, respectively, at P and T. Activity coefficients for the aqueous species are assumed to be unity, which implies a standard state of unit activity at 1 molal concentration of the species in a solution of ionic strength equal to that of the experiments. For this reason, application of the results should be limited to solutions which have total chloride concentrations in the range represented by the experiments. The goal of the experiments described here was to determine the values of n for the most dominant tin-chloride species and the values of the corresponding equilibrium constants, $K_{1,n}$ and $K_{2,n}$.

The experimental setup, shown in Fig. 1, consists of 45 mg reagent grade SnO_2 with 60–80 μl HCl solution, between 0.18 and 3.3 molal, in a sealed gold metal tube (4.0 mm I.D., 4.2 mm O.D, 25 mm long). The oxygen buffer, either hematite-magnetite (HM) or nickel-nickel oxide (NNO), is contained within an inner, sealed platinum metal tube (2.8 mm I.D., 3.0 mm O.D., 15 mm long) together with 25 μl distilled water. This variation of the standard double-tube arrangement (see EUGSTER and WONES, 1962) provides a larger solution volume for analysis and a quicker extraction. Each run was held at temperature and 1.5 kb pressure for between 14 and 90 days in a 30 cm long cold-seal pressure vessel with a 30 cm long rapid-quench extension (see EUGSTER et al., 1987). To effect reversals, run temperature was approached from below in some runs, and from above in the others. For approach from above, the runs were first held at 100°C higher than the final run temperature for 1 day before being brought down to the run temperature.

After a run was quenched, the capsule was cleaned, weighed, and centrifuged at 14,000 G for 40 seconds. It was placed in a glovebox purged with water-saturated argon to prevent oxidation and evaporation. The fluid was extracted and collected on a teflon watchglass. Aliquots of 10 or 20 μl were diluted in volumetric flasks to 5 or 10 ml with 1.2

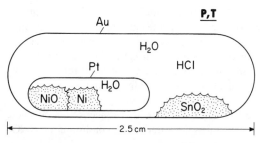

FIG. 1. Schematic diagram of the experimental setup for cassiterite solubility experiments.

N HCl for the Sn analysis, and with 10 to 50 ml distilled water for the chloride analysis. The remaining fluid was used to measure pH with a combination microelectrode while still in the glovebox. Chloride concentrations were measured with a Buchler chloridometer and tin with an AA spectrophotometer and graphite furnace. Excellent precision was achieved in the tin analysis by using a stabilized temperature platform, adding 500 ppm $NiNO_3$ to the solution, and adding 10% H_2 to the purge gas (SLAVIN et al., 1981; RAYSON and HOLCOMBE, 1982). The estimated uncertainty in the tin and chloride determinations is estimated to be ±5% based on the number of dilutions and the precision of the measurements. A check on gold concentrations revealed no detectable gold in solution. Results of the successful experiments are listed in Table 1.

INTERPRETATION OF DATA: REGRESSION PROCEDURES

From the measurements of total chloride, total tin, and pH at 25°C, the oxidation state of aqueous tin can be calculated. In these acid solutions, the contribution of OH^- can be neglected, thus, the room-temperature charge balance equation becomes

$$XSn_{total} + (10)^{-pH_Q} = Cl_{total}, \qquad (5)$$

where X is the oxidation state of aqueous tin, Sn_{total} and Cl_{total} are the tin and chloride concentrations measured after quench, and pH_Q is the pH of the quenched solution. The largest uncertainty in calculating X from Equation (5) arises from the pH measurement (see Table 1), because of the extremely small fluid volume and possible poisoning of the microelectrode by tin ions. Nevertheless, the average value of X at all temperatures is 2.18 ± 0.24 for runs buffered by NNO, and 3.63 ± 0.77 for runs buffered by HM. Due to the uncertainty associated with this determination, and the lack of a clear relationship between the calculated values of X and Cl_{total}, we assume that only one oxidation state of aqueous tin is present in significant quantities. Note that a mixture of oxidation states is impossible in *both* the HM and NNO runs due to the many orders of magnitude difference in f_{O_2} between these two buffers. Thus, it appears that aqueous tin is bivalent and quadruvalent under the conditions of NNO and HM, respectively.

Because of the use of rapid-quench techniques and the extraction of the fluid under an inert atmosphere, we assume that X is also the oxidation state of tin in solution at P and T. Consequently, once X is known, the electrical neutrality equation at P and T can be expressed as

$$(H^+)^{P,T} + \sum_n [X - n](SnCl_n^{X-n})^{P,T} = (Cl^-)^{P,T},$$

$$(6)$$

where the summation is over all species, $SnCl_n^{X-n}$, that exist in solution at significant concentrations. The tin and chloride mass balance at P and T are given by

$$Sn_{total} = \sum_n (SnCl_n^{X-n})^{P,T}, \quad (7)$$

and

$$Cl_{total} = (Cl^-)^{P,T} + (HCl^0)^{P,T}$$
$$+ \sum_n n(SnCl_n^{X-n})^{P,T}. \quad (8)$$

In addition, the dissociation of HCl^0 at P and T yields the following expression:

$$K_{HCl} = \frac{(H^+)^{P,T}(Cl^-)^{P,T}}{(HCl^0)^{P,T}}, \quad (9)$$

(assuming unit activity coefficients) where the values for K_{HCl} are from FRANTZ and MARSHALL (1984). Finally, the dissociation reaction for the nth tin chloride species,

$$SnCl_n^{X-n} \rightleftarrows SnCl_{n-1}^{X-n+1} + Cl^-, \quad (10,n)$$

provides the relation,

$$K_{10,n} = \frac{(SnCl_{n-1}^{X-n+1})^{P,T}(Cl^-)^{P,T}}{(SnCl_n^{X-n})^{P,T}}. \quad (11,n)$$

One equation of this form can be written for each two consecutive tin-chloride species that are proposed to exist in solution. In other words, if a total of m consecutive species are proposed, then there will be m − 1 independent equations of the form of Equation (11,n). Together with Equations (6),

Table 1. Conditions and analytical results of cassiterite solubility experiments

Run #	Temp (°C)	Buf.	Dur. (days)	App. dir.	Cl_T (m)	Sn_T (m)	pH_Q	X
40	400	NNO	86	↓	0.654	0.210	0.890	2.50
41	400	NNO	95	↑	1.21	0.435	0.655	2.27
44	400	NNO	86	↓	2.03	0.815	0.658	2.22
45	400	NNO	36	↑	0.642	0.211	0.830	2.34
46	400	NNO	36	↑	2.08	0.793	0.374	2.09
47	400	NNO	35	↑	0.286	0.0790	1.028	2.43
48	400	NNO	30	↓	0.291	0.0910	1.083	2.29
23	500	NNO	41	↑	1.956	0.771	0.503	2.13
24	500	NNO	35	↑	1.854	0.773	0.548	2.03
25	500	NNO	21	↑	0.191	0.0710	1.440	2.18
28	500	NNO	18	↓	0.610	0.222	1.261	2.50
29	500	NNO	22	↑	1.14	0.461	0.875	2.18
49	500	NNO	32	↑	0.696	0.257	0.938	2.26
50	500	NNO	27	↓	1.19	0.511	0.804	2.02
51	500	NNO	27	↓	0.200	0.0794	1.384	2.00
62	500	NNO	19	↑	4.82	2.01	≤0	≤1.9
76	500	NNO	54	↑	3.23	1.27	0.250	2.10
77	500	NNO	44	↓	3.13	1.33	0.245	1.93
70	500	NNO	33	↑	0.305	0.0586	0.872	2.92
72	500	NNO	33	↑	1.25	0.318	0.354	2.54
73	500	NNO	22	↓	1.21	0.329	0.360	2.35
74	600	NNO	33	↑	4.45	1.84	0.160	2.04
75	600	NNO	23	↓	4.36	1.75	N.D.	—
84	600	NNO	43	↑	0.302	0.599	0.793	2.36
85	600	NNO	34	↓	1.11	0.303	0.258	1.83
100	600	NNO	15	↓	0.614	0.141	0.531	2.27
101	600	NNO	15	↓	2.49	0.800	0.092	2.10
90	700	NNO	9	↑	1.21	0.261	0.242	2.44
91	700	NNO	9	↑	4.45	1.49	≤0	≤2.3
53	500	HM	29	↑	1.15	0.225	0.372	3.21
64	500	HM	36	↑	0.174	0.0195	1.021	4.05
66	500	HM	36	↑	0.794	0.142	0.398	2.78
67	500	HM	30	↓	1.02	0.172	N.D.	—
69	500	HM	30	↓	3.01	0.652	N.D.	—
82	600	HM	43	↑	2.39	0.320	≤0	≤4.3
86	600	HM	10	↑	0.300	0.0107	N.D.	—
87	600	HM	10	↑	1.20	0.120	0.180	4.48
88	600	HM	10	↑	4.64	0.682	≤0	≤5.3

Oxygen buffer (Buf.): NNO = nickel-nickel oxide; HM = hematite-magnetite. Dur. = number of days at temperature. App. dir. = direction of approach to temperature. Sn_T and Cl_T are the total measured concentrations of tin and chloride. pH_Q is the pH measured after quench. N.D. = not determined. ≤0 = off scale at zero pH. X is the valence of aqueous tin from Equation (5).

(7), (8), and (9), there will be a total of m + 3 equations describing the conditions of each individual run.

In this set of equations, however, there are 2m + 2 unknowns for each run: $(Cl^-)^{P,T}$, $(H^+)^{P,T}$, $(HCl^0)^{P,T}$, $(SnCl_{n_1}^{X-n_1})^{P,T}, \ldots, (SnCl_{n_m}^{X-n_m})^{P,T}$, and $K_{10,n_1+1}, \ldots, K_{10,n_m}$. A set of runs at the same pressure and temperature must then be considered as a whole, and the equations solved by nonlinear least squares regression. Since the unknown K_{10}'s are constant for a set of runs at a given P and T, the regression is expressed by m − 1 equations of the form

$$\frac{d\sigma^2}{d(\log K_{10,n})} = 0 \quad (12)$$

which expresses the condition of least squares best fit (BEVINGTON, 1969; DRAPER and SMITH, 1981). Using the case where X = 2 as an example, the standard deviation, σ, is defined by

$$\sigma^2 = \frac{1}{R} \sum_{i=1}^{R} [\log K_{1,n}^i - \log \bar{K}_{1,n}]^2, \quad (13)$$

where $K_{1,n}^i$ is the value of the equilibrium constant for Reaction (1) as determined by run i, $\bar{K}_{1,n}$ is the mean value for all runs at the same pressure and temperature, and R is the total number of runs under that set of conditions. The standard deviation is the same if it is defined in terms of $\log K_{1,n}$ (as in Equation 13), $\log K_{1,n-1}$ or $\log Sn_{total}$. Thus, the standard deviation is independent of m and can be compared between trials of the calculation where different species, and different numbers of species, are considered. With equations of the form of (11,n) and (12), and Equations (6), (7), (8), and (9), there are 2m + 2 equations, and the system is fully defined.

This regression analysis can be expressed as

$$\log K_{1,n}^i = f(K_{10,n}, Sn_{total}^i, Cl_{total}^i), \quad (14)$$

with $\log K_{1,n}$ (or $\log K_{2,n}$ if X = 4) as the response variable, $K_{10,n}$ as the regression parameter, and Sn_{total}^i and Cl_{total}^i as the predictor variables from each individual run i. The function, f, represents the combination of Equations (3 or 4), (6), (7), (8), (9), and (11,n), but due to the nature of this set of equations, it cannot be expressed explicitly as a simple regression equation. The analysis can be performed, however, by repeatedly solving the equations (function f) for different values of $K_{10,n}$ until a minimum is found in σ. The solution where σ is at a minimum will yield the best-fit values for the unknowns.

INTERPRETATION OF DATA: INITIAL SPECIES SELECTION

In order to minimize computational difficulty and to avoid over-fitting the data, it is helpful to initially identify the most important species in this system under the temperature and pressure conditions of interest. To do this, Equations (6) through (12) are simplified by assuming that only a single species is present (i.e., m = 1) in a set of runs. The equations are solved for the set of runs considering each possible species, one at a time, and the fits of the results are compared.

With only one species, Reaction (10) is undefined as are Equations (11,n) and (12). Equations (6), (7), (8), and (9) define the system completely for the unknowns $(H^+)^{P,T}$, $(Cl^-)^{P,T}$, $(HCl^0)^{P,T}$, and

the concentration of the single tin-chloride species, $(SnCl_n^{X-n})^{P,T}$. The best value of n is found by minimizing the standard deviation as a function of the species chosen, or n. This will be illustrated graphically here. Equations (3) and (4) can be rearranged to the following form:

$$\log\left[\frac{(SnCl_n^{2-n})f_{O_2}^{1/2}}{(H^+)^2}\right] = n \log(Cl^-) + \log K_{1,n},$$

$$(15)$$

and

$$\log\left[\frac{(SnCl_n^{4-n})}{(H^+)^4}\right] = n \log(Cl^-) + \log K_{2,n}, \quad (16)$$

where all terms are at P and T. Equations (15) and (16) describe straight lines of slope n. Figure 2 shows the run data for 400°C, buffered by NNO (where X = 2) where Equations (6) through (9) are solved assuming the presence of Sn^{2+} (a), $SnCl^+$ (b), and $SnCl_2^0$ (c). A line of slope n is positioned by using $\log \bar{K}_{1,n}$ as the y-intercept. Figure 2(a) shows that the data do not fit the model of Sn^{2+} (n = 0) as the dominant tin-chloride species. In Fig. 2(b), the trend of the points is slighly steeper than the reference line, and in Fig. 2(c), it is slightly shallower. Thus, the mean value of n is between 1 and 2 and it is concluded that significant amounts of $SnCl^+$ and $SnCl_2^0$ are present in solution at 400°C. This is also indicated by the standard deviations which are approximately equal for n = 1 and n = 2, and lower than for n = 0.

The same conclusions can be drawn for the data at 500°C, NNO (Fig. 3) and 600°C, NNO (Fig. 4), although the standard deviations show a slight preference for n = 1 at 500°C and n = 2 at 600°C. Thus, a final model must include both $SnCl^+$ and $SnCl_2^0$ at these temperatures as well. At 700°C, NNO (Fig. 5), the only reasonable fit of the two data points is for n = 2. The ratio of Cl_{total} to Sn_{total} for most of these runs does not permit n = 3.

Under the conditions of HM (where X = 4) at 500 and 600°C, Figs. 6 and 7 indicate that $SnCl_3^+$ is dominant.

INTERPRETATION OF DATA: FINAL SPECIATION CALCULATIONS

Knowing that $SnCl^+$ and $SnCl_2^0$ are the species of interest at 400 to 700°C and NNO, both can be considered simultaneously. The results will yield $K_{1,1}$ for the reaction

$$SnO_2 + 2H^+ + Cl^- \rightleftharpoons SnCl^+ + H_2O + \frac{1}{2}O_2$$

$$(1,1)$$

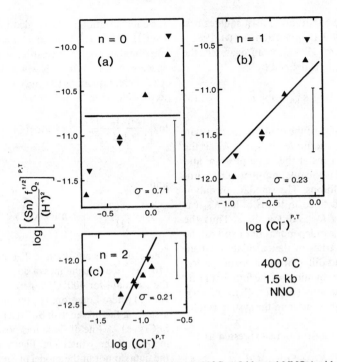

FIG. 2. Experimental cassiterite solubility results at 400°C, 1.5 kb, and NNO (oxidation state of tin is 2). Results for each run are calculated by Equations (6), (7), (8), and (9) and are plotted according to Equation (15). Only one tin-chloride species is considered to be present: Sn^{2+} (n = 0) in (a); $SnCl^+$ (n = 1) in (b); and $SnCl_2^0$ (n = 2) in (c). The total concentration of tin is represented by (Sn). The slope of the line in each diagram conforms to the chosen value of n. Triangles point in the direction of approach to equilibrium (from higher or lower temperature). The standard deviation, σ, is defined by Equation (13). The average estimated experimental error bar is shown on the right-hand side of each diagram and is based on the estimated 5% uncertainty in the tin and chloride measurements (the error in the horizontal direction is smaller than the size of the symbols).

and $K_{1,2}$ for the reaction

$$SnO_2 + 2H^+ + 2Cl^- \rightleftarrows SnCl_2^0 + H_2O + \tfrac{1}{2}O_2.$$
$$(1,2)$$

Equations (6), (7), (8), and (9) are written with $SnCl^+$ and $SnCl_2^0$ as the aqueous tin species and Equation (11,n) takes the form of

$$K_{10,2} = \frac{(SnCl^+)^{P,T}(Cl^-)^{P,T}}{(SnCl_2^0)^{P,T}}. \qquad (11,2)$$

Combined, these 5 equations can be solved for the unknowns of an individual run at P and T, $(SnCl^+)^{P,T}$, $(SnCl_2^0)^{P,T}$, $(H^+)^{P,T}$, $(Cl^-)^{P,T}$, $(HCl^0)^{P,T}$, provided that a value is assumed for $\log K_{10,2}$.

For a set of runs, the procedure is to assume a value of $\log K_{10,2}$, solve the set of equations for each run, and calculate σ by Equation (13). This is done repeatedly using different values of $\log K_{10,2}$ until a minimum is found in σ, and Equation (12) is satisfied. The minimum indicates the best fit of the data to the two-species model.

The regression analysis is illustrated in Fig. 8 where σ is plotted against different assumed values of $\log K_{10,2}$. At 400, 500, and 600°C, the calculations yield a distinct minimum in σ at $\log K_{10,2} = -0.45$, 0.16, and -1.35, respectively. This confirms that, statistically, the best fit of these experimental data is for a model where a combination of both $SnCl^+$ and $SnCl_2^0$ is present. As the assumed value of $\log K_{10,2}$ is made larger or smaller, σ increases rapidly before finally leveling off at the value obtained when only one species is assumed to be present ($SnCl^+$ or $SnCl_2^0$, respectively). For the 700°C data, σ decreases rapidly as the assumed value of $\log K_{10,2}$ is decreased to about -3.75. At that point, σ nearly reaches its lowest value; approximately that obtained when only $SnCl_2^0$ was assumed to be present. As the assumed value of $\log K_{10,2}$ is made smaller, σ continues to decrease, but only very slightly, while the resulting $\log K_{1,2}$ remains constant. Thus, at 700°C, it can be concluded that $SnCl_2^0$ is dominant in this concentration range and that the maximum amount of $SnCl^+$ compatible with the data is determined by $\log K_{1,1} \le \log K_{1,2} + (-3.75)$.

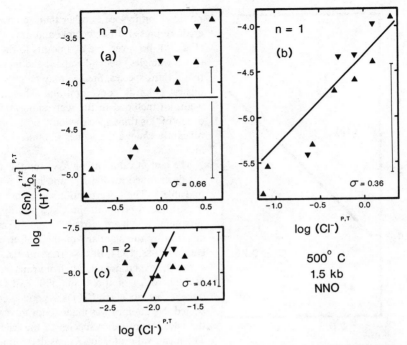

FIG. 3. Experimental cassiterite solubility results at 500°C, 1.5 kb, and NNO (oxidation state of tin is 2). See Fig. 2 caption.

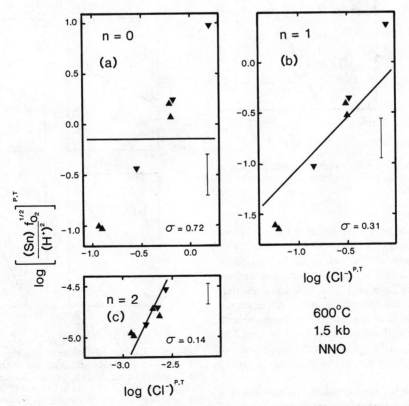

FIG. 4. Experimental cassiterite solubility results at 600°C, 1.5 kb, and NNO (oxidation state of tin is 2). See Fig. 2 caption.

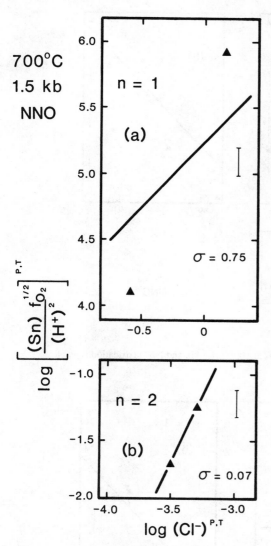

FIG. 5. Experimental cassiterite solubility results at 700°C, 1.5 kb, and NNO (oxidation state of tin is 2). See Fig. 2 caption.

The best-fit results are listed in Table 2. These are illustrated in Fig. 9 for 400°C (a), 500°C (b), 600°C (c), and 700°C (d), at 1.5 kb and NNO. Again, as in Figs. 2 through 7, the solubility diagram as defined by Equation (15) is used where the data points represent the total solubility of cassiterite calculated for each run. The bold curve is the best-fit curve through the points and is determined by both $\log \bar{K}_{1,1}$ and $\log \bar{K}_{1,2}$. It is the sum of the two lighter lines shown for the individual species, $SnCl^+$ and $SnCl_2^0$.

Figure 9(a, b, and d) shows that there is no strong systematic deviation of the points from the mean at 400, 500, and 700°C. This is an indication that

no tin-chloride species other than the two considered are present in detectable amounts. At 600°C (Fig. 9c), however, the two points to the far right (open triangles) which lie above the mean suggest that a third species, $SnCl_3^-$, may be present at the highest chloride concentrations. These two runs were not included in the regression analysis. Addition of this third species would probably not significantly change the values obtained for $\log K_{1,1}$ and $\log K_{1,2}$.

The best-fit values of $\log K_{10,2}$ and the mean values determined for $\log K_{1,1}$ and $\log K_{1,2}$ are listed in Table 3. The uncertainties shown are based on the estimated errors in the Cl_{total} and Sn_{total} measurements which are carried through the entire calculation by the standard method of error propagation (REES, 1984; BEVINGTON, 1969).

For the experimental data from runs buffered by HM where X = 4 at 500 and 600°C, the fit to the one-species model ($SnCl_3^+$) is as good as can be expected. No attempt was made to improve the fit by the addition of other species to the calculations. The mean values for $\log K_{2,3}$ as determined in the previous section are also given in Table 3.

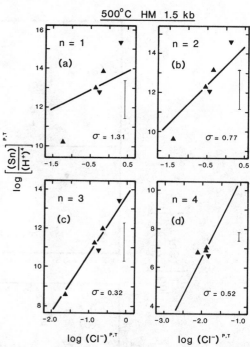

FIG. 6. Experimental cassiterite solubility results at 500°C, 1.5 kb, and HM (oxidation state of tin is 4). See Fig. 2 caption for explanation. One species is considered to be present: $SnCl^{3+}$ (n = 1) in (a); $SnCl_2^{2+}$ (n = 2) in (b); $SnCl_3^+$ (n = 3) in (c); and $SnCl_4^0$ (n = 4) in (d). In one of the runs (#69) no solution could be obtained for n = 4.

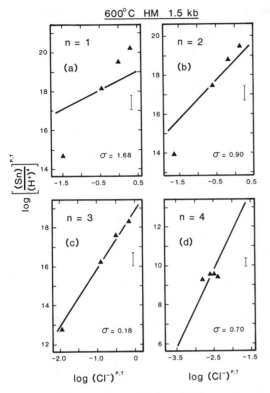

FIG. 7. Experimental cassiterite solubility results at 600°C, 1.5 kb, and HM (oxidation state of tin is 4). See captions for Figs. 2 and 6.

THERMODYNAMIC PARAMETERS

The equilibrium constants determined in this study can be represented as a function of temperature by an equation of the form

$$\log K = a + \frac{b}{T} + \frac{c}{T^2} \qquad (17)$$

where the coefficients a, b, and c are determined by least-squares regression of the data in Table 3 and T is in Kelvin. The coefficients for Reactions 1,1; 1,2; 2,3; and 10,2 are given in Table 4.

In Fig. 10, $\log K$ is plotted against $1/T(K)$ for Reactions 1,1; 1,2; and 10,2. The uncertainties in $\log K$ are less than or equal to the size of the symbols. The regression curve as defined by Equation (17) is also shown. The non-unique solutions for $\log K_{1,1}$ and $\log K_{10,2}$ at 700°C are the maximum possible values. These were not used in determining the regression curves but are included in Fig. 10 to show that they are compatible with the extrapolations of the lower temperature data.

For comparison, the data of PATTERSON et al. (1981) and JACKSON and HELGESON (1985a) at

350°C and pressure equal to the saturated vapor pressure of water (P_{sat}) are also plotted in Fig. 10. Although the difference in pressure makes this comparison only approximate, there is reasonable agreement in the value for $\log K_{10,2}$. Shown also is the value for $\log K_{1,2}$ at 500°C, 1.0 kb from KOVALENKO et al. (1986) which shows reasonable agreement as well.

The results for $\log K_{10,2}$ as a function of temperature can be used to derive information on the thermodynamics of complex dissociation. The slope of the lower curve in Fig. 10 is related to the standard enthalpy of reaction, $\Delta H^0_{10,2}$, by the equation

$$\frac{d \log K_{10,2}}{d(1/T)} = \frac{-\Delta H^0_{10,2}}{(2.303)R}, \qquad (18)$$

where R is the gas constant and T is in Kelvin. The entropy of reaction, $\Delta S^0_{10,2}$, is given by

$$\Delta S^0_{10,2} = \frac{\Delta H^0_{10,2}}{T} + (2.303)R \log K_{10,2}. \qquad (19)$$

The resulting values of $\Delta H^0_{10,2}$ and $\Delta S^0_{10,2}$ are listed in Table 5. It can be seen that $\Delta H^0_{10,2}$ and $\Delta S^0_{10,2}$ are both positive at 400°C and negative at temperatures of 500°C and above. There is a systematic decrease in both $\Delta H^0_{10,2}$ and $\Delta S^0_{10,2}$ with increasing temperature from 400 to 700°C. As discussed by SEWARD (1981), this reflects the changing nature of the solvent, the hydration of the charged species, and the chemical bonding between the metal ion and the ligand. The positive enthalpy at 400°C is indicative of predominantly covalent bonding. The charged species are not strongly solvated under these conditions and relatively little water is involved in the reaction. Thus, a small positive entropy is caused by an increase in the number of particles and an increased vibrational and rotational energy in the system. At 500°C and above, the increasingly negative entropy values are indicative of an increase in the electrostatic (ionic) character of the bonding with temperature. The negative $\Delta S^0_{10,2}$ values indicate that what solvation water *is* involved in the reaction is derived from an increasingly disordered solvent.

GEOLOGIC APPLICATIONS

Data presented here can be used to predict the oxidation state of aqueous tin, the nature of the aqueous tin-chloride complexes in natural environments, the solubility of cassiterite in ore fluids, and mechanisms likely to cause cassiterite precipitation. The physical and chemical conditions existing at the time of cassiterite deposition, such as f_{O_2}, T, pH, and salinity, can be estimated from mineralogical, fluid inclusion, and isotopic studies.

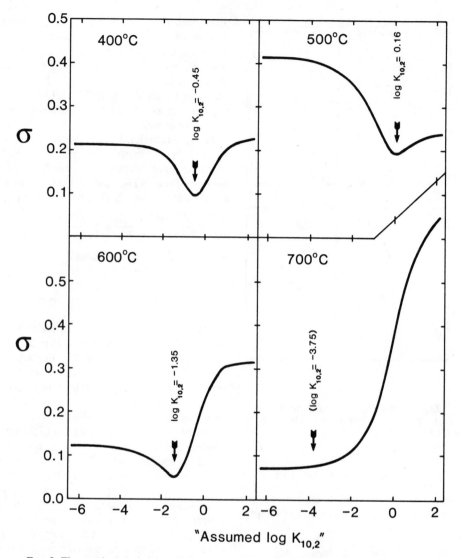

FIG. 8. The standard deviation as defined by Equation (13) plotted against assumed values for log $K_{10,2}$. A minimum in the curve is where Equation (12) is satisfied and indicates the value of log $K_{10,2}$ which best fits the experimental data. No minimum is found at 700°C, but the largest permissible value of log $K_{10,2}$ is indicated (see text).

Acidic conditions are indicated by the common occurrence of acid-alteration assemblages in cassiterite deposits, such as greisenization of granitic host-rock, alteration of feldspars to muscovite, alteration of biotite to muscovite or chloride in schists, and the formation of skarns in carbonate rock. These alterations set an upper limit on the pH of the ore-forming fluids at below neutral near the site of deposition, although the fluids could be more acidic at their source. Individual studies of fluid inclusions and phase relations indicate that mineralizing fluids had pH values in the range of 3.9 to 5.4 at the site

of ore deposition (PATTERSON *et al.*, 1981; JACKSON and HELGESON, 1985b).

Oxygen fugacities of the ore-forming fluids at Renison Bell, Tasmania were estimated by PATTERSON *et al.* (1981) using CO_2/CH_4 ratios in fluid inclusions, and at Mt. Lindsay, Tasmania by KWAK (1983) using a rare occurrence of iron-titanium oxides. They found values between those defined by the NNO and QFM buffers. Topaz rhyolites associated with small cassiterite deposits in the western United States have f_{O_2} values approximately equal to those defined by the QFM buffer (CHRISTIANSEN

Table 2. Best-fit results of cassiterite solubility data

Run	$\log(\text{SnCl}^+)$	$\log(\text{SnCl}_2^0)$	$\log(\text{Cl}^-)$	$\log(\text{H}^+)$	$\log K_{1,1}^i$	$\log K_{1,2}^i$
			400°C, NNO ($\log f_{\text{O}_2} = -27.572$)			
40	−0.858	−1.146	−0.738	−1.355	−11.196	−10.746
41	−0.618	−0.713	−0.545	−1.356	−11.147	−10.697
44	−0.424	−0.358	−0.383	−1.429	−10.970	−10.520
45	−0.855	−1.146	−0.741	−1.379	−11.141	−10.691
46	−0.436	−0.370	−0.384	−1.336	−11.165	−10.715
47	−1.210	−1.761	−1.001	−1.418	−11.160	−10.710
48	−1.151	−1.690	−0.989	−1.496	−10.957	−10.507
			500°C, NNO ($\log f_{\text{O}_2} = -22.796$)			
23	−0.255	−0.669	−0.254	−3.323	−4.752	−4.912
24	−0.254	−0.667	−0.253	−3.452	−4.494	−4.654
25	−1.169	−2.495	−1.166	−3.346	−4.709	−4.869
28	−0.709	−1.577	−0.708	−3.267	−4.865	−5.025
29	−0.435	−1.030	−0.435	−3.429	−4.540	−4.700
49	−0.652	−1.464	−0.651	−3.283	−4.832	−4.992
50	−0.398	−0.955	−0.397	−3.569	−4.260	−4.420
51	−1.122	−2.403	−1.120	−3.462	−4.476	−4.636
62	0.053	−0.055	0.053	−3.343	−4.711	−4.871
76	−0.090	−0.340	−0.090	−3.265	−4.868	−5.028
77	−0.075	−0.310	−0.075	−3.447	−4.505	−4.665
			600°C, NNO ($\log f_{\text{O}_2} = -19.113$)			
70	−1.475	−1.600	−1.475	−4.315	−0.928	0.422
72	−1.005	−0.659	−1.005	−4.270	−1.016	0.334
73	−0.996	−0.642	−0.996	−4.325	−0.906	0.444
84	−1.469	−1.587	−1.468	−4.334	−0.890	0.460
85	−1.017	−0.684	−1.017	−4.346	−0.864	0.486
100	−1.221	−1.092	−1.221	−4.321	−0.915	0.435
101	−0.775	−0.199	−0.775	−4.340	−0.877	0.473
			700°C, NNO ($\log f_{\text{O}_2} = -16.187$)			
90	(−2.173)	−0.595	−2.172	−4.825	(1.555)	5.305
91	(−1.791)	0.169	−1.790	−4.878	(1.662)	5.412

Run	$\log(\text{SnCl}_3^+)$	$\log(\text{Cl}^-)$	$\log(\text{H}^+)$	$\log K_{2,3}^i$
		500°C, HM		
53	−0.648	−0.646	−3.151	13.892
64	−1.710	−1.653	−2.567	13.516
66	−0.848	−0.845	−2.993	13.660
67	−0.764	−0.761	−2.912	13.165
69	−0.186	−0.185	−3.398	13.959
		600°C, HM		
82	−0.495	−0.495	−4.520	19.069
86	−1.971	−1.962	−3.688	18.666
87	−0.921	−0.921	−4.283	18.973
88	−0.166	−0.166	−4.613	18.782

All concentrations are for the species at 1.5 kb and T. Parentheses indicate maximum permitted values at 700°C (see text).

et al., 1986). HAAPALA and KINNUNEN (1982) summarized published data on fluid inclusions in cassiterite from tin deposits around the world. They found that the most common values for homoge-

nization temperatures were in the range of 300 to 400°C, but some were as high as 500°C and a few were as low as 250°C. They point out that when corrected for pressure, the depositional tempera-

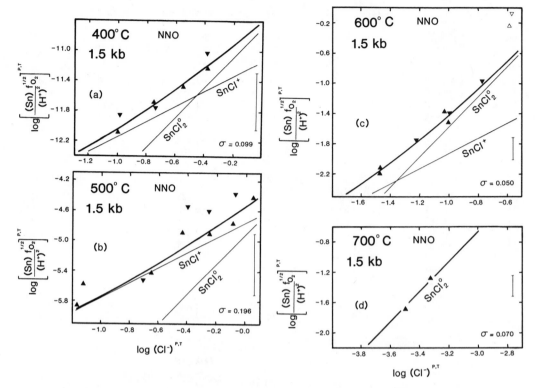

FIG. 9. Experimental cassiterite solubility at 400 (a), 500 (b), 600 (c), and 700°C (d), all at 1.5 kb and NNO (oxidation state of tin is 2). Each point represents the best-fit solution of Equations (6), (7), (8), (9), (11,2), and (12) as discussed in the text and Fig. (8). The standard deviation, σ, is defined by Equation (13). The average estimated error bar is shown on the right-hand side of each diagram and is based on the estimated 5% uncertainty in the tin and chloride measurements (uncertainty in the horizontal direction is smaller than the size of the symbols). See text for explanation of the two runs at 600°C represented by open symbols.

tures could be 100°C or more higher than the homogenization temperatures.

Salinities of ore-forming fluids as determined by fluid inclusion studies of cassiterite deposits are variable, ranging from about 1 m NaCl equivalent to as high as 8 m (HAAPALA and KINNUNEN, 1982; ROEDDER, 1984). Most commonly, however, during the stage of cassiterite formation, salinities were in the range of 1 to 2 m (KELLY and RYE, 1979;

COLLINS, 1981; PATTERSON et al., 1981; ZAW and DAW, 1983; JACKSON and HELGESON, 1985b; JACKSON et al., 1982; NORMAN and TRANGCOTCHASAN, 1982). Estimates of fluid pressures range from 100 to 1800 bars (HAAPALA and KINNUNEN, 1982).

To obtain the relationship between stannous and stannic chloride complexes in hydrothermal fluids, Reactions (1,1) and (1,2) can be combined with Reaction (2,3) to yield the following:

$$SnCl^+ + 2Cl^- + 2H^+ + \tfrac{1}{2}O_2 \rightleftarrows SnCl_3^+ + H_2O$$
(20)

and

$$SnCl_2^0 + Cl^- + 2H^+ + \tfrac{1}{2}O_2 \rightleftarrows SnCl_3^+ + H_2O.$$
(21)

The equilibrium constants for these reactions can be obtained by combining the data listed in Table 4:

Table 3. Best-fit equilibrium constants for Reactions (1,1), (1,2), (10,2), and (2,3)

T (°C)	400	500	600	700
$\log K_{1,1}$	−11.14	−4.76	−0.93	(1.58)
	±0.15	±0.21	±0.06	(±0.09)
$\log K_{1,2}$	−10.69	−4.92	0.42	5.33
	±0.16	±0.22	±0.07	±0.10
$\log K_{10,2}$	−0.45	0.16	−1.35	(−3.75)
$\log K_{2,3}$	—	13.64	18.87	—
		±0.25	±0.16	

Table 4. Coefficients for Equation (17) for Reactions (1,1), (1,2), (10,2), and (2,3)

Reaction	a	b	c
(1,1)	−3.217	2.640×10^4	-2.143×10^7
(1,2)	70.080	-8.246×10^4	1.891×10^7
(10,2)	−70.812	1.054×10^5	-3.904×10^7
(2,3)	59.064	-3.511×10^4	—

$$\log K_{20} = 62.28 - \frac{6.15 \times 10^4}{T} + \frac{2.14 \times 10^7}{T^2}$$

$$(22)$$

and

$$\log K_{21} = -11.02 + \frac{4.74 \times 10^4}{T} - \frac{1.89 \times 10^7}{T^2}.$$

$$(23)$$

Equations (22) and (23) are only valid at 500 and 600°C because Reaction (2,3) was only determined at these temperatures. The relation between stannous and stannic chloride species depends on chloride ion concentration, pH, and f_{O_2}. To evaluate the relative importance of these two oxidation states of aqueous tin in natural fluids, Equations (22) and (23) were used to plot the boundary between stannous chloride dominance and stannic chloride dominance. This is shown in Fig. 11 as a function of f_{O_2} and T for a 2 m Cl_{total} fluid in equilibrium with the assemblage K-feldspar + muscovite + quartz. This assemblage is commonly associated with cassiterite deposits, and at a given KCl concentration, it fixes the pH of the fluid. For reference, the oxygen buffer curves for HM and QFM are also shown. The boundary between stannous and stannic dominance lies well above the region of f_{O_2} typical of cassiterite-bearing fossil hydrothermal systems. Thus, only the stannous species are expected to be found in such systems. This conclusion holds even if the pH is as much as 1.5 units lower. The species $SnCl^+$ and $SnCl_2^0$ then, are the tin-chloride species of importance under most natural hydrothermal conditions.

The total solubility of cassiterite as a function of temperature is plotted in Fig. 12 at the oxygen buffers QFM, NNO, HM. Again, it was assumed that a fluid of 2 m Cl_{total} is in equilibrium with the acid buffer assemblage K-feldspar-muscovite-quartz. The two species, $SnCl^+$ and $SnCl_2^0$, make up the total tin concentration in the fluid. Figure 12 shows that low f_{O_2} and high temperature are required for significant concentrations of tin in solution. For instance, at 500°C and QFM, the tin concentration is only 1 ppm. If the pH is decreased by 1 unit, that

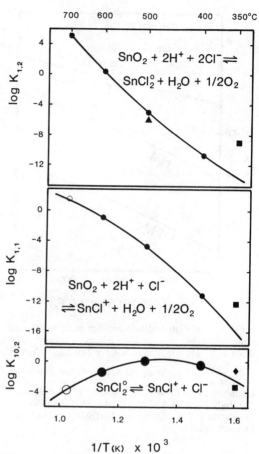

FIG. 10. Plot of log K vs. $1/T(K)$ for Reactions (1,1), (1,2), and (10,2). Round symbols are from this study (experimental) where $P = 1.5$ kb. The open round symbols represent a non-unique solution of the experimental data at 700°C as discussed in the text. The curves are fit to the solid round symbols only. The diamond is from the data of PATTERSON et al. (1981) and the squares are from JACKSON and HELGESON (1985a). Both are theoretical and for total pressure equal to the saturated vapor pressure of water. The triangle is from the experimental data of KOVALENKO et al. (1986) at 1.0 kb.

concentration is obtained at 400°C, QFM. That, however, is probably the lowest pH obtainable in this temperature range. Therefore, for the transport

Table 5. Enthalpy and entropy of Reaction (10,2)

T (°C)	ΔH^0 (k J)	ΔS^0 (k J/K)
400	203.7	0.294
500	−83.56	−0.105
600	−305.0	−0.375
700	(−481.0)	(−0.566)

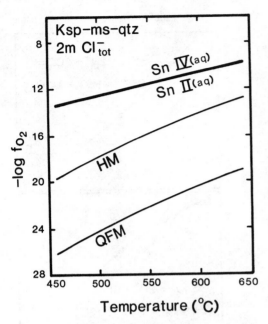

FIG. 11. Calculated relation between stannous (II) and stannic (IV) chloride complexes in natural hydrothermal fluids at 1.5 kb. Calculation is for a 2 m Cl_{total} solution with $\frac{K_{total}}{Na_{total}} = 0.15$ in equilibrium with the acid-buffer assemblage, K-feldspar-muscovite-quartz. Data are from this study, GUNTER and EUGSTER (1980), FRANTZ and MARSHALL (1984), QUIST and MARSHALL (1968), and FRANCK (1956).

Using the data presented here, the mechanisms responsible for cassiterite deposition can be quantitatively evaluated. Accepting $SnCl^+$ and $SnCl_2^0$ as the dominant species, precipitation is given by the reverse of Reaction (1),

$$SnCl_n^{2-n} + \tfrac{1}{2}O_2 + H_2O \rightarrow SnO_2 + 2H^+ + nCl^-$$
(24)

Reaction (24) can be driven to the right by a drop in temperature, an increase in f_{O_2} and/or pH, and probably also a change in pressure, although the pressure dependency was not evaluated. For the conditions depicted in Fig. 12, for instance, a drop of 50°C can decrease the tin concentration in solution by as much as an order of magnitude. Similarly, a one unit increase in pH, all other parameters held constant, forces cassiterite to become supersaturated by two orders of magnitude. An increase in pH is most likely caused by interaction of the acidic ore fluids with the country rock, such as carbonates at Dachang (S.E. China) and Renison Bell (Tasmania) or schists at Panasqueira (Portugal). Acid neutralization by carbonates occurs through the reaction

$$CaCO_3 + H^+ \rightarrow Ca^{2+} + HCO_3^-, \quad (25)$$

and can be monitored by the increased Ca^{2+} contents of fluid inclusions (KWAK and TAN, 1981a,b). In schists and granites, H^+ can be consumed

of 1 ppm Sn or more in solution, it is necessary to have temperatures in excess of 400°C, pH at or below that of the K-feldspar-muscovite-quartz assemblage, and oxygen fugacity in the area of QFM-NNO. A decrease in temperature, increase in pH, or increase in oxygen fugacity can cause the deposition of cassiterite.

The solubilities of cassiterite and magnetite are compared in Fig. 13 where the data for magnetite are from CHOU and EUGSTER (1977). Although the solubilities of these two minerals are dependent on the same parameters, Fig. 13 shows that cassiterite solubility has a stronger dependence on oxygen fugacity and temperature than does magnetite solubility. At QFM, for instance, iron concentration decreases by about 4 orders of magnitude from 700 to 400°C whereas the tin concentration decreases by about 8 orders of magnitude at the same pH and chloride ion concentration. Consequently, at 500°C or lower, f_{O_2} near NNO, and free chloride ion concentration of 0.1 molar or more, Fe solutes greatly dominate Sn solutes in a fluid in equilibrium with magnetite + cassiterite.

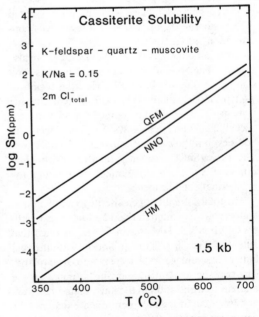

FIG. 12. Total solubility of cassiterite in natural hydrothermal fluids at 1.5 kb as a function temperature and f_{O_2}. Fluid is the same as in Fig. 11.

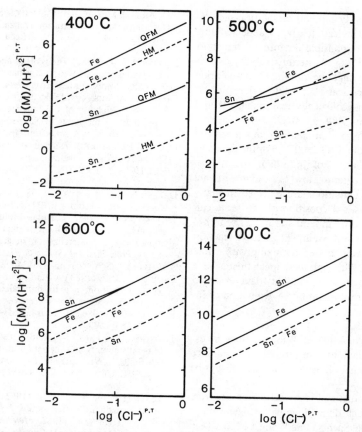

FIG. 13. Solubility diagram for magnetite and cassiterite at 400 to 700°C. Magnetite solubility is at 2 kb from the data of CHOU and EUGSTER (1977). Cassiterite solubility is at 1.5 kb from this study. Solid curves are calculated at QFM and dashed curves are at HM. (M) represents the total concentration of either Sn or Fe. The concentrations of H^+ and Cl^- are for the free ions at P and T.

through the alteration of feldspars to muscovite by the reaction (HEMLEY, 1959),

$$3KAlSi_3O_8 + 2H^+ \rightarrow KAl_3Si_3O_{10}(OH)_2$$
$$+ 6SiO_2 + 2K^+. \quad (26)$$

Indeed, cassiterite-quartz veins frequently are lined with coarse muscovite selvages.

A decrease in chloride molality also can initiate cassiterite deposition and it is most easily accomplished by mixing of ore fluids with dilute, meteoric waters. Such mixing has been inferred for many hydrothermal systems from stable isotope studies (see for instance PATTERSON et al., 1981; and JACKSON et al., 1982).

Since the oxidation state of tin is 2 in solution and 4 in cassiterite, oxidation is an essential requirement for cassiterite deposition. In principle, an increase in f_{O_2} can be accomplished by a differential loss of hydrogen, but as pointed out elsewhere (EUGSTER, 1986), coupled redox reactions without a net transfer of oxygen are probably more effective. Simultaneous precipitation of cassiterite and arsenopyrite, and of cassiterite and sulfides are examples (HEINRICH and EADINGTON, 1986; EUGSTER, 1986):

$$3SnCl_2^0 + 2H_3AsO_3^0 + 2FeCl_2^0 + 2H_2S^0 \rightarrow$$
$$3SnO_2 + 2FeAsS + 10HCl^0, \quad (27)$$

and

$$7SnCl_2^0 + FeCl_2^0 + 2H_2SO_4^0 + 6H_2O \rightarrow$$
$$7SnO_2 + FeS_2 + 16HCl^0. \quad (28)$$

The acidity produced by Reactions (27) and (28) can be consumed by reactions such as (25) and (26).

CONCLUSIONS

Solubility experiments in supercritical HCl solutions can provide the information necessary to identify the dominant aqueous species and to formulate and quantify mineral dissolution-precipitation reactions. This leads to a better understanding of the processes responsible for mineral transport and deposition. From cassiterite solubility experiments we found that the dominant tin-chloride species in supercritical HCl solutions are $SnCl^+$, and $SnCl_2^0$ under geologically reasonable pH and f_{O_2} conditions from 400 to 700°C, 1.5 kb. The species, $SnCl_3^-$, has also been identified at a lower pH and higher f_{O_2} from 500 to 600°C. For the transport of significant quantities of tin in hydrothermal solutions, high temperature, low f_{O_2}, and low pH are required. These conditions are reflected in many natural tin deposits. Deposition of cassiterite can be initiated by a drop in temperature, a rise in pH through reaction with wallrock, a redox reaction with coupled precipitation, or mixing with dilute meteoric waters. Although these mechanisms have been suggested previously, solubility data presented in this paper permit quantitative evaluation of individual processes.

Acknowledgements—G. A. Wilson wishes to express his gratitude for the inspiration and guidance of Professor Hans P. Eugster. Hans will always be remembered.

Many conceptual and technical difficulties were overcome through discussions with Eugene Ilton, Sigudur Gislason, Sheng-Yuan Wang, I-Ming Chou, Terry Seward, Phil Candela, and Scott Wood. Dimitri Sverjensky is also gratefully acknowledged for sharing with us his insight. This manuscript was greatly improved thanks to constructive reviews by Phil Candela, I-Ming Chou, and Jeremy Fein. This work represents part of the senior author's Ph.D. dissertation at The Johns Hopkins University. Support was provided by National Science Foundation grants EAR 82-06177 and EAR 84-11050.

REFERENCES

BAUMANN L., STEMPROK M., TISCHENDORF G. and ZOUBEK V. (1974) Metallogeny of tin and tungsten in the Krusne Hory-Erzgebirge. Geol. Survey, Prague. 66p.

BEVINGTON P. R. (1969) *Data Reduction and Error Analysis for the Physical Sciences.* 336 pp. McGraw-Hill, New York.

BRAY C. J. and SPOONER E. T. C. (1983) Sheeted vein Sn-W mineralization and greisenization associated with economic kaolinitization, Goonbarrow china clay pit, St. Austell, Cornwall, England: geologic relationships and geochronology. *Econ. Geol.* **78,** 1064–1089.

CHOU I-M. and EUGSTER H. P. (1977) Solubility of magnetite in supercritical chloride solutions. *Amer. J. Sci.* **277,** 1296–1314.

CHRISTIANSEN E. H., SHERIDAN M. J. and BURT D. M. (1986) The geology and geochemistry of Cenozoic topaz rhyolites from the western United States. *Geol. Soc. Amer. Spec. Paper* **205.**

CLARK A. H., PALMA V. V., ARCHIBALD D. A., FARRAR E., ARENAS M. J. F. and ROBERTSON R. C. R. (1983) Occurrence and age of tin mineralization in the Cordillera Oriental, Southern Peru. *Econ. Geol.* **78,** 514–520.

COLLINS P. L. F. (1981) The geology and genesis of the Cleveland tin deposit, western Tasmania: fluid inclusion and stable isotope studies. *Econ. Geol.* **76,** 365–392.

CRERAR D. A., SUSAK N. J., BORSIK M. and SCHWARTZ S. (1978) Solubility of the buffer assemblage pyrite + pyrrhotite + magnetite in NaCl solutions from 200 to 300°C. *Geochim. Cosmochim. Acta* **42,** 1427–1437.

DADZE T. P., SOROKHIN V. I. and NEKRASOV I. Y. (1982) Solubility of SnO_2 in water and in aqueous solutions of HCl, HCl + KCl, and HNO_3 at 200 to 400°C and 101.3 Mpa. *Geochem. Internat.* **18,** 142–152.

DRAPER N. R. and SMITH H. (1981) *Applied Regression Analysis.* (second edition). 709 pp. John Wiley & Sons.

EADINGTON P. J. (1982) Calculated solubilities of cassiterite in high temperature hydrothermal brines, and some applications to mineralization in granitic rocks and skarns. *Proc. First Int. Symp. on Hydrothermal Reactions* (ed. S. SOMIYA), pp. 335–345. Tokyo Inst. Tech.

EADINGTON P. J. (1983) A fluid inclusion investigation of ore formation in a tin-mineralized granite, New England, New South Wales. *Econ. Geol.* **78,** 1204–1221.

EUGSTER H. P. (1986) Minerals in hot water. *Amer. Mineral.* **71,** 655–673.

EUGSTER H. P., CHOU I-M. and WILSON G. A. (1987) Mineral solubility and speciation in supercritical chloride fluids. In *Hydrothermal Experimental Techniques* (eds. G. C. ULMER and H. L. BARNES), pp. 1–19. John Wiley & Sons.

EUGSTER H. P. and WONES D. R. (1962) Stability relations of the ferruginous biotite, annite. *J. Petrol.* **3,** 82–125.

EVANS A. M. (EDITOR) (1982) *Metallization Associated with Acid Magmatism.* 385 pp. John Wiley & Sons.

FRANCK E. U. (1956) Hochverdichteter Wasserdampf III. Ionendissoziation von KCl, KOH, und H_2O in überkritischem Wasser. *Z. Phys. Chem.* **8,** 192–206.

FRANTZ J. D. and MARSHALL W. L. (1984) Electrical conductances and ionization constants of salts, acids, and bases in supercritical aqueous fluids: I. Hydrochloric acid from 100 to 700°C and at pressures to 4000 bars. *Amer. J. Sci.* **284,** 651–667.

GUNTER W. D. and EUGSTER H. P. (1980) Mica-feldspar equilibria in supercritical alkali chloride solutions. *Contrib. Miner. Petrol.* **75,** 235–250.

HAAPALA I. and KINNUNEN K. (1982) Fluid inclusion evidence on the genesis of tin deposits. In *Metallization Associated with Acid Magmatism* (Ed. A. M. EVANS), pp. 101–110. John Wiley & Sons.

HALLS C. (EDITOR) (1985) *High Heat Production Granites, Hydrothermal Circulation and Ore Genesis.* Institute of Mining and Metallurgy. London. 593 pp.

HARTLY F. R., BURGESS C. and ALCOCK R. M. (1980) *Solution Equilibria.* 361 pp. Ellis Horwood Limited, Halstead Press.

HEINRICH C. A. and EADINGTON P. J. (1986) Thermodynamic predictions of the hydrothermal chemistry of arsenic, and their significance for the paragenetic sequence of some cassiterite-arsenopyrite-base metal sulfide deposits. *Econ. Geol.* **81,** 511–529.

HEMLEY J. J. (1959) Some mineralogical equilibria in the system K_2O-Al_2O_3-SiO_2-H_2O. *Amer. J. Sci.* **257,** 241–270.

ISHIHARA S. and TAKENOUCHI S. (EDITORS) (1980) *Granite Magmatism and Related Mineralization.* Japan Soc. Mining Geol. Spec. Issue **8,** Tokyo, 247p.

JACKSON K. J. and HELGESON H. C. (1985a) Chemical and thermodynamic constraints on the hydrothermal transport and deposition of tin: I. Calculations of the solubility of cassiterite at high pressures and temperatures. *Geochim. Cosmochim. Acta* **49**, 1–22.

JACKSON K. J. and HELGESON H. C. (1985b) Chemical and thermodynamic constraints on the hydrothermal transport and deposition of tin: II. Interpretation of phase relations in the southeast Asian tin belt. *Econ. Geol.* **80**, 1365–1378.

JACKSON N. L., HALLIDAY A. N., SHEPPARD S. M. F. and MITCHELL J. G. (1982) Hydrothermal activity in the St. Just mining district, Cornwall, England. In *Metallization Associated with Acid Magmatism* (ed. A. M. EVANS), pp. 137–179. John Wiley & Sons.

JOHANSSON L. (1970) Iteration procedures in solution chemistry with special reference to solubility measurements. *Acta Chemica. Scandinavica* **24**, 1572–1578.

KELLY W. C. and RYE R. O. (1979) Geologic, fluid inclusion, and stable isotope studies of the tin-tungsten deposits of Panasqueira, Portugal. *Econ. Geol.* **74**, 1721–1819.

KELLY W. C. and TURNEAURE F. S. (1970) Mineralogy, paragenesis and geothermometry of the tin and tungsten deposits of the Eastern Andes, Bolivia. *Econ. Geol.* **65**, 609–680.

KLINSTOVA A. P. and BARSUKOV V. L. (1973) Solubility of cassiterite in water and in aqueous NaOH solutions at elevated temperatures. *Geochem. Int.* **10**, 540–547.

KLINTSOVA A. P., BARSUKOV V. L., SHEMARYKINA T. P. and KHODAKOVSKIY I. L. (1975) Measurement of the stability constants for Sn(IV) hydrofluoride complexes. *Geochem. Int.* **12**, no. 2, 207–215.

KOVALENKO, N. I., RYZHENKO B. N., BARSUKOV V. L., KLINTSOVA A. P., VELYUKHANOVA T. K., VOLYNETS M. P. and KITAYEVA L. P. (1986) The solubility of cassiterite in HCl and HCl + NaCl (KCl) solutions at 500°C and 1000 atm under fixed redox conditions. *Geochem. Int.* **23**, No. 7, 1–16.

KWAK T. A. P. (1983) The geology and geochemistry of the zoned, Sn-W-F-Be skarns at Mt. Lindsay, Tasmania, Australia. *Econ. Geol.* **78**, 1440–1465.

KWAK T. A. P. (1987) *W -Sn Skarn Deposits and Related Metamorphic Skarns and Granitoids.* 452 pp. Elsevier.

KWAK T. A. P. and TAN T. H. (1981a) The geochemistry of zoning in skarn minerals at the King Island (Dolphin) mine. *Econ. Geol.* **76**, 468–497.

KWAK T. A. P. and TAN T. H. (1981b) The importance of CaCl₂ in fluid composition trends—evidence from the King Island (Dolphin) skarn deposit. *Econ. Geol.* **76**, 955–960.

NEKRASOV I. Y. and LADZE T. P. (1973) Solubility of cassiterite in silicic chloride solutions at 300°C and 400°C. *Doklady Akad. Nauk SSSR* **213**, 145–147.

NORMAN D. I. and TRANGCOTCHASAN Y. (1982) Mineralization and fluid inclusion study of the Yod Nam tin mine, southern Thailand. In *Metallization Associated with Acid Magmatism* (Ed. A. M. EVANS), pp. 261–272. John Wiley & Sons.

PABALAN R. T. (1986) Solubility of cassiterite (SnO₂) in NaCl solutions from 200°C–350°C, with geologic applications. Ph.D. Dissertation. The Pennsylvania State University, University Park, Pennsylvania.

PATTERSON D. J., OHMOTO H. and SOLOMON M. (1981) Geologic setting and genesis of cassiterite-sulfide mineralization at Renison Bell, western Tasmania. *Econ. Geol.* **76**, 393–438.

PETERSON E. U. (1986) Tin in volcanogenic massive sulfide deposits: An example from the Geco mine, Manitouwadge District, Ontario, Canada. *Econ. Geol.* **81**, 323–342.

PUCHNER C. C. (1986) Geology, alteration and mineralization of the Kougarok Sn deposit, Seward Peninsula, Alaska. *Econ. Geol.* **81**, 1775–1794.

QUIST A. S. and MARSHALL W. L. (1968) Electrical conductances of aqueous sodium chloride solutions from 0 to 800°C and at pressures to 4000 bars. *J. Phys. Chem.* **72**, 684–703.

RAYSON G. D. and HOLCOMBE J. A. (1982) Tin atom formation in a graphite furnace atomizer. *Anal. Chim. Acta* **136**, 249–260.

REES C. E. (1984) Error propagation calculations. *Geochim. Cosmochim. Acta* **48**, 2309–2311.

ROEDDER E. (1984) *Fluid Inclusions. Mineral. Soc. Amer., Rev. in Mineral.* **12**, 644p.

ROSSOTTI F. J. C. and ROSSOTTI H. (1961) *The Determination of Stability Constants.* 425 pp. McGraw-Hill.

SEWARD T. M. (1973) Thio complexes of gold and the transport of gold in hydrothermal ore solutions. *Geochim. Cosmochim. Acta* **37**, 379–399.

SEWARD T. M. (1976) The stability of chloride complexes of silver in hydrothermal solutions up to 350°C. *Geochim. Cosmochim. Acta* **40**, 1329–1341.

SEWARD T. M. (1981) Metal complex formation in aqueous solutions at elevated temperatures and pressures. In *Chemistry and Geochemistry of Solutions at High Temperatures and Pressures* (eds. D. T. RICKARD and F. E. WICKMAN), *Phys. and Chem. of the Earth* **13 & 14**, 113–132.

SILLITOE R. H., HALLS C. and GRANT J. N. (1975) Porphyry tin deposits in Bolivia. *Econ. Geol.* **70**, 913–927.

SLAVIN W., MANNING D. C. and CARNICK G. R. (1981) The stabilized temperature platform furnace. *Atomic Spectrosc.* **2**, 137–145.

SUN S-S. and EADINGTON P. J. (1987) Oxygen isotope evidence for the mixing of magmatic and meteoric waters during tin mineralization in the Mole Granite, New South Wales, Australia. *Econ. Geol.* **82**, 43–52.

TAYLOR R. G. (1979) *Geology of Tin Deposits.* 543 pp. Elsevier.

VON GRUENEWALDT G. and STRYDOM J. H. (1985) Geochemical distribution patterns surrounding tin-bearing pipes and the origin of the mineralizing fluids at the Zaaiplaats tin mine, Potgietersrus District. *Econ. Geol.* **80**, 1201–1211.

WILSON G. A. (1986) Cassiterite solubility and metal-chloride speciation in supercritical solutions. Ph.D. Dissertation. The Johns Hopkins University, Baltimore, Maryland.

WILSON G. A. and EUGSTER H. P. (1984) Cassiterite solubility and metal-chloride speciation in supercritical solutions (abstr.). *Geol. Soc. Amer. Ann. Meeting Abstr. Prog.* **16**, 696.

WOOD S. A. and CRERAR D. A. (1985) A numerical method for obtaining multiple linear regression parameters with physically realistic signs and magnitudes: Applications to the determination of equilibrium constants from solubility data. *Geochim. Cosmochim. Acta* **49**, 165–172.

ZHANG Z. and LI X. (1981) Studies on mineralization and composition of Dachang ore field, Guangxi Province, China. *Geochimica* **1**, 74–86.

ZAW U. K. and DAW K. M. T. (1983) A note on a fluid inclusion study of tin-tungsten mineralization at Mawchi Mine, Kayah State, Burma. *Econ. Geol.* **78**, 530–534.

Part D.
Diagenetic Systems

Fluid-Mineral Interactions: A Tribute to H. P. Eugster
© The Geochemical Society, Special Publication No. 2, 1990
Editors: R. J. Spencer and I-Ming Chou

Clay-carbonate reactions in the Venture area, Scotian Shelf, Nova Scotia, Canada

IAN HUTCHEON

Department of Geology and Geophysics, The University of Calgary, Calgary, Alberta, T2N 1N4 Canada

Abstract—The sandstones of the over pressured zone of the Venture Field at 4500 m depth show abundant authigenic chlorite that can be interpreted to have formed by a reaction between kaolinite and ankerite. The stability of this mineral reaction in a H_2O-CO_2-NaCl fluid can be determined from mineral compositions, thermodynamic data and the solubility of CO_2 in saline fluids. The predicted equilibrium temperatures are approximately 160°C, which is in good agreement with measured temperatures. The measurement of ^{13}C for CO_2 and calcite in the gas shows many samples to be in isotopic equilibrium at reservoir temperature, providing additional evidence that the CO_2 in the gas was derived from calcite dissolution.

The reaction mechanism that produces CO_2 in this and other sedimentary basins and in geothermal wells is not defined by the equilibrium treatment of the stability of clay carbonate reactions. However examination of temperature-pCO_2 trends reported by other authors suggests that silicate hydrolysis may be the source of hydrogen ions that cause calcite dissolution. This hypothesis can be tested by examining representative fluid compositions and imposing equilibrium with the appropriate mineral assemblage to determine the value of pCO_2 in equilibrium with silicates and carbonates. Calculated values for pCO_2 for quartzose sediments containing kaolinite, illite and calcite, used to model pCO_2 trends in the Gulf Coast, show good agreement with observed data. Reactions with albite, calcite and analcime, used to model pCO_2 trends in Iceland, do not show as reasonable agreement with measured trends of pCO_2. While not demonstrating unequivocally that CO_2 at deeper levels in sedimentary basins results from dissolution of carbonates, driven by silicate hydrolysis, the evidence to support this hypothesis is strong.

INTRODUCTION

CARBON DIOXIDE is present in natural gases in many sedimentary basins. Studies of the Gulf Coast and North Sea (LUNDEGARD *et al.*, 1984; LUNDEGARD and LAND, 1986; SMITH and EHRENBERG, 1989) have shown that the partial pressure of CO_2 increases with increasing temperature and, therefore, depth. The continuous increase observed in these two basins suggests that the process responsible for the accumulation of CO_2 at greater depths is progressive and HUTCHEON and ABERCROMBIE (1989a), by considering the analyses of produced waters from steam flood pilots, have suggested that the hydrolysis of silicates produces hydrogen ions which are consumed by the dissolution of calcite, releasing CO_2. HUTCHEON *et al.* (1990) have shown that the isotopic composition of CO_2, generated in steam flood pilots, is consistent with an origin from dissolution of calcite. In the Venture Field, there is evidence for the reaction between clays and carbonates and the purpose of this paper is: to describe the conditions under which this reaction might have occurred; to present some supporting isotopic evidence for carbonates being the source of CO_2; and to propose a reaction mechanism that will produce CO_2 and adequately describe the pCO_2-temperature relationships noted by SMITH and EHRENBERG (1989).

The potential for reactions between clay minerals and carbonates during late diagenesis and low grade metamorphism was recognized by ZEN (1959). The reaction of kaolinite and dolomite (or ankerite) to form chlorite was described in the Salton Sea hydrothermal system by MUFFLER and WHITE (1969). Reactions between silicates and carbonate to produce CO_2 have been proposed by HUTCHEON *et al.* (1980), MCDOWELL and PACES (1985) and SMITH and EHRENBERG (1989). HUTCHEON *et al.* (1980) described the potential of univariant reactions between silicates and carbonates to buffer the CO_2 content of an aqueous fluid to the solubility surface between H_2O and CO_2 and, potentially, to produce a CO_2-rich vapour once the solubility surface was reached; however, a reaction mechanism was not proposed. For a fluid containing only H_2O and CO_2, and using pure magnesium end-member chlorite (clinochlore), dolomite and calcite, they calculated the intersection between the kaolinite-dolomite-chlorite mineral reaction and the solubility surface to be at approximately 200°C at 100 MPa (1 Kbar). In general, mineral reactions during diagenesis and low grade metamorphism take place in a saline aqueous fluid and the minerals are not pure end-members. To produce applicable estimates of the stability of the kaolinite-dolomite-chlorite reaction relative to the H_2O-CO_2 miscibility surface, the calculations should incorporate the effect of NaCl on the solubility of CO_2 and the effect of variations in the chemical compositions of chlorite and carbonates.

In this paper, evidence is presented for the reaction of kaolinite and ankerite (or dolomite) to produce chlorite, iron-bearing calcite and CO_2 in the Venture area of the Scotian Shelf (Fig. 1). Data from natural gas wells have been used for estimates of fluid salinity and electron microprobe analyses have provided the composition of the minerals. The stability of chlorite, relative to the H_2O-CO_2 solubility surface in an NaCl-bearing fluid, has been calculated from these compositions and thermochemical data. The possibility that reactions between silicates and carbonates may be a source of CO_2 is examined using the isotopic compositions of the carbonates and CO_2.

The Venture area is an ideal location to examine mineral reactions during diagenesis at higher temperatures. Gas is contained in highly over pressured sandstones (up to 120 MPa) of the Missisauga Formation and temperatures may be in excess of 160°C, depending on the depth. The formation fluids are saline (approximately 20 wt. percent NaCl equivalent) and the gas contains CH_4 and CO_2. Drilling for natural gas in the area has provided rock samples from cored boreholes, water salinities and gas samples for compositional and isotopic analyses.

GEOLOGICAL SETTING

The locations of wells studied are shown in Fig. 1. Not all wells have been cored and most of the samples discussed in this paper are from the Missisauga Formation and were obtained from Venture B-13, B-43, H-22 and B-52, Arcadia J-16, Olympia A-12, Citnalta I-59, Bluenose G-47 and 47A, South Venture O-59, West Venture C-62 and Thebaud I-93.

The stratigraphy of the Scotian Shelf was first presented by MCIVER (1972) and modified by GIVEN (1977). Biostratigraphy was presented by GRADSTEIN et al. (1975). The study by GIVEN (1977) incorporated lithofacies descriptions (PIPER, 1975; SWIFT et al., 1975) and paleogeographic and structural studies (HAWORTH, 1975; PARSONS, 1975; HOWIE and BARSS, 1975; JANSA and WADE, 1975; SMITH, 1975). KEEN (1979, 1983) and KEEN and LEWIS (1982) delineated potential zones of organic maturity on the Scotian Shelf from thermal data. Organic maturation studies have been completed by CASSOU et al. (1977), POWELL and SNOWDON (1979), PURCELL et al. (1979), POWELL

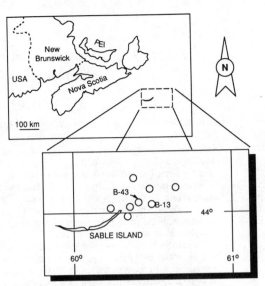

FIG. 1. General location map showing the position of the Venture field relative to the Eastern Coast of Canada and the locations of wells from which rock and fluid samples were obtained in this study.

(1982) and ISSLER (1984). The only diagenetic studies available are by HERB (1975), which documents diagenetic cements in the Nova Scotia Group sandstones of the Missisauga and Logan Canyon Formations, and NOGUERA URREA (1987), which examines the diagenetic history of over pressure in the Missisauga and Mic Mac Formations. Neither study examines the physicochemical conditions of the mineral reactions observed in the over pressured zone, nor is the relationship between mineral reactions and CO_2 discussed.

The main depositional centre of the Scotian Basin underlies the eastern part of the Scotian Shelf and consists of a wedge of Mesozoic to Cenozoic sediments up to 12 km thick in the vicinity of Sable Island. Triassic and Lower Jurassic redbeds are overlain by Middle Jurassic to Lower Cretaceous deltaic sediments and shelf carbonates. The Verrill Canyon Formation shales are suggested to be the source rock for the gas accumulation at the Venture structure (POWELL, 1982).

Depth relationship of temperature and pressure

Pressures in the Venture area are significantly higher than the hydrostatic pressure ("over pressured"). Fig. 2 shows the variation of pressure and temperature with depth in well B-43 obtained from drill stem test data provided by Mobil Canada. Pressures in this well reach 110 MPa (1100 bars) at 5500 m depth and increase significantly below 4500

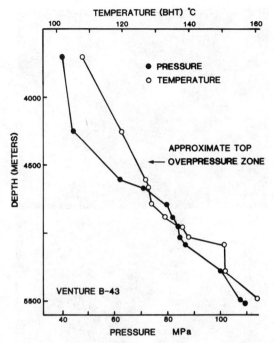

FIG. 2. Depth *vs.* temperature and pressure for well B-43 in the Venture area.

m, the top of the "over pressured" zone. Rock and fluid samples are available over the range from 4000–5500 m and temperatures over this interval, also estimated from drilling data, range from 110 to 160°C. JANSA and NOGUERA URREA (1989) note that the main over pressure zone is in sandstones and that the shales show normal compaction trends and attribute the over pressure to the low geothermal gradient, formation of diagenetic seals, generation of gas and release of water from later stages of shale diagenesis. ISSLER (1984) shows temperatures ranging up to approximately 160°C at 4800 m depth on the Scotian Shelf. Vitrinite reflectance has been measured in samples obtained from wells on the Scotian Shelf by CASSOU *et al.* (1977) and DAVIES and AVERY (1984), who show a marked rate of increase of reflectance with depth in the over pressured zone. The depth to the over pressured zone is variable between wells, but is generally in the range of 4500 m within the Venture area. Over the entire Scotian Shelf the distribution of over pressures is variable.

DIAGENETIC MINERALS, COMPOSITION AND DISTRIBUTION

Observation of the early diagenetic sequence is limited because core samples are available only at deeper intervals. The over pressured zone has been cored in many of the wells in the Venture area and petrographic relationships relevant to the clay-carbonate reaction are described on the basis of samples from these wells.

X-ray diffraction analyses of the less than 2 μm size fraction from sandstones indicates that 14 Å-chlorite is the most abundant clay mineral in the over pressured zone (Fig. 3), ranging up to 98% by weight of the clay size fraction. Above this zone, less than 10% of the clay-size fraction is chlorite; with kaolinite and illite being the most abundant minerals. No expanding clays were observed in the clay-size fraction of the sandstones at these depths. HUTCHEON and NAHNYBIDA (1990) have observed smectite in the shales at depths up to 2300 m and report the details of the analytical procedures used to identify clay minerals in these rocks. The abrupt increase in chlorite content in the over pressured zone can be confirmed by thin section and scanning electron microscope (SEM) examination.

Thin sections were examined in detail but the relevant petrographic features are best summarized by secondary and backscattered electron photomicrographs. Authigenic minerals observed with the SEM include pyrite, quartz overgrowths, illite, kaolinite, chlorite, calcite, ankerite, rutile (?) and sphene. Chlorite is observed lining and filling pores. Quartz overgrowths enclose the pore lining chlorite (Fig. 4a), separating it from the pore filling chlorite. This is interpreted to represent two stages of chlorite

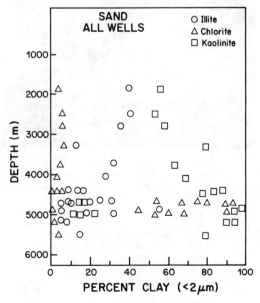

FIG. 3. Clay mineralogy of the less than 2 μm size fraction of sandstones.

FIG. 4. Petrography of samples examined by scanning electron microscopy. a. Quartz overgrowths enclose pore lining chlorite. The detailed section on the right shows that euhedral quartz has grown over the chlorite rims. The chlorite is interpreted to be an early diagenetic event in this context. (4959 m depth) b. Blocky authigenic kaolinite has well developed crystal faces and the morphology suggests the dickite polytype (4431 m depth). c. Although kaolinite is abundant in some samples, at 4440 m, just above the over pressured zone, the edges of kaolinite crystals are frayed, interpreted to indicate reaction of the kaolinite. d. Fine authigenic chlorite plates coat rhombs of iron calcite at 4955 m depth. This texture is interpreted to be evidence for the chlorite reaction. e. Coarse chlorite rosettes, up to 20 mm, fill pores and are interpreted to be late diagenetic chlorite formed by the clay-carbonate reaction (4959 m depth) f. Backscatter electron micrograph of kaolinite, ankerite and chlorite. Note that ankerite and kaolinite are not observed in contact (4962 m depth).

growth, one early and one later in the paragenetic sequence. At 4430 m, above the over pressured zone in B-43, blocky kaolinite, possibly the dickite variety, (Fig. 4b) is observed. At 4440 m, the kaolinite cleavage flakes are frayed (Fig. 4c), interpreted as possible dissolution of kaolinite. Authigenic pore filling chlorite is present at 4955 m as fine platelets that coat authigenic calcite (Fig. 4d) and at slightly greater depths (4960 m) abundant rosettes of pore filling chlorite are observed (Fig. 4e) to be intergrown with calcite that contains some iron. Ankerite appears above this interval as a replacement of plagioclase feldspar. In the zone with abundant pore-filling chlorite, ankerite is rare. In samples that contain kaolinite, ankerite and chlorite, kaolinite and chlorite are not observed in contact with ankerite (Fig. 4f). These textures are interpreted to indicate a reaction, similar to that documented by HUTCHEON *et al.* (1980), of ankerite and kaolinite to produce calcite and chlorite.

Using the Fe and Mg end-member thermodynamic components, this reaction can be described by the equilibria:

$$5\ CaMg(CO_3)_2 + Al_2Si_2O_5(OH)_4 + SiO_2$$
$$+ 2\ H_2O \rightleftarrows Mg_5Al_2Si_3O_{10}(OH)_8$$
$$+ 5\ CaCO_3 + 5\ CO_2 \quad (1)$$
$$5\ FeCO_3 + Al_2Si_2O_5(OH)_4 + SiO_2$$
$$+ 2\ H_2O \rightleftarrows Fe_5Al_2Si_3O_{10}(OH)_8 + 5\ CO_2. \quad (2)$$

Thermodynamic data are available for all the components in reactions (1) and (2) in HELGESON *et al.* (1978) and WALSHE (1986). Mineral compositions and a model for the dependence of activity on composition are required to examine the phase relationships of reactions (1) and (2) in CO_2-H_2O-NaCl fluids. Compositions of chlorite, calcite and ankerite have been measured by electron microprobe. The compositions measured were used to calculate the activities of the Fe and Mg end-members in chlorite and the activity of $FeCO_3$ in calcite, assuming an ideal solid solution.

Electron microprobe analyses

Mineral compositions were measured on an ARL SEMQ electron microprobe at the University of Calgary using mineral standards of known composition and wavelength dispersive spectrometers. A JOEL 733 electron microprobe at CANMET in Ottawa was also used and analyses were by energy dispersive spectrometer using mineral standards. Points for analyses on the ARL were located by transmitted light optics. In general, operating conditions were 15 kV and 20 nanoamps beam current, although these were varied depending on the stability of the material being analysed. Backscatter electron images, obtained on the JOEL 733, were found to be useful to identify grains suitable for analyses of the fine clay mineral particles. Because the intensity of the backscatter image is proportional to the mean atomic number of the substance under the beam, fine clay plates, with a high mean atomic number and oriented nearly vertically in the low mean atomic number mounting material (epoxy), were relatively easy to recognize from their relative brightness. Although edge effects are a problem with analysis of fine particles, grains as small as a few μm could be analysed and yielded reasonable totals for chlorite (Table 1). The epoxy mounting medium contains no elements common to chlorite and thus the X-rays generated from the epoxy should not substantially affect the analyses.

Composition of chlorite

A number of chlorite grains in various samples were analysed, and the chlorite compositions from wells B-43, B-13, B-52 and J-16 are reported in Table 1. The calculations in this paper are based on the compositions in Table 1. The chlorites are generally iron rich with minor amounts of magnesium, between approximately 3.7 and 6.2 wt.%. There are minor to trace amounts of Na_2O, K_2O, CaO, TiO_2 and MnO. The chlorite grains are relatively homogeneous at the scale of microprobe analyses and the backscatter photomicrographs do not show any bright patches or other indications of inhomogeneous compositions. The observation of fractured samples in reflected electron images shows that some authigenic chlorites have fine fibres of illite growing on them. These might be undetectable using back scatter electrons and could contribute to the K_2O content of the chlorite. Calcite was also observed to be intergrown with chlorite and this may account for the CaO observed in chlorite. Fine grains of authigenic TiO_2, presumably remaining from the destruction of detrital biotite and muscovite, were observed and may account for some of the TiO_2 in the analyses of chlorite and other authigenic phases.

Composition of carbonates

Calcite, siderite, dolomite and ankerite can all be recognized in the samples examined. Ankerite replaces plagioclase feldspar and ranges from 30–50 mol% of the dolomite end-member. Late, pore filling calcite, associated with the pore filling chlorite

Table 1. Chemical compositions (wt.%) of authigenic chlorite in from the Venture Field determined by electron microprobe

Well	Depth (m)	Mineral	No. points	Na₂O	K₂O	CaO	Al₂O₃	SiO₂	TiO₂	FeO	MgO	MnO	SrO	Total*
B-13	4951.0	chlorite	(5)	0.0	0.06	0.0	23.12	26.05	0.0	33.47	6.22	0.09	nd	89.01
B-13	4951.0	calcite	(4)	0.0	nd	54.05	0.0	nd	nd	1.16	0.40	0.79	0.04	56.44
B-13	4956.0	chlorite	(3)	0.0	0.32	0.10	23.97	28.21	0.09	29.85	6.37	0.12	nd	89.03
B-13	4956.0	calcite	(6)	0.0	nd	53.74	0.01	nd	nd	1.07	0.34	0.69	0.04	55.89
B-43	4432.0	ankerite	(5)	nd	nd	31.32	0.0	nd	nd	16.39	8.25	0.97	nd	56.93
B-43	4670.0	ankerite	(4)	nd	nd	31.23	0.0	nd	nd	15.29	8.37	1.33	nd	56.22
B-43	4876.0	ankerite	(3)	nd	nd	31.94	0.0	nd	nd	14.06	9.15	1.10	nd	56.25
B-43	4876.0	calcite	(4)	nd	nd	54.09	0.0	nd	nd	0.80	0.19	0.03	nd	55.11
B-43	4877.0	calcite	(3)	0.10	nd	53.88	0.15	nd	nd	1.21	0.38	0.49	nd	56.21
B-43	4877.0	ankerite	(6)	0.04	nd	30.95	0.11	nd	nd	15.00	8.89	1.27	nd	56.26
B-43	4877.0	chlorite	(3)	0.0	0.11	0.09	23.65	28.97	0.56	30.31	5.69	0.1	nd	89.48
B-43	4956.65	chlorite	(7)	0.0	0.33	0.30	22.99	24.05	0.56	32.15	5.12	0.69	nd	86.19
B-43	4956.65	chlorite	(7)	0.0	0.33	0.30	22.99	24.05	0.56	32.15	5.12	0.69	nd	86.19
B-43	4958.65	chlorite	(18)	0.26	0.08	0.12	22.04	23.71	0.38	33.07	4.84	0.28	nd	84.78
B-43	4958.65	illite	(6)	0.06	9.34	0.12	32.68	49.73	0.21	1.57	0.04	0.56	nd	94.31
B-43	4958.65	albite	(3)	8.72	0.12	0.0	20.23	69.64	0.09	0.16	0.0	0.0	nd	98.96
B-43	4959.	chlorite	(7)	0.02	0.51	0.11	23.93	27.69	nd	30.86	5.76	0.09	nd	89.48
B-43	4959.0	calcite	(4)	0.12	nd	53.63	0.01	nd	nd	0.86	0.50	0.50	nd	55.62
B-43	4432.0	albite	(6)	11.23	0.08	0.82	20.01	67.40	nd	0.04	0.0	nd	nd	99.58
B-43	4670.0	albite	(4)	11.85	0.06	0.03	19.39	67.83	nd	0.18	0.0	nd	nd	99.34
B-43	4876.0	albite	(4)	11.68	0.08	0.08	19.25	68.28	nd	0.03	0.0	nd	nd	99.40
B-43	4958.83	chlorite	(7)	0.0	0.11	0.0	22.5	23.33	0.29	32.65	4.92	0.3	nd	84.10
B-43	4958.83	ankerite	(2)	0.0	0.46	30.51	0.0	0.0	0.0	13.76	9.55	1.38	nd	55.66
B-43	4958.83	albite	(3)	10.10	0.00	0.0	19.24	69.93	0.0	0.0	0.0	0.67	nd	99.94
B-43	4963.0	albite	(6)	11.39	0.26	0.21	19.55	68.73	nd	0.17	0.0	nd	nd	100.31
J-16	5159.0	chlorite	(3)	0.35	0.21	0.20	22.37	21.85	0.25	34.91	4.24	0.29	nd	84.67
J-16	5159.0	ankerite	(3)	0.24	0.39	29.66	0.12	0.33	0.08	17.83	7.11	0.72	nd	56.48
J-16	5159.0	albite	(3)	8.88	0.12	0.08	19.39	70.22	0.18	0.0	0.0	0.0	nd	98.87
J-16	5159.0	siderite	(2)	0.0	0.0	1.95	0.0	0.19	0.0	51.14	7.19	1.76	nd	62.23
J-16	5168.0	chlorite	(13)	0.55	0.25	0.23	23.77	24.46	0.34	36.07	3.66	0.42	nd	89.75
J-16	5168.0	siderite	(6)	0.0	0.0	2.37	0.0	0.91	0.0	49.54	6.68	3.26	nd	62.76
J-16	5168.0	albite	(3)	8.63	0.0	0.12	19.33	72.53	0.21	0.63	0.0	0.0	nd	62.76
B-52	4711.0	chlorite	(2)	0.0	0.15	0.0	13.65	24.78	1.40	35.97	6.20	0.36	nd	82.50
B-52	5125.0	chlorite	(4)	nd	0.34	0.24	25.33	27.53	nd	30.82	4.67	0.07	nd	89.00
B-52	5125.0	ankerite	(5)	0.0	nd	31.02	0.06	nd	nd	15.02	8.61	1.27	0.04	56.02
B-52	5276.26	chlorite	(3)	0.37	0.12	0.0	23.03	23.70	0.27	34.22	4.99	0.38	nd	86.45
B-52	5276.26	calcite	(3)	0.0	0.42	54.44	0.0	0.0	0.0	1.34	0.62	0.32	nd	57.14

* Totals do not include volatiles such as H_2O and CO_2.

and interpreted to be a reaction product, contains approximately 3 to 4 mol% $FeCO_3$. Detrital carbonate grains are variable in abundance in most samples and care was taken during electron microprobe analyses to select only grains that were obviously authigenic. Some spot analyses of detrital calcite grains showed them to have similar compositions to the cement, suggesting that recrystallization of all calcites has taken place. Analyses of selected carbonates are reported in Table 1.

Isotopic compositions

Isotopic compositions of CH_4, CO_2 and calcite were measured in the Stable Isotope Laboratory in the Department of Physics at The University of

Calgary. The data collected are presented in Table 2. The compositions of gases in wells O-51, I-93, B-13, B-43, B-52, G-47 and H-22 are from KEN-DALL et al. (1989). Calcite compositions and gas compositions from C-62 and C-74 were measured in this study. The $\delta^{13}C$ of CH_4 ranges from -38 to $-45‰$ (PDB), reaching the most enriched $\delta^{13}C$ in the over pressured zone (Fig. 5). If the data from each well are observed separately, there is also a tendency for the methane to be more enriched in ^{13}C with depth. The $\delta^{13}C$ values for CO_2 range from -2 to $-12‰$ (PDB), also reaching the most enriched values in the over pressured zone (Fig. 6). The CO_2 from C-62 and B-52 becomes more enriched with depth, as was observed for methane. A plot of $\delta^{13}CH_4$ vs. $\delta^{13}CO_2$ shows the data from all wells

Table 2. Isotopic composition of calcite and CO_2 from the Venture field. The compositions of CO_2 for wells O-51, I-93, B-13, B-43, G-47 and H-22 are from KENDALL *et al.* (1989)

Well	Depth (m)	Temp (°C)	$\delta^{13}CO_2$	$\delta^{13}C$ calcite	Depth (m)	Temp (°C)	$\delta^{13}CO_2$	$\delta^{13}C$ calcite
C-74	3876.5	110	−5.9	−1.56	3876.5	110		−5.05
C-74	3876.5	110		−5.05	3876.5	110		−3.05
C-74	3922	111		−1.35	3922	111		−3.48
C-74	4314.5	121	−6.53	−7.21	4413	123	−3.71	−6.38
C-74	3876.5	110	−5.9	−1.56	3876.5	110		−5.05
C-74	3876.5	110		−3.05	3922	111		−1.35
C-74	3922	111		−3.48	4314.5	121	−6.53	−7.21
C-74	4413	123	−3.71	−6.38	4413		−3.49	
C-74	4514.5	123	−3.12	−4.61	4689.5	130	−3.86	−5.28
C-74	4754.5	131	−2.61	−5.6				
C-62	5021.5	138		17.63	5021.5	138		15.27
C-62	4926.5	136	−10.06	−6.02	4742	131	−12.64	−5.04
C-62	4742		−6.16		4742		−11.62	
O-51	4371	122		−8.48				
I-93	3916	111		−6.67	3928.5	112		2.65
I-93	3998.5	113		−4.12	4086.5	115		−2.69
I-93	4326	121		−0.39				
B-13	4694.7	130		−0.72	4696	130		−1.11
B-13	4699	130		−10.2	4699.5	130		−1.24
B-13	4700	131		−1.1	4700.5	131		−0.94
B-13	4721	131		−4.88	4725.7	131		−9.77
B-13	4727	131		−3.59	4730.9	131		−5.67
B-13	4731	131		−6.75	4732	131		−5.05
B-13	4733	131		−4.49				
B-43	4431	124		−8.85	4435	124		−27.39
B-43	4436	124		−22.5	4437	124		−7.25
B-43	4439	124		−10.73	4440	124		−8.98
B-43	4443	124		−3.26	4670	129		−7.27
B-43	4671	129		−10.23	4676	129		−4.37
B-43	4951	136		−6.3	4966	137		−3.73
B-52	4719	131	−8.51	−5.28	4719	130	−8.51	−4.47
B-52	4967.5		−11.95		5033.5	138	−7.7	−2.91
B-52	5045.5		−7.82		5072.5		−10.21	
B-52	5290	144	−3.86	−3.23	5290	145	−3.86	−4.24
B-52	5290	145	−3.86	−2.43	5728.5		−1.81	
B-52	5802		−3.22					
G-47	4583	127	−5.3	2.79	5230	143	−1.69	1.91
H-22	5022	138	−8.29	−5.27	5022	138	−8.29	−6.21
H-22	5022	138	−8.29	−2.19				
J-16	4869.5	135	−3.56	−1.89	4896.5	135	−3.56	−3.82
J-16	5170	142	−3.92	−0.76	5170	142	−6.52	−3.83

tends to be limited by an $a_c = 1.04$ (Fig. 7) line from WHITICAR *et al.* (1986).

PHASE RELATIONSHIPS OF MINERALS IN H_2O-CO_2-NaCl FLUIDS

The stability of chlorite relative to kaolinite and carbonate (ankerite or dolomite) can be considered by an examination of the phase relationships among these minerals in a CO_2-H_2O-NaCl fluid. Previous studies (TROMMSDORFF and SKIPPEN, 1986) have described the influence of univariant mineral reactions on the compositions of coexisting fluids in metamorphic rocks. They conclude that mineral reactions involving CO_2 and H_2O can buffer the fluid composition to increasing NaCl and CO_2 contents, whereas dehydration reactions tend to buffer the fluid to more H_2O-rich compositions. These conclusions are correct if the dehydration reaction has the "normal" topology in *P-T* space, that is, the equilibrium temperature increases as water pressure increases. The reaction of kaolinite to illite is interpreted to be important during diagenesis in the Venture area and in other sedimentary basins and can be written as a dehydration reaction. It does, however, have the reverse topology, that is the equilibrium temperature decreases as pressure increases. Reactions of this type also tend to coexist

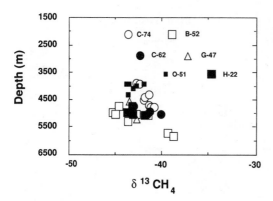

FIG. 5. $\delta^{13}C$ of CH_4 vs. depth in meters. Data for wells other than C-62 and C-74 are from KENDALL et al. (1989).

FIG. 7. $\delta^{13}C$ of CO_2 vs. $\delta^{13}C$ of CH_4 for all wells with data for both gases. The line labelled $a_c = 1.04$ is from WHITICAR et al. (1986).

with a fluid that increases in NaCl and CO_2 with increasing temperature.

Clay-carbonate reactions, specifically with chlorite, have been postulated or described by various authors (ZEN, 1959; MUFFLER and WHITE, 1969; HUTCHEON et al., 1980; McDOWELL and PACES, 1985) but the mineral compositions required to more accurately determine their stability have not been measured. Further, the phase relationships in a CO_2-H_2O-NaCl fluid, particularly at the limit of CO_2 miscibility in the aqueous phase, have not been determined.

The data used in the following calculations includes the thermodynamic properties of the minerals, mainly from HELGESON et al. (1978), with Fe-chlorite data from WALSHE (1986). The solubility of CO_2 in H_2O-NaCl fluids was obtained from BOWERS and HELGESON (1983), ELLIS and GOLDING (1963) and GEHRIG (1980). Using the same methods as HUTCHEON et al. (1980), the activity

coefficients for mixing of CO_2 and H_2O were obtained on the pseudo-binary at 0, 6 and 20 wt% NaCl using a two-constant Margules mixing model. The activity coefficients for H_2O in the liquid phase and CO_2 in the vapour phase at the conditions of interest were found to be close to unity and not to have a large effect on the calculated stability and have therefore been ignored in the calculations.

The activities of the components in reactions (1) and (2) were calculated using an ideal solution model and the compositions of the appropriate chlorite and carbonate. The stabilities of the reactions are shown on a pseudo-binary T-X_{CO_2} diagram (Fig. 8), at constant pressure, relative to the solubility of CO_2 in 0.0, 6.0 and 20.0 weight percent NaCl solutions. The reaction equilibria that include corrections for the mineral compositions are la-

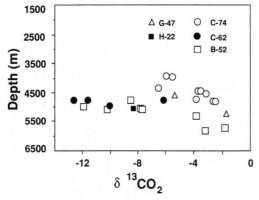

FIG. 6. $\delta^{13}C$ of CO_2 vs. depth in meters. Data for from KENDALL et al. (1989) were used in addition to data from C-62 and C-74.

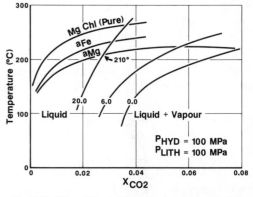

FIG. 8. Stability of Fe- and Mg-chlorite reactions in CO_2-H_2O-NaCl fluid at constant pressure. Note that the intersection of the end member reactions, adjusted for the activities of chlorite and calcite (curves labelled "a_{Mg}" and "a_{Fe}") intersect the 20 weight percent NaCl curve for CO_2 solubility at very similar temperatures.

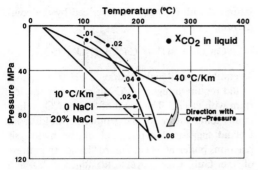

FIG. 9. The trace of the intersections of the kaolinite-carbonate-chlorite reaction with the solubility surface for CO_2 in a CO_2-H_2O-NaCl fluid can be shown as functions of pressure and temperature. The solid lines with the dots are 0 and 20 weight percent NaCl. The numbers beside each dot are the mole fraction of CO_2 in equilibrium at the intersection of the mineral reaction with the CO_2 solubility surface. Note that X_{CO_2}, and thus pCO_2, increases with increasing depth and temperature.

belled on the diagram and intersect the 20 wt% NaCl solubility curve at temperatures between 160 and 170°C, appropriate for a depth of 5000 m or greater in the B-43 well (Fig. 2). Examination of salinities from drill stem tests and interpreted from well logs suggest this is an appropriate salinity for formation waters in the Venture field.

By repeating the calculations at a number of different pressures it is possible to find the temperature of intersections of reactions such as (1) and (2) with the CO_2 solubility surface in a CO_2-H_2O-NaCl fluid. These can then be presented as a P-T diagram with the locus of the intersections for various NaCl contents and the mole fraction of CO_2 noted along the trace of the intersection. Fig. 9 shows such a diagram and a number of features are evident upon examination. Firstly, the evolution of a fluid coexisting with the reactions as written has the appropriate sense of compositional evolution. That is, the fluid becomes enriched in NaCl and CO_2 with increasing temperature and pressure. This does not imply that the solubility of CO_2 increases with increasing NaCl content because the pressure is also increasing. Over the range of pressures, temperatures and NaCl contents examined here it is apparent that increases in pressure increase the solubility of CO_2 and are sufficient to overcome the decrease in CO_2 solubility imposed by higher NaCl contents. Superimposed on the diagram are geothermal gradients for 10 and 40°C/km. The trace of the intersection between the CO_2 solubility surface and the mineral reaction first intersects the geothermal gradient at a temperature of about 140–150°C for a solution with 0.0 wt% NaCl. This is the minimum temperature at which

the kaolinite-carbonate to calcite reaction will be in equilibrium on the CO_2 solubility surface.

Using the fractionation factors for ^{13}C between CO_2 and calcite and temperature estimates, it is possible to calculate the expected isotopic composition of CO_2 from the measured calcite composition and compare it to the measured CO_2 composition. Temperature at various depths has been measured by Mobil Oil for the wells in this study. The isotopic compositions of calcite was determined as part of this study and the isotopic compositions for CO_2 in the gas that were not available from KENDALL et al. (1989) were measured, if samples were available. The results of these calculations are shown in Fig. 10. Some of the points are not in good agreement with the predicted value expected if the calcite were in isotopic equilibrium with CO_2 in the gas. Some of these samples, shown as shaded points in Fig. 10, have considerable amounts of detrital carbonate.

BUFFERING

The calculation of the stability of clay-carbonate reactions does not define the reaction mechanism that produces the CO_2 observed in sedimentary basins. It is possible that other reactions, such as those involving smectite (GUNTER and BIRD, 1988) or illite (HUTCHEON et al., 1980) and mixed volatiles would be stable at lower temperatures at the same pressure, and intersect the solubility surface of CO_2 in H_2O-CO_2-NaCl mixtures at lower temperatures. Reactions involving smectite or illite may be candidates to produce CO_2 by calcite dissolution during the diagenesis of shales.

MCDOWELL and PACES (1985) note that reaction among silicates and carbonates takes place in an

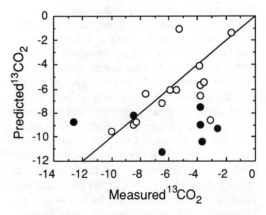

FIG. 10. The comparison of measured and expected compositions of CO_2. The shaded symbols represent samples with primarily detrital carbonate.

aqueous liquid and involves aqueous ions such as HCO_3^- and Ca^{2+}. Without measured concentrations of ions in solution it is not possible to calculate the stability of such reactions in a CO_2-H_2O-NaCl fluid. Once the aqueous phase is saturated with CO_2, further dissolution of carbonate will produce CO_2 directly into the vapour phase. Consideration of different reactions raises the question of how carbonates are dissolved and thus the question of how pH is buffered during diagenesis. SURDAM et al. (1989) have suggested that the organic acids are the most effective pH buffers over a considerable range of diagenetic temperatures. LUNDEGARD and LAND (1989) have examined the carbonate equilibria in the presence of organic acids and changing pCO_2. ABERCROMBIE (1989) examined buffering capacities for the carbonate minerals, dissolved carbonate and silicate minerals and found that hydrolysis of silicates has the greatest buffer capacity. His calculations did not consider aqueous organic species. HUTCHEON and ABERCROMBIE (1989a) show that, for most of the pH range, acetate solutions of 3000 mg/l have lower buffering capacity than a reaction between albite and smectite. GUNTER and BIRD (1988) have observed the production of CO_2 during hydrothermal experiments and suggest that silicate hydrolysis reactions are the source of protons for calcite dissolution. HUTCHEON and ABERCROMBIE (1989a,b) also present evidence from field data that aluminous silicates, such as kaolinite, function as "acids" as temperature increases during diagenesis. They show that the Na^+/H^+ and K^+/H^+ activity ratios of the fluids are in metastable equilibrium with silicate minerals observed in thermal flooding of heavy oil reservoirs, but that the $Ca^{2+}/(H^+)^2$ activity ratio was determined by calcite dissolution.

SMITH and EHRENBERG (1989) show a trend of temperature and pCO_2 for the Gulf Coast and North Sea and suggest the trend may be the result of hydrolysis reactions among silicates causing dissolution of carbonates. ARNÓRSSON et al. (1983a,b) have studied the compositions of fluids from geothermal wells and springs in Iceland and show fluid saturation with respect to silicates such as albite, smectite and analcime. They report a similar trend of pCO_2 vs. temperature for geothermal wells in basaltic rocks from Iceland. HUTCHEON and ABERCROMBIE (1989a) note that the data of ARNÓRSSON et al. (1983b) show a lower pCO_2 at any particular temperature than the data shown by SMITH and EHRENBERG (1989) for sedimentary rocks from the Gulf Coast.

The dissolution of calcite in an aqueous solution saturated with CO_2 can be described by the equilibrium:

$$CaCO_3 + 2H^+ \rightleftarrows Ca^{2+} + H_2O + CO_2. \quad (3)$$

At equilibrium, a higher imposed aH^+ (lower pH) would require a higher pCO_2 to maintain equilibrium with calcite. The trend of lower pCO_2 values at any temperature for the geothermal fluids from Iceland suggests that potential silicate hydrolysis reactions buffering pH are more "basic" than those of the Gulf Coast. ABERCROMBIE (1988) and HUTCHEON and ABERCROMBIE (1989b) have concluded from samples of hot fluids recovered from volcaniclastic rocks that these rocks are in metastable equilibrium with analcime and smectite. HUTCHEON et al. (1988) sampled waters from quartz-rich reservoirs and volcaniclastic reservoirs and determined that the fluids in quartz rich reservoirs were more "acidic" and in metastable equilibrium with kaolinite-smectite and the fluids from the volcaniclastic reservoirs were more "basic" and in equilibrium with smectite-analcime. Examination of the phase relationships for Na-Al-Si-O-H minerals shows that, in general, the smectite-analcime reaction is stable at values of Na^+/H^+ activity approximately two orders of magnitude (depending on temperature) higher than the kaolinite-smectite reaction. At similar Na^+ activities, the Na^+/H^+ activity ratio can only change significantly by changes in aH^+.

The calcite dissolution reaction (3), as written above, has a value of pCO_2 that depends on the activity of Ca^{2+} and H^+ in solution. Typically activities of H^+ will be 10^{-5} to 10^{-7} and the activity of Ca^{2+} will be approximately 10^{-2}, or three to five orders of magnitude greater. For reaction (3) the equilibrium constant is:

$$K = \frac{[aCa^{2+}][fCO_2]}{[aH^+]^2}.$$

Inspection of the equilibrium constant indicates that very small changes in aH^+, which is a squared term and a very small number compared to aCa^{2+}, will cause relatively large changes in fCO_2. For example, a one order of magnitude change in aH^+ will be reflected by a two order of magnitude change in log fCO_2 and, therefore, in log pCO_2. Fig. 11 shows the temperature-pCO_2 trends for fluids from sediments of the Gulf Coast (SMITH and EHRENBERG, 1989) and fluids from geothermal wells in basaltic rocks from Iceland (ARNÓRSSON et al., 1983b). The Gulf Coast rocks would be expected to have more acidic pH (higher aH^+) values imposed by reactions among kaolinite and smectite or illite, whereas the

Icelandic fluids would (for similar $a\text{Na}^+$) be expected to have more basic pH (lower $a\text{H}^+$) imposed by equilibria between smectite, analcime and albite.

Rocks in the Gulf Coast are reported to contain kaolinite and calcite as authigenic phases. To test the proposition that CO_2 partial pressures are determined by calcite dissolution regulated by silicate hydrolysis, water analyses were selected from the Gulf Coast (KHARAKA et al., 1985) and Iceland (ARNÓRSSON et al., 1983a). The analyses from the Gulf Coast were processed using EQ3 (WOLERY, 1983) with the conditions that the activity of Al^{3+} be set by equilibrium with kaolinite, $a\text{H}^+$ be set by equilibrium with illite and $a\text{HCO}_3^-$ be set by calcite. Illite was included in the calculations as muscovite and the resulting $p\text{CO}_2$ adjusted by considering the activity of muscovite in illite to be 0.1. The effect of varying muscovite activity in illite from 1.0 to 0.1 has a minimal effect on the calculations. The program was executed at temperatures between 25 and 250°C under these conditions, providing calculated values for the fugacity of CO_2. The calculated fugacity of CO_2 was used with fugacity coefficients (ANGUS et al., 1976), assuming a geothermal gradient of 30°C/km and a fluid pressure gradient of 105 bars per kilometer, to determine the value of $p\text{CO}_2$ at each temperature. These values were plotted as a line on Fig. 11 that is in reasonable agreement with the data points measured by SMITH and EHRENBERG (1989). EQ3 calculates high temperature aqueous reactions for pressures determined by the boiling curve for pure water and clearly the formation waters of the Gulf Coast have high salin-

ities. The partial pressure of CO_2 has been calculated only from equilibrium with calcite and all other variables are essentially imposed by the mineral assemblage chosen, thus the salinity effect should not be significant.

Attempts to saturate different water analyses from Iceland with smectite, calcite and albite were only successful up to 200°C (Fig. 11). Above this temperature EQ3 failed to execute successfully. The thermodynamic data for smectite and analcime are probably not of as good quality as the other data in the EQ3 data base and this may explain the lack of success in achieving numerical solutions to the problem. The data of ARNÓRSSON et al. (1983a,b) show that in the rocks they examined there are two different mineral assemblages, smectite-albite below 200°C and chlorite-epidote without smectite above 200°C. Fig. 11 shows a break in the trend of $p\text{CO}_2$ vs. temperature at approximately 200°C that may be related to this change in mineral assemblages. The data in Fig. 10 was used to calculate the trend of log $p\text{CO}_2$ vs. temperature for the kaolinite-carbonate-chlorite reaction in the Venture Field. The line passes through the higher temperature data reported by ARNÓRSSON et al. (1983b) in the temperature range reported for stable chlorite in the rocks from Iceland.

The trend of temperature with $p\text{CO}_2$ can be explained for both sets of data by considering schematically the change in position of the intersection of reaction (3) with limit of miscibility of CO_2 in a H_2O-NaCl aqueous solution (Fig. 12). At P_1 the intersection of the mineral reaction and the CO_2

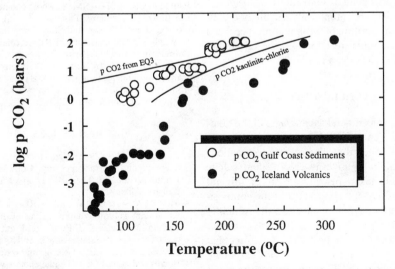

FIG. 11. Temperature-$p\text{CO}_2$ trends from Iceland and Gulf Coast. The line that passes through the Gulf Coast data was calculated using EQ3 and assuming equilibrium between the fluid, calcite, kaolinite, illite and CO_2.

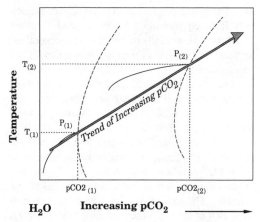

FIG. 12. Schematic representation of intersections of mineral reactions with the CO_2 solubility surface as a function of pCO_2 and temperature at two pressures, P_1 and P_2.

solubility surface is at a particular temperature, T_1. At some higher pressure, P_2, the solubility of CO_2 in the aqueous phase will be higher and there will be a point of intersection of reaction (3) at some higher temperature, T_2. If aH^+ is buffered by a relatively "acidic" assemblage, such as kaolinite-illite, the corresponding pCO_2 at any temperature will be higher than if aH^+ is buffered by a relatively "basic" mineral assemblage such as illite-K-feldspar or smectite-albite. It is likely that the locus of intersections of the mineral reactions with the solubility surface for most silicate hydrolysis reactions in rocks containing calcite (or other carbonates) will produce a relatively linear trend of temperature $vs.$ $\log pCO_2$. Fig. 12 also describes graphically the method used to calculate the position of the intersection between the chlorite-kaolinite reaction and the miscibility surface for CO_2 as shown in Fig. 9.

CONCLUSIONS

The authigenic mineral assemblages of the sandstones in the Venture Field show textural and petrographic evidence that clay-carbonate reactions involving kaolinite, ankerite chlorite and calcite probably have taken place. Analysis of mineral compositions and the application of the composition, with thermodynamic data and the solubility of CO_2 in H_2O-NaCl aqueous fluids shows that equilibria (1) and (2) can be used to determine that equilibrium temperatures for the clay-carbonate reaction are about 160°C, which is near the upper limit of reasonable temperatures for the over pressured zone of the Venture Field. The isotopic com-

position of calcite and CO_2 is consistent with an origin for the CO_2 from a clay-carbonate reaction.

The metastable equilibrium stability of chlorite, kaolinite and carbonates does not shed any light on the reaction mechanism. The suggestion that silicate hydrolysis controls the pH of the aqueous phase has been tested by considering the compositions of waters, buffered by silicate hydrolysis reactions, and calculating the values of pCO_2 that would result at any temperature. The calculated pCO_2 agrees reasonably well with measured trends for fluids from quartzose sedimentary rocks from the Gulf Coast and less well with geothermal fluids from basaltic volcanic rocks in Iceland suggesting that the trends of pCO_2 observed in sedimentary basins and hydrothermal fluids at temperatures in the range of 300°C, or lower, are controlled by silicate-carbonate mineral reactions.

Acknowledgements—This work was completed while the author was on a Senior Industrial Fellowship sponsored by NSERC (The Natural Sciences and Engineering Research Council, Canada), Shell Canada Limited and The University of Calgary. I am in debt to all three organizations for their support. Most of the data were collected with the cooperation of Mobil Oil Canada over an extended period between 1982 and 1986. Jo van Elsberg, Catherine Chaplin and Dave Blair, all formerly of Mobil, among many others at Mobil Oil Canada, were extremely helpful and cooperative with sampling core, interpreting data and providing moral support and discussions. The isotopic analyses were done in the Physics Department in the NSERC supported laboratory of H. R. Krouse at The University of Calgary and I am, as always, grateful for their assistance. Cynthia Nahnybida and Steve Machemer did some of the electron microprobe analyses, the remainder were done by the author under the guidance of Paul Mainwaring of CANMET, whose patience with probe initiations is legendary. Discussions with Hugh Abercrombie, Glenn Wilson and Steve Ehrenberg have been very helpful. The assistance of Ritsue Imakiire and Bill Bourcier in getting EQ3/6 running was much appreciated. Critical comments by John Bloch improved the final manuscript. Financial support for this project was received from Mobil Canada, a research contract with the Geological Survey of Canada and NSERC.

REFERENCES

ABERCROMBIE H. J. (1988) Water-rock interaction during diagnesis and thermal recovery, Cold Lake, Alberta. Unpublished Ph.D. Thesis. The University of Calgary, Calgary, Alberta.
ANGUS S., ARMSTRONG B. and DE RUECK K. (1976) International Thermodynamic Tables of the Fluid State Carbon Dioxide, Vol. 3. Pergamon Press.
ARNÓRSSON S., GUNNLAUGSSON E. and SVAVARSSON H. (1983a) The chemistry of geothermal waters in Iceland. II. Mineral equilibria and independent variables controlling water compositions. *Geochim. Cosmochim. Acta* **47**, 547–566.
ARNÓRSSON S., GUNNLAUGSSON E. and SVAVARSSON H.

(1983b) The chemistry of geothermal waters in Iceland. III. Chemical geothermometry in geothermal investigations. *Geochim. Cosmochim. Acta* **47**, 567–577.

BOWERS T. S. and HELGESON H. C. (1983) Calculation of thermodynamic and geochemical consequences of nonideal mixing in the system H₂O-CO₂-NaCl on phase relations in geologic systems: Equation of state for H₂O-CO₂-NaCl fluids at high pressures and temperatures. *Geochim. Cosmochim. Acta* **47**, 1247–1275.

CASSOU A., CONNAN J. and PORTHAULT B. (1977) Relations between maturation of organic matter and geothermal effect, as exemplified in Canadian east coast offshore wells. *Bull. Canadian Petrol. Geol.* **25**, 174–194.

DAVIES E. H. and AVERY M. P. (1984) A system for vitrinite reflectance analysis on dispersed organic matter for offshore Eastern Canada. *Geol. Surv. Canada Paper* 84-1A, 367–372.

ELLIS A. J. and GOLDING R. M. (1963) The solubility of carbon dioxide above 100°C in water and in sodium chloride solutions. *Amer. J. Sci.* **261**, 47–60.

GEHRIG M. (1980) Phasengleichgewichte und PVT-Daten ternarer Mishungen aus Wasser, Kohlendioxid und Natriumchlorid bis 3 Kbar und 550C. Unpublished Ph.D. Dissertation. Karlsruhe University, Germany.

GIVEN M. M. (1977) Mesozoic and early Cenozoic geology of offshore Nova Scotia. *Bull. Canadian Petrol. Geol.* **25**, 63–91.

GRADSTEIN F. M., WILLIAMS G. L., JENKINS W. A. M. and ASCOLI P. (1975) Mesozoic and Cenozoic stratigraphy of the Atlantic continental margin, Eastern Canada. *Canadian Soc. Petrol. Geol. Mem.* **4**, 103–131.

GUNTER W. G. and BIRD G. W. (1988) CO₂ production in tar sand reservoirs under in situ steam temperatures: reactive calcite dissolution. *Chem. Geol.* **70**, 301–311.

HAWORTH R. T. (1975) The development of Atlantic Canada as a result of continental collision—evidence from offshore gravity and magnetic data. *Canadian Soc. Petrol. Geol. Mem.* **4**, 59–77.

HELGESON H. C., DELANY J. M., NESBITT H. W. and BIRD D. K. (1978) Summary and critique of the thermodynamic properties of rock-forming minerals. *Amer. J. Sci.* **278A**, 1–229.

HERB G. (1975) Diagenesis of deeply buried sandstones on the Scotian Shelf. Ph.D. Thesis. Dalhousie University, Dartmouth, Nova Scotia.

HOWIE R. D. and BARSS M. S. (1975) Paleogeography and sedimentation in the upper Paleozoic, Eastern Canada. *Canadian Soc. Petrol. Geol. Mem.* **4**, 45–57.

HUTCHEON I. and NAHNYBIDA C. G. (1990) Clay mineralogy and diagenesis of sandstones and shales from the Venture Region, Nova Scotia, Canada. (in preparation).

HUTCHEON I., ABERCROMBIE H. and KROUSE H. R. (1990) Inorganic origin of carbon dioxide production in low temperature thermal recovery of bitumen: Chemical and isotopic evidence. *Geochim. Cosmochim. Acta* (in press).

HUTCHEON I. and ABERCROMBIE H. (1989a) Carbon dioxide in clastic rocks and silicate hydrolysis. *Geology* (submitted).

HUTCHEON I. and ABERCROMBIE H. (1989b) Fluid-rock interactions in thermal recovery of bitumen, Tucker Lake Pilot, Cold Lake, Alberta. *Amer. Assoc. Petrol. Geol.* Special volume on reservoir prediction (submitted).

HUTCHEON I., ABERCROMBIE H., SHEVALIER M. and

NAHNYBIDA C. (1988) A comparison of formation reactivity in quartz-rich and quartz-poor reservoirs during steam assisted recovery. *Fourth International UNITAR/UNDP Conference on Heavy Crude and Tar Sands.* Paper 235. Edmonton.

HUTCHEON I., OLDERSHAW A. and GHENT E. (1980) Diagenesis of Cretaceous sandstones of the Kootenay Formation at Elk Valley (southeastern British Columbia) and Mt. Allan (southwestern Alberta). *Geochim. Cosmochim. Acta* **44**, 1425–1435.

ISSLER D. R. (1984) Calculation of organic maturation levels for offshore eastern Canada,—Implications for general application of Lopatin's method. *Canadian J. Earth Sci.* **21**, 477–488.

JANSA L. F. and WADE J. A. (1975) Paleogeography and sedimentation in the Mesozoic and Cenozoic, southeastern Canada. *Canadian Soc. Petrol. Geol. Mem.* **4**, 103–131.

JANSA L. F. and NOGUERA URREA V. H. (1989) Geology and genesis of over pressured sandstone reservoirs in the Venture gas field, Offshore Nova Scotia, Canada (abstr.). *Annual Convention Amer. Assoc. Petrol. Geol.,* San Antonio, Texas.

KEEN C. E. (1979) Thermal history and subsidence of rifted continental margins—evidence from wells on the Nova Scotia and Labrador shelves. *Canadian J. Earth Sci.* **16**, 505–522.

KEEN C. E. (1983) Salt diapirs and thermal maturity: Scotian Basin. *Bull. Canadian Petrol. Geol.* **31**, 101–108.

KEEN C. E. and LEWIS T. (1982) Measured radiometric heat production in sediments from the continental margin of eastern North America: Implications for petroleum generation. *Bull. Amer. Assoc. Petrol. Geol.* **66**, 1402–1407.

KENDALL S., KROUSE H. R. and ALTEBAUEMER F. (1989) Isotopic evidence for the generation and migration of gas, Sable sub-basin, Scotian Shelf, Canada. *Bull. Canadian Petrol. Geol.* (submitted).

KHARAKA Y. K., HULL R. W. and CAROTHERS W. W. (1985) Water-rock interactions in sedimentary basins. In *Relationship of Organic Matter and Mineral Diagenesis* (ed. D. L. GAUTIER). Soc. Econ. Paleontol. Mineral. Short Course 17.

LUNDEGARD P. D., LAND L. S. and GALLOWAY W. E. (1984) Problem of secondary porosity: Frio Formation (Oligocene), Texas Gulf Coast. *Geology* **12**, 399–402.

LUNDEGARD P. D. and LAND L. S. (1986) Carbon dioxide and organic acids: their role in porosity enhancement and diagenesis of the Texas Gulf Coast. *Soc. Econ. Paleontol. Mineral. Spec. Publ.* **38**, 129–146.

LUNDEGARD P. D. and LAND L. S. (1989) Carbonate equilibria and pH buffering by organic acids—response to changes in pCO₂. *Chem. Geol.* **74**, 277–287.

MCDOWELL S. D. and PACES J. B. (1985) Carbonate alteration minerals in the Salton Sea geothermal system, California, USA. *Mineral. Mag.* **49**, 469–479.

MCIVER N. L. (1972) Cenozoic and Mesozoic stratigraphy of the Nova Scotia shelf, *Canadian J. Earth Sci.* **9**, 54–69.

MUFFLER L. P. J. and WHITE D. E. (1969) Active metamorphism of Upper Cenozoic sediments in the Salton Sea geothermal field and the Salton Trough, Southeastern California. *Bull. Geol. Soc. Amer.* **80**, 157–182.

NOGUERA URREA V. H. (1987) Geology and diagenetic history of over pressured reservoirs in the lower Mississauga-Mic Mac Formations of the Venture gas field,

Scotian Shelf, Nova Scotia. Unpublished M.Sc. Thesis. Dalhousie University, Dartmouth, Nova Scotia.

PARSONS M. G. (1975) The geology of the Laurentian fan and the Scotian rise. *Canadian Soc. Petrol. Geol. Mem.* **4,** 155–167.

PIPER D. J. W. (1975) Late quaternary deep water sedimentation off Nova Scotia and western Grand banks. *Canadian Soc. Petrol. Geol. Mem.* **4,** 195–204.

POWELL T. G. and SNOWDON L. R. (1979) Geochemistry of crude oils and condensates from the Scotian Basin, offshore eastern Canada. *Bull. Canadian Petrol. Geol.* **27,** 453–466.

POWELL T. G. (1982) Petroleum geochemistry of the Verrill Canyon formation: a source for Scotian shelf hydrocarbons. *Bull. Canadian Petrol. Geol.* **30,** 167–179.

PURCELL L. P., RASHID M. A. and HARDY I. A. (1979) Geochemical characteristics of sedimentary rocks in the Scotian Basin. *Bull. Amer. Assoc. Petrol. Geol.* **63,** 87–105.

SMITH H. A. (1975) Geology of the West Sable structure. *Bull. Canadian Petrol. Geol.* **23,** 109–130.

SMITH J. T. and EHRENBERG S. N. (1989) Correlation of carbon dioxide abundance with temperature in clastic hydrocarbon reservoirs: Relationship to inorganic chemical equilibrium. *Marine Petrol. Geol.* **6,** 129–135.

SURDAM R. C., CROSSEY L. G., HAGEN E. S. and HEASLER H. P. (1989) Organic-inorganic interactions and sandstone diagenesis. *Bull. Amer. Assoc. Petrol. Geol.* **73,** 1–23.

SWIFT J. H., SWITZER R. W. and TURNBULL W. F. (1975) The Cretaceous Petrel Limestone of the Grand Banks, Newfoundland. *Canadian Soc. Petrol. Geol. Mem.* **4,** 181–194.

TROMMSDORFF V. and SKIPPEN G. B. (1986) Vapour loss ("boiling") as a mechanism for fluid evolution in metamorphic rocks. *Contrib. Mineral. Petrol.* **94,** 317–322.

WALSHE J. L. (1986) A six-component chlorite solid solution model and the conditions of chlorite formation in hydrothermal and geothermal systems. *Econ. Geol.* **81,** 681–703.

WHITICAR M. J., FABER E. and SCHOELL M. (1986) Biogenic methane formation in marine and freshwater environments: CO_2 reduction *vs.* acetate fermentation—Isotope evidence. *Geochim. Cosmochim. Acta* **50,** 693–709.

WOLERY T. J. (1983) EQ3NR A program for geochemical aqueous speciation-solubility calculations: User's guide and documentation. Lawrence Livermore Laboratory Report UCRL-53414.

ZEN E-AN (1959) Clay mineral-carbonate relations in sedimentary rocks. *Amer. J. Sci.* **257,** 29–43.

Fluid-Mineral Interactions: A Tribute to H. P. Eugster
© The Geochemical Society, Special Publication No. 2, 1990
Editors: R. J. Spencer and I-Ming Chou

Normative analysis of groundwaters from the Rustler Formation associated with the Waste Isolation Pilot Plant (WIPP), southeastern New Mexico

MARC W. BODINE, JR.[†] and BLAIR F. JONES

U.S. Geological Survey, MS 432 National Center, Reston, Virginia 22092, U.S.A.

Abstract—Salt norms and simple salt assemblages have been calculated for 124 water analyses from the three aquifers of the Ochoan Series of the Rustler Formation of Permian age (plus four analyses from overlying strata) at and near the Waste Isolation Pilot Plant (WIPP) in southeastern New Mexico. The results indicate that Rustler Formation waters range from hypersaline, primitive-diagenetic fluids, probably synsedimentary, to meteoric recharge waters. The distribution of normative halite in the aquifer fluids is compatible with zones that describe the extent of halite dissolution in the Rustler Formation. Recharge waters have entered the Rustler aquifers through overlying strata and/or laterally. Their norms reflect negligible to extensive resolution of Rustler or Permian Age Salado Formation halite, pervasive resolution of calcium sulfate in the Rustler Formation throughout the region, and silicate hydrolysis of framework minerals in clastic sediments within and above the Rustler.

The salt norms indicate (1) small, apparently isolated areas of the Culebra and Magenta Dolomite aquifers in the Rustler Formation contain primitive-diagenetic fluid residuals because of very low formational transmissivity; (2) within most areas of the Rustler Formation that are not overlain by halite, extensive infiltration of recharge waters into the Magenta, Culebra, and Rustler-Salado contact zone aquifers is correlative with normative sulfate content; (3) much of the halite resolution effected by Rustler-Salado contact zone waters has taken place in uppermost Salado Formation halites; (4) mixtures of primitive-diagenetic solutes with alkali sulfate-bearing recharge can produce NaCl-rich norms without halite dissolution; and (5) normative alkali sulfate content can be related to the ground-water flux from overlying continental (siliciclastic) sediment.

The large variation in hydrologic properties of the extensively studied Culebra Dolomite aquifer is emphasized by the diversity in normative results. In addition, the flow field simulations presented by DAVIES (1989) are reflected in the areal distribution of normative compositions. Interpretation of the norms is consistent with the north-to-south flow direction and velocities in the model, and is compatible with paleo-recharge from the Nash Draw area to the north and west of the WIPP site.

INTRODUCTION

AN UNDERSTANDING of the evolution and hydrodynamics of ground waters in salt beds at the Waste Isolation Pilot Plant (WIPP) in the Delaware Basin in southeastern New Mexico is essential for developing the technology to assure effective containment of long-lived nuclear waste. The groundwaters occur in several geologic horizons above and below the waste facility and exhibit a wide range of compositions, origins, and flow regimes that are currently being investigated by a variety of techniques in several laboratories.

We propose that recasting a water's chemical composition into a "salt norm" (BODINE and JONES, 1986) is a useful diagnostic tool in this effort. The salt norm is the idealized equilibrium assemblage of mineral salts at surficial conditions that is quantitatively equivalent to the solute concentrations in the water. The salt norm can be envisioned as the assemblage of salts that would have precipitated from the evaporated water sample and would be in equilibrium with the last vestage of remaining water.

The normative assemblage yields a diagnostic chemical-mineralogical characterization of the water, aids in the interpretation of the origin of the water's solutes, is indicative of the character of water-rock interaction in subsurface environments, and may contribute to determining the evolutionary path of the water chemistry.

In this report, after a brief review of the calculation of the salt norm and the criteria for its interpretation, we present salt norms for 124 analyses of groundwaters from the Rustler Formation and 4 analyses of groundwaters from overlying strata from the WIPP-site region and discuss their use in characterizing the waters and indicating their origin and evolution.

THE SALT NORM

Presenting a water's dissolved constituents as neutral salt abundances is not new. In the nineteenth century it was the accepted practice to report water compositions in terms of abundances of simple salts. As HEM (1970) pointed out, this practice predated acceptance of the Arrhenius concept of dissociated ions, and although it also attempted to express water compositions in terms of the salts produced upon evaporation, it was actually more closely related to classical gravimetric analytical procedures

---

[†] Deceased.

with minimal regard to salt speciation and association. It is not surprising then, that this form of expressing water composition diminished in the twentieth century with only occasional use for specific purposes in the more recent literature either by forming assemblages of simple salts (RANKAMA and SAHAMA, 1950, p. 318) or attempts at forming reasonable associations of mineral salts (LAMBERT, 1978).

The striking differences in salt mineral assemblages in marine and nonmarine evaporite deposits are clearly indicative of both lithologic origin and subsequent geochemical evolution for the dissolved constituents (EUGSTER and JONES, 1979; EUGSTER et al., 1980). Thus it is informative to reconstruct solute content of a natural water into the equilibrium salt assemblage to be obtained if the water evaporated to dryness under earth surface conditions.

The salt norm is analogous to the CIPW norm (CROSS et al., 1902) which is an idealized equilibrium assemblage of igneous minerals that is calculated from the rock's chemical composition. The CIPW norm has proved useful in igneous petrology not only for characterizing and classifying igneous rocks but for providing invaluable quantitative data for rigorous interpretation of the origin and evolution of an igneous complex. We suggest similar advantages for the salt norm in characterizing natural waters and interpreting their chemical evolution. Characterization of a water composition as a salt norm is more detailed and more suggestive of solute origin and subsequent interaction than major cation-anion predominance that is currently the most commonly utilized system of hydrochemical classification.

Calculation of the salt norm

The salt norm is calculated with the computer program SNORM (BODINE and JONES, 1986) and this discussion will be limited to the program's major features. The program computes the norm by quantitatively assigning up to 18 solutes (Table 1) into the equilibrium assemblage of salts obtained from a list of 63 normative salts (Table 2). SNORM computes the normative assemblage directly from the solute matrix without proceeding along a succession of reaction paths. The salt norm characterizes a natural water's composition and indirectly indicates solute source

Table 1. Solutes in water analyses assigned to normative salts

	Cations		Anions	
Major	Mg^{2+}	Na^+	Cl^-	HCO_3^-
	Ca^{2+}	K^+	SO_4^{2+}	CO_3^{2-}
Minor	Li^+	Sr^{2+}	Fl^-	NO_3^-
	NH_4^+	Ba^{2+}	$*Br^-$	$†B$
			$*I^-$	PO_4^{3-}

* Assigned as solid solution to chloride normative salts.

† Recast as borate with assignment of a preliminary charge of $-^7/_{12}$ per boron, that is, the artificial borate ion $(BO_{1.79})^{-7/12}$; if feasible, reassigned the charge per boron appropriate for the borate stoichiometry of the borate salt in the normative assemblage (see BODINE and JONES, 1986).

Table 2. Chemical formula of the normative salts

Ammonia niter	NH_4NO_3
Anhydrite	$CaSO_4$
Antarcticite*	$CaCl_2 \cdot 6H_2O$
Aphthitalite*	$K_3Na(SO_4)_2$
Arcanite*	K_2SO_4
Barite	$BaSO_4$
Bischofite*	$MgCl_2 \cdot 6H_2O$
Bloedite*	$Na_2Mg(SO_4)_2 \cdot H_2O$
Borax	$Na_2B_4O_7 \cdot 10H_2O^+$
Burkeite*	$Na_6CO_3(SO_4)_2$
Calcite*	$CaCO_3$
Carnallite*	$KMgCl_3 \cdot 6H_2O$
Celestite	$SrSO_4$
Dolomite*	$CaMg(CO_3)_2$
Epsomite*	$MgSO_4 \cdot 7H_2O$
Fluorapatite	$Ca_5(PO_4)_3F$
Fluorite	CaF_2
Glauberite*	$Na_2Ca(SO_4)$
Gypsum	$CaSO_4 \cdot 2H_2O$
Halite*	$NaCl$
Hydroxyapatite	$Ca_5(PO_4)OH†$
Indirite	$Mg_2B_6O_{11} \cdot 15H_2O**$
Inyoite	$Ca_2B_6O_{11} \cdot 13H_2O**$
Kainite*	$KMgClSO_4 \cdot 3H_2O$
Kalicinite*	$KHCO_3$
Kieserite*	$MgSO_4 \cdot H_2O$
Leonite*	$K_2Mg(SO_4)_2 \cdot 4H_2O$
Magnesite*	$MgCO_3$
Mascagnite*	$(NH_4)_2SO_4$
Mirabilite	$Na_2SO_4 \cdot 10H_2O$
Niter	KNO_3
Nitrobarite	$Ba(NO_3)_2$
Nitrocalcite	$Ca(NO_3)_2$
Nitromagnesite	$Mg(NO_3)_2$
Picromerite*	$K_2Mg(SO_4)_2 \cdot 6H_2O$
Pirssonite*	$Na_2Ca(CO_3)_2 \cdot 2H_2O$
Polyhalite*	$K_2Ca_2Mg(SO_4)_4 \cdot 2H_2O$
Salammoniac	NH_4Cl
Sellaite	MgF_2
Soda niter	$NaNO_3$
Strontionite	$SrCO_3$
Sylvite*	KCl
Syngenite*	$K_2Ca(SO_4)_2 \cdot 2H_2O$
Tachyhydrite*	$Mg_2CaCl_6 \cdot 12H_2O$
Teschemacherite	NH_4HCO_3
Thenardite*	Na_2SO_4
Trona*	$Na_3H(CO_3)_2 \cdot 2H_2O$
Ulexite	$NaCaB_5O_9 \cdot 8H_2O**$
Villiaumite	NaF
Wagnerite	Mg_2PO_4F
Witherite	$BaCO_3$
—	$BaCl_2 \cdot H_2O$
—	$BaCl_2 \cdot 2H_2O$
—	$LiCl \cdot H_2O$
—	Li_2CO_3
—	LiF
—	$LiNO_3 \cdot 3H_2O$
—	$Li_2SO_4 \cdot H_2O$
—	$Mg_3(PO_4)_2$
—	Na_3PO_4
—	$SrCl_2 \cdot 2H_2O$
—	$SrCl_2 \cdot 6H_2O$
—	$Sr(NO_3)_2$

*Diagnostic major cation and anion salt (gypsum and mirabilite omitted because their respective hydration equilibria preclude chloride or nitrate in the water).

† Normative hydroxyapatite calculated as $Ca_{4.75}(PO_4)_{3.17}$ in order to preserve charge balance.

** Stoichiometry of borate salts modified if norm calculated with average borate charge of $-^7/_{12}$ per boron (see BODINE and JONES, 1986).

and chemical evolution of the water; no direct information on reaction paths or extent of mineral saturation for an individual water sample is given by SNORM.

Solutes

The solutes (Table 1) were selected to comply with two criteria: (1) the solute, at least occasionally, occurs in natural waters in more than negligible (trace) concentrations; and (2) the solute forms its normative salt(s) through direct combination with other solutes in the analysis without interacting with the aqueous environment or requiring any chemical or charge modification. Thus, neither pH nor redox enter directly into the SNORM calculations. Only boron, which is reported as the element in conventional water analyses, fails to comply with these criteria. In SNORM boron is recalculated as borate and contributes to the total anion charge of the analysis.

Total cation equivalency must equal total anion equivalency in order to assign all solutes to neutral salt species. This is accomplished in SNORM by proportioning all cation and anion concentrations in the analysis to yield neutral charge balance. The deviation of the cation and anion proportionation factors from unity is a measure of the completeness and accuracy of the water sample analysis.

Normative salts

SNORM forms the salt norm from a list of 63 permissable normative salts (Table 2). Most of the salts occur as minerals and are thermodynamically stable at 25°C, 1 bar total pressure, and atmospheric partial pressure of carbon dioxide ($10^{-3.5}$ bars) at the water activity imposed by the salt assemblage.

After SNORM establishes the maximum number of phases in the assemblage with the Gibbs phase rule, the salt norm is calculated by assigning the solutes to the single assemblage that quantitatively consumes all solutes and contains no unstable salt associations. The unstable salt associations have been determined with free energy minimization reactions. Where thermodynamic data were lacking, mineral associations in surficial assemblages and analogy with relations in similar equilibria were used to estimate unstable associations.

Interpretation of the salt norm

Characterization and interpretation of the salt norm relies almost exclusively on the major cation-major anion salts in the norm (those salts denoted with an asterisk in Table 2) although salts of minor solutes are included in the norm. Throughout this report we express normative salt relative abundances as their *anhydrous weight percentages* in the salt norm. Anhydrous weight percentage of the normative salt excludes waters of hydration in the salt's formula from the salt's weight in calculating relative abundance and such normative abundances more closely reflect solute proportions.

Because the salt norm relates water composition to the relative abundances of both cations and anions, one or more of the key salts in the salt norm more fully characterizes water composition than does cation or anion predominance. Thus, the designation of a water sample as a "tachyhydrite water" is more explicit than its designation as a "chloride-rich water." Even more importantly, the

norm, both qualitatively and quantitatively, is indicative of solute source and the character of water-rock interactions that occurred during the water's evolution.

To that end BODINE and JONES (1986) developed a diagnostic chart (Fig. 1) relating the occurrence of normative salts to solute source and chemical evolution of the water. Three major categories of solute sources are recognized: a meteoric source in which the waters derive their solutes through surficial weathering processes; a marine source that, with or without modification, reflects a dominant sea water origin; and a diagenetic source in which the solutes reflect substantial rock-water interaction. In Table 3 we have compiled a representative group of salt norms from BODINE and JONES (1986) that illustrate the major features of Fig. 1.

Meteoric norms

Norms for meteoric waters characteristically contain alkali-bearing sulfate or carbonate normative salts (Fig. 1). Dissolution of soluble minerals (such as calcite and other carbonates, gypsum, and halite), contamination by anthropogenic constituents, or introduction of aerosol often contribute an appreciable fraction of the solutes in meteoric waters. However, it is the hydrolysis reactions involved in the weathering of silicate rocks that generate the most diagnostic features in the norms of meteoric water. Alkali-bearing carbonate salts in the norm reflect atmospheric carbon dioxide-water interaction forming carbonic acid and accompanying hydrolysis of silicate minerals. Alternatively, alkali-bearing sulfate salts in the norm reflect sulfide mineral oxidation forming sulfuric acid with consequent hydrolysis of silicate minerals.

Salt speciation and abundance in the norm frequently permits more detailed interpretation (Fig. 1). Although the norms in Table 3 include but four of the many examples comparing the character of the norm of meteoric waters with the weathering processes and host lithologies that are in BODINE and JONES (1986), they illustrate some important diagnostic features. The trona-burkeite abundance in the Arizona norm (Table 3, no. 1) illustrates extensive carbonic acid hydrolysis during the weathering of an Na-rich silicate host. Anhydrite-epsomite-bloedite in the Sudbury norm (Table 3, no. 3) reflect sulfuric acid hydrolysis of mafic rocks, whereas thenardite-glauberite in the Virginia norm (Table 3, no. 2) characterize similar weathering of a granite rock. The abundant halite in the Yilgarn norm (Table 3, no. 4) most likely reflects halite dissolution that practically overwhelms but does not obliterate the indicators (polyhalite-leonite) of sulfuric acid hydrolysis.

Marine norms

Sea water has a characteristic halite-bischofite-kieserite-carnallite-anhydrite norm (Table 3, no. 5) with the magnesium chloride association as represented by the bischofite-carnallite pair in the norm as the single unique diagnostic feature. It is difficult to attribute this association ultimately to any other than a marine source, though there may have been subsequent continental recycling (BODINE and JONES, 1986).

Only a few subsurface waters (other than those in the immediate vicinity of a marine source) yield norms quantitatively similar to that for sea water. This is not unex-

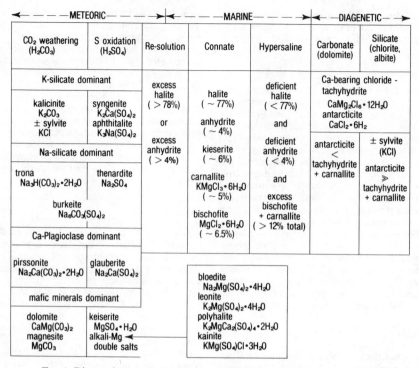

FIG. 1. Diagnostic chart for normative salts from BODINE and JONES (1986).

pected. Mixing, continental cycling, and water-rock interaction, including resolution of marine evaporite salts, can be expected not only to produce quantitative variations among the sea water normative minerals but to qualitatively modify the normative mineralogy, as in many cases reviewed by BODINE and JONES (1986). As one example of the latter, a connate marine groundwater mixing with a meteoric water bearing normative alkali sulfates can eliminate bischofite and generate sylvite or kainite in the norm as magnesium becomes associated with sulfate rather than chloride. The presence of polyhalite in an otherwise marine-like norm provides an even stronger indication of marine-meteoric water mixing.

Surface brines that have evolved through fractional precipitation of calcium sulfate, halite, or even the bitter salts yield norms qualitatively identical to sea water but quantitatively distinct (Fig. 1). The salt norm of a Black Sea evaporite brine sampled at the early states of magnesium sulfate deposition (Table 3, no. 6) illustrates the pronounced relative enrichment of bischofite, carnallite, and kieserite in the norm with a corresponding decrease of halite. Subsurface waters from a marine evaporite source are subject to similar modifications as subsurface marine waters; however, elimination of the bischofite or bischofite-carnallite signature from the norm becomes less likely.

Diagenetic norms

The presence of calcium-bearing chloride minerals, tachyhydrite and antarcticite (Table 2), in the salt norm is indicative of solute diagenesis (Fig. 1). Nearly all such norms in BODINE and JONES (1986) were obtained from subsurface fluids associated with deep sedimentary basins, highly saline strata, or crystalline shield areas, and are most

readily interpreted as evolving through the exchange of magnesium and sodium in solution for calcium in the solid phase. Because of the relatively low solubility of calcium sulfate, the abundance of normative sulfate salts becomes vanishingly small as calcium-bearing chlorides increase.

In carbonate rocks the exchange of magnesium in solution for calcium in the solid phase primarily reflects dolomitization. The extent of the exchange is limited by the Ca/Mg concentration ratio for the calcite-dolomite equilibrium. Although affected by aqueous calcium and magnesium ion activity coefficients, temperature, and mineralogy of the carbonate species, the Ca/Mg ratio, reflected in the norm by the antarcticite to tachyhydrite ratio, is not expected to greatly exceed unity (maximum of about 4?) due to dolomitization alone. For example, the norm of a marine evaporite brine associated with extensive recent dolomitization from the north Sinai (Table 3, no. 7) contains abundant tachyhydrite with only minor antarcticite.

Alternatively, in silicate rocks the exchange of magnesium and sodium in solution for calcium in plagioclase and Ca-bearing mafic minerals, reflecting albitization and chloritization, can proceed much farther than similar exchanges accompanying dolomitization, particularly because of the decrease in solubility of magnesium silicate at moderately elevated temperatures. For example, magnesium salts are nearly absent in the halite-antarcticite norm of a Sudbury mine water in crystalline rocks (Table 3, no. 8).

Graphic representation of the salt norm

A simplified representation of the salt norm for graphic display on maps has led to the definition of the "simple-

Table 3. Examples of salt norms (anhydrous weight percent) of waters representing various paths of solute evolution selected from BODINE and JONES (1986)

	1	2	3	4	5	6	7	8
TDS	477 mg/kg	905 mg/kg	4,280 mg/l	227,000 mg/l	35,150 mg/kg	309,200 mg/kg	275,500 mg/kg	241,000 mg/l
Salt norm								
Calcite	—	7.4	—	—	—	—	—	—
Dolomite	0.9	4.5	—	—	—	—	<0.1	—
Magnesite	0.4	—	3.4	<0.1	0.3	—	—	—
Trona	60.2	—	—	—	—	—	—	—
Burkeite	31.4	—	—	—	—	—	—	—
Anhydrite	—	—	40.8	—	4.0	—	0.2	0.1
Kieserite	—	—	—	—	6.1	26.1	—	—
Epsomite	—	—	28.6	4.7	—	—	—	—
Polyhalite	—	—	4.5	2.2	—	—	—	—
Leonite	—	—	—	2.5	—	—	—	—
Syngenite	—	4.7	—	—	—	—	—	—
Bloedite	—	—	18.8	—	—	—	—	—
Glauberite	—	24.9	—	—	—	—	—	—
Aphthitalite	0.3	—	—	—	—	—	—	—
Thenardite	—	52.9	—	—	—	—	—	—
Kainite	—	—	—	1.8	—	—	—	—
Halite	6.5	4.7	3.7	88.7	78.1	30.8	35.0	19.0
Sylvite	—	—	—	—	—	—	—	0.1
Carnallite	—	—	—	—	4.9	19.3	0.1	<0.1
Bischofite	—	—	—	—	6.5	23.7	—	—
Tachyhydrite	—	—	—	—	—	—	58.9	—
Antarcticite	—	—	—	—	—	—	5.8	79.7

1. Meteoric groundwater evolved chiefly through carbonic acid hydrolysis of sodium-rich silicate rocks, Navaho Sandstone, Mexican Water, Arizona (WHITE *et al.*, 1963).

2. Meteoric groundwater evolved chiefly through sulfuric acid hydrolysis of Paleozoic granite, Chester, Virginia (WHITE *et al.*, 1963).

3. Meteoric groundwater evolved chiefly through sulfuric acid hydrolysis of mafic igneous rocks, Sudbury area, Ontario (FRAPE *et al.*, 1984).

4. Meteoric groundwater evolved chiefly through resolution of marine solutes (halite) diluting sulfuric acid hydrolysis effects, Yilgarn Block, Western Australia (MANN, 1983).

5. Normal sea water (RILEY and CHESTER, 1971).

6. Marine evaporite brine at initial deposition of magnesium sulfate, Black Sea pond, Ukraine (ZHEREBTSOVA and VOLKOVA, 1966).

7. Diagenetic water resulting chiefly from dolomitization of calcarenite, Hayareah Sabka, Bardawil Lagoon region, north Sinai (LEVY, 1977).

8. Diagenetic water most likely resulting chiefly from albitization and chloritization of crystalline rocks, North Mine, Sudbury District, Ontario (FRAPE and FRITZ, 1982).

salt assemblage" (BODINE and JONES, 1986). The simple-salt assemblage is constructed by SNORM from the salt norm through separation of each major-ion normative salt into its simple-salt components (all salts with minor solutes are omitted). For example, one mole of polyhalite (Table 2) in the norm is recalculated as 2 moles of $CaSO_4$, 1 mole of K_2SO_4, and 1 mole of $MgSO_4$. The abundance of each simple salt from its respective normative-salt sources is summed and normalized to its weight and molar percentage in the simple-salt assemblage. Each simple salt is defined with an anion charge of -2 to maintain consistency throughout an array of simple salts; this results in molar units of the alkali chlorides being expressed as Na_2Cl_2 and K_2Cl_2. Twelve different simple salts are recognized: the carbonates (bicarbonate in normative salts is recast as carbonate), the sulfates, and the chlorides (chloride includes

any bromine and iodine as if in solid solution) of each major cation, including calcium, magnesium, potassium, and sodium.

Graphic display of the sample salt abundances is formatted as a rose diagram in which the circle is divided into twelve 30 degree segments with each segment representing a specific simple salt (Fig. 2A). We have arranged the simple salts on the diagram with the sulfates occupying the upper third of the circle, the carbonates occupying the lower left third, and the chlorides occupying the lower right third. The three anion groups are separated by radial boundary lines on the rose diagram. The salts within each anion group are sequentially arranged in terms of their respective relative solubilities. For example, calcium sulfate, the least soluble of the sulfates, occupies the rightmost segment in the sulfate area and is followed from right to

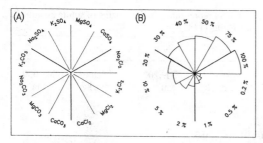

FIG. 2. The simple-salt rose diagram: A. Location of simple-salt segments on the rose diagram. B. Relative scale of segment radii as a function of the cube root of the salt's weight or mole fraction in the assemblage.

left by Mg_2SO_4, then K_2SO_4, with Na_2SO_4, the most soluble of the sulfates, in the leftmost segment of the sulfate group.

The radius of each segment of the rose is a function of that salt's relative molar or weight abundance in the simple-salt assemblage. Because the more diagnostic salts commonly have low abundance we have adopted the convention of equating segment radius to the cube root of the salt's fraction in the assemblage (Fig. 2B), recognizing that small-to-moderate differences in abundance above 25 percent will not be readily apparent.

Fig. 3 contains the simple-salt rose diagrams illustrating the variations of the salt norms for the array of waters in Table 3. Most full salt norm features that characterize these waters are preserved in the simple-salt assemblage.

THE WIPP SITE

The Waste Isolation Pilot Plant (WIPP) site is located 40 kilometers (25 miles) east of Carlsbad, New Mexico in an area of Eddy County referred to as Los Medanos (Fig. 4). On this and all ensuing maps the site is identified by the distinctive Zone IV boundary that formerly defined the exclusion area controlled by the Department of Energy (POWERS et al., 1978). The site is in the northern part of the Delaware Basin about 10 kilometers (6 miles) south of the buried Capitan reef front. The site area has little relief and, in general, slopes gently westward. To the west of the site is the escarpment of Livingstone Ridge, which marks the eastern margin of Nash Draw, a pronounced, but broad valley trending south-southwest. The area is drained by the Pecos River, a perennial southerly flowing stream 16 kilometers (10 miles) west-southwest of the WIPP site boundary at its closest point. The land surface is partially blanketed with sand dunes and is sparsely vegetated with a northern Chihuahuan desert flora.

Geologic setting, Delaware Basin

The geology of the Delaware Basin has been extensively studied over the past half century. In ad-

dition to the recent geologic investigations of the WIPP-site area, the Delaware Basin has long been a focus of geologic interest because of its abundant fluid hydrocarbon resources, its thick and somewhat unusual marine evaporite succession containing the nation's major potash deposits, and its striking geologic features, particularly the complex reef systems culminating with the imposing Capitan reef that is so well exposed in the Guadalupe Mountains along the western margin of the basin.

In this report we rely heavily on data from POWERS et al. (1978), MERCER (1983), and DAVIES (1989) for our summary of WIPP-site geology and hydrology.

The Delaware Basin is a north–south elongated structural basin in western Texas and southeastern New Mexico that is 220–260 kilometers (135–160 miles) long and 120–160 kilometers (75–100 miles) wide. The stratigraphic relations are illustrated in Fig. 5. Tectonic activity other than downwarping and subsidence has been minor within the northern part of the basin giving it an intercratonic character. Subsidence of the basin began in early Pennsylvanian and continued to the end of the Permian resulting in maximum structural relief of about 6,000 meters (20,000 feet). Normal marine sediments were deposited throughout much of its depositional history, but in the latest Permian (Ochoan) Series the basin was restricted, and a thick pile of evaporite salts and associated clastics accumulated. At the WIPP site about 4,500 meters (15,000 feet) of Pennsylvanian and Permian sediments were deposited, of which the upper 1,200 meters (4,000 feet) are the Ochoan evaporite units.

The Delaware Basin has been emergent since the

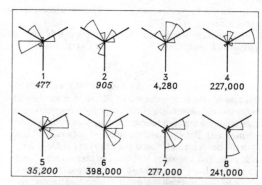

FIG. 3. Simple-salt rose diagrams of representative waters in Table 3. Identity of the segments is from Fig. 2A, segment radius proportional to the cube root of the salt's weight fraction is from Fig. 2B, and labels under each rose are the sample identification and total dissolved solids in the sample as mg/L or, if in italics, mg/kg; these specifications apply to all subsequent rose diagrams.

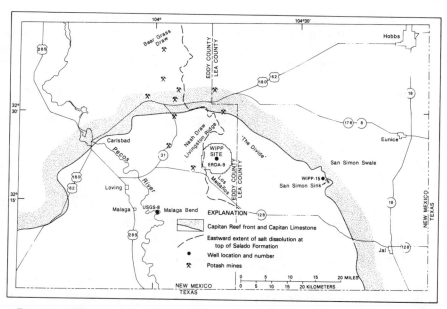

FIG. 4. Location map of the WIPP site and relevant geographic features in the northern part of the Delaware Basin, southeastern New Mexico, showing the Capitan reef front, locations of nearby potash mines, and the eastward boundary of any apparent salt dissolution in the uppermost Salado Formation from POWERS *et al.*, 1978.

Permian with only two relatively short Pre-Pleistocene depositional episodes. These sediments, Late Triassic and Miocene-Pliocene in age, are exclusively terrestrial with a total (present day) thickness of less than 150 meters (500 feet) in the WIPP-site vicinity. A discontinuous veneer of Pleistocene bolson-type deposits in channels and depressions, a Pleistocene calcrete, and nearly omnipresent recent

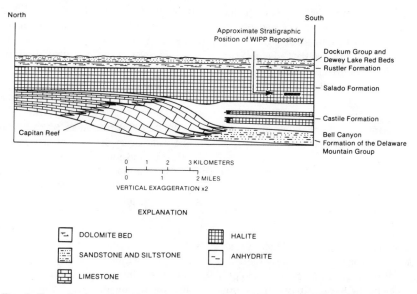

FIG. 5. Generalized stratigraphic section (from DAVIES, 1989) showing the relationships among the Guadalupian Series reef complex, the fore-reef Bell Canyon Formation of the Delaware Mountain Group (stippled), and the later Ochoan Series evaporite rocks in the Delaware Basin; along the northern and eastern segments of the Capitan reef, the Salado Formation and younger Ochoan Series rocks overlap the reef.

dune sands with minor alluvium and small playa deposits complete the depositional history of the basin.

The basin was gently tilted one or more times after the Late Triassic, resulting in a net regional dip of about 2 degrees to the east. Thus, the Capitan reef front on the western margin of the basin is well exposed and there is a west-to-east sequence of Pre-Pleistocene north–south trending outcrop belts of successively younger strata extending across the basin. Along the eastern rim of the reef the Permian Capitan Limestone is overlain by nearly a kilometer of Pre-Pleistocene strata.

Salt flowage in the Ochoan strata has resulted in the development of salt anticlines with extension and fracturing of the overlying more brittle strata at various localities throughout the basin. Dissolution and removal of salt is common throughout the basin. This process has resulted in sagging and collapse of overlying beds with concomitant fracturing and brecciation of the more brittle lithologies. Near-surface dissolution sink holes, discontinuous drainage patterns, and other geomorphic features of karst processes have developed (BACHMAN, 1985). The most conspicuous result of such processes is the formation of Nash Draw, the distinct valley to the west of the WIPP site.

Stratigraphic-geohydrologic framework of Permian units

The stratigraphic interval that is related to the WIPP facility extends from the upper Guadalupian Series water-bearing rocks immediately below the basal evaporites upward through the Ochoan Series and younger rocks. A stratigraphic summary based on core from the ERDA-9 drill hole (Fig. 4) is given in Table 4.

Inasmuch as the focus of the effort was on fluids in the Rustler Formation, only the strata above and immediately below this unit will be considered here. The description emphasizes hydrologic properties discussed in the reports of MERCER (1983) and DAVIES (1989).

Salado Formation

This formation, consisting chiefly of halite (Table 4), contains no evidence of regionally circulating water. Most drill-stem tests in the Salado yield permeability values below the sensitivity of the procedures. Some slightly higher permeabilities can be associated with anhydrite and polyhalite beds and with clay seams. Small isolated pockets of brine have been encountered during mining and drilling, as well as nonflammable pressurized gas pockets ("blowouts"). In addition, inclusion fluids are abundant in the salts throughout the formation, and in underground workings slowly accumulating brine "seeps" on the floors and "drips" from the backs are common. The only recognized groundwater circulation associated with the Salado Formation occurs at the contact with the overlying Rustler Formation.

West of WIPP site the Salado Formation is close to the land surface, and most of the soluble salts have been removed by dissolution (Fig. 4). Solution features such as sinks and karst mounds are common in the west, and within Nash Draw the uppermost Salado salts have been dissolved.

Rustler Formation

At least three well-defined aquifers occur in the Rustler Formation: the Rustler-Salado contact zone (see above), the Culebra Dolomite member, and the Magenta Dolomite member (Table 4). These are separated by varying proportions of halite, anhydrite or gypsum, and clastic beds. Halite can account for as much as 50 meters (160 feet) or more of the formation's total thickness and occurs above and below both the Culebra and Magenta aquifers. Halite dissolution in the Rustler Formation (Fig. 6) affects progressively deeper stratigraphic horizons from west to east. Only at the eastern boundary of the WIPP site is halite present throughout the formation. In the center of the site halite is present only below the Culebra aquifer. No halite is in the Rustler Formation at the western site boundary. Except in areas of obvious dissolution, such as associated with the Pecos River Valley or Nash Draw, or zones affected by secondary fracturing, there appears to be little hydrologic connection among the three aquifers.

DAVIES (1989) references several authors providing geologic evidence for halite dissolution in the Rustler and underlying Salado Formations in the vicinity of the WIPP site, and to the west. This evidence is primarily the spatial distribution of halite beds (Fig. 6) and the stratigraphic correlation of halite strata at depth in the east with mudstone beds near the land surface in the west. Further, the spatial correlation between halite dissolution and permeability in the Rustler led MERCER (1983) and others to underline the apparent causative link of the deformation accompanying dissolution with generation of secondary permeability in the Rustler Formation.

HOLT and POWERS (1988) have recently concluded that lateral variation in halite occurrence in

Table 4. Summary of the stratigraphy at the WIPP site (POWERS *et al.,* 1978)

Formation		Thickness meters (feet)
	RECENT	
Surficial sands	Blanket sands; chiefly dune sand, some alluvium and playa deposits.	0–30 (0–100)
	PLEISTOCENE	
Mescalero caliche	Hard, white crystalline caliche (limestone) crust.	0–10 (0–35)
Gatuña Formation	Pale reddish brown, fine grained, friable sandstone.	
	UPPER TRIASSIC	
Santa Rosa Sandstone	Pale red to gray, crossbedded, arkosic nonmarine medium- to coarse-grained friable sandstone; preserved only in eastern half of site.	0–75 (0–250)
	PERMIAN-OCHOAN	
Dewey Lake Redbeds	Uniform dark red-brown marine mudstone and siltstone with interbedded very fine-grained sandstone; thins westward.	30–170 (100–550)
Rustler Formation[1]	Gray, gypsiferous anhydrite with siltstone interbeds in upper part; reddish-brown siltstone or very fine silty sandstone in lower part, halitic throughout. Contains two dolomite marker beds: Magenta in upper part and Culebra in lower part. Thickens eastward due to increasing content of undissolved rock salt.	85–130 (275–425)
Salado Formation[2]	Mainly rock salt (85–90%) with minor interbedded anhydrite, polyhalite, and clayey to silty clastics. Trace of potash minerals in the McNutt interval. Minor interbeds are thin and occur in complexly alternating sequences. Thickest nonhalite bed is the Cowden anhydrite. Multiple anhydrite beds are most common immediately below the Cowden and immediately above the base of the formation.	535–610 (1750–2000)
Castile Formation[3]	Thick massive units of finely laminated ("varved") anhydrite-calcite alternating with thick halite units containing thinly interbedded anhydrite. Top anhydrite unit lacks calcite interlaminations.	380$^\pm$ (1250$^\pm$)
	PERMIAN-GUADALUPIAN	
Bell Canyon Formation[4]	Mostly light gray fine-grained sandstone with varying amount of silty and shaley interbeds and impurities. Contains considerable limestone interbeds and lime-rich intervals. Fore-reef facies of the Capitan reef system.	300$^\pm$ (1000$^\pm$)

[1] In the ERDA-9 drill hole the Rustler Formation is 94.5 meters (310 feet) thick; the Magenta Dolomite, 7.3 meters (24 feet) thick, occurs 17.7 meters (58 feet) below the top of the formation, and the Culebra Dolomite, 7.9 meters (26 feet) thick, occurs 36.6 meters (120 feet) above the base of the formation.

[2] In the ERDA-9 drill hole the Salado Formation is 602 meters (1976 feet) thick; with 156 meters (512 feet) of upper member, 112.5 meters (369 feet) of McNutt potash interval, and 334 meters (1095 feet) of lower member. The Cowden anhydrite bed, 5.5 meters (18 feet) thick, occurs 84.7 meters (278 feet) above the base of the formation in the lower member.

[3] In the WIPP-site vicinity the following approximate thicknesses have been assigned to the major lithologic intervals from the base to the top of the Castile Formation: Anhydrite I—75 meters (250 feet); Halite I—100 meters (330 feet); Anhydrite II—30 meters (100 feet); Halite II—65 meters (210 feet); and the upper anhydrite interval, presumably Anhydrites III and IV (halite III is absent)—110 meters (360 feet).

[4] On approaching the margin of the basin the Bell Canyon Formation interfingers with and grades into the Capitan Limestone, the thick, up to 600 meters (2000 feet) reef lithology that virtually encircles the basin. The Capitan Limestone is a light colored, fossiliferous and vuggy limestone, dolomite, and carbonate breccia. The belt of Capitan Limestone reaches 16–23 kilometers (10–14 miles) in width along the Northwestern Shelf and is always at least 6 times as broad as thick.

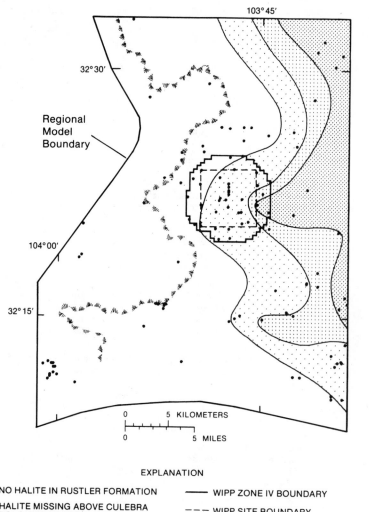

FIG. 6. Areal distribution of halite in the Rustler Formation of the WIPP site and vicinity, from MERCER (1983) and SNYDER (1985).

the Rustler formation is more the result of facies changes and syndepositional dissolution than later ground-water circulation. As DAVIES (1989) points out, if these were the only processes operative (particularly syndepositional dissolution versus later ground-water effects), other unknown mechanisms would have to be called on to produce several orders of magnitude of secondary permeability variation in the Culebra.

Certainly evaporite dissolution and related secondary processes have been active in Nash Draw, where subsidence has produced a marked valley. This feature trends north–south and has two important eastwardly trending reentrants, one to the north of the WIPP site and one to the south.

The *Rustler-Salado contact aquifer* at the WIPP site consists of a clayey dissolution residuum containing fragmental gypsum and some mudstone or siltstone. Salt associated magnesium clays of the Salado Formation have been described by BODINE (1978). The residuum thickens markedly to the west and in some exposures along the Pecos River involves the entire Salado Formation. To the east of the site the residuum grades into and interfingers with its original clayey halite lithology.

Two flow regimes have been postulated for the Rustler-Salado Formation contact aquifer. To the west of the WIPP site (Fig. 7) is a region of relatively high transmissivity that was first described by ROB-INSON and LANG (1938) and referred to as the "brine aquifer." They suggested it begins with a recharge area some 40 kilometers (25 miles) north-northwest of the northern WIPP-site boundary in the Bear Grass Draw area, flows southerly into the Nash Draw area, and turns southwest to discharge into the Pecos River at Malaga Bend (Fig. 7). The area of greatest permeability is associated with the residuum of the Nash Draw area, where flow is primarily through fractures and intergranular pore spaces. MERCER (1983) has confirmed the earlier reports of a well developed Rustler-Salado contact zone and brine under artesian conditions in the central and northern parts of Nash Draw. Drilling in support of the site investigations also has extended the eastern border of the "brine aquifer" across the western boundary of the WIPP site.

The second regime lies to the east of the "brine aquifer" in the WIPP-site area. Here transmissivities are low and flow rates are extremely slow if not stagnant. In this area the Rustler-Salado contact residuum is less well developed or absent, and halite in the lower part of the Rustler Formation suggests that little or no meteoric water has infiltrated directly into the Rustler-Salado Formation contact

aquifer. What flow exists is primarily along bedding planes at the formational contact and in the lowermost part of the Rustler Formation. Transmissivities range from greater than 0.1 foot squared per day to the west in the "brine aquifer" to less than 0.001 foot squared per day in the east.

The *Culebra Dolomite aquifer* (Table 4) is a vuggy finely crystalline dolomite that is the most persistent and productive aquifer in the WIPP-site area. Water flows through fractures in the dolomite and is confined by the overlying thick anhydrite or gypsum and the underlying clay and anhydrite. As with the Rustler-Salado contact aquifer, the principal recharge probably has come from the north, and principal discharge is at Malaga Bend. The flow path is generally southerly in the Nash Draw area.

The extent of structural deformation of the Culebra Dolomite, and thus its hydraulic properties, are related to the degree of dissolution of the enclosing evaporite strata, which varies considerably throughout the WIPP-site vicinity. Where the Rustler Formation has been subjected to considerable anhydrite/gypsum dissolution, as in Nash Draw and vicinity, the Culebra aquifer has been fractured and broken into large disconnected blocks with steep, erratic dips. In contrast, to the east where the Dewey Lake Redbeds overlie the Rustler formation, halite dissolution is negligible and fracturing within the Culebra aquifer is minimal. Within the WIPP-site, for example, the variable extent of salt dissolution in the Rustler Formation (Fig. 6) and resultant fracturing of the Culebra aquifer yield transmissivities that range from 0.001 to 140 feet squared per day (MERCER, 1983). Because of the directional differences in hydraulic conductivity imposed by the fracture flow (anisotropy), local flow paths are difficult to determine. An extensive program of multiscale flow modelling of the Culebra aquifer system, coupled with site-specific tracer and aquifer tests, has been carried out and is continuing (DAVIES, 1989; LAPPIN and HUNTER, 1989).

The *Magenta Dolomite aquifer* (Table 4), the uppermost aquifer in the Rustler Formation, is a clastic carbonate unit containing thin laminae of anhydrite. The Magenta is a confined aquifer except where the dolomite and its underlying strata are severely fractured. Flow in the aquifer characteristically occurs along thin silt or silty dolomite beds and, in places, along bedding planes. In the Nash Draw area, however, fracture flow is dominant. Recharge of the Magenta is most likely to the north, as with the underlying Rustler aquifers, but it is supplemented in other nearby areas where the aquifer is near the surface. Based on potentiome-

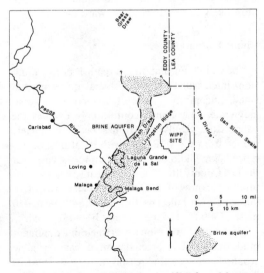

FIG. 7. Location of the "brine aquifer" from MERCER (1983).

tric-surface data, MERCER (1983) suggested that another recharge area occurs east of the WIPP site.

The development of Nash Draw had a profound effect upon the geohydrology of the Magenta Dolomite. Within the central part of the draw and continuing southwestward to the Malaga Bend, the Magenta has been removed by erosion. The dolomite is exposed along the western rim of the draw, but it is weathered and not saturated; the Magenta, along with its underlying anhydrite beds, are highly fractured. Furthermore, immediately east of the draw and west of the WIPP site the aquifer is also unsaturated. Because of the fracturing in the underlying anhydrite, the absence of water in several wells immediately east of Nash Draw, and the trend of the potentiometric surface, it is likely that Magenta waters drain into the underlying fractured anhydrites and clastics. They may also enter the Culebra aquifer in the Nash Draw area for eventual discharge into the Malaga Bend.

Dewey Lake Redbeds

The Dewey Lake Redbeds (Table 4) overlie the Rustler Formation throughout the WIPP site but are absent in Nash Draw to the west, except for occasional erratic blocks preserved in collapse structures. The unit consists of alternating thin-bedded siltstones and mudstones with occasional lenticular bodies of fine-grained sandstone; the siltstones and mudstones in the subsurface contain pervasive veins of selenite gypsum.

Because of their generally low permeability the Dewey Lake Redbeds contain little water, although during drilling in the WIPP-site area minor zones of saturation (moist well cuttings), mostly in the lenticular sands, were encountered. Groundwater movement in the Dewey Lake is highly restricted because of the lenticular geometry of the sand "aquifers." There is little indication of hydrologic connection between the Dewey Lake Redbeds and the underlying Rustler Formation.

Stratigraphic-geohydrologic framework of younger rocks

The Santa Rosa Sandstone and Chinle Formations (Triassic) unconformably overly the Ochoan strata. McGOWEN et al. (1979) suggested that the Triassic rocks in western Texas and eastern New Mexico cannot be correlated with the Santa Rosa and Chinle Formations and should be collectively referred to as the Dockum Group in the Delaware Basin (Fig. 5). However, we have here retained the formational designations that conform with pre-

vious hydrogeologic reports for the WIPP site (see, for example, POWERS et al., 1978; BACHMAN, 1980; MERCER, 1983). The Miocene-Pliocene Ogallala Formation unconformably overlies the Triassic strata. This fluvial sand and gravel unit is absent at the WIPP site but is exposed at, and east of, the Divide (Fig. 4). Above the Ogallala is the Pleistocene Gatuna Formation which occurs throughout the region as discontinuous bolson-type deposits, and the dense Mescalero caliche. Recent surficial sands, chiefly dunes, blanket the entire WIPP-site area.

Santa Rosa Sandstone

The Santa Rosa Sandstone (Table 4) is preserved in the eastern half of the WIPP site. West of ERDA-9 (Fig. 4) the unit has been completely removed by erosion. Its large-scale trough-type cross bedding, lack of sorting, arkosic composition, and grain angularity suggest rapid deposition in a fluvial system draining a predominantly crystalline terrain.

Although the Santa Rosa is an important aquifer in southwestern Lea County, only a small quantity of water has been found in the unit at the WIPP site. This occurs in a single drill hole in the lowermost sandstone interval immediately overlying the Dewey Lake Redbeds. The water was under water-table conditions but an attempt to test the zone failed other than collecting a sample for chemical analysis. The Santa Rosa is presumably recharged by precipitation in areas where the unit is directly overlain by permeable Cenozoic sediments. The water moves downward through the formation to the contact with the Dewey Lake aquitard, and then most likely moves down dip toward the east.

Chinle Formation

The Chinle Formation is a reddish-brown mudstone intercalated with greenish-gray mudstone and small, uncommon lenses of sandstone and conglomerate. The Chinle conformably overlies the Santa Rosa Sandstone. It has been completely eroded from the WIPP site with its nearest occurrence being 8 kilometers (5 miles) to the east near the Lea County line (Fig. 4) where it reaches a maximum thickness of 30 meters (100 feet). About 18 kilometers (11 miles) northeast of the site up to 250 meters (800 feet) of the unit has been observed.

The Chinle Formation's overwhelming dominant mudstone lithology precludes more than negligible groundwater. For this reason, and because of its distance from the site, Chinle waters are not included in this report.

Cenozoic deposits

The Miocene-Pliocene Ogallala sands and gravels, although forming one of the major regional aquifers to the east, occur only as isolated erosional remnants in the northern Delaware Basin. The closest of these to the WIPP site are along the Divide to the east (Fig. 4) and are not water bearing. The veneer of Pleistocene sediments that forms the surface of the WIPP-site area only rarely contains groundwater and no water has been found in these sediments at the site. Only some distance away from the site, such as along the Pecos River, are more than sporadic occurrences and minor quantities of groundwater present. Although the surficial sands and underlying Gatuna Formation in the WIPP-site area are characteristically permeable, they act principally as conduits for downward percolating waters rather than as reservoirs. In contrast, the dense Mescalero Caliche generally acts as a barrier. Except for a single water sample from the thick Pleistocene fill in the San Simon sink (Fig. 4, site WIPP-15) no information is available on water compositions from these units. Stream gravels of the Gatuna Formation found north of the WIPP site boundary indicate that Nash Draw was the location of a major stream during middle Pleistocene (BACHMAN, 1985). Groundwater circulation associated with the stream apparently caused considerable dissolution and subsidence in both the Rustler and upper part of the Salado Formation, and provided a major mechanism responsible for the formation of Nash Draw.

Gypsiferous spring deposits on the eastern margin of Nash Draw indicate that by the late Pleistocene Nash Draw was a well developed valley, and that dissolution was still active at that time (BACHMAN, 1985). These deposits at an elevation well above the present water table illustrate that the late Pleistocene ground water system contained more water than the present day flow system, reflecting wetter climatic conditions.

Regional hydrology

With particular attention to the Culebra Dolomite member of the Rustler Formation, DAVIES (1989) has presented a comprehensive series of analyses of groundwater flow in the vicinity of the WIPP site including variable-density flow simulations (Fig. 8). This work has developed the regional flow field from simulation both with and without consideration of vertical flux. It indicates that flow velocities are relatively fast west of the WIPP site and east and northeast of the site. In the transition

zone between the two extremes, velocities are highly variable. The simulations also indicate that up to 25 percent of total inflow to the Culebra could enter vertically, but most of the influx must be in the western part of the transition zone adjacent to Nash Draw.

DAVIES (1989) has also presented a simple cross-sectional model for flow system drainage across the WIPP site to Nash Draw following recharge during a past pluvial period. This model suggests that the system could sustain flow from purely transient drainage since the last glacial maximum. The model also indicates that the observed underpressuring of the Culebra in the vicinity of the WIPP site is most likely the hydrodynamic result of the Culebra having a relatively high hydraulic conductivity, and being well connected to its discharge area, but poorly connected to sources of recharge.

Water samples

We have selected 128 chemical analyses of water from 42 sites (Table 5) representing 77 horizon-site sampling points. Multiple analyses from the same site and horizon are either replicate data from different investigators for the water sample, or data for samples collected at different times.

Site locations

Almost all sites that provided data for our study are located on Fig. 9, using base map coordinates from RICHEY (1989). Many of the wells were drilled in support of the WIPP-site geologic and hydrologic characterization studies (the DOE, H, P, and WIPP groups of wells in Table 5). A few were drilled in support of the GNOME project (an underground nuclear test) and one (USGS-8; Fig. 4) was drilled as part of the Malaga Bend Salinity Alleviation project. The remainder were drilled for a variety of other purposes ranging from petroleum and potash exploration to domestic water supply. A few samples came from sites without exact coordinates (Table 6).

Chemical analyses

Analytical data for these waters are from a wide variety of sources. Most have been published, principally in reports concerned with site characterization studies for the WIPP project. Unpublished analyses include those compiled for us by Karen Robinson and Steven Lambert of Sandia Laboratories for this study. Others are from the U.S. Geological Survey's Water Resources Division either in older Technical Files or more recent data from the

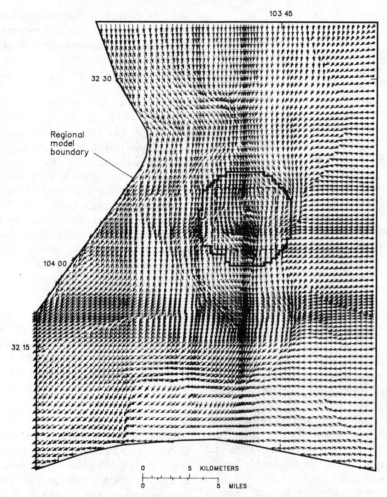

FIG. 8. Direction and magnitude of ground water flow produced by the baseline simulation model of DAVIES (1989).

Denver Central Laboratory. In addition, 34 new analyses by S. L. Rettig (U.S. Geological Survey, Reston), heretofore unpublished, are included. Most of these are replicate samples previously analyzed by the Denver Central Laboratory and reported in MERCER (1983); these new analyses generally yield improved ionic balances and include analytical data for bromide.

RESULTS

Tables 7 through 11 report our SNORM-calculated salt norm and simple-salt assemblages as well as the original analytical data for each of the 128 chemical analyses. The tables are arranged stratigraphically from the Rustler-Salado contact waters

(Table 7), discharge waters from the Rustler Formation in the Pecos River area (Table 8), waters from the Culebra Dolomite (Table 9) and Magenta Dolomite (Table 10) aquifers in the Rustler Formation, and four waters from younger strata (Table 11).

Each table identifies the site of sample collection, and if known, reports sample collection date, sample density, and sample pH. The chemical analysis, excluding solutes not forming normative salts, is reported with its source cited in the footnote. Those analyses used for plotting on maps and diagrams with simple-salt rose are marked with an asterisk. The total dissolved solids (TDS), rounded to three significant figures, is the sum from SNORM of the solutes reported, and is neither the residue weight

Table 5. Sample site abbreviations, site names, location data, and stratigraphic horizons sampled, with number of analyses

		Location data	Stratigraphic horizon					
			1	2	3	4	5	6
ANDER-1	Anderson Lake (surface)	Table 6	1					
ANDER-2	Anderson Lake (10 ft. d)	Table 6	1					
D-1	DOE-1 well	Fig. 9		1				
D-2	DOE-2 well	Fig. 9		1				
ENGEL	Engel well	Fig. 9		1				
GNOME	GNOME shaft	Fig. 9		1				
GNOME-1	GNOME USGS-1 well	Fig. 9		1				
GNOME-4	GNOME USGS-4 well	Fig. 9		1				
GNOME-5	GNOME USGS-5 well	Fig. 9	1		1			
GNOME-8	GNOME USGS-8 well	Fig. 9		1				
GRANDE	Laguna Grande de la Sal	Table 6	1					
H-1	H-1 (Hydrology hole)	Fig. 9	1	1	1			
H-2	H-2 (Hydrology hole)	Fig. 9	1	2	1			
H-3	H-3 (Hydrology hole)	Fig. 9	1	3	1			
H-4	H-4 (Hydrology hole)	Fig. 9	2	4	1			
H-5	H-5 (Hydrology hole)	Fig. 9	2	4	1		1	
H-6	H-6 (Hydrology hole)	Fig. 9	1	3	1			
H-7	H-7 (Hydrology hole)	Fig. 9	2	3				
H-8	H-8 (Hydrology hole)	Fig. 9	2	3	2			
H-9	H-9 (Hydrology hole)	Fig. 9	2	2	2			
H-10	H-10 (Hydrology hole)	Fig. 9	2	1	1			
H-11	H-11 Hydrology hole)	Fig. 9		1				
H-12	H-12 (Hydrology hole)	Fig. 9		1				
INDIAN	Indian well	Fig. 9		1				
P-14	P-14 (Potash hole)	Fig. 9	2	2				
P-15	P-15 (Potash hole)	Fig. 9	1	1				
P-17	P-17 (Potash hole)	Fig. 9	1	2				
P-18	P-18 (Potash hole)	Fig. 9	1	1				
SOUTH	South well	Fig. 9		1				
SURPRISE	Surprise spring	Table 6	2					
TWIN	Twin Pasture well	Table 6					1	
TWO-MILE	Two-mile mill well	Fig. 9		1				
USGS-8	USGS-8 well	Fig. 4	3					
W-15	WIPP-15 well	Fig. 4						1
W-25	WIPP-25 well	Fig. 9	2	4	3			
W-26	WIPP-26 well	Fig. 9	3	3				
W-27	WIPP-27 well	Fig. 9	2	2	2			
W-28	WIPP-28 well	Fig. 9	2	2				
W-29	WIPP-29 well	Fig. 9	2	3				
W-30	WIPP-30 well	Fig. 9	2	2	1			
WALKER	Walker well	Fig. 9				1		
WINDMILL	Little windmill well	Fig. 9		2				

1. Rustler Formation: Rustler-Salado contact.
2. Rustler Formation: Culebra aquifer.
3. Rustler Formation: Magenta aquifer.
4. Dewey Lake Redbeds.
5. Santa Rosa Sandstone.
6. Pleistocene collapse fill.

after drying, nor necessarily the value reported in the cited reference. The Cl/Br ratio is reported as the weight ratio.

The cation–anion charge balance for each analysis, recorded as "balance (+/−)", is calculated by SNORM and is the ratio of the sum of cation equivalencies to the sum of anion equivalencies with elemental boron calculated as the borate anion. In general, the greater the balance departs from unity, the less reliable the analytical data and the ensuing SNORM results. Except for the possibility of offsetting analytical errors, we are confident of the salt

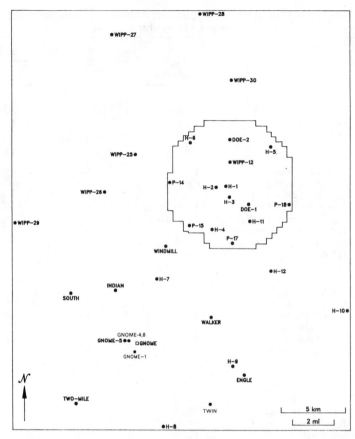

FIG. 9. Water sample locations within and near the WIPP site; solid circles are well sites and open squares are mine shafts. Locations are from RICHEY, 1989, DAVIES, 1989, and LAPPIN and HUNTER, 1989.

norm if the balance is within 1.00 ± 0.02, but feel that balances greater than 1.05 or less than 0.95 may give misleading results, particularly for speciation and abundance of salts with quantities of less than 5 percent in the normative assemblages.

The salt norm for each analysis is reported in full. The major salt abundances, that is, those salts composed solely of major solutes (Table 1), are listed in the body of the table, whereas those for salts containing one or more minor solutes are reported in the table's footnotes. The simple-salt assemblage for each analysis, often plotted on maps or illustrations as simple-salt rose diagrams, is also listed. Normative salt abundance is reported in anhydrous weight percentage and simple salt abundance in weight percentage of their respective assemblages.

We have not reported all replicate analyses. For example, we have excluded many of the analyses reported by MERCER (1983). They are either nearly identical to S. L. Rettig's data reported here or yield charge balances that deviate markedly from unity. Furthermore, none of the analyses in MERCER (1983) report bromide concentrations. We have also excluded the analyses, chiefly of waters from the Culebra aquifer, that were reported by UHLAND and RANDALL (1986) and UHLAND et al. (1987). These data yield charge balances that frequently deviate appreciably from unity and are replicated by other data.

DISCUSSION

Fluids from the three aquifers in the Rustler Formation (Tables 7, 9, and 10) and from surface and near-surface sites in the discharge area of Rustler waters (Table 8) exhibit a wide range of salinities and salt norms. The variation is attributed to multiple origins of the waters and their solutes. Retention of connate or primitive brines, diagenetic pro-

Table 6. Approximate locations of water-sample sites lacking map coordinates

ANDERSON	Waters from Anderson Lake in the vicinity of Malaga Bend on the Pecos River and USGS-8 (Fig. 4).
GRANDE	Water samples from Laguna Grande de la Sal approximately 18 kilometers (12 miles) west-southwest of the WIPP site (Fig. 7).
SURPRISE	Water from Surprise spring immediately north of Laguna Grande de la Sal (Fig. 7).
TWIN	Twin Pastures Well (also called Twin well) located (Fig. 9) approximately 19 kilometers (12 miles) south of the WIPP site (UHLAND and RANDALL, 1986).

cesses within the Rustler Formation, influx of younger meteoric weathering waters, and dissolution of evaporite minerals within the formation all played a role in the evolution of these fluids. We also include data for 4 waters in units that overlie the Rustler Formation (Table 11) as potential examples of downward percolating waters that may have recharged the Rustler Formation.

Seawater evaporation salt norms

For discussion of norms for waters from the Rustler Formation we will refer to salt norms calculated for progressively evaporating sea water brines (Table 12). The first group, Group (A) with waters A-1 through A-7, are brine compositions determined from 25°C experimental salt solubility studies, chiefly in the Na-K-Mg-Cl-SO$_4$-H$_2$O system, that are compiled and summarized in BRAITSCH (1971). These experiments exclude the precipitation of polyhalite or other Ca-bearing sulfate salts other than gypsum/anhydrite through initial halite formation. Beyond initial halite saturation calcium in the brine is arbitrarily omitted. The second group, Group (B) with waters B-1 through B-7, are the Ca-Mg-Na-K-Cl-SO$_4$-H$_2$O computer-simulated brine compositions for 25°C from EUGSTER *et al.* (1980), beginning with sea water composition. The data show equilibrium precipitation throughout the succession; thus, for example, primary polyhalite, rarely, if ever, observed in natural primary assemblages, begins to crystallize in the model during precipitation of the halite facies. The program of EUGSTER *et al.* (1980) continuously re-equilibrates the brine composition with salts already crystallized with the result that no transitory salts, such as gypsum, polyhalite, or epsomite, remain in the assem-

blage after the brine composition leaves the stability fields for these phases.

The differences between the two sets of data in Table 12 reflect the inclusion of polyhalite in the calculations by EUGSTER *et al.* (1980). Reaching the polyhalite stability field within the halite facies results not only in direct precipitation of polyhalite on continued evaporation, but also in replacing most of the previously precipitated anhydrite with polyhalite. This sufficiently reduces sulfate, potassium, and magnesium concentrations in the brine so that neither bloedite nor kainite saturation is attained. Thus, when the epsomite field is reached in Group (B), carnallite in the salt norm of the coexisting brine (B-5) is only 6.86 percent of the assemblage, rather than the 23.6 percent (A-4) coexisting with initial epsomite in Group (A). Carnallite is at a maximum (25.6 percent) in the norm for the brine at initial kainite precipitation (A-5) in Group (A), whereas in Group (B) maximum carnallite (15.0 percent) in the norm is reached well before initial bitter salt deposition at initial stability of polyhalite (B-4).

In Table 13 we summarize the salt norms and Cl/Br weight ratios from two successions of marine brine ponds constructed for the commercial retrieval of halite. The first of these, a succession of pans at the Secovlje salt works near Portoroz, Yugoslavia, was sampled and analyzed (HERRMANN *et al.*, 1973) only through the early third of the halite facies of the marine succession. The second, the Morton Bahama salt works on Great Inagua Island in the Bahamas, was sampled and analyzed (McCAFFREY *et al.*, 1987) through more than half of the halite facies. Furthermore, three samples of the most concentrated Bahamian brines were further evaporated in the laboratory through the remaining interval of the halite facies, the magnesium sulfate facies, and into the potash salt facies. Unfortunately, the mineralogy of the bitter salts was not determined. It also appears that some addition of water, possibly from water vapor sorption by the experimental brines with very low water activity, diluted some solutions between their separation from the precipitated salts and their analysis; the decrease in salinity from no. 36-4 to no. 39-6 and, in particular, from no. 39-6 to no. 39-4 (Table 13) can only be attributed to dilution.

In each set of analyses, the Cl/Br ratio between seawater and at initial halite saturation (Seawater through Pan 13d and W-64 through W-46 in Table 13) remains at about 300. The ratio begins decreasing markedly with increasing halite precipitation to about 50 at the initial precipitation of bitter salts

M. W. Bodine, Jr. and B. F. Jones

Table 7. Chemical analysis, salt norm (anhydrous weight percent),

	1*	2*	3*	4	5*
Site	H-1	H-2	H-3	H-4	H-4
Date collected	2/23/77	2/23/77	2/23/77	3/16/79	3/16/79
Density (g/cc)	—	—	—	—	1.212
pH	7.2	5.9	7.6	—	3.00
Chemical analysis	mg/L	mg/L	mg/L	mg/L	mg/kg
Mg	30,000.	25,000.	25,000.	27,000.	20,600.
Ca	13,000.	9,200.	18,000.	8,300.	7,140.
Sr	—	—	—	—	229.
Li	—	—	—	—	16.
Na	56,000.	66,000.	59,000.	66,000.	56,400.
K	17,000.	9,100.	14,000.	8,600.	6,850.
F	—	—	—	1.7	
Cl	210,000.	200,000.	210,000.	210,000.	167,000.
Br	—	—	—	—	1,355.
NO$_3$	1.3	4.9	3.4	1.2	—
HCO$_3$	675.	199.	467.	1.0	—
SO$_4$	520.	1,300.	370.	1,400.	594.
B	110.	150.	1.9	360.	167.
TDS	327,000.	311,000.	327,000.	322,000.	260,000.
Balance (+/−)	1.006	0.989	0.990	0.959	0.995
Cl/Br	—	—	—	—	123.
Salt norm					
Dolomite	0.2	<0.1	0.1	<0.1	—
Magnesite	—	—	—	—	—
Anhydrite	0.2	0.6	0.2	0.6	0.3
Kieserite	—	—	—	—	—
Epsomite	—	—	—	—	—
Polyhalite	—	—	—	—	—
Leonite	—	—	—	—	—
Bloedite	—	—	—	—	—
Kainite	—	—	—	—	—
Halite	43.3	54.3	46.2	53.5	55.6
Sylvite	—	—	—	—	—
Carnallite	22.5	12.8	18.7	11.9	11.5
Bischofite	5.0	11.5	—	16.0	12.5
Tachyhydrite	28.7	20.5	31.1	17.5	19.5
Antarcticite	—	—	3.7	—	—
Simple salts					
CaCO$_3$	0.1	<0.1	0.1	<0.1	—
MgCO$_3$	0.1	<0.1	0.1	<0.1	—
CaSO$_4$	0.2	0.6	0.2	0.6	0.3
MgSO$_4$	—	—	—	—	—
K$_2$SO$_4$	—	—	—	—	—
Na$_2$SO$_4$	—	—	—	—	—
Na$_2$Cl$_2$	43.4	54.4	46.2	53.8	55.9
K$_2$Cl$_2$	9.9	5.6	8.2	5.3	5.1
MgCl$_2$	35.7	31.7	30.1	33.9	31.5
CaCl$_2$	10.6	7.6	15.2	6.5	7.2

* Simple-salt rose plotted on subsequent maps and diagrams.
Source of analysis; minor normative salts (anhydrous weight percent):
1. MERCER (1983). Inyoite—0.2, nitrocalcite—0.0005.
2. MERCER (1983). Inyoite—0.2, nitrocalcite—0.002.
3. MERCER (1983). Inyoite—0.003, nitrocalcite—0.001.
4. MERCER (1983). Inyoite—0.5, fluorite—0.001, nitrocalcite—0.0005.
5. Rettig (this report). Inyoite—0.3, SrCl$_2$ · 2H$_2$O—0.2, LiCl · H$_2$O—0.04.

and simple salts (weight percent) of Rustler-Solado contact waters

6 H-5 5/16/79 — —	7* H-5 5/16/79 1.263 4.34	8* H-6 4/9/79 1.209 2.91	9 H-7 3/20/80 1.048 6.80	10* H-7 3/20/80 1.047 7.13	11 H-8 9/6/80 — 7.60	12* H-8 9/6/80 1.085 7.22
mg/L	mg/kg	mg/kg	mg/L	mg/kg	mg/L	mg/kg
82,000.	63,300.	16,500.	910.	788.	430.	392.
2,100.	1,580.	3,470.	2,600.	2,340.	1,200.	1,110.
—	26.	83.	—	42.	—	18.
—	18.	1.7	—	0.3	—	—
14,000.	12,400.	68,800.	22,000.	21,500.	46,000.	43,700.
21,000.	14,100.	6,450.	210.	164.	660.	350.
—	—	—	0.8	—	0.4	—
290,000.	207,000.	166,000.	41,000.	37,800.	70,000.	66,400.
—	3,060.	907.	—	12.	—	8.3
—	—	—	0.1	—	—	—
300.	—	—	—	47.	—	—
2,000.	1,090.	1,220.	2,900.	2,710.	5,300.	4,700.
67.	194.	117.	3.1	25.	1.3	3.4
412,000.	303,000.	264,000.	69,600.	65,400.	124,000.	117,000.
0.972	1.052	0.993	0.959	0.998	1.014	1.015
—	68.	183.	—	3,150.	—	8,000.
0.1	—	—	—	0.1	—	—
—	—	—	—	—	—	—
0.7	0.5	0.7	5.8	5.9	1.4	2.2
—	—	—	—	—	0.9	—
—	—	—	—	—	—	1.1
—	—	—	—	—	3.8	2.2
—	—	—	—	—	—	0.3
—	—	—	—	—	—	—
8.8	10.1	66.4	82.5	83.6	93.8	94.3
—	—	—	—	—	—	—
22.6	19.6	10.7	1.3	1.1	<0.1	—
65.6	67.3	13.6	—	—	—	—
2.2	2.1	8.1	7.1	6.5	—	—
—	—	—	3.2	2.6	—	—
<0.1	—	—	—	<0.1	—	—
<0.1	—	—	—	<0.1	—	—
0.7	0.5	0.7	5.8	5.9	3.3	3.2
—	—	—	—	—	1.7	1.6
—	—	—	—	—	1.2	0.7
—	—	—	—	—	—	0.2
8.8	10.1	66.9	82.5	83.9	93.8	94.3
9.9	8.7	4.7	0.6	0.5	<0.1	—
79.7	79.9	24.7	5.3	4.7	<0.1	—
0.8	0.8	3.0	5.9	5.0	—	—

* Simple-salt rose plotted on subsequent maps and diagrams.
Source of analysis; minor normative salts (anhydrous weight percent):
  6. MERCER (1983). Inyoite—0.08.
  7. Rettig (this report). Inyoite—0.3, $LiCl \cdot H_2O$—0.04, $SrCl_2 \cdot 2H_2O$—0.02.
  8. Rettig (this report). Inyoite—0.2, $SrCl_2 \cdot 2H_2O$—0.06, $LiCl \cdot H_2O$—0.004.
  9. MERCER (1983). Inyoite—0.02, fluorite—0.002, nitrocalcite—0.0002.
  10. Rettig (this report). Inyoite—0.2, $SrCl_2 \cdot 2H_2O$—0.1, $LiCl \cdot H_2O$—0.003.
  11. MERCER (1983). Indirite—0.005, sellaite—0.0005.
  12. Rettig (this report). Celestite—0.03, indirite—0.01.

Table 7. (Continued)

	13	14*	15	16*	17
Site	H-9	H-9	H-10	H-10	P-14
Date collected	5/20/80	5/20/80	5/19/80	5/19/80	2/24/77
Density (g/cc)	1.202	1.203	1.198	1.202	—
pH	7.0	6.21	6.3	5.74	7.2
Chemical analysis	mg/L	mg/kg	mg/L	mg/kg	mg/L
Mg	870.	707.	11,000.	8,400.	1,200.
Ca	1,300.	1,120.	1,500.	1,120.	570.
Sr	—	15.	—	42.	—
Li	—	3.3	—	3.7	—
Na	130,000.	102,000.	100,000.	86,500.	120,000.
K	1,200.	790.	4,000.	2,950.	1,300.
F	0.1	—	0.7	—	—
Cl	190,000.	155,000.	190,000.	155,000.	180,000.
Br	—	79.	—	832.	—
NO_3	4.9	—	3.7	—	1.5
HCO_3	—	—	—	—	222.
SO_4	2,600.	3,990.	3,300.	3,830.	10,000.
B	19.	20.	120.	97.	1.7
TDS	326,000.	264,000.	310,000.	259,000.	313,000.
Balance (+/−)	1.075	1.027	0.999	1.027	1.017
Cl/Br	—	1,960.	—	186.	—
Salt norm					
Dolomite	—	—	—	—	—
Magnesite	—	—	—	—	<0.1
Anhydrite	1.2	1.2	1.5	1.4	—
Kieserite	—	0.5	—	0.6	—
Epsomite	—	—	—	—	0.1
Polyhalite	—	0.4	—	—	1.3
Leonite	—	—	—	—	0.9
Bloedite	—	—	—	—	2.3
Kainite	—	—	—	—	—
Halite	97.0	96.8	82.0	83.7	95.4
Sylvite	—	—	—	—	—
Carnallite	1.5	1.0	5.6	4.9	—
Bischofite	—	—	10.7	9.2	—
Tachyhydrite	0.2	—	—	—	—
Antarcticite	<0.1	—	—	—	—
Simple salts					
$CaCO_3$	—	—	—	—	—
$MgCO_3$	—	—	—	—	<0.1
$CaSO_4$	1.2	1.4	1.5	1.5	0.6
$MgSO_4$	—	0.6	—	0.6	1.8
K_2SO_4	—	0.1	—	—	0.9
Na_2SO_4	—	—	—	—	1.2
Na_2Cl_2	97.1	96.9	82.1	83.9	95.4
K_2Cl_2	0.7	0.5	2.5	2.1	—
$MgCl_2$	1.0	0.6	13.9	12.0	—
$CaCl_2$	0.1	—	—	—	—

* Simple-salt rose plotted on subsequent maps and diagrams.

Source of analysis; minor normative salts (anhydrous weight percent):

13. MERCER (1983). Inyoite—0.03, nitrocalcite—0.002, fluorite—0.00006.
14. Rettig (this report). Indirite—0.03, celestite—0.01, $Li_2SO_4 \cdot H_2O$—0.01.
15. MERCER (1983). Inyoite—0.2, indirite—0.03, nitrocalcite—0.002, fluorite—0.0005.
16. Rettig (this report). Indirite—0.2, celestite—0.03, $Li_2SO_4 \cdot H_2O$—0.01.
17. MERCER (1983). Indirite—0.002, niter—0.0008.

Table 7 (Continued)

18*	19	20*	21*	22*	23	24
P-14	P-15	P-17	P-18	WIPP-25	WIPP-25	WIPP-26
2/4/80	4/3/79	5/11/79	5/20/80	3/19/80	7/17/80	3/18/80
1.126	1.042	1.193	1.266	1.171	—	1.078
3.55	7.98	6.36	5.35	7.14	7.4	8.5
mg/kg	mg/kg	mg/kg	mg/kg	mg/kg	mg/L	mg/L
941.	240.	29,300.	41,100.	2,390.	3,260.	1,300.
1,190.	864.	14,400.	22,400.	572.	560.	2,700.
20.	30.	380.	55.	9.4	11.3	—
—	—	1.0	8.5	0.8	1.58	—
70,200.	21,900.	20,100.	15,200.	78,600.	122,800.	52,000.
1,380.	1,250.	9,640.	11,800.	2,050.	3,330.	1,000.
111,000.	34,400.	152,000.	199,000.	127,000.	192,000.	88,000.
—	28.	2,725.	3,120.	41.	51.	—
—	—	—	—	—	—	0.2
—	152.	—	—	105.	130.	—
7,640.	3,220.	478.	197.	9,380.	12,400.	7,600.
42.	3.4	184.	161.	40.	40.9	30.
192,000.	62,100.	229,000.	293,000.	220,000.	335,000.	153,000.
0.980	1.007	0.981	0.974	0.980	1.009	0.957
—	1,230.	56.	64.	3,100.	3,760.	—
—	0.2	—	—	<0.1	<0.1	—
<0.1	2.1	0.3	0.1	—	—	6.2
1.1	—	—	—	2.8	1.9	0.7
—	—	—	—	1.9	1.2	—
4.2	5.4	—	—	—	—	—
—	—	—	—	—	—	—
—	—	—	—	1.4	2.8	—
0.1	—	—	—	—	—	—
93.9	89.4	22.7	13.6	92.0	92.9	88.9
—	2.1	—	—	—	—	—
0.4	0.7	18.6	18.0	1.7	1.2	2.9
—	—	11.1	9.5	—	—	1.2
—	—	46.7	58.5	—	—	—
—	0.2	—	—	<0.1	<0.1	—
2.1	4.7	0.3	0.1	—	—	—
2.1	1.2	—	—	0.9	0.6	6.2
1.4	1.7	—	—	4.1	3.9	0.7
—	—	—	—	0.6	0.4	—
94.0	89.5	22.8	13.6	92.1	92.9	89.0
0.2	2.4	8.2	7.9	1.3	1.6	1.3
0.2	0.4	51.4	56.8	1.0	0.6	2.8
—	—	17.3	21.6	—	—	—

* Simple-salt rose plotted on subsequent maps and diagrams.

Source of analysis; minor normative salts (anhydrous weight percent):

18. Rettig (this report). Indirite—0.1, celestite—0.02.
19. Rettig (this report). Celestite—0.1, indirite—0.03.
20. Rettig (this report). Inyoite—0.4, $SrCl_2 \cdot 2H_2O$—0.3, $LiCl \cdot H_2O$—0.03.
21. Rettig (this report). Inyoite—0.3, $SrCl_2 \cdot 2H_2O$—0.04, $LiCl \cdot H_2O$—0.02.
22. Rettig (this report). Indirite—0.08, celestite—0.009, $Li_2SO_4 \cdot H_2O$—0.003.
23. Bendix (unpub.). Indirite—0.06, celestite—0.007, $Li_2SO_4 \cdot H_2O$—0.004.
24. MERCER (1983). Indirite—0.09, nitromagnesite—0.0002.

Table 7 (Continued)

	25	26*	27	28	29*
Site	WIPP-26	WIPP-26	WIPP-27	WIPP-27	WIPP-28
Date collected	3/18/80	7/23/80	5/21/80	7/24/80	3/20/80
Density (g/cc)	1.107	—	1.204	1.073	1.138
pH	8.03	7.7	7.43	7.31	6.72
Chemical analysis	mg/kg	mg/L	mg/kg	mg/kg	mg/kg
Mg	1,170.	1,660.	1,040.	1,680.	2,070.
Ca	1,310.	1,410.	1,160.	1,070.	615.
Sr	20.	27.	10.	21.	11.
Li	1.0	1.15	1.2	2.0	0.9
Na	50,100.	68,600.	102,000.	32,100.	65,000.
K	814.	1,200.	2,570.	1,580.	2,070.
F					—
Cl	82,400.	108,000.	154,000.	52,400.	102,000.
Br	24.	19.3	51.	16.	36.
NO_3	—	—	—	—	—
HCO_3	123.	270.	—	—	260.
SO_4	6,730.	7,480.	5,190.	8,670.	11,000.
B	43.	31.8	13.	16.	54.
TDS	143,000.	189,000.	266,000.	97,600.	183,000.
Balance (+/−)	0.957	1.004	1.043	0.982	0.990
Cl/Br	3,430.	5,600.	3,020.	3,280.	2,830.
Salt norm					
Dolomite	—	—	—	—	—
Magnesite	<0.1	0.1	—	—	0.1
Anhydrite	2.6	2.1	0.1	0.6	—
Kieserite	2.4	2.4	—	4.9	3.0
Epsomite	—	—	—	—	2.4
Polyhalite	1.2	0.8	2.8	6.5	1.4
Leonite	—	—	—	—	2.2
Bloedite	—	—	—	—	—
Kainite	—	—	—	—	—
Halite	91.7	92.2	95.0	84.6	90.8
Sylvite	—	—	0.3	—	—
Carnallite	1.8	2.3	1.8	3.2	—
Bischofite	—	—	—	—	—
Tachyhydrite	—	—	—	—	—
Antarcticite	—	—	—	—	—
Simple salts					
$CaCO_3$	—	—	—	—	—
$MgCO_3$	<0.1	0.1	—	—	0.1
$CaSO_4$	3.2	2.5	1.4	3.8	1.2
$MgSO_4$	2.7	2.5	0.6	6.3	5.4
K_2SO_4	0.4	0.3	0.9	2.0	1.6
Na_2SO_4	—	—	—	—	—
Na_2Cl_2	91.8	92.3	95.0	84.7	91.0
K_2Cl_2	0.8	1.0	1.1	1.4	0.8
$MgCl_2$	1.0	1.3	1.0	1.8	—
$CaCl_2$	—	—	—	—	—

* Simple-salt rose plotted on subsequent maps and diagrams.
Source of analysis; minor normative salts (anhydrous weight percent):
25. Rettig (this report). Indirite—0.1, celestite—0.03, $Li_2SO_4 \cdot H_2O$—0.006.
26. Bendix (unpub.). Indirite—0.08, celestite—0.03, $Li_2SO_4 \cdot H_2O$—0.005.
27. Rettig (this report). Indirite—0.02, celestite—0.008, $Li_2SO_4 \cdot H_2O$—0.004.
28. Rettig (this report)—"recollect" (shallower in aquifer). Indirite—0.07, celestite—0.05, $Li_2SO_4 \cdot H_2O$—0.02.
29. Rettig (this report). Indirite—0.1, celestite—0.01, $Li_2SO_4 \cdot H_2O$—0.004.

Table 7 (Continued)

30 WIPP-28 7/31/80 1.180 7.00	31* WIPP-29 3/18/80 1.067 6.86	32 WIPP-29 7/24/80 — 7.20	33 WIPP-30 3/19/80 1.200 6.23	34* WIPP-30 7/17/80 — 7.50	35* GNOME-5 11/27/61 1.200 7.0
mg/L	mg/kg	mg/L	mg/kg	mg/L	mg/kg
3,400.	1,930.	2,320.	1,830.	2,770.	2,710.
605.	872.	1,080.	758.	955.	424.
11.6	17.	21.	12.	18.4	33.
1.82	0.9	1.34	—	0.72	2.82
97,100.	29,500.	36,100.	92,500.	120,600.	94,700.
4,300.	842.	1,480.	1,330.	2,180.	2,090.
—	—	—	—	—	4.2
155,000.	45,200.	58,000.	150,000.	192,000.	156,000.
29.	12.	12.5	72.	78.	5.5
—	—	—	—	—	0.4
170.	154.	200.	—	620.	18.
16,700.	11,300.	12,000.	6,300.	7,390.	540.
46.2	45.	19.9	79.	81.6	—
277,000.	89,900.	111,000.	253,000.	327,000.	257,000.
0.983	0.995	0.981	0.972	0.999	1.001
5,340.	3,770.	4,640.	2,080.	2,460.	28,400.
—	—	—	—	—	<0.1
<0.1	0.1	0.1	—	0.1	—
—	<0.1	<0.1	0.8	0.9	0.3
—	—	7.5	2.0	1.9	—
2.0	7.9	—	—	—	—
1.6	6.8	6.9	0.4	0.1	—
1.2	—	—	—	—	—
—	2.2	—	—	—	—
5.2	—	—	—	—	—
90.0	82.7	83.6	94.6	94.0	93.8
—	—	—	—	—	—
—	—	1.8	2.1	2.8	3.5
—	—	—	—	—	1.8
—	—	—	—	—	0.6
—	—	—	—	—	<0.1
<0.1	0.1	0.1	—	0.1	<0.1
0.7	3.3	3.3	1.0	1.0	0.3
6.0	10.3	9.0	2.1	1.9	—
1.2	2.1	2.1	0.1	<0.1	—
—	1.2	—	—	—	—
90.0	82.9	83.7	94.7	94.1	93.8
2.0	—	0.8	0.9	1.2	1.6
—	—	1.0	1.2	1.6	4.1
—	—	—	—	—	0.2

* Simple-salt rose plotted on subsequent maps and diagrams.
Source of analysis; minor normative salts (anhydrous weight percent):
30. Bendix (unpub.). Indirite—0.07, celestite—0.009, $Li_2SO_4 \cdot H_2O$—0.005.
31. Rettig (this report). Indirite—0.2, celestite—0.04, $Li_2SO_4 \cdot H_2O$—0.008.
32. Bendix (unpub.). Indirite—0.08, celestite—0.04, $Li_2SO_4 \cdot H_2O$—0.01.
33. Rettig (this report). Indirite—0.1, celestite—0.01.
34. Bendix (unpub.). Indirite—0.1, celestite—0.01, $Li_2SO_4 \cdot H_2O$—0.002.
35. U.S.G.S. (unpub. Tech files). $SrCl_2 \cdot 2H_2O$—0.02, $LiCl \cdot H_2O$—0.007, fluorite—0.003, $Sr(NO_3)_2$—0.0003.

Table 8. Chemical analysis, salt norm (anhydrous weight percent), and simple salts (weight percent) of

Site	1 USGS-8	2* USGS-8	3 USGS-8
Date collected	2/25/54	1/17/64	5/1/64
Density (g/cc)	1.207	1.209	1.214
pH	—	—	—
Chemical analysis	mg/L	mg/L	mg/L
Mg	2,020.	2,750.	3,540.
Ca	411.	550.	436.
Sr	—	—	—
Li	—	—	—
Na	101,000.	120,000.	119,000.
K	3,310.	4,500.	6,300.
Cl	158,000.	187,000.	187,000.
Br	—	—	—
HCO_3	78.	102.	146.
SO_4	10,200.	13,100.	15,300.
B	—	15.	19.
TDS	275,000.	328,000.	332,000.
Balance (+/−)	0.999	1.007	1.010
Cl/Br	—	—	—
Salt norm			
Dolomite	—	—	—
Magnesite	<0.1	<0.1	<0.1
Anhydrite	—	—	—
Epsomite	0.5	0.3	—
Leonite	0.9	0.5	0.1
Kieserite	—	—	—
Polyhalite	1.1	1.2	0.9
Kainite	4.1	5.4	8.1
Halite	93.4	92.6	90.7
Sylvite	—	—	0.3
Carnallite	—	—	—
Bischofite	—	—	—
Tachyhydrite	—	—	—
Simple salts			
$CaCO_3$	—	—	—
$MgCO_3$	<0.1	<0.1	<0.1
$CaSO_4$	0.5	0.6	0.4
$MgSO_4$	3.6	4.1	5.2
K_2SO_4	0.8	0.6	0.3
Na_2Cl_2	93.4	92.6	90.7
K_2Cl_2	1.6	2.1	3.3
$MgCl_2$	—	—	—
$CaCl_2$	—	—	—

* Simple-salt rose plotted on subsequent maps and diagrams.
Source of analysis; minor normative salts (anhydrous weight percent):
1. KUNKLER (1980).
2. KUNKLER (1980). Indirite—0.02.
3. KUNKLER (1980). Indirite—0.03.

(between no. 40-4 and no. 36-4) and further drops to 34 in their most fractionated brine (no. 39-4) precipitating potash salts. Salt norms for the two successions (Table 13) and their corresponding salt precipitates are similar to the succession calculated from experimental data (A-1 through A-7 in Table 12) indicating that polyhalite was not forming in either succession of pans.

Validity of rustler water samples

LAMBERT and HARVEY (1987), as part of their stable isotope study, have designated certain water samples from drill sites of questionable validity based on drilling records and properties of the sampled fluids. Coupled with our own evaluation of fluid properties and site locations, samples from one waters discharged in the Malaga Bend area (1–5), Laguna Grande de Sal (6), and Surprise Spring (7–8)

4 ANDERSON 4/20/66 1.214 —	5 ANDERSON 4/20/66 1.215 —	6 GRANDE 9/81 1.213 7.40	7 SURPRISE 1938 — —	8 SURPRISE 1975 — —
mg/L	mg/L	mg/kg	mg/L	mg/L
3,790.	3,820.	5,610.	304.	2,200.
480.	428.	376.	76.6	2,000.
—	—	24.	1.0	6.0
—	—	7.5	—	4.0
118,000.	118,000.	87,400.	1,175.	14,400.
6,920.	6,930.	15,700.	128.	580.
187,000.	187,000.	157,000.	3,080.	30,000.
—	—	73.2	—	16.
158.	164.	—	150.	170.
17,100.	16,900.	12,900.	2,610.	3,470.
29.	28.	29.	—	—
334,000.	333,000.	279,000.	8,210.	52,900.
1.002	1.003	0.997	0.819	1.003
—	—	2,144.	—	1,875.
—	—	—	—	0.2
<0.1	<0.1	—	1.2	—
—	—	—	36.7	9.3
—	—	—	—	—
0.6	0.5	—	—	—
—	—	—	5.3	—
1.0	0.9	1.0	—	—
8.2	8.4	8.0	—	—
89.8	89.9	79.7	42.1	69.2
0.2	0.3	4.5	—	—
—	—	6.8	7.8	4.8
—	—	—	6.9	8.8
—	—	—	—	7.4
—	—	—	—	0.1
<0.1	<0.1	—	1.2	0.1
0.5	0.4	0.5	36.7	9.3
5.5	5.6	5.1	5.3	—
0.7	0.6	0.3	—	—
89.9	89.9	79.8	42.1	69.4
3.4	3.5	10.5	3.4	2.1
—	—	3.8	11.2	16.2
—	—	—	—	2.7

* Simple-salt rose plotted on subsequent maps and diagrams.
Source of analysis; minor normative salts (anhydrous weight percent):
4. Lake surface water. KUNKLER (1980). Indirite—0.04.
5. Water at 10 foot depth. KUNKLER (1980). Indirite—0.04.
6. Rettig (this report). Indirite—0.05, $Li_2SO_4 \cdot H_2O$—0.02, celestite—0.02.
7. U.S.G.S. (unpublished). Celestite—0.03.
8. U.S.G.S. (unpublished). $SrCl_2 \cdot 2H_2O$—0.2, $LiCl \cdot H_2O$—0.05.

or more of the Rustler horizons at three sites were eliminated from further consideration in this discussion (Table 14).

Five samples in Table 14 questioned by LAMBERT and HARVEY (1987) were retained for our discussion, as their salt norms and Cl/Br ratios (when available) suggest only negligible contamination. Furthermore, we have added to the list three samples from two sites where recent anthropogenic activity, chiefly the release of waters from nearby potash mining, have locally influenced water composition.

M. W. Bodine, Jr. and B. F. Jones

Table 9. Chemical analysis, salt norm (anhydrous weight

	1*	2	3*	4	5*	6
Site	H-1	H-2	H-2	H-3	H-3	H-3
Date collected	6/2/76	2/22/77	4/21/86	3/17/77	6/11/84	2/4/85
Density (g/cc)	—	—	1.009	—	1.035	—
pH	7.6	8.4	8.0	7.4	7.4	7.4
Chemical analysis	mg/L	mg/L	mg/L	mg/L	mg/L	mg/L
Mg	280.	160.	167.	670.	829.	783.
Ca	780.	690.	743.	1,500.	1,550.	1,470.
Sr	—	—	9.54	—	22.6	30.5
Li	—	—	0.225	—	0.53	0.4
Na	9,400.	2,100.	3,570.	19,000.	17,400.	18,000.
K	190.	91.	93.5	630.	495.	425.
F	5.1	2.0	2.24	0.5	2.09	1.94
Cl	12,000.	2,800.	5,310.	29,600.	29,500.	30,300.
Br	—	—	5.6	—	28.8	25.8
I	—	—	0.081	—	0.133	0.138
NO₃	—	0.04	—	0.3	—	—
HCO₃	105.	59.	60.	115.	52.0	52.0
CO₃	—	5.0	—	—	—	—
SO₄	7,400.	3,000.	2,980.	5,700.	5,130.	4,820.
B	2.4	9.5	10.4	20.	30.	26.3
TDS	30,200.	8,920.	13,000.	57,200.	55,000.	55,900.
Balance (+/−)	0.962	0.987	0.977	1.017	0.973	0.974
Cl/Br	—	—	948.	—	1,020.	1,170.
Salt norm						
Dolomite	—	—	—	—	—	—
Magnesite	0.2	0.5	0.3	0.10	0.1	0.1
Anhydrite	—	19.9	17.2	6.8	9.7	9.1
Kieserite	—	—	—	3.0	2.9	2.6
Epsomite	—	—	0.4	—	—	—
Polyhalite	—	7.5	5.3	4.3	—	—
Bloedite	9.4	14.3	9.1	—	—	—
Glauberite	18.4	5.7	—	—	—	—
Aphthitalite	1.8	—	—	—	—	—
Thenardite	5.2	—	—	—	—	—
Syngenite	—	—	—	—	—	—
Kainite	—	—	—	—	—	—
Halite	64.8	51.6	67.1	83.5	81.8	83.2
Sylvite	—	—	—	—	—	—
Carnallite	—	—	—	2.2	4.0	3.4
Bischofite	—	—	—	—	1.3	1.4
Tachyhydrite	—	—	—	—	—	—
Antarcticite	—	—	—	—	—	—
Simple salts						
CaCO₃	—	—	—	—	—	—
MgCO₃	0.2	0.5	0.3	0.11	0.1	0.6
CaSO₄	9.0	26.4	19.9	8.8	9.8	9.1
MgSO₄	4.3	8.2	5.7	3.9	2.9	2.6
K₂SO₄	1.4	2.3	1.6	1.3	—	—
Na₂SO₄	20.2	10.7	5.0	—	—	—
Na₂Cl₂	64.8	51.8	67.5	83.7	82.0	83.4
K₂Cl₂	—	—	—	1.0	1.8	1.5
MgCl₂	—	—	—	1.2	3.5	3.3
CaCl₂	—	—	—	—	—	—

* Simple-salt rose plotted on subsequent maps and diagrams.
Source of analysis; minor normative salts (anhydrous weight percent):
1. MERCER (1983). Borax—0.04, sellaite—0.03.
2. MERCER (1983). Ulexite—0.5, sellaite—0.04, niter—0.0007.
3. Bendix (unpub.). Indirite—0.4, celestite—0.2, sellaite—0.03, Li₂SO₄ · H₂O—0.01.
4. MERCER (1983). Indirite—0.2, sellaite—0.001, niter—0.0009.
5. Bendix (unpub.). Indirite—0.2, celestite—0.09, Li₂SO₄ · H₂O—0.008, sellaite—0.006.
6. Bendix (unpub.). Indirite—0.2, celestite—0.1, Li₂SO₄ · H₂O—0.006, sellaite—0.006.

percent), and simple salts (weight percent) of Culebra waters

7 H-4 12/14/78 — 7.6	8 H-4 5/29/81 1.010 8.0	9 H-4 8/10/84 1.010 7.8	10* H-4 7/20/85 1.015 7.7	11 H-5 12/19/78 — 6.8	12 H-5 6/1/81 1.100 7.9	13 H-5 10/15/81 1.100 7.9
mg/L	mg/L	mg/L	mg/L	mg/L	mg/L	mg/L
430.	455.	505.	427.	1,900.	2,140.	2,150.
180.	700.	698.	691.	360.	1,710.	1,720.
—	14.	17.8	14.3	—	31.6	31.3
—	0.39	0.49	0.40	—	0.77	0.77
5,800.	6,080.	6,150.	5,850.	53,000.	52,400.	52,300.
180.	215.	222.	210.	1,400.	1,290.	1,300.
1.9	—	2.13	—	1.4	—	—
7,500.	7,980.	7,950.	7,480.	86,000.	89,500.	89,500.
—	41.5	47.7	42.6	—	62.5	64.
—	—	0.226	—	—	—	—
0.09	—	—	—	0.04	—	—
59.	71.	75.	70.	41.	80.	86.
—	—	—	—	—	—	—
4,000.	6,230.	5,700.	5,520.	810.	7,360.	7,570.
19.	17.9	19.8	14.1	36.	33.2	35.2
18,200.	21,800.	21,400.	20,300.	144,000.	155,000.	155,000.
1.015	0.959	1.014	1.007	1.029	0.963	0.958
—	192.	167.	176.	—	1,430.	1,400.
—	—	—	—	<0.1	—	—
0.2	0.2	0.2	0.2	—	<0.1	—
—	2.0	5.5	3.3	0.8	3.0	<0.1
—	—	—	—	—	2.5	3.9
—	—	—	—	—	—	2.6
0.5	7.3	7.5	7.5	—	—	—
24.3	19.0	21.0	18.4	—	—	—
2.5	11.2	3.5	9.1	—	—	—
—	—	—	—	—	—	—
3.6	—	—	—	—	—	—
—	—	—	—	—	—	—
68.3	59.6	61.7	61.0	92.2	88.4	88.3
—	—	—	—	—	—	—
—	—	—	—	4.2	3.7	3.8
—	—	—	—	2.7	1.4	1.4
—	—	—	—	—	—	—
—	—	—	—	—	—	—
—	—	—	—	<0.1	—	—
0.2	0.2	0.2	0.2	<0.1	<0.1	<0.1
3.1	11.1	10.8	11.4	0.8	3.9	3.9
11.3	10.4	11.3	10.1	—	2.5	2.6
2.2	2.3	2.3	2.3	—	—	—
14.5	16.1	13.2	14.7	—	—	—
68.7	60.0	62.1	61.3	92.3	88.5	88.4
—	—	—	—	1.8	1.6	1.7
—	—	—	—	5.0	3.5	3.5
—	—	—	—	—	—	—

* Simple-salt rose plotted on subsequent maps and diagrams.
Source of analysis; minor normative salts (anhydrous weight percent):
  7. MERCER (1983). Ulexite—0.5, sellaite—0.02, niter—0.0008.
  8. Bendix (unpub.). Ulexite—0.4, celestite—0.1, $Li_2SO_4 \cdot H_2O$—0.02.
  9. Bendix (unpub.). Ulexite—0.5, celestite—0.2, $Li_2SO_4 \cdot H_2O$—0.02, sellaite—0.02.
 10. Bendix (unpub.). Ulexite—0.3, celestite—0.1, $Li_2SO_4 \cdot H_2O$—0.02.
 11. MERCER (1983). Indirite—0.1, inyoite—0.01, fluorite—0.002, nitrocalcite—0.00004.
 12. Bendix (unpub.). Indirite—0.1, celestite—0.04, $Li_2SO_4 \cdot H_2O$—0.004.
 13. Bendix (unpub.). Indirite 0.1, celestite—0.04, $Li_2SO_4 \cdot H_2O$—0.004.

Table 9 (Continued)

	14*	15	16	17*	18	19
Site	H-5	H-6	H-6	H-6	H-7	H-7
Date collected	8/27/85	12/20/78	5/2/81	9/15/85	3/20/80	3/20/87
Density (g/cc)	1.104	—	1.040	1.042	1.001	1.000
pH	7.4	7.3	7.0	6.9	7.0	7.65
Chemical analysis	mg/L	mg/L	mg/L	mg/L	mg/L	mg/kg
Mg	2,170.	970.	1,080.	1,040.	130.	130.
Ca	1,700.	1,200.	2,150.	2,040.	590.	530.
Sr	29.3	—	32.0	30.4	—	12.
Li	0.81	—	0.44	0.45	—	0.1
Na	54,100.	18,000.	18,600.	18,000.	210.	238.
K	1,350.	500.	450.	375.	1.4	11.
F	1.95	1.5	—	1.9	1.4	—
Cl	85,400.	28,000.	33,000.	32,330.	350.	337.
Br	48.6	—	34.3	34.0	—	0.76
I	0.191	—	—	0.096	—	—
NO3	—	0.09	—	—	1.8	—
HCO3	49.	—	96.	94.	—	130.
CO3	—	—	—	—	—	—
SO4	7,840.	3,800.	3,980.	3,570.	1,900.	1,960.
B	33.7	9.5	10.9	10.4	0.78	1.9
TDS	153,000.	52,500.	59,400.	58,000.	3,190.	3,350.
Balance (+/−)	1.030	1.076	1.001	0.992	0.995	0.915
Cl/Br	1,760.	—	962.	951.	—	443.
Salt norm						
Dolomite	—	—	0.1	0.1	—	—
Magnesite	<0.1	—	—	—	—	2.7
Anhydrite	3.7	7.4	9.5	8.8	63.1	57.1
Kieserite	3.2	2.8	—	—	18.8	—
Epsomite	—	—	—	—	—	13.0
Polyhalite	—	—	—	—	—	2.6
nBloedite	—	—	—	—	—	7.1
Glauberite	—	—	—	—	—	—
Aphthitalite	—	—	—	—	—	—
Thenardite	—	—	—	—	—	—
Syngenite	—	—	—	—	—	—
Kainite	—	—	—	—	—	—
Halite	88.4	83.3	79.6	80.0	16.8	16.4
Sylvite	—	—	—	—	0.2	—
Carnallite	3.8	4.0	3.3	2.9	0.9	—
Bischofite	0.7	2.5	1.6	1.0	—	—
Tachyhydrite	—	—	5.8	7.0	—	—
Antarcticite	—	—	—	—	—	—
Simple salts						
CaCO3	—	—	0.1	0.1	—	—
MgCO3	<0.1	—	0.1	0.1	—	2.7
CaSO4	3.7	7.4	9.5	8.8	63.3	59.0
MgSO4	3.2	2.8	—	—	18.8	17.0
K2SO4	—	—	—	—	—	0.8
Na2SO4	—	—	—	—	—	3.9
Na2Cl2	88.6	83.4	79.7	80.2	16.9	16.6
K2Cl2	1.7	1.7	1.5	1.3	0.1	—
MgCl2	2.9	4.7	7.1	7.1	1.0	—
CaCl2	—	—	2.1	2.6	—	—

* Simple-salt rose plotted on subsequent maps and diagrams.
Source of analysis; minor normative salts (anhydrous weight percent):
14. Bendix (unpub.). Indirite—0.1, celestite—0.04, Li$_2$SO$_4$·H$_2$O—0.004, sellaite—0.002.
15. MERCER (1983). Indirite—0.08, sellaite—0.005, nitromagnesite—0.0002.
16. Bendix (unpub.). Inyoite—0.09, SrCl$_2$·2H$_2$O—0.1, LiCl·H$_2$O—0.005.
17. Bendix (unpub.). Inyoite—0.09, SrCl$_2$·2H$_2$O—0.1, LiCl·H$_2$O—0.005, fluorite—0.007.
18. MERCER (1983). Indirite—0.1, nitromagnesite—0.07, sellaite—0.07.
19. Rettig (this report). Celestite—0.8, indirite—0.3, Li$_2$SO$_4$·H$_2$O—0.03.

Table 9 (Continued)

20* H-7 3/26/86 1.001 7.2	21 H-8 2/11/80 1.000 7.3	22 H-8 2/11/80 1.000 7.61	23* H-8 1/22/86 1.002 7.6	24 H-9 2/5/80 1.000 7.57	25* H-9 11/14/85 1.002 7.4	26* H-10 3/21/80 1.044 8.01
mg/L	mg/L	mg/kg	mg/L	mg/kg	mg/L	mg/kg
130.	170.	150.	157.	150.	137.	785.
587.	570.	530.	548.	580.	590.	1,580.
8.51	—	7.0	6.92	11.	7.5	19.
0.1	—	0.1	0.115	0.2	0.175	0.6
207.	82.	81.	55.1	220.	146.	20,600.
7.0	4.7	4.3	3.83	8.	6.85	479.
1.51	2.4	—	2.45	—	3.31	—
320.	57.0	60.	30.5	327.	194.	33,600.
0.57	—	0.18	0.085	0.58	.24	12.
0.0525	—	—	0.145	—	0.11	—
—	4.2	—	—	110.	110.	50.
120.	—	76.	95.	—	—	—
—	—	—	—	—	—	—
1,850.	2,000.	2,050.	1,950.	2,070.	1,903.	4,950.
0.755	0.58	0.89	0.485	1.1	0.63	29.
3,230.	2,890.	2,960.	2,850.	3,480.	3,100.	62,100.
0.995	1.061	0.933	0.994	0.947	1.007	0.999
561.	—	333.	359.	564.	808.	2,800.
—	—	—	—	—	—	—
2.6	—	1.8	2.4	2.2	2.5	0.1
62.3	63.9	64.3	66.2	58.9	64.7	8.6
—	—	—	—	—	—	2.3
16.0	23.7	19.7	20.6	17.4	16.2	—
1.6	0.5	1.1	1.0	1.8	1.6	—
0.1	8.1	9.28	7.2	3.3	3.6	—
—	—	—	—	—	—	—
—	—	—	—	—	—	—
—	—	—	—	—	—	—
—	—	—	—	—	—	—
16.6	3.3	3.3	1.8	15.5	10.5	84.3
—	—	—	—	—	—	—
—	—	—	—	—	—	3.4
—	—	—	—	—	—	1.1
—	—	—	—	—	—	—
—	—	—	—	—	—	—
—	—	—	—	—	—	—
2.6	—	1.8	2.4	2.2	2.5	0.1
63.6	64.5	65.3	67.2	60.3	66.1	8.7
16.5	27.7	24.3	24.4	19.5	18.3	2.3
0.5	0.1	0.3	0.3	0.5	0.5	—
0.1	4.4	5.0	4.0	1.8	2.0	—
16.7	3.3	3.4	1.8	15.7	10.6	84.6
—	—	—	—	—	—	1.5
—	—	—	—	—	—	3.0
—	—	—	—	—	—	—

* Simple-salt rose plotted on subsequent maps and diagrams.
Source of analysis; minor normative salts (anhydrous weight percent):
20. Bendix (unpub.). Celestite—0.6, indirite—0.1, sellaite—0.078, $Li_2SO_4 \cdot H_2O$—0.03.
21. MERCER (1983). Niter—0.2, indirite—0.09, sellaite—0.1.
22. Rettig (this report). Celestite—0.5, indirite—0.1, $Li_2SO_4 \cdot H_2O$—0.03.
23. Bendix (unpub.). Celestite—0.5, indirite—0.08, sellaite—0.1, $Li_2SO_4 \cdot H_2O$—0.03.
24. Rettig (this report). Celestite—0.7, indirite—0.1, $Li_2SO_4 \cdot H_2O$—0.05.
25. Bendix (unpub.). Celestite—0.5, indirite—0.09, sellaite—0.2, $Li_2SO_4 \cdot H_2O$—0.05.
26. Rettig (this report). Indirite—0.2, celestite—0.6, $Li_2SO_4 \cdot H_2O$—0.008.

Table 9 (Continued)

	27*	28*	29	30	31*	32
Site	H-11	H-12	ENGEL	P-14	P-14	P-15
Date collected	6/3/85	8/8/85	3/5/85	3/14/77	2/26/86	5/10/77
Density (g/cc)	1.090	1.096	—	—	1.019	—
pH	7.2	7.2	7.4	6.0	6.8	—
Chemical analysis	mg/L	mg/L	mg/L	mg/L	mg/L	mg/L
Mg	1,320.	1,980.	152.	760.	840.	63.
Ca	1,700.	1,760.	588.	3,100.	3,510.	770.
Sr	24.7	30.6	8.36	—	50.8	—
Li	0.615	1.15	0.17	—	0.275	—
Na	40,400.	49,200.	200.	7,600.	4,360.	6,900.
K	943.	1,270.	5.6	600.	37.9	1,700.
F	—	—	2.8	0.9	1.69	1.2
Cl	65,900.	79,000.	231.	20,000.	14,470.	11,000.
Br	47.4	76.	0.27	—	72.	—
I	—	—	0.116	—	0.415	—
NO$_3$	55.	53.	110.	0.04	—	0.2
HCO$_3$	—	—	—	357.	110.	63.
CO$_3$	—	—	—	—	—	24.
SO$_4$	7,180.	7,210.	1,990.	1,400.	1,590.	3,200.
B	31.7	39.3	0.97	0.7	0.72	4.7
TDS	118,000.	141,000.	3,290.	33,800.	25,000.	23,700.
Balance (+/−)	0.983	1.018	1.019	0.940	0.983	1.022
Cl/Br	1,390.	1,040.	856.	—	201.	—
Salt norm						
Dolomite	—	—	—	0.8	0.3	—
Magnesite	<0.1	0.026	2.4	—	—	0.3
Anhydrite	4.8	4.20	60.3	5.8	9.0	2.4
Kieserite	3.1	2.71	—	—	—	—
Epsomite	—	—	15.1	—	—	—
Polyhalite	0.3	—	1.2	—	—	3.8
Bloedite	—	—	8.2	—	—	—
Glauberite	—	—	—	—	—	—
Aphthitalite	—	—	—	—	—	—
Thenardite	—	—	—	—	—	—
Syngenite	—	—	—	—	—	15.1
Kainite	—	—	—	—	—	—
Halite	88.3	88.0	11.8	59.7	45.0	73.0
Sylvite	—	—	—	—	—	5.3
Carnallite	3.3	3.88	—	8.1	0.7	—
Bischofite	—	1.02	—	—	—	—
Tachyhydrite	—	—	—	6.8	20.3	—
Antarcticite	—	—	—	18.9	24.4	—
Simple salts						
CaCO$_3$	—	—	—	0.4	0.2	—
MgCO$_3$	<0.1	0.26	2.4	0.4	0.2	0.3
CaSO$_4$	5.0	4.21	61.4	5.8	9.0	10.9
MgSO$_4$	3.1	2.72	19.4	—	—	0.8
K$_2$SO$_4$	0.1	—	0.4	—	—	9.7
Na$_2$SO$_4$	—	—	4.5	—	—	—
Na$_2$Cl$_2$	88.4	88.1	11.9	59.7	45.1	73.1
K$_2$Cl$_2$	1.5	1.71	—	3.5	0.3	5.3
MgCl$_2$	1.9	3.20	—	8.8	13.2	—
CaCl$_2$	—	—	—	21.4	32.0	—

* Simple-salt rose plotted on subsequent maps and diagrams.
Source of analysis; minor normative salts (anhydrous weight percent):
27. Bendix (unpub.). Indirite—0.1, celestite—0.04, Li$_2$SO$_4 \cdot$ H$_2$O—0.004.
28. Bendix (unpub.). Indirite—0.1, celestite—0.05, Li$_2$SO$_4 \cdot$ H$_2$O—0.006.
29. Bendix (unpub.). Celestite—0.5, indirite—0.1, sellaite—0.1, Li$_2$SO$_4 \cdot$ H$_2$O—0.04.
30. MERCER (1983). Inyoite—0.01, fluorite—0.005, nitrocalcite—0.0002.
31. Bendix (unpub.). SrCl$_2 \cdot$ 2H$_2$O—0.3, inyoite—0.01, fluorite—0.01, LiCl $\cdot$ H$_2$O—0.007.
32. MERCER (1983). Ulexite—0.1, sellaite—0.008, niter—0.001.

Table 9 (Continued)

33 P-17 5/10/77	34* P-17 3/17/86	35* P-18 5/10/77	36 WIPP-25 8/14/80	37 WIPP-25 8/14/80	38* WIPP-25 8/20/80	39 WIPP-25 2/12/86
—	1.065	—	1.014	1.011	1.010	1.010
7.4	7.5	7.2	7.3	7.11	6.9	7.2
mg/L	mg/L	mg/L	mg/L	mg/kg	mg/L	mg/L
1,600.	1,460.	16,000.	250.	247.	260.	315.
1,700.	1,620.	5,600.	920.	890.	905.	1,140.
—	29.1	—	—	12.	12.	16.6
—	0.87	—	—	0.2	0.2	0.22
30,000.	28,300.	9,200.	5,100.	5,300.	3,160.	3,180.
120.	782.	6,200.	0.9	69.	73.5	102.
1.5	1.87	1.2	1.4	—	—	1.68
54,000.	48,220.	80,000.	8,300.	8,020.	5,250.	6,320.
—	71.5	—	—	2.4	2.6	3.35
—	0.175	—	—	—	—	0.042
0.3	—	3.6	3.0	—	—	—
77.	61.	310.	—	—	210.	130.
—	—	—	—	—	—	—
5,000.	6,020.	980.	2,400.	2,570.	2,500.	2,380.
1.7	37.5	100.	1.9	1.8	1.52	1.67
92,500.	86,600.	118,000.	17,000.	17,100.	12,400.	13,600.
0.936	0.978	0.942	1.014	1.063	1.013	0.974
—	674.	—	—	3,340.	2,020.	1,890.
—	—	0.2	—	—	—	0.7
0.1	<0.1	—	—	—	1.2	—
6.5	6.5	1.2	18.2	17.0	24.9	24.7
0.9	2.9	—	1.7	4.1	3.5	—
—	—	—	—	—	—	—
—	—	—	—	—	—	—
—	—	—	—	—	—	—
—	—	—	—	—	—	—
—	—	—	—	—	—	—
—	—	—	—	—	—	—
86.0	84.4	20.6	75.7	75.8	65.0	60.8
—	—	—	—	—	—	—
0.6	4.0	23.7	<0.1	1.7	2.6	3.3
6.0	2.0	20.6	4.3	1.2	2.6	1.7
—	—	33.4	—	—	—	8.4
—	—	0.1	—	—	—	0.4
0.1	<0.1	0.1	—	—	1.2	0.3
6.5	6.5	1.2	18.3	17.0	24.9	24.8
0.9	2.9	—	1.7	4.1	3.5	—
—	—	—	—	—	—	—
86.0	84.6	20.6	75.7	75.9	65.1	61.0
0.3	1.8	10.4	<0.1	0.7	1.1	1.5
6.3	4.3	55.2	4.3	2.2	4.1	8.9
—	—	12.3	—	—	—	3.1

* Simple-salt rose plotted on subsequent maps and diagrams.
Source of analysis; minor normative salts (anhydrous weight percent):
33. MERCER (1983). Indirite—0.008, sellaite—0.003, nitromagnesite—0.0004.
34. Bendix (unpub.). Indirite—0.2, celestite—0.07, $Li_2SO_4 \cdot H_2O$—0.008, sellaite—0.004.
35. MERCER (1983). Inyoite—0.4, nitrocalcite—0.004, fluorite—0.002.
36. MERCER (1983). Indirite—0.05, nitromagnesite—0.02, sellaite—0.01.
37. Rettig (this report). Celestite—0.1, indirite—0.05, $Li_2SO_4 \cdot H_2O$—0.009.
38. Bendix (unpub.). Celestite—0.2, indirite—0.06, $Li_2SO_4 \cdot H_2O$—0.01.
39. Bendix (unpub.). $SrCl_2 \cdot 2H_2O$—0.2, inyoite—0.06, fluorite—0.03, $LiCl \cdot H_2O$—0.01.

Table 9 (Continued)

Site	40* WIPP-26	41 WIPP-26	42 WIPP-26	43 WIPP-27	44 WIPP-27	45 WIPP-28
Date collected	8/18/80	8/24/80	11/25/85	8/22/80	9/3/80	8/21/80
Density (g/cc)	1.009	1.005	1.012	1.092	1.090	1.041
pH	7.34	6.9	7.1	6.63	6.3	6.80
Chemical analysis	mg/kg	mg/L	mg/L	mg/kg	mg/L	mg/kg
Mg	362.	355.	380.	1,740.	1,900.	624.
Ca	1,190.	1,240.	1,340.	2,700.	3,210.	1,130.
Sr	19.	16.8	19.5	44.	50.9	17.
Li	0.2	0.24	0.23	—	0.33	—
Na	3,770.	3,620.	4,220.	38,500.	39,200.	19,900.
K	149.	170.	343.	7,140.	8,060.	346.
F	—	—	1.73	—	—	—
Cl	7,260.	7,200.	8,770.	73,300.	78,500.	32,400.
Br	3.0	3.2	3.9	27.	28.3	1.0
I	—	—	0.0695	—	—	—
NO_3	—	—	—	—	—	—
HCO_3	—	140.	120.	—	120.	—
CO_3	—	—	—	—	—	—
SO_4	2,380.	2,480.	2,420.	3,300.	3,830.	3,360.
B	1.8	1.45	1.65	2.1	2.3	2.9
TDS	15,100.	15,200.	17,600.	127,000.	135,000.	57,800.
Balance (+/−)	1.012	0.985	0.970	1.000	0.971	0.999
Cl/Br	2,420.	2,250.	2,250.	2,710.	2,770.	32,400.
Salt norm						
Dolomite	—	0.7	0.5	—	0.1	—
Magnesite	—	—	—	—	—	—
Anhydrite	22.4	23.1	19.3	3.7	4.0	6.6
Kieserite	—	—	—	—	—	1.4
Epsomite	—	—	—	—	—	—
Polyhalite	—	—	—	—	—	—
Bloedite	—	—	—	—	—	—
Glauberite	—	—	—	—	—	—
Aphthitalite	—	—	—	—	—	—
Thenardite	—	—	—	—	—	—
Syngenite	—	—	—	—	—	—
Kainite	—	—	—	—	—	—
Halite	62.8	61.3	62.3	77.2	75.3	87.6
Sylvite	—	—	—	6.5	7.2	—
Carnallite	4.2	4.9	8.7	9.6	10.0	2.6
Bischofite	—	—	—	—	—	1.7
Tachyhydrite	9.0	9.7	5.6	—	—	—
Antarcticite	1.2	<0.1	3.4	2.9	3.4	—
Simple salts						
$CaCO_3$	—	0.4	0.3	—	0.1	—
$MgCO_3$	—	0.3	0.2	—	—	—
$CaSO_4$	22.4	23.1	19.4	3.7	4.0	6.7
$MgSO_4$	—	—	—	—	—	1.4
K_2SO_4	—	—	—	—	—	—
Na_2SO_4	—	—	—	—	—	—
Na_2Cl_2	63.0	61.5	62.5	77.3	75.3	87.7
K_2Cl_2	1.9	2.2	3.8	10.8	11.6	1.1
$MgCl_2$	9.3	8.9	8.40	5.4	5.6	3.1
$CaCl_2$	3.3	3.6	5.4	2.9	3.4	—

* Simple-salt rose plotted on subsequent maps and diagrams.
Source of analysis; minor normative salts (anhydrous weight percent):
40. Rettig (this report). $SrCl_2 \cdot 2H_2O$—0.2, inyoite—0.06, $LiCl \cdot H_2O$—0.008.
41. Bendix (unpub.). $SrCl_2 \cdot 2H_2O$—0.2, inyoite—0.05, $LiCl \cdot H_2O$—0.01.
42. Bendix (unpub.). $SrCl_2 \cdot 2H_2O$—0.2, inyoite—0.05, fluorite—0.002, $LiCl \cdot H_2O$—0.008.
43. Rettig (this report). $SrCl_2 \cdot 2H_2O$—0.06, inyoite—0.008.
44. Bendix (unpub.). $SrCl_2 \cdot 2H_2O$—0.07, inyoite—0.008, $LiCl \cdot H_2O$—0.002.
45. Rettig (this report). Celestite—0.06, indirite—0.02.

Table 9 (Continued)

46* WIPP-28 9/11/80 1.030 6.5	47 WIPP-29 8/20/80 1.174 6.88	48 WIPP-29 8/28/80 1.160 6.1	49 WIPP-29 12/14/85 1.216 5.9	50* WIPP-30 8/13/80 1.070 6.36	51 WIPP-30 9/6/80 1.020 8.8
mg/L	mg/kg	mg/L	mg/L	mg/kg	mg/L
555.	3,750.	5,480.	6,500.	804.	460.
1,180.	6,980.	950.	413.	1,010.	1,140.
15.5	20.	28.8	12.8	15.	17.9
0.3	1.0	0.78	0.7	—	0.27
15,200.	67,600.	71,400.	94,900.	35,500.	8,570.
485.	1,270.	15,600.	23,300.	888.	255.
—	—	—	4.59	—	—
24,800.	123,000.	138,000.	178,800.	56,500.	14,600.
7.2	44.	45.	61.	30.	10.5
—	—	—	0.385	—	—
—	—	—	—	—	40.
—	—	210.	160.	—	17.
—	—	—	—	5,050.	4,120.
4,380.	1,120.	14,000.	20,000.	—	—
5.83	—	4.4	5.17	18.	6.09
46,600.	219,000.	246,000.	324,000.	99,800.	29,200.
0.984	0.975	0.956	0.967	0.991	0.950
3,440.	2,800.	3,070.	2,930.	1,880.	1,390.
—	—	—	—	—	—
—	—	0.1	<0.1	—	0.2
7.7	—	—	—	1.9	13.5
3.1	—	—	—	1.9	5.0
2.0	2.3	2.8	0.9	3.2	0.3
—	—	—	—	—	—
—	—	—	—	—	—
—	—	—	—	—	—
—	—	—	—	—	—
—	7.1	7.8	11.1	—	—
83.7	79.7	76.0	76.0	90.9	77.0
—	5.6	5.1	7.6	—	—
3.3	5.3	8.6	4.3	2.0	3.7
—	—	—	—	—	—
—	—	—	—	—	—
—	—	—	—	—	—
—	—	0.1	<0.1	—	0.2
8.7	1.1	1.4	0.4	3.5	13.7
3.5	4.9	5.2	7.0	2.6	5.1
0.6	0.7	0.9	0.3	1.0	0.1
—	—	—	—	—	—
83.8	79.7	76.0	76.0	91.0	77.2
1.5	10.6	11.7	13.8	0.9	1.6
1.9	3.0	4.8	2.4	1.1	2.1
—	—	—	—	—	—

* Simple-salt rose plotted on subsequent maps and diagrams.
Source of analysis; minor normative salts (anhydrous weight percent):
46. Bendix (unpub.). Indirite—0.06, celestite—0.07, $Li_2SO_4 \cdot H_2O$—0.005.
47. Rettig (this report). Celestite—0.02, $Li_2SO_4 \cdot H_2O$—0.004.
48. Bendix (unpub.). Celestite—0.03, indirite—0.008, $Li_2SO_4 \cdot H_2O$—0.003.
49. Bendix (unpub.). Indirite—0.007, celestite—0.009, sellaite—0.002, $Li_2SO_4 \cdot H_2O$—0.002.
50. Rettig (this report). Indirite—0.08, celestite—0.03.
51. Bendix (unpub.). Indirite—0.09, celestite—0.1, $Li_2SO_4 \cdot H_2O$—0.008.

Table 9 (Continued)

Site	52 GNOME SHAFT	53* GNOME-1	54 GNOME-4	55 GNOME-8	56* INDIAN
Date collected	12/10/60	8/18/60	12/5/61	1/27/63	1/22/63
Density (g/cc)	—	—	—	—	—
pH	7.7	7.6	7.5	7.1	7.6
Chemical analysis	mg/kg	mg/kg	mg/kg	mg/kg	mg/kg
Mg	153.	146.	134.	155.	169.
Ca	597.	608.	644.	624.	624.
Sr	—	—	8.8	1.0	1.1
Li	—	—	0.1	0.25	0.27
Na	555.	520.	640.	630.	315.
K	17.	11.	16.	27.	8.6
F	1.9	0.3	1.0	2.4	2.1
Cl	828.	770.	948.	1,190.	533.
Br	—	—	—	—	—
I	—	—	—	—	—
NO$_3$	11.	7.8	10.	7.6	4.2
HCO$_3$	106.	114.	114.	108.	193.
CO$_3$	—	—	—	—	—
SO$_4$	2,040.	1,960.	1,950.	2,050.	1,950.
B	12.	—	0.8	—	—
TDS	4,320.	4,140.	4,470.	4,800.	3,810.
Balance (+/−)	0.981	1.011	1.031	0.923	1.005
Cl/Br	—	—	—	—	—
Salt norm					
Dolomite	—	—	—	—	—
Magnesite	1.7	1.9	1.8	1.53	3.6
Anhydrite	47.3	49.7	47.9	47.1	56.9
Kieserite	—	—	—	10.7	15.1
Epsomite	12.7	13.6	11.4	—	—
Polyhalite	1.8	1.1	1.5	—	—
Bloedite	3.8	2.1	0.8	—	—
Glauberite	—	—	—	—	—
Aphthitalite	—	—	—	—	—
Thenardite	—	—	—	—	—
Syngenite	—	—	—	—	—
Kainite	—	—	—	—	—
Halite	31.8	31.2	35.8	35.6	21.5
Sylvite	—	—	—	—	—
Carnallite	—	—	—	2.61	1.0
Bischofite	—	—	—	1.60	0.9
Tachyhydrite	—	—	—	—	—
Antarcticite	—	—	—	—	—
Simple salts					
CaCO$_3$	—	—	—	—	—
MgCO$_3$	1.7	2.0	1.8	1.55	3.6
CaSO$_4$	48.6	50.4	49.0	47.5	57.4
MgSO$_4$	15.0	14.8	12.2	10.8	15.3
K$_2$SO$_4$	0.5	0.3	0.5	—	—
Na$_2$SO$_4$	2.1	1.2	0.4	—	—
Na$_2$Cl$_2$	32.1	31.3	36.1	35.9	21.7
K$_2$Cl$_2$	—	—	—	1.15	0.4
MgCl$_2$	—	—	—	3.08	1.5
CaCl$_2$	—	—	—	—	—

* Simple-salt rose plotted on subsequent maps and diagrams.
Source of analysis; minor normative salts (anhydrous weight percent):
52. U.S.G.S. (unpub. tech file). Niter—0.4, wagnerite—0.4, Mg$_3$(PO$_4$)$_2$—0.08.
53. U.S.G.S. (unpub. tech file). Niter—0.3, sellaite—0.01.
54. U.S.G.S. (unpub. tech file). Celestite—0.4, niter—0.4, sellaite—0.03, wagnerite—0.03, Li$_2$SO$_4 \cdot$ H$_2$O—0.02.
55. U.S.G.S. (unpub. tech file). Celestite—0.5, nitromagnesite—0.2, sellaite—0.08, Li$_2$SO$_4 \cdot$ H$_2$O—0.04.
56. U.S.G.S. (unpub. tech file). Celestite—0.6, nitromagnesite—0.1, sellaite—0.09, Li$_2$SO$_4 \cdot$ H$_2$O—0.06.

Table 9 (Continued)

57* DOE-1 4/24/85 1.090 7.1	58* DOE-2 3/11/85 1.040 7.0	59* TWO-MILE 8/8/62 — 6.7	60* WINDMILL 9/14/61 — 8.0	61 WINDMILL 8/8/62 — 6.8	62* SOUTH 8/8/62 — 7.4
mg/L	mg/L	mg/kg	mg/kg	mg/kg	mg/kg
1,610.	1,060.	177.	139.	138.	90.
1,730.	1,960.	630.	564.	580.	589.
26.2	37.7	14.	—	10.	10.
0.635	0.47	0.32	—	0.32	0.2
45,800.	18,400.	1,410.	525.	500.	24.
1,100.	410.	25.	23.	28.	4.8
—	1.69	2.0	—	3.0	1.4
73,600.	34,600.	2,200.	515.	588.	18.
56.	33.5	—	—	—	—
—	0.225	—	—	—	—
—	—	2.5	—	2.6	9.0
45.	67.	66.	108.	113.	196.
—	—	—	—	—	—
7,350.	3,950.	2,200.	2,290.	2,080.	1,660.
36.6	15.6	—	—	—	—
131,000.	60,500.	6,730.	4,160.	4,040.	2,600.
1.004	0.940	0.993	0.985	1.018	0.993
1,310.	1,030.	—	—	—	—
—	0.1	—	—	—	—
<0.1	—	0.7	1.8	2.0	5.4
4.5	9.0	31.1	45.2	46.6	80.4
3.0	—	11.5	—	—	—
—	—	—	0.8	5.6	8.8
—	—	2.1	4.1	4.7	—
—	—	—	27.5	15.8	2.7
—	—	—	—	—	—
—	—	—	—	—	—
—	—	—	—	—	—
88.4	80.4	53.8	20.6	24.5	1.2
—	—	—	—	—	—
3.6	3.1	0.3	—	—	—
0.2	2.3	—	—	—	—
—	4.9	—	—	—	—
—	—	—	—	—	—
—	<0.1	—	—	—	—
<0.1	<0.1	0.7	1.8	2.0	5.5
4.5	9.1	32.3	47.1	49.2	81.6
3.1	—	12.0	14.3	14.0	10.2
—	—	0.7	1.3	1.5	—
—	—	—	14.9	8.7	1.5
88.6	80.6	54.1	20.6	24.7	1.2
1.6	1.4	0.1	—	—	—
2.3	7.1	0.1	—	—	—
—	1.8	—	—	—	—

* Simple-salt rose plotted on subsequent maps and diagrams.
Source of analysis; minor normative salts (anhydrous weight percent):
57. Bendix (unpub.). Indirite—0.1, celestite—0.04, $Li_2SO_4 \cdot H_2O$—0.004.
58. Bendix (unpub.). Inyoite—0.1, $SrCl_2 \cdot 2H_2O$—0.1, $LiCl \cdot H_2O$—0.005, fluorite—0.006.
59. U.S.G.S. (unpub. tech file). Celestite—0.4, niter—0.06, sellaite—0.049, $Li_2SO_4 \cdot H_2O$—0.03.
60. U.S.G.S. (unpub. tech file).
61. U.S.G.S. (unpub. tech file). Celestite—0.5, sellaite—0.1, niter—0.1, $Li_2SO_4 \cdot H_2O$—0.06.
62. U.S.G.S. (unpub. tech file). Celestite—0.8, niter—0.5, sellaite—0.09, $Li_2SO_4 \cdot H_2O$—0.064, soda niter—0.07.

Table 10. Chemical analysis, salt norm (anhydrous weight

	1*	2*	3	4*
Site	H-1	H-2	H-3	H-3
Date collected	6/4/76	2/22/77	5/10/77	7/1/85
Density (g/cc)	—	—	—	1.006
pH	7.4	8.6	8.0	8.0
Chemical analysis	mg/L	mg/ml	mg/L	mg/L
Mg	270.	170.	480.	292.
Ca	890.	820.	1,200.	1,000.
Sr	—	—	—	17.3
Li	—	—	—	0.32
Na	5,700.	2,700.	9,300.	1,520.
K	70.	81.	250.	34.5
F	2.8	—	1.8	2.43
Cl	8,000.	4,100.	15,000.	3,360.
Br	—	—	—	5.85
I	—	—	—	1.2
NO_3	—	0.2	0.4	—
HCO_3	92.	74.	51.	46.
SO_4	3,900.	2,400.	3,400.	2,310.
B	2.2	0.22	13.	2.05
TDS	18,900.	10,400.	29,700.	8,590.
Balance (+/−)	1.025	1.045	1.030	0.982
Cl/Br	—	—	—	574.
Salt norm				
Dolomite	—	—	—	0.4
Magnesite	0.3	0.5	0.1	—
Anhydrite	14.5	25.1	13.5	38.0
Kieserite	—	5.4	2.6	—
Epsomite	0.6	—	—	—
Polyhalite	2.7	2.5	—	—
Bloedite	11.4	—	—	—
Glauberite	—	—	—	—
Aphthitalite	—	—	—	—
Thenardite	—	—	—	—
Syngenite	—	—	—	—
Halite	70.5	64.7	78.1	45.6
Sylvite	—	—	—	—
Carnallite	—	1.8	3.6	1.8
Bischofite	—	—	1.9	10.0
Tachyhydrite	—	—	—	3.6
Antarcticite	—	—	—	—
Simple salts				
$CaCO_3$	—	—	—	0.2
$MgCO_3$	0.3	0.5	0.1	0.2
$CaSO_4$	15.8	26.3	13.5	38.2
$MgSO_4$	6.4	5.9	2.6	—
K_2SO_4	0.8	0.8	—	—
Na_2SO_4	6.2	—	—	—
Na_2Cl_2	70.5	64.7	78.3	45.9
K_2Cl_2	—	0.8	1.6	0.8
$MgCl_2$	—	1.0	3.9	13.4
$CaCl_2$	—	—	—	1.4

* Simple-salt rose plotted on subsequent maps and diagrams.
Source of analysis; minor normative salts (anhydrous weight percent):
1. MERCER (1983). Indirite—0.5, sellaite—0.03.
2. MERCER (1983). Indirite—0.01, niter—0.003.
3. MERCER (1983). Indirite—0.2, sellaite—0.01, nitromagnesite—0.002.
4. Bendix (unpub.). $SrCl_2 \cdot 2H_2O$—0.4, inyoite—0.1, fluorite—0.06, $LiCl \cdot H_2O$—0.02.

percent), and simple salts (weight percent) of Magenta waters

5* H-4 12/14/78 — 8.0	6* H-5 12/14/78 — 7.8	7* H-6 12/20/78 — 7.3	8 H-8 2/12/80 1.006 9.3	9* H-8 2/12/80 1.004 7.8
mg/L	mg/L	mg/L	mg/L	mg/kg
410.	170.	160.	17.	15.
210.	240.	520.	870.	845.
—	—	—	—	20.
—	—	—	—	0.2
7,000.	1,500.	1,100.	2,400.	2,300.
130.	53.	46.	84.	75.
2.5	2.8	1.4	0.7	—
7,500.	880.	1,200.	3,500.	3,520.
—	—	—	—	0.82
—	—	—	—	—
0.04	0.04	0.1	0.3	34.
63.	50.	51.	—	2,490.
7,000.	3,200.	2,700.	2,100.	6.9
13.	11.	2.5	3.1	9,310.
22,300.	6,110.	5,780.	8,980.	0.959
0.980	0.996	0.967	1.061	4,290.
—	—	—	—	—
—	—	—	—	—
0.2	0.6	0.6	—	0.3
—	—	23.6	29.3	26.0
—	—	—	—	—
—	—	5.9	4.1	2.2
19.4	28.0	25.9	—	—
6.6	27.3	9.8	—	7.2
1.7	2.5	—	—	—
16.9	17.0	—	—	—
—	—	—	0.6	2.1
55.0	23.7	34.0	65.4	61.4
—	—	—	0.3	—
—	—	—	—	—
—	—	—	—	—
—	—	—	—	—
—	—	—	—	—
0.2	0.6	0.6	—	0.3
3.3	13.5	31.3	31.7	31.8
8.9	13.0	13.2	0.9	0.5
1.3	2.0	1.8	1.6	1.9
31.2	47.1	19.0	—	3.7
55.1	23.9	34.1	65.5	61.9
—	—	—	0.3	—
—	—	—	—	—
—	—	—	—	—

* Simple-salt rose plotted on subsequent maps and diagrams.
Source of analysis; minor normative salts (anhydrous weight percent):
5. MERCER (1983). Borax—0.3, sellaite—0.02, niter—0.0003.
6. MERCER (1983). Borax—0.8, sellaite—0.08, niter—0.001.
7. MERCER (1983). Ulexite—0.2, sellaite—0.04, niter—0.003.
8. MERCER (1983). Ulexite—0.2, sellaite—0.01, niter—0.006.
9. Rettig (this report). Ulexite—0.4, celestite—0.5, $Li_2SO_4 \cdot H_2O$—0.02.

Table 10 (Continued)

	10*	11	12*	13
Site	H-9	H-9	H-10	WIPP-25
Date collected	2/5/82	2/5/82	3/21/80	9/4/80
Density (g/cc)	1.003	1.000	1.171	1.010
pH	8.5	7.2	4.70	7.5
Chemical analysis	mg/L	mg/kg	mg/kg	mg/L
Mg				
Ca	170.	170.	2,130.	240.
Sr	550.	535.	2,310.	910.
Li	—	15.	38.	—
Na	—	0.3	5.	—
K	800.	800.	78,600.	3,100.
F	28.	26.	307.	0.8
Cl	1.8	—	—	1.5
Br	750.	750.	136,000.	5,600.
I	—	2.5	158.	—
NO$_3$	0.09	—	—	—
HCO$_3$	—	42.	—	2.8
SO$_4$	2,700.	2,760.	2,680.	1,900.
B	2.6	2.5	22.	1.9
TDS	5,000.	5,100.	222,000.	11,800.
Balance (+/−)	0.992	0.963	0.955	1.012
Cl/Br	—	300.	861.	—
Salt norm				
Dolomite	—	—	—	—
Magnesite	—	0.565	—	—
Anhydrite	34.6	32.4	1.7	23.0
Kieserite	—	—	—	—
Epsomite	—	—	—	—
Polyhalite	4.1	3.80	—	—
Bloedite	34.7	33.5	—	—
Glauberite	1.7	4.75	—	—
Aphthitalite	—	—	—	—
Thenardite	—	—	—	—
Syngenite	—	—	—	—
Halite	24.6	24.1	92.5	66.5
Sylvite	—	—	—	—
Carnallite	—	—	0.6	<0.1
Bischofite	—	—	0.8	3.8
Tachyhydrite	—	—	4.2	6.6
Antarcticite	—	—	—	—
Simple salts				
CaCO$_3$	—	—	—	—
MgCO$_3$	—	0.570	—	—
CaSO$_4$	37.6	36.9	1.7	23.0
MgSO$_4$	16.8	16.3	—	—
K$_2$SO$_4$	1.3	1.18	—	—
Na$_2$SO$_4$	19.7	20.8	—	—
Na$_2$Cl$_2$	24.7	24.3	92.6	66.6
K$_2$Cl$_2$	—	—	0.3	<0.1
MgCl$_2$	—	—	3.9	8.0
CaCl$_2$	—	—	1.6	2.4

* Simple-salt rose plotted on subsequent maps and diagrams.
Source of analysis; minor normative salts (anhydrous weight percent):
10. MERCER (1983). Ulexite—0.3, sellaite—0.06, niter—0.003.
11. Rettig (this report). Celestite—0.6, ulexite—0.2, Li$_2$SO$_4$ · H$_2$O—0.05.
12. Rettig (this report). Inyoite—0.05, SrCl$_2$ · 2H$_2$O—0.03, LiCl · H$_2$O—0.01.
13. MERCER (1983). Inyoite—0.08, nitrocalcite—0.03, fluorite—0.03.

Table 10 (Continued)

14* WIPP-25 9/4/80 1.007 7.41	15 WIPP-25 9/17/80 1.004 6.9	16 WIPP-27 7/24/80 1.095 6.66	17 WIPP-27 9/25/80 1.090 6.3	18* WIPP-30 9/24/80 1.012 8.2	19* GNOME-5 11/15/61 — 7.6
mg/kg	mg/L	mg/kg	mg/L	mg/kg	mg/kg
248.	260.	1,690.	2,100.	203.	122.
893.	905.	3,290.	3,660.	731.	648.
12.	12.	50.	58.6	15.	—
0.2	0.2	—	0.34	0.2	—
3,340.	2,910.	38,900.	43,200.	5,930.	150.
65.	71.5	7,030.	8,090.	119.	8.3
—	—	—	—	—	1.5
5,460.	5,250.	75,500.	85,200.	7,980.	250.
2.1	2.5	26.	28.3	2.1	—
—	—	—	—	—	13.
—	180.	—	210.	—	103.
2,480.	2,490.	2,830.	3,410.	3,660.	1,940.
1.4	1.54	4.9	2.32	10.	0.72
12,500.	12,100.	129,000.	146,000.	18,700.	3,240.
1.032	0.963	0.994	0.986	1.042	0.993
2,600.	2,100.	2,900.	3,010.	3,800.	—
—	—	—	0.1	—	—
—	1.0	—	—	—	2.2
23.8	26.3	3.1	3.3	7.3	69.4
4.0	2.3	—	—	—	14.7
—	—	—	—	4.5	—
—	—	—	—	9.4	—
—	—	—	—	6.9	—
—	—	—	—	—	—
—	—	—	—	—	—
66.6	63.3	76.7	76.0	71.5	11.7
—	—	6.4	6.2	—	—
2.2	2.7	9.2	10.1	—	1.1
3.2	4.2	—	—	—	—
—	—	—	—	—	—
—	—	4.5	4.2	—	—
—	—	—	0.1	—	—
—	1.0	—	—	—	2.3
23.8	26.4	3.1	3.3	12.9	69.9
4.0	2.3	—	—	5.3	14.8
—	—	—	—	1.4	—
—	—	—	—	8.6	—
66.7	63.4	76.8	76.0	71.8	11.8
1.0	1.2	10.4	10.7	—	0.5
4.5	5.7	5.1	5.7	—	0.6
—	—	4.5	4.2	—	—

* Simple-salt rose plotted on subsequent maps and diagrams.
Source of analysis; minor normative salts (anhydrous weight percent):
14. Rettig (this report). Celestite—0.2, indirite—0.05, $Li_2SO_4 \cdot H_2O$—0.01.
15. Bendix (unpub.). Celestite—0.2, indirite—0.06, $Li_2SO_4 \cdot H_2O$—0.01.
16. Rettig (this report). $SrCl_2 \cdot 2H_2O$—0.07, inyoite—0.02.
17. Bendix (unpub.). $SrCl_2 \cdot 2H_2O$—0.07, inyoite—0.008, $LiCl \cdot H_2O$—0.001.
18. Rettig (this report). Ulexite—0.3, celestite—0.2, $Li_2SO_4 \cdot H_2O$—0.008.
19. U.S.G.S. (unpub. tech file). Soda niter—0.4, nitromagnesite—0.1, indirite—0.1, sellaite—0.08.

Table 11. Chemical analysis, salt norm (anhydrous weight percent), and simple salts (weight percent) of Dewey Lake (1, 2), Santa Rosa (3), and Quaternary sediment (4) waters

Site	1 WALKER WELL	2 TWIN PASTURE	3 H-5	4 WIPP-15
Date collected	7/31/62	1/30/86	5/24/78	3/12/79
Density (g/cc)	—	0.998	—	—
pH	7.1	7.7	—	—
Chemical analysis	**mg/kg**	**mg/L**	**mg/L**	**mg/L**
Mg	145.	22.5	51.	81.
Ca	613.	80.4	56.	35.
Sr	8.4	1.06	—	—
Ba	—	—	—	0.2
Li	0.19	—	—	—
Na	140.	25.4	280.	300.
K	3.6	3.85	25.	19.
F	2.0	0.58	1.2	2.0
Cl	325.	44.1	120.	93.
Br	—	0.17	—	—
NO_3	8.0	—	1.6	—
HCO_3	134.	230.	240.	988.
SO_4	1,790.	75.1	530.	170.
B	—	0.13	0.89	1.0
TDS	3,170.	483.	1,310.	1,690.
Balance (+/−)	1.001	1.070	1.072	0.975
Cl/Br	—	259.	—	—
Salt norm				
Calcite	—	7.2	—	—
Dolomite	—	42.0	0.1	13.1
Magnesite	3.0	—	14.2	16.8
Trona	—	—	—	30.7
Burkeite	—	—	—	22.2
Anhydrite	67.1	29.7	—	—
Kieserite	12.6	—	—	—
Glauberite	—	—	30.3	—
Aphthitalite	—	—	5.5	4.4
Thenardite	—	—	32.0	—
Halite	11.5	16.9	17.1	12.1
Sylvite	—	0.9	—	—
Carnallite	0.5	2.3	—	—
Bischofite	4.3	—	—	—
Simple salts				
$CaCO_3$	—	30.3	0.1	7.5
$MgCO_3$	3.0	19.4	14.4	24.2
Na_2CO_3	—	—	—	33.7
$CaSO_4$	67.8	30.0	15.0	—
$MgSO_4$	12.7	—	—	—
K_2SO_4	—	—	4.3	3.6
Na_2SO_4	—	—	49.1	18.1
Na_2Cl_2	11.6	17.0	17.2	12.9
K_2Cl_2	0.2	1.9	—	—
$MgCl_2$	4.6	1.3	—	—

Source of analysis; minor normative salts (anhydrous weight percent):
1. U.S.G.S. (unpub. tech file). Celestite—0.6, nitromagnesite—0.3, sellaite—0.1, $Li_2SO_4 \cdot H_2O$—0.05.
2. Bendix (unpub.). $SrCl_2 \cdot 2H_2O$—0.5, fluorite—0.3, inyoite—0.2.
3. MERCER (1983). Borax—0.4, niter—0.2, fluorite—0.2.
4. GONZOLES (1981). Borax—0.4, villiaumite—0.4, barite—0.03.

Characterization of Rustler Formation waters

The salt norms for the Rustler Formation fluids readily fit four water types: (1) highly concentrated primitive or primary brines, most of which have been involved in diagenetic exchange of magnesium for calcium, and thus termed *primitive-diagenetic* waters; (2) dilute *alkali-bearing carbonate recharge*

Table 12. Salt norms of synthetic isothermal (25°C) evaporating sea water: (A) Experimental data compiled and summarized in BRAITSCH (1971); and (B) Computer-simulated Ca-Mg-Na-K-Cl-SO₄ brines (EUGSTER *et al.*, 1980)

(A)	A-1	A-2	A-3	A-4	A-5	A-6	A-7
TDS (mg/kg)	106,200	273,800	346,600	321,700	328,000	331,900	359,900
Percent water remaining	29.8	9.12	1.66	1.53	1.37	0.82	0.61
Normative salts							
Anhydrite	3.53	0.56	—	—	—	—	—
Kieserite	6.56	6.75	26.5	27.4	24.3	8.41	1.11
Halite	78.4	80.9	27.2	14.5	12.5	3.18	1.08
Carnallite	4.82	4.95	19.4	23.6	25.6	6.60	0.78
Bischofite	6.66	6.88	26.9	34.6	37.6	81.8	97.0

(B)	B-1	B-2	B-3	B-4	B-5	B-6	B-7
TDS (mg/kg)	35,400	181,600	275,200	295,300	308,400	331,700	357,000
Percent water remaining	100.0	15.58	9.16	2.45	1.34	0.84	0.41
Normative salts							
Anhydrite	3.50	1.50	0.328	0.071	0.024	0.013	0.017
Kieserite	6.59	6.85	8.88	21.6	22.1	8.94	1.36
Halite	78.5	80.0	80.7	39.1	13.1	3.40	0.980
Carnallite	4.79	4.93	4.90	15.0	6.86	6.62	0.657
Bischofite	6.64	6.69	5.24	24.3	57.9	81.0	97.0

A-1. At initial gypsum.
A-2. At initial halite.
A-3. At initial bloedite.
A-4. At initial epsomite.
A-5. At initial kainite.
A-6. At initial carnallite.
A-7. At bischofite saturation.

B-1. Sea water.
B-2. During gypsum precipitation.
B-3. Shortly after initial halite.
B-4. Shortly after initial polyhalite.
B-5. Shortly after initial epsomite.
B-6. Shortly after initial carnallite.
B-7. At bischofite saturation.

Table 13. Chloride-bromide ratio (weight) and salt norm (anhydrous weight percent) of marine brines from commercial seawater evaporation pan successions from analyses of Yugoslavian brines (HERRMAN *et al.*, 1973) and Bahamian brines (MCCAFFREY *et al.*, 1987)

	Evaporating pans at the Sečovlje salt works, Portorož, Yugoslavia					
Sample†	Seawater	Pan 11	Pan 13a	Pan 13d	Pan 13e	Pan 13i
Cl/Br	295	309	326	305	180	140
TDS (mg/kg)	34,500	124,000	170,000	256,000	274,000	274,000
Normative salts						
Anhydrite	4.0	3.8	2.0	0.5	0.2	0.2
Kieserite	6.3	6.5	8.1	7.4	13.4	15.1
Halite	76.4	76.5	78.2	78.7	61.3	51.5
Carnallite	5.2	5.0	4.9	5.1	7.7	10.2
Bischofite	8.1	8.2	6.8	8.3	17.4	23.0

	Evaporating pans at the Morton Bahamas salt works, Great Inagua Island, Bahamas								
Sample†	W-64	W-56	W-46	W-43	W-39	40-4	36-4*	39-6*	39-4*
Cl/Br	283	295	286	214	80	53	48	43	34
TDS (mg/L)	36,900	125,000	318,000	331,000	374,000	422,000	449,000	414,000	309,000
Normative salts									
Anhydrite	3.6	4.2	0.2	0.1	**	**	**	**	**
Kieserite	6.5	6.4	6.8	9.0	22.3	30.2	32.5	17.7	13.0
Halite	78.8	77.8	80.8	76.9	39.2	14.1	8.1	5.2	2.8
Carnallite	5.1	5.2	5.3	6.7	16.2	22.6	25.0	20.5	6.1
Bischofite	6.0	6.4	6.9	7.2	22.2	33.0	34.3	56.6	78.0

* Evaporated in laboratory with starting brine from halite precipitating pans.
** Calcium not determined.

† Sample description and salt precipitating: Seawater; Pans 11 and 13a, carbonate; Pan 13d, gypsum; Pans 13e and 13i, halite; W-64 (seawater); W-56 (pan R2A); W-46 (pan R9), gypsum; W-43 (pan Y3-Middle), W-39 (pan D3-Middle), and 40-4 (pan F4), halite; 36-4 (pan C4-Middle), magnesium sulfate; and 39-6 (pan D3-Middle) and 39-4 (pan D3-Middle), potash salt(s).

Table 14. Samples from sites in the Rustler Formation of questionable validity

Site/horizon	LAMBERT and HARVEY (1987)	This report
H-1 Culebra	Very low yield, and possible mixing with Rustler-Salado contact fluids.	Salt norm (Table 9, #1) suggests no or little mixing with Rustler-Salado fluid (Table 7, #1); norm even more alkali sulfate-rich than neighboring (H-2 and H-3) Culebra norms (Table 9, #2–6). Retained for this discussion.
P-15 R-S contact	Unit not productive; likely contaminated with Culebra fluids; unexpectedly low salinity.	Normative potassium sulfate (Table 9, #19) and low salinity abnormal. Omitted from this discussion.
P-15 Culebra	Similarities to Rustler-Salado water suggests contamination.	Concur. Omitted from this discussion.
P-17 Culebra	Packer leak suspected.	Salinities, norms, and Cl/Br different (Table 9, #32 and Table 7, #19); little mixing indicated. Retained for this discussion.
P-18 R-S contact and Culebra	Neither very productive; each possibly contaminated with well bore fluids.	Substantial salinity difference (Table 7, #21 and Table 9, #35); location for preserving primitive norms appropriate. Retained for this discussion.
WIPP-27 R-S contact, Culebra, and Magenta	Dropping water level and pump rate, and high iron content suggest prolonged contact with pipe (R-S). Culebra and Magenta not questioned.	Concur (R-S); very high potassium (Culebra and Magenta fluids (Table 9, #44 and Table 10, #16 and 17) suggests Mississippi mine outfall. Omitted from this discussion.
WIPP-29 Culebra	Not questioned.	Abnormally high salinity and high abundance of normative potassium salts (Table 9, #47–49) suggest IMC mine outfall. Omitted from this discussion.
WIPP-30 Magenta	No stability of field measurements.	May be due to continued dissolution along flow path. Retained for this discussion.

waters that obtained their solutes chiefly through carbonic acid hydrolysis of detrital silicate minerals in the weathering environment, such as the feldspars of the Gatuna Formation (Table 4); (3) dilute *sulfate-rich recharge* waters in which meteoric weathering waters were modified by interaction with Rustler anhydrite along a relatively halite-free flow path; and (4) variably concentrated *halite-rich resolution* brines in which meteoric waters were modified by interaction not only with Rustler anhydrite but also with Rustler and uppermost Salado halite.

Primitive-diagenetic waters

The primitive endmember brine is chloride-rich with appreciable quantities of normative carnallite and bischofite coexisting with halite and anhydrite-kieserite. The presence of anhydrite and halite intervals throughout the Rustler Formation, indicative of penesaline to hypersaline depositional conditions, requires connate water compositions that vary from normal seawater to that approaching bitter salt saturation, that is, brine compositions from A-1 to A-3 and B-1 to B-5 from the Yugoslavian salt pans in Table 12, and from samples W-64 to 40-4 from the Bahamian salt pans in Table 13. For these marine-evaporite-brine sequences halite abundance in the salt norm never exceeds 81 percent and is as low as 15–25 percent after halite precipitation. At the same time, kieserite abundance is always appreciably greater than that of anhydrite, and bischofite is always greater in abundance than carnallite (Tables 12 and 13). Primitive brines are also characterized by low Cl/Br ratios; seawater ($\sim$300) is about the maximum value, and values as low as 50 accompany halite precipitation (Table 13).

None of the fluids in the three Rustler Formation aquifers illustrate a truly pure, primary, primitive hypersaline brine. To be sure, some of the Rustler-Salado contact waters are hypersaline, contain normative halite abundance equal to or less than that in the seawater norm, exhibit the characteristic normative carnallite-bischofite association, and have a Cl/Br ratio of $\sim$300 or less. However, their norms contain no kieserite and always tachyhydrite, indicating that these brines have been modified to a varying degree by diagenetic fluid-rock reaction.

According to LAMBERT (pers. com., 1987), less productive horizons of wells that have been extensively treated with process fluids during development may derive solute $CaCl_2$ from the leaching of concrete behind the casing; $CaCl_2$ is a commonly used additive to cement mixes in the well service industry. This might be the case with P-18 Rustler/Salado (Table 7, no. 21) and Culebra samples (Table 9, no. 35).

The extent of diagenetic exchange through dolomitization can be reflected in the norm by loss of kieserite and appearance of tachyhydrite, with concomitant decrease in normative bischofite. The loss of normative kieserite ($MgSO_4$) can be expressed by

$$Mg^{2+} + SO_4^{2-} + 2CaCO_3$$
$$\text{brine} \qquad \text{calcite}$$

$$= CaMg(CO_3)_2 + CaSO_4 \qquad (1)$$
$$\text{dolomite} \qquad \text{anhydrite}$$
$$\text{(or gypsum)}$$

wherein calcium released by dolomitization is precipitated as calcium sulfate. After sulfate concentration is reduced and normative kieserite is eliminated the further exchange of calcium for magnesium through the reaction

$$Mg^{2+} + 2CaCO_3 = Ca^{2+} + CaMg(CO_3)_2 \qquad (2)$$

results in normative tachyhydrite (Mg_2CaCl_6), or eventually antarcticite ($CaCl_2$) at the expense of bischofite. With increased calcium concentration, additional loss of sulfate is required to maintain saturation with gypsum or anhydrite.

An example of a primitive hypersaline Rustler-Salado contact brine showing limited diagenesis is H-5 (Table 7, nos. 6 and 7); only enough Ca-Mg exchange has occurred to eliminate kieserite and produce minor tachyhydrite in the norm. In contrast, there has been sufficient Ca-Mg exchange in the Rustler-Salado contact brine from P-17 (Table 7, no. 20) for tachyhydrite to dominate the norm with only minor bischofite remaining. For both of these brines, anhydrite in the norm is low (0.1–0.5 percent), carnallite abundance is moderate (18.0–22.6 percent), and alkaline-earth chloride salts (68.0–69.4 percent) dominate the assemblages. When compared to the norms in Tables 12 and 13, these Rustler-Salado contact brines suggest a primary seawater evaporite brine that had progressed to the end of halite facies or possibly into the early stages of bitter salt deposition (though the P-17 fluid appears somewhat diluted).

Among Culebra fluids only the brine from P-18 (Table 9, no. 35) appears to be sufficiently concentrated to be classed primitive-diagenetic; its norm is tachyhydrite-rich with coexisting carnallite-halite-bishofite-anhydrite. Unfortunately no bromide determination was made. Except for salinity, the brine is very similar to that from the Rustler-Salado contact below (Table 7, no. 21), and thus might reflect contamination due to minimal purging of low productivity units (MERCER, 1983).

Alkali-bearing carbonate recharge

The other water types, the recharge waters, have a meteoric origin and entered the Rustler Formation by lateral migration and/or downward percolation from overlying strata. The solutes in these waters can be expected to reflect the extent and character of weathering reactions with framework minerals in the host rock and other mineral-rock interaction in the aquifer. Their isotopic compositions are consistent with a purely meteoric origin for their water molecules (LAMBERT and HARVEY, 1987).

The alkali-bearing water type gains solutes through dissolution of carbonate minerals and carbonic acid hydrolysis of framework silicates. A representative silicate hydrolysis reaction involving the albite component of plagioclase feldspar, for example, can be written as follows:

$$2NaAlSi_3O_8 + H_2O + 2H_2CO_3$$
$$\text{albite}$$

$$= Al_2Si_2O_5(OH)_4 + 4SiO_2 + 2Na^+ + 2HCO_3^-$$
$$\text{kaolinite} \qquad \text{quartz}$$
$$(3)$$

in which a kaolin-type mineral and quartz (or hydrous silica) are formed and sodium and bicarbonate are added to the solution. Similar reactions can be written for the hydrolysis of potash feldspar or mafic silicate minerals yielding dissolved potassium or magnesium with the bicarbonate in solution. Such reactions generate salt norms with alkali-bearing carbonate/bicarbonate salts, as exemplified by the norm for a water from Pleistocene sediment, the Gatuna Formation, deposited in a collapse structure at WIPP-15 (Table 11, no. 4) east of the site (Fig. 4). The presence of burkeite (Table 2) in the norm suggests either some sulfuric acid hydrolysis of silicates accompanying the oxidation of sulfide minerals or, more likely, dissolution of small quantities of calcium sulfate, most likely secondary gypsum, with subsequent precipitation of calcite. As might be expected the WIPP-15 fluid contains only a minor quantity of chloride, all as normative halite. The chloride most likely came from dissolution of atmospheric aerosol particles.

Sulfate-rich recharge water

Interaction of the alkali-bearing carbonate-type recharge water with Rustler Formation rocks that contained anhydrite, but from which halite had been dissolved would produce the *sulfate-rich recharge* type. With a carbonate-rich, sulfate-poor water, such as the WIPP-15 water (Table 11, no. 4), the principal interaction would be dissolution of anhydrite and precipitation of calcite through the reaction

$$2HCO_3^- + CaSO_4$$
$$= CaCO_3 + CO_2 + H_2O + SO_4^{2-} \quad (4)$$

that decreases carbonate salts in the norm and replaces them with sulfate salts. Because anhydrite/gypsum is considerably more soluble than calcite, calcium concentration in the fluid will be enhanced as dissolved carbonate is depressed.

Where only minor silicate hydrolysis has occurred in the weathering environment, anhydrite dissolution will give rise to salt norms with strikingly high abundances of anhydrite. Examples of this include the H-7, H-8, H-9, INDIAN, and SOUTH waters from the Culebra aquifer (Table 9, nos. 18–20, 21–23, 24–25, 56, and 62, respectively) and the GNOME-5 water from the Magenta aquifer (Table 10, no. 19). All contain more than 55 percent normative anhydrite, with the norm for Culebra water from SOUTH (Table 9, no. 62) containing a maximum observed normative anhydrite of 80 percent. These waters contain only small quantities of alkali-bearing sulfate salts in the norm.

Alternatively, if silicate hydrolysis in the weathering zone has been appreciable, such as for the WIPP-15 water (Table 11, no. 4), the dissolution of anhydrite in the Rustler Formation results in the replacement of carbonate in the water by sulfate and, rather than producing norms with abundant anhydrite, yields a normative sulfate salt assemblage with a large proportion of alkali sulfate. Extreme examples of this are the norms for the H-1 Culebra water (Table 9, no. 1) and the H-4 and H-5 Magenta water (Table 10, nos. 5 and 6) that contain normative aphthitalite and thenardite coexisting with glauberite and bloedite.

Halite-rich recharge water

The *halite-rich recharge* type of water is characterized by normative halite that substantially exceeds the 78 percent of the seawater norm (Table 3, no. 5, Table 13, Seawater and no. W-64). Examples of Rustler Formation recharge waters in which halite dissolution apparently played a dominant role include H-8, H-9, P-14, WIPP-25, WIPP-27, and GNOME-5 Rustler-Salado contact brines (Table 7, nos. 12, 13–14, 18, 23, 27–28, and 35, respectively), H-5, H-11, H-12, WIPP-30, and DOE-1 Culebra fluids (Table 9, nos. 11–14, 27, 28, 50, and 57), and the H-10 Magenta brine (Table 10, no. 12). The norms contain more than 85 percent halite (several more than 90 percent) and, as expected, these waters have Cl/Br ratios that range from 810 to 8,000, where measured. Cl/Br ratios far greater than 300, the approximate seawater value, reflect resolution of the characteristically low-bromine halites initially precipitated from marine evaporate brines.

Hypersaline halite-rich mixed recharge-primitive waters

Halite dissolution alone cannot destroy a meteoric signature. In other words, the absence of alkaline-earth chlorides and the presence of normative alkali-bearing sulfate salts in the norm may be preserved even though extensive halite dissolution occurs. This latter feature of the norm should only be diluted, not destroyed, during halite dissolution by recharge waters. Yet the above cited norms nearly always contain alkaline-earth-bearing chloride salts. Furthermore, sodium-bearing sulfate normative salts, except in one analysis of H-8 (Table 7, nos. 11–12), are absent from these assemblages and the only commonly occurring, alkali sulfate-bearing, normative salt is polyhalite. The presence of these characteristics is attributed to mixing of solutes picked up during halite dissolution into the recharge waters. Such dissolution releases intergranular and intragranular (fluid inclusion) solutes, which are added to the recharge water. Thus, brines from the Rustler Formation that are rich in halite all contain solutes characteristic of primitive fluids to a variable degree.

Mixing a sulfate-rich recharge water containing normative alkali-bearing sulfates with a primitive water containing normative bischofite and carnallite or a primitive-diagenetic fluid containing normative tachyhydrite and carnallite (with or without bischofite or antarcticite) produces considerable variation in the mixed-water norm. These brines are all highly saline; nearly all contain greater than 100,000 mg/l dissolved solids and many, particularly those from the Rustler-Salado contact zone, contain more than 200,000 mg/l.

Significant changes in the composition and salt norm of mixed hypersaline fluids can be produced by effective titration. For example, the alkali sulfate in the recharge waters mixing with the calcium chloride component of a primitive-diagenetic brine can result in precipitation of secondary calcium sulfate. This can be the origin of gypsum veins and pods in post-Salado strata. Similarly, effective titration of a normative alkali sulfate in the recharge water by normative magnesium and potassium chloride of a primitive brine can result in either direct precipitation of polyhalite or the replacement of primary anhydrite by polyhalite. The formation of polyhalite by fluid mixing may explain the origin of the polyhalite that SNYDER (1985) describes in some cores of the Tamarisk Member between the Magenta and Culebra aquifers. If the mixed fluid is taken to undersaturation, the components generated by the titration, halite and/or alkaline-earth sulfate, will appear in the salt norm. Only if all the other normative chloride salts of the primitive water are converted to halite can a sodium-bearing sulfate remain in the norm. The single example of this in the Rustler waters is one of two analyses of H-8 from the Rustler-Salado contact (Table 7, no. 12) in which the norm contains a very small fraction (0.3 percent) of bloedite with halite (94 percent) and the sulfates anhydrite-epsomite-polyhalite (5.5 percent). It is suggested that only a small increment of primitive brine mixed with the halite-dissolving recharge water; the extremely high Cl/Br ratio (8,000) in H-8 supports this interpretation.

Besides eliminating all normative sodium-bearing sulfate salts, a larger proportion of primitive brine in the mixed fluid will exchange potassium with calcium or magnesium to produce alkaline-earth sulfate and potassium chloride until all antarcticite, tachyhydrite, or bischofite are removed from the norm. However, carnallite-polyhalite is a stable salt-pair, and is the only instance in which the simple salts K_2SO_4-$MgCl_2$ coexist. Carnallite-polyhalite in the salt norm, or the K_2SO_4-$MgCl_2$ salt pair in the simple-salt assemblage, thus appears diagnostic of mixing marine-derived and meteoric waters. Many of the Rustler halite-dissolution fluids have this association in the norm. For example, four of the six Rustler-Salado contact brines and two of the five Culebra brines cited above contain the carnallite-polyhalite association in their norms.

Even larger proportions of primitive or primitive-diagenetic brines in the mixed fluid can result in retention of alkaline-earth chloride salts in the norm. The only indications of these fluids' mixed origins are the increased abundance of halite in the norm and a Cl/Br ratio in excess of 300. Halite-enriched, alkaline-earth chloride norms can also be attributed to a primitive-diagenetic fluid in which halite-resolution brines from gypsum dewatering have mixed with the primitive brine. Examples of

such waters include H-7 and GNOME-5 from the Rustler-Salado contact zone (Table 7, nos. 9–10 and 35), H-3, H-5, H-6, H-10, H-12, WIPP-28, DOE-1, and DOE-2 from the Culebra aquifer (Table 9, nos. 4–5, 11–14, 15–17, 26, 28, 45–46, 57, and 58), and H-10 from the Magenta aquifer (Table 10, no. 12).

Less saline mixed recharge-primitive waters

A number of samples from the Culebra and Magenta aquifers that suggest mixing of sulfate-rich recharge with primitive or primitive-diagenetic waters are relatively dilute (generally less than 25,000 mg/l total dissolved solids). These waters have dissolved negligible to moderate amounts of halite, and are confined to the Culebra and Magenta aquifers because of omnipresent salt underlying the Rustler-Salado contact. Many of these waters yield norms that are difficult to interpret unequivocally, particularly those with relatively low normative chloride salt content.

For example, H-4 from the Culebra aquifer (Table 9, nos. 7–10), with approximately 20,000 mg/l total salinity, contains normative alkali-bearing sulfate salts and about 60 percent normative halite. The water appears to be essentially a recharge fluid. However, the Cl/Br ratio ranges from 167 to 192, which is too low to account for by halite resolution alone. Thus the normative halite concentration must result from mixing with primitive brine. This requires that the recharge water had sufficient normative alkali sulfate to eliminate the normative alkaline-earth chloride of the primitive brine and still not destroy the alkali-bearing sulfate signature of the recharge water's norm. In this way, the apparent conflict between the normative assemblage and the Cl/Br ratio can be reconciled.

With other examples, further contradictions develop. Samples from the Culebra in hole P-14 (Table 9, nos. 30–31) with total salinites of 34,000 mg/l (no. 30) and 25,000 mg/l (no. 31) yield salt norms with 60 and 45 percent halite (nos. 30 and 31, respectively). The available Cl/Br ratio is 201 (no. 31). Thus the primitive-diagenetic brine characteristics are preserved both in the salt norm and in the bromide data; dilution is apparently the result of either mixing with very dilute recharge water or with earlier diagenetic fluids from gypsum dewatering. Its water molecules, however, are isotopically meteoric (LAMBERT and HARVEY, 1987). In addition, WIPP-26 from the Culebra aquifer, with total salinity varying from 15,000 mg/l to nearly 18,000 mg/l in three analyses (Table 9, nos. 40–42), contains abundant tachyhydrite and antarcticite (9.7

to 10.2 percent) in the salt norm. This fluid, however, has a Cl/Br ratio that ranges from 2,250 to 2,420. Again, the water molecules are isotopically meteoric (LAMBERT and HARVEY, 1987).

These three similar moderately saline fluids from the same stratigraphic horizon show both internal and comparative inconsistencies: one (H-4) with a characteristic halite-dissolution recharge salt norm but a bromide content that suggests a primitive fluid; a second (P-14) with a strong primitive-diagenetic signature in both the salt norm and the bromide quantity; and a third (WIPP-26) with a salt norm that suggests a primitive-diagenetic fluid but a Cl/Br ratio and normative anhydrite that suggests extensive evaporite resolution by a recharge brine. We can only suggest that the apparent inconsistencies are related to the extent and character of silicate hydrolysis and evaporite resolution in the recharge water, which in turn, probably reflects permeability and flow path in strata overlying the Culebra. For the three sites discussed here, the post-Culebra de-

tritus becomes thinner from H-4 westward, fractures are more likely close to the Nash Draw scarp (P-14), and extensive fracturing occurs in the Draw itself (WIPP-26); all of which are consistent with the nature of the mixtures proposed.

Areal distribution of salt norms from Rustler Formation fluids

The areal distribution of salt norms for waters from the three Rustler aquifers is shown by maps with simple-salt rose diagrams representing the salt norms of the water composition at each location (Figs. 10, 11, and 12). Because of the closely spaced boreholes within the WIPP-site boundary (Fig. 9), smaller scale diagrams are presented for a few samples which are located only approximately in order to avoid overlapping rose plots. We excluded from the maps samples that we regard of questionable validity or severely perturbed by anthropogenic activity (Table 14).

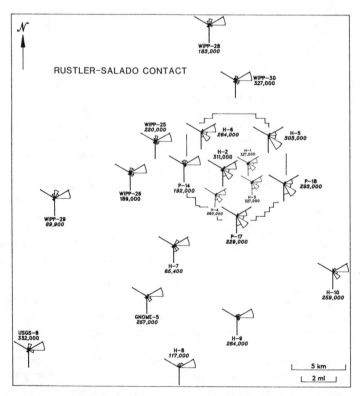

FIG. 10. Map of the WIPP-site area with simple-salt rose diagrams at sites (Fig. 9) of analyzed waters from the Rustler-Salado contact aquifer (the sample numbers plotted are identified in Tables 7 and 8 with an asterisk). For most of the rose diagrams, the intersection of axes is plotted exactly at the sample site. To avoid overlap, a few diagrams have been made smaller and plotted near the corresponding sample sites, given exactly in Fig. 9. The criteria for interpreting rose diagrams are presented in Figs. 2 and 3.

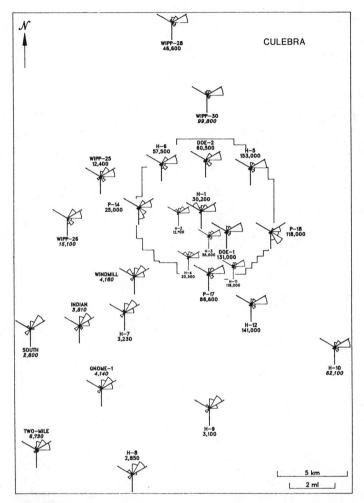

FIG. 11. Map of the WIPP-site area with simple-salt rose diagrams at sites (Fig. 9) of analyzed waters from the Culebra aquifer of the Rustler Formation (the sample numbers plotted are identified in Table 9 with an asterisk). For most of the rose diagrams, the intersection of axes is plotted exactly at the sample site. To avoid overlap, a few diagrams have been made smaller and plotted near the corresponding sample sites, given exactly in Fig. 9. The criteria for interpreting rose diagrams are presented in Figs. 2 and 3.

Rustler-Salado contact zone waters

The salt norms and total salinities of the Rustler-Salado contact fluids reflect zonation of chloride-rich brines east to west from a primitive-diagenetic type, through mixed primitive-recharge fluids, to waters dominated by halite resolution recharge.

Primitive-diagenetic normative assemblages characterize brine from the eastern central portion of the WIPP-site area, including H-5 and H-6 to the north and west, H-1, H-2, H-3, and P-18 in the central and eastern areas and P-17 and H-4 to the south (Fig. 10). Each Rustler-Salado contact brine within this area is hypersaline, contains less than

70 percent normative halite, and moderate to high quantities of normative tachyhydrite (2 to 31 percent) and bishofite (5 to 67 percent). The sum of normative bischofite and tachyhydrite always greatly exceeds carnallite, and the Cl/Br ratio ranges from 56 to 183. These fluids appear to be modified from brines that could have accompanied deposition of the halite or even of the initial bitter salts in the uppermost Salado Formation. The deuterium and oxygen-18 content for these brines also deviate significantly from the meteoric water relationship. The projection of a trend in isotope values and normative composition for these sample sites suggests an increasing primitive component eastward.

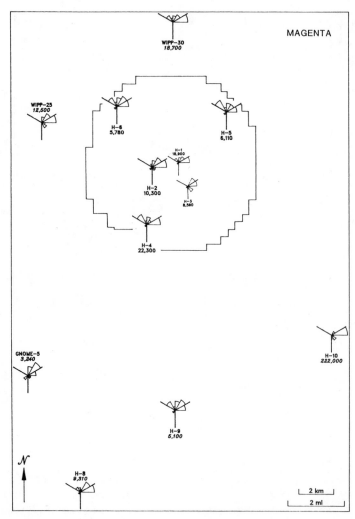

FIG. 12. Map of the WIPP-site area with simple-salt rose diagrams at sites (Fig. 9) of analyzed waters from the Magenta aquifer of the Rustler Formation (the sample numbers plotted are identified in Table 10 with an asterisk). For most of the rose diagrams, the intersection of axes is plotted exactly at the sample site. To avoid overlap, a few diagrams have been made smaller and plotted near the corresponding sample sites, given exactly in Fig. 9. The criteria for interpreting rose diagrams are presented in Figs. 2 and 3.

South and west of the WIPP site are Rustler-Salado contact brines characterized by moderate to high salinity, and a primitive-diagenetic, normative salt assemblage (anhydrite, carnallite, bischofite, generally tachyhydrite, and little or no kieserite), but with a pronounced excess of halite (84 to 94 percent) over the seawater norm. This strongly suggests mixing of recharge waters that dissolved substantial halite with a primitive-diagenetic fluid. This group is represented by H-7, GNOME-5, and H-10 (Fig. 10). The extent of apparent diagenesis among these waters is highly variable. Tachyhydrite

in the norm ranges from none in H-10 and 0.6 percent in GNOME-5 to 6.5 percent (with 2.6 percent anarcticite) in H-7 (Table 7, nos. 15, 35, and 9, respectively). This range can develop through two disparate processes. One is a great variation in the exchange of solute calcium for magnesium accompanying dolomitization (note the Bardawil brines, LEVY, 1977, as cited by BODINE and JONES, 1986). The second is calcium precipitation as gypsum when a calcium-rich brine (primitive-diagenetic) mixes with an alkali sulfate-rich fluid (recharge). We suspect both processes played a role. Alternatively, one or more of these fluids may have evolved solely as primitive-diagenetic brines in which gypsum dehydration provided fluid for halite resolution, and dolomitization eliminated kieserite and added tachyhydrite to the norm.

The brine of H-10 (Table 7, nos. 15 and 16) is an enigma. It has high normative halite that characterizes halite resolution, but has a Cl/Br ratio of only 186, suggestive of a primitive fluid. Our only suggestion is that a low chloride sodium-rich fluid, such as an alkaline evaporite water from an ephermeral surface playa, mixed with a calcium chloride-bearing, primitive-diagenetic brine with resultant precipitation of calcium and increase in normative halite.

To the north, west, and south of the WIPP site there is a progressive increase in the proportion of halite-resolution recharge water mixed with primitive-diagenetic waters. In the Rustler-Salado contact brines at WIPP-30, WIPP-25, P-14, WIPP-26, and H-9 (Fig. 10) normative halite (83–95 percent) coexists with normative polyhalite-carnallite (Table 7, nos. 33–34, 22–23, 17–18, 24–46, and 13–14, respectively). Such an assemblage is, as discussed above, good evidence for mixing of primitive-marine and recharge waters. All of these fluids have a high Cl/Br ratio (greater than 1,900) which along with the high normative halite abundance, indicates considerable halite resolution by the recharge water component. In the sample sites furthest north, west, and south of WIPP site at WIPP-28, WIPP-29, USGS-8, and H-8 (Table 7, nos. 29–30, nos. 31–32; Table 8, nos. 1–3; Table 7, nos. 11–12, respectively) all alkaline earth-bearing chlorides are absent from the norms and at WIPP-29 and H-8 small amounts of normative alkali-bearing sulfates present an added recharge signature. The normative halite abundance (84–94 percent) and high Cl/Br (greater than 2,800) indicate an extensive proportion of resolution halite in this group of waters.

It is not surprising that the Rustler-Salado contact fluids are highly saline; none contain less than 60,000 mg/kg total dissolved solids. This cannot be explained by halite dissolution in the overlying Rustler Formation for waters west of the halite dissolution front (Fig. 6). We envision recharge waters percolating downward or laterally through the Rustler Formation attaining their sulfate-rich character by interaction with Rustler anhydrite but, for the most part, not becoming chloride-rich until reaching the Rustler-Salado contact interval. Thereupon the recharge fluids dissolve Salado Formation halite that floors the Rustler-Salado contact interval and mix with any available uppermost Salado primitive-

diagenetic brines and their solutes. Presumably uncontaminated primitive-diagenetic fluids are only preserved where tight masses of halite remain in the overlying Rustler Formation so as to inhibit downward percolation into the Rustler-Salado contact interval.

In Table 8 we have listed analyses and salt norms of Rustler and Salado waters from the "brine aquifer" that discharge along the Pecos River or have been retrieved by drilling. The geographic relationships between the "brine aquifer" (the best developed water-bearing zone associated with the Rustler-Salado contact), Nash Draw, Laguna Grande de la Sal (including Surprise Spring at its north end) and Malaga Bend, the discharge point for brines entering the Pecos River are shown in Fig. 7. These waters are highly saline and, except for greater concentrations of potassium, yield norms that are similar to the mixed halite-resolution recharge and primitive-diagenetic norms that are characteristic of the Rustler-Salado contact zone. We attribute the high potassium content, reflected in normative kainite and leonite (Table 8, nos. 1–5) in the Malaga Bend area, or in normative sylvite-carnallite-polyhalite in the Laguna Grande de Sal water (Table 8, no. 6), to contamination with outfall from the potash mining facilities to the north and east. The substantial increase in salinity of Surprise Spring water from 1938 to 1975 (Table 8, nos. 6 and 7) is most likely also the result of contamination from the potash operations. The deuterium and oxygen-18 values for the latest Surprise Spring sample is consistent with partial evaporation of waters imported for mine processing (LAMBERT and HARVEY, 1987).

Culebra waters

Salt norms and total salinites present an entirely different range of features for fluids from the Culebra aquifer. Salinities rarely exceed 100,000 mg/l and are frequently less than 10,000 mg/l. The salt norms are often dominated by sulfates and sodium-bearing sulfate salts are common. Furthermore, neither the salt norms nor the total salinities of these waters (Fig. 11) yield as distinct an areal pattern as do those for the Rustler-Salado contact brines.

In contrast to the results for Rustler-Salado contact fluids, only two Culebra norms can be readily interpreted as dominantly primitive-diagenetic. The sample from P-18 (Table 9, no. 35), comes from the eastern margin of the WIPP site (Fig. 11), and, based on the absence of extensive halite dissolution in the overlying strata of the Rustler Formation (Fig. 6), its normative character is not unexpected, despite problems with small sample noted previously. Although no bromide data were obtained for this fluid, the abundances of tachyhydrite, bischofite, and carnallite (33, 21, and 24 percent respectively) are characteristic of a well-developed primitive-diagenetic assemblage. The norm of the other primitive-diagenetic Culebra fluid, P-14 (Table 9, no. 31) is somewhat different in that antarcticite, tachyhydrite, and carnallite (24, 20, and 1 percent, respectively; we doubt the validity of the low potassium concentration) constitute the alkaline earth-bearing chloride salts. The brine also has a Cl/Br ratio of 201, a primitive-diagenetic value. According to LAMBERT and HARVEY (1987), however, the isotopic composition of the water is clearly meteoric. Such a fluid in such a locality, characterized by high permeability, a meteoric isotope composition, and isolation from other primitive-diagenetic solute signatures, is decidely anomalous.

Two features of the chemistry of these Culebra waters are noteworthy. First, both brines are well undersaturated with halite or any other chloride salts; the sample from P-18 has a total salinity of 118,000 mg/l and that from P-14 has only 25,000 mg/l. Either these waters were severely diluted with recharge waters that underwent negligible reaction in the weathering regime, or the dilution occurred much earlier as a result of diagenetic dewatering of the primary bedded gypsum. The meteoric isotope signatures of both samples make the latter result very unlikely.

The second feature, restricted to the P-14 sample, is the markedly high Ca/Mg ratio when compared with the several primitive-diagenetic Rustler-Salado contact brines or the Culebra P-18 sample. In the P-14 sample, normative antarcticite is not only present, but its abundance is greater than that of tachyhydrite. Normative antarcticite does not occur in the associated Rustler-Salado contact brine or in the P-18 fluid from the Culebra. This may be the result of extensive dolomitization (or Mg-phyllosilicate authigenesis) in or near the Culebra aquifer by the P-14 fluid. This contrasts with the absence of bedded carbonate in the Rustler-Salado contact interval, and therefore, less opportunity for calcium-magnesium exchange between the fluid and the host rock, resulting in lower calcium concentrations in the fluid. The P-18 fluid from the Culebra contains no normative antarcticite and may not have been in contact with appropriate materials in or near the aquifer to effect such extensive exchange. The areal uniformity and carbon-13 content of both Culebra and Magenta dolomites suggest that they evolved not from alteration of limestone, but as evaporative dolomites. In contrast, evidence of dedolomitization

in the presence of Rustler-like meteoric water has been found in the Magenta at WIPP-33, between the WIPP site and Nash Draw (LAMBERT and HARVEY, 1987).

Eight Culebra samples from H-3, H-5, H-6, H-10, H-12, P-17, DOE-1, and DOE-2 (Table 9, nos. 4–6, 11–14, 15–17, 26, 28, 33–34, 57, and 58) give salt norms qualitatively indicative of primitive-diagenetic origin. However, the normative halite abundances (80–88 percent) and Cl/Br ratios (674–2,800) significantly exceed the seawater values. Each is also quite saline (55,000–153,000 mg/l). Three alternative origins can be ascribed to these waters. The first is that a primitive-diagenetic fluid mixed with a halite-resolution recharge water. The second is that, like the origin proposed for P-18, early gypsum dehydration water diluted a primitive halite-precipitating brine, but in this case, effected considerable halite resolution. The third is that the original primitive water was undersaturated with halite, and dissolved halite after migrating to or into a neighboring bedded halite. The relatively high amount of normative carnallite, as contrasted with bischofite-tachyhydrite abundance, in most of these waters (all but H-6 and DOE-2) favors the first alternative, that is, mixing of the primitive-diagenetic brine with a halite-resolution recharge water. During the mixing process, alkali sulfate salts in the recharge water and abundant calcium in the primitive-diagenetic water interact to precipitate secondary gypsum and to decrease the normative alkaline-earth chlorides in the fluid. It appears plausible that all samples but H-6 and DOE-2 involved mixing with recharge waters, whereas the latter two brines evolved through an earlier diagenetic process of halite resolution accompanying diagenetic gypsum dewatering. However, the stable-isotope data indicate that all these waters are meteoric in origin (LAMBERT and HARVEY, 1987).

Two Culebra samples from WIPP-25 and WIPP-26 (Table 9, nos. 36–39 and 40–42) have normative halite quantities significantly less than seawater (63–65 percent), but have Cl/Br ratios well in excess (2,000–4,000) of the seawater value. Furthermore, they contain abundant normative alkaline-earth sulfate salts, chiefly anhydrite with minor kieserite in WIPP-25, and are unusually dilute (12,000–15,000 mg/l) for waters with carnallite-bischofite norms. These characteristics suggest substantial simple dissolution of calcium sulfate (and perhaps some dolomite) followed by dilution of primitive-diagenetic fluids. Isotopically, these waters are also of meteoric origin (LAMBERT and HARVEY, 1987).

The remaining Culebra waters with total salinity greater than 10,000 mg/l are H-1, H-4, H-11, WIPP-

28, and WIPP-30 (Table 9, nos. 1, 7–10, 27, 45–46, and 50–51). Three of these, H-11, WIPP-28, and WIPP-30, with 47,000 to 118,000 mg/l contain the diagnostic carnallite-polyhalite association indicating mixing of primitive-diagenetic fluids with sulfate-resolution recharge water. The high halite contents in the norms of these three waters (84–91 percent) and the high Cl/Br (1,400–3,400) point toward some halite resolution by the recharge waters. H-3, on the other hand, is less saline (30,000 mg/l) and contains normative thenardite and aphthitalite with 68 percent normative halite. This water is a typical anhydrite-resolution recharge water that has lost calcium to mixing with sodium carbonate derived from silicate hydrolysis, and has also dissolved minor amounts of halite. H-4, with 61 percent normative halite coexisting with the bloedite-glauberite association, also has an anhydrite-resolution recharge signature. The fluid's Cl/Br ratio is only 176, suggesting mixing of a primitive-diagenetic fluid with an anhydrite-resolution recharge water that contained negligible chloride. Again, the water molecules appear isotopically meteoric (LAMBERT and HARVEY, 1987).

Of the 13 Culebra sites with dilute waters, 10 (H-2, H-7, H-8, H-9, ENGEL, GNOME SHAFT, GNOME-1, GNOME-4, WINDMILL, and SOUTH in Table 9, nos. 2–3, 18–20, 23–23, 24–25, 29, 52, 53, 54, 60–61, and 62) yield norms containing sodium-bearing sulfates, either bloedite or the bloedite-glauberite pair, and halite abundances from 1.2 to 67 percent. The TWO-MILE (Table 9, no. 59) well, Culebra water, contains a polyhalite-carnallite association with 54 percent halite in the norm. Two wells (GNOME-8 and INDIAN in Table 9, nos. 55 and 56) yield Culebra waters with a normative carnallite-bischofite association and 36 and 22 percent halite, respectively. All 13 fluids must have been sodium carbonate-bearing recharge waters that have interacted with Rustler anhydrite, deposited secondary calcite, and, for those with greater than 15–25 percent chloride salts in the norm, redissolved minor remnants of Rustler halite.

Two of these 13 sites (H-8 and SOUTH) yielded waters with exceptionally low chloride content, less than 4 percent halite in otherwise chloride salt-free norms, suggesting recharge waters did not encounter any primitive-diagenetic fluid, nor participate in halite resolution accompanying migration into the Culebra aquifer. The very high abundance of anhydrite in each norm strongly suggests extensive dissolution of Rustler anhydrite.

The conclusions from the area distribution of the norms of the Culebra waters (Fig. 11) are (1) that

the distribution of salt norms is compatible with the flow field presented by DAVIES (1989), and (2) that the Culebra Dolomite aquifer has extremely variable hydrologic properties. The basis for the first conclusion is that the one primitive-diagenetic, but somewhat diluted brine with a dominantly tachyhydrite-bischofite-carnallite norm and a very low Cl/Br ratio (P-18) occurs where the Rustler Formation is the least permeable; bedded halite is preserved above both the Culebra and Magenta members. P-18 is surrounded by sampling sites where the Culebra waters are moderately saline with norms yielding abundant halite (greater than 80 percent) coexisting with the carnallite-bischofite assemblage, with (H-6 and DOE-2), or without tachyhydrite (H-3, H-5, H-10, H-12, P-17, and DOE-1). These associations suggest mixing with anhydrite-halite-resolution recharge waters, while preserving the alkaline-earth chloride salt(s) in the normative assemblage. Two Culebra brines in Nash Draw north of the WIPP site (WIPP-28 and WIPP-30) contain normative polyhalite-carnallite, indicating a larger fraction of anhydrite-halite-resolution recharge water mixing with primitive-diagenetic fluids. Finally, the large area of dilute Culebra water with alkali sulfate-bearing norms south and west of the WIPP site clearly reflects (1) the paucity of halite above the Culebra, (2) significant infiltration of recharge waters, and (3) relatively rapid flow to the southwest. The normative and solute concentration patterns in Fig. 11 actually fit well with DAVIES (1989) flow-field simulation without vertical influx.

The variability of the Culebra aquifer's hydrologic properties (conclusion 2, above) is illustrated by the great diversity of normative characteristics for its fluids, such as the unusual preservation of a severely diluted primitive-diagenetic brine at P-14 and its apparent mixing to the west with anhydrite-halite-resolution recharge or halite-resolution gypsum-dehydration waters at WIPP-25 and WIPP-26. It is not at all clear what hydrologic conditions have preserved such normative compositions in an area where all Rustler halite is missing (Fig. 6), and where the primitive-diagenetic brine (P-14) is clearly separated by unequivocal resolution waters at H-1, H-2, and H-3 from the nearest primitive-diagenetic brine (P-18). It is also not clear what hydrologic features allow a closely spaced group of sites (H-1, H-2, and H-3) to yield waters of such vastly different salinities (8,920 to 55,000 mg/l) and normative character (thenardite-aphthitalite association vs. carnallite-bischofite association). Certainly hydrologic connections must be minimal. Evidence for significant vertical infiltration is given in the norms of dilute waters in the southwest quadrant of Fig.

11. These norms are relatively sulfate-rich and chloride-poor, but range in normative mineralogy from carnallite-bischofite (INDIAN) to abundant bloedite (WINDMILL). This strongly suggests recharge of the aquifer by infiltration of weathering waters. Local variations in the extent of silicate hydrolysis, interaction with Rustler anhydrite, and dissolution of isolated remnants of Rustler halite, rather than evolution along a well-defined flow path, best explain the random variations of the salt norms in this area.

Magenta waters

Waters from the Magenta member of the Rustler Formation were obtained at far fewer sites than either the Rustler-Salado contact-zone brines or the Culebra fluids because the Magenta horizon is along the eastern margin of Nash Draw, or removed by erosion within Nash Draw and throughout much of the area southwest of the WIPP and GNOME sites (Fig. 9).

No Magenta waters have solutes indicating a predominantly primitive origin. The only highly saline water (222,000 mg/kg) is H-10 (Table 10, no. 12) where the norm has tachyhydrite, bischofite, and carnallite, but also 93 percent halite. The Cl/Br ratio is 860. This is a typical mixture of a halite-rich recharge fluid and a primitive-diagenetic brine. Two other much more dilute waters from H-3 and WIPP-25 (Table 10, nos. 4 and 14) with 8,600 and 12,500 mg/l total dissolved solids, have qualitatively primitive norms characterized by bischofite, carnallite, anhydrite, and either tachyhydrite (H-3) or kieserite (WIPP-25), but in both the Cl/Br ratio is high (574 and 2,600, respectively). The low halite abundance is due to high anhydrite content, rather than abundant alkaline earth-bearing chloride. Thus, each appears to be a mixture of an anhydrite-halite resolution recharge water with a primitive-diagenetic fluid.

Two other wells (H-2 and GNOME-5; Table 10, nos. 2 and 19) have Magenta waters with slight primitive or primitive-diagenetic normative characteristics. The sample from H-2 with a total salinity of 10,000 mg/l contains small quantities of carnallite and polyhalite and 65 percent halite in the norm. The GNOME-5 fluid with 3,200 mg/l total dissolved solids contains 1.1 percent carnallite, 15 percent kieserite, only 12 percent halite, but 69 percent anhydrite in the norm. Bromide content was not determined in either water. Both waters are very dilute so that it is difficult to be certain of any primitive-diagenetic contribution.

The remaining 7 Magenta waters show substantial quantities of sodium-bearing alkali salts in the norm and are clearly dominated by anhydrite-resolution recharge waters with some halite resolution. In two samples, H-4 and H-5 (Table 10, nos. 5 and 6) silicate hydrolysis was sufficient to yield considerable thenardite (17 percent in both) and some aphthitalite (1.7 and 2.5 percent, respectively) in their norms. The moderate amounts of normative halite (55 and 24 percent, respectively) and low salinities (22,000 and 6,100 mg/l) indicate moderate to minor halite resolution; unfortunately bromide data for both H-4 and H-5 are lacking. The water from 4 of the remaining 5 sites (H-1, H-6, H-9, and WIPP-30 in Table 10, nos. 1, 7, 10–11 and 18) contains abundant normative bloedite (9–35 percent), glauberite and anhydrite. H-8 (Table 10, no. 9), on the other hand, contains the glauberite-syngenite-polyhalite assemblage of alkali-bearing sulfate salts. Salinities in these waters range from 5,000 to 19,000 mg/l with normative halite varying from 25 to 72 percent. Bromide determinations are reported for three of the waters (H-8, H-9, and WIPP-30) with Cl/Br ratios of 4,300, 300 and 3,800, respectively. While the value for H-9 appears anomalous, the low normative halite abundance (25 percent) and the low salinity (5,000 mg/l) suggest that the Cl/Br ratio probably reflects generation in the weathering environment. As expected, all Magenta waters analyzed for deuterium and ^{18}O gave meteoric values (LAMBERT and HARVEY, 1987).

The areal distribution of normative data for fluids from the Magenta aquifer (Fig. 12) locates the only highly saline water (H-10) to southeast of the WIPP site in an area in which bedded halite presumably remains in the strata overlying the Magenta dolomite (Fig. 7). This water is a mixed halite-resolution recharge and primitive-diagenetic fluid. No dominantly primitive-diagenetic Magenta fluid has been observed within the WIPP-site area.

The remaining Magenta waters are relatively dilute (less than 23,000 mg/l) and represent variable but lesser proportions of primitive-diagenetic solutes as compared to anhydrite-dissolution recharge solutes. Halite resolution is negligible or minor. As with the Culebra fluids, no definitive areal pattern occurs in the norms of these dilute Magenta waters, and much recharge is probably from vertical infiltration into the aquifer. For example, the norms for the closely spaced cluster of wells, H-1, H-2, and H-3 within the WIPP site, show considerable variation (Table 10, nos. 1, 2, and 3–4, and Fig. 12). H-3 has a strong primitive-diagenetic signature (normative tachyhydrite-bischofite), with a lesser fraction of anhydrite-resolution recharge. H-2 shows a greater fraction of anhydrite-resolution recharge fluid (a polyhalite-carnallite association), whereas H-1, with normative bloedite-anhydrite, is characteristic of a water that dissolved minor amounts of halite and is dominated by anhydrite-resolution recharge solutes.

Other variations in the norms of Magenta fluids within the WIPP-site area, particularly the isolated occurrence of the WIPP-25 water with its $MgCl_2$-$CaSO_4$ character and evidence of halite resolution (Table 10, nos. 13–15 and Fig. 12) further document the variability of the aquifer. Note that this norm and individual solute values are very similar to those of the underlying Culebra sample (Table 9, no. 38). Taken with very similar isotopic composition (LAMBERT and HARVEY, 1987) and hydraulic properties (MERCER, 1983), this indicates that the Magenta and Culebra aquifers are interconnected at WIPP-25. This is one of the few places where there is such an interconnection. Here DAVIES (1989) has demonstrated the dissipation of a large pressure differential between the Magenta and the Culebra, which results from dissolution and fracture-induced hydraulic conductivity related to, but east of, Nash Draw.

Salt norm variations among Rustler Formation aquifers

Because the distribution of Rustler Formation salt norms in the WIPP-site area, particularly the Culebra and Magenta norms, indicate highly variable aquifer properties and hydrologic connections with overlying strata, variations in salt norms were examined among the three aquifers at each site where more than one Rustler Formation horizon was sampled. These comparisons are shown on Fig. 13 for eighteen sampling localities.

The general features are well illustrated. Most sites show a progressive drop in fluid salinities upward from the Rustler-Salado contact zone through the Culebra aquifer and into the Magenta Dolomite. Secondly, a progression upward in the norms at most sites represents the succession from primitive-diagenetic to anhydrite-resolution recharge with minor halite resolution characteristics. Waters from H-5 (Fig. 13), for example, illustrate both trends over broad ranges. The Rustler-Salado contact sample from this well is hypersaline (303,000 mg/l) and is dominated by primitive-diagenetic solutes yielding a bischofite-tachyhydrite salt norm with minor halite. The overlying Culebra water is substantially less saline, (153,000 mg/l), with a bis-

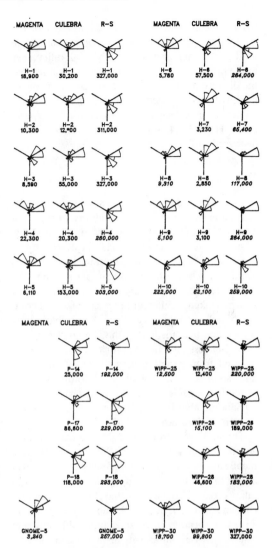

FIG. 13. Simple-salt rose diagrams comparing waters from the Magenta, Culebra, and Rustler-Salado Formation contact aquifers for those sites in the WIPP-site area with analyzed samples from more than one horizon (the samples plotted are identified in Tables 7, 9, and 11 with an asterisk). The criteria for interpreting rose diagrams are presented in Figs. 2 and 3.

chofite-carnallite association and 88 percent halite in a norm that is diagnostic of mixed halite-dissolving primitive-recharge fluid. The uppermost Magenta water, on the other hand, is dilute (6,100 mg/l), with the thenardite-aphthitalite pair and 23.7 percent halite in the norm. This is characteristic of an anhydrite-resolution-recharge fluid that has picked up little primitive brine, and has dissolved only minor amounts of halite.

Although about half the Rustler waters do not

strictly follow both these trends, a number of the anomalies, particularly when comparing dilute fluids, are minor, and reflect only small differences in the proportion of recharge-primitive solutes, halite dissolution, and/or extent and character of silicate hydrolysis effected by the recharge fluid. Those sequences that deviate markedly from the general trend, H-10, P-14, WIPP-25, and WIPP-26, were discussed earlier.

Salt norm variations with time

LAMBERT and ROBINSON (1984) discussed in detail the extended chemical instability in the sampling of brines from the drill holes in Nash Draw. A significant portion of this chemical change can exceed analytical error and is most readily attributed to mixing of brines from different parts of the sampling interval. Thus a comparison of salt norms for analyses taken at different times can highlight compositional changes suggestive of variable solute source and perhaps flow path. The great majority of and largest time interval for these comparisons resulted from recent systematic recollection and analysis of brines in the Culebra Dolomite Member of the Rustler Formation that were provided to us by Sandia Laboratories (ROBINSON and LAMBERT, written communication, 1987).

Three samples from the Rustler-Salado contact aquifer in Nash Draw (WIPP-25, 29, and 30) were recollected and analyzed in different laboratories at a four month interval with significant increases in salinity, but without an essential change in chemical character (Table 7, nos. 22–23, nos. 31–32, and nos. 33–34, respectively). A substantial dilution and sulfate increase contrast with an increase in alkaline-earth chloride in the only example of recollection and analysis after a long interval (slightly over 8 years) in the Magenta dolomite aquifer (H-3 in Table 10, nos. 3–4). These results suggest increased mixing of gypsum dissolution with residual solutes such as found in the Culebra at P-14.

Compositional variation in the Culebra brines with time of collection can be separated into four basic types. These types assist in categorizing variation in chemical composition with degree of evaporite dissolution or leaching with respect to area, stratigraphic interval, and/or time. A shift in the normative assemblage toward increased alkaline-earth chloride indicates a greater fraction of primitive-diagenetic brine, and characterizes the slight change in TDS over 4 years in H-6 (Table 9, nos. 15–16), over 6 years in WIPP-25 (Table 9, nos. 38–39), and over 5 years in WIPP-26 (Table 9, nos.

41–42). The largest shift of this type is seen in P-14 over 9 years (Table 9, nos. 30–31; we are somewhat puzzled by and suspicious of the data for potassium, only 37.9 mg/l, reported for the 2/26/86 sample, no. 31), but the total salinity actually decreased by 25 percent. This decrease might reflect a smaller proportion of halite-resolution brine. A second type of normative compositional change is the increase of NaCl, probably associated with an increase in salinity, as illustrated by H-2 collection over a 9-year interval (Table 9, no. 2–3). This may be related either to simple resolution of halite or to titration of an alkaline-earth chloride component of the brine by sodium carbonate, with precipitation of secondary calcite. A change in the salt norm toward increased alkaline-earth sulfate alone, the third type, can accompany either a slight increase in salinity, as represented by H-4 and H-5 (Table 9, nos. 7–8 and nos. 11–12), or a slight decrease in salinity, as shown by H-9, and P-17 (Table 9, nos. 24–25 and nos. 33–34), and probably reflects gypsum resolution or titration of a calcium chloride component by sodium sulfate, with precipitation of gypsum (or anhydrite). Finally, an increase in normative alkali sulfate should be expected to accompany further dilution of a Culebra fluid by sodium carbonate-bearing recharge waters, such as the WIPP-15 water from Pleistocene sediment (Table 11, no. 4), and additional precipitation of secondary calcite. The only slight, but rather inconclusive indication of this process is given in the comparison of the H-9 analyses (Table 9, nos. 24–25) separated by a 5 year interval.

In conclusion, however, it should be re-emphasized that the changes in chemical composition for well samples over extended intervals of time are just as likely to be due to sampling techniques and well condition as to the time itself.

SUMMARY AND CONCLUSIONS

The salt norms and derivative simple-salt assemblages calculated for 124 water analyses from the three aquifer units of the Ochoan (Permian) Rustler Formation (plus 4 analyses from overlying strata) at or near the Waste Isolation Pilot Plant (WIPP) in Southeastern New Mexico indicate that Rustler waters range from hypersaline, marine-like, primitive-diagenetic fluids to dilute meteoric recharge waters. The chemical compositions of these waters can be attributed to variable mixing of solutes from the following sources: 1) the intergranular or fluid inclusion retention of primary seawater-evaporative bitterns; 2) diagenetic reaction of brines with precursor carbonate or silicate sediments, principally the exchange of Ca for Mg; 3) influx of meteoric water, with and without a significant solute-weathering component; and 4) dissolution of evaporite minerals, primarily halite and anhydrite.

The Rustler Formation fluids can be classified according to principal normative component into four basic types, generally in order of decreasing salinity and depth: 1) primitive-diagenetic fluids, characterized by alkaline-earth chloride; 2) NaCl-rich resolution brines; 3) calcium sulfate-dominated recharge waters; 4) alkali carbonate-bearing recharge waters. The recharge waters have entered the Rustler aquifers through overlying strata or laterally. Their norms reflect negligible to extensive resolution of Rustler or Salado halite, pervasive resolution of calcium sulfate in the Rustler Formation throughout the region, and silicate hydrolysis of framework minerals in clastic sediments both within and overlying the Rustler.

The primitive-diagenetic brines are most prevalent in the contact zone of the Rustler with the underlying Salado halite, particularly for samples from wells east of the central part of the WIPP site where transmissivities and flow rates in the Rustler-Salado contact zone are very low (MERCER, 1983). In fact, there is a general zonation of high chloride brines from the most alkaline-earth rich in the east to the most sodium-rich (including some sulfate) in the west. The norms of the primitive-diagenetic fluids in the Rustler are similar to those for marine hypersaline bitterns, and Cl/Br ratios are appropriately low, but tachyhydrite ($CaMg_2Cl_6$) is present rather than kieserite ($MgSO_4$). End member examples are from H-5 for maximum $MgCl_2$, and from P-18 for the most $CaCl_2$, at salinities in excess of 290 g/l. For brines from the Culebra Dolomite aquifer a minor $CaCl_2$ component is restricted to norms from four wells all within the WIPP site, though a small amount of $CaCl_2$ is also present in the norm of H-10 Magenta to the southeast of the site.

Halite resolution brines are the most common type of fluid in the Rustler Formation of the region around the WIPP site. About 70 percent of the samples from the Rustler-Salado contact zone and 30 percent from the Culebra Dolomite are dominated by an NaCl component. The abundance of normative halite in the aquifer samples is compatible, for the most part, with zones describing the extent of apparent halite dissolution observed in the principal evaporite-bearing intervals of the Rustler Formation. In some sample norms (*e.g.*, H-9 and WIPP-30 from the Rustler-Salado contact) halite exceeds 90 percent of the normative salts,

and the Cl/Br ratio is five times the seawater value, as expected from the dissolution of bromine-poor halite. Most of the samples from the Culebra aquifer in the eastern part of the region appear to be variable mixtures of these resolution brines with fluids of primitive-diagenetic origin. Simple halite dissolution in itself cannot destroy the normative signatures of primitive-diagenetic brines (alkaline-earth chloride) or weathering waters (alkali sulfo-carbonate). This is seen in a number of the NaCl dominated brines, which very generally fall into two groups with salinities exceeding 100 g/l or less than 25 g/l. The former typically contain some residual primitive-diagenetic indicator (usually $MgCl_2$) and the latter commonly have excess sulfate. Complex mixtures of primitive-diagenetic brine with alkali sulfate-bearing recharge can produce halite-rich norms without halite dissolution, but they are distinguished by low Cl/Br ratios. Primitive-diagenetic residuals apparently have remained in isolated pockets where formational transmissivities are low.

In that part of the WIPP site region where halite is thin, absent or has been essentially removed by dissolution, such as in Nash Draw and vicinity, the further solution of anhydrite or gypsum leads to ground-water norms dominated by calcium sulfate. Culebra aquifer samples from Nash Draw appear to result from extensive dilution of primitive-diagenetic residuals by $CaSO_4$ dissolution waters. Norms with more than 50 percent $CaSO_4$ are common for samples from the Culebra aquifer to the south and west of the WIPP site area, where the most dilute of the Culebra waters are found. As with halite, the simple dissolving of anhydrite or gypsum will not obliterate a primitive-diagenetic or weathering solute-signature, but the mixing of alkaline-earth chloride and alkali sulfate components will generate added Ca and Mg sulfates in the norm. Similarly, Na or Mg carbonate from silicate hydrolysis mixed with dissolved $CaSO_4$ will result in normative Na-Mg sulfates, as in the case for some central WIPP-site Culebra, and more especially, a number of Magenta aquifer waters. The norm for H-8 Culebra exceeds 90 percent alkaline-earth sulfate, whereas the sample of the Culebra water from the Windmill well south of the WIPP site has mixed sulfates comprising 80 percent of the norm, almost matched by the 75 percent for H-9 Magenta. Indeed, an increased sodium sulfate component appears to correlate with a greater continental sediment fraction upward in the stratigraphic section. The maximum Na_2SO_4 percentage is in the norm for a sample from the Triassic Santa Rosa Formation (H-5). For the WIPP-15 water from Cenozoic deposits, straightforward feldspar and mafic mineral weathering is seen in the dominant Na-Mg carbonate components in the norm.

Nearly all the salt norms for waters from the shallowest of the Rustler Formation aquifers, the Magenta Dolomite, illustrate the dominance of $CaSO_4$ dissolution. The Magenta is dry at several locations in north central Nash Draw and has been removed by erosion to the south. Normative indications of halite dissolution are low in most samples, except for the most saline Magenta fluid, as might be expected, from the eastern edge of the area. Many of the Magenta norms contain sodium sulfate salts, which appear to correlate with the presence of overlying continental (siliciclastic) strata. In two uncontaminated Magenta norms, a small CaMgCl primitive-diagenetic signature remains, and even in the two norms from wells in the overlying Triassic Dewey Lake Redbeds a $MgCl_2$ residuum is present.

The large variation in hydrologic properties of the Culebra Dolomite, the most extensively considered of the Rustler Formation aquifers, is emphasized by the diversity in normative results. In addition, the flow field simulations presented by DAVIES (1989) are reflected in the areal distribution of normative compositions. The increase in flow velocities of the model from east to west across the region fits the distribution of concentrated primitive-diagenetic brines in the eastern part of area, halite-dominated waters thru the central part of the region, and dilute $CaSO_4$-type solutions to the west and south. For the WIPP site in detail, the highest flow velocity in the northwest, the high permeability zone in the center-south of the site, and the complex, relatively high flow conditions in the area some distance south and west of the site are in agreement with the normative interpretations. The predominant north to south flow direction (or even northwest to southeast, south of the WIPP site) is clearly consistent with normative inference. This is also compatible with the idea of paleo-recharge to the Rustler aquifers from the Nash Draw area (LAMBERT and HARVEY, 1987). More equivocal is the amount of vertical influx first correlated with the alkali sulfate content of some of the norms. The small shifts in flow direction predicted by DAVIES (1989) for a vertical flux ranging from negligible to a maximum of 25 percent of the total ground-water flow cannot be resolved with existing normative data.

Acknowledgements—We greatly appreciate Shirley L. Rettig for the careful analysis of many of these waters. We also thank Peter Davies and Steven Richey for providing us with the WIPP boundary and sampling site coordinates that are plotted on the many maps as well as for their willing help in locating analytical data and their enthusiastic interest and support of this work. We are very grateful to Steven Lambert, Malcom Siegel, Jerry Mercer, Karen Robinson, and Carol Stein of the Sandia Laboratories in Albuquerque for providing us with unpublished data and considerable discussion. We thank George Garcia for several of the illustrations. Walter Dean and Peter Davies reviewed the manuscript and made many helpful suggestions. Marge Shapira helped very much with manuscript preparation.

REFERENCES

ANDERSON R. Y. (1978) Deep dissolution of salt, northern Delaware Basin, New Mexico. Rept. to Sandia Laboratories, Albuquerque, NM, 106p.

ANDERSON R. Y. (1983) Deformation-dissolution potential of bedded salt, Waste Isolation Pilot Plant site, Delaware Basin, NM. Fifth International Sympos. Scientific Basis of Radioactive Waste Management, Berlin, (West) Germany, 7–10 June, 1982, 10p.

BACHMAN G. O. (1980) Regional geology and Cenozoic history of the Pecos region, southeastern New Mexico. U.S. Geol. Surv. Open-File Rept. 80-1099.

BACHMAN G. O. (1985) Assessment of near-surface dissolution at and near the Waste Isolation Pilot Plant (WIPP), southeastern New Mexico. Sandia Labs. Rept. SAND 86-7078, 33p.

BODINE M. W. JR. (1978) Clay-mineral assemblages from drill core of Ochoan evaporites, Eddy County, New Mexico. In *Geology and mineral deposits of Ochoan rocks in Delaware Basin and adjacent areas.* (ed. G. S. AUSTIN), pp. 21–31. Mexico Bur. Mines Min. Res. Circ. 159.

BODINE M. W. JR. and JONES B. F. (1986) The salt norm: A quantitative chemical-mineralogical characterization of natural waters. U.S. Geol. Surv. Water Res. Inv. Rept. 86-4086.

BRAITSCH O. (1971) *Salt Deposits, Their Origin and Composition.* 297pp. Springer-Verlag.

CROSS C. W., IDDINGS J. P., PIRSSON L. V. and WASHINGTON H. S. (1902) A quantitative chemico-mineralogical classification and nomenclature of igneous rocks. *J. Geol.* **10**, 555–690.

DAVIES P. B. (1989) Variable-density ground-water flow and paleohydrology in the Waste Isolation Pilot Plant (WIPP) region, Southeastern New Mexico. U.S. Geol. Surv. Open-File Rept. 88-490.

EUGSTER H. P., HARVIE C. E. and WEARE J. H. (1980) Mineral equilibria in the six-component sea water system, Na-K-Mg-SO_4-Cl-H_2O, at 25°C. *Geochim. Cosmochim. Acta* **44**, 1335–1348.

EUGSTER H. P. and JONES B. F. (1979) Behavior of major solutes during closed-basin brine evolution. *Amer. J. Sci.* **279**, 609–631.

FRAPE S. K. and FRITZ P. (1982) The chemistry and isotope composition of saline groundwaters from the Sudbury Basin, Ontario. *Canadian J. Earth Sci.* **19**, 645–661.

FRAPE S. K., FRITZ P. and MCNUTT R. H. (1984) Water-rock interaction and chemistry of groundwaters from the Canadian Shield. *Geochim. Cosmochim. Acta* **48**, 1617–1627.

GONZALEZ D. D. (1981) Hydrological data. In *Basic data report for drillhole WIPP 15 (Waste Isolation Pilot Plant—WIPP)*. Sandia National Labs, SAND 79-0274 24–29.

HEM J. D. (1970) Study and interpretation of the chemical characteristics of natural waters. U.S. Geol. Surv. Water-Supply Paper 1473.

HERRMANN A. G., KNAKE D., SCHNEIDER J. and PETERS H. (1973) Geochemistry of modern seawater and brines from salt pans. Main components and bromine distribution. *Contrib. Mineral. Petrol.* **40**, 1–24.

HOLT R. M. and POWERS D. W. (1988) Facies variability and post-depositional alteration within the Rustler Formation in the vicinity of the Waste Isolation Pilot Plant, southeastern New Mexico. U.S. Department of Energy Report DOE/WIPP-88-004.

KUNKLER J. L. (1980) Evaluation of the Malaga Bend salinity alleviation project. U.S. Geol. Surv. Open-File Rept. 80-1111.

LAMBERT S. J. (1978) Geochemistry of Delaware Basin ground waters. In *Geology and mineral deposits of Ochoan rocks in Delaware Basin and adjacent areas.* (ed. G. S. AUSTIN), pp. 33–38. New Mexico Bur. Mines Min. Res. Circ. 159.

LAMBERT S. J. and HARVEY D. M. (1987) Stable-isotope geochemistry of groundwaters in the Delaware Basin of southeastern New Mexico. Sandia National Labs., SAND 87-0138, 258p.

LAMBERT S. J. and ROBINSON K. L. (1984) Field geochemical studies of groundwaters in Nash Draw, southeastern New Mexico. Sandia National Labs., SAND 83-1122, 38p.

LAPPIN A. R. and HUNTER R. L. eds. (1989) Systems analysis, long-term radionuclide transport, and dose assessments, Waste Isolation Pilot Plant (WIPP), southeastern New Mexico; March, 1989. Sandia National Labs., SAND 89-0462, 632p.

LEVY Y. (1977) The origin and evolution of brine in coastal sabkhas, northern Sinai. *J. Sed. Petrol.* **47**, 451–462.

MANN A. W. (1983) Hydrogeochemistry and weathering on the Yilgarn Block, Western Australia—ferrolysis and heavy metals in continental brines. *Geochim. Cosmochim. Acta* **47**, 181–190.

McCAFFREY M. A., LAZAR B. and HOLLAND H. D. (1987) The evaporation path of seawater and the coprecipitation of Br^- and K^+ with halite. *J. Sed. Petrol.* **57**, 928–937.

McGOWEN J. H., GRANATA G. E. and SENI S. J. (1979) Depositional framework of the lower Dockum Group (Triassic) Texas Panhandle. University of Texas Bur. Econ. Geol., Rept Inv. 97.

MERCER J. W. (1983) Geohydrology of the proposed Waste Isolation Pilot Plant site, Los Medanos area, southeastern New Mexico. U.S. Geol. Surv. Water Res. Inv. Rept. 83-4016.

POWERS D. W., LAMBERT S. J., SHAFFER S. E., HILL L. R. and WEART W. D. (editors) (1978) Geological characterization report, Waste Isolation Pilot Plant (WIPP) site, southeastern New Mexico. Sandia National Labs. Contractor Rept., SAND 78-1596, 2 vols.

RANKAMA K. and SAHAMA T. G. (1950) *Geochemistry.* 912pp. University of Chicago Press.

RICHEY S. F. (1989) Geologic and hydrologic data for the Rustler Formation near the Waste Isolation Pilot Plant, southeastern New Mexico. U.S. Geol. Surv. Open-File Rept. 89-32.

RILEY J. P. and CHESTER R. (1971) *Introduction to Marine Chemistry.* 465pp. Academic Press.

ROBINSON T. W. and LANG W. B. (1938) Geology and ground-water conditions of the Pecos River valley in the vicinity of Laguna Grande de la Sal, New Mexico. In *12th and 13th Biennial Reports.* New Mexico State Engineer, p. 77–100.

SNYDER R. P. (1985) Dissolution of halite and gypsum, and hydration of anhydrite to gypsum, Rustler Formation, in the vicinity of the Waste Isolation Pilot Plant, southeastern New Mexico. U.S. Geol. Surv. Open-File Rept. 85-229.

UHLAND D. W. and RANDALL W. S. (1986) 1986 annual water quality data report for the Waste Isolation Pilot Plant. Technology Development Dept., Waste Isolation Pilot Plant Project, Westinghouse Electric Corporation, Carlsbad, New Mexico, DOE-WIPP-86-006.

UHLAND D. W., RANDALL W. S. and CARRASCO R. C. (1987) 1987 annual water quality data report for the Waste Isolation Pilot Plant. Engineering and Repository Technology, Westinghouse Electric Corp., Carlsbad, New Mexico, DOE-WIPP-87-006.

WHITE D. E., HEM J. D. and WARING G. A. (1963) Chemical composition of subsurface waters. In *Data of Geochemistry* [6th edition]. (ed M. FLEISCHER), U.S. Geol. Surv. Prof. Paper 440-F.

ZHEREBTSOVA I. K. and VOLKOVA N. N. (1966) Experimental study of behavior of trace elements in the process of natural solar evaporation of Black Sea water and Sasyk-Sivash brine. *Geochem. Int.* **3**, 656–670.

Fluid-Mineral Interactions: A Tribute to H. P. Eugster
© The Geochemical Society, Special Publication No. 2, 1990
Editors: R. J. Spencer and I-Ming Chou

Thermochemistry of the formation of fossil fuels

MOTOAKI SATO

U.S. Geological Survey, MS 959, Reston, Virginia 22092, U.S.A.

Abstract—Photosynthesis liberates O_2 from a mixture of CO_2 and H_2O in forming biological substances. Formation of fossil fuels is also associated with progressive depletion of oxygen from biological substances. The two processes are thus sequential in terms of deoxygenation. To evaluate thermochemical relations of the deoxygenation processes in a simple manner, the concept of deoxygenation free energy G_{deox} and deoxygenation enthalpy H_{deox} (defined as Gibbs energy and enthalpy, respectively, of reaction to release unit mole of O_2 in forming a compound $C_xH_yN_zO_w$ from CO_2, H_2O, and N_2) are computed for over 150 compounds. If compound A changes to compound B by dehydration and/or decarboxylation, and A has a higher G_{deox} value than B, energy is released and the transformation can occur spontaneously; likewise heat is released if A has a higher H_{deox}. The computed results indicate that energy input is required only for the initial stage of photosynthesis, and progressive deoxygenation and aromatic condensation during diagenesis should decrease free energy and release heat. Catagenesis and metagenesis are also most probably spontaneous and exothermic because both low C-number paraffinic hydrocarbons and polycondensed aromatic kerogens have the lowest G_{deox} and H_{deox} values among various homologous groups even at room temperature. The ultimate products of the spontaneous evolution of fossil fuels are methane and graphite. The role of increased temperature associated with deeper burial is not to change the position of equilibrium, but simply to increase the reaction rate exponentially so that the extremely slow reactions would proceed in geologic time.

INTRODUCTION

SOLAR ENERGY IS chemically stored, *via* photosynthesis, as an unstable coexistence of free oxygen and organic matter near the Earth's surface. When organisms die, most of the biological matter is aerobically degraded to gases and water-soluble compounds and is recycled to the atmosphere and hydrosphere. Only about 0.1% escapes chemical and microbial oxidation within the first few centimeters of burial and undergoes anaerobic transformation to fossil fuels in sediments. Because of accumulation over geologic periods of time, however, the total amount of fossilized organic matter is estimated to be orders of magnitude (about 2000 times according to WELTE, 1970) greater than the live biomass, warranting a thorough understanding of the nature of the transformation process.

Much biogeochemical work was devoted to the molecular identification and quantification of the organic substances in relation to geologic factors (basin structure, burial depth, age, temperature, etc.) to deduce the rate and chemical mechanisms of the degradation of biological matter and transformation to fossil fuels (BREGER, 1963; TISSOT and WELTE, 1978; HUNT, 1979). Consequently, recent investigators have arrived at the deduction that the maturation process occurs spontaneously in stages when the temperature increases due to burial along a normal geothermal gradient. The present work is an attempt to provide an explicit thermochemical basis for this deduction.

BACKGROUND

Photosynthesis may be viewed as a natural process in which photosynthetic organisms liberate oxygen from a mixture of mainly H_2O and CO_2, and some N and S compounds, to form biological substances. Transformation of biological remains to fossil fuels may also be viewed as a process in which oxygen is removed from biological substances. Graphic presentations by OWEN (1948), VAN KREVELEN (1961), and others show at a glance that O as well as N and S are depleted progressively from biological substances (*e.g.*, carbohydrates, proteins, lipids, and lignin) during their transformation to fossil fuels (*e.g.*, kerogen, petroleum hydrocarbons, natural gas, and coal). In this sense, photosynthesis and fossil fuel formation are natural sequential processes in which the compounds of the C-H-O-(N-S) system are deoxygenated.

As caloric values of organic compounds generally increase with increasing degree of deoxygenation, some earlier investigators searched for the sources of the energy thought to be necessary for the transformation, analogous to the solar energy input in photosynthesis. Bacteriological processes, radioactivity, heat, and pressure have been studied as the possible sources of the transformation energy (BREGER, 1963). Except perhaps for the effect of heat in thermal metamorphism of coal to anthracite, however, the search for these energy sources had remained uncertain.

More recent studies in the field and laboratory (summarized by TISSOT and WELTE, 1978; HUNT,

1979), have indicated that most biological compounds undergo rapid and extensive anaerobic chemical changes within several hundred meters of burial, where the temperature remains within the biologically active range (<50°C). Diagenetic processes, both chemical and microbial, transform reactive biological substances to more inert, insoluble polymers such as humic acids and kerogen. Further burial with accompanying temperature rise (to about 150–200°C) causes thermal degradation of kerogen to petroleum-range hydrocarbons, a process called catagenesis. Burial deeper than several kilometers (>200°C) results in metagenesis and metamorphism, in which methane and other gaseous hydrocarbons and carbonized residues are formed.

In the above scheme of fossil fuel evolution, temperature seems to play a controlling role. Because of the inevitable association of the depth of burial with increased time, temperature, and pressure, it is difficult to clearly determine if energy input as heat is indeed necessary, or if the late stages of transformation are merely hindered by high activation energies (mechanism) and require higher temperatures to proceed even given geologic periods of time. The ambiguities are especially strong in the discussions of thermal degradation, and "cracking" of geopolymers.

THERMOCHEMICAL CONSIDERATIONS

Evaluation of free energy changes for reactions involving natural organic substances is handicapped by the fact that too many complex and sometimes ill-defined compounds are involved, and also many reaction paths are still insufficiently understood. The approach described below, however, bypasses the complexities arising from uncertainties in reaction paths and outlines energy relations among many organic compounds. The approach is based on the assumption that the deoxygenation during maturation of fossil fuels proceeds via loss of H_2O and CO_2, a model well documented for coalification (VAN KREVELEN, 1963) and kerogen evolution (TISSOT and WELTE, 1978, p. 149).

The principle of the approach is that when one organic compound can theoretically be obtained from another by releasing H_2O and/or CO_2 (and N_2 if nitrogen is involved; sulfur is not considered here as it is a minor component), the free energy balance in the overall reaction, however complex the actual mechanism may be, is directly related to the difference between the Gibbs free energies required to make the two compounds from a mixture of H_2O, CO_2, and N_2 upon releasing 1 mole of free oxygen (O_2).

For example, the formation of ethanol (C_2H_6O) from glucose ($C_6H_{12}O_6$) at 25°C and 1 atm:

$$C_6H_{12}O_6 = 2\ C_2H_6O + 2\ CO_2,$$

$$\Delta G_r = -226.44\ \text{kJ} \quad (1)$$

(where ΔG_r is the Gibbs free energy of reaction, and kJ is kilojoule) may be regarded as being composed of two reactions, each of which releases 1 mole of O_2, in opposite directions:

$$CO_2 + H_2O = (1/6)\ C_6H_{12}O_6 + O_2,$$

$$\Delta G_r = 479.82\ \text{kJ} \quad (2)$$

$$(2/3)\ CO_2 + H_2O = (1/3)\ C_2H_6O + O_2,$$

$$\Delta G_r = 442.08\ \text{kJ}. \quad (3)$$

It is easy to see that $(1/6)\ \Delta G_r\ (1) = \Delta G_r\ (3) - \Delta G_r\ (2)$. Heat is known to be generated during the fermentation of carbohydrates. Reaction (1) can occur spontaneously as indicated by the negative free energy of reaction, although the presence of enzymes in malt increases the reaction rates multifold. The spontaneity of the conversion of glucose to ethanol is also indicated by the fact that $\Delta G_r\ (3) < \Delta G_r\ (2)$.

As seen above, whether one organic compound should spontaneously be converted to another, upon losing H_2O, CO_2, and/or N_2, could easily be surveyed by calculating the Gibbs free energy necessary to remove 1 mole of O_2 in forming each compound from CO_2, H_2O, and N_2 (if needed), and by comparing the computed free-energy values. Such a free energy of reaction may be defined as *deoxygenation free energy*, G_{deox}, and can be obtained from the Gibbs free energy of reaction for the general reaction:

$$x\ CO_2 + \frac{y}{2}\ H_2O + \frac{z}{2}\ N_2$$

$$= C_xH_yN_zO_w + \left(\frac{4x + y - 2w}{4}\right) O_2 \quad (4)$$

as

$$G_{\text{deox}} = \frac{4\Delta G_r(4)}{(4x + y - 2w)} \quad (5)$$

and

$$\Delta G_r(4) = \Delta G_f^0(C_xH_yN_zO_w)$$

$$- x\Delta G_f^0(CO_2) - \frac{y}{2}\Delta G_f^0(H_2O) \quad (6)$$

where ΔG_f^0 is the Gibbs energy of formation for each chemical species involved at 1 atm pressure and at a specified temperature (25°C in this paper unless otherwise noted). If compound A changes to compound B via dehydration, and/or decarboxylation, the transformation can proceed spontaneously with a release of energy when A has a higher G_{deox}.

Similarly, *deoxygenation enthalpy, H_{deox}*, may be defined on the basis of equation (4) as

$$H_{deox} = \frac{4\Delta H_r(4)}{(4x + y - 2w)} \quad (7)$$

where the enthalpy of the reaction can be obtained from the enthalpy of formation, ΔH_f^0, of the chemical species involved as:

$$\Delta H_r(4) = \Delta H_f^0(C_xH_yN_zO_w)$$

$$- x \Delta H_f^0(CO_2) - \frac{y}{2} \Delta H_f^0(H_2O). \quad (8)$$

The availability of Gibbs energy of formation data for complex solid and liquid organic compounds is rather limited, and an estimation is often difficult, particularly for uncharacterized geopolymers. In contrast, enthalpy of formation data or heat of combustion (H_c) data are available for a large number of organic substances in fair to good accuracy. In fact, combustion calorimetry constitutes the basis for obtaining thermochemical data for the majority of organic substances. It may be noted here that reaction (4) is exactly reverse of the combustion reaction, and that $H_c = \Delta H_r(4)$. Hence,

$$H_{deox} = \frac{4H_c}{(4x + y - 2w)}. \quad (9)$$

If compound A changes to compound B via dehydration and/or decarboxylation, the transformation reaction is exothermic when A has a higher H_{deox}. In view of complexity of estimating G_{deox} values for complex compounds as discussed later, the H_{deox} relation may be substituted for the G_{deox} relation in certain cases.

In summary, if compound A changes to compound B by dehydration and/or decarboxylation, and A has a higher G_{deox} value than B, energy is released and the transformation can occur spontaneously; heat is released if A has a higher H_{deox} than B.

In order to relate the G_{deox} or H_{deox} values of various compounds to the degree of deoxygenation, which is an index of maturation of fossil fuels, another quantity called *deoxygenation quotient, Q_{deox}*, is introduced. The deoxygenation quotient indicates the fraction of oxygen atoms removed from the total oxygen atoms present in the starting mixture of CO_2

and H_2O (and N_2) in forming a compound in the C-H-O-(N) system:

$$Q_{deox} = \frac{4x + y - 2w}{4x + y} \quad (10)$$

where x, y, and w are defined in equation (4). Note that nitrogen as N_2 is not involved in determining Q_{deox} (it may if incorporated in the deoxygenation equation as NH_3 and/or NO_2, but for simplicity we disregard this possibility for now). The Q_{deox} for glucose, for example, is 0.667, and that for ethanol is 0.857, according to the above definition.

The G_{deox} and Q_{deox} values have been computed for some 150 common organic compounds, for which the Gibbs free energy data are available at 25°C (STULL *et al.*, 1969) and listed in Table 1. Some of these values have been plotted on a G_{deox} *vs.* Q_{deox} diagram (deoxygenation free energy diagram) in Fig. 1. Also, the G_{deox} values of some hydrocarbons (all $Q_{deox} = 1$) are plotted against the atomic H/C ratio in Fig. 2.

Figures 1 and 2 show many interesting facts regarding the thermochemical relations among the common organic compounds. Notable points are listed below.

(a) Among the oxygenous homologous series, fatty acids have the lowest G_{deox} for a given Q_{deox}, followed by amino acids, proteins, alcohols, carbohydrates, aldehydes, and finally by ethers, in ascending order.

(b) Within each oxygenous homologous series, the G_{deox} decreases regularly with increasing C number, and the slope of the decrease against Q_{deox} also decreases regularly, approaching zero at $Q_{deox} = 1$. The differences in G_{deox} among various oxygenous homologous series also decrease as Q_{deox} approaches 1.

(c) Among hydrocarbons, the G_{deox} of the olefinic series decreases regularly (allowing some reversals within the limits of uncertainties in the thermochemical data) with increasing C number, whereas that of the paraffinic series has the reverse trend. The cycloparaffinic series roughly follows the trend of the olefinic series, but the pattern is highly irregular.

(d) When the C number increases, the G_{deox} values of most common nonaromatic homologous series converge to a single value, which is about 424.7 kJ at 25°C (this value will be referred to as "the convergence value" of the G_{deox}). The G_{deox} values of aromatic rings uniquely diverge from the convergence value as the C number increases, decreasing toward the graphite value (394.4 kJ) as the degree of condensation increases.

274 M. Sato

Table 1. A list of H_{deox} (deoxygenation enthalpy) and G_{deox} (deoxygenation free energy), and Q_{deox} (deoxygenation quotient) values of common compounds of the C-H-N-O system. For hydrocarbons, atomic H/C ratio is given instead of Q_{deox} (all 1.00)

SECTION A

Name	Formula	State	Q_{deox}	H_{deox} (kJ)	G_{deox} (kJ)
Fatty acids					
Formic acid	CH_2O_2	(l)	.333	509.17	540.24
Acetic acid	$C_2H_4O_2$	(l)	.667	437.30	436.89
Butyric acid	$C_4H_8O_2$	(l)	.833	436.70	429.72
Valeric acid	$C_5H_{10}O_2$	(l)	.867	436.57	428.57
Palmitic acid	$C_{16}H_{32}O_2$	(s)	.958	433.78	425.61
Polybasic and hydroxy acids					
Oxalic acid	$C_2H_2O_4$	(s)	.200	485.82	649.61
Lactic acid	$C_3H_6O_3$	(s)	.667	447.98	457.27
Succinic acid	$C_4H_6O_4$	(s)	.636	425.90	440.48
Pyruvic acid	$C_3H_4O_3$	(l)	.625	467.07	477.66
Fumaric acid, trans.	$C_4H_4O_4$	(s)	.600	444.88	466.08
Maleic acid	$C_4H_4O_4$	(s)	.600	451.70	473.57
Citric acid · H_2O	$C_6H_{10}O_8$	(s)	.529	471.63	499.84
Fatty alcohols					
Methanol	CH_4O	(l)	.750	484.41	468.36
Ethanol	C_2H_6O	(l)	.857	455.85	442.07
Propanol	C_3H_8O	(l)	.900	448.74	435.83
Butanol	$C_4H_{10}O$	(l)	.923	446.23	433.73
Pentanol	$C_5H_{12}O$	(l)	.938	443.28	431.18
Hexanol	$C_6H_{14}O$	(l)	.947	442.50	430.56
Heptanol	$C_7H_{16}O$	(l)	.955	442.14	430.11
Hexadecanol	$C_{16}H_{34}O$	(s)	.980	436.29	426.79
Polyhydric and other alcohols					
Ethylene glycol	$C_2H_6O_2$	(l)	.714	475.84	470.78
Glycerol	$C_3H_8O_3$	(l)	.700	472.93	472.82
Erythritol	$C_4H_{10}O_4$	(s)	.692	465.06	472.67
Pentaerythritol	$C_5H_{16}O_4$	(s)	.778	476.43	465.08
Fufuryl alcohol	$C_5H_6O_2$	(l)	.846	463.40	459.87
Pentamethylene glycerol	$C_5H_{12}O_2$	(l)	.875	463.26	451.82
Cyclopentanol	$C_5H_{10}O$	(l)	.933	442.37	432.86
Cyclohexanol	$C_6H_{12}O$	(l)	.944	438.57	430.13
Aldehydes					
Formaldehyde	CH_2O	(g)	.667	563.45	521.66
Acetaldehyde	C_2H_4O	(g)	.833	476.93	451.94
Propionaldehyde	C_3H_6O	(g)	.889	461.50	441.07
Butylaldehyde	C_4H_8O	(l)	.917	450.67	437.65
Heptanal	$C_7H_{14}O$	(l)	.952	444.37	432.04
Furfural	$C_5H_4O_2$	(l)	.833	468.85	466.77
Ketones, esters					
Aceton	C_3H_6O	(l)	.889	447.48	434.83
2-Butanone	C_4H_8O	(l)	.917	444.38	431.80
2-Pentanone	$C_5H_{10}O$	(l)	.933	442.39	430.06
2-Octanone	$C_8H_{16}O$	(l)	.958	439.15	427.15
Ketene	C_2H_2O	(g)	.800	505.88	482.83

SECTION A

Name	Formula	State	Q_{deox}	H_{deox} (kJ)	G_{deox} (kJ)
Furan	C_4H_4O	(l)	.900	462.97	456.03
Ethyl acetate	$C_4H_8O_2$	(l)	.833	447.67	438.72
Ethers					
Methyl ether	C_2H_6O	(g)	.857	486.82	462.47
Ethyl methyl ether	C_3H_8O	(g)	.900	468.32	447.61
Ethyl ether	$C_4H_{10}O$	(l)	.923	453.95	440.09
Isopropyl ether	$C_6H_{14}O$	(l)	.947	445.63	433.43
Carbohydrates					
Glucose, D	$C_6H_{12}O_6$	(s)	.667	466.94	479.82
Alpha-Galactose, D	$C_6H_{12}O_6$	(s)	.667	465.12	478.20
Sorbose, L	$C_6H_{12}O_6$	(s)	.667	467.58	480.03
Beta-Lactose	$C_{12}H_{22}O_{11}$	(s)	.686	469.13	481.23
Sucrose	$C_{12}H_{22}O_{11}$	(s)	.686	470.35	483.09
Alpha-Lactose · H_2O	$C_{12}H_{24}O_{12}$	(s)	.667	474.40	487.45
Beta-Maltose · H_2O	$C_{12}H_{24}O_{11}$	(s)	.667	474.40	487.45
Galactitol	$C_6H_{14}O_6$	(s)	.684	463.86	473.24
Manitol, D	$C_6H_{14}O_6$	(s)	.684	465.33	474.52
Xylose, D	$C_5H_{10}O_5$	(s)	.667	469.47	484.35
Peptides—amino acids, proteins, etc.					
Glycine	$C_2H_5NO_2$	(s)	.692	428.61	446.25
Alanine, D, L	$C_3H_7NO_2$	(s)	.789	431.30	437.67
Valine, L	$C_5H_{11}NO_2$	(s)	.871	432.84	432.22
Leucine, L	$C_6H_{13}NO_2$	(s)	.892	432.99	430.42
Isoleucine, L	$C_6H_{13}NO_2$	(s)	.892	434.05	431.62
Phenylalanine, L	$C_9H_{11}NO_2$	(s)	.915	432.25	431.86
Serine, L	$C_3H_7NO_3$	(s)	.684	447.57	462.81
Tyrosine, L	$C_9H_{11}NO_3$	(s)	.872	433.38	435.93
Tryptophan, L	$C_{11}H_{12}N_2O_2$	(s)	.929	432.96	434.00
Arginine, D	$C_6H_{14}N_4O_2$	(s)	.895	439.80	445.45
Creatinine	$C_4H_7N_3O$	(s)	.913	444.94	453.06
Aspertic acid, L	$C_4H_7NO_4$	(s)	.652	426.96	447.33
Glutamic acid, L	$C_5H_9NO_4$	(s)	.724	427.45	439.62
Hippuric acid	$C_9H_9NO_3$	(s)	.867	432.71	435.61
Alantoin	$C_4H_6N_4O_3$	(s)	.727	428.49	460.72
Aspargine, L	$C_4H_8N_2O_3$	(s)	.750	428.41	443.57
Glycylglycine	$C_4H_8N_2O_3$	(s)	.750	430.25	452.38
Alanylglycine, L	$C_5H_{10}N_2O_3$	(s)	.800	428.38	437.54
Leucylglycine, D, L	$C_8H_{16}N_2O_3$	(s)	.875	435.76	436.51
Hippurylglycine	$C_{11}H_{12}N_2O_4$	(s)	.857	434.22	438.85
Others					
Graphite	C	(s)	1.00	393.51	394.38
Carbon monoxide	CO	(g)	.500	565.93	514.21
Carbon dioxide	CO_2	(g)	0.00	0	0
Hydrogen	H_2	(g)	1.00	571.68	474.38
Water	H_2O	(l)	0.00	0	0
Hydrocyanic acid	HCN	(g)	1.00	533.57	506.48
Ammonia	NH_3	(g)	1.00	510.76	452.85
Hydrazine	N_2H_4	(g)	1.00	666.86	632.91
Urea	CH_4N_2O	(s)	.750	421.34	447.74

Table 1. (Continued)

SECTION B

Name	Formula	State	H/C	H_{deox} (kJ)	G_{deox} (kJ)
Paraffins					
Methane	CH_4	(g)	4.00	445.17	408.97
Ethane	C_2H_6	(g)	3.00	445.67	419.26
Propane	C_3H_8	(g)	2.67	444.00	421.69
Butane	C_4H_{10}	(g)	2.50	439.32	422.84
Pentane	C_5H_{12}	(l)	2.40	438.67	423.20
Hexane	C_6H_{14}	(l)	2.33	438.22	423.40
Heptane	C_7H_{16}	(l)	2.29	437.90	423.57
Octane	C_8H_{18}	(l)	2.25	437.65	423.70
Nonane	C_9H_{20}	(l)	2.22	437.46	423.79
Decane	$C_{10}H_{22}$	(l)	2.20	437.31	423.88
Dodecane	$C_{12}H_{26}$	(l)	2.17	437.10	424.04
Hexadecane	$C_{16}H_{34}$	(l)	2.13	436.69	424.10
Tetracosane	$C_{24}H_{50}$	(s)	2.08	434.98	424.70
Dotriacontane	$C_{32}H_{66}$	(s)	2.06	434.12	423.98
Olefins					
Ethylene	C_2H_4	(g)	2.00	470.33	443.76
Propene	C_3H_6	(g)	2.00	457.43	434.99
1-Butene	C_4H_8	(g)	2.00	452.87	432.93
1-Pentene	C_5H_{10}	(l)	2.00	446.67	431.48
1-Hexene	C_6H_{12}	(l)	2.00	444.85	430.32
1-Heptene	C_7H_{14}	(l)	2.00	443.57	429.51
1-Octene	C_8H_{16}	(l)	2.00	442.60	428.89
1-Decene	$C_{10}H_{20}$	(l)	2.00	441.26	428.05
1-Hexadecene	$C_{16}H_{32}$	(l)	2.00	439.41	426.97
Cycloparaffins and other saturated alicyclic hydrocarbons					
Cyclopropane	C_3H_6	(g)	2.00	464.74	444.25
Cyclobutane	C_4H_8	(g)	2.00	453.41	439.62
Cyclopentane	C_5H_{10}	(l)	2.00	438.78	425.90
Cyclohexane	C_6H_{12}	(l)	2.00	435.54	424.55
Cycloheptane	C_7H_{14}	(l)	2.00	437.84	426.20
Cyclooctane	C_8H_{16}	(l)	2.00	438.80	427.54
Methylcyclopentane	C_6H_{12}	(l)	2.00	437.52	424.55
Ethylcyclopentane	C_7H_{14}	(l)	2.00	437.32	424.60
Propylcyclopentane	C_8H_{16}	(l)	2.00	437.13	424.65
Bytylcyclopentane	C_9H_{18}	(l)	2.00	435.35	423.73
Methylcyclohexane	C_7H_{14}	(l)	2.00	434.78	422.99

SECTION B

Name	Formula	State	H/C	H_{deox} (kJ)	G_{deox} (kJ)
Ethylcyclohexane	C_8H_{16}	(l)	2.00	435.21	423.47
Propylcyclohexane	C_9H_{18}	(l)	2.00	435.24	423.59
Butylcyclohexane	$C_{10}H_{20}$	(l)	2.00	435.35	423.73
Cyclohexylheptane	$C_{13}H_{26}$	(l)	2.00	434.76	423.17
Cyclopentyldecane	$C_{15}H_{30}$	(l)	2.00	436.62	424.73
Cyclohexyldecane	$C_{16}H_{32}$	(l)	2.00	435.46	423.99
Diolefins and acetylides					
Allene	C_3H_4	(g)	1.33	486.08	464.98
1,3-Butadiene	C_4H_6	(l)	1.50	457.62	443.40
Isoprene	C_5H_8	(l)	1.60	451.46	437.99
1,4-Pentadiene	C_5H_8	(l)	1.60	455.21	441.13
Acetylene	C_2H_2	(g)	1.00	519.83	494.06
Propyne	C_3H_4	(g)	1.33	484.41	462.99
1-Bytyne	C_4H_6	(l)	1.50	467.81	453.26
2-Butyne	C_4H_6	(l)	1.50	463.89	449.89
Aromatic hydrocarbons					
Benzene	C_6H_6	(l)	1.00	435.68	426.96
Naphthalene	$C_{10}H_8$	(s)	0.80	429.71	424.47
Anthracene	$C_{14}H_{10}$	(s)	0.71	428.33	423.82
Phenanthrene	$C_{14}H_{10}$	(s)	0.71	427.54	422.95
Fluoranthene	$C_{16}H_{10}$	(s)	0.63	427.84	423.73
Pyrene	$C_{16}H_{10}$	(s)	0.63	423.78	419.76
Naphthacene	$C_{18}H_{12}$	(s)	0.67	426.53	422.90
Perylene	$C_{20}H_{12}$	(s)	0.60	424.69	420.96
Toluene	C_7H_8	(l)	1.14	434.43	424.80
Ethylbenzene	C_8H_{10}	(l)	1.25	434.74	424.83
Propylbenzene	C_9H_{12}	(l)	1.33	434.85	424.74
m-Xylene	C_8H_{10}	(l)	1.25	433.51	423.68
Mesitylenebenzene	C_9H_{12}	(l)	1.33	432.76	423.03
Tetramethylbenzene	$C_{10}H_{14}$	(s)	1.40	430.18	421.63
Pentamethylbenzene	$C_{11}H_{16}$	(s)	1.45	432.00	422.87
Hexamethylbenzene	$C_{12}H_{18}$	(s)	1.50	432.16	423.32
Biphenyl	$C_{12}H_{10}$	(s)	0.83	431.16	425.71
Diphenylethane	$C_{14}H_{14}$	(l)	1.00	431.94	424.39
Bibenzyl	$C_{14}H_{14}$	(l)	1.00	432.08	425.65
Styrene	C_8H_8	(l)	1.00	439.53	430.62
Stilbene, trans.	$C_{14}H_{12}$	(s)	0.86	432.89	427.18

(e) Among unpolymerized species, carbohydrates have exceptionally high G_{deox} for the number of carbon atoms. They are exceeded only by a handful of compounds such as oxalic, formic, citric acids and formaldehyde.

Calculations now in progress for G_{deox} values at 127°C (400 K) and 227°C (500 K) also indicate that the above relations observed for 25°C remain unchanged. Except for condensed aromatic compounds, the G_{deox} values converge to 424.3 kJ with increasing carbon number at 127°C.

ESTIMATION OF DEOXYGENATION PARAMETERS FOR NATURAL POLYMERS

Certain organic substances have the ability to form macromolecules upon polymerization. Compared to the respective monomers, these polymers are much less water-soluble and inherently slow to react, and thus have a better chance of being preserved in sediments upon resisting microbial and chemical oxidation. In fact they are the dominant precursors of fossil fuels. Biopolymers such as cellulose, lignin, resins, and proteins are used as the structural material to support free-standing weights by land plants. Upon burial these polymers undergo extensive aromatic condensation to kerogen, which makes up more than 90% of the total organic matter in the Earth's crust (HUNT, 1979). For thermochemical consideration of fossil fuel formation, therefore, it is important to estimate the deoxygenation parameters for these polymers, although only rough approximations are possible because (1) the entropy change associated with polymerization is generally difficult to accurately assess, and (2) most of these polymers are compositionally variable within a range.

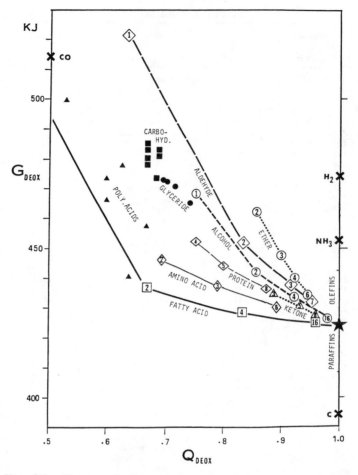

FIG. 1. Plots of G_{deox} (deoxygenation free energy) versus Q_{deox} (deoxygenation quotient) for some compounds of the C-H-N-O system. Homologous series are identified by labelled connecting lines. Arabic numerals indicate the number of carbon atoms in these homologues. Some other groups are also identified with labels, where possible. Inorganic compounds are individually labelled. Except for labelling, hydrocarbons are excluded in this diagram. The star indicates the convergence value of the homologous series.

Many methods have been proposed for estimating thermochemical properties of organic compounds (reviewed by JANZ, 1967) including the method of VAN KREVELEN and CHERMIN (1951) for Gibbs energy of formation based on group contributions. However, most of these methods are derived for the ideal gas state, and, hence, are applicable only to gaseous compounds and condensed compounds with known heat of vaporization and measurable vapor pressures. Biopolymers and geopolymers have negligible vapor pressures at room temperature and decompose if heated. So in some cases it is preferable to use the heat of combustion data and rely on the similarity of H_{deox} relation to the G_{deox} relation as the first approximation.

If heat of combustion is available only for the monomer or major components of a complex polymer, and particularly if the monomer is gas or liquid, there may be a significant change in entropy associated with polymerization. The entropy of polymerization may be ignored if the transformation is from an amorphous solid to a crystalline state. For example, the transformation of amorphous Se to linear macromolecule trigonal Se (a simple model case of polymerization) decreases entropy by 6.77 J/mol·K at 25°C according to GAUER et al. (1981), which corresponds to a polymerization energy of about 2 kJ per mole of the monomer compound.

Cellulose

Cellulose is the basic structural material for higher plants. It is a highly polymerized polysaccharide consisting of several thousands glucose (cellbiose) units in chains. Upon hydration it produces glucose. The chemical formula, $(C_6H_{10}O_5)_n$, gives Q_{deox} of 0.706. The heat of combustion of cellulose at 25°C is given as 2812.4 kJ/mole (COLBERT et al., 1981), which gives H_{deox} as 468.7 kJ according to Equation (9). The H_c value gives ΔH_f^0 of cellulose as −977.88 kJ/mole. The S^0 for cellulose is estimated at 205 J/mole·K by subtracting the entropy of macrochain polymerization (6.77 J/mole·K for Se) from the entropy of glucose (212.13 J/mole·K, STULL et al., 1969, p. 680). The entropy change in the substitution of 2(-OH) by (-O-) linkage in a solid polyhydroxy compound may be neglected (JANZ, 1967, Table 4.1). This gives ΔG_f^0 for cellulose as −681.5 kJ/mole and G_{deox} as 478.5 kJ. The G_{deox} value is 0.7 kJ lower than that of glucose. The reversal in the order of H_{deox} and that of G_{deox} between the two compounds is due to the entropy of liquid H_2O liberated in the dehydration-polymerization.

Lignin

Lignin is a collective term for amorphous aromatic macromolecular substances, which cement cellulose in land plants, and is an important precursor to coal. It is basically constructed from phenyl-propane derivatives, such as coniferyl alcohol, sinapyl alcohol, and p-coumaryl alcohol, that are biosynthesized from carbohydrates (SARKANEN and LUDWIG, 1971; KIRK et al., 1980). The polymerization of these monomer units are associated with replacement of two end hydroxyls by an ether linkage. The atomic composition ranges typically from $(C_{11}H_{12}O_3)_n$ to $(C_9H_8O)_n$, depending on the average number of the methoxyl (-OCH₃) groups attached to the phenol ring, giving a Q_{deox} range of 0.893 to 0.955. A compound that is closest to these monomers in structure, and for which the heat of combustion or enthalpy of formation is known, is isoeugenol $CH_3CH:CH \cdot C_6H_3(OCH_3)OH$. To make a chain polymer of coniferyl lignin from isoeugenol, it is required that one each of the (C-H) and (C-OH) bonds are replaced by one (C-O-C) bond. Using the bond contributions for heats of combustion of liquids (applicable to amorphous solids) given by LAIDLER (1956), and 5345.5 kJ for the H_c of isoeugenol, the H_c and H_{deox} for the coniferyl lignin is computed as 5170 kJ/mole and 450 kJ, respectively.

The variations in H_{deox} by addition or subtraction of a methoxyl group is estimated at about 5 kJ per mole of the lignin unit. This is made on the basis of the difference in ΔH_f^0 between 4-hydroxy-m-anisolaldehyde (vanillin) $C_6H_3(OH)(OCH_3)CHO$ and p-hydroxybenzaldehyde $C_6H_4(OH)CHO$ (STULL et

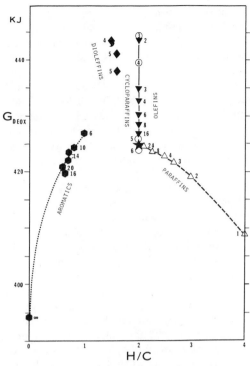

FIG. 2. Plots of G_{deox} versus H/C atomic ratio for common hydrocarbons. Arabic numerals indicate the number of carbon atoms. For paraffinic and olefinic series, only those of straight chains are plotted. Graphite is considered to be the product of infinite aromatic condensation, and included in the aromatic series. The star indicates the convergence value of the oxygenous homologous series shown in Fig. 1.

al., 1969), which is 168.6 kJ/mole (510.7 kJ in terms of H_c between the two), and applying equation (7) to the chemical formula of the lignin monomers. Ignoring the small differences in the entropy term, the approximate H_{deox} value for the lignin group ranges from 455 kJ (for dimethoxy sinapyl unit, Q_{deox} 0.893) to 445 kJ (for p-coumaryl unit, Q_{deox} 0.955). Other more complex and random configurations would probably average out to this range.

There is no simple reliable way of estimating the G_{deox} of lignins because of their complex structures.

Humic matter

This is an operational term applied to polymerized, water-insoluble brown to black substances formed, under restricted oxidative conditions in soil and also in young sediments in swamps and other subaqueous environments, from the residue of biological matter unused by microorganisms. Humic substances are soluble in dilute alkali solutions, but only partially so in dilute acid solutions. The acid soluble fraction is called "fulvic acids" and have a range of composition of C: 30–40%, H: 6–8%, O: 45–55%, N: 4.5–5.5% in weight. The acid-insoluble fraction is called "humic acids" and have the compositional range of C: 50–55%, H: 5.5–6.5%, O: 30–35%, N: 1–4% (GALIMOV, 1980), and atomic ratio range of H/C: 0.5–1.5, and O/C: 0.2–0.5 (TISSOT and WELTE, 1978). Fulvic acids tend to disappear early during the diagenesis. On their way to becoming protokerogen and eventually to kerogen, humic acids are increasingly depleted of oxygen, and become more aromatic.

Molecular characterization of humic acids is very incomplete. A general consensus is that the biological precursors are dominantly lignins and tannins which are widespread in plants. Probably proteins and perhaps carbohydrates are also involved. Laboratory studies indicate that both aromatic and aliphatic/alicyclic substances are incorporated. Functional groups detected are carboxyls, carbonyls, phenolic and enolic hydroxyls, and amides in the order of abundance. Methoxyls are present only in low concentrations. For example, a typical humic acid from peat described by CHRISTMAN and OGLESBY (1971) contains C: 58.35%, H: 4.97%, O: 32.2%, N: 2.62% in weight, and the following functional groups in millimoles/gram: 0.2 methoxyl, 2.9 phenolic hydroxyl, 2.2 enolic hydroxyl, 5.5 carbonyl, and 8.6 carboxyl. The functional groups in total take up 63.4% of the dried, ash-free humic acid, the balance presumably being hydrocarbon chains and rings. None of the proposed structural models (reviewed by SWAIN, 1963; CHRISTMAN and OGLESBY, 1971; HUC, 1980; GALIMOV, 1980) appears to accommodate the abundant carboxyls and carbonyls, and, at the same time, conform with an approximate C:H:O ratio of 2:2:1 indicated by the above weight composition. The alicyclic quinones could account for the carbonyls but have too high O/C ratios. Ketonic bridges between H-rich structures may resolve this problem. Such a unit may be constructed, for example, from one molecule of quinic acid $(OH)_4 \cdot C_6H_7 \cdot COOH$ and one molecule of protocatechuic acid $C_6H_3(OH)_2COOH$, bridged by a ketonic or ether bond (Fig. 3a or 3b). A ketonic bond of this type is found in benzoin $C_6H_5 \cdot CHOH \cdot CO \cdot C_6H_5$. Quinic acid is an alicyclic polyhydroxy acid derived from carbohydrate and, upon dehydration, changes to aromatic protocatechuic acid (CONNANT, 1936, p. 458). Protocatechuic acid is also formed by oxidation of the p-coumaryl alcohol unit of lignin. Such units may be linked together by ether, ketonic, or ester (SCHNITZER and NEYROND, 1975) bonds upon releasing H_2O.

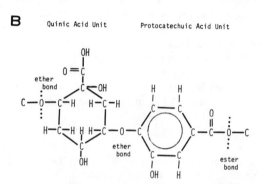

FIG. 3. Suggested models of humic acids that could be derived from lignin and cellulose, and versatile in the aliphatic vs. aromatic ratio; (A) a model in which the aliphatic and aromatic groups are bridged by a ketonic bond, and (B) that bridged by an ether bond.

In view of the uncertainties of the chemical structure of humic acids, direct estimation of G_{deox} values of humic acids is not attempted. Instead, a list of chemicals that have been identified upon degradation of humic matter, or proposed to be likely components on theoretical grounds, are listed, together with their H_{deox} (and G_{deox} in rare cases where Gibbs energy of formation is given in the literature) in Table 2. Nitrogenous compounds are excluded for simplicity. Compounds with methoxyls are eliminated in view of the low abundance of this group. Needless to say, compounds without heat of combustion or enthalpy of formation data are excluded. Sample calculations show that G_{deox} or H_{deox} values are very similar among isomers when they have more than several C atoms. Only the most stable isomer is listed when H_{deox} values vary by less than 1 kJ among the isomers. The compound name is given by a structural name in such a case. Alicyclic quinoids are grouped together with aromatic polyhydroxy compounds because of the close relationship existing between the two groups.

The Q_{deox} values for humic acids inferred from

the Van Krevelen (H/C vs. O/C) diagrams (Fig. II. 2.9 in TISSOT and WELTE, 1978; Fig. 7-6 in HUNT, 1979) are in the range of 0.778 to 0.927, and a typical value is 0.88. The range is richer in oxygen than the above model lignins.

Kerogen

This term was originally applied to the organic matter in oil shales that produced waxy oils (keros = wax in Greek) upon heating, but the term has been broadened in recent years to include all the disseminated organic matter of sedimentary rocks insoluble in nonoxidizing acids, bases, and organic solvents regardless of capacity for oil distillation.

On the basis of microscopic characterization, kerogens are classified into sapropelic and humic. Sapropelic kerogens originate in decomposition and polymerization products of lipid-rich biological remains such as spores and algae and deposited in subaquatic muds, usually under oxygen-poor conditions, and have relatively high H/C ratios (1.3–1.7, HUNT, 1979). Oil shales and boghead coals mature from organic-rich sapropelic deposits and correspond to Type I of the three kerogen evolution types recognized by TISSOT et al. (1974). The molecular sizes are dominantly in the range of C_{15} to C_{40} and have average chemical composition ranging from $C_{40}H_{68}O_5$ (Q_{deox} 0.956) to $C_{40}H_{54}O_1$ (Q_{deox} 0.991) on a C_{40} basis. Q_{deox} increases as the depth of burial increases. This kind of H-rich composition requires predominance of saturated hydrocarbons of aliphatic/alicyclic groups and lesser amount (in terms of carbon numbers) of aromatic hydrocarbons. The oxygenous functional groups are scarce, and consist of ester and ether linkages, hydroxyls, carboxyls, and carbonyls in varying proportions (VITOROVIC, 1980). The G_{deox} values are probably somewhere between those of palmitic acid (Q_{deox} 0.958, G_{deox} 425.6 kJ, Table 1) and the convergence value (Q_{deox} 1.00, G_{deox} 424.7 kJ) discussed earlier.

Humic kerogens correspond to Type III of TISSOT et al. (1974), and include humic coals. Atomic ratios on C_{40} basis vary approximately from $C_{40}H_{32}O_{12}$ (Q_{deox} 0.875) to $C_{40}H_{28}O_4$ (Q_{deox} 0.957) as they evolve during diagenesis. This type has the lowest H content among the three types. The compositional boundary between humic kerogen and humic acids is not well defined and the two overlap substantially in the Q_{deox} range. In contrast to the aliphatic-rich Type I, this type is characterized by a high proportion of aromatic carbons. Chemical and physicochemical evidence suggest that aromatic carbon atoms increase from about 70% in hard brown coals to over 90% in anthracites (TISSOT and

WELTE, 1978, p. 210). Also the formation of polycyclic aromatic rings is suggested (several to tens of rings). The main functional groups are ketonic and carboxyl groups, but ester group is missing (ROBIN et al., 1977). Similar to humic acids, Gibbs energy data are available only for a small number of solid or liquid compounds which tend to be low in C-number. H_{deox} can be estimated from polycyclic aromatic compounds with carboxyl and carbonyl groups in Table 2.

DISCUSSIONS

The significance of the G_{deox} versus Q_{deox} relations shown in Fig. 1 regarding the transformation of plant matter to fossil fuels becomes apparent when one notes that typical photosynthetic products have higher G_{deox} and lower Q_{deox} values compared with fossil fuel substances. A higher G_{deox} means, by definition, that more energy is chemically stored in liberating a unit quantity of free oxygen from a mixture of H_2O, CO_2, and N_2, and also that a substance could be deoxygenated spontaneously by loss of CO_2 and/or H_2O to produce another substance of a lower G_{deox}. This fact clearly has an important bearing on the flow of energy in the biosphere as discussed below.

Photosynthesis

The higher G_{deox} values for primary photosynthetic products indicate that photosynthetic organisms chemically capture as much solar energy as possible at the initial stage of photosynthesis by utilizing the thermochemical nature of the C-H-(N)-O system with regard to deoxygenation. When starting out from CO_2, H_2O, and N_2 of ultimate volcanic origin, the most effective chemical storage of solar energy by deoxygenation can be achieved by forming compounds with highest G_{deox} values, which turn out to be compounds of low Q_{deox} values. In terms of G_{deox}, oxalic acid (Q_{deox} 0.200, G_{deox} 649.6 kJ), formic acid (0.333, 554.7 kJ), formaldehyde (0.667, 521.7 kJ), and citric acid (0.529, 499.8 kJ) lead other compounds. The formation of formaldehyde, followed by its polymerization to monosaccharides, has been suggested as the initial stage of photosynthesis by many investigators (CONANT, 1936, p. 561). This path is energetically workable. Carbohydrates (G_{deox} 464 to 487 kJ) can form from formaldehyde spontaneously with release of energy as indicated by the higher G_{deox} of the latter. The organic acids mentioned are also found in plants. These acids will change to saccharides spontaneously if catalyzed by suitable enzymes, as exemplified by the ripening of citric fruits.

Table 2. Deoxygenation parameters of compounds relevant to humic matter

Name	Formula	State	Q_{deox}	H_{deox}	G_{deox}
		Biopolymers			
Cellulose	$(C_6H_{10}O_5)_n$	(s)	.706	468.7	478.5
Lignin (conyferyl)	$(C_{10}H_{10}O_2)_n$	(s)	.920	450.0	—
		Aromatic acids, hydroxyacids, anhydrides			
Benzoic acid	$C_7H_6O_2$	(s)	.882	430.26	430.27
p-Hydroxylbenzoic acid	$C_7H_6O_3$	(s)	.824	432.51	436.52
Salicylic acid	$C_7H_6O_3$	(s)	.824	432.39	436.31
Phthalic acid	$C_8H_6O_4$	(s)	.789	429.78	436.68
Phthalic anhydride	$C_8H_4O_3$	(s)	.833	434.54	439.79
Protocatechuic acid	$C_7H_6O_4$	(s)	.765	435.78	—
Gallic acid	$C_7H_6O_5$	(s)	.706	441.97	—
2,4-Cresotic acid	$C_8H_8O_3$	(s)	.850	432.41	—
Phenylacetic acid	$C_8H_8O_2$	(s)	.900	432.80	—
Phenoxyacetic acid	$C_8H_8O_3$	(s)	.850	444.72	—
p-Toluic acid	$C_8H_8O_2$	(s)	.900	429.13	—
p-Anisic acid	$C_8H_8O_3$	(s)	.850	440.68	—
Piperonylic acid	$C_8H_6O_4$	(s)	.789	448.55	—
Atropic acid	$C_9H_8O_2$	(s)	.909	437.32	—
Cinnamic acid, trans	$C_9H_8O_2$	(s)	.909	434.77	—
Hydrocinnamic acid	$C_9H_{10}O_2$	(s)	.913	432.68	—
p-Cumaric acid	$C_9H_8O_3$	(s)	.864	437.39	—
Uvitic acid	$C_9H_8O_4$	(s)	.818	431.84	—
Trimesic acid	$C_9H_6O_6$	(s)	.714	428.16	—
Phenylpropiol acid	$C_9H_6O_2$	(s)	.905	449.76	—
p-Isopropylbenzoic acid	$C_{10}H_{12}O_2$	(s)	.923	432.03	—
		Aromatic hydroxyl derivatives, aldehydes			
Phenol	C_6H_6O	(s)	.933	436.22	432.49
Benzyl alcohol	C_7H_8O	(l)	.944	439.63	433.17
o-Cresol	C_7H_8O	(s)	.944	434.53	—
o-Hydroxybenzyl alcohol	$C_7H_8O_2$	(s)	.889	442.15	—
p-Ethylphenol	$C_8H_{10}O$	(s)	.952	435.25	—
Thymol	$C_{10}H_{14}O$	(s)	.963	434.79	—
Isoeugenol	$C_{10}H_{12}O_2$	(l)	.923	445.84	—
Pyrocatecohol	$C_6H_6O_2$	(s)	.867	441.16	—
Hydroquinone	$C_6H_6O_2$	(s)	.867	438.83	—
p-Benzoquinone	$C_6H_4O_2$	(s)	.857	457.65	459.17
Phloroglucinol	$C_6H_6O_3$	(s)	.800	430.78	—
2-Naphthol	$C_{10}H_8O$	(s)	.958	430.76	—
1,4-Naphthaquinone	$C_{10}H_6O_2$	(s)	.913	438.94	—
Benzaldehyde	C_7H_6O	(l)	.941	440.37	—
o-Anisaldehyde	C_8H_8O	(l)	.900	447.72	—
Cinnamaldehyde	C_9H_8O	(l)	.955	443.63	—
Phenylpropiol aldehyde	C_9H_6O	(l)	.952	452.37	—
		Aromatic ethers, ketones			
Anisole	C_7H_8O	(l)	.944	444.50	—
Phenetol	$C_8H_{10}O$	(l)	.952	441.57	—
p-Dimethoxybenzene	$C_8H_{10}O_2$	(s)	.905	447.24	—
m-Methylanisole	$C_8H_{10}O$	(l)	.952	441.99	—
p-Allylanisole	$C_{10}H_{12}O$	(l)	.962	447.12	—
Methylisoeugenol	$C_{11}H_{14}O_2$	(l)	.931	448.95	—
Diphenyl ether	$C_{12}H_{10}O$	(s)	.966	437.11	433.05
Acetophenone	C_8H_8O	(l)	.950	436.73	430.19
Banzophenone	$C_{13}H_{10}O$	(l)	.968	435.12	—
Benzoion	$C_{14}H_{12}O_2$	(s)	.941	435.99	—

Table 2. (Continued)

Name	Formula	State	Q_{deox}	H_{deox}	G_{deox}
		Others			
Cumen	C_9H_{12}	(l)	1.00	434.61	424.74
p-Cymen	$C_{10}H_{14}$	(l)	1.00	433.90	424.00
Decahydronaphthalene, trans	$C_{10}H_{18}$	(l)	1.00	432.89	423.19
2-Methylnaphthalene, trans	$C_{11}H_{10}$	(s)	1.00	429.82	423.46
Inositol	$C_6H_{12}O_6$	(s)	.667	461.98	—
p-Cymen-2,5-diol	$C_{10}H_{14}O_2$	(s)	.926	437.85	—
Borneol	$C_{10}H_{18}O$	(s)	.966	438.26	—
Terpinol	$C_{10}H_{18}O$	(s)	.966	439.25	—
Campholic acid	$C_{10}H_{16}O_2$	(s)	.929	439.59	—
Cyclohexane carboxylic acid	$C_7H_{12}O_2$	(s)	.900	434.23	—
Diphenyl carbonate	$C_{13}H_{10}O_3$	(s)	.903	438.81	438.36
Eicosanoic acid	$C_{20}H_{40}O_2$	(s)	.967	433.59	—
Phenylurea	$C_7H_8N_2O$	(s)	.944	433.27	—
Quinolin	C_9H_7N	(l)	1.00	436.40	432.03

The monosaccharides and glycerol (0.700, 472.8 kJ), once synthesized, become the primary ingredients for a variety of biological substances of higher Q_{deox} values. As long as the biosynthesis proceeds *via* loss of H_2O and/or CO_2, the process is most probably spontaneous, because both CO_2 and H_2O are far more stable than the parental substance. Structural material (*e.g.,* cellulose, lignin, waxes, resins) and energy storage material (*e.g.,* starch, fats, oils) are probably produced from the primary ingredients without net addition of energy. In the presence of NH_3 (G_{deox} 452.8 kJ), which is found in volcanic condensates as NH_4Cl (WHITE and WARING, 1963), proteins may also be synthesized from these ingredients.

Transformation to fossil fuels

Fossil fuel substances, such as natural gas (methane), petroleum (paraffins, cycloparaffins, aromatic hydrocarbons), coal (polycyclic aromatic compounds), and oil shale (Type I kerogen), all have higher Q_{deox} and lower G_{deox} values than do primary photosynthetic products. Methane has the lowest G_{deox} (409.0 kJ) among organic substances. Most hydrocarbons found in petroleum in nature have G_{deox} lower than 427.2 kJ. The only known exception appears to be cyclobutane (439.2 kJ), which was reportedly found in a trace quantity (0.001%) in a Venezuelan oil (WHITEHEAD and BREGER, 1963, p. 252). Kerogens in oil shale and boghead coal (Type I) probably have G_{deox} values between 424.7 kJ and 425.6 kJ as discussed earlier.

There may be slight reversals in Q_{deox} and G_{deox} when humic acids form from lignin because of addition of oxygen during oxidative degradation. In the absence of definitive chemical characterization of humic acids and entropy data for lignins, this point is very uncertain. The portion of humic acid derived from cellulose and other carbohydrates have definitely lower G_{deox} values than the precursor substances. The formation of quinic acid from carbohydrates and its subsequent transformation to protocatechuic acid by dehydration and oxidation mentioned earlier may be cited as an example.

Type III humic kerogens and coals, ranging in Q_{deox} from 0.875 to 0.957, probably cluster around 430 kJ in G_{deox} as discussed earlier. With the decrease of functional groups as burial depth increases, it is probable that G_{deox} falls below 430 kJ in bituminous coal, and below 420 kJ in anthracitic coal judging from that of perylene $C_{20}H_{12}$ (421 kJ). The aromatic polycondensation ultimately results in the formation of graphite (394.4 kJ).

In the overall deoxygenation energy relation of fossil fuels, a significant point is that the G_{deox} values of fossil fuel substances are very close to or lower than the convergence G_{deox} value (424.7 kJ) for non-aromatic compounds mentioned earlier. This fact appears to indicate that no net input of energy has taken place during the fossilization of biological remains and the evolutionary history of fossil fuels. If energy had been added, in net, during the processes, some fossil-fuel substances would show substantially higher G_{deox} values than the convergence G_{deox} value. It is well known that thermal cracking of petroleum produces olefins (443.5 to 426.8 kJ). Hydrogen (H_2, 474.4 kJ) is scarce in natural gas. It may be argued that H_2 escapes readily because of its high buoyancy and mobility, but the accumulation of helium, which is equally mobile, in natural gas reservoirs does not support this argument. In the presence of unsaturated organic compounds, H_2, if ever produced, is expected to be consumed readily for hydrogenation of the unsaturated compounds. It is more probable, however, that so-called hydrogenation reactions are caused by intra- or inter-molecular proton transfer and that H_2 partial pressure never reaches a significant level during the transformation and maturation.

Chemical changes of organic sedimentary deposits are rapid at the beginning of burial and soon reach a plateau stage when the carbon content reaches about 85 weight percent (Q_{deox} about 0.96) (BREGER, 1963). This is where the mode of energy release changes from deoxygenation to hydrogen disproportionation, and a sharp bend in the kerogen evolution lines occurs in the Van Krevelen diagram. The H/C ratio begins to decrease rapidly at this point. With most of the oxygen depleted (the atomic O/C < 0.1), the dominant form of the energy release shifts to simultaneous polycondensation of aromatic rings and production of methane and, in case of aliphatic Type I kerogens, other light paraffinic hydrocarbons. As polycondensation continues, H atoms are liberated from condensing alicyclic, ketonic, and benzene rings and spontaneously partition to form saturated light (C_8 or smaller) hydrocarbons. These compounds all have G_{deox} values lower than the convergence G_{deox}. This stage is catagenesis.

The ultimate state of energy release without oxidation is the graphite (393.4 kJ) and methane (409.0 kJ) combination. This is the extreme result of the spontaneous hydrogen partioning, a process that operates in metagenesis. Methane escapes easily so that graphite becomes the ultimate *in situ* product of spontaneous evolution of organic matter in an anaerobic environment.

The large positive G_{deox} values for all organic substances of the C-H-N-O system indicate that isolation from free oxygen is essential for fossil-fuel formation. The Gibbs free energies of reaction evolved in the direct oxidation by free oxygen are at least an order of magnitude greater than those in the deoxygenation and hydrogen disproportionation reactions. An access to free oxygen would change the course of reactions, ultimately oxidizing the whole organic matter back to the pre-photosynthetic material, *i.e.*, H_2O, CO_2, and N_2.

Higher temperatures (up to a few hundred °C) do not appear to change the energy relations, but they surely would increase the reaction rates, which increase exponentially following the Arrhenius equation, $k = Ae^{-Ea/RT}$, where Ea is the activation energy and k the rate constant. Because the spontaneity of evolution of fossil fuels is indicated even

at room temperature, the role of increasing temperature associated with increasing depth of burial is to increase, by orders of magnitude, the rates of reactions which would otherwise not proceed detectibly even in geologic time scale. The activation energy for the polycyclic aromatic condensation to cause hydrogen disproportionation must be large, requiring catagenic temperatures (100 to 150°C) for the process to proceed in geologic time. This, however, does not necessarily imply that a heat input is essential for advanced stages of fossil fuel evolution. A large mass of organic sediment could create its own thermal environment, if heat is trapped by impervious overlying sediments, as the transformation reactions are generally exothermic. Also, adiabatic compression of fluids may elevate the temperature in a large sedimentary basin such as the Gulf of Mexico (JONES, 1970).

A totally confined environment, on the other hand, may not favor evolution of fossil fuels. The mass action law indicates that, at equilibrium, the removal of the reaction products facilitates for a reaction to proceed. The removal is particularly important when the free energy of reaction is small, as in the late stages of fossil fuel evolution. Geological settings in which fluids (H_2O, CO_2, CH_4, and other light hydrocarbons) can escape through porous media or fractures, or be absorbed by reactions with lithic material (such as unhydrated volcanic ash), would accelerate the evolutionary process.

In conclusion, I would like to emphasize that this work is a preliminary attempt to obtain a broad picture of the significance of the thermochemical relationship of organic substances to the process of fossil-fuel formation. Thermochemical data for important natural substances such as lignin, humic acids, and kerogens are incomplete and only a very broad framework can be constructed. Also, computations at elevated temperatures and pressures are yet to be completed. Hopefully a better framework will be built in future.

Acknowledgement—This work was seeded in my mind when the late Wilmot H. Bradley recruited me to study the algal sediment at Mud Lake, Florida, in 1967 and asked me if algal sediment could be converted to oil shale under normal burial conditions. It seems to me that the present paper is appropriate to this memorial volume, because the late Hans Eugster was interested in the thermochemistry of the C-H-O system and he had a close friendship with W. H. Bradley. I deeply cherish the memory of my friendship with both of them, two scientists of extraordinary caliber and visions. I thank the late Irving A. Breger for challenging discussions in the early days of this work. I also thank Richard A. Robie for his help and comments. Rama K. Kotra, William H. Orem, and Everett L.

Shock reviewed the current version. I am grateful for their constructive criticisms.

REFERENCES

BREGER I. A. (1963) Classification of naturally occurring carbonaceous substances. In *Organic Geochemistry* (ed. I. A. BREGER), pp. 50–86. *Inter. Ser. Mono. Earth Sci. 16.* MacMillan.

CHRISTMAN R. F. and OGLESBY R. T. (1971) Microbiological degradation and the formation of humus. In *Lignins* (eds. K. V. SARKANEN and C. H. LUDWIG), pp. 769–795. Wiley-Interscience.

COLBERT J. C., HE XIHENG and KIRKLIN D. R. (1981) Entropy of microcrystalline cellulose. *J. Res. National Bur. Standards* **86,** 655–659.

CONNANT J. B. (1936) *The Chemistry of Organic Compounds.* MacMillan.

GALIMOV E. M. (1980) C^{13}/C^{12} in kerogen. In *Kerogen* (ed. B. DURAND), pp. 271–299. Editions Technip. Imprimerie Bayensaine. Paris.

GAUER U., SHU H.-C., MEHTA A. and WUNDERLICH B. (1981) Heat capacity and other thermodynamic properties of linear macromolecules I, selenium. *J. Phy. Chem. Ref. Data* **10,** 89–110.

HUC A. Y. (1980) Origin and formation of organic matter in recent sediments and its relation to kerogen. In *Kerogen* (ed. B. DURAND), pp. 445–474. Editions Technip. Imprimerie Bayensaine. Paris.

HUNT J. M. (1979) *Petroleum Geochemistry and Geology.* W. H. Freeman.

JANZ G. J. (1967) *Thermodynamic Properties of Organic Compounds.* Academic Press.

JONES P. H. (1970) Geothermal resources of the northern Gulf of Mexico basin. *Geothermics,* Special Issue 2, Pt. 1, 14–26.

KIRK T. K., HIGUCHI T. and CHANG H.-M. (eds.) (1980) *Lignin Biodegradation: Microbiology, Chemistry, and Potential Applications.* Vol. I; Vol. II. CRC Press.

LAIDLER K. J. (1956) A system of molecular thermochemistry for organic gases and liquids. *Canadian J. Chem.* **34,** 626–628.

OWEN L. (1948) The origin and evolution of petroleum. *Petrol. Times* **52,** 1052–1054.

ROBIN P. L., ROUXHERT P. G. and DURAND B. (1977) Caracterisation de kerogenes et de leu evolution pa spectroscopie infrarouge. In *Advances in Organic Chemistry, 1975* (eds. R. COMPRES and J. GONI), pp. 693–716. Enadimsa. Madrid.

SARKANEN K. V. and LUDWIG C. H. (eds.) (1971) *Lignins-Occurrence, Formation, Structure and Reactions.* Karl Freudenberg Symposium Volume. Wiley-Interscience.

SCHNITZER M. and NEYROND J. A. (1975) Alkanes and fatty acids in humic substances. *Fuel* **54,** 17–19.

STULL D. R., WESTRUM JR., E. F. and SINKE G. C. (1969) *The Chemical Thermodynamics of Organic Compound.* John Wiley.

SWAIN F. M. (1963) Geochemistry of Humus. In *Organic Geochemistry* (ed. I. A. BREGER), pp. 87–147. *Inter. Ser. Mono. Earth Sci. 16.* MacMillan.

TISSOT B., DURAND B., ESPITALIE J. and CONLAZ A. (1974) Influence of nature and diagenesis of organic matter in formation of petroleum. *Amer. Assoc. Petrol. Geol. Bull.* **58,** 499–506.

TISSOT B. P. and WELTE D. H. (1978) *Petroleum Formation and Occurrence.* Springer-Verlag. (2nd ed. 1984).

VAN KREVELEN D. W. (1961) *COAL.* Elsevier.

VAN KREVELEN D. W. (1963) Geochemistry of coal. In *Organic Geochemistry* (ed. I. A. BREGER), pp. 183–247. *Inter. Ser. Mono. Earth Sci. 16.* MacMillan.

VAN KREVELEN D. W. and CHERMIN H. A. G. (1951) Estimation of the free enthalpy (Gibbs free energy) of formation of organic compounds from group contributions. *Chem. Eng. Sci.* **1,** 66–80.

VITROVICIC D. (1980) Structure elucidation of kerogen by chemical methods. In *Kerogen,* pp. 301–338. (ed. B. DURAND). Editions Technip. Imprimerie Bayensaine. Paris.

WELTE D. H. (1970) Organischer Kohlenstoff und die Entwicklung der Photosynthesese auf der Erde. *Naturwissenschaften* **57,** 17–23.

WHITE D. E. and WARING G. A. (1963) *Data of Geochemistry* (Sixth Ed.), Chapter K, *Volcanic Emanations.* U.S. Geol. Surv. Prof. Paper 440-K.

WHITEHEAD W. L. and BREGER I. A. (1963) Geochemistry of petroleum. In *Organic Geochemistry* (ed. I. A. BREGER), pp. 248–332. *Inter. Ser. Earth Sci. 16.* MacMillan.

Fluid-Mineral Interactions: A Tribute to H. P. Eugster
© The Geochemical Society, Special Publication No. 2, 1990
Editors: R. J. Spencer and I-Ming Chou

Paleohydrogeology of the Colorado Plateau—background and conceptual models

RICHARD F. SANFORD

Mail Stop 905, U.S. Geological Survey, Denver Federal Center, Denver, CO 80225, U.S.A.

Abstract—Tectonic, stratigraphic, paleo-climatic, and hydrologic data are compiled for reconstructing regional paleogroundwater flow through the Phanerozoic in the Colorado Plateau. Following the early Paleozoic, which was characterized by a stable shelf environment, the Colorado Plateau underwent four major tectonic-sedimentary cycles: Pennsylvanian-Permian, Triassic, Jurassic-early Cretaceous, and late Cretaceous-Tertiary. Each cycle consisted of marine deposition, followed by uplift and fluvial-lacustrine-eolian deposition, followed by uplift and erosion. Seven principal aquifer systems and eight principal confining units controlled groundwater flow during these four cycles. During the same period, northward drift of the North American plate took the Colorado Plateau region from tropical equatorial through temperate mid-latitude climatic zones. These climatic changes profoundly affected groundwater flow and chemistry. Uplift of mountain ranges, principally the Pennsylvanian-Permian ancestral Rocky Mountains, the Triassic-Jurassic Mogollon highlands, and the Tertiary Rocky Mountains, profoundly affected climatic conditions and favored topographically controlled or gravity-driven groundwater flow. Repeated transgressions of shallow epeiric seas caused burial and compaction. Interfaces between fluids of contrasting composition formed as environments alternated among marine, evaporitic marginal marine, sabkha, fluvial, fresh lacustrine, saline lacustrine, and non-depositional environments. Mixing of fluids at such interfaces may have caused precipitation of uranium, vanadium, and copper. Variations in groundwater flow through time probably also influenced the distribution of hydrocarbon resources.

INTRODUCTION

THE COLORADO PLATEAU physiographic province of the western United States (Fig. 1) contains reserves of oil and gas, uranium, vanadium, copper, carbon dioxide, and evaporite (saline) minerals. Although the formation of these resources was controlled by the movement of groundwater, little is known about the paleohydrogeology. Diagenetic and mineralogical studies that have hydrogeologic implications have focused on parts of the San Juan, Paradox, and Henry Mountains basins in the Colorado Plateau (*e.g.* KELLER, 1962; NORTHROP, 1982; BELL, 1983, 1986; BREIT, 1986; TURNER-PETERSON and FISHMAN, 1986; HANSLEY, 1989). Regional hydrologic studies typically are limited to the present environment (*e.g.* BERRY, 1959; JOBIN, 1962; HANSHAW and HILL, 1969; LYFORD *et al.*, 1980; FREETHEY and CORDY, 1989; GELDON, 1989). As a step toward a synthesis of geologic and hydrologic data into a comprehensive paleohydrogeologic model, this paper presents a summary of the most important data and a conceptual model for groundwater flow through time in the Colorado Plateau. The data constitute input for quantitative hydrologic models that will test the speculative scenarios presented here (SANFORD, 1982, 1989).

The conceptual models are constructed by combining principles and data from observed stratigraphy, inferred paleoclimates and paleoenvironments, measured and derived hydrologic properties

of rocks, hydrologic properties of modern sediment analogs, and groundwater flow in modern analog systems. First, the basin is reconstructed using modern stratigraphic thicknesses. The paleotopographic base level is restored using modern slopes in similar environments combined with paleo-slope indicators such as current directions and grain-size distribution. Then major factors affecting groundwater flow are evaluated based on basic hydrologic principles and modern groundwater systems (*e.g.* HUBBERT, 1940; FREEZE and CHERRY, 1979). Factors considered are gravity-driven flow down the topographic slope, outcrop pattern, evaporation of surface water, evapotranspiration by phreatophytes, compaction of sediments and accompanying overpressure, decompression due to erosional unloading, permeability and hydraulic conductivity variations, composition and density contrasts, temperature gradients, and fluid mixing. The flow paths are then constructed to be consistent with these factors. Finally, the flow paths are compared with mineral alteration and deposition patterns. In many cases, opposing hydrologic forces complicate the qualitative prediction of flow paths, and quantitative modeling is essential.

Mixing of different types of groundwater is given particular attention because of its potential to cause diagenetic changes and form certain types of ore deposits (BACK and HANSHAW, 1965; RUNNELS, 1969; PLUMMER, 1975). Because the thermodynamic saturation index is rarely a linear function of mixing between two fluids, precipitation can occur even when both end-member fluids are originally undersaturated with respect to a particular mineral (BACK and HANSHAW, 1965). Types of ore deposits linked to fluid interfaces include tabular-type uranium deposits, which will be used as an example in this paper.

Oil and gas migration and accumulation are also controlled by groundwater, and the location of hydrocarbons today may be as much a function of the history of groundwater flow as it is of the present flow (HUBBERT, 1953). Petroleum and natural gas are not specifically addressed in this paper, but the framework provided may be useful to petroleum geologists.

TECTONIC CYCLES AND GROUND-WATER CONTROLS

The Phanerozoic history of the Colorado Plateau was characterized by four major tectonic cycles: Pennsylvanian-Permian, Triassic, Jurassic, and Cretaceous-Tertiary. Each cycle lasted from 64 to 100 million years and consisted of marine-marginal marine deposition, fluvial-lacustrine-eolian deposition, and erosion (SANFORD, 1983; SANFORD, in GRANGER et al., 1988). Many fluctuations occurred within individual cycles (Fig. 2). A diagram of deposition rate is shown instead of a stratigraphic column in order to show duration of events in proper proportion rather than relative thicknesses of stratigraphic units. Periods of erosion or non-deposition are just as important for understanding groundwater and spanned as much geologic time in the Colorado Plateau as periods of deposition.

Periods of deposition involve loading and expulsion of connate pore water that moves upward and outward from deeper parts of the basins (Fig. 3a) (MAGARA, 1976). The greater density of solids compared to groundwater, including saline groundwater, forces pore water upward with respect to the framework grains. Flow due to compaction is relatively slow and takes place over long periods of time in intracratonic basins (BETHKE, 1985). In the Colorado Plateau, Upper Cretaceous marine shales and abundant fluvial mudstones of all ages would be sources of expelled pore water. Fluvial, marine, and eolian sandstones compact less than mudstones, but can still expel large quantities of pore water. Compaction has the greatest influence on groundwater movement during rapid marine deposition, when topographic effects are minimal.

During fluvial deposition in general, gravity-driven flow probably outweighs compaction (BETHKE, 1985). Gravity-driven flow results from topographic variations and is characterized by flow from elevated areas toward topographic depressions (FREEZE and WITHERSPOON, 1967; HITCHON, 1969). Permeability variations in recharge areas tend to focus flow into aquifers; in discharge areas flow tends to be upward and out of aquifers (GARVEN and FREEZE, 1984). Lakes may be discharge areas or recharge areas, but large permanent lakes and lakes close to base level are nearly always discharge areas (WINTER, 1976; FREEZE and CHERRY, 1979, pp. 226-229).

Periods of nondeposition or erosion may have little compaction but much gravity-driven flow (Fig. 3b). In the Colorado Plateau, compaction due to tectonic compression was probably insignificant, except in the extreme northwest part of the region during the Sevier orogeny. During uplift and erosional downcutting, dilute meteoric water mixes with and may eventually flush out saline formation water in aquifers, although less permeable or very deep units may retain older water (ISSAR, 1981; DOMENICO and ROBBINS, 1985). In arid environments, much of the section above base level (e.g. sea level) may be underpressured or even unsaturated, and flow may be minimal (ORR and KREITLER, 1985; FREETHEY and CORDY, 1989).

Uplift and associated erosion also cause decompression, fracturing, and dissolution. Decompression can lower the hydrostatic head and tend to draw fluid into the region of unloading (NEUZIL and POLLOCK, 1983). Fracturing and dissolution of highly soluble minerals, such as halite and calcite, increases the hydraulic conductivity in uplifted areas (JOHNSON, K. S., 1981; FREETHEY and CORDY, 1989).

Pore water, originally sea water or meteoric water, is typically saline at depth owing to salt and carbonate dissolution, chemical equilibration with sediments, or membrane filtration (e.g. BREDEHOEFT et al., 1963; HANSHAW and COPLEN, 1973; KREITLER, 1979). In all phases of the tectonic cycle, salinity variations in the fluid tend to cause the groundwater to move toward a state of gravitational equilibrium or steady state: when both fluids are stationary, the denser fluid lies below less dense fluid, and the interface is horizontal; when the upper, less dense fluid is flowing, the interface slopes upward in the direction of flow, (HUBBERT, 1940). Typically, salinity increases along the flow path from the basin margin to center, and a lens or wedge of fresh water rests upon saline water below (Fig. 3c). The meeting of dilute meteoric water moving down dip with saline pore water from below often creates a brine-fresh water interface (PAYNE, 1968, 1970,

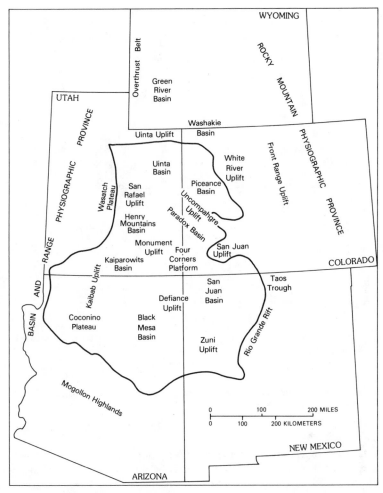

FIG. 1. Map of the Colorado Plateau physiographic province and surrounding area showing major structural features. From GROSE, 1972; GREEN, in GRANGER *et al.*, 1988.

1972, 1975; KREITLER, 1979; KREITLER *et al.*, 1977). Also, the meeting of meteoric, down-dip flow with intruding sea water or playa lake water often forms an interface (PERLMUTTER *et al.*, 1959; COOPER *et al.*, 1964; MCLEAN, 1970, 1975; CUSTODIO, 1981; KISHI *et al.*, 1982; MAGARITZ and LUZIER, 1985; RANDAZZO and BLOOM, 1985). Aquifers may be well mixed, whereas confining units may exhibit large compositional variations. Whenever the configuration of the basin changes due to tectonic forces or sedimentation, the relative flow rates of the different water types changes, or the groundwater changes density due to chemical effects, the bodies of groundwater and the positions of the interfaces must readjust to reestablish gravitational equilibrium. Salinity can also affect fluid flow by changing fluid viscosity and hydraulic conductivity.

Gradients in salinity across a low permeability layer can cause migration of water from the fresh water side toward the brine in response to the chemical potential gradient. This osmotic effect can explain certain abrupt changes in hydraulic head in the San Juan and Paradox basins (BERRY, 1959; HANSHAW and HILL, 1969). However, later workers found no osmotic membrane effects in the Paradox basin (THACKSTON *et al.*, 1981).

Temperature may affect groundwater flow at depth. The upward flux of geothermal energy tends to result in hotter, less dense water below cooler, denser water. Free convection may result, if the temperature effect overcomes the compositional effect on density (WOODING, 1962; WOOD and HEWETT, 1982). Compaction-driven flow is probably so slow that a normal geothermal gradient is maintained (BETHKE, 1985), but gravity-driven flow may

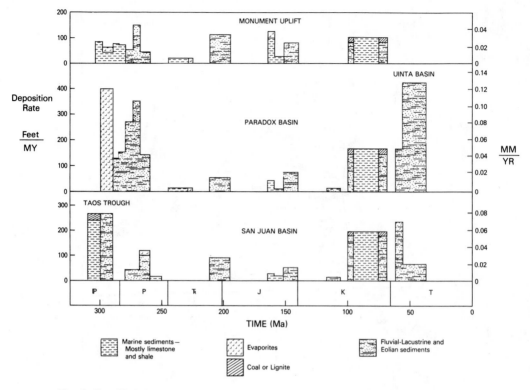

FIG. 2. Simplified deposition rate versus time for selected parts of the Colorado Plateau. Data primarily from MALLORY, 1972.

be large enough that the temperature is lowered in area of recharge and raised in areas of discharge (GARVEN and FREEZE, 1984). Because the salinity of groundwater in the Colorado Plateau ranges from fresh (<1000 mg/l TDS) to as much as 400,000 mg/l (*e.g.* HANSHAW and HILL, 1969), density variations due to salinity would typically outweigh those due to temperature, as suggested also for the Gulf Coast basin (RANGANATHAN and HANOR, 1988). However, temperature may be a major driving force locally in compositionally homogeneous aquifers (WOOD and HEWETT, 1982). Temperature also affects fluid viscosity and hydraulic conductivity.

The sequence of events determines which compositionally distinct types of pore water will mix and interact. Marine transgressions can be expected to cause sea water to mix with and displace interstitial fresh water and to cause the gravitationally unstable configuration of sea water overlying fresh water. Prograding fluvial deposits encroaching on regressing seas may lead to interstitial fresh water overlying sea water. When marine regressions occur during arid periods having evaporite deposition, fresh water can be expected to overlie hypersaline brine. Young fluvial sediments typically contain

fresh groundwater, but associated lacustrine deposits may be either fresh or saline. Fluvial-lacustrine deposits therefore can have mixing of fresh and saline water in the near-surface as well as mixing of these types with other types at depth.

TECTONIC HISTORY AND PALEOCLIMATES

Plate-tectonic reconstructions and analysis of climatically sensitive deposits, such as coal and evaporites, show that the North American plate was drifting northward and rotating counterclockwise during most of the Paleozoic (*e.g.*, PARRISH *et al.*, 1982). A stable shelf environment existed in the Colorado Plateau region from Cambrian to Mississippian, while carbonates and clean sands were deposited (MALLORY, 1972; FOUCH and MAGATHAN, 1980). Gentle upwarping during late Mississippian and early Pennsylvanian caused karst formation in the Mississippian limestones. During the early Pennsylvanian, the Colorado Plateau was just north of the equator, and a shallow epeiric sea lay to the east. Warm trade winds from the northeast and cross-equatorial monsoon winds from the south provided a humid environment in coastal areas and

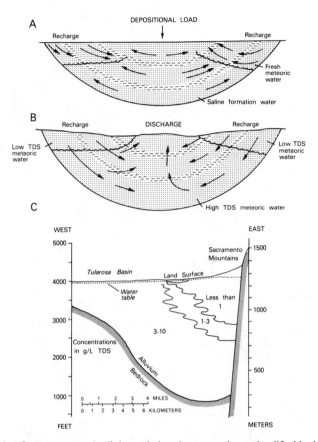

FIG. 3. Typical flow patterns and salinity variations in compacting and uplifted basins. A) Young basin undergoing active deposition and compaction. From KREITLER, 1979; B) Mature basin dominated by gravity-driven ground water flow. From KREITLER, 1979; C) Continental basin in arid climate showing water table and fresh water-brine interface. From McLEAN, 1975.

inland as shown by "terra rosa" soils developed on exposed Mississippian carbonates (LESSENTINE, 1965) and by Morrowan lignite deposits in the Taos trough (CASEY, 1980). The environment was probably like that of east-facing (windward) coasts near the equator today, for example, in Indonesia.

By middle Pennsylvanian, the ancestral Rocky Mountains had created an extensive rain shadow inland where vast evaporite deposits began to form (Fig. 4a). By the Triassic, when the highlands were substantially reduced by erosion, the Carboniferous seas had retreated, and the vast Pangean continent lay to the northeast. A monsoon climate prevailed from the late Permian until the latest Triassic (Fig. 4c) (PARRISH et al., 1982; DUBIEL, 1989). In winter months, cool, dry continental air masses would have come from the interior to the northeast. During the summer, the landmass would have drawn air from the Pacific Ocean. In the latest Triassic, conditions again became arid. The rising Mogollon highlands may have blocked the moist air from the southwest.

Also, the Colorado Plateau region drifted north from equatorial latitudes into the trade-wind zone where winds tend to be from the northeast.

During the Triassic and Jurassic the ancestral Rocky Mountains were eroded to base level while the Mogollon volcanic arc and highlands rose to great height (Fig. 4d). Numerous transgressions and regressions of the sea from the northwest alternated with periods of dominantly fluvial-lacustrine-eolian sedimentation and non-deposition. Overall, a dominantly marine-marginal marine-evaporite period was followed by a dominantly fluvial-lacustrine-eolian phase in both the Triassic and Jurassic.

When the Four Corners had drifted to about 40°N, the Colorado Plateau region entered the temperate zone with dominantly westerly winds, cooler temperatures, and increased rainfall (Fig. 4e) (PARRISH et al., 1982). In late Cretaceous, coal beds were widespread. An extensive epeiric sea that transgressed from the northeast covered much of the Colorado Plateau.

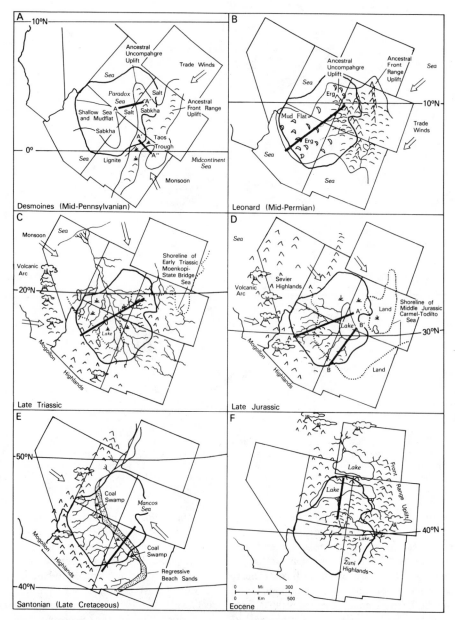

FIG. 4. Paleogeography and paleoclimate of the Colorado Plateau for selected periods in the Phanerozoic. Data from BLAKEY and GUBITOSA, 1983; PETERSON and HITE, 1969; MALLORY, 1972; BLAKEY, 1974; MACK *et al.,* 1979; CAMPBELL, 1980; CASEY, 1980; HECKEL, 1980; PETERSON, 1980; BLAKEY, 1980; PARRISH *et al.,* 1982; PARRISH and PETERSON, 1988; PETERSON, 1988a; DUBIEL, 1989.

Paleocene and Eocene uplift led to locally arid conditions controlled by orographic effects (Fig. 4f). Coal accumulation nearly ceased by the Paleocene (TREMAINE *et al.,* 1981), and alkaline-saline lakes became extensive by the Eocene (EUGSTER and HARDIE, 1978; JOHNSON and KEIGHIN, 1981). Late Tertiary uplift led to conditions much like those today.

HYDROLOGIC PROPERTIES AND HYDROSTRATIGRAPHIC UNITS

The wide range of sediment types and degrees of lithification in the sedimentary basins of the Colorado Plateau has resulted in extreme variation in hydrologic parameters. Horizontal hydraulic conductivity can range from 10^{-3} m/s for clean, uncemented littoral and eolian sand to 10^{-12} m/s for compacted, unfractured shale (FREEZE and CHERRY, 1979, p. 29). The actual range is probably less in general, because sandstones are commonly cemented by calcite or silica, lowering their hydraulic conductivity, and shales are commonly fractured and faulted, raising their hydraulic conductivity. Vertical hydraulic conductivity measurements are not available for most confining units in the Colorado Plateau, but estimates suggest modern values in the range 10^{-13}–10^{-10} m/s (FRENZEL and LYFORD, 1982).

Similarly, porosity is highly variable. In general, it decreases with increasing depth and lithostatic pressure. Argillaceous sediments may be deposited with 70–80% porosity but may only have 25–30% porosity after burial to 1000 m; sand typically has 35–45% porosity at the surface but 25–35% at 1000 m (HANOR, 1979; BOND and KOMINZ, 1984; BALDWIN and BUTLER, 1985). Most compaction data is from marine basins; sediments deposited subaerially may have only 40–50% porosity at the surface. Shales release relatively more of their pore water near the surface and are mostly compacted at relatively shallow depth, whereas sands compact gradually over a greater depth.

For analysis of groundwater flow, lithostratigraphic units are grouped into hydrostratigraphic units. Hydrostratigraphic units are parts of or groups of lithostratigraphic units that are similar in hydrologic properties and act as a single hydrologic unit. They typically do not correspond to individual formations or time-stratigraphic units. Aquifers are hydrostratigraphic units that have high transmissivity on a regional scale. Confining units are those that have generally low transmissivity.

Rocks of the Colorado Plateau can be grouped into seven aquifer systems and eight confining units (Table 1). Lithology and type of pore water included in the original sediment are shown for each lithostratigraphic unit. Permeabilities have been converted to hydraulic conductivities in meters per second (m/s) for pure water at 20°C. The highest hydraulic conductivities reflect fracturing at shallow depths (FREETHEY and CORDY, 1989; GELDON, 1989). Because this fracture permeability formed mainly during Tertiary uplift, values toward the lower end of the range probably should be used for pre-Tertiary basin reconstructions. Modern sediments that may serve as analogs for ancient environments have hydraulic conductivities comparable to the more permeable fractured sandstones.

The main aquifers are the Mississippian karstic limestones, Pennsylvanian-Permian eolian sandstones, Triassic-Jurassic eolian sandstones, Middle-Upper Jurassic eolian sandstones, and three Cretaceous fluvial-marine sandstones. Shales, siltstones, evaporites, and massive limestones form the confining units.

Faults have been variously interpreted as conduits and as barriers to groundwater flow (HANSHAW and HILL, 1969; THACKSTON et al., 1981; HUNTOON, 1983; FREETHEY and CORDY, 1989).

GROUNDWATER CHEMISTRY

The past composition of groundwater can be inferred from analyses of existing water in the rocks of interest, analyses of water in analog modern environments, analyses of the original fluid trapped as inclusions in minerals, and thermodynamic stabilities of diagenetic minerals. Groundwater from the Colorado Plateau today indicates the types of water to be expected in the past when arid and semiarid climates prevailed, and when fresh water dissolved soluble constituents from the sediments. Modern analogs can suggest the types of water trapped in sediments when deposited. Fluid inclusions can provide evidence of paleo-temperatures and salinities, but usually only for atypically coarse-grained or high-temperature deposits. Thermodynamic calculations place constraints on fluid compositions, but kinetic effects and experimental uncertainty are limiting factors.

Modern groundwater in the Colorado Plateau ranges from fresh (<1000 mg/L TDS) to saline (>35,000 mg/L) (HANSHAW and HILL, 1969; IORNS et al., 1965; PRICE and ARNOW, 1974; THACKSTON et al., 1981; FREETHEY et al., 1984; HOOD and PATTERSON, 1984; WARNER et al., 1985; FREETHEY and CORDY, 1989; GELDON, 1989). Fresh water occurs at shallow depth near recharge areas. Dissolved solids increase with depth away from recharge areas (Fig. 3). The deeper basins typically contain highly saline to briny water; for example, more than 300,000 mg/L TDS occurs locally in the Leadville and Redwall Limestones (HANSHAW and HILL, 1969; GELDON, 1989), and 400,000 mg/L TDS is in the Paradox Member of the Hermosa Formation (THACKSTON et al., 1981).

Major solutes in groundwater of the Colorado Plateau are sodium, calcium, magnesium, potassium, chloride, bicarbonate, carbonate, sulfate, and silica (FREETHEY and CORDY, 1989; GELDON, 1989). These constituents can usually be related to dissolution of soluble minerals such as calcite, gypsum, halite, sylvite, dolomite, and unstable silicates. The most common water types are calcium bicarbonate and sodium chloride. Major dissolved constituents in fresh water are typically calcium and bicarbonate. Brines (>35,000 mg/L) are commonly

Table 1. Primary lithostratigraphic and hydrostratigraphic units of the Colorado Plateau

Age	Primary lithostratigraphic units*	Lithologies	Original pore water	Hydrostratigraphic unit**	Hydraulic conductivity (m/s)†
Cambrian-Mississippian	Ignacio	Quartzite	Sea water	Lower Paleozoic aquifer	10^{-7} to 60
	Elbert	Dolomite, shale, sandstone	Sea water		
	Redwall	Limestone, orthoquartzite	Sea water		
	Leadville	Limestone, orthoquartzite	Sea water		
Pennsylvanian-Permian	Supai (lower)	Shale	Fresh-saline	Upper Paleozoic confining unit	10^{-6} to 3
	Molas	Paleosol	Fresh		
	Pinkerton Trail	Marine carbonate, shale	Sea water		
	Paradox	Halite, gypsum, shale	Hypersaline brine		
	Honaker Trail	Marine carbonate, shale	Sea water		
	Cutler	Arkose	Fresh		
	Halgaito	Shale	Fresh-saline		
	Abo	Shale, sandstone	Fresh-saline		
	Hermit	Shale	Fresh-saline		
	Organ Rock	Shale	Fresh-saline		
	Yeso	Shale, sandstone, carbonate	Fresh-saline		
Pennsylvanian-Permian	Supai (upper)	Eolian sandstone	Fresh-saline	Upper Paleozoic aquifer	10^{-5} to 60
	Weber	Eolian sandstone	Fresh-saline		
	Elephant Canyon	Marine limestone, sandstone	Sea water		
	Cedar Mesa	Eolian sandstone	Fresh-saline		
	De Chelly	Eolian sandstone	Fresh-saline		
	White Rim	Eolian sandstone	Fresh-saline		
	Coconino	Eolian sandstone	Fresh-saline		
	Glorieta	Eolian sandstone	Fresh-saline		
	Toroweap	Marine limestone, sandstone	Sea water		
	Kaibab	Marine limestone	Sea water		
	San Andres	Marine limestone	Sea water		
Triassic	Moenkopi	Marine siltstone, limestone, gypsum	Sea water, hypersaline brine	Triassic confining unit (Fluvial sandstone locally transmissive)	10^{-6} to 10^{-4}
	Chinle (lower)	Fluvial-lacustrine sandstone, mudstone	Fresh		
	Chinle (upper)	Bentonitic lacustrine mudstone, sandstone	Fresh-alkaline saline		
Triassic-Jurassic	Wingate	Eolian sandstone	Fresh-saline	Glenn Canyon Group aquifer	10^{-5} to 10^{-3} (Wingate)
	Kayenta	Fluvial sandstone	Fresh		10^{-3} to 0.1 (Navajo)
	Navajo	Eolian sandstone	Fresh-saline		
	Nugget	Eolian sandstone	Fresh-saline		
	Page	Eolian sandstone	Fresh-saline		
Middle-Jurassic	Carmel	Marine siltstone, limestone	Sea water-hypersaline brine	Carmel confining unit	N.A.
	Entrada (lower)	Silty sandstone	Fresh-saline		
Middle-Late Jurassic	Entrada (upper)	Eolian sandstone	Fresh-saline	Entrada-Morrison aquifer	10^{-5} to 10^{-2}
	Cow Springs	Eolian sandstone	Fresh-saline		
	Bluff	Eolian sandstone	Fresh-saline		
	Salt Wash	Fluvial sandstone, mudstone	Fresh		
	Recapture	Fluvial sandstone, mudstone	Fresh-saline		
	Westwater Canyon	Fluvial sandstone, mudstone	Fresh		

Table 1. (Continued)

Age	Primary lithostratigraphic units*	Lithologies	Original pore water	Hydrostratigraphic unit**	Hydraulic conductivity (m/s)†
Middle-Late Jurassic	Curtis	Sandstone, mudstone, gypsum	Sea water-brine	Summerville-Curtis confining unit	N.A.
	Summerville	Mudstone, sandstone	Sea water-brine		
	Wanakah	Sandstone, siltstone, limestone, gypsum	Sea water-brine		
	Tidwell	Mudstone, gypsum	Sea water-brine		
Late Jurassic	Brushy Basin	Bentonitic lacustrine mudstone, sandstone	Fresh-alkaline saline	Brushy Basin confining unit	N.A.
Early-Late Cretaceous	Burro Canyon	Fluvial sandstone	Fresh	Dakota aquifer	10^{-3} to 10^{-2}
	Dakota	Fluvial-marine sandstone	Fresh-sea water		
Late Cretaceous	Mancos	Marine shale	Sea water	Mancos confining unit	N.A.
Late Cretaceous	Gallup	Fluvial-marine sandstone	Fresh-sea water	Mesaverde aquifer	0.01 to 1.0
	Mesaverde	Fluvial-marine sandstone	Fresh-sea water		
	Crevasse Canyon	Fluvial-marine sandstone	Fresh-sea water		
	Point Lookout	Coastal or marine sandstone	Fresh-sea water		
	Cliff House	Coastal or marine sandstone	Fresh-sea water		
Late Cretaceous	Lewis	Marine shale	Sea water	Lewis confining unit	N.A.
Late Cretaceous	Pictured Cliffs	Coastal or marine sandstone	Fresh-sea water	Pictured Cliffs aquifer	0.01 to 1.0
Late Cretaceous-Tertiary	Fruitland	Sandstone, shale, coal	Fresh-sea water	Tertiary confining unit	10^{-11} to 10^{-6} (Fluvial beds and fracturing enhance permeability locally)
	Kirtland	Fluvial-lacustrine shale	Fresh		
	Animas	Conglomerate, arkose, shale	Fresh		
	Nacimiento	Shale and arkose	Fresh		
	San Jose	Fluvial shale, sandstone	Fresh		
	Currant Creek	Conglomerate, sandstone, shale	Fresh		
	North Horn	Shale, limestone, sandstone	Fresh		
	Wasatch	Fluvial shale, sandstone	Fresh		
	Green River	Lacustrine shale, marl, evaporite	Fresh-alkaline-saline		
Holocene		Playa lake salt crust			10^{-3}
		Beach and dune sand			10^{-5} to 10^{-3}
		Fluvial sand			10^{-6} to 10^{-3}
		Clay			10^{-12} to 10^{-5}

Sources of data: JOBIN, 1962; WIT, 1967; COOLEY et al., 1969; HANSHAW and HILL, 1969; PRYOR, 1973; TURK et al., 1973; HOOD, 1976; FREEZE and CHERRY, 1979; ETHRIDGE et al., 1980a and b; LYFORD et al., 1980; THACKSTON et al., 1981; BLAKEY et al., 1988; CONDON and HUFFMAN, 1988; PETERSON, 1988b; FREETHEY and CORDY, 1989; GELDON, 1989.

* Listed in approximate stratigraphic order from oldest to youngest.

** Hydrostratigraphic units are named for purposes of discussion in this paper and are not official U.S. Geological Survey designations.

† N.A. stands for "not available."

sodium chloride-type. Intermediate saline water with TDS from 1000 to 35,000 mg/L has combinations of calcium, magnesium, sodium, sulfate, and bicarbonate as dominant ions.

Modern analogs to ancient surface water are sea water, river water, fresh lake water, and saline lake water. Variations in certain components of sea water, such as sulfur isotope ratios, can be useful in determining fluid sources (*e.g.* HOLSER and KAPLAN, 1966). Lake water varies from fresh to saline, oxidizing to reducing. Lake bottom sediments typically are reducing. Lake water compositions are highly variable and depend largely on the composition of rocks in the drainage basin (EUGSTER, 1980).

Alteration minerals reveal the chemistry of paleolake and surface water. Kaolinite typically forms from high flux of low-TDS water that may be acidic from the decomposition of organic matter. Analcime, potassium feldspar, and albite may indicate alkaline-saline alteration in a closed basin. Certain saline minerals such as gypsum, halite, and sylvite suggest arid environments where lake or sea water has evaporated; thick-bedded deposits of these minerals, as in the Paradox basin, suggest a restricted arm of the sea (HITE, 1968).

Fluid inclusions have been used to determine the temperature and salinity of breccia-pipe-hosted silver-base metal-uranium deposits in northern Arizona (WENRICH and SUTPHIN, 1989). Fluid inclusions yielded temperatures and salinities of fluids associated with veins of carbonate, sulfate, and copper-silver sulfide minerals in the Paradox Basin (MORRISON and PARRY, 1986).

GROUNDWATER FLOW THROUGH TIME

From Cambrian through Mississippian, stable marine shelf carbonates and sandstones accumulated on the flanks of the transcontinental arch which extended in a northeast-southwest direction through New Mexico and eastern Colorado (Figs. 4a and 5). Cambrian, Devonian, and Mississippian strata make up most of the section and thicken to the north and west away from the arch into Utah, Arizona, and Nevada. In late Mississippian and early Pennsylvanian, upwarping and regression of the sea exposed the limestones to weathering, karst formation, and soil development in a warm humid environment near the equator.

Groundwater would have consisted of connate sea water in submarine sediments and fresh water in exposed areas. The karst features may have formed on land and at the sea water-fresh water interface. In similar settings today, Florida and the Yucatan Peninsula, limestone is dissolved and dolomite is precipitated (WARD and HALLEY, 1985; BACK *et al.*, 1986).

After the lengthy stable period characterized by the marine shelf environment ended, the Colorado Plateau experienced four major cycles of marine deposition, subaerial deposition, and erosion.

Cycle 1: Pennsylvanian to Permian

In Morrowan time, shallow epeiric seas transgressed across the Mississippian limestone surface, and the ancestral Rocky Mountains began to emerge (Fig. 4a and 5). Winds from the east created a lush, humid environment, as indicated by lignite deposits in the Taos trough (CASEY, 1980). An arid climate prevailed on the lee side where vast evaporite deposits formed (OHLEN and McINTYRE, 1965; BAARS *et al.*, 1967; PETERSON and HITE, 1969).

On the windward side, in the Taos trough, ample precipitation kept the water table near the ground surface, and debris shed from the emerging highlands was largely saturated with fresh water. In low-lying coastal areas, fresh to brackish water was locally acidified by decomposing organic matter. A mixing zone of brackish water between the fresh- and sea-water masses extended down and landward due to sea water intrusion. Because sediment loading was greatest adjacent to the highlands, expelled pore water tended to move updip and seaward toward the shallower parts of the trough (Fig. 5).

On the leeward side of the highlands, for example in the Paradox basin, sparse precipitation quickly percolated down through coarse clastic material of the alluvial fans. Because of the low recharge rate, high transmissivity, and steep slopes, the water table was probably only slightly above sea level. Discharge was at the toe of the alluvial fans and at the shoreline of the sea owing to decrease in topographic slope, decrease in grain size and transmissivity, and the presence of a salt water wedge. Gravity flow through the fan became increasingly more saline and alkaline due to evapotranspiration by phreatophytes, evaporative pumping by capillary action, and mixing with connate brine. Evaporation from shoreline pans or sabkhas drew brine from the basin shoreward. Compaction was greatest in muds interbedded with evaporites where deposition was most voluminous. Most of the water trapped during deposition of the evaporites and interbedded clay was probably lost very near the surface because of the high compressibility of the material; porosities of 1-2 percent are typical of evaporites at relatively shallow depths (GEVANTMAN, 1981). Fluid in the core of the Paradox Member evaporites probably

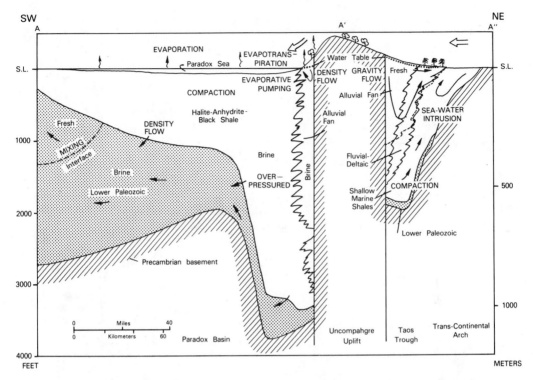

FIG. 5. Paleo ground water flow in Desmoinesian time (Middle Pennsylvanian). Line of section is shown in Fig. 4a; note offset in section at A′. S.L. stands for sea level. (Stratigraphy from OHLEN and MCINTYRE, 1965; BAARS *et al.,* 1967; PETERSON and HITE, 1969; HITE and CATER, 1972; SUTHERLAND, 1972; CASEY, 1980.

experienced overpressures, as observed today (SPENCER, 1975; THACKSTON *et al.,* 1981). Any groundwater that was forced out of the clastic sediments at the base of the section by compaction, density differences, or topographic forces probably moved through the lower Paleozoic aquifer rather than through the evaporites. Topography and compaction tended to force this deep groundwater up and out; however, the increased density due to salt dissolution tended to keep the most dense fluid in the deepest part of the basin. Any fresh water that had been trapped in the Mississippian aquifer from its previous subaerial environment would have been flushed out and replaced with brine at this time, as suggested by high-TDS water in the Leadville and Redwall Limestones today (HANSHAW and HILL, 1969; THACKSTON *et al.,* 1981; GELDON, 1989).

Toward the close of the Pennsylvanian, the shelf carbonate facies encroached on the evaporite facies, as subsidence of the Paradox basin and uplift of the Uncompahgre highlands slowed temporarily. In the Wolfcampian and Leonardian (early Permian), uplift and subsidence resumed at a greater rate, and large volumes of clastic material, fluvial arkosic sand and silt, and eolian sand and dust, spread southwestward from the Uncompahgre highlands over the Colorado Plateau (Fig. 4b and 6). Heterogeneous arkosic material next to the uplift grades into and intertongues with the fine-grained silt and eolian sand facies to the southwest. Toward the close of the early Permian, the sea, which had regressed far to the northwest, encroached on the western and southern parts of the Colorado Plateau.

Like west-facing shoreline areas in the trade-wind belt today, such as northern Chile, Namibia, and northwestern Australia, the climate was probably extremely arid. Precipitation mainly fell in the mountains. The water table at higher elevations probably was far below the ground surface due to low recharge, steep slopes, and high permeability of the coarse fan material. Evaporation and evapotranspiration would cause the fresh meteoric water to become alkaline and saline as it percolated through the fan and basin-fill sediments (*e.g.* MCLEAN, 1975; CARLISLE *et al.,* 1978; BRIOT, 1983). Ubiquitous calcite cement suggests that calcrete formed downstream at lower elevations as it does today in similar environments.

Where shallow groundwater and sea water met at the shoreline, a mixing zone would develop. Sea

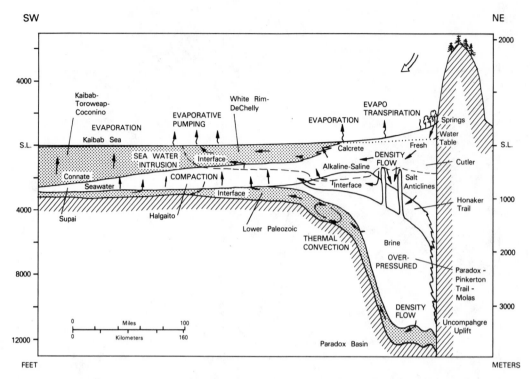

Fig. 6. Paleo ground water flow in Leonardian time (early Permian). Line of section is shown in Fig. 4b. S.L. stands for sea level. (Stratigraphy from BAARS, 1962; RASCOE and BAARS, 1972; CAMPBELL, 1980; PETERSON, 1980; and references cited for Fig. 5). Selected lithostratigraphic units labeled.

water is typically drawn landward by evaporation in similar sabkha environments. Because the shallow groundwater may have been less dense than sea water, the sea water probably intruded landward at depth and formed a sloping interface within the permeable sands and carbonate sediments.

Compaction continued to expel connate water from the deepest part of the basin adjacent to the Uncompahgre block. The black shale beds of the Paradox Member probably were a source of brine. Algal material and bacterial activity would lower the Eh of the brine. Expelled low-Eh brine then migrated into aquifers above and below. Brine expelled into the underlying Mississippian aquifer tended to accumulate in the deepest parts of the basin or to move updip toward the shoreline to the southwest and into thicker parts of the aquifer to the northwest (normal to the section of Fig. 6). In the core of the Paradox Member evaporites, overpressure probably developed as sediment loading occurred. Above the Paradox Member, the locally transmissive Honaker Trail Member of the Hermosa Formation was a conduit for expelled water of compaction from black shales in the Paradox and from clastic sediments in the upper part of the section.

Growth of salt anticlines began shortly after Paradox deposition (Fig. 4c of CATER, 1972). The bounding faults may have been favorable conduits for fluid flow. If descending fresh water encountered the salt, dissolution of salt and density-driven flow might have been important, as they are in the Gulf Coast today (e.g. RANGANATHAN and HANOR, 1988). Interfaces formed between connate brine, sea water, and fresh water in different parts of the basin. Gypsum, dolomite, and other minerals are evidence for such interfaces (RAUP, 1982; RANDAZZO and BLOOM, 1985). These interfaces would also be favorable zones for precipitation of copper, silver, and related metals. Gravity stratification of groundwater probably was taking place in the deepest Paleozoic sediments during the Pennsylvanian and Permian, and a stably stratified configuration may have persisted through the Mesozoic while these sediments remained deeply buried.

Cycle 2: Triassic

Tectonic uplift and deposition waned in the late Permian and earliest Triassic, and the topography became subdued. Hydrologically, conditions were like those in the early Permian except that com-

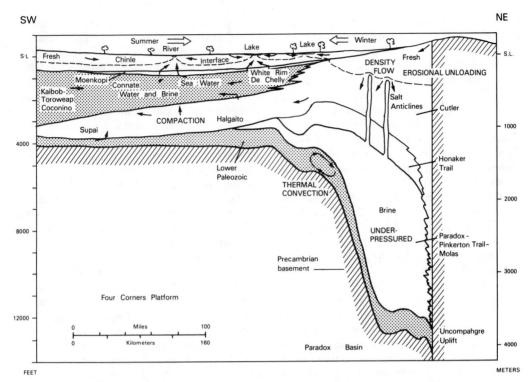

FIG. 7. Paleo ground water flow in Petrified Forest time (late Triassic). Line of section is shown in Fig. 4c. S.L. stands for sea level. (Stratigraphy from BLAKEY and GUBITOSA, 1983; BLAKEY, 1974; DUBIEL, 1987a, 1987b, 1989; and references cited for Fig. 6). Selected lithostratigraphic units labeled.

paction-driven expulsion of pore water ceased as active deposition came to an end. Groundwater flow was dominantly gravity-driven. Fresh water displaced brine in upper parts of the Permian section in the Paradox basin. Locally, especially in the Paradox basin, salt dissolution and resulting density-driven flow may have been important. Very little erosion at the top surface of the Kaibab Limestone and the Coconino and De Chelly Sandstones indicates a very gentle gradient over most of the Colorado Plateau area (BAARS, 1962). The lack of compaction, the gentle topographic gradient, and the presence of gravity stratification may have allowed evolved sea water at depth to occupy pore spaces of Permian sediments into Triassic time, although the upper part would have been recharged with fresh meteoric water.

During the early Triassic, the Moenkopi Formation was laid down during several episodes of regression and transgression of shallow epeiric seas (BLAKEY, 1974). The seas invaded from the northwest and formed a restricted bay surrounded by the Mogollon highlands, which were beginning to rise in the southwest, the Uncompahgre highlands, now largely eroded to the northeast, and the Defiance

uplift, a subdued uplift to the southeast. Interstitial pore water trapped during deposition was dominantly sea water with pockets of hypersaline brine from marginal-marine evaporites and pockets of fresh water in areas of fluvial, subaerial deposition.

In late Triassic, a fluvial-deltaic-lacustrine system prograded over the marine deposits (Fig. 4c and 7) (BLAKEY and GUBITOSA, 1983; DUBIEL, 1987a, 1987b). An aggrading river system flowed northwestward toward the sea. A warm monsoon climate prevailed in which dry winters alternated with wet summers. Rising base level caused the formation of local but extensive lakes, deltas, and marshes. In these areas, green mudstones, well-preserved organic matter, and worm burrows indicate a water table close to the surface; away from these areas, red strata with poorly preserved organic matter and no worm burrows indicate a locally deep water table (DUBIEL, 1989). Abundant bentonitic beds indicate significant deposition of volcanic ash from a magmatic arc to the southwest.

Seasonally abundant precipitation provided fresh meteoric groundwater in the near-surface environment. Regional surface and groundwater flow was from highland areas marginal to the depositional basin toward the sea to the northwest. Complex local flow systems probably developed on the undulating ground surface, and topographic lows tended to be local discharge areas.

Topographic flow and compaction forced connate water upward from the Moenkopi and other underlying marine sediments. As deposition proceeded, fresh water in lower parts of the Chinle was replaced mainly by evolved sea water from below. The interface between fresh and saline water moved upwards as the newly deposited sediments subsided. Expelled saline water collected in areas of more active subsidence where compaction was greater and where downwarps tended to pool the denser fluid.

Erosional unloading could have had major effects on the regional groundwater flow (NEUZIL and POLLOCK, 1983). From late Permian through early Jurassic, the Uncompahgre highlands generally were eroding. As both the Precambrian core of the uplift and the upper strata of arkosic debris were stripped off, the release of stress caused underpressuring at depth and tended to draw fluid in from areas of higher pore pressure where erosional unloading was slower or where deposition was taking place. Groundwater in the permeable sandstones flowed relatively quickly in response to this change in stress, but less permeable units, particularly the Paradox evaporites remained underpressured for a significant length of time. At the same time, the depositional basin in the central and northwest parts of the Colorado Plateau was compacting and subsiding as the Moenkopi and Chinle were deposited. Deep ground-water expelled from beneath the central Colorado Plateau therefore tended to flow toward the areas of maximum decompression, i.e., toward the Paradox basin. Determining whether this force overcame the opposing regional topographic slope would require a quantitative analysis.

Two types of ore deposits probably formed about this time (210 ± 10 Ma); sandstone-hosted, tabular uranium and breccia-pipe-hosted uranium-copper-silver-base metal deposits (WENRICH, 1985; LUDWIG et al., 1986; GRANGER et al., 1988; LUDWIG and SIMMONS, 1988). The role of groundwater was critical to the formation of both types, but exactly how is still uncertain.

Tabular uranium deposits in fluvial sandstones of the Chinle commonly are associated with complexly interbedded low-energy sediments and green lacustrine mudstones (LUPE, 1977; DUBIEL, 1983). Geometry and mineralogy of this deposit type in general (both Triassic and Jurassic) suggest formation at the interface between brine and fresh water (FISCHER, 1947; GRANGER and WARREN, 1981; NORTHROP, 1982; GRANGER and SANTOS, 1986).

A Triassic fresh water-brine interface first formed when fluvial sediments containing fresh interstitial water were deposited over marine sediments containing sea water and locally hypersaline brine. Gradually, the interface rose in response to sedimentation and subsidence. Topographic relief and compaction combined to force the brine toward topographic depressions, defined by thicker accumulation of sediments, lacustrine and generally finer-grained deposits, reducing conditions, and high water table. At the same time, fresh meteoric water was moving toward topographic lows as surface flow and groundwater. Abundant volcanic ash enriched this fresh water in uranium (ZIELINSKI, 1983). The two solutions met in the subsurface, and flow paths converged beneath the topographic depressions. Because discharge areas typically are much smaller than recharge areas (FREEZE and WITHERSPOON, 1967), and because fresh-water lenses narrow toward the discharge areas (HUBBERT, 1940), uranium-bearing groundwater and brine from depth were focused and were forced together near the discharge areas. Mixing of the two solutions caused uranium to precipitate (SWANSON et al., 1966; ANDREWS, 1981). The topographic depression controlled the location of groundwater discharge and mixing, and the focusing of flow lines allowed a deposit, rather than disseminated low-grade concentrations, to form.

Breccia-pipe deposits are most abundant in the southwest part of the Colorado Plateau in northern Arizona (WENRICH, 1985). Fluid inclusions in the early carbonate and base-metal stages of mineralization have homogenization temperatures of 80–173°C and salinities typically more than 18 weight per cent equivalent NaCl (WENRICH and SUTPHIN, 1989). Groundwater having such salinities was probably widespread at depth in the Colorado Plateau, but a source for the high temperatures is more difficult to identify. Heated groundwater may have been associated with igneous activity in the magmatic are south and west of the Colorado Plateau, or it may have come from a deep basin. One source for a basinal brine of this temperature in the Colorado Plateau would have been the Paradox basin. The present reconstruction (Fig. 7) shows a maximum depth of 4000 m, which suggests temperatures of 100–150°C given a normal geothermal gradient (e.g. HANOR, 1979). However, the great distance from the Paradox basin to northern Arizona, the cooling that would have resulted from transport, and the unfavorable topographic slope in the Triassic-early Jurassic make the Paradox basin an unlikely source. A similar argument can be made for the Oquirrh basin on the northwest margin of the

Colorado Plateau. Thus the mineralizing fluid probably was not a basinal brine under a normal geothermal gradient. A magmatic heat source related to the magmatic arc to the southwest was a more likely source, because the transport distance would have been smaller and the topographic slope would have been more favorable.

Cycle 3: Jurassic to early Cretaceous

The climate changed from monsoon to arid in the latest Triassic, and arid conditions prevailed until the late Jurassic (CRAIG et al., 1955; GREEN, 1975, 1980; BLAKEY et al., 1983; KOCUREK and DOTT, 1983; DUBIEL, 1989). Sand seas or ergs covered much of the Colorado Plateau from latest Triassic to early Jurassic. The eolian Wingate and Navajo Sandstones were deposited and later became one of the principal aquifers in the region. Tectonically, the area was relatively quiet, and the topographic relief probably remained low. The central part of the Colorado Plateau was a topographic basin surrounded on three sides by the remnants of highlands that existed from the Permian through Triassic. The magmatic arc to the southwest was intermittently active through the latest Jurassic and early Cretaceous.

During the Middle Jurassic, the shallow restricted Carmel sea invaded from the northwest and covered much of the western and central Colorado Plateau region, while erg and coastal sabkha environments dominated in the southeastern part (BLAKEY et al., 1983; KOCUREK and DOTT, 1983; CONDON and HUFFMAN, 1988; PETERSON, 1988b). Clastic, eolian, and evaporitic sediments were deposited.

Regression of the Carmel sea was accompanied by progradation of the Entrada erg. Another transgression of the sea from the northwest then deposited marine and evaporitic marginal-marine sediments of the Curtis Formation and Todilto Limestone Member of the Wanakah Formation. Transgressions and regressions resulted in complex intertonguing of marine, marginal-marine, evaporite, and sabkha facies characteristic of an arid climate from Middle to latest Jurassic.

Given the arid climate of the early Jurassic, shallow groundwater in the Colorado Plateau probably ranged from fresh to saline. Fresh meteoric water recharged in the highlands around the basin. As it moved toward lowland areas, groundwater may have become more saline and alkaline due to evaporation, evapotranspiration, and reaction with detrital material. Groundwater was mainly saline in the central, topographically low part of the area.

During transgressions of the Carmel and Curtis seas, groundwater flow in subaerial environments remained like that in the early Jurassic, but groundwater in strata beneath and adjacent to the seas was altered by sea-water intrusion and by compaction-driven flow upward and outward from deeper parts of the basin. Tidal and supratidal areas were dominated by landward flow driven by evaporation as in modern sabkha environments. Surficial fresh water tended to be displaced by sea water and hypersaline brine with each marine transgression, and fresh water tended to displace sea water and hypersaline brine as the seas retreated. With continued deposition and subsidence, however, fresh water lenses tended to remain on top of the denser saline groundwater, and eventually the entire column of sediments deposited from early Jurassic through middle late Jurassic was saturated by saline groundwater.

Uranium deposits in the Todilto Limestone are considered to have formed syngenetically in a sabkha environment where landward-migrating, low Eh-high pH sea water interacted with seaward-migrating, high Eh-low pH terrestrial water (RENFRO, 1974; RAWSON, 1980a, 1980b).

In the latest Jurassic, continued northward movement of the North American plate began to produce a cooler, less arid climate. Eolian sands, gypsum, and salt casts in the lower members of the Morrison Formation (GREEN, 1975; BLAKEY et al., 1988; PETERSON, 1988b) suggest similarity to earlier arid conditions rather than to later Morrison conditions (GREEN, 1975), although the later fluvial sedimentation may have been the result of increased uplift in the source area rather a change in climate (FRED PETERSON, U.S.G.S., personal communication, 1989).

By the late Jurassic, the Uncompahgre uplift had eroded to base level so that sedimentation reached across and beyond to the present Central Plains region. The volcanic arc to the west and southwest of the Colorado Plateau resumed production of vast quantities of ash that are best preserved in lacustrine mudstones of the Brushy Basin Member at the top of the Morrison Formation. Because of lower base level to the northeast and rising highlands to the southwest, the axis of maximum deposition and subsidence migrated from the southwest Colorado Plateau in the Triassic to the central part by latest Jurassic.

The Upper Jurassic Morrison Formation consists of fluvial conglomerates and sandstones from sources in the southwest and south that grade into lacustrine mudstones in the Paradox and San Juan basins (Fig. 4d and 8) (CRAIG et al., 1955; BELL, 1983, 1986; TURNER-PETERSON, 1985). Mineralogy

300   R. F. Sanford

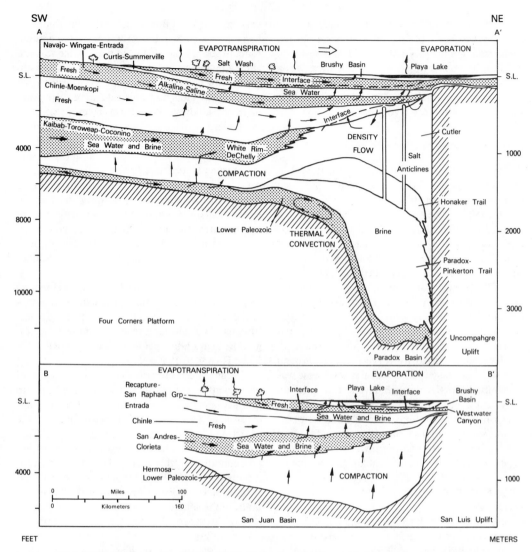

FIG. 8. Paleo ground water flow in Brushy Basin time (late Jurassic). Lines of section are shown in Fig. 4d. S.L. stands for sea level. (Stratigraphy from CRAIG et al., 1955; PETERSON et al., 1965; CONDON et al., 1984; CONDON and PETERSON, 1986; CONDON and HUFFMAN, 1988; PETERSON, 1988b; and references cited for Fig. 7). Selected lithostratigraphic units labeled.

and facies in the Brushy Basin Member suggest that it was locally a playa lake. Carbonized logs and humic material in the fluvial sandstones indicate that the climate was humid enough for plant growth.

Because of the considerable uranium resources hosted by the fluvial sandstones of the Morrison Formation, much has been written about it over the past 40 years. Redistributed-type deposits are generally thought to be roll-type deposits that formed in the Tertiary by remobilization of older tabular-type deposits. However, there is still no general agreement on how Upper Jurassic-Lower Cretaceous tabular-type deposits formed. The fol-

lowing hydrologic model is believed to be consistent with the observed mineral parageneses and facies, but a complete discussion of ore formation is beyond the scope of this paper.

In response to the regional topographic slope, groundwater moved generally from the southwest and south toward the northeast (Fig. 8). Water that recharged in the highlands to the southwest preferentially flowed through fluvial sandstones of the Morrison Formation, the Middle Jurassic eolian sandstones, the Upper Permian sandstones and limestones, and the lower Paleozoic aquifer system. During downdip flow, evaporation near the surface and reaction with detrital materials at depth increased the alkalinity and salinity of the groundwater. Where these aquifers pinch out and grade into less-permeable finer-grained material, the northeastward flow was forced upward. At the same time, sedimentation and the resulting compaction forced connate water up and outward from sediments in the deeper parts of the basins. Faults, especially those associated with salt anticlines in the Paradox basin, were especially favorable conduits. Due to dissolution of evaporites and reaction with detrital material, this groundwater was highly saline, possibly as much as 400,000 mg/l TDS, based on modern analyses. As these two types of groundwater approached the discharge areas, they displaced and forced upward pore water ahead of them. Consequently, connate saline water trapped in Middle to Upper Jurassic marine and evaporitic marginal-marine and/or lacustrine sediments (Curtis and Summerville Formations and Tidwell and Recapture Members of the Morrison Formation) tended to move upward and displace fresh water in the base of the Morrison sandstones. The resulting interface sloped upward toward the basin in the direction of flow.

Discharge areas were not only where aquifers pinch out, but also in topographically low areas where fine-grained lacustrine sediments accumulated. The decrease in slope and the finer-grained sediments combined to favor discharge at lake margins. Deeper, regional flow that passed under the lake ultimately was forced up to the surface upon encountering the buried Precambrian blocks of the Uncompahgre and San Luis uplifts.

A third source of groundwater was the connate alkaline-saline brine trapped in the lacustrine facies of the Brushy Basin muds. Compaction tended to drive this water upward, but its higher density relative to fresh water, especially in the center of the playa, might have tended to cause it to displace fresh water outward and downward (TURNER-PETERSON, 1985; TURNER-PETERSON and FISHMAN, 1986).

Downward flow through the base of the Brushy Basin Member encountered upward gravity- and compaction-driven brine. If the upward-flowing fluid were a denser brine, either hypersaline connate brine from marginal-marine evaporites, or a saline brine from dissolution of evaporites, the lighter lacustrine alkaline-saline brine would be displaced upward.

If the upward-flowing groundwater were less dense, the forces due to topographic elevation and compaction would still tend to force the deeper groundwater upward, because the small volume of pore water in the Brushy Basin muds would be overwhelmed by the much greater volume of water from below. An estimate of the relative volumes of pore water from a column one meter square and the thickness of the sedimentary section deep can be made using observed porosity-depth curves (*e.g.* HANOR, 1979; BOND and KOMINZ, 1984) and stratigraphic thicknesses. Given a thickness for the Brushy Basin of 100 m, a porosity of 50% at the top, and a porosity of 15% at the bottom, the amount of water expelled per meter of burial is about 100 m × (50%–15%) or 35 m. (Units are volume of fluid in m^3 expelled per unit area in m^2).

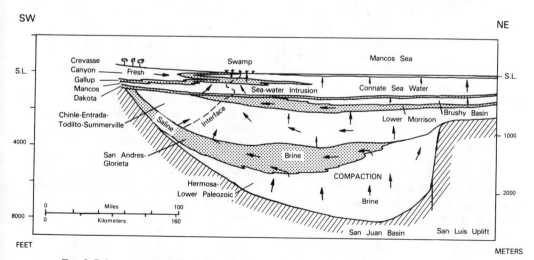

FIG. 9. Paleo ground water flow in Santonian time (late Cretaceous). Lines of sections are shown in Fig. 4e. S.L. stands for sea level. (Stratigraphy from MOLENAAR, 1983; and references cited for Fig. 8). Selected lithostratigraphic units labeled.

Given a thickness for the Paleozoic and Mesozoic rocks beneath the Brushy Basin of 3600 m, a porosity of 40% at the top, and a porosity of 20% at the base, the corresponding amount of pore water expelled per meter of burial is about 3600 m × (40%–20%) or 720 m, 20 times that from the Brushy Basin. The porosities selected are deliberately chosen to maximize the contribution from the Brushy Basin, and thus favor the hypothesis that pore water from the Brush Basin is significant. Even with this bias, the pore water from the Brushy Basin is small compared to that from the underlying sediments. Despite the Brushy Basin muds having greater porosity than the lithified sediments below, the much greater volume of these sediments can yield many times more pore water from compaction.

Alteration patterns indicate that playa lake water locally may have moved downward into the underlying fluvial sandstones. In the San Juan basin, alteration typical of the central playa lake generally extends only into the upper part of the underlying Westwater Canyon Member of the Morrison Formation, except in the central part of the paleo-playa lake (HANSLEY, 1989). In the Paradox basin, the playa lake facies are confined to the upper part of the Brushy Basin Member and do not extend down into the Salt Wash Member of the Morrison Formation (KELLER, 1962; TURNER-PETERSON, 1987).

The available evidence suggests that the interface between playa lake and deeper pore water was within the Brushy Basin Member in the Paradox basin, was in the upper part of the Westwater Canyon Member on the margins of the San Juan basin, and was near the base of the Westwater Canyon Member in the center of the San Juan basin. The difference in interface location between the Paradox and San Juan basins can be explained by the greater thickness of underlying Paleozoic and Mesozoic sediments and thus the greater volume of upward moving pore fluid in the Paradox basin.

Saline pore water expelled from depth near the margins of the playa lake and flanks of the basin encountered fresh or slightly alkaline shallow meteoric groundwater flowing down dip in the fluvial sandstones of the Salt Wash and Westwater Canyon Members. A fresh-water lens formed on top of the brine like that in coastal aquifers and continental basins today. Present-day interfaces are marked by deposition of dolomite, humic matter, and metals (SWANSON and PALACAS, 1965; SWANSON et al., 1966; MAGARITZ and LUZIER, 1985; RANDAZZO and BLOOM, 1985). Similarly, in the Salt Wash and the Westwater Canyon Members, dolomite, humic matter, uranium, and metals may mark the former

position of this fresh water-brine interface (GRANGER et al., 1961; GRANGER et al., 1980; ADAMS and SAUCIER, 1981; NORTHROP, 1982). The zone of iron-titanium oxide destruction, which has been attributed to mildly alkaline relatively fresh shallow groundwater (ADAMS et al., 1974; REYNOLDS et al., 1986; TURNER-PETERSON and FISHMAN, 1986), may define the shape and position of the fresh water lens. The shape of this zone is irregular but remarkably like that of present-day fresh-water lenses. The slope of the tabular uranium deposits, stratigraphically upward toward the basin, is also consistent with observed brine interface today.

As the Brushy Basin and overlying Cretaceous strata were deposited, the saline groundwater beneath the Morrison eventually rose and displaced all of the fresh water in the fluvial sandstones. The interface occupied progressively higher positions in the stratigraphic column. Uranium deposits in the Jackpile Sandstone Member at the top of the Morrison Formation (ADAMS et al., 1978), and in the base of the Upper Cretaceous Dakota Sandstone (GREEN, 1980) may be relicts of progressively higher positions of the interface.

In the early Cretaceous, moderate uplift over most of the Colorado Plateau, especially in the southern part, led to beveling of exposed units and only minor deposition of fluvial sediments. The topographic slope was to the northeast as in the late Jurassic. Fresh water recharged at the outcrop and displaced shallow saline and alkaline-saline groundwater. Widespread kaolinite alteration beneath the Dakota Formation suggests that organic acid-bearing, oxidizing fresh water percolated downward prior to Dakota deposition (GREEN, 1980; ADAMS and SAUCIER, 1981).

Cycle 4: late Cretaceous to early Tertiary

Late Cretaceous transgressions and regressions of the sea from the northeast accompanied the most voluminous and rapid depositional event since the uplift of the ancestral Rocky Mountains in the late Pennsylvanian and Permian (Fig. 4e and 9). Locally, rapid deposition continued into the Paleocene and Eocene. Absence of evaporites and abundance of coal beds in the late Cretaceous indicate that the climate had fully changed from arid subtropical to humid temperate.

As marine sediments were deposited over fluvial deposits in the late Cretaceous, gravity-driven flow became less important, and compaction-driven flow increased. Expelled connate sea water from the marine sediments moved upward through the compacting muds and also downward into the Dakota

Sandstone aquifer and then landward. Fresh water in sandstones of the Morrison and Dakota may have been replaced either by connate sea water from above or saline pore water from depth. At the maximum extent of marine transgression, when most of the Colorado Plateau was beneath the sea, compaction-driven flow tended to drive pore fluid west and southwest toward the shoreline.

The deepest and highest temperature fluids in the Colorado Plateau were expelled at this time (excluding local systems associated with Tertiary igneous intrusions). Temperatures recorded by vitrinite reflectance in Upper Cretaceous rocks of the San Juan Basin (HANSLEY, 1989) are highest over the thickest accumulations of Paleozoic and Mesozoic sedimentary rocks, which were the major sources of deeper pore fluid. Conversion of smectite to illite and the formation of chlorite may also have been the result of late Cretaceous deposition and compaction. In the Westwater Canyon Member, illite and chlorite distribution and isotopic relations suggest a warm fluid from depth (WHITNEY, 1986; WHITNEY and NORTHROP, 1987) as does the etching of detrital garnets (HANSLEY, 1987). Post-ore calcite, barite, dolomite, and copper minerals in the Morrison Formation in the Paradox basin suggest a warm brine from the Pennsylvanian evaporites at depth (BREIT, 1986). The timing and temperature relations of these diagenetic features are consistent with maximum pore water expulsion during the late Cretaceous.

Rapid deposition and low-permeability sediments may have caused over-pressuring, pressure in excess of hydrostatic pressure, during the late Cretaceous. A calculation suggests that the Mancos and Lewis Shales were probably the only clastic units in the Colorado Plateau that might have had excess pore pressure due to compaction. (The Paradox evaporites may also have had excess pressure.) The theory of BREDEHOEFT and HANSHAW (1968) with the following parameters (MOLENAAR, 1983; BREDEHOEFT and HANSHAW, 1968):

	Mancos Shale	Lewis Shale
Thickness (m)	690	745
Elapsed time (my)	13	7
Specific storage (m^{-1})	3×10^{-3}	3×10^{-3}
Hydraulic conductivity (m/s)	10^{-11}	10^{-11}
Density of sediment (gm/cm^3)	1.9	1.9

yields:

Calculated heads at base of unit (m)		
Hydrostatic	690	745
Lithostatic	1311	1416
Excess over hydrostatic	62	148

The results show that pore pressure might be 10 and 22 percent, respectively, of the difference between hydrostatic and lithostatic pressure at the close of deposition. Lower values are more likely, however, because fracture permeability could raise the hydraulic conductivity by several orders of magnitude and because some pore water would escape through basal and interfingering aquifers. No other shale or mudstone, such as the Petrified Forest Member of the Chile Formation or the Brushy Basin Member of the Morrison Formation, is likely to have had significant excess pore pressure.

Uplift during the latest Cretaceous and early Tertiary caused progradation of fluvial, lacustrine, and palludal sediments over Upper Cretaceous marine deposits (OSMOND, 1965; PETERSON et al., 1965; QUIGLEY, 1965; JOHNSON, 1985). A deep and widespread kaolinite alteration zone at the beveled top of Cretaceous rocks indicates a hiatus in deposition and a period of soil formation (JOHNSON and MAY, 1980; JOHNSON, 1985). Largely isolated continental basins generally received fluvial sediments early in their evolution and later received lacustrine sediments in the central parts of the basins and intertonguing fluvial sediments around the margins. In the Uinta and Piceance basins, where the most voluminous lower Tertiary sediments were deposited, early fluvial sedimentation was followed by fresh water lacustrine sedimentation. More fluvial sediments filled the basin, then a new fresh water lake formed. During the Eocene, salinity gradually increased while the lake expanded. Finally, vast amounts of volcanic detritus were deposited in a fluvial-deltaic complex that filled in the lake (JOHNSON, 1985).

In the basins, the complex transgressions and regressions of lakes and the episodic influxes of fluvial sediments created a complex hydrologic system (Fig. 4f and 10). The early fluvial and fresh water lacustrine stages were characterized by fresh water runoff and near-surface groundwater that probably left the basin. Increased alkalinity and salinity may indicate a change to locally more arid conditions, but could simply be a result of a rising outlet (JOHNSON, 1985). In either case, later lake sediments contained alkaline-saline pore water instead of fresh water. Fresh-water lenses probably overlay brine at the lake margins.

Topographic and compaction-driven flow would

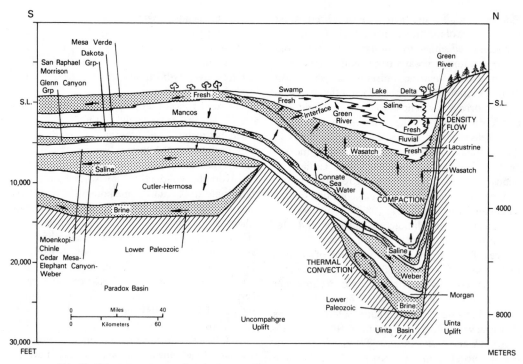

FIG. 10. Paleo ground water flow in Eocene time. Line of section is shown in Fig. 4f. S.L. stands for sea level. (Stratigraphy from OSMOND, 1965; QUIGLEY, 1965; JOHNSON, R. C., 1981; JOHNSON, 1985; and references cited for Fig. 8). Selected lithostratigraphic units labeled.

have first brought up relatively fresh water from the underlying fluvial sediments such as the Paleocene and Eocene Wasatch and Upper Cretaceous Mesaverde Formations. Within the lake sediments, an unstable interface formed between alkaline-saline brine on top and fresh water below. The combination of gravity head and compaction counteracted the density difference, as discussed for the Jurassic Brushy Basin lake, except that the denser brine above fresh water tended to favor downward migration. The relative volumes of pore water suggests that the interface gradually moved upward, but mineral alteration patterns may also indicate flow directions.

The depositional and hydrologic setting during Green River time was similar to that during Triassic Petrified Forest and Jurassic Brushy Basin time; lacustrine sediments containing volcanic material overlay fluvial channel sandstones. However, there are no significant tabular uranium deposits of Tertiary age. The volcanic rocks contained enough uranium to have been a source and were present during lacustrine deposition. The major difference between the Tertiary environment and the Mesozoic environments is the absence of a shallow interface between fresh water above and brine below

in the Tertiary. Because of the climatic change in the Cretaceous, the late Cretaceous marginal marine environments were characterized by coal swamps with fresh to brackish, organic-rich, acidic water instead of evaporites with interstitial hypersaline brine. The Tertiary lake therefore overlay some 3000 m of fluvial sediments originally with fresh pore water grading down into marginal marine sediments originally with fresh to brackish water. The sediments immediately beneath the lake would have contained the freshest water. There were no evaporites or marine sediments just below the lacustrine sediments and none at depth in hydrologic continuity. Even though the shallow fresh water percolating down through volcanic-rich sediments probably would have carried ample uranium, there was no interface with brine below to cause precipitation of uranium. Instead, roll-type deposits formed locally, as discussed below.

Areas of substantial uplift, exposure, and erosion increased sporadically throughout the Cenozoic. Sedimentary units that had been buried for all or most of their history were for the first time uplifted above sea level. Dense brine that may have resided in the deepest parts of the basin for a long time could now be drained owing to the deep incision of rivers. Sediments that previously had experienced small fluctuations in groundwater flow due to gentle tilting, downwarping, uplift in marginal areas, and so on, now were raised far above sea level. Regional flow began to assume the directions now taken by surface drainage and groundwater in the Colorado River basin. The total volume of flow increased dramatically as topographic relief increased to thousands of meters. Large volumes of rock, particularly the upper units at higher elevations, became underpressured or unsaturated and subject to oxidation for the first time. While depositional basins dominated in pre-Tertiary time, downward-moving fluid was nearly matched by upward-moving fluid; after the Colorado Plateau region rose, in latest Cretaceous and Tertiary, much of the recharge simply passed downward through the sedimentary section and discharged at lower elevations.

For uranium resources, the interaction of the oxidizing environment with preexisting reducing conditions was important in altering existing deposits and forming new ones. In the San Juan basin, tabular uranium deposits were partly destroyed and new roll-type deposits formed by the downdip flow of shallow, oxidizing groundwater. In Wyoming, roll-type deposits formed from uranium leached from fresh volcanic rock or recently exposed granitic basement. As pyrite and organic matter in the fluvial sandstones was oxidized, the groundwater progressively became less oxidized until hexavalent uranium was reduced and precipitated as UO_2. In the Uravan mineral belt, in the northeastern Paradox basin, carnotite deposits formed at the outcrop by oxidation of tabular uranium deposits.

Igneous activity took place in the Colorado Plateau and vicinity from late Cretaceous through Cenozoic time (STEVEN et al., 1972). Upper Cretaceous and Paleocene intrusive bodies were emplaced along a northeast-trending zone from northeastern Arizona to north-central Colorado. Most of the exposed igneous rocks are Eocene and Miocene. Intrusive and extrusive rocks of this period are widely distributed in and around the Colorado Plateau and are remnants of a much larger volcanic field now largely removed by erosion. Upper Cenozoic volcanics were emplaced around the margins of the Colorado Plateau.

Regional flow of groundwater during each of these volcanic periods was disturbed by the local topographic and thermal effects of the intrusions. Uplift associated with magma intrusion caused groundwater flow away from the intrusions toward topographic lows. The locally elevated temperature set up thermal convection. Flow was upward where groundwater was heated next to the intrusions and downward as groundwater cooled away from the intrusions. Deposition of carbonate, sulfate, and copper-silver-sulfide minerals associated with faults in the Paradox Basin may have been related to these late Cretaceous and Cenozoic intrusions (MORRISON and PARRY, 1986).

SUMMARY

Groundwater in the Colorado Plateau during most of the Phanerozoic must have been significantly different in some respects from that observed today. In pre-Tertiary, before widespread uplift, saline groundwater was dominant, and fresh water was limited to the shallow subsurface of areas elevated moderately above sea level. The past environment most like today's may have been the early Permian, when high relief and arid conditions prevailed in the Colorado Plateau. For most of the Phanerozoic, however, the Colorado Plateau consisted of basins that probably were saturated with saline water and had much less total flow than today owing to low relief, lack of dissection, and density stratification.

Mixing of dissimilar groundwater types is probably a common phenomenon, and in some cases can be related to known diagenetic and ore-forming processes. This reconstruction of paleo groundwater in the Colorado Plateau indicates where such mixing zones may have occurred and may be useful as an exploration tool for concentrations of uranium, vanadium, and copper. The occurrence of tabular-type uranium deposits is closely related to the presence of evaporitic rocks below the fluvial sandstones and to the development of a fresh water-brine interface.

Diagenetic patterns can also be better understood when the entire flow system is examined. Paleotemperatures and alteration patterns of illite, chlorite, and anhydrite deposition and garnet etching in the San Juan basin, for example, can be related to upward flow of warm saline fluid from below during compaction in the late Cretaceous. In the Uravan area, sulfur isotopes and late-stage chlorite are consistent with the predicted groundwater flow upward from underlying evaporites.

The conceptual models presented in this paper illustrate only the broadest outline for ancient groundwater flow in the Colorado Plateau. Each of the time periods that were important for the accumulation of economic resources should be studied in greater detail with quantitative models. Quantitative models of groundwater in the San Juan basin during late Jurassic to early Cretaceous are in progress (SANFORD, 1989). More work is needed to evaluate hydrologic properties in the past.

Acknowledgements—I am ultimately indebted to the late Hans Eugster of Johns Hopkins University for this study because he gave me a permanent appreciation of the importance of the fluid phase in geologic phenomena. In his typical multidisciplinary fashion, he integrated the mineralogy, fluid chemistry, sedimentology, and hydrology in pioneering studies of evaporites and saline lakes. The Colorado Plateau attracted him especially because of the Tertiary Green River Formation. Thanks to his influence, I am able to appreciate better the complex interactions of fluids, sediments, evaporites, and diagenetic minerals of the Colorado Plateau. Grant Garven, Charles Kreitler, C. M. Molenaar, and Warren Wood reviewed the manuscript and made many constructive suggestions.

REFERENCES

ADAMS S. S. and SAUCIER A. E. (1981) Geology and recognition criteria for uraniferous humate deposits, Grants uranium region, New Mexico. *Nat. Uranium Res. Eval.*, U.S. Dept. of Energy, GJBX-2(81), 226p.

ADAMS S. S., CURTIS H. S. and HAFEN P. L. (1974) Alteration of detrital magnetite-ilmenite in continental sandstones of the Morrison Formation, New Mexico. In *Formation of Uranium Deposits*, pp. 219–253. Intl. Atomic Energy Agency, Vienna.

ADAMS S. S., CURTIS H. S., HAFEN R. L. and NEJAD H. S. (1978) Interpretation of postdepositional processes related to the formation and destruction of the Jackpile-Paguate uranium deposit, northwest New Mexico. *Econ. Geol.* **73**, 1635–1654.

ANDREWS S. A. (1981) Physical modeling of controls on primary uranium ore deposition in alluvial fans. M.S. Thesis, Colorado State University, Fort Collins, 83p.

BAARS D. L. (1962) Permian system of Colorado Plateau. *Amer. Assoc. Petrol. Geol. Bull.* **46**, 149–218.

BAARS D. L., PARKER J. W. and CHRONIC J. (1967) Revised stratigraphic nomenclature of Pennsylvanian system, Paradox Basin. *Amer. Assoc. Petrol. Geol. Bull.* **51**, 393–403.

BACK W. and HANSHAW B. B. (1965) Chemical geohydrology. In *Advances in Hydrosci.*, (ed. V. T. CHOW), pp. 49–109. Academic Press, New York.

BACK W., HANSHAW B. B., HERMAN J. S. and VAN DRIEL J. N. (1986) Differential dissolution of a Pleistocene reef in the groundwater mixing zone of coastal Yucatan, New Mexico. *Geology* **14**, 137–140.

BALDWIN B. and BUTLER C. O. (1985) Compaction curves. *Amer. Assoc. Petrol. Geol. Bull.* **69**, 622–626.

BELL T. E. (1983) Deposition and diagenesis of the Brushy Basin and upper Westwater Canyon Members of the Morrison Formation in northwest New Mexico and its relationship to uranium mineralization. Ph.D. thesis, University of California, Berkeley, 101p.

BELL T. E. (1986) Deposition and diagenesis of the Brushy Basin Member and upper part of the Westwater Canyon Member of the Morrison Formation, San Juan Basin, New Mexico. In *A Basin Analysis Case Study: The Morrison Formation, Grants Uranium Region, New Mexico*, (eds. C. E. TURNER-PETERSON E. S. SANTOS and N. S. FISHMAN), *Studies in Geol.* 22, pp. 77–91. Amer. Assoc. Petrol. Geol.

BERRY F. A. F. (1959) Hydrodynamics and geochemistry of the Jurassic and Cretaceous systems in the San Juan basin, northwestern New Mexico and southwestern Colorado. Ph.D. thesis, Stanford University, 192p.

BETHKE C. M. (1985) A numerical model of compaction-driven groundwater flow and heat transfer and its application to the paleohydrology of intracratonic sedimentary basins. *J. Geophys. Res.* **90**, 6817–6828.

BLAKEY R. C. (1974) Stratigraphic and depositional analysis of the Moenkopi Formation, southeastern Utah. *Utah Geol. and Min. Surv. Bull.* **104**, 81p.

BLAKEY R. C. (1980) Pennsylvanian and Early Permian paleogeography, southern Colorado Plateau and vicinity. In *Paleozoic Paleogeography of the West-Central United States,* (eds. T. D. FOUCH and E. R. MAGATHAN), *Rocky Mountain Paleogeog. Symp. 1,* pp. 239–257. Rocky Mountain Sec., Soc. Econ. Paleont. Mineral.

BLAKEY R. C. and GUBITOSA R. (1983) Late Triassic paleogeography and depositional history of the Chinle Formation, southern Utah and northern Arizona. In *Mesozoic Paleogeography of the west-central United States,* (eds. M. W. REYNOLDS and E. D. DOLLY), *Rocky Mountain Paleogeog. Symp. 2,* pp. 57–77. Rocky Mountain Sec., Soc. Econ. Paleont. Mineral.

BLAKEY R. C., PETERSON F., CAPUTO M. V., GEESAMAN R. C. and VOORHEES B. J. (1983) Paleogeography of Middle Jurassic continental, shoreline, and shallow marine sedimentation, southern Utah. In *Mesozoic Paleogeography of the west-central United States,* (eds. M. W. REYNOLDS and E. D. DOLLY), *Rocky Mountain Paleogeog. Symp. 2,* pp. 77–100. Rocky Mountain Sec., Soc. Econ. Paleont. Mineral.

BLAKEY R. C., PETERSON F. and KOCUREK G. (1988) Synthesis of late Paleozoic and Mesozoic eolian deposits of the western interior of the United States. *Sed. Geol.* **56**, 3–125.

BOND G. C. and KOMINZ M. A. (1984) Construction of tectonic subsidence curves for the early Paleozoic miogeocline, southern Canadian Rocky Mountains: Implications for subsidence mechanisms, age of breakup, and crustal thinning. *Geol. Soc. Amer. Bull.* **95**, 155–173.

BREDEHOEFT J. D., BLYTH C. R., WHITE W. A. and MAXEY G. B. (1963) Possible mechanism for concentration of brines in subsurface formations. *Amer. Assoc. of Petrol.* **47**, 257–269.

BREDEHOEFT J. D. and HANSHAW B. B. (1968) On the maintenance of anomalous fluid pressures: I. Thick sedimentary sequences. *Geol. Soc. Amer. Bull.* **79**, 1097–1106.

BREIT G. N. (1986) Geochemical study of authigenic minerals in the Salt Wash Member of the Morrison Formation, Slick Rock district, San Miguel County, Colorado. Ph.D. thesis, Colorado School of Mines, Golden, 267p.

BRIOT P. (1983) L'Environnement hydrogéochimique du calcrete uranifère de Yeelirrie (Australie Occidentale). *Mineralium Deposita* **18**, 191–206.

CAMPBELL J. A. (1980) Lower Permian depositional systems and Wolfcampian paleogeography, Uncompahgre basin, eastern Utah and southwestern Colorado. In *Paleozoic Paleogeography of the West-Central United States,* (eds. T. D. FOUCH and E. R. MAGATHAN), *Rocky Mountain Paleogeog. Symp. 1,* pp. 321–340. Rocky Mountain Sec., Soc. Econ. Paleont. Mineral.

CARLISLE D., MERIFELD P. M., ORME A. R., KOHL M. S., KOLKER O. and LUNT O. R. (1978) The distribution of calcretes and gypcretes in southwestern United States and their uranium favorability based on a study of deposits in Western Australia and South West Africa (Namibia). U.S. Dept. Energy, Subcontract Number 76-022-E, 274p.

CASEY J. M. (1980) Depositional systems and paleogeographic evolution of the Late Paleozoic Taos Trough, northern New Mexico. In *Paleozoic Paleogeography of the West-Central United States*, (eds. T. D. FOUCH and E. R. MAGATHAN), *Rocky Mountain Paleogeog. Symp. 1*, pp. 181–196. Rocky Mountain Sec., Soc. Econ. Paleont. Mineral.

CATER F. W. (1972) Salt anticlines within the Paradox basin. In *Geologic Atlas of the Rocky Mountain Region*, (ed. W. W. MALLORY), pp. 137–138. Rocky Mountain Assoc. Geol.

CONDON S. M., FRANCZYK K. J. and BUSH A. L. (1984) Geologic map of the Piedra River Wilderness Study Area, Archuleta and Hinsdale Counties, Colorado. U.S. Geol. Surv. Misc. Field Studies Map MF-1630-B, scale: 1:50,000.

CONDON S. M. and HUFFMAN A. C. JR. (1988) Revisions to stratigraphic nomenclature of Jurassic and Cretaceous rocks of the Colorado Plateau. U.S. Geol. Surv. Bull. 1633-A, 1–12.

CONDON S. M. and PETERSON F. (1986) Stratigraphy of Middle and Upper Jurassic rocks of the San Juan Basin: Historical perspective, current ideas, and remaining problems. In *A Basin Analysis Case Study: The Morrison Formation, Grants Uranium Region, New Mexico*, (eds. C. E. TURNER-PETERSON, E. S. SANTOS and N. S. FISHMAN), *Studies in Geol. 22*, pp. 7–26. Amer. Assoc. Petrol. Geol.

COOLEY M. E., HARSHBARGER J. W., AKERS J. P. and HARDT W. F. (1969) Regional hydrogeology of the Navajo and Hopi Indian Reservations, Arizona, New Mexico, and Utah. U.S. Geol. Surv. Prof. Paper 521-A, 61p.

COOPER H. H. JR., KOHOUT F. A., HENRY H. R. and GLOVER R. E. (1964) Sea water in coastal aquifers. U.S. Geol. Surv. Water-Supply Paper 1613-C, 84p.

CRAIG L. C., HOLMES C. N., CADIGAN R. A., FREEMAN V. L., MULLENS T. E. and WEIR G. W. (1955) Stratigraphy of the Morrison and related formations, Colorado Plateau Region, a preliminary report. U.S. Geol. Surv. Bull. 1009-E, 168p.

CUSTODIO E. (1981) Sea water encroachment in the Llobregat and Besós areas, near Barcelona (Catalonia, Spain). *International Hydrological Programme*, Seventh Salt Water Intrusion Meeting, Uppsala, Sweden, pp. 120–171.

DOMENICO P. A. and ROBBINS G. A. (1985) The displacement of connate waters from aquifers. *Geol. Soc. Amer. Bull.* **96**, 328–335.

DUBIEL R. F. (1983) Sedimentology of the lower part of the Upper Triassic Chinle Formation and its relationship to uranium deposits, White Canyon area, southeastern Utah. U.S. Geol. Surv. Open-File Report 83-459, 48p.

DUBIEL R. F. (1987a) Sedimentology and new fossil occurrences of the Upper Triassic Chinle Formation, southeastern Utah. *Four Corners Geol. Soc. Guidebook*, 10th Field Conference, Cataract Canyon, pp. 99–107.

DUBIEL R. F. (1987b) Sedimentology of the Upper Triassic Chinle Formation, southeastern Utah: Paleoclimatic implications. *J. Arizona-Nevada Acad. Sci.* **22**, 35–45.

DUBIEL R. F. (1989) Depositional and climatic setting of the Upper Triassic Chinle Formation, Colorado Plateau. In *Dawn of the Age of Dinosaurs in the American Southwest*, (eds. S. G. LUCAS and A. P. HUNT), New Mexico Museum of Natural History, Albuquerque, pp. 171–187.

ETHRIDGE F. G., ORTIZ N. V., GRANGER H. C., FERENCHAK J. A. and SUNADA D. K. (1980a) Effects of

groundwater flow on the origin of Colorado Plateau-type uranium deposits. *New Mexico Bur. Mines and Min. Res., Mem.* **38**, 98–106.

ETHRIDGE F. G., ORTIZ N. V., SUNADA D. K. and TYLER N. (1980b). Laboratory, field, and computer flow study of the origin of Colorado Plateau type uranium deposits. U.S. Geol. Surv. Open-File Rept. 80-805, 81p.

EUGSTER H. P. (1980) Geochemistry of evaporitic lacustrine deposits. *Annual Reviews Earth Planetary Sci.* **8**, 35–63.

EUGSTER H. P. and HARDIE L. A. (1978) Saline lakes. In *Lakes—Chemistry, Geology, Physics*, (ed. A. LERMAN), pp. 237–293. Springer-Verlag.

FISCHER R. P. (1947) Deposits of vanadium-bearing sandstone. In *Mineral Resources of Colorado* (ed. J. W. VANDERWILT), pp. 451–456. State of Colorado Mineral Resources Board, Denver.

FOUCH T. D. and MAGATHAN E. R. (EDS.) (1980) *Paleozoic Paleogeography of the West-Central United States. Rocky Mountain Paleogeography Symposium 1*. Rocky Mountain Sec., Soc. Econ. Paleont. Mineral.

FREETHEY G. W. and CORDY G. E. (1989) Geohydrology of Mesozoic rocks in the upper Colorado River Basin—Excluding the San Juan Basin—in Arizona, Colorado, New Mexico, Utah and Wyoming. U.S. Geol. Surv. Prof. Paper 1411-C. (in press).

FREETHEY G. W., KIMBALL B. A., WILBERG D. E. and HOOD J. W. (1984) General hydrogeology of the aquifers of Mesozoic age, upper Colorado River Basin—excluding the San Juan Basin—Colorado, Utah, Wyoming, and Arizona. U.S. Geol. Surv. Open-File Rept. 84-716, 11p.

FREEZE R. A. and CHERRY J. A. (1979) Groundwater. Prentice-Hall, Englewood Cliffs, 604p.

FREEZE R. A. and WITHERSPOON P. A. (1967) Theoretical analysis of groundwater flow. 2. Effect of water-table configuration and subsurface permeability variation. *Water Resources Res.* **3**, 623–634.

FRENZEL P. F. and LYFORD F. P. (1982) Estimates of vertical hydraulic conductivity and regional groundwater flow rates in rocks of Jurassic and Cretaceous age, San Juan Basin, New Mexico and Colorado. U.S. Geol. Surv. Water-Resources Investig. 82-4015, 36p.

GARVEN G. and FREEZE R. A. (1984) Theoretical analysis of the role of groundwater flow in the genesis of stratabound ore deposits. I. Mathematical and numerical model. *Amer. J. Sci.* **284**, 1085–1124.

GELDON A. L. (1989) Hydrologic properties and flow systems of the Paleozoic rocks in the upper Colorado Basin, excluding the San Juan Basin, in Arizona, Colorado, Utah, and Wyoming. U.S. Geol. Surv. Prof. Paper 1411-B. (in press).

GEVANTMAN L. H. (ED.) (1981) Physical properties data for rock salt. *U.S. Dept. Commerce/Nat. Bur. Standards NBS Monograph* **167**, pp. 3–43.

GRANGER H. C., FINCH W. I., BROMFIELD C. S., DUVAL J. S., GRAUCH V. J. S., GREEN M. J., HILLS F. A., PETERSON F., PIERSON C. T., SANFORD R. F., SPIRAKIS C. S. and WAHL R. R. (1988) The Colorado Plateau uranium province, USA. In *Recognition of Uranium Provinces*, pp. 157–193. International Atomic Energy Agency, Vienna.

GRANGER H. C., FINCH W. I., KIRK A. R. and THADEN R. E. (1980) Research on uranium resource models, a progress report: Part III—Genetic-geologic model for tabular humate uranium deposits, Grants Mineral Belt,

San Juan Basin, New Mexico. U.S. Geol. Surv. Open-File Rept. 80-2018-C, 58p.

GRANGER H. C. and SANTOS E. S. (1986) Geology and ore deposits of the Section 23 mine, Ambrosia Lake district, New Mexico. In *A Basin Analysis Case Study: The Morrison Formation, Grants Uranium Region, New Mexico,* (eds. C. E. TURNER-PETERSON, E. S. SANTOS and N. S. FISHMAN), *Studies in Geol. 22,* pp. 185-210. Amer. Assoc. Petrol. Geol.

GRANGER H. C., SANTOS E. S., DEAN B. G. and MOORE F. B. (1961) Sandstone-type uranium deposits at Ambrosia Lake, New Mexico—An interim report. *Econ. Geol.* **56,** 1179-1210.

GRANGER H. C. and WARREN C. G. (1981) Genetic implications of the geochemistry of vanadium-uranium deposits in the Colorado Plateau region. (abstr.) Amer. Assoc. Petrol. Geol., Rocky Mountain Section, 29th Annual Meeting p. 24.

GREEN M. W. (1975) Paleodepositional units in Upper Jurassic rocks in the Gallup-Laguna uranium area, New Mexico. U.S. Geol. Surv. Open-File Rept. 75-610, 13p.

GREEN M. W. (1980) Disconformities in Grants mineral belt and their relationship to uranium deposits. *New Mexico Bur. Mines and Min. Res., Mem.* **38,** 70-85.

GROSS L. T. (1972) Tectonics. In *Geologic Atlas of the Rocky Mountain Region,* (ed. W. W. MALLORY), pp. 35-44. Rocky Mountain Assoc. Geol.

HANOR J. S. (1979) The sedimentary genesis of hydrothermal fluids. In *Geochemistry of Hydrothermal Ore Deposits, Second Edition,* (ed. H. L. BARNES), pp. 137-172. Wiley-InterSci., New York.

HANSHAW B. B. and COPLEN T. B. (1973) Ultrafiltration by a compacted clay membrane—II. Sodium ion exclusion at various ionic strengths. *Geochim. Cosmochim. Acta* **37,** 2311-2327.

HANSHAW B. B. and HILL G. A. (1969) Geochemistry and hydrodynamics of the Paradox Basin region, Utah, Colorado and New Mexico. *Chem. Geol.* **4,** 263-294.

HANSLEY P. L. (1987) Petrologic and experimental evidence for the etching of garnets by organic acids in the Upper Jurassic Morrison Formation, northwestern New Mexico. *J. Sed. Petrol.* **57,** 666-681.

HANSLEY P. L. (1989) Regional diagenesis of sandstones in the Upper Jurassic Morrison Formation, San Juan Basin, New Mexico and Colorado: geologic, chemical and kinetic constraints. *U.S. Geol. Surv. Bull. 1808H.* (in press).

HECKEL P. H. (1980) Paleogeography of eustatic model for deposition of midcontinent Upper Pennsylvanian cyclothems. In *Paleozoic Paleogeography of the West-Central United States,* (eds. T. D. FOUCH and E. R. MAGATHAN), *Rocky Mountain Paleogeog. Symp. 1,* pp. 197-215. Rocky Mountain Sec., Soc. Econ. Paleont. Mineral.

HITCHON B. (1969) Fluid flow in the western Canada sedimentary basin. *Water Resources Res.* **5,** 186-195.

HITE R. J. (1968) Salt deposits of the Paradox basin, southeastern Utah and southwestern Colorado. *Guidebook to the Geology of the Paradox basin, Intermountain Assoc. Petrol. Geol. 9th Ann. Field Conf.,* pp. 221-225.

HITE R. J. and CATER F. W. (1972) Pennsylvanian rocks and salt anticlines, Paradox basin, Utah and Colorado. In *Geologic Atlas of the Rocky Mountain Region,* (ed. W. W. MALLORY), pp. 133-138. Rocky Mountain Assoc. Geol.

HOLSER W. T. and KAPLAN I. R. (1966) Isotopic geochemistry of sedimentary sulfates. *Chem. Geol.* **1,** 93-135.

HOOD J. W. (1976) Characteristics of aquifers in the northern Uinta basin area, Utah and Colorado. State of Utah Dept. Nat. Res., Tech. Pub. No. 53, 71p.

HOOD J. W. and PATTERSON D. J. (1984) Bedrock aquifers in the northern San Rafael Swell area, Utah, with special emphasis on the Navajo sandstone. State of Utah Dept. Nat. Res. Tech. Pub. No. 78, 128p.

HUBBERT M. K. (1940) The theory of groundwater motion. *J. Geol.* **48,** 785-944.

HUBBERT M. K. (1953) Entrapment of petroleum under hydrodynamic conditions. *Bull. Amer. Assoc. Petrol. Geol.* **37,** 1954-2026.

HUNTOON P. W. (1983) Fault severing of aquifers and other geologically controlled permeability contrasts in the basin-mountain interface, and the implications for groundwater recharge to and development from the major artesian basins of Wyoming. Wyoming Water Research Center, University of Wyoming, Research Project Technical Completion Report (A-034-WYO), Agreement No. 14-34-0001-2154, pp. 1-16.

IORNS W. V., HEMBREE C. H. and OAKLAND G. L. (1965) Water resources of the upper Colorado River basin—Technical report. U.S. Geol. Surv. Prof. Paper 441, 370p.

ISSAR A. (1981) The rate of flushing as a major factor in determining the chemistry of water in fossil aquifers in southern Israel. *J. Hydrology* **54,** 285-296.

JOBIN D. A. (1962) Relation of the transmissive character of the sedimentary rocks of the Colorado Plateau to distribution of uranium deposits. *U.S. Geol. Surv. Bull. 1124,* 151p.

JOHNSON K. S. (1981) Dissolution of salt on the east flank of the Permian basin in the southwestern U.S.A. *J. Hydrology* **54,** 75-93.

JOHNSON R. C. (1981) Stratigraphic evidence for a deep Eocene Lake Uinta, Piceance Creek Basin, Colorado. *Geology* **9,** 55-62.

JOHNSON R. C. (1985) Early Cenozoic history of the Uinta and piceance creek basins, Utah and Colorado, with special reference to the development of eocene Lake Uinta. In *Cenozoic Paleogeography of the West-Central United States,* (ed. R. M. FLORES and S. S. KAPLAN), pp. 247-276. Soc. Econ. Paleont. and Mineral., Rocky Mountain Section.

JOHNSON R. C. and KEIGHIN C. W. (1981) Cretaceous and Tertiary history and resources of the Piceance Creek basin, western Colorado. *New Mexico Geol. Soc., Guidebook, 32nd Field Conference,* Western Slope Colorado, pp. 199-210.

JOHNSON R. C. and MAY F. (1980) A study of the Cretaceous-Tertiary unconformity in the Piceance Creek Basin, Colorado: The underlying Ohio Creek Formation (Upper Cretaceous) redefined as a member of the Hunter Canyon or Mesaverde Formation. *U.S. Geol. Surv. Bull. 1482-B.* 27p.

KELLER W. D. (1962) Clay minerals in the Morrison Formation of the Colorado Plateau. *U.S. Geol. Surv. Bull. 1150,* 90p.

KISHI Y., FUKUO Y., KAKINUMA T. and IFUKU M. (1982) The regional steady interface between fresh water and salt water in a coastal aquifer. *J. Hydrology* **58,** 63-82.

KORCUREK G. and DOTT R. H. JR. (1983) Jurassic paleogeography and paleoclimate of the central and southern Rocky Mountains region. In *Mesozoic Paleogeography of the west-central United States,* (eds. M. W. REYNOLDS and E. D. DOLLY), *Rocky Mountain*

Paleogeog. Symp. 2, pp. 101–116. Rocky Mountain Sec., Soc. Econ. Paleont. Mineral.

KREITLER C. W. (1979) Groundwater hydrology of depositional systems. In *Depositional and Groundwate Flow Systems in the Exploration for Uranium,* (eds. W. E. GALLOWAY, C. W. KREITLER and J. H. MCGOWEN), pp. 118–176. Bureau Econ. Geol., University of Texas, Austin.

KREITLER C. W., GUEVARA E., GRANATA G. and MCKALIPS D. (1977) Hydrogeology of Gulf Coast aquifers, Houston-Galveston area, Texas. *Gulf Coast Assoc. Geol. Socs. Trans., 27.*

LESSENTINE R. H. (1965) Kaiparowits and Black Mesa Basins: stratigraphic synthesis. *Amer. Assoc. Petrol. Geol. Bull.* **49,** 1997–2019.

LUDWIG K. R., RASMUSSEN J. D. and SIMMONS K. R. (1986) Age of uranium ores in collapse-breccia pipes in the Grand Canyon area, northern Arizona. (abstr.) *Geol. Soc. Amer., Rocky Mountain Section Meeting, Abstracts with Programs* **18,** 392.

LUDWIG K. R. and SIMMONS K. R. (1988) Progress in U/Pb isotope studies of collapse-breccia pipes in the Grand Canyon region, northern Arizona. (abstr.) *Geol. Soc. Amer., Annual Abstracts with Programs Meeting* **20,** A139.

LUPE R. (1977) Depositional environments as a guide to uranium mineralization in the Chinle Formation, San Rafael Swell, Utah. *J. Res. U.S. Geol. Surv.* **5,** 365–372.

LYFORD F. P., FRENZEL P. F. and STONE W. J. (1980) Preliminary estimates of effects of uranium-mine dewatering on water levels, San Juan Basin. *New Mexico Bur. Mines and Min. Res., Memoir 38,* 320–333.

MACK G. H., SUTTNER L. J. and JENNINGS J. R. (1979) Permo-Pennsylvanian climatic trends in the ancestral Rocky Mountains. *Four Corners Geol. Soc. Guidebook, 9th Field Conference,* Permianland, pp. 7–12.

MAGARA, K. (1976) Water expulsion from clastic sediments during compaction—directions and volumes. *Amer. Assoc. Petrol. Geol.* **60,** 543–553.

MAGARITZ, MORDECKAI and LUZIER J. E. (1985) Water-rock interactions and seawater-freshwater mixing effects in the coastal dunes aquifer, Coos Bay, Oregon. *Geochim. Cosmochim. Acta* **49,** 2515–2525.

MALLORY W. W. (EDITOR) (1972) *Geologic atlas of the Rocky Mountain region.* Rocky Mountain Assoc. Geol., Denver, 331p.

MCLEAN J. S. (1970) Saline groundwater resources of the Tularosa basin, New Mexico. U.S. Geol. Surv. Res. and Devel. Prog. Rept. No. 561, 128p.

MCLEAN J. S. (1975) Saline groundwater in the Tularosa Basin, New Mexico. *New Mexico Geol. Soc. Guidebook, 26th Field Conference,* Las Cruces Country, p. 237–238.

MOLENAAR C. M. (1983) Major depositional cycles and regional correlations of Upper Cretaceous rocks, southern Colorado Plateau and adjacent areas. In *Mesozoic Paleogeography of the west-central United States,* (eds. M. W. REYNOLDS and E. D. DOLLY), *Rocky Mountain Paleogeog. Symp. 2,* pp. 201–224. Rocky Mountain Sec., Soc. Econ. Paleont. Mineral.

MORRISON S. J. and PARRY W. T. (1986) Carbonate-sulfate veins associated with copper ore deposits from saline basin brines, Lisbon Valley, Utah: Fluid inclusion and isotopic evidence. *Econ. Geol.* **81,** 1853–1866.

NEUZIL C. E. and POLLOCK D. W. (1983) Erosional unloading and fluid pressures in hydraulically "tight" rocks. *J. Geol.* **91,** 179–193.

NORTHROP H. R. (1982) Origin of tabular-type vanadium-uranium deposits in the Henry structural basin, Utah. Ph.D. Thesis, Colorado School of Mines, Golden, 194p.

OHLEN H. R. and MCINTYRE L. B. (1965) Stratigraphy and tectonic features of Paradox basin, Four Corners area. *Amer. Assoc. Petrol. Geol. Bull.* **49,** 2020–2040.

ORR E. D. and KREITLER C. W. (1985) Interpretation of pressure-depth data from confined underpressured aquifers exemplified by the deep-basin brine aquifer, Palo Duro basin, Texas. *Water Resources Res.* **21,** 533–544.

OSMOND J. C. (1965) Geologic history of site of Uinta Basin, Utah. *Amer. Assoc. Petrol. Geol. Bull.* **49,** 1957–1973.

PARRISH J. T. and PETERSON F. (1988) Wind directions predicted from global circulation models and wind directions determined from eolian sandstones of the western United States—a comparison. *Sediment. Geol.* **56,** 261–282.

PARRISH J. T., ZIEGLER A. M. and SCOTESE C. R. (1982) Rainfall patterns and the distribution of coals and evaporites in the Mesozoic and Cenozoic. *Palaeogeog., Palaeoclim., Palaeoecol.* **40,** 67–101.

PAYNE J. N. (1968) Hydrologic significance of the lithofacies of the Sparta Sand in Arkansas, Louisiana, Mississippi, and Texas. U.S. Geol. Surv. Prof. Paper 569-A, 17p.

PAYNE J. N. (1970) Geohydrologic significance of the lithofacies of the Cockfield Formation of Louisiana and Mississippi and of the Yegua Formation of Texas. U.S. Geol. Surv. Prof. Paper 569-B, 14p.

PAYNE J. N. (1972) Significance of lithofacies of the Cane River Formation or equivalents of Arkansas, Louisiana, Mississippi, and Texas. U.S. Geol. Surv. Prof. Paper 569-C, 17p.

PAYNE J. N. (1975) Hydrologic significance of the lithofacies of the Carrizo Sand of Arkansas, Louisiana, and Texas, and the Meridian Sand of Mississippi. U.S. Geol. Surv. Prof. Paper 569-D, 11p.

PERLMUTTER N. M., GERAGHTY J. J. and UPSON J. E. (1959) The relation between fresh and salty groundwater in southern Nassau and southeastern Queens counties, Long Island, New York. *Econ. Geol.* **54,** 416–435.

PETERSON F. (1988a) Pennsylvanian to Jurassic eolian transportation systems in the western United States. *Sediment. Geol.* **56,** 207–260.

PETERSON F. (1988b) Stratigraphy and nomenclature of Middle and Upper Jurassic Rocks, Western Colorado Plateau, Utah and Arizona. *U.S. Geol. Surv. Bull. 1633-B,* 13–56.

PETERSON J. A. (1980) Permian paleogeography and sedimentary provinces, west central United States. In *Paleozoic Paleogeography of the West-Central United States,* (eds. T. D. FOUCH and E. R. MAGATHAN), *Rocky Mountain Paleogeog. Symp. 1,* pp. 271–292. Rocky Mountain Sec., Soc. Econ. Paleont. Mineral.

PETERSON J. A. and HITE R. J. (1969) Pennsylvanian evaporite-Carbonate cycles and their relation to petroleum occurrence, southern Rocky Mountains. *Amer. Assoc. Petrol. Geol. Bull.* **53,** 884–908.

PETERSON J. A., LOLEIT A. J., SPENCER C. W. and ULLRICH R. A. (1965) Sedimentary history and economic geology of San Juan Basin. *Amer. Assoc. Petrol. Geol. Bull.* **49,** 2076–2119.

PLUMMER L. N. (1975) Mixing of sea water with calcium carbonate groundwater. *Geol. Soc. Amer., Mem. 142,* 219–236.

PRICE D. and ARNOW T. (1974) Summary appraisals of the nation's groundwater resources—Upper Colorado region. U.S. Geol. Surv. Prof. Paper 813-C, 40p.

PRYOR W. A. (1973) Permeability-porosity patterns and variations in some Holocene sand bodies. *Amer. Assoc. Petrol. Geol. Bull.* **57**, 162–189.

QUIGLEY M. D. (1965) Geologic history of Piceance Creek-Eagle Basins. *Amer. Assoc. Petrol. Geol. Bull.* **49**, 1974–1996.

RANDAZZO A. F. and BLOOM J. I. (1985) Mineralogical changes along the freshwater/saltwater interface of a modern aquifer. *Sediment. Geol.* **43**, 219–239.

RANGANATHAN V. and HANOR J. S. (1988) Density-driven groundwater flow near salt domes. *Chem. Geol.* **74**, 173–188.

RASCOE B. JR. and BAARS D. L. (1972) Permian systems. In *Geologic Atlas of the Rocky Mountain Region,* (ed. W. W. MALLORY), pp. 143–165. Rocky Mountain Assoc. Geol.

RAUP O. B. (1982) Gypsum precipitation by mixing seawater brines. *Amer. Assoc. Petrol. Geol. Bull.* **66**, 363–367.

RAWSON R. R. (1980a) Uranium in the Jurassic Todilto Limestone of New Mexico—An example of a sabkha-like deposit. In *Uranium in Sedimentary Rocks: Application of the Facies Concept to Exploration,* (ed. C. E. TURNER-PETERSON), Short Course Notes, pp. 127–147. Rocky Mountain Sec., Soc. Econ. Paleont. Mineral.

RAWSON R. R. (1980b) Uranium in Todilto Limestone (Jurassic) of New Mexico—Example of a sabkha-like deposit. *New Mexico Bur. Mines and Min. Res., Mem. 38,* 304–312.

RENFRO A. R. (1974) Genesis of evaporite-associated stratiform metalliferous deposits—A sabkha process. *Econ. Geol.* **69**, 33–45.

REYNOLDS R. L., FISHMAN N. S., SCOTT J. H. and HUDSON M. R. (1986) Iron-titanium oxide minerals and magnetic susceptibility anomalies in the Mariano Lake-Lake Valley cores—Constraints on conditions of uranium mineralization in the Morrison Formation, San Juan Basin, New Mexico, In *A Basin Analysis Case Study: The Morrison Formation, Grants Uranium Region, New Mexico,* (eds. C. E. TURNER-PETERSON, E. S. SANTOS and N. S. FISHMAN), *Studies in Geol. 22,* pp. 303–313. Amer. Assoc. Petrol. Geol.

RUNNELS D. D. (1969) Diagenesis, chemical sediments, and the mixing of natural waters. *J. Sediment. Petrol.* **39**, 1188–1201.

SANFORD R. F. (1982) Preliminary model of regional Mesozoic groundwater flow and uranium deposition in the Colorado Plateau. *Geology* **10**, 348–352.

SANFORD R. F. (1983) Reconstruction of paleo-flow systems: Applications to sedimentary uranium deposits. (abstr.) *Geol. Soc. Amer., Annual Meeting, Abstracts with Programs* **15**, 677.

SANFORD R. F. (1989) Late Jurassic paleohydrology of the San Juan basin: simulated effects of variable transmissivity and lake levels. (abstr.). *Amer. Assoc. Petrol. Geol., 1989 Rocky Mountain Section Meeting, Abstracts with Programs,* p. 90.

SPENCER C. W. (1975) Petroleum geology of east-central Utah and suggested approaches to exploration. *Four Corners Geol. Soc. Guidebook, 8th Field Conf.,* Canyonlands, p. 263–276.

STEVEN T. A., SMEDES H. W., PROSTKA H. J., LIPMAN P. W. and CHRISTIANSEN R. L. (1972) Upper Cretaceous and Cenozoic igneous rocks. In *Geologic Atlas of the Rocky Mountain Region,* (ed. W. W. MALLORY), pp. 229–232. Rocky Mountain Assoc. Geol.

SUTHERLAND P. K. (1972) Pennsylvanian stratigraphy, southern Sangre De Cristo Mountains, New Mexico. In *Geologic Atlas of the Rocky Mountain Region,* (ed. W. W. MALLORY), pp. 139–142. Rocky Mountain Assoc. Geol.

SWANSON V. E., FROST I. C., RADER L. F. JR. and HUFFMAN CLAUDE JR. (1966) Metal sorption by northwest Florida humate. U.S. Geol. Surv. Prof. Paper 550-C, C174–C177.

SWANSON V. E. and PALACAS (1965) Humate in coastal sands of northwest Florida. *U.S. Geol. Surv. Bull. 1214-B,* 29p.

THACKSTON J. W., MCCULLEY B. L. and PRESLO L. M. (1981) Groundwater circulation in the western Paradox Basin, Utah. *Rocky Mountain Assoc. Geol. 1981 Field Conference,* pp. 201–225.

TREMAIN C. M., BORECK D. L. and KELSO B. S. (1981) Methane in Cretaceous and Paleocene coals of western Colorado. *New Mexico Geol. Soc., Guidebook, 32nd Field Conference,* Western Slope Colorado, pp. 241–248.

TURK L. J., DAVIS S. N. and BINGHAM C. P. (1973) Hydrogeology of lacustrine sediments, Bonneville Salt Flats, Utah. *Econ. Geol.* **68**, 65–78.

TURNER-PETERSON C. E. (1985) Lacustrine-humate model for primary uranium ore deposits, Grants Uranium Region, New Mexico. *Amer. Assoc. Petrol. Geol. Bull.* **69**, 1999–2020.

TURNER-PETERSON C. E. (1987) Sedimentology of the Westwater Canyon and Brushy Basin Members, Upper Jurassic Morrison Formation, Colorado Plateau, and relationship to uranium mineralization. Ph.D. Thesis, University of Colorado, Boulder, 169p.

TURNER-PETERSON C. E. and FISHMAN N. S. (1986) Geologic synthesis and genetic models for uranium mineralization in the Morrison Formation, Grants uranium region, New Mexico. In *A Basin Analysis Case Study: The Morrison Formation, Grants Uranium Region, New Mexico,* (eds. C. E. TURNER-PETERSON, F. S. SANTOS and N. S. FISHMAN), *Studies in Geol. 22,* pp. 357–388. Amer. Assoc. Petrol. Geol.

WARD W. C. and HALLEY R. B. (1985) Dolomitization in a mixing zone of near-seawater composition, Late Pleistocene, northeastern Yucatan Peninsula. *J. Sediment. Petrol.* **55**, 407–420.

WARNER J. W., HEIMES F. J. and MIDDELBURG R. F. (1985) Groundwater contribution to the salinity of the upper Colorado River Basin. U.S. Geol. Surv. Water-Resources Investig. Rept. 84-4198, 113p.

WENRICH K. J. (1985) Mineralization of breccia pipes in northern Arizona. *Econ. Geol.* **80**, 1722–1735.

WENRICH K. J. and SUTPHIN H. B. (1989) Lithotectonic setting necessary for formation of a uranium-rich, solution-collapse breccia-pipe province, Grand Canyon region, Arizona. U.S. Geol. Surv. Open-File Rept. 89-0173, 33p.

WHITNEY C. G. (1986) Petrology of clay minerals in the subsurface Morrison Formation near Crownpoint, southern San Juan Basin, New Mexico: An interim report. In *A Basin Analysis Case Study: The Morrison Formation, Grants Uranium Region, New Mexico,* (eds. C. E. TURNER-PETERSON, E. S. SANTOS and N. S. FISHMAN), *Studies in Geol. 22,* pp. 315–329. Amer. Assoc. Petrol. Geol.

WHITNEY C. G. and NORTHROP H. R. (1987) Diagenesis

and fluid flow in the San Juan Basin, New Mexico—Regional zonation in the mineralogy and stable isotope composition of clay minerals in sandstone. *Amer. J. Sci.* **287**, 353–382.

WINTER T. C. (1976) Numerical simulation analysis of the interaction of lakes and groundwater. U.S. Geol. Surv. Prof. Paper 1001, 45p.

WIT K. E. (1967) Apparatus for measuring hydraulic conductivity of undisturbed soil samples. In *Permeability and Capillarity of Soils.* Amer. Soc. Test. Mat., Spec. Tech. Pub. No. 417, 72–83.

WOOD J. R. and HEWITT T. A. (1982) Fluid convection and mass transfer in porous sandstones—a theoretical model. *Geochim. Cosmochim. Acta* **46**, 1707–1713.

WOODING R. A. (1962) Convection in a saturated porous medium at large Rayleigh number or Peclet number. *J. Fluid Mechanics* **15**, 527–546.

ZIELINSKI R. A. (1983) Tuffaceous sediments as source rocks for uranium: A case study of the White River Formation, Wyoming. *J. Geochem. Expl.* **18**, 285–306.

Part E.
Surface Environments

Fluid-Mineral Interactions: A Tribute to H. P. Eugster
© The Geochemical Society, Special Publication No. 2, 1990
Editors: R. J. Spencer and I-Ming Chou

Trace metal geochemistry of Walker, Mono, and Great Salt Lakes

JOSEPH L. DOMAGALSKI,* and HANS P. EUGSTER†

Department of Earth and Planetary Sciences, The Johns Hopkins University, Baltimore, MD 21218, U.S.A.

and

BLAIR F. JONES

United States Geological Survey, M.S. 432, Reston, VA 22092, U.S.A.

Abstract—Recent sediments were collected from three Basin and Range closed basin lakes (Walker Lake, NV, Mono Lake, CA, and Great Salt Lake, UT) to evaluate the response of a suite of trace metals (Fe, Mn, Cu, Pb, Zn, Cd, Co, Mo, V) to diagenetic processes and assess reactivity, mobility, mineral transformations and precipitation, and organic complexation. Sediments and pore fluids, representing up to 1000 years of deposition, were analyzed for brine and metal composition, pH, alkalinity, and H_2S. Sequential extractions of metals from solids were valuable in locating core horizons where metals are mobilized or precipitated and in the interpretation of pore-fluid metal profiles. Sedimentary metals initially associated with Fe-Mn oxides and organic matter respond to changes in biologically induced redox conditions by forming sulfide solids and humic acid complexes. Walker Lake metals are tightly bound near the sediment-water interface by oxides which only slowly dissolve in the anoxic sediments. At 50–60 cm of sediment depth, dissolution of oxides results in the most extensive metal redistribution. At Great Salt Lake sedimenting particles are entrained within the dense bottom brines where maximum organic decomposition and H_2S production occurs. Pyrite formation occurs very close to the sediment-water interface and renders most metals (Mo, V, and Cd excluded) insoluble except to hot concentrated acids. Mono Lake metals show maximum reactivity to extracting solutions just below the sediment-water interface. Very high pore-fluid sulfide levels (0.02 M) immobilize metals through the formation of metastable Fe monosulfides but metal reactivity is high as pyrite formation is hindered in the high pH environment. The Great Salt Lake is the only system where metal immobilization, through the formation of pyrite, was observed to have gone to completion.

INTRODUCTION

THE FATE of trace metals in sedimentary basins is dependent on the speciation of the metals during deposition and subsequent reactions during diagenesis. Previous work (JENNE, 1977, TUREKIAN, 1977, BALISTRIERI and MURRAY, 1981, 1982, 1983, 1984) has shown that trace metals, in oxygenated low temperature environments, are associated with reactive particles such as amorphous to crystalline oxides of Fe and Mn, organic coatings on suspended particles, and biological material. More recent work (JACOBS et al., 1985; BOULEGUE et al., 1982; LORD and CHURCH, 1983; KREMLING 1983; LUTHER et al., 1982) has been directed towards the chemistry of trace metals across redox boundaries where the reductive dissolution of metal oxides and anaerobic degradation of organic matter is occurring. The release of trace metals to brines occurs during these processes. The major problems affecting trace metals in these environments include solute speciation with respect to chloride, bisulfide, polysulfide, and

dissolved organic ligands, and the nature of the solid phases controlling metal solubility. It has been suggested that trace metals are not in equilibrium with pure metal sulfides during low temperature diagenesis but instead are involved in co-precipitation reactions, possibly with Fe sulfides (JACOBS et al., 1985). It is necessary to understand the solid phase associations of trace metals and the dissolved speciation so that thermodynamic models of metal redistribution and ore-forming processes can be formulated.

This study examines the solution and solid phase chemistry of Fe, Mn, Cu, Pb, Zn, Co, Cd, Mo, and V in the lake waters and Recent sediments of closed basin lakes of the Basin and Range province. Two alkaline (Na-CO_3-Cl-SO_4) systems, Walker Lake, NV, and Mono Lake, CA, and one non-alkaline (Na-Mg-Cl-SO_4) system, Great Salt Lake, UT, were studied. These lakes offer unique opportunities for investigating the effects of brine composition and evolution on redox chemistry, organic matter diagenesis, and sulfide precipitation, and their influence on trace metal distribution. Both Mono and the Great Salt Lake were chemically stratified with anoxic bottom waters during this study whereas

* *Present address:* United States Geological Survey, 2800 Cottage Way, Sacramento, CA 95825, U.S.A.

† Deceased.

Walker Lake was not stratified and had an oxygenated sediment-water interface. Alkaline lake basins have been proposed by EUGSTER (1985) as potential sites of ore forming processes. In addition, the processes occurring here can be contrasted with the more widely studied marine systems. This study examines the initial reactions occurring during trace metal deposition.

STUDY AREA

The lakes chosen for this study lie within the Basin and Range physiographic province of North America. Field work took place during the summers of 1984, 1985, and 1986. The area is characterized by north-south trending mountain ranges many of which are bounded by fault scarps. The ranges are separated by valleys which are filled with Cenozoic continental deposits. The lakes are perennial closed basin systems with no drainage outlets. They are remnants of much larger glacial lakes which had very high stands during the Pleistocene. A map of the present lakes and Pleistocene basins is shown in Fig. 1. The lake level chronology for these basins has been reported by BENSON (1978), THOMPSON *et al.* (1986), BENSON and THOMPSON (1987), BRADBURY *et al.* (1988), and SPENCER *et al.* (1984). These studies have shown that a rapid drop in lake level affected all three basins about 13,000 years ago in response to waning glacial climates.

Walker Lake

Walker Lake is one of the remnants of pluvial Lake Lahontan. It is bounded by the Wassuk Range on the west and the Gillis Range on the east. The Wassuk Range is composed of volcanic and granitic rocks whereas the Gillis Range is composed of intermediate to felsic Triassic volcanics and Mesozoic granitic rocks. The Walker River is the major source of fresh water with the drainage originating within the Sierra Nevada Mountains. The inflow water has carbonate alkalinity exceeding the concentration of alkaline earth ions. Evaporative concentration leads to a $Na-CO_3-Cl-SO_4$ brine composition (EUGSTER and HARDIE, 1978). The lake is at an early stage of brine evolution; its ionic strength is 0.17 and its pH is 9.2. The lake is monomictic, with a fall overturn following summer thermal stratification.

Evidence for past desiccation or extreme low stands of Walker Lake are recorded in the pore fluids. A region designated as the zone of salt flux (SPENCER, 1977) is shown on the map in Fig. 2a. This is the hydrographic low of the basin where concentrated brines and/or salts were deposited. A

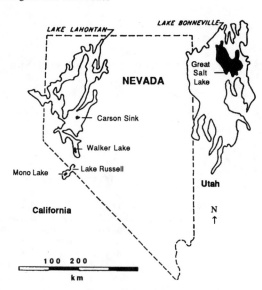

FIG. 1. Study area map showing locations of present lakes and Pleistocene precursors.

diffusive flux of these dissolved salts supplies solutes to the modern lake. The most recent desiccation or extreme low stand ended about 2500 years ago (BRADBURY *et al.*, 1988). The lake filled rapidly after this time to close to the present level. The desiccation may have been related to either climatic factors or diversion of the Walker River. Two sampling sites were chosen for Walker Lake and their locations are shown in Fig. 2a. Site WL2, at 34.5 m of depth, is at the hydrographic low, within the area of salt flux. Site WL1, at 33 m of depth, is outside of the area of salt flux.

Mono Lake

Mono Lake is the modern remnant of pluvial Lake Russell and is located at the base of the east facing escarpment of the Sierra Nevada Mountains. Three sides of the lake are flanked by Cenozoic volcanics. Metasedimentary rocks, lower Paleozoic in age, which were intruded by Mesozoic plutonic rocks are found on the western side. Changes in historic lake levels before 1940 reflect the general nature of drought and wet periods. Currently, the lake is severely affected by diversion of inflow streams to the Los Angeles Aqueduct. The lake level has fallen over 12 m since diversion began in 1941. A 2.5 meter rise in lake level occurred in the early 1980's in response to two winters with heavy precipitation (Los Angeles Department of Water and Power Briefing Document, March, 1984). The influx in fresh water resulted in chemical stratification of the water column. The stratification broke down

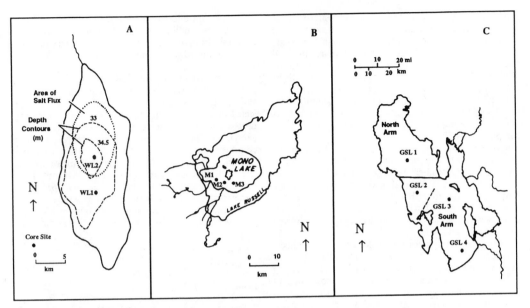

FIG. 2. Locations of sample sites and lake morphology: a) Walker Lake; b) Mono Lake; c) Great Salt Lake: The dashed line marks the location of a sub-aqueous carbonate ridge. The solid line marks the location of the railroad causeway.

and the entire water column mixed in November, 1988 in response to two consecutive drought years (A. MAEST, personal communication, 1989).

The inflow to Mono Lake is Sierra Nevada streams. The inflow water has carbonate alkalinity in excess of the concentration of alkaline earths, so a Na-CO$_3$-Cl-SO$_4$ brine forms upon evaporative concentration. Mono Lake waters are in an advanced state of brine evolution; its pH is 9.65. Three sampling sites were chosen for sediment coring (Fig. 2b). They are located in a region of a thick accumulation of mud as reported in a study of Mono Lake bathymetry (SCHOLL et al., 1967). The water depths range from 30 to 33 m.

Great Salt Lake

The Great Salt Lake is the modern remnant of pluvial Lake Bonneville. The lake occupies three interconnecting fault bounded depressions or grabens which are flanked by two parallel horsts and crossed by two other discontinuous and partly submerged horsts (STOKES, 1980).

The lake has been divided into a North and South Arm by a railroad causeway. In addition, a sub-aqueous carbonate ridge separates the lake into a North and South Basin (SPENCER et al., 1984) (Fig. 2c). Three rivers supply most of the dilute input with all of the inflow occurring in the South Arm. The North Arm receives water from rain, runoff

and a few hydrothermal springs. Brine concentrations differ in the two arms of the lake; the most concentrated waters occur in the North Arm. The South Arm is chemically stratified with the most concentrated brines occurring in the bottom waters of the North Basin region of the South Arm. In addition, a deep brine occurs in the southern basin of the lake. Sampling sites, shown in Fig. 2c, were chosen to coincide with these brine bodies. Coring was attempted but was not possible in the North Arm due to a layer of halite at the sediment-water interface. The successful sampling locations include one North Basin site and two South Basin sites. The southernmost site will hereafter be referred to as the South Basin deep site, after SPENCER (1982) and SPENCER et al. (1985), to distinguish it from the more northerly South Basin site. The water depths ranged from 12.5 to 13 m.

The concentration of alkaline earths, of the inflow water, is close to the amount of carbonate alkalinity. Neither calcium nor carbonate species increase in concentration during evaporative concentration. The resulting brine is a Na-Mg-Cl-SO$_4$ system with low buffer capacity relative to Walker and Mono Lakes.

METHODS

Field work

Lake water samples were obtained with a plastic Kemmerer type sampler. Temperature was measured on sample splits with an Orion temperature probe. Water samples were then placed in polyethylene bottles and stored on ice until filtration. All sample bottles were cleaned by soaking for several days in a 0.2 N solution of Ultrapure nitric and hydrochloric acid (Alfa Chemical Co.) followed by rinsing with ultrapure water from a Millipore water system. Samples were also taken, with the water sampler, for pH, alkalinity, and sulfide measurements. The remaining water was filtered through all plastic Nuclepore filter holders which had been cleaned by a similar procedure as the sample bottles. The filter holder was equipped with 0.2 micrometer Nuclepore filters and connected to a plastic syringe. For the Great Salt Lake water column 0.45 micrometer Millipore filters were used. Filtered waters from the Great Salt and Walker Lakes were preserved with 65% Ultrapure HNO$_3$ to a pH between 1 to 2. This was not possible with Mono Lake waters because of the high alkalinity which would require too large a quantity of acid to effect a drop in pH.

Lake sediment samples were collected by two methods. Cores, up to one meter in length, were collected with a gravity corer. This consisted of a cylinder for holding polycarbonate tubing and a plunger which closes on contact with sediment and retains the solids by vacuum. Upon retrieval the cores were capped and stored in an upright position.

Solid sampling of the sediment-water interface was supplemented by the use of an Eckman dredge. Undisturbed samples of the top 20 to 25 cm of sediment were taken from the dredge with either a cut 60 cc syringe, modified to act as a mini piston corer, or with polycarbonate tubing.

Within hours after collection, pore fluids were recovered from the cores. The tubing was cut in 5 to 10 cm sections and teflon filter holders, containing 0.2 micrometer Nuclepore filters, were installed in each end. A squeezing device, which employs a teflon piston, compresses the sediment. This causes fluid to flow through the filters and into tygon tubing with final collection in plastic syringes. The entire squeezing configuration is closed to the atmosphere, excluding any contact with oxygen. This was the same type of pore-fluid squeezer described in SPENCER *et al.* (1985). Subsamples of pore fluid were taken for measurements of pH, alkalinity, and sulfide. The remaining water was placed in polyethylene bottles. Walker and Great Salt Lake pore fluids were acidified to pH 1 to 2 with Ultrapure HNO$_3$ and the Mono Lake pore fluids were stored at the natural pH (between 9.2 to 9.6). The squeezed sediment was double wrapped in plastic ziplock bags, double wrapped in plastic, and taped shut. The samples were refrigerated until subsequent analyses. The Eckman sub-samples were extruded in two to three cm sections in a nitrogen filled glove bag, and placed in 50 cc centrifuge tubes. The tightly sealed tubes were spun at 5500 rpm. The recovered pore fluid was filtered and saved for metal analyses. The tubes were re-sealed under nitrogen and refrigerated for sequential and humic acid trace metal extractions.

Field analytical methods

The following methods were used on lake and interstitial fluids. pH measurements were made on samples at room temperature (25°C). An Orion Research Model 231 portable pH/mv/temperature meter was used with a combination electrode. The meter was calibrated with pH 7 phosphate and pH 10 carbonate buffers.

Alkalinity titrations, using standardized sulfuric acid,

were made on diluted samples from Walker and Mono Lakes and on undiluted samples from the Great Salt Lake using the Orion pH meter. The end-points of the titrations were determined with a Gran plot (STUMM and MORGAN 1981).

The methylene blue method of LINDSAY and BAEDECKER (1988) was used for the determination of total sulfide (H$_2$S, HS$^-$, and S^{2-}). An 8 ml water sample, collected in a syringe from the squeezing device, was connected to a three way stopcock along with another syringe containing 2 ml of the spectrophotometric reagents ferric chloride and N,N,-dimethyl-*p*-phenylene diamine. After a 20 minute development time in the dark the sample was analyzed, with a spectrophotometer, at 670 nm. Using this colorimetric method, the reagents and spectrophotometer were calibrated in the laboratory with a standardized sulfide solution. In the field, samples of lake water were placed in a syringe soon after collection and analyzed in an identical manner.

Laboratory methods

Brines

Sodium, K, and Mg were analyzed from diluted samples by atomic absorption spectrophotometry. Since the samples were highly diluted, salt matrix affects were reduced, if not eliminated, and concentrations were determined from standard curves. Calcium concentrations tended to be low in the alkaline waters and therefore the samples could not be diluted to avoid matrix effects. As a result, the method of standard additions had to be used for quantification.

Chloride and sulfate were determined on diluted samples using a Dionex Ion Chromatograph with a standard curve prepared from peak area measurements.

Dissolved organic carbon (DOC) was determined with a Dohrman DC-80 Organic Carbon Analyzer. The measurement is based on oxidation of organic carbon to carbon dioxide by ultra-violet light in the presence of potassium persulfate plus oxygen. The evolved CO$_2$ was analyzed by infra-red detection. Prior to oxidation the sample was acidified and purged with nitrogen to remove dissolved CO$_2$.

Dissolved Fe, Mn, Mo, V, Co, and Cu were determined by direct injection of diluted samples into graphite tubes for analysis by flameless atomic absorption spectrophotometry using a Perkin-Elmer HGA-500 programmer interfaced to a Perkin-Elmer 4000 atomic absorption spectrophotometer. Quantification was accomplished by the method of standard additions. For the analysis of Mn and Co 50 micrograms of magnesium nitrate were co-injected as a matrix modifier. Fe, Mo, V, and Co were analyzed with standard graphite tubes whereas the L'vov platform type were used for Mn and Cu (SLAVIN *et al.*, 1983). High purity grade argon was used as the purge gas. Background correction was accomplished with either a tungsten or a deuterium lamp, depending on the wavelength. Dissolved Zn was analyzed by flame atomic absorption spectrophotometry. Quantification was by the method of standard additions.

Dissolved Pb and Cd were analyzed by differential pulse anodic stripping voltammetry. The working electrode is a hanging mercury drop and a silver/silver chloride electrode is the reference. The instrument was an EG & G Princeton Applied Research Polarographic Analyzer/Stripping Voltammeter. The plating potential was −1.20 volts. The purging time was 4 minutes followed by a deposition time of 30 seconds and an equilibration time of 15 seconds. The differential pulse height was 50 mv and the anodic scan was from -1.20 volts to $+0.15$ volts at a rate of 2 mv per second. Data is collected on a chart recorder and quantification is by a standard curve of peak height.

Gel permeation chromatography was used to separate complexed metals from "non-complexed" ions. The eluate from the column was analyzed by flame or flameless atomic absorption spectrophotometry. Complete details of the gel permeation procedure are described in the section on the humic acid bound metal extraction.

Sediment analyses

Total Organic Carbon (TOC). Sediment samples were dried at 70°C in an oven. A 3 to 5 g sample is placed in 10% HCl until all carbonates have dissolved. The sample was then centrifuged, washed with distilled/de-ionized water, re-centrifuged, dried, and weighed. The dry sample is analyzed by combustion on a LECO carbon analyzer. Vanadium pentoxide is added to the sample to aid combustion. The per cent organic carbon is reported relative to a standard and on the dry weight of the sample prior to carbonate removal.

Total sedimentary metals. Fe, Mn, Cu, Zn, Pb, Co, Cd, and Mo analyses were performed by X-Ray Assay Laboratories, Ltd. of Ontario, Canada, employing total acid dissolution of the sediment, followed by flame atomic absorption for quantification. Vanadium was analyzed by a fusion recovery technique.

Aqua regia extractable metals. This is defined for this study as those metals which can be solubilized by treatment with hot aqua regia. This quantity of metal is assumed to be the maximum amount which is subject to re-mobilization by diagenetic processes. The extracting solution is 2.6% Ultrapure nitric acid plus 2.4% Ultrapure hydrochloric acid. A 0.5 g aliquot of oven dried sediment was placed in a polyethylene bottle with 10 ml of aqua regia and baked for one hour at 75°C. The solution was shaken for one hour and filtered through a 0.4 micrometer Nuclepore filter in an all-plastic holder. The sediment was washed with ultrapure water and the wash was filtered, combined with the aqua regia solution and diluted to 25 ml. The metals were analyzed by atomic absorption spectrophotometry.

Sequential extractions. These extractions were performed on aliquots of wet sediment, 2 to 3 g equivalent dry weight, according to the methods of TESSIER *et al.* (1979) with some modifications. The dry weight is determined on a separate aliquot. The sample is placed in a 50 ml centrifuge tube with 20 ml of extraction solution under a nitrogen atmosphere. All extraction solutions, except for hydrogen peroxide, are purged with nitrogen before injection to remove dissolved oxygen. At the end of the extraction step the solution is recovered by centrifugation, the sediment is washed with nitrogen purged ultrapure water, and the wash water is combined with the extraction solution. The extracting steps, times, and chemicals used are as follows: Solution 1. Millipore ultrapure water wash, 15 mls; Solution 2. Neutral cation exchanger, 1 M NH_4CH_3COO, pH = 7.15, prepared from reagent grade ammonium hydroxide (Fisher Scientific) and Ultrapure acetic acid (Alfa Chemical Co.), extraction time = 2 hours; Solution 3. Acidic cation exchanger, 0.5 M $NaCH_3COO$, pH = 4.85, prepared from Ultrapure acetic acid and Ultrapure sodium hydroxide (Alfa Chemical Co.), extraction

time = 4 hours; Solution 4. Reducing Solution, 0.5 M hydroxylamine hydrochloride in 0.01 N nitric acid, prepared from 99+% $NH_2OH \cdot HCl$ (Alfa Chemical Co.) and Ultrapure nitric acid (Alfa Chemical Co.), extraction time = 4 hours; Solution 5. Oxidizing Solution, 30% hydrogen peroxide adjusted to pH 2.5 with nitric acid, prepared from reagent grade hydrogen peroxide (Baker Chemical Co.) and Ultrapure nitric acid. 15 ml of solution are introduced to a sample until foaming ceases. This usually takes several days. After oxidation, the sediment is extracted overnight with 0.5 M ammonium acetate, pH = 2.2. This is prepared from Ultrapure acetic acid and Suprapur ammonium chloride (MCB Chemical Co.). The ammonium acetate step is necessary to extract metals which may have adsorbed onto oxidized phases after the peroxide treatment. Analysis of all the extracting solutions showed that they contained less than one part per billion of interferring metals.

The purpose of the selective extraction experiments is to determine what compositions of solutions are capable of mobilizing metals from the sediments. This provides an indication of the strength of metal bonding in the solid phases and may provide some clues as to the solid phase residence of the trace metals. The ultimate goal is to predict metal mobilization and precipitation interactions during diagenesis. Knowledge of detailed solid phase composition is necessary to model metal mobilization thermodynamically. Extractions were carried out on two separate aliquots. A sequential extraction was made of the first aliquot and an extraction for humic acid bound metals was made of the second. Ideally, the neutral exchanger extracts metals from exchangeable positions on clay or other minerals and organic matter, the acidic cation exchanger attacks carbonate minerals, the reducing solution attacks Mn oxides and labile Fe oxides, and the oxidizing solution attacks labile sulfides and labile organic matter. Labile sulfides and organic matter are solids which completely dissolve during the time frame of the experiment.

There has been much discussion of the selectivity and utility of these methods for the interpretation of metal solid phase residence. See, for example JONES and BOWSER (1978), VAN VALIN and MORSE (1982), RETTIG and JONES (1986), and MARTIN *et al.* (1987). The major problems cited include possible dissolution of non-target phases and/or re-adsorption of trace metals by other solids. Since the extractions involve pH and redox changes trace metals may be re-distributed to different solids during the course of the experiment. The peroxide extraction can generate acidity which may mobilize metals. Oxidation of labile metal sulfides will generate protons in a manner analagous to the generation of acid from pyrite-containing mine tailings as described in STUMM and MORGAN (1981). Therefore, more than one possible interpretation may be obtained from extraction data. The extraction results and pore fluid data are used together to interpret the most probable reactions and solid phase compositions affecting trace metal diagenesis for the sedimentary interval under consideration. One core from each lake was chosen for these extractions.

Humic acid bound metals. This extraction is performed on a separate aliquot of sediment. The sample size for Mono and Walker Lakes was 2 to 4 g whereas for the Great Salt Lake up to eleven g, dry weight, were extracted. The wet sediment was placed in 50 ml polycarbonate centrifuge tubes under a nitrogen atmosphere. To this, 20 mls of nitrogen purged 0.1 N Ultrapure NaOH were added and the reaction was allowed to proceed overnight. Both humic and fulvic acids are recovered by this extraction.

The basic extraction may also result in the dissolution of fine clays. The fluid was recovered by centrifugation and all fluid transfers were performed under a nitrogen atmosphere. The extraction was repeated with fresh NaOH. Five extractions were necessary to recover the majority of humic substances. The combined extracts were diluted to a known volume and analyzed for metals. Prior to instrumental analysis an aliquot of the fluid was spun on a high speed Eppendorf microcentrifuge to remove suspended particles.

Gel permeation chromatography was used to separate the humic acid complexed metals from the free metals. The chromatographic medium consisted of a 10 mm internal diameter column packed to a height of 19 cm with Sephadex G-25 packing material, bead size of 50–150 micrometers. The mobile phase consisted of 0.01 M Tris buffer, pH = 8.0. The pH was adjusted with Ultrapure HCl. The buffer was dissolved in 0.01 M Ultrapure NaCl (Alfa Chemical Co.). The elution volume of a nonretained solute, or high molecular weight organic matter, was determined with Blue Dextran (molecular weight = 2,000,000) whereas the elution volume was determined with standard solutions. The humic acids are colored and their progress through the column can be determined visually. The colored humic acids eluted at the same volume as the Blue Dextran. However, fulvic acids are recovered by the sodium hydroxide extraction along with other organic compounds with a wide variety of molecular weights. There are likely other organic compounds present which elute after the colored material. Therefore, the elution volume of Blue Dextran is used, operationally, to define the separation between metals complexed to high molecular weight organic compounds and uncomplexed metals. Some of the metals eluting after the elution volume of Blue Dextran may also be organically complexed. Chromatographic fractions were collected manually and analyzed by flame or flameless atomic absorption spectrophotometry.

RESULTS AND DISCUSSION

Walker Lake brine chemistry

The Walker Lake water column is alkaline, with a pH of 9.2. The concentration of the major elements is given in Table 1. The DOC concentrations of the lake water are moderately high and range from 27 to 30 mg/l. Thermal stratification of the lake waters occurs from late spring to summer with hypolimnetic oxygen depletion occurring during late summer. At all other times of the year oxygen reaches the bottom sediments. A redox gradient occurs very near to the sediment-water interface. Just above the sediment-water interface the pe is presumably controlled by the O_2/H_2O system, at a value near 12 (STUMM and MORGAN, 1981). Just below the sediment-water interface at site WL1, the pe, as calculated from the SO_4^{2-}/HS^- system, is between −4.75 to −5.0. At site WL2, the corresponding pe is −5.1 throughout the core. The water column, with a pH of 9.2, is a favorable environment for the adsorption of cationic trace metals onto Fe

Table 1. Composition of Walker Lake water (samples collected: July, 1984)

Species	Concentration (mg/l)
Na^+	3000.0
K^+	153.0
Ca^{2+}	9.4
Mg^{2+}	121.0
Li^+	1.0
Sr^{2+}	2.3
Cl^-	2080.0
SO_4^{2-}	2172.0
F^-	16.5
SiO_2	1.2
Alkalinity as milliequivalents/l:	44.8
pH: 9.2	

oxides and organic matter. The values of pe, from both sites, suggest that Fe and Mn oxides should start dissolving below the sediment-water interface (STUMM and MORGAN, 1981).

The waters are supersaturated with respect to calcite. The saturation index varied from 1.3 near the surface to 1.1 at depth during the summer sampling. For aragonite the saturation index varied from 1.1 near the surface to 1.0 at depth. The primary carbonate mineral accumulating at the sediment-water interface is monohydrocalcite.

The pore-fluid profiles of Na^+, Cl^-, SO_4^{2-}, sulfide, alkalinity, pH, and DOC, shown in Fig. 3, reveal that the two coring locations have different brine compositions. The most concentrated waters occur in the hydrographic low of the basin where brines or salts were deposited during past desiccation events. These brines are now diffusing to the surface. Sulfur chemistry is different at the two sites. Sulfate reduction goes to completion near the bottom of core WL1 but abundant sulfate is present at the bottom of core WL2 from upward diffusion of brines. Sulfide production is affected by sulfate concentrations at the two sites. Sulfide concentration reaches a maximum near the bottom of core WL1. At site WL2 the concentrations are higher and sulfide continues to form past the point of maximum core penetration. The production is limited only by the availability of labile organic matter. The increase in pore-fluid alkalinity, relative to the lake water, at site WL1 is driven by sulfate reduction, whereas at site WL2 the increase is driven by both sulfate reduction and the diffusive flux coming from buried evaporites not present at depth at site WL1. Despite the increases in alkalinity the pore-fluid pH drops at both locations. This is due to the release of carbon dioxide and organic acids from decomposing organic matter. The pH drops

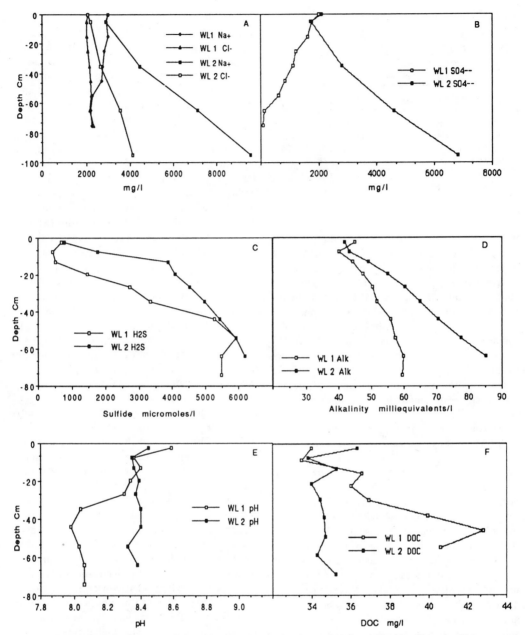

FIG. 3. Walker Lake pore-fluid major element geochemistry: a) Na$^+$ and Cl$^-$; b) sulfate; c) sulfide; d) alkalinity; e) pH; f) DOC.

by over one half unit across the sediment-water interface. It is buffered at values near 8.4 at site WL2 and at a value near 8.1 at site WL1. The DOC of the pore fluids increases relative to the water column with concentrations between 34 to 43 mg/l.

The water column metal concentrations are given in Table 2. Fe and Mn concentrations are low which is consistent with the oxygenated conditions. Mo-

lybdenum has the highest concentration of all metals. This is due to its presence as molybdate, MoO_4^{2-}, which has high solubility in an oxidizing alkaline brine.

Significant changes in dissolved metal concentrations occur just below the sediment-water interface. The trace metal profiles are shown in Fig. 4. Iron concentrations are higher than the overlying

Table 2. Walker Lake water column. Trace elements and
dissolved organic carbon

Depth (m)	Metal concentration: micrograms/liter Dissolved organic carbon concentration: milligrams/liter								
	Fe	Mn	Cu	Zn	Cd	Co	Mo	V	DOC
1.0	2.9	1.7	0.8	41.0	N.A.*	N.D.**	N.A.	7.3	30.5
5.0	N.A.	9.0	N.A.	16.6	1.9	N.D	269	6.7	29.0
15.0	3.7	4.7	0.7	11.3	0.4	0.4	334	8.3	26.7
25.0	5.3	4.7	0.4	10.2	0.4	1.0	321	8.3	27.2

* N.A. Not analyzed.
** N.D. Not detected.

water with the mobilization due, in part, to reductive dissolution of Fe oxide and decomposition of algal debris. The difference in Fe concentrations at the two coring locations could be due to solubility control by sulfide as the two concentrations are inversely correlated, but solubility calculations indicate that Fe is not in equilibrium with pure solid sulfide phases. The ion activity products for the Fe-S system of site WL2 are shown in Fig. 4b. These were calculated with WATEQ4F (BALL *et al.*, 1987). Supersaturation with respect to amorphous Fe sulfide, mackinawite, and pyrite occurs throughout the core.

Mn concentrations increase with depth indicating the presence of a dissolving phase. Copper concentrations are lowest near the sediment-water interface. There is some indication of mobilization in the central region of the core and a trend towards decreasing concentration lower. A mobilization of Zn occurs at 5 to 8 cm below the sediment-water interface. Zn concentrations tend to stabilize below that depth. The Co profiles are substantially different at the two sites with higher concentrations at site WL2. This indicates either a greater rate of Co release from dissolving solids at this site or, alternatively, stabilization of dissolved Co by some ligand. The dissolved Pb profile for site WL2 is considered tentative as this element was analyzed by differential pulse anodic stripping voltammetry and the peak position was shifted in a negative direction from that of the standard. This can occur if the metal is strongly complexed to a ligand. A mobilization of Mo and V is indicated between 10 to 20 cm of core. Both metals tend to decrease in concentration below that depth. Cadmium is undetectable in the pore fluids.

Walker Lake sedimentary chemistry

The mineralogy of Walker Lake sediments has been described by SPENCER (1977). His analyses

showed that the sediments are composed of 45 to 49% clay minerals, from 9 to 16% quartz, from 7 to 12% feldspars, and about 20% carbonate minerals. The clay minerals are of the smectite-illite type with smectite dominating. The carbonate mineral which is currently accumulating at Walker Lake is monohydrocalcite. This is a rare mineral which precipitates as trigonal bipyramidal crystals from a gel-like precursor. It is not known what conditions at Walker Lake are responsible for the precipitation of this mineral. SPENCER (1977) showed that it is slowly replaced by low magnesium calcite within the sediments. This forms in response to a drop in the pore-fluid magnesium to calcium ratio. Magnesium is lost from the pore fluids due to uptake by clay minerals (SPENCER, 1977). An Fe-oxide gel is present just under the sediment-water interface. The gel is thin, about 1 to 2 mm thick, and bright red. It probably originates from upward diffusion of reduced Fe which is oxidized and precipitated in the region of higher pore-fluid pe near the sediment-water interface. The gel is of major significance for trace metals as it is freshly precipitated and capable of adsorbing metals out of solution. It is near the sediment-water interface where the pH is close to 9, a favorable environment for trace-metal adsorption (BENJAMIN and LECKIE, 1981).

The concentrations of aqua regia extractable metals and total sedimentary metals are given in Table 3. The profiles for Mo and V are shown in Fig. 5. The shape of these profiles suggest that diagenetic mobilization or re-distribution of these two metals has occurred. These profiles are similar to that described for solid phase Mn in Lake Michigan sediments (ROBBINS and CALLENDER, 1975). A diagenetic model was proposed by those workers to explain this type of re-distribution. The mechanism for this mobilization, in Walker Lake sediments, is discussed later in this paper. The other metals show no apparent sedimentary concentration profiles which could be due to anthropogenic or diagenetic re-distribution for the length of the core. Iron, Mn, Cu, and Zn show somewhat higher concentrations near the bottom of the core. The concentrations of Cu, Zn, Pb, and Co are very close to the values of the "average black shale" of VINE and TOURTELOT (1970). Molybdenum is above this value in the enriched zone whereas V only approaches the value near the enriched zone.

Total sedimentary organic carbon is shown in Fig. 6. The profiles show that concentrations tend to decrease down to the 45 cm horizon with slight increases from 45 to 65 cm. Three horizons (0 to 16 cm, 23 to 35 cm, and 49 to 59 cm) were analyzed for humic acid concentrations. In these horizons,

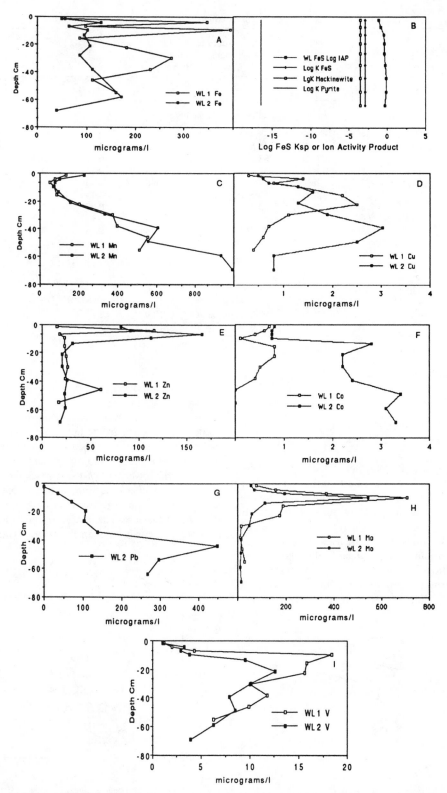

FIG. 4. Walker Lake pore-fluid trace metal concentrations and iron sulfide activity product profiles:
a) Fe; b) FeS ion activity product and saturation indices for mackinawite, amorphous FeS and pyrite;
c) Mn; d) Cu; e) Zn; f) Co; g) Pb; h) Mo; i) V.

Table 3. Walker Lake sediments. Total extractable and total sedimentary metals

	Iron concentrations: milligrams/gram								
	All other metals: micrograms/gram								
	Core location: WL2								
Depth (cm)	Fe	Mn	Cu	Zn	Pb	Co	Cd	Mo	V
				Total extractable metals					
2.5	10.6	543	28.0	38.9	11.5	4.6	0.1	25.3	62.2
7.5	9.7	533	26.5	35.8	10.0	6.3	0.1	27.4	61.6
13.5	9.9	557	27.3	33.2	11.4	7.4	0.12	24.1	85.3
21.0	11.3	640	35.2	34.0	10.0	8.3	0.16	5.3	70.0
29.5	8.0	535	26.7	31.7	9.2	6.0	0.12	7.9	49.5
39.0	10.5	587	28.6	39.3	10.2	9.0	0.15	2.7	35.7
49.0	14.7	538	26.4	42.8	7.9	4.8	0.07	1.8	29.2
59.0	16.1	660	34.3	52.4	8.7	6.1	0.1	1.7	38.4
69.0	13.2	641	32.8	43.5	10.1	7.4	0.15	3.0	36.2
				Total sedimentary metals					
2.5	38.0	670	44.0	81.0	16.0	16.0	<1	44.0	110
19.5	33.0	670	43.0	68.0	16.0	13.0	<1	63.0	140
34.5	40.0	630	49.0	84.0	12.0	15.0	<1	17.0	98
46.5	43.0	740	49.0	92.0	20.0	17.0	<1	9.0	94
62.0	42.0	820	51.0	87.0	14.0	16.0	<1	13.0	88

humic acids account for 6.4, 1.2, and 7.8% of the TOC, respectively.

Walker Lake sequential and humic acid extractions

Site WL2 was chosen for sequential extraction experiments. The results for Fe and Mn are shown in Fig. 7. Iron was unreactive to distilled/de-ionized water and neutral exchanger but was extracted by the acid cation exchanger and the reducing and oxidizing solutions. The most reactive zones for the reducing solution are at 30 cm and the bottom two core sections. The amount of sequential extractable Fe is greatest near the bottom of the core where 10 to 11% of the aqua regia extractable is recovered.

This region of the core shows the greatest amount of reactivity. Above 30 cm the oxidizing solution tends to extract the most Fe whereas below 30 cm the reducing solution extracts the most. The peroxide extractable Fe in the upper part of the core is probably derived from algal debris and sulfides, whereas the reducing solution attacks dissolving oxides in the lower part of the core. Some of the peroxide extractable Fe is from FeS as the pore fluids are supersaturated with respect to this mineral throughout the core. The concentration of humic-acid bound Fe is greatest in the bottom two core

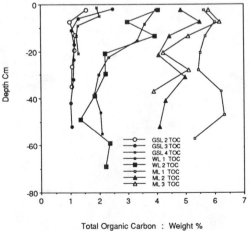

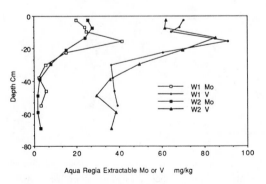

FIG. 5. Aqua regia extractable Mo and V from Walker Lake sediments.

FIG. 6. Sedimentary total organic carbon at Walker, Mono and Great Salt Lakes.

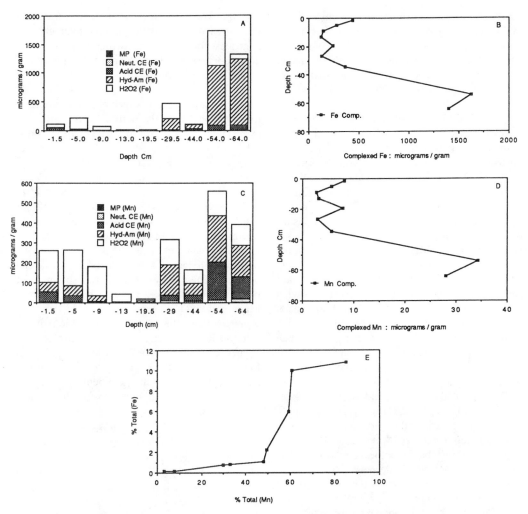

FIG. 7. Sequential and humic acid extraction results for Fe and Mn from Walker Lake sediments; The patterns represent results for each extraction (MP: Millipore water; Neut. CE: neutral cation exchanger; Acid CE: acidic cation exchanger; Hyd-Am: hydroxylamine hydrochloride or the reducing solution; H_2O_2: hydrogen peroxide or the oxidizing solution): a) Fe sequential extraction; The largest stack represents 4.2% of total sedimentary Fe; b) Fe humic acid extraction; The greatest amount of complexed Fe is 4% of the total; c) Mn sequential extraction; The largest stack represents 73% of total sedimentary Mn; d) Mn humic acid extraction; The greatest amount of complexed Mn is 4.5% of the total; e) comparison of Fe and Mn reactivity to sequential extraction solutions.

sections. In these horizons about 10% of the aqua regia extractable Fe is present as humic acid complexes.

Manganese was also unreactive to the distilled/deionized water and neutral cation exchanger extractions. Approximately 30% of the aqua regia extractable Mn was recovered with hydrogen peroxide on samples from the sediment-water interface. Sequential extractable Mn decreased to the 20 cm horizon and then tended to increase to the bottom of the core. The largest amount of Mn, 80% of the aqua regia extractable, was recovered at 55 cm. Similar to the situation with Fe, hydrogen peroxide

extracted the most in the upper part of the core; hydroxylamine hydrochloride extracted more in the lower part. This may be accounted for by oxidation of algal debris in the upper part of the core and attack of dissolving oxides below. The humic acid bound Mn profile is very similar in shape to that for Fe. However, less than 5% of the aqua regia extractable Mn is complexed to humic acid. The acid cation exchanger also extracts a significant amount of Mn. This may be from dissolution of carbonates or, alternatively, from the exchange of Mn from active surface sites by proton exchange.

A comparison of the percent of Fe and Mn recovered by sequential extraction solutions is shown in Fig. 7e. The figure shows that Fe reactivity is low until the point where about 50% of the Mn is recovered. The data is consistent with the redox chemistry of these elements. Manganese oxides decompose at a higher pe than Fe. Therefore, its reactivity will precede that of Fe down a sedimentary redox gradient. The pore-fluid profiles for Mn show increasing concentrations with depth, and reactivity to the reducing solution increases below the 20 cm horizon indicating that the dissolving phase is an oxide. The Mn oxides are dissolving prior to the Fe oxides whose reductive dissolution is initiated near the 50 cm horizon. Dissolved Mn^{2+} is re-adsorbed to Fe oxides at and above this horizon. This is indicated by the humic acid results which show the greatest amount of Mn and Fe complexation in the same horizons. Apparently, Fe oxides outcompete humic acid for adsorption of reduced Mn. Manganese is released from the surfaces of dissolving Fe oxide and some is adsorbed onto humic acid. The reduced Fe is either complexed to humic acid or precipitates as FeS. Pyrite does not precipitate because the pore fluids must first be undersaturated with respect to FeS before this can occur according to MORSE et al., 1987.

The sequential and humic acid extraction profiles for Cu, Zn, Pb, Co, and Cd are shown in Fig. 8. Copper is mobilized by the oxidizing solution. The greatest amount recovered was close to 10% of the aqua regia extractable Cu at the 5 cm horizon. The sequential extractable Cu decreases to the 20 cm horizon and then increases to the bottom of the core. A greater amount of Cu is complexed to humic acid. Close to 20% of the aqua regia extractable Cu is complexed near the 20 cm horizon and at the bottom of the core. The solid phase transformations of Cu are difficult to interpret. The only reactivity is to the oxidizing solution and the humic acid extraction. This is typical for Cu extractions and has been observed in a number of other sedimentary samples, as discussed by MARTIN et al., 1987. Some Cu apparently is associated with oxide phases. Pore-fluid Cu concentrations are lowest at the sediment-water interface near the Fe oxide gel. Dissolved Cu concentrations are also generally low throughout the core. Copper extractability and humic acid complexation increase below 20 cm which might be attributed to the liberation of Cu from dissolving oxides.

The sequential extraction chemistry of Zn is dominated by two horizons. The first is near the 5 cm horizon where the greatest amount, about 12% of the aqua regia extractable, is recovered. The pore-fluid Zn concentrations are also highest at this depth. All solutions extract some Zn at the 5 cm horizon. The acid cation exchanger extracts the most. Reactivity to hydroxylamine hydrochloride is greatest for the bottom two core sections. Hydrogen peroxide extracts the most Zn at the 5 and 55 cm horizons. The humic acid bound Zn profile is very similar to those for Fe and Mn. About 11 to 12% of the aqua regia extractable Zn is complexed to humic acid near the bottom of the core. The extractability of Zn with the oxidizing solution in the upper part of the core suggests that Zn is released from decomposing algal debris or sulfides. It is known that phytoplankton accumulate Zn (VINOGRADOVA and KOVALSKIY, 1962; BOSTROME et al., 1974). Iron oxides also adsorb Zn out of solution (BENJAMIN and LECKIE, 1981). These may be the two major sources of sedimentary Zn. Reactivity to the acid cation exchanger may be due to desorption from the active sites of metal oxides or clay minerals. There is evidence for adsorption of Zn onto Fe oxides. This is indicated by the relatively lower pore-fluid concentrations at the sediment-water interface, the release of Zn by the reducing solution in the lower two core sections, and the greatest complexation of Zn by humic acid in the lower two core sections. The extraction results suggest a number of possible sedimentary residences of Zn which include carbonates, oxides, humic acid, and sulfides.

Lead is extractable with the reducing and oxidizing solutions. The oxidizer always extracts the most. The greatest reactivity to the reducing solution is at and below 55 cm, which is similar to Fe. The horizon of greatest reactivity is at 55 cm where close to 30% of the aqua regia extractable is recovered. The shape of the humic acid complexed Pb profile is very similar to that for Fe, Mn, and Zn. This suggests that some Pb is transferred to humic acid following the dissolution of Fe oxide. The greatest percentage of complexed Pb, about 10% of the aqua regia extractable, is at the bottom two core horizons. The sequential and humic acid extractions suggest that Pb is affected by the presence of Fe oxides throughout the core. Oxides have the dominant effect on Pb geochemistry in this sedimentary interval. Although the experiments suggest that some Pb is transferred to humic acid during oxide dissolution, it is not possible to infer the nature of other authigenic Pb phases. As the oxidizing solution extracts the greatest amount of Pb it is possible that precipitation of Pb sulfide or co-precipitation with other metal sulfides occurs.

The sequential extraction of Co is dominated by reactivity to the oxidizing solution. The reactivity is high in the upper 5 cm, decreases at the 20 cm horizon, and then tends to increase at the bottom of the core. The acid cation exchanger extracts some Co in the upper 5 cm and the reducing solution extracts some in the lower two core sections. Between 10 to 15% of the aqua regia extractable Co is complexed to humic acid throughout the core except for the 55 cm horizon where close to 50% is complexed. The geochemistry of Co is probably controlled by Fe and Mn oxides throughout the length of the core as suggested by the extraction profiles. This is indicated by the greatest reactivity to the reducing solution and the highest amount of humic acid complexation occurring near the bottom of the core. The reactivity to peroxide in the upper 10 cm is probably due to oxidation of algal debris or sulfides. Solubility calculations suggest that a Co sulfide should form in these sediments.

Cadmium is extractable only by the oxidizing solution. In most core sections the oxidizing solution quantitatively extracts the aqua regia extractable Cd. Cadmium is not complexed to humic acid. The Cd extracted by peroxide is either bound to non-humic organic matter or to sulfides. There is no evidence for Cd adsorption onto oxides within these sediments. Results from laboratory experiments (BENJAMIN and LECKIE, 1981) show that Cd is less reactive to adsorption onto oxides as compared to the other cationic metals in this study. Cadmium is probably outcompeted for adsorption sites on these oxides and precipitates as a discrete sulfide.

The extraction profiles for Mo and V are shown in Fig. 9. Molybdenum is quantitatively recovered by sequential extraction reagents from Walker Lake sediments, except for the sediment-water interface. The highest concentrations are recovered at the 10 cm horizon due to diagenetic redistribution. Less than 20% of the aqua regia extractable Mo is recovered at the sediment-water interface. Below, there is an easily extractable form which is released by distilled/de-ionized water and the neutral cation exchanger. The acidic cation exchanger and the reducing solution extract progressively smaller amounts. Hydrogen peroxide extracts the more refractory Mo in amounts comparable to the first two extractants. Molybdenum is also effectively extracted by sodium hydroxide. However, there is no association of Mo with humic acid. Humic acid can be quantitatively separated from Mo by gel permeation chromatography. This is consistent with the results of DISNAR (1981) who showed that acidic conditions, below pH 3, are required for the adsorption of Mo onto sedimentary organic matter.

The pore-fluid and selective extraction results show that Mo solubility is quite low at the sediment-water interface. Only sodium hydroxide, at pH 13, is capable of extracting Mo from this horizon. In oxidizing streams Mo has been found in association with Fe oxides (KABACK and RUNNELLS, 1980) whereas for reducing sediments, such as the Black Sea, Mo precipitates as a discrete sulfide or in association with Fe monosulfides (KOROLOV, 1958; VOLKOV and FOMINA, 1974; BERTINE, 1972). Molybdenum may be adsorbed to the Fe oxide gel near the sediment-water interface. However, there is no evidence for adsorption of Mo onto oxides below this region.

The most likely mechanism for the sedimentary Mo re-distribution involves redox chemistry. Within the enriched zone Mo can be transported as the oxidized molybdate ion. Sufficient sulfide is present below to immobilize the metal, possibly by co-precipitation with Fe monosulfide. The Mo sulfide which forms is stable only in the presence of dissolved sulfide. It can be extracted by the anoxic sodium hydroxide extraction, and is partially extractable with other reagents. The peroxide extractable Mo in the deeper part of the core is probably associated with Fe monosulfides which can be dissolved with this reagent. The highest solubility of Mo is within the horizon below the Fe oxide gel and above the region of Mo sulfide precipitation.

Vanadium is also affected by diagenetic re-distribution at Walker Lake. The mechanism has some similarities to that for Mo but the redox speciation of V results in different solid phase associations. Only the reducing and oxidizing solutions extract this metal. The percent of V recovered by the sequential extraction relative to the aqua regia extractable is shown in Fig. 9c. The greatest extractability occurs just above the sedimentary horizon of highest total V. Extractability tends to decrease below this horizon. Unlike Mo, V does have an association with organic matter. The bottom two core horizons have the greatest amount of humic acid complexed V, about 10% of the aqua regia extractable. The shape of the humic acid bound V profile is similar to those already shown for Fe, Mn, Zn, and Pb.

In alkaline, oxidizing environments dissolved V is present as some type of an oxyanion such as HVO_4^{2-} (HEM, 1977). Under reducing conditions and at the pH of Walker Lake pore fluids, V should be present as the oxycation, $V(OH)_2^+$ (HEM, 1977). The transition occurs somewhere below 10 cm. The oxycation readily adsorbs onto oxides at the pH of these pore fluids and is also complexed by humic acid or other organic compounds. The association of V with Fe oxides below 10 cm is supported by the similarity in the shape of the humic acid complexed metal profiles for these two metals. The highest concentrations of humic acid bound Fe and V occur near the bottom of the core due to reductive

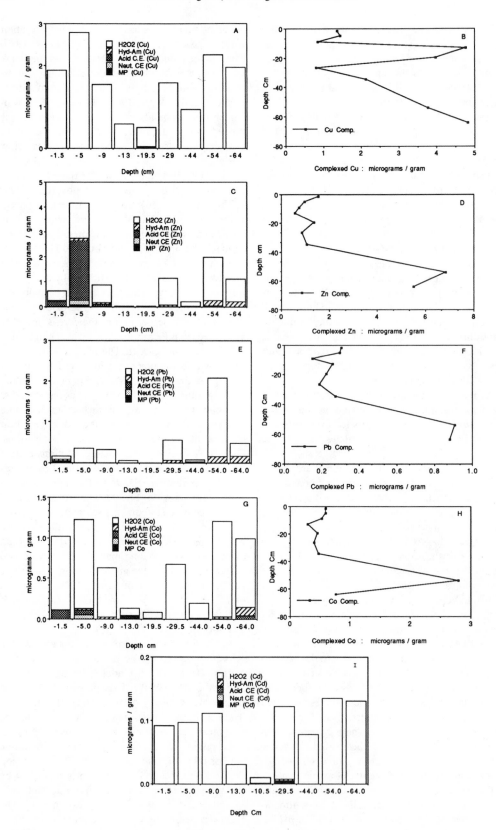

dissolution of Fe oxides and subsequent adsorption of the two reduced metals. The decrease in V reactivity to sequential extraction solutions below 10 cm of sediment is due to reduction and adsorption onto Fe oxide or organic matter. Like Mo, V has low pore-fluid concentration and low extractability near the sediment-water interface. This may also be due to adsorption on the freshly precipitated Fe oxide gel. The sedimentary re-distribution is a result of the diffuse redox gradient in Walker Lake sediments. Vanadium has highest solubility in the pore fluids close to the 10 cm horizon. The high solubility implies that V is present as an oxyanion in this region. Below this horizon V is immobilized by reduction and is adsorbed onto Fe oxides or is complexed by organic matter. Vanadium is mobilized at and below the 55 cm horizon due to reductive dissolution of Fe oxides.

Walker Lake summary

Walker Lake is a monomictic alkaline closed basin lake at an early stage of brine evolution. The sediment-water interface is oxygenated throughout most of the year except during late summer thermal stratification. The combination of oxygenated waters and high pH results in very low trace metal concentrations throughout the water column due to the scavenging of fine particulate oxides and organic matter. Changes in water chemistry occur across the sediment-water interface, which affect trace metals. This includes a drop in pH, due to carbon dioxide release from decomposing organic matter, and the production of hydrogen sulfide. The two sampling sites differ in that WL2 is at the hydrographic low of the basin coincident with buried brines and salts from past desiccation events. These brines are now diffusing to the sediment-water interface. Upward diffusion of dissolved carbonate species results in a higher buffering capacity of the pore-fluid brine relative to site WL1. Sulfate reduction goes to completion at site WL1, near 55 cm of sediment, but adequate sulfate is available at the same horizon of WL2 for further sulfide production. The abundant sulfate at site WL2 is in part from brine diffusion.

The trace metal chemistry in Walker Lake sediments is influenced by the presence of Fe and Mn oxides and hydrogen sulfide, which are present throughout the core. The reductive dissolution of Mn oxides begins before Fe oxides. Some adsorption of Mn^{2+} onto Fe oxide occurs after the dissolution of Mn oxide. Evidence for the presence of metal oxides and their influence on trace metals was provided by the following data: the reactivity of Fe to the reducing solution was greatest at the bottom of the core; a concentration increase in humic acid bound Fe coincides with the same core horizons as increases in the concentration of humic acid bound trace metals (Mn, Zn, Pb, Co, Cu, V). The results suggest the transfer of Fe from dissolving oxides to humic acid. Trace metals, initially bound to Fe or Mn oxides, are adsorbed on the humic material as they are released to the pore fluids by reductive dissolution. The humic acids are therefore secondary sinks of metals, following the reductive dissolution of metal oxides. Other secondary sinks include sulfides. The Walker Lake pore fluids are supersaturated with respect to Fe monosulfides which probably precipitate throughout this sedimentary interval. The degree of supersaturation is about two log units which inhibits the formation of pyrite. Solubility calculations suggest that the pore fluids may be supersaturated with respect to the sulfides of Cu^+, Zn^{2+}, Pb^{2+}, and Co^{2+}. However, there is insufficient information on the pore-fluid speciation of these metals to determine if they are in equilibrium with pure sulfide phases. Co-precipitation with Fe monosulfides is possible but the extraction experiments show no conclusive proof of this.

In addition to Fe monosulfides there is precipitation of diagenetically mobilized Fe oxide at the sediment-water interface. The source of this Fe is most likely algal debris as the Fe oxides are not labile to reduction in this region. This is supported by the sequential extraction data which showed that hydrogen peroxide released the most Fe near the sediment-water interface. The Fe released from decaying algal debris either precipitates as a sulfide or migrates to the sediment-water interface where oxidation occurs. The presence of this oxide layer near the sediment-water interface has important impacts on dissolved trace metals. It is a recently precipitated hydrous Fe oxide which can adsorb trace metals.

FIG. 8. Sequential and humic acid extraction results for Cu, Zn, Pb, Co, and Cd from Walker Lake sediments: a) Cu sequential extraction; The largest stack represents 6.4% of total sedimentary Cu; b) Cu humic acid extraction; The greatest amount of complexed Cu is 11% of the total; c) Zn sequential extraction; The largest stack represents 5.1% of total sedimentary Zn; d) Zn humic acid extraction; The greatest amount of complexed Zn is 8.4% of the total; e) Pb sequential extraction; The largest stack represents 14.8% of total sedimentary Pb; f) Pb humic acid extraction; The greatest amount of complexed Pb is 6.5% of the total; g) Co sequential extraction; The largest stack represents 7.5% of total sedimentary Co; h) Co humic acid extraction; The greatest amount of complexed Co is 17.4% of the total; i) Cd sequential extraction; Most stacks represent 100% of aqua regia extractable Cd.

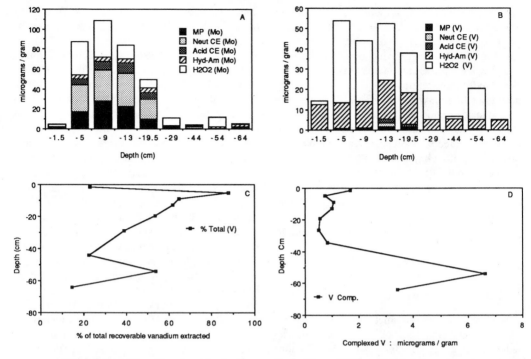

FIG. 9. Sequential and humic acid extraction results for Mo and V from Walker Lake sediments: a) Mo sequential extraction; Most stacks represents 100% of total sedimentary Mo; b) V sequential extraction; Most stacks represent 50% of total sedimentary V; c) % of total extractable V recovered by sequential extraction; d) V humic acid extraction; The greatest amount of complexed V is 7% of the total.

Furthermore, it is located near the sediment-water interface at a pH near 9 which is ideal for the adsorption of metal cations. Consistent with this, the pore-fluid trace metal concentrations tend to be lowest near this gel, relative to other core horizons.

The geochemistry of Cd is distinct from that of Fe, Mn, Cu, Zn, Pb, and Co. The evidence for this is with the sequential extraction data, in particular, the percent Cd extracted relative to the aqua regia extractable. For most core horizons Cd is quantitatively extractable with hydrogen peroxide. Cadmium is probably precipitating as a discrete sulfide. Cadmium is less reactive to adsorption onto Fe and Mn oxides relative to the other trace metals (BEN-JAMIN and LECKIE, 1981). In the sulfidic environment Cd is most likely extracted from oxide surfaces and precipitated as a sulfide. There is no evidence of humic acid complexation of Cd.

The oxyanion forming elements, Mo and V, have a sedimentary geochemistry based on redox conditions which results in diagenetic re-distribution. The redox chemistry of Walker Lake sediments is ideal for this mobilization as a diffuse gradient of pe, from oxygen saturation to Mn oxide dissolution to Fe oxide dissolution, occurs over an interval of 1 meter of sediment. Both Mo and V migrate to

the upper region of sediment in the oxyanion form. The oxyanions are resistant to adsorption and hence migrate to sedimentary environments of higher pe. Molybdenum is immobilized by sulfide precipitation. Vanadium is immobilized by reduction to an oxycation adsorbed onto oxide surfaces or complexed by organic matter.

Sequential extraction procedures recover various amounts of trace elements from Walker Lake sediments. The highest recoveries (100% relative to the aqua regia extractable) are for Mo and Cd. The lowest recoveries (less than 1% to 24%) are for Fe, Zn, Cu, and Pb. This is probably caused by the presence of Fe oxides throughout the core. The recovery of cobalt is higher in most horizons relative to Fe, Zn, Cu, and Pb, but is always lower than Mn. Manganese recovery is much higher, up to 85% of the aqua regia extractable, which is probably related to the dissolution of labile oxides. Humic acids are a significant sink for Fe, Zn, Co, V, and Pb in the lower part of the core and a significant sink for Cu throughout. Up to 20% of the aqua regia extractable Cu is present as humic acid complexes. Humic acids account for 10 to 15% of the aqua regia extractable Fe, Zn, Pb, and V. Cobalt complexation is highest at the 55 cm horizon where it accounts for 47% of the aqua regia extractable. Humic acids are an insignificant to minor sink for Mn. The greatest amount of complexation is 5% of the aqua regia extractable Mn at the bottom of the core.

Deeper cores, from both sites, are required to determine the diagenesis of metals associated with unstable solids, such as Fe monosulfides and humic acids. A comparison of sulfide mineralogy and organic matter, as trace metal sinks, could be made in the high and low sulfate environments.

Mono Lake brine chemistry

Mono Lake contains a $Na-CO_3-Cl-SO_4$ water with a pH near 9.65. A chemocline was present at 15 to 18 m of depth during the summers of 1985 and 1986. The deepest part of the lake sampled was 32 m. The ionic strength of the mixolimnion was 1.6 and that of the bottom brines was 1.9. The concentrations of the major elements in the water column are listed in Table 4 and the water column and pore-fluid profiles of sodium, alkalinity, sulfate, sulfide, and DOC are shown in Fig. 10. The lake is in an advanced state of brine evolution. The high alkalinity is an effective pH buffer and the high pH provides a favorable environment for trace metal adsorption onto Fe and Mn oxides.

Sulfate behaves non-conservatively in the water column. The concentration increases across the chemocline but appears to drop near the bottom of the lake (Fig. 10). The drop can be attributed to the initiation of sulfide production. Sulfide concentration in the bottom brine was 182.7 micromoles/l at 25 m of depth and 423.2 micromoles/l at 32 m of depth during this sampling. At a pH of 9.65 almost all of this sulfide is in the HS^- form. At a pH above 9 the bottom brines are also an ideal environment for the formation of polysulfide species which are effective ligands for trace metals (BOULEGUE and MICHARD, 1978). The calculated pe of the bottom brine, from the SO_4^{2-}/HS^- redox couple, is −6.25 which is sufficiently negative to result in the reductive dissolution of Mn and Fe oxides. The lake waters contain high levels of DOC. The concentrations are listed in Table 4. The high pH and evaporative concentration are probably most responsible for the high DOC levels. The high pH waters are probably effective in hydrolyzing algal debris and thus contributing to the DOC. Algal productivity is high and has been measured at 340 to 550 g C/m^2/year (JELLISON and MELACK, 1988). The lake water DOC may affect the chemistry of trace metals but this was not investigated in this study.

The pore-fluid concentrations of major elements decrease with depth as illustrated by the Na^+ profile

Table 4. Mono Lake water column. Concentration of major elements and dissolved organic carbon

	Ion concentration: milligrams/liter Dissolved organic carbon concentration: milligrams/liter Alkalinity: milliequivalents/liter							
Depth (m)	Na^+	K^+	Mg^{2+}	Ca^{2+}	Cl^-	SO_4^{2-}	ALK	DOC
1	30660	1610	38.9	10.3	9430	5040	630	N.A.*
5	30660	1640	38.9	6.7	9570	5300	634	78.9
10	30780	1640	39.3	5.5	9550	5170	625	79.9
15	30420	1630	38.8	6.1	9600	5250	632	88.4
20	30900	1650	39.6	6.1	9830	5240	645	79.1
25	32720	1760	41.3	5.7	10270	5640	691	87.6
30	34410	1840	43.2	5.2	10910	6010	709	86.4
34	35140	1890	43.7	5.8	11210	5880	719	89.4

* N.A. Not analyzed.

(Fig. 10b). Alkalinity and chloride have the same trend. The lake and pore-fluid brines are out of equilibrium. The rapid evaporative drawdown of the lake has resulted in more concentrated brines in the water column relative to the pore fluids. Despite the drop in alkalinity the pore fluids are well buffered against changes in pH. The pH drops only slightly in the sediments from 9.65 to 9.24. Pore-fluid sulfate decreases in concentration more markedly than the other major solutes because of reduction. The sulfide profiles are shown in Fig. 10f. The sulfate:sulfide ratio is shown in Fig. 10g. A polynomial equation derived from this data predicts that sulfate reduction should go to completion at a depth near one m of sediment. The calculated pe of the pore fluids, based on the SO_4^{2-}/HS^- ratio, is −6.37 at the sediment-water interface and decreases to −6.16 at 65 cm.

The pore fluids have high concentrations of DOC. The concentrations at sites ML1 and ML2 are similar, but site ML3 is lower. The DOC of the pore fluids is composed of hydrophilic acids which are capable of complexing metals (DOMAGALSKI, 1988). These acids have a structure which is similar to Mono Lake humic acids as determined by ^{13}C nuclear magnetic resonance spectroscopy (DOMAGALSKI, 1988). This was the only lake for which a sufficient amount of pore-fluid DOC was recovered for this type of analysis. Although the DOC of Walker Lake pore fluids was not investigated by NMR spectroscopy it is likely to be structurally similar to the DOC of Mono Lake. Both Walker and Mono Lakes have similar humic acid NMR spectra and the DOC in both lakes is derived from the decomposition of algal material in an anoxic environment (DOMAGALSKI, 1988). The molecular weight of the Mono Lake pore-fluid DOC is at least greater than 500 atomic mass units. Most of the

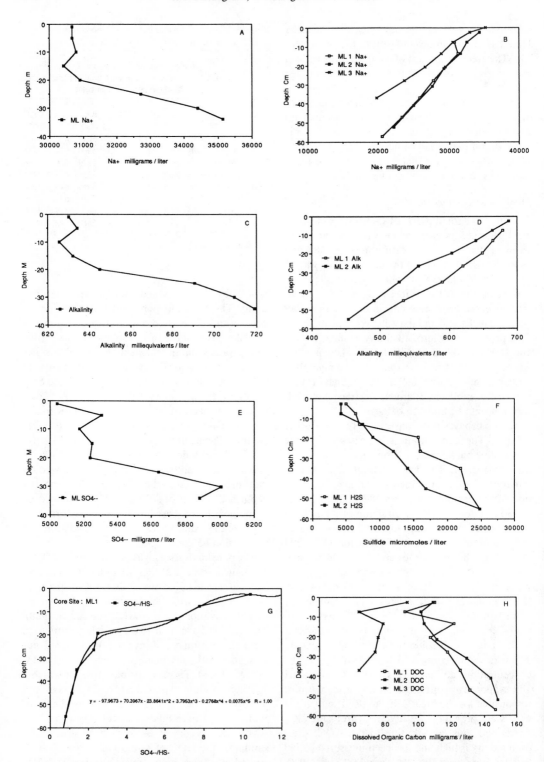

FIG. 10. Mono Lake water column and pore-fluid major element geochemistry: a) water column Na$^+$; b) pore-fluid Na$^+$; c) water column alkalinity; d) pore-fluid alkalinity; e) water column sulfate; f) pore-fluid sulfide; g) pore-fluid SO$_4^{--}$/HS$^-$ ratio; h) pore-fluid dissolved organic carbon.

Table 5. Mono Lake water column. Trace metals

Depth (m)	Metal concentration: micrograms/liter								
	Fe	Mn	Cu	Zn	Co	Mo	V	Cd	Pb
5	210	38.0	7.1	17.3	0.5	49.2	16.3	0.5	N.D.*
10	695	52.5	9.5	16.8	0.3	82.0	17.9	N.A.**	N.D.
15	833	65.9	8.9	15.5	0.9	77.0	N.A.	0.5	N.D.
20	526	40.6	6.9	16.8	0.7	84.0	17.5	N.A.	N.D.
25	258	60.5	5.0	18.0	N.D.	80.1	20.6	0.5	N.D.
30	153	63.0	6.4	16.8	N.D.	83.9	22.9	0.5	N.D.
34	129	71.6	7.6	16.8	N.D.	85.0	22.4	N.A.	N.D.

* N.D. Not detected.
** N.A. Not analyzed.

DOC was recovered by ultra-filtration using a membrane with this nominal cut-off.

The water column trace metal concentrations are given in Table 5. Iron concentration is highest near the chemocline and decreases with depth due to precipitation of FeS. The water column and pore-fluid trace metal profiles are shown in Fig. 11. Solubility calculations by WATEQ4F (BALL et al., 1987) show that supersaturation with respect to all Fe sulfide polymorphs occurs in the bottom brines. The concentration of Fe continues to decrease across the sediment-water interface. The pore-fluid Fe profile is shown in Fig. 11c along with the concentration of Fe complexed to DOC. The uncomplexed Fe concentrations were used in WATEQ4F for solubility calculations. The results, graphically shown in Fig. 11d, show that supersaturation with respect to amorphous FeS, mackinawite, and pyrite occurs throughout the core. These calculations assume no solution complexing of Fe with bisulfide ions.

Manganese concentrations are also greatest in the anoxic bottom waters (Fig. 11b), but there is no evidence of precipitation. Manganese concentrations drop rapidly across the sediment-water interface. Only a small amount of the dissolved Mn is complexed to DOC (Fig. 11e). The bottom brines are undersaturated with respect to MnS. The log saturation index at 32 m of depth is -0.5. Across the sediment-water interface the saturation index increases to 0.6. Below 5 cm the ion activity products are close to the Ksp. The saturation indices were calculated using a pKsp of 0.4 (JACOBS and EMERSON, 1982). Authigenic Mn sulfides are rare in modern sediments and their presence has only been documented in the sediments of the Baltic (BOESEN and POSTMA, 1988).

Copper concentration is greatest near the chemocline, decreases below, and then increases in the bottom brine. Zinc concentrations are relatively uniform with depth. Lake water Cd concentrations

are at detection limits. Lead concentration is below the detection limit of 0.5 micrograms/l. Cobalt concentrations are less than one microgram/l above the chemocline and undetectable below.

The pore-fluid Cu profile indicates mobilization in the upper 10 cm of sediment (Fig. 11f). Organic complexes of Cu increase in importance below this horizon. Zinc concentrations are also greatest just below the sediment-water interface. Lead is not detected in the pore fluids. Pore-fluid Cd concentrations are low but clearly detectable below 30 cm (Fig. 11i). This element was measured by differential pulse anodic stripping voltammetry and the sample peak was displaced from that of the pure standard by -40.0 mv throughout the pore fluids. This is taken as evidence that Cd is complexed to a polymolecular ligand in the pore fluids.

Cobalt pore-fluid profiles are similar at sites ML1 and ML2 but the concentrations are lower at site ML3 (Fig. 11h). Although Co was undetected in the sulfidic bottom brines of the lake, the Co concentrations tend to increase with the sulfide content of the pore fluids. It was not possible to determine if Co is complexed to DOC or other ligands. The pore-fluid profiles for the three sites suggest that Co and DOC concentrations are correlated. This can be seen by comparing the Co and DOC profiles in Figs. 10h and 11h. Solubility calculations for Cu^+, Zn^{2+}, Co^{2+}, and Cd^{2+} suggest extreme supersaturation with respect to metal sulfides. However the speciation of these metals in the pore fluids has not been defined. Possible ligands include polysulfide species and organic compounds.

Lake water V concentrations increase at the chemocline but there is little apparent mobilization in the pore fluids (Figs. 11k and 11l). The highest concentrations are at sites ML1 and ML2, which also have higher DOC relative to site ML3. This is similar to the profiles discussed for Co. In reducing waters V is present as an oxycation which can complex with hydrophilic acids. Molybdenum concentrations are lowest near the top of the water column with relatively higher and constant concentrations occurring at the bottom of the lake. A mobilization of Mo occurs in the upper 10 cm of sediment (Fig. 11j).

Mono Lake sedimentary chemistry

Mono Lake sediments are dominated by smectite clay minerals. Feldspars, derived from the Sierra Nevada Range, are also of importance. Aragonite is present in minor concentrations. The cores are black throughout, due to the presence of Fe monosulfides.

The concentration of aqua regia extractable metals and total metals are given in Table 6. Copper,

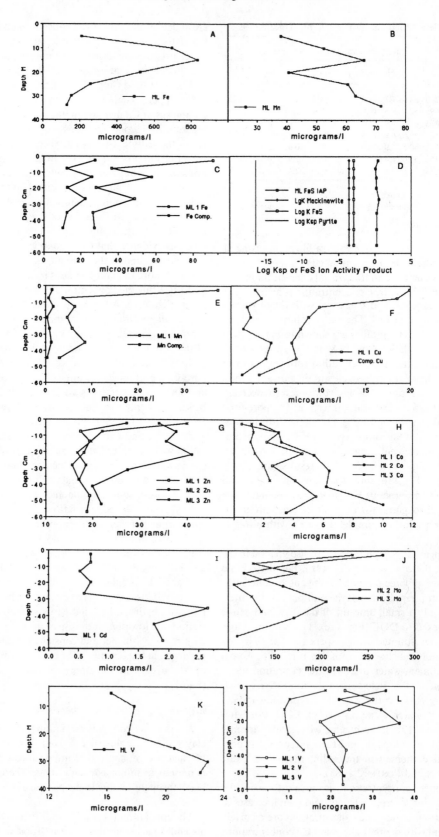

Co, Zn, and V concentrations are below that of the "average black shale" of VINE and TOURTELOT (1970). Lead is close to that value and Mo is above. The total metal concentrations are higher than the corresponding aqua regia extractable. There is no apparent evidence for diagenetic re-distribution or anthropogenic enrichment of these metals.

The weight per cent of total organic carbon for the three coring sites is shown in Fig. 6. The highest concentrations are at site ML1. This may be due to its proximity to inflow streams which supply nutrients to the lake. Three horizons were analyzed for humic acid concentrations (0 to 16 cm, 23 to 30 cm, and 40 to 50 cm). Humic acids account for 8.4, 7.5, and 7.4% of the TOC in these horizons.

Mono Lake sequential and humic acid extractions

The extraction results for Fe and Mn are shown in Fig. 12. Iron can be extracted from Mono Lake sediments by the acidic cation exchanger, the reducing solution, and the oxidizing solution. In the upper 7 cm of core, the reducing solution releases the most Fe. Below, the oxidizing solution tends to release the most. There is very little Fe mobilized by the reducing solution between 13 to 20 cm. Below this horizon the amount mobilized by this reagent increases. This is probably due to crystalline oxides which only slowly dissolve in the anoxic environment. The most reactive zone for Fe is between 5 to 10 cm where close to 25% of the aqua regia extractable is recovered by sequential extraction solutions. The most unreactive zone is between 13 to 20 cm. Below this horizon the extractable Fe tends to be constant at values close to 10% of the aqua regia extractable. The amount of Fe complexed to humic acid is very low in Mono Lake sediments. In only two core sections does the amount exceed 1.0%. This suggests that the Fe released by the oxidizing solution is not organic but more likely is from a labile sulfide.

Manganese is extractable by the same three reagents. The maximum amount of reducible Mn is at the 9 cm horizon. The amounts extracted at and below the 13 cm horizon are always lower than those from above this horizon. The profiles for the acidic cation exchanger and reducing solution have a similar configuration. Throughout the core the oxidizing solution releases the most Mn. The greatest

amount of Mn extracted by these reagents is at 9 cm where 43.4% of the aqua regia extractable was recovered. Below this horizon there is no significant change in reactivity and close to 25% of the aqua regia extractable was mobilized by selective sequential extractions.

The amount of Mn complexed to humic acid is extremely low in Mono Lake sediments. The highest amount is only 0.7% of the aqua regia extractable between 15 to 20 cm of depth. As with Fe, this suggests that the metal extracted with the oxidizing solution is not organically complexed but more likely is a sulfide.

The water chemistry data shows that the reductive dissolution of Fe and Mn oxides is initiated near the chemocline and the reactivity to hydroxylamine hydrochloride suggests that the greatest amount of reducible metal is present in the upper ten cm of sediment. Iron sulfide precipitation occurs in the water column and continues in the sediments as the oxides are reduced. Solubility calculations with WATEQ4F (BALL et al., 1987) show that the bottom brines are supersaturated with respect to Fe monosulfides. This supersaturation continues through the sediments. Pyrite does not form in these sediments despite high supersaturation. Pyrite was not detected in X-ray diffraction patterns from

Table 6. Mono Lake sediments. Total extractable and total sedimentary metals

Depth (cm)	Iron concentrations: milligrams/gram All other metals: micrograms/gram Core location: ML1								
	Fe	Mn	Cu	Zn	Pb	Co	Cd	Mo	V
	Total extractable metals								
2.5	12.3	318	21.6	53.3	15.0	5.7	0.21	14.3	32.1
7.5	13.2	326	21.3	58.5	10.5	4.6	0.26	17.8	37.0
13.5	9.9	293	17.8	48.6	15.0	4.5	0.33	14.4	33.8
20.5	9.5	338	15.7	44.6	12.1	4.4	0.27	10.7	38.5
28.0	13.2	392	22.9	58.8	13.4	5.0	0.35	18.6	39.9
37.0	10.9	331	18.1	53.3	9.8	5.3	0.27	18.3	41.2
47.0	9.4	318	15.4	44.2	9.0	4.4	0.29	27.8	34.1
57.0	12.0	362	22.8	52.3	7.6	7.4	0.30	35.3	31.4
	Total sedimentary metals								
28.0	26.0	530	31.0	86.0	18.0	11.0	<1	33.0	68.0
57.0	22.0	500	27.0	74.0	14.0	9.0	<1	61.0	66.0

FIG. 11. Mono Lake water column and pore-fluid trace metal concentrations: a) water column Fe; b) water column Mn; c) pore-fluid Fe, total and complexed by DOC; d) pore-fluid FeS ion activity products and saturation indices for amorphous FeS, mackinawite, and pyrite; e) pore-fluid Mn, total and complexed by DOC; f) pore-fluid Cu; g) pore-fluid Zn; h) pore-fluid Co; i) pore-fluid Cd; j) pore-fluid Mo; k) water column V; l) pore-fluid V.

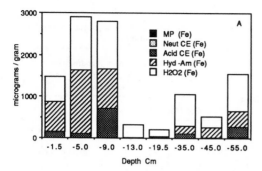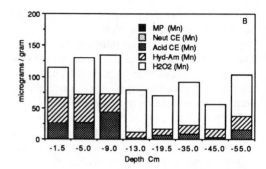

FIG. 12. Sequential extraction results for Fe and Mn from Mono Lake sediments: a) Fe; The largest stack represents 12% of total sedimentary Fe; b) Mn; The largest stack represents 26% of total sedimentary Mn.

mineral grains separated from the bulk sediment by bromoform extraction. Apparently the initial precipitation of FeS is kinetically favored. All other factors for pyrite formation are ideal in these sediments. Elemental sulfur, required for pyrite precipitation, should be abundant in the sediments because of reduction reactions involving HS^- and metal oxides. The supersaturation with respect to metastable Fe monosulfides prevents the formation of pyrite. Direct precipitation of pyrite requires either dissolution of FeS or undersaturation of the pore fluids with respect to these minerals (RICKARD, 1975). Supersaturation in Mono Lake pore fluids is driven by high pH and sulfide levels. Humic acid is a minor sink of Fe in these sediments. The primary Fe reactions include the reductive dissolution of oxides in the lake waters and upper ten cm of sediment, and the precipitation of FeS. The reactivity of Fe to the acidic cation exchanger indicates that carbonate minerals may also be a secondary sink of Fe. However, this Fe may have been mobilized from other solids by proton exchange.

Manganese oxide dissolution is initiated below the chemocline, but there is no evidence of precipitation in this region. There is evidence of rapid precipitation directly across the sediment-water interface. Like Fe, Mn oxide reductive dissolution is greatest in the upper ten cm of sediment. Solubility calculations indicate that the lake bottom brines are undersaturated with respect to MnS. Directly below the sediment-water interface the pore fluids are supersaturated with respect to this mineral. These results suggest that a discrete MnS phase controls the solubility of this metal in Mono Lake sediments.

The extraction profiles for Cu, Zn, Pb, Co, and Cd are shown in Fig. 13. Copper was reactive to hydrogen peroxide. The greatest amount is extracted just below the sediment-water interface where 61%

of the aqua regia extractable is recovered. This amount decreases to a relatively constant 20% at depth. In most horizons humic acid complexes account for 2 to 4% of the aqua regia extractable. In two horizons the complexed Cu accounted for 7 and 9.5% of the aqua regia extractable. The sequential extraction results and pore-fluid data for Cu suggest that a mobilization occurs in the top ten cm of sediment. The amount extracted is relatively constant below this horizon. The most likely reactions for Cu include release from dissolving oxides and algal debris and precipitation of a sulfide or humic acid complex. Since humic acid complexation is minor, sulfides represent the major sedimentary sink. The sequential extraction experiment cannot distinguish the type of sulfide forming.

Hydrogen peroxide extracts the most Zn in the upper 10 cm of sediment. Below 15 cm the relative reactivities of the acidic cation exchanger and the reducing and oxidizing solutions are similar. Reactivity to the neutral exchanger increases up to 9 cm after which it drops to a constant value. Reactivity to the acidic cation exchanger is also greatest in this region. The highest reactivity is in the upper 10 cm where 25% of the aqua regia extractable is recovered by sequential extraction solutions. Below this horizon the recovered amount is constant at values close to 10%. Humic acid complexes account for 3 to 5% of the aqua regia extractable Zn throughout the core.

The chemistry of Zn at Mono Lake is complicated. The metal is mobilized in the pore fluids of the upper few cm of sediment. The pore-fluid concentrations do not change below. The pore fluids are supersaturated with respect to Zn sulfides, unless some unknown ligands are complexing the metal in solution. Some of the peroxide extractable Zn in the upper 10 cm of sediment may be from organic matter of planktonic origin, but the extracted metal

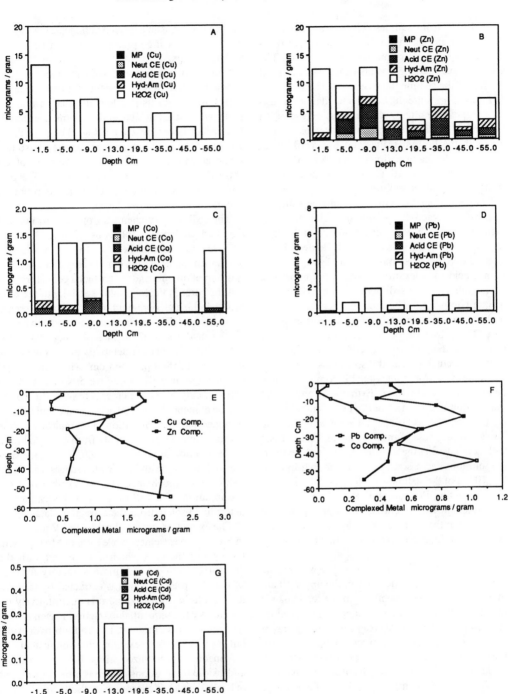

FIG. 13. Sequential and humic acid extraction results for Cu, Zn, Pb, Co, and Cd from Mono Lake sediments: a) Cu sequential extraction; The largest stack represents 46% of total sedimentary Cu; b) Zn sequential extraction; The largest stack represents 16% of total sedimentary Zn; c) Co sequential extraction; The largest stack represents 16% of total sedimentary Co; d) Pb sequential extraction; The largest stack represents 40% of total sedimentary Pb; e) Cu and Zn humic acid extractions; The greatest amount of complexed Cu is 7.5% of the total; The greatest amount of complexed Zn is 2.5% of the total; f) Co and Pb humic acid extractions; The greatest amount of complexed Pb is 6.5% of the total; The greatest amount of complexed Co is 9.5% of the total; g) Cd sequential extraction; Most stacks represent 100% of the aqua regia extractable Cd.

may also be from sulfides. There are other labile forms of Zn present. Reactivity to the reducing solution suggests adsorption onto metal oxides. The acidic cation exchanger is also effective, except at the top of the sedimentary column. The pathways of Zn diagenesis are difficult to elucidate from this data. A possible interpretation is the release of Zn from metal oxides and organic matter of planktonic origin in the bottom brines and upper 10 cm of sediment. The major sedimentary sink is some type of sulfide, either an authigenic Zn sulfide or coprecipitated with other sulfides.

Lead is extractable with the oxidizing solution. The greatest reactivity occurs just below the sediment-water interface. In this region 41% of the aqua regia extractable Pb was recovered. Below this horizon, the recoverable amount varies from 3 to 20%. Humic acid complexation increases in importance with depth. Near the sediment-water interface there is little complexed. The amount increases to 10% of the aqua regia extractable near the bottom of the core. The extraction profile for Pb is very similar to Cu with the greatest reactivity to the oxidizing solution occurring near the sediment-water interface. The extraction is releasing Pb from non-humic organic matter and sulfides. Extraction from sulfides may be dominant source of Pb. Lead is rapidly precipitated as there is no detectable metal in the pore fluids. Humic acid complexation of Pb increases with depth despite the fact that reactivity to extracting solutions decreases. The oxidizing solution, below 10 cm of the sediment-water interface, is probably extracting Pb from sulfides. The per cent of Pb extractable by sequential extraction is positively correlated to that of Fe which suggests that Pb may be involved in coprecipitation with Fe monosulfides.

The majority of reactive Co is released by the oxidizing solution, particularly in the upper 10 cm of core. There is also reactivity to the reducing solution and the acid cation exchanger in this region. Reactivity to the latter two reagents is largely absent below the 10 cm horizon except at the bottom of the core where the acid cation exchanger extracts a small amount. The highest percentage of aqua regia extractable metal, complexed to Mono Lake humic acid, is for Co. The complexes account for 10 to 25% of the aqua regia extractable Co. Cobalt has decreasing reactivity to the reducing solution in the upper 10 cm of core, which indicates that some metal is adsorbed on the dissolving metal oxides. The amount of humic acid bound Co increases down to the 10 cm horizon as adsorption onto metal oxides decreases in importance. The primary associations for Co are with organic matter and sulfides. It is not possible to tell if authigenic Co sulfides are forming, or if co-precipitation with Fe monosulfides occurs.

Cadmium is extractable with the oxidizing solution, but there was no evidence for complexation with humic acid. Its profile is unusual in that the metal was not extractable at the sediment-water interface but is very reactive below. The pore fluids are supersaturated with respect to the Cd sulfide phase greenockite. There is no correlation of Cd extraction chemistry with other elements. The precipitation of a discrete Cd sulfide phase probably occurs.

The sedimentary chemistry of Cu, Zn, Pb, and Co is affected by the following factors: 1) the reductive dissolution of Fe and Mn oxides, which is greatest in the upper ten cm of core; 2) the decomposition of organic material, which occurs throughout the core; 3) the high sulfide activity of the pore fluids; 4) the high pore-fluid pH; and 5) the presence of humic acid. The dissolution of Fe and Mn oxides results in the loss of major adsorbing phases for these elements. The extractability of these elements is greatest in the upper ten cm, where the oxides are undergoing active reductive dissolution. Some extractability of these elements may be due to the release of metals from algal debris in the initial stages of anaerobic decomposition. The majority of the extracted metals are probably from labile sulfides. The extraction profiles for Cu, Zn, Pb, Co, are similar to those for Fe and Mn which suggests a related sedimentary geochemistry. It is possible that coprecipitation with FeS occurs but this cannot be determined solely from the extraction data.

The extraction profiles for Mo and V are shown in Fig. 14. An extremely labile form of Mo is present throughout the core which is recovered with the first three extractions. A more refractory form of Mo is also present which is extracted by the oxidizing solution. Molybdenum is quantitatively recovered by the sequential extraction reagents. Molybdenum is also mobilized by sodium hydroxide, but there is no complexation with humic acid.

Vanadium is mobilized primarily by the reducing and oxidizing solutions. There is a minor amount of labile V which is released with distilled/de-ionized water and the neutral cation exchanger. There is no change throughout the sediments in the amount extracted by the reducing solution but the amount of peroxide extractable V increases with depth. In most core sections about 50% of the aqua regia extractable V is recovered by sequential extraction reagents. There is some V complexed to humic acid but the amount is very low, less than 1% of the aqua regia extractable.

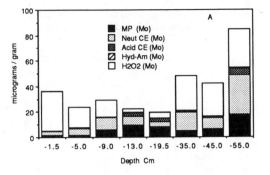
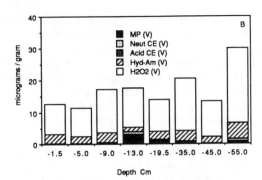

FIG. 14. Sequential extraction results for Mo and V from Mono Lake sediments: a) Mo; Most stacks represent 100% of total sedimentary Mo; b) V; Most stacks represent 45% of total sedimentary V.

There is no evidence of sedimentary re-distribution of V and Mo in Mono Lake sediments. Since reducing conditions occur throughout the sediments there is no chance for re-distribution. A gradient of oxidizing to reducing conditions is required. Vanadium is present in these sediments as the reduced oxycation, $V(OH)_2^+$, according to the pe and pH conditions. The oxycation will readily adsorb onto sedimentary particles or be complexed by organic matter. Molybdenum has been shown to be associated with sulfides in anoxic environments (VOLKOV and FOMINA, 1974; BERTINE, 1972). It has been suggested (BERTINE, 1972) that Mo co-precipitates with Fe monosulfides. Molybdenum is easily mobilized from anoxic sediments and is only immobile in the presence of sulfide. Molybdenum solubility is probably controlled by sulfide precipitation throughout Mono Lake sediments.

Mono Lake summary

Mono Lake is a saline alkaline lake, pH = 9.65, with anoxic bottom waters due to the presence, during 1985 and 1986, of a chemocline near 15 m of depth. The presence of an oxycline/chemocline results in the initiation of the reductive dissolution of Fe and Mn oxides and of sulfate reduction in the bottom brines. Maximum dissolution of labile sedimentary oxides (Fe, Mn) occurs in the top ten cm of sediment and refractory oxides are reduced below. Sulfide production occurs throughout the length of core in response to high total sedimentary organic carbon (5–7%), which is labile with respect to bacterial metabolism, and abundant sulfate. The Fe and Mn oxides are replaced primarily by metal sulfides following reductive dissolution.

Supersaturation with respect to Fe monosulfide is initiated in the water column. Across the sediment-water interface the degree of supersaturation increases and stabilizes at about 3 log units. The pore-fluid supersaturation with respect to Fe sulfide is driven by the positive sulfide gradient and the high pH, which is buffered above 9 for the length of the core. The lake water brines are not supersaturated with respect to MnS, but the pore fluids are. The Fe monosulfide minerals are metastable with respect to pyrite. Pyrite formation is inhibited because of the pore-fluid supersaturation with respect to Fe monosulfide.

Sequential extraction procedures tend to recover higher amounts of some trace metals from Mono Lake sediments relative to Walker Lake sediments. This is particularly the case for Fe, Zn, Cu, Co, and Pb (2 to 61% relative to the aqua regia extractable). This is probably related to the dissolution of Fe and Mn oxides in the upper part of the core and the precipitation of these elements into extractable phases. Manganese extractability was less from Mono Lake sediments relative to Walker Lake sediments (between 18 to 43% relative to the aqua regia extractable). This is probably related to the initiation of Mn oxide dissolution in the Mono Lake water column, as opposed to the sediments at Walker Lake, and its possible precipitation as an authigenic sulfide at Mono Lake. Similar to the situation at Walker Lake, Cd and Mo can be quantitatively extracted from Mono Lake sediments. The extraction of V is relatively constant in Mono Lake sediments because only one oxidation state is possible. Humic acids, although abundant in the Mono Lake sediments (8% of the TOC), are an insignificant sink of Fe and in general account for less than 1% of the aqua regia extractable. Cobalt is most reactive to humic acid complexation. Between 10 to 25% of the aqua regia extractable is complexed to humic acid. Copper is the next most reactive element. The greatest complexation of Cu by humic acid is close to 10% of the aqua regia extractable, but is closer to 2 to 4% in most core sections.

The group of elements Fe, Mn, Cu, Zn, Pb, and Co are related in the sense that their extraction profiles are similar. In all cases the greatest extractability occurs in the upper part of the core with a trend of constant reactivity below. These elements are associated with metal oxides, algal debris, and sulfides upon sedimentation and are released to the pore fluids as these phases are degraded. Sulfide phases are the dominant sink below 10 cm. The similarity of the extraction profiles suggests that co-precipitation with Fe monosulfides occurs but this would have to be proven by experimentation. An alterative interpretation is that this group of metals is mobilized during the peroxide extraction due to the generation of acid by oxidation of FeS. In this case the metals are not geochemically linked with FeS by co-precipitation, but rather, their geochemical mobility is linked to the oxidation of FeS. Manganese may be present as a discrete sulfide as suggested by the ion activity products and the extraction profiles. Cadmium probably also precipitates as a discrete sulfide, as suggested by the extraction results.

Molybdenum and V are not involved in diagenetic sedimentary re-distribution at Mono Lake as was the case at Walker Lake. This is due to the reducing conditions occurring throughout the sedimentary column and the anoxic bottom brines. There is no opportunity for these two elements to migrate along a redox gradient as at Walker Lake. Molybdenum precipitates as a sulfide. Vanadium is reduced to the oxycation species and is adsorbed onto organic matter and residual oxides. High levels of DOC, up to 150 mg/l, occur in the Mono Lake pore fluids. This material is a ligand for Cu, Fe, and Mn. The pore-fluid data suggests that it may also be a ligand for Co and V. This is indicated by the positive correlation among Co and V concentrations and the amount of DOC at the three sites.

The metal geochemistry at Mono Lake sediments is therefore dominated by the reductive dissolution of metal oxides in the top ten cm and metastable equilibrium with respect to Fe sulfide. Trace metals are extracted by the oxidizing solution which degrades the Fe monosulfides. Deeper cores are required to determine how and where metal stabilizing reactions occur.

Great Salt Lake brine chemistry

The Great Salt Lake is a meromictic hypersaline Na-Mg-Cl-SO$_4$ brine of low alkalinity. The major element water-column composition for site GSL3

Table 7. Great Salt Lake water column. Composition at site GSL3. Samples collected 9-86 (STURM, 1986)

Depth (m)	Concentration (g/l)					
	Na$^+$	Mg^{2+}	K$^+$	Ca^{2+}	Cl$^-$	SO$_4^{2-}$
0.15	15.5	1.82	1.20	0.12	27.7	3.48
1.5	15.6	1.68	1.25	0.12	28.4	3.48
3.1	15.7	1.79	1.25	0.13	28.4	3.44
4.6	15.5	1.78	1.20	0.13	28.4	3.46
6.1	15.7	1.80	1.20	0.13	28.4	3.41
7.6	22.0	2.25	1.80	0.15	38.3	4.53
9.1	45.5	4.95	2.88	0.25	80.1	9.36
10.6	56.5	6.02	3.60	0.28	100.6	12.2
12.2	57.7	6.46	3.71	0.30	104.4	12.7
13.0	58.5	6.65	3.75	0.31	105.9	13.9

is given in Table 7. The bottom waters have similar compositions but varying densities. The most concentrated brines occur in the North Basin. The ionic strength of the surface waters of the South Arm is close to 1.0. The ionic strength of the bottom waters of the North Basin of the South Arm is 4.5 whereas the South Basin bottom brines have an ionic strength of 3.0, and those of the South Basin deep site have an ionic strength of 3.6. These measurements were taken in September, 1986. All waters are undersaturated with respect to halite and gypsum but are close to saturation with aragonite (SPENCER et al., 1985). Water column sulfate concentrations range from 3.5 to 13.9 g/l in the South Basin and from 3.4 to 20.2 g/l in the North Basin.

The maximum concentration of solutes related to microbiological activity and organic decomposition, such as alkalinity, DOC, and sulfide, are found in the bottom brines. These species tend to decrease in concentration across the sediment-water interface. The concentrations of these species are dependent upon the decomposition of organic matter by anaerobic processes with sulfate as the terminal electron acceptor. Fig. 15 shows the water column and pore-fluid concentration profiles for alkalinity, pH, DOC, and sulfide for 1986. Total alkalinity doubles in concentration from the surface to the bottom brines. The increase is due to sulfate and metal oxide reduction which consumes protons. Although total alkalinity increases, the pH of the water column decreases. This is due to the release of carbon dioxide and organic acids from decomposing organic matter. The profile of Fig. 15c suggests that there are no major alkalinity producing reactions in the sediments. The pH also decreases across the sediment-water interface. The values stabilize at 6.5 and 6.6 at the North and South Basin sites. Sulfide is not present in the surface brines.

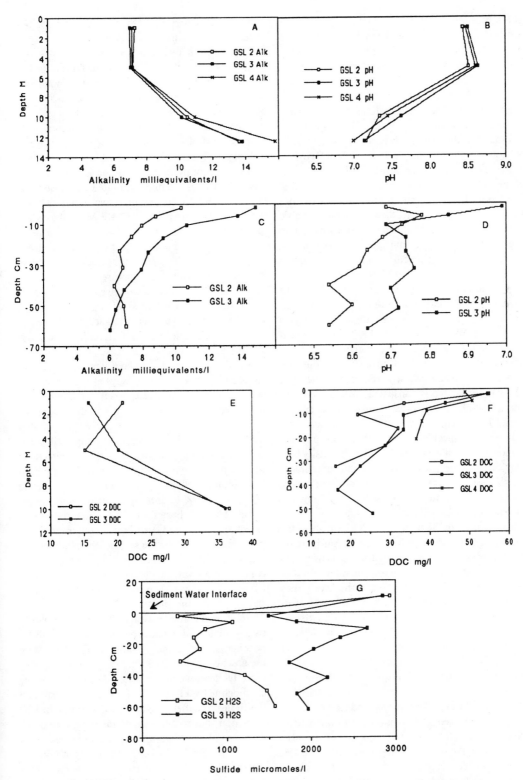

FIG. 15. Water column and pore-fluid geochemistry of Great Salt Lake: a) water column alkalinity;
b) water column pH; c) pore-fluid alkalinity; d) pore-fluid pH; e) water column DOC; f) pore-fluid
DOC; g) bottom brine and pore-fluid hydrogen sulfide.

The total sulfide bottom brine concentrations are close to 3 millimolar. H_2S is the major sulfide species below 10 m in the lake and throughout the pore fluids. The pe of the brines is just slightly more negative than -3.38 as calculated from the SO_4^{2-}/HS^- ratio and pH.

Pore-fluid sulfide concentrations are low relative to the bottom brines of the lake. The drop in concentration in the pore fluids is not due to Fe sulfide precipitation. The reductive dissolution of Fe oxides is greatest in the bottom brines of the lake and is of much less importance in the sediments. Sulfate is also not a limiting factor; pore fluids contain from 10 to 30 g SO_4^{2-}/l. Instead, sedimentary organic matter is refractory and degrades at a slower rate relative to the organic matter of the lake brines. The amount of organic carbon currently accumulating in Great Salt Lake sediments is lower than what would be expected for a productive lake with a shallow oxic mixolimnion and a reducing monimolimnion. The TOC of the surficial sediments is 2.5% for the South Basin site and 1.5% for the North Basin site. These concentrations decrease and stabilize at values between 1.0 to 1.1% for either site at 5 to 8 cm below the sediment-water interface. The data presented is consistent with an interpretation that planktonic organic matter is largely recycled to carbon dioxide within the anoxic water column. The mechanism is entrainment and decomposition of algal debris in the high density bottom brines. The process is most effective at the North Basin site which has the most dense fluids. The lowest accumulation of sedimentary organic carbon and the lowest pore-fluid sulfide concentrations occur there. Variations in DOC are also consistent with this interpretation. DOC can increase in concentration in anoxic environments from microbial metabolism of labile organic matter by sulfate reduction. DOC is highest in the lake brines and upper 5 to 10 cm of sediment but decreases in concentration below. Since the organic matter of the sediments degrades at a much lower rate, relative to the organic matter suspended in the bottom brines, DOC production and concentration are much lower in the pore fluids relative to the bottom brine.

Iron and Mn concentrations increase across the chemocline by reductive dissolution of the oxides. Fine-grained oxides of these two metals are also entrained in the bottom brines. Bisulfide ion is the most probable electron donor for reductive dissolution. The lake and pore-fluid profiles are shown in Fig. 16. The concentration profiles are similar to those for DOC. This is due, in part, to entrainment of metal oxide particles in the anoxic bottom brines where reductive dissolution occurs.

Precipitation reactions also affect Fe and Mn. Solubility calculations show that the bottom brines and pore fluids of the upper 10 cm of sediment are supersaturated with respect to mackinawite and pyrite (Fig. 16e). Precipitation of a FeS phase probably occurs in the bottom brines, which accumulates as a layer of black particles at the sediment-water interface. Pyrite forms below this zone and the mechanism of pyrite precipitation is discussed below. The waters are undersaturated with respect to all Mn sulfide phases by over two orders of magnitude and therefore no discrete MnS phases are precipitating.

The concentration profiles of most cationic trace metals show no definite trends. The water column and pore-fluid concentrations are given in Table 8. Dissolved Cu differs at the three sites with the concentrations either decreasing, increasing, or remaining more or less unchanged across the chemocline and sediment-water interface. The Cu pore-fluid concentrations do not vary significantly from those of the lake water. Lake water Zn concentrations also vary in different ways at the three sites. A concentration increase is noted in the upper few cm of sediment with decreasing concentrations below. Cobalt is present in the surface waters but is undetectable in the bottom brines suggesting that rapid precipitation occurs. Cobalt is undetectable in the upper 12 cm of sediment. A maximum concentration is observed near 17 cm with decreasing amounts below. Cadmium is not detected in the Great Salt Lake brines. Lead was analyzed by differential pulse anodic stripping voltammetry. A peak was detected for the surface brines which had the same half wave potential of the standard. The concentration was 2.5 micrograms/l. A peak, shifted by -150 mv, was detected in the bottom brine sample. The concentration was 15 micrograms/l. The shift in potential indicates that Pb is complexed by a strong ligand. These peaks completely disappear across the sediment-water interface indicating the absence of dissolved Pb.

Molybdenum concentrations decrease across the chemocline by a factor of 4 for both the North and South Basins. After a slight mobilization in the surficial pore fluids, Mo concentrations continue to decrease. Water column V concentrations show contrasting behavior in the North and South Basins. An increase with depth is observed in the North Basin and a decrease in the South. The South Basin pore-fluid concentrations are always higher than those of the North.

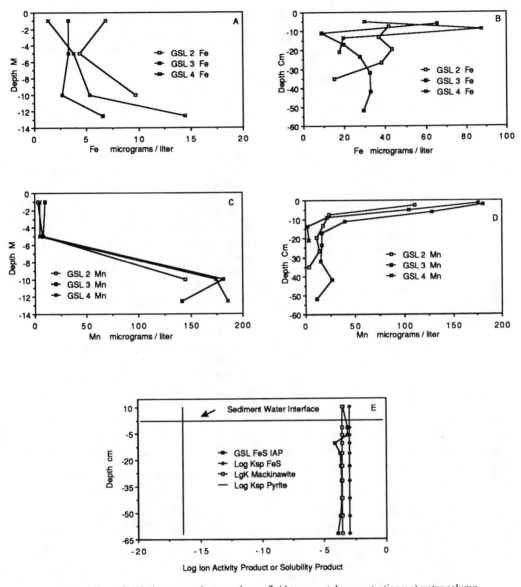

FIG. 16. Great Salt Lake water column and pore-fluid trace metal concentrations: a) water column Fe; b) pore-fluid Fe; c) water column Mn; d) pore-fluid Mn; e) water column and pore-fluid FeS ion activity products and saturation indices for amorphous FeS, mackinawite, and pyrite.

Great Salt Lake sedimentary chemistry

The mineralogy of Great Salt Lake sediments has been described by SPENCER (1982) and SPENCER *et al.* (1984) as part of their study on the lake level chronology of the Bonneville Basin. The sediments under study in this report were deposited after the fall from the Bonneville level about 13,000 years ago (SPENCER's, 1982, Unit 1). The sediments are pelleted sand. The mineralogy is quite uniform and consists of 15% quartz, 25% aragonite, 10% calcite,

and up to 50% clay minerals. The clays are predominantly smectite. Minor amounts of feldspars and sulfides are present (SPENCER *et al.*, 1984). The sediments are black for the upper 5 cm and gray below. Iron monosulfides are responsible for the black color at the sediment-water interface.

The profiles for total sedimentary organic carbon are shown in Fig. 6. The surficial sediments contain the most organic carbon with the highest amounts in the South Basin. Both basins have similar amounts below 10 cm. Two horizons were analyzed

Table 8. Great Salt Lake water column and pore fluids. Dissolved trace metals

Lake waters
Depth: meters
Metal concentration: micrograms/liter

Depth	Cu	Zn	Co	Pb	Cd	Mo	V
			Site GSL2				
1.0	6.5	8.7	N.A.*	N.A.	N.A.	23.6	3.1
5.0	10.2	14.1	N.A.	N.A.	N.A.	27.8	3.8
10.0	3.2	21.1	N.A.	N.A.	N.A.	9.4	8.7
			Site GSL3				
1.0	4.9	13.3	1.2	3.4	N.D.**	30.4	4.0
5.0	3.2	10.9	0.5	N.A.	N.D.	27.3	4.0
10.0	3.6	10.3	N.D.	10.0	N.D.	8.2	2.1
12.5	6.1	11.0	N.D.	12.5	N.D.	4.9	1.7
			Site GSL4				
1.0	5.1	12.8	N.A.	N.A.	N.A.	N.A.	N.A.
5.0	3.7	8.0	N.A.	N.A.	N.A.	N.A.	N.A.
10.0	8.6	11.4	N.A.	N.A.	N.A.	N.A.	N.A.
12.5	10.4	18.9	N.A.	N.A.	N.A.	N.A.	N.A.

Pore fluids
Depth: centimeters

Depth	Cu	Zn	Co	Pb	Cd	Mo	V
			Site GSL2				
2.5	5.6	21.7	N.A.	N.A.	N.A.	7.6	1.2
7.5	5.5	20.0	N.A.	N.A.	N.A.	3.5	1.1
13.0	8.4	18.8	N.A.	N.A.	N.A.	3.9	0.8
19.5	10.5	19.8	N.A.	N.A.	N.A.	11.2	0.8
26.5	12.7	18.8	N.A.	N.A.	N.A.	3.7	1.4
35.0	5.6	14.9	N.A.	N.A.	N.A.	4.8	1.2
			Site GSL 3				
2.0	9.7	32.0	N.D.	N.A.	N.A.	12.9	3.6
6.0	9.2	24.0	N.D.	N.D.	N.D.	9.9	1.5
11.0	5.9	22.3	N.D.	N.D.	N.D.	8.6	2.2
17.0	7.2	26.5	2.8	N.A.	N.A.	8.1	3.3
23.5	8.6	20.6	1.4	N.A.	N.A.	3.6	2.4
32.0	9.3	17.1	0.9	N.D.	N.D.	6.1	3.8
42.0	6.4	13.8	0.9	N.A.	N.A.	2.9	2.2
52.0	6.0	14.2	0.5	N.A.	N.A.	2.3	2.0

for humic acid concentrations (0 to 8 cm and 13 to 20 cm). Humic acids account for 0.7 and 4.3% of the TOC in these horizons.

The concentrations of total metals and aqua regia extractable metals are given in Table 9. There are no total metal analyses for the bottom brine particles due to the small sample size. Immediately apparent is the surface enrichment in Cu, Pb, Zn, Cd, and Mo. This is due to runoff from a large open pit Cu mine in the nearby Oquirrh Mountains. Despite the anthropogenic enrichment, the dissolved concentrations of these metals in the pore fluids are not unusually high. The pre-industrial total sedimentary concentrations of all metals are less than that of the average black shale except for Mo.

Great Salt Lake sequential and humic acid extractions

Sequential extractions of Great Salt Lake sediment included particles suspended in the bottom brines as well as the sedimentary material.

The sequential extraction profiles for Fe and Mn are shown in Figs. 17a and 17b. Iron is extractable from the particles suspended in the bottom brines by both the reducing and oxidizing solutions. The reactivity decreases considerably across the sediment-water interface, and the oxidizing solution is the only reagent which releases any Fe. Iron is not extractable below 5 cm. There is no humic acid complexed Fe in the upper 8 cm. Below this horizon, between 3.5 to 6 ppm of Fe are complexed, which accounts for, at most, 0.2% of the aqua regia extractable Fe.

The greatest amount of Mn is extractable from particles suspended in the bottom brines. The acidic cation exchanger extracts the most Mn followed by the reducing solution. The neutral cation exchanger

Table 9. Great Salt Lake sediments. Total extractable and total sedimentary metals

Metal concentration: Micrograms/gram
Core location: GSL3

Depth (cm)	Fe	Mn	Cu	Zn	Pb	Co	Cd	Mo	V
			Total extractable metals						
L.P.*	3000	120	130	160	36	1.5	0.06	20.1	6.7
2.0	3330	180	180	110	110	2.5	3.4	29.4	12.5
6.0	4400	180	41	66	54	2.6	0.7	4.5	11.8
11.0	3600	210	7.1	27	3.4	1.7	0.02	3.9	11.4
17.0	4300	240	7.8	29	5.6	2.4	0.09	3.3	11.4
23.5	3900	250	7.1	29	4.3	2.2	0.02	2.3	10.8
32.0	3500	230	6.2	29	4.7	2.4	0.15	1.9	10.4
42.0	3800	220	7.5	25	3.6	2.3	0.10	1.5	11.5
52.0	2700	180	4.6	26	3.6	1.9	0.08	0.2	8.7
			Total sedimentary metals						
2.5	11000	180	210	120	110	6.0	3.0	61.0	30
10.5	14000	240	24	67	14	6.0	<1	10.0	34
23.5	15000	280	11	54	6	6.0	<1	8.0	30
52.0	15000	250	11	59	6	6.0	<1	7.0	32

* L.P. refers to particles suspended in the bottom brines of the lake.

and hydrogen peroxide extract similar amounts. Nearly all of the aqua regia extractable Mn is recovered by these procedures. Across the sediment-water interface the reactivity is much less and hydrogen peroxide is the most efficient extractant, followed by the reducing solution. Below 5 cm of the sediment-water interface the reactivity does not change. Small amounts are released by the oxidizing and reducing solutions and the acidic cation exchanger. There is no Mn complexed to humic acid in these sediments.

Manganese and Fe undergo reductive dissolution below the chemocline which results in a concentration increase of dissolved metals. The suspended metal oxides can be extracted by the reducing solution. The pore-fluid metal concentrations decrease with depth except for a mobilization of Fe in the top few cm. Of most significance is the reversal in reactivity across the sediment-water interface of solid phase Fe and Mn with respect to the extracting reagents. Up to 50% of the lake water particulate Fe and 90% of the Mn is extractable. This decreases to about 5% across the sediment-water interface. The Fe extracted from the surficial sediment is released by hydrogen peroxide and below this horizon there is no reactivity. The reactivity of Mn with respect to the sequential extraction procedures is constant at 5% of the aqua regia extractable below 5 cm of sediment.

It is inferred that sedimentary Fe is stabilized by the formation of pyrite. This is supported by SPENCER et al. (1984), who found that 50% of the sedimentary sulfur, in this interval, is pyritic. The formation of pyrite involves the following reactions in the water column and across the sediment-water interface: sulfate reduction producing sulfide ions; reduction of Fe oxides, by sulfide, to dissolved ferrous Fe; oxidation of sulfide ions by oxygen at the chemocline or by metal oxides in the bottom brines and sediments to provide elemental sulfur; reaction of the elemental sulfur with HS^- to provide polysulfides; precipitation of Fe monosulfide, and final production of pyrite by one of two mechanisms. Pyrite may form by direct precipitation following dissolution of the Fe monosulfide. Alternatively, the Fe monosulfide may react with residual oxide minerals to form greigite which would then alter to pyrite. These reactions have been described from experiments and field observations by BERNER (1970), RICKARD (1974, 1975), and PYZIK and SOMMER (1981) and were reviewed by MORSE et al. (1987). Direct precipitation, following the dissolution of Fe monosulfide, is probably the most important pathway for the Great Salt Lake. The critical requirement is the dissolution of the Fe monosulfide. The

lake bottom brines are close to equilibrium with respect to mackinawite, but the pore fluids, below 10 cm of the sediment-water interface, are undersaturated. All other conditions for pyrite formation are ideal. Sulfate reduction occurs and elemental sulfur can form in the bottom brines, or at the sediment-water interface, by reaction of metal oxides with sulfide ions. The decrease in pore-fluid sulfide, dissolved Fe, and pH results in undersaturation of the pore fluids with respect to Fe monosulfides. While this occurs, all fluids are supersaturated with respect to pyrite. Direct precipitation of pyrite has been observed in salt marsh sediments where sulfate reduction is rapid, organic matter concentrations are high, and pH is low (HOWARTH, 1979; HOWARTH and TEAL, 1979; LUTHER et al., 1982; GIBLIN and HOWARTH, 1984).

The water-column reactions are dominated by Fe monosulfide formation due to rapid kinetics and supersaturation. A decrease in pore-fluid pH and H_2S allows for the dissolution of FeS phases as the pore fluids become either undersaturated, or close to equilibrium, with respect to these minerals. The environment is also favorable for elemental sulfur production and the result is pyrite precipitation. The pyrite formation reactions go to completion in the upper 5 to 10 cm of sediment. This is indicated by the unreactivity of Fe to the extracting solutions below 5 cm and the change in sediment color from black to gray. Pyrite is not attacked by hydrogen peroxide but the metastable Fe sulfides are. These, and organic bound Fe in algal debris, account for the peroxide extractable Fe in the top 5 cm of sediment. Humic acid bound Fe is low at the Great Salt Lake. Maximum humic acid concentration occurs below the zone of pyrite formation. Pyrite is stable in anoxic sediments and therefore complexation is low.

The reactions for Mn are dissolution of oxides in the bottom brines and upper few cm of sediment, release from algal debris, and precipitation of a stable phase and an extractable phase. It has been shown that pyrite is capable of incorporating Mn into its structure (JACOBS et al., 1985). These same authors showed that in low temperature marine systems metals do not appear to be in equilibrium with pure sulfide phases. It is proposed that this process also occurs at the Great Salt Lake and that it represents the major mechanism of Mn immobilization in Great Salt Lake sediments. There is a small amount of peroxide extractable Mn below 5 cm of the sediment-water interface. The solid phase association of peroxide extractable Mn cannot be inferred from these experiments.

The extraction profiles for Cu, Zn, Pb, Co, and

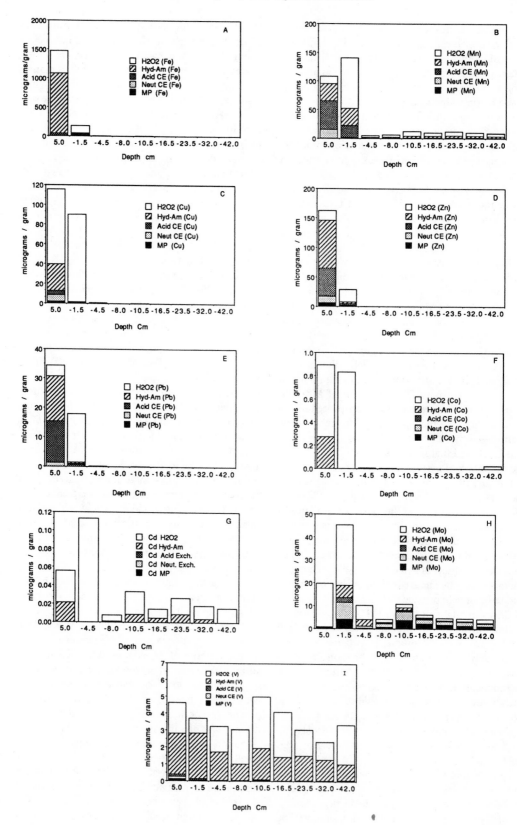

Cd are shown in Figs. 17c–17g. The shape of the profiles for Cu, Zn, Pb, and Cd are affected by the anthropogenic enrichment at the sediment-water interface. The greatest quantity of Cu, Zn, Pb, and Co, relative to the aqua regia extractable, is recovered from particles suspended in the bottom brines. Hydrogen peroxide and the reducing solution are most effective for Cu and Co. The reducing solution and the acidic cation exchanger are most effective for Zn and Pb. Most of the aqua regia extractable metals are recovered in this region. Across the sediment-water interface the reactivity changes. There is a small residual reactivity to hydrogen peroxide near the sediment-water interface. Below this horizon the particles are not reactive to the sequential extraction procedures.

Copper is the most reactive metal to humic acid complexation in these sediments. Below 10 cm of sediment, 4.5% of the aqua regia extractable Cu is complexed. Between 0 and 10 cm, 1 to 2% is complexed. There is a small amount of Zn which is complexed, but the amount is always well below 1% of the aqua regia extractable. No Pb or Co is complexed by humic acid.

Most of the aqua regia extractable Cd is recovered from the bottom brine particles by the oxidizing solution. There is also slight reactivity to the reducing solution. Most of the sedimentary Cd is extractable by hydrogen peroxide. No Cd is complexed to humic acid.

The elements Fe, Mn, Cu, Zn, Pb, and Co have similar reactivity to sequential extraction reagents. It is inferred that the trace elements are stabilized by pyrite formation. These elements may also have been associated with the Fe monosulfide precursors, and this association is preserved during conversion of the monosulfide to pyrite. The result is a significant decrease in reactivity across the sediment-water interface. The most compelling evidence for the co-precipitation of these elements with pyrite is provided by the sequential extraction data.

The bottom brine solid phase association of Cu, Zn, Pb, Co, and Cd can be with either metal oxides, organic matter, and sulfides. This is suggested by the high reactivity to both the reducing and oxidizing solutions. The oxides and labile organic matter are largely destroyed in the bottom brines and surficial sediments. The small reactivity of these elements at the sediment-water interface is probably due to their association with Fe monosulfide prior to pyrite formation.

Humic acids are an insignificant sink of trace metals because pyrite forms above the horizon of maximum humic acid concentration. Only Cu has any appreciable reactivity to the Great Salt Lake humic acids. The incorporation of the trace metals into pyrite is apparently an extremely stable configuration and prevents organic complexation from taking place.

Cadmium chemistry at Great Salt Lake is different from the other metals. Only a small amount of Cd is associated with metal oxides in the bottom brines and sediments. BENJAMIN and LECKIE (1981) showed that pH levels above 8 are required for quantitative removal of Cd from solution by adsorption onto Fe oxide. Therefore, adsorption should occur in the surface brines, but desorption should occur below the chemocline. The only extracting reagent which is effective for Cd is the oxidizing solution. There are several core horizons where Cd recovery is 100% of the aqua regia extractable. Cadmium can be extracted with hydrogen peroxide while iron cannot. Thus, Cd is probably not co-precipitated with pyrite, but instead precipitates as a discrete sulfide phase.

The extraction profiles for Mo and V are shown in Figs. 17h and 17i. Molybdenum is highly reactive to sequential extraction reagents throughout the core. The Mo associated with water column particles is most reactive to hydrogen peroxide. Molybdenum of the surficial sediments is also extractable with hydrogen peroxide but a significant residual fraction is also present. Below 5 cm of the sediment-water interface, a labile fraction is present which is extractable with distilled/de-ionized water and the neutral cation exchanger. There is also a hydrogen peroxide extractable fraction present. The only residual Mo was found in the surficial sediments. It has already been demonstrated (Table 9) that these sediments are enriched in Mo due to runoff from an open pit Cu mine. The residual Mo in this region may be present in stable phases from this run-off. There is no Mo complexed by humic acid in these sediments.

FIG. 17. Sequential extraction results for Fe, Mn, Cu, Zn, Pb, Co, Cd, Mo, and V from Great Salt Lake sediments: a) Fe; The largest stack represents 50% of aqua regia extractable Fe; b) Mn; The largest stack represents 77% of total sedimentary Mn; c) Cu; The largest stack represents 86% of aqua regia extractable Cu; d) Zn; The largest stack represents 94% of aqua regia extractable Zn; e) Pb; The largest stack represents 96% of aqua regia extractable Pb; f) Co; The largest stack represents 60% of aqua regia extractable Co; g) Cd; Most stacks represent 100% of total sedimentary Cd; h) Mo; The largest stack represents 75% of total sedimentary Mo.; Other stacks represent 60 to 100% of the total; i) V; Most stacks represent 14.7% of total sedimentary V.

The data for Mo suggests that precipitation of a discrete Mo sulfide phase occurs in the bottom lake brines and surficial sediments. As discussed, Mo is associated with Fe monosulfides in anoxic waters but is released during the transformation to pyrite (KOROLOV, 1958). BERTINE (1972) showed for anoxic marine systems that 70% of the sedimentary Mo is associated with Fe sulfides. Bertine suggested that precipitation of a discrete Mo sulfide occurs. The data for this study shows that Mo is removed from solution below the chemocline. In this region Mo is taken out of solution as either a discrete Mo sulfide or adsorbed onto the surface of Fe monosulfide. Adsorption onto metal oxides is unlikely since reductive dissolution is taking place. There is a slight increase in dissolved Mo just below the sediment-water interface with decreasing concentrations below. This mobilization is probably due to the release of Mo from the surface of Fe monosulfide minerals which are dissolving in this region. The extraction of Mo from the water column particles is accomplished with hydrogen peroxide but none of this material is associated with organic matter, as indicated by gel permeation chromatography, which suggests that the metal is being released from sulfides. Sedimentary Mo is highly labile and can be extracted quantitatively below 5 cm of sediment with a variety of reagents including hydrogen peroxide. Molybdenum can be extracted with hydrogen peroxide whereas iron cannot. Thus, there is no association of Mo with pyrite. The data supports the concept of a discrete Mo sulfide.

Vanadium reactivity is quite constant below the sediment-water interface. The water column particles are more reactive than the sediments but have less total V. Reactivity to the reducing solution and hydrogen peroxide is similar. About 75% of the aqua regia extractable V is recovered from the water column particles whereas 25 to 45% of the aqua regia extractable is recovered from the sedimentary column. There is no V complexed by humic acid in these sediments.

The extraction of V is dominated by the reducing and oxidizing solutions and there is no change in reactivity throughout the core. The hydrogen peroxide extractable form of V may be an organic complex. According to the Eh-pH diagram of HEM (1977) the V species, consistent with the conditions at the Great Salt Lake bottom waters and pore fluids, should be $V(OH)_2^+$. This species can be complexed by organic matter.

The chemistry of Mo and V in Great Salt Lake are considerably different. Only 40 to 50% of the aqua regia extractable V is labile to sequential extraction solutions, whereas Mo is quantitatively ex-

tractable. Only the reducing and oxidizing solutions are effective for V whereas for Mo the milder reagents, such as distilled water and the neutral cation exchanger, are also effective. Vanadium is therefore present in more stable forms relative to Mo. Although the peroxide extractable form of V is probably an organic complex, the form extracted by the reducing solution is unknown.

Great Salt Lake summary

At the Great Salt Lake a combination of physical, chemical, and biological factors result in the stabilization of the metals Fe, Mn, Cu, Zn, Pb, and Co, by pyrite formation, in the bottom brines and upper few cm of sediment. Fine-grained material, such as algal debris and metal oxides, are entrained in the dense bottom brines. Here, organic matter degradation, sulfide production, and metal oxide reductive dissolution are greatest. Associated trace metals are released from the metal oxides and precipitate in some sulfide phase. The drop in pH across the chemocline is due to the release of carbon dioxide from decaying organic matter and poor buffering of the brine. The pH continues to drop across the sediment-water interface. Sulfide concentrations are lower in the pore fluids relative to the lake and as a result Fe monosulfide dissolves and is replaced by pyrite. This reaction goes to completion in a narrow zone just below the sediment-water interface. The geochemical conditions are such that euhedral pyrite is expected to form by direct precipitation following the dissolution of Fe monosulfide. Pyrite is formed from ferrous Fe, sulfide, and polysulfide ions. The low alkalinity helps this process by allowing pH to drop in concert with falling sulfide levels.

The selective extraction results indicate that Mn, Zn, Cu, Pb, and Co co-precipitate with Fe in pyrite whereas Mo and Cd form discrete sulfides. Vanadium is most likely present as an oxycation which is complexed to organic matter or adsorbed onto sedimentary material. Furthermore, maximum humic acid production occurs below the region of pyrite formation and as a result the Great Salt Lake humic acids have extremely low trace metal content.

Two conclusions can be drawn for the Great Salt Lake system. The first is that pyrite formation and the co-precipitation of Mn, Cu, Zn, Pb, and Co result in the immobilization of this group along with Fe. This group of metals is extremely unreactive to the sequential extraction solutions and therefore is expected to be resistant to any further mobilization during diagenesis unless a change in pore-fluid chemistry occurs which is capable of dissolving pyrite. Although Mo and Cd precipitate as sulfides, they are extractable. This indicates that they do not co-precipitate with pyrite, but are subject to mobilization, particularly by oxidizing fluids. Molybdenum appears to be immobile only in the presence of dissolved sulfide. Vanadium can also be mobilized by oxidizing conditions, particularly if it undergoes an oxidation state change to the oxyanion or if the organic matter to which it is complexed is degraded. The second conclusion is that despite high surface water algal productivity, a relatively narrow oxygenated zone of lake waters, and intensely reducing bottom brines, the accumulation of organic carbon in Great Salt Lake sediments is not very high with concentrations barely above 1% by weight. A stratified water column with very dense bottom brines and high sulfate concentrations results in the recycling of organic carbon to carbon dioxide in the water column with low sedimentary accumulation. The degradation of algal debris in the bottom brine is rapid since the reactions are not limited by diffusion of fresh oxidant to the algal surfaces and the concentration of sulfate is very high. Therefore surface productivity is not the only consideration in predicting the accumulation of organic carbon in sedimentary basins. Physical and chemical factors of the water column are important. This study shows that organic carbon can be effectively mineralized to carbon dioxide in an anoxic stratified system if an excess of sulfate is present.

CONCLUSIONS AND SUMMARY

Local conditions and brine composition impact the chemistry of trace metals associated with reactive particles such as metal oxides and organic matter in closed basin lakes. Brine composition and concentration affect: 1) the stratification of the water column, which defines a sharp redox transition and may also affect the sedimentation rate of fine-grained particles; 2) the sulfate concentration, which affects anaerobic organic matter degradation, sulfide production, and metal sulfide precipitation; 3) the total alkalinity, which affects buffering capacity and pH; 4) the location of organic matter degradation and the oxidants involved, which affects pH, alkalinity, sulfide and DOC concentrations; and 5) the location of the reductive dissolution of metal oxides, which affects trace metal adsorption. The lakes of this study are in different phases of brine evolution and the effects of two distinct brine types (alkaline and non-alkaline) on trace metal geochemistry were observed. The geochemistry of trace metals is affected by these compositional differences. The early diagenesis of trace metals in these three

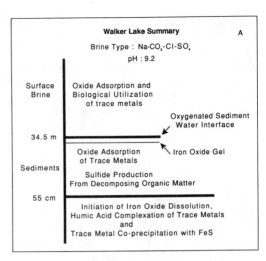

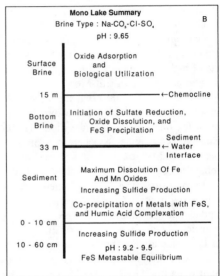

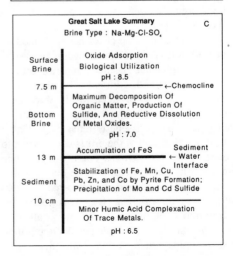

FIG. 18. Schematic representation of trace metal diagenesis: a) Walker Lake; b) Mono Lake; c) Great Salt Lake.

lakes is shown schematically in Figs. 18a, 18b and 18c. A chemically stratified water column at Mono and Great Salt Lakes results in the initiation of the reductive dissolution of Fe and Mn oxides in the bottom brines. This reaction goes to completion in the water column at the Great Salt Lake due to high brine density which results in the entrainment of metal oxides. At Mono Lake, reductive dissolution of Fe and Mn oxides is of greatest importance in the upper ten cm of sediment due to rapid deposition. The reductive dissolution of metal oxides and the release of trace metals occurs in a deeper sedimentary horizon at Walker Lake in as much as the water column is not stratified and oxygen reaches the bottom sediments. Differences in sulfide concentrations in the alkaline lake sediments result in different relative proportions of metal sulfide precipitation and humic acid complexation. Lower sulfide concentrations at Walker Lake result in a much higher amount of humic acid complexation of trace metals, relative to Mono Lake. This occurs in spite of higher sedimentary organic carbon and humic acid concentrations at Mono Lake. It was shown (DOMAGALSKI, 1988) that the structure of humic acids in both Walker and Mono Lakes is favorable for trace metal complexation. The dependence of trace metal geochemistry on brine composition and brine evolution suggests that the solid phase association of metals will be highly variable over long time frames. Brine chemistry can change rapidly in these lakes due to changes in inflow and climate. Stratification of the water column is a transient phenomenon. The stratification of Mono Lake has already broken down and trace metal geochemistry will adjust to an oxygenated sediment-water interface. Trace metal geochemistry will therefore only rarely assume steady state conditions.

One of the major differences between alkaline and non-alkaline lakes, as determined by this study, is the type of authigenic Fe sulfide minerals forming in the lake waters and sediment. The pH profiles and Fe sulfide activity products for the three lakes are shown in Fig. 19. Pyrite is prevented from forming during early diagenesis in the alkaline lakes due to supersaturation of the brines with respect to Fe monosulfide. This assumes that Fe monosulfide formation is favored, kinetically, over that of pyrite. The subsequent formation of pyrite requires dissolution of Fe monosulfide (RICKARD, 1975). Supersaturation with respect to Fe monosulfide is caused by high pH, due to buffered alkaline brines, and high sulfide concentrations. All other factors for pyrite formation are ideal. At the Great Salt Lake, poor buffering capacity of the brine results

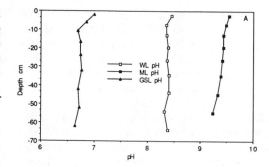

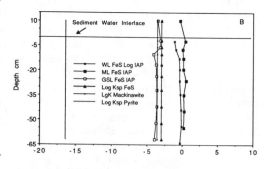

FIG. 19. Inter-basinal comparison of pore-fluid pH and FeS ion activity products: a) pH comparison; b) FeS ion activity product comparison.

in a pH drop of two units from the surface waters to pore fluids. The drop in pH is driven by degradation of organic matter. Sedimentary pyrite forms rapidly at the Great Salt Lake due to decreasing pore-fluid pH and sulfide concentrations which lead to the dissolution of Fe monosulfides which precipitated in the bottom brines of the lake. The metastable Fe monosulfides persist in the upper meter of Mono and Walker Lake sediments.

Iron sulfide chemistry has an important impact on the trace metals Mn, Cu, Pb, Zn, and Co. These elements have zero to very low extractability from Great Salt Lake sediments due to co-precipitation with pyrite. They will only be mobilized, during diagenesis, by the dissolution of this stable phase. This requires oxidation. In contrast, these metals are highly extractable from Mono Lake sediments, particularly with hydrogen peroxide. Metal mobilization, to form an ore-forming fluid, can occur in alkaline sediments by a change in pore-fluid chemistry resulting in the degradation of FeS. The mobilization of cationic trace metals does not require oxidation if trace metals co-precipitate with FeS. Decreases in either sulfide concentration or pH allow for FeS dissolution. If sedimentary conditions for pyrite formation are not appropriate, following the dissolution of FeS, organic matter may be an important sink of Fe and those metals associated with the monosulfide. Pyrite formation requires the presence of elemental sulfur (BERNER, 1970). Elemental sulfur may be generated by oxidation of bisulfide if suitable oxidants, such as metal oxides, are available (PYZIK and SOMMER, 1981). Alternatively, FeS was shown to be easily oxidized and this reaction can contribute acidity to an ore-forming fluid, thereby increasing the solubility of cationic metals. EUGSTER (1985) suggested that alkaline basins may be potential sites of ore-forming processes. He proposed that metals can be transported as hydroxy complexes in the high pH environment and classified such deposits as the Green River Type. This work shows that metals, in alkaline sediments, are not present in stable phases during early diagenesis. Mobilization of metals, to form an ore-forming fluid, is therefore possible under the right set of conditions.

Pyrite formation, during early diagenesis in marine and non-marine systems, has been discussed by BERNER (1984) and reviewed by MORSE et al. (1987). Pyrite in marine sediments frequently has a framboidal texture which has been linked to a series of reactions involving Fe monosulfide and greigite precursors (SWEENEY and KAPLAN, 1973). The rapid formation of pyrite in salt marshes has been reported by HOWARTH (1979). The rapid precipitation of pyrite in salt marshes is attributed to conditions which favor undersaturation of Fe monosulfides. The Great Salt Lake, and systems like it, with highly dense bottom brines represents a unique pyrite forming environment. The bottom brine, with high sulfate, Fe oxide, and organic matter is a favorable environment for Fe monosulfide formation. Reactive organic matter is degraded and Fe minerals are reduced in the bottom brines. The Fe monosulfide dissolves directly across the sediment water interface and is replaced by pyrite. The locations of Fe monosulfide and pyrite precipitation are, therefore, separate. Although pyrite textures were not identified in this study, the conditions are right for the formation of euhedral crystals.

The geochemistry of Mo and V involves redox chemistry, precipitation of Mo sulfide, and adsorption of reduced V species onto metal oxides or sedimentary organic matter. Diagenetic mobilization of these two elements occurs at Walker Lake due to the presence of a diffuse redox transition zone within the sediments. In contrast, a sharp redox gradient occurs at the chemocline in meromictic systems and prevents any chance of sedimentary diagenetic re-distribution from occurring.

The Mono Lake brines and sediments provide an excellent natural laboratory for the study of low temperature metal-ligand interactions. The ligands include high molecular weight organic acids and also inorganic species. In particular, polysulfide ions should be abundant in the pore fluids due to high pH conditions which favor their production. Future work for trace metals should be directed towards the study of dissolved organic ligands, polysulfide ligands, the surface chemistry of Fe monosulfide minerals, and the metal sulfide mineralogy in deeper sedimentary sections. Future work at Walker Lake should be directed at comparing the effects of high and low pore-fluid sulfate concentrations on trace metal-organic carbon interactions and Fe sulfide mineralogy in deeper sedimentary horizons than those of this study. The two sites described in this study are ideal for this type of investigation.

Acknowledgements—This work was part of a Ph.D. dissertation by Joseph Domagalski at the Johns Hopkins University, Dept. of Earth and Planetary Sciences. The authors gratefully acknowledge the Utah Geological and Mineral Survey and the Utah Department of Parks who made sampling on the Great Salt Lake possible. Sampling at Mono Lake was made possible by Jungle Laboratories, of Cibolo, TX, and at Walker Lake by Lloyd Thompson, of Hawthorne, NV, both of whom provided boat transportation. We gratefully acknowledge reviews by Mary Jo Baedecker and Edward Callender who provided valuable comments on the manuscript. This work was supported by the National Science Foundation (Grant: EAR-87-06384). Field work was supported by the Geological Society of America, Sigma Xi, and grants from the Johns Hopkins University.

REFERENCES

BALISTRIERI L. S. and MURRAY J. W. (1981) The surface chemistry of goethite (αFeOOH) in major ion seawater. *Amer. J. Sci.* **281**, 788–806.

BALISTRIERI L. S. and MURRAY J. W. (1982) The adsorption of Cu, Pb, Zn, and Cd on goethite from major ion seawater. *Geochim. Cosmochim. Acta* **46**, 1253–1265.

BALISTRIERI L. S. and MURRAY J. W. (1983) Metal-solid interactions in the marine environment: Estimating apparent equilibrium binding constants. *Geochim. Cosmochim. Acta* **47**, 1091–1098.

BALISTRIERI L. S. and MURRAY J. W. (1984) Marine scavenging: Trace metal adsorption by interfacial sediment from MANOP Site H. *Geochim. Cosmochim. Acta* **48**, 921–929.

BALL J. W., NORDSTROM D. K. and ZACHMANN D. W. (1987) WATEQ4F—A personal computer fortran translation of the geochemical model WATEQ2 with revised data base. *U.S. Geol. Surv. Open-File Rep.* 87-50.

BENJAMIN M. M. and LECKIE J. O. (1981) Multiple-site adsorption of Cd, Cu, Zn, and Pb on amorphous iron oxyhydroxide. *J. Colloid Inter. Sci.* **79**, 209–221.

BENSON L. V. (1978) Fluctuation in the level of pluvial

Lake Lahontan during the last 40,000 years. *Quat. Res.* **9**, 300–318.

BENSON L. V. and THOMPSON R. S. (1987) Lake-level variation in the Lahontan Basin for the last 50,000 years. *Quat. Res.* **28**, 69–85.

BERNER R. A. (1970) Sedimentary pyrite formation. *Amer. J. Sci.* **268**, 1–23.

BERNER R. A. (1984) Sedimentary pyrite formation: An update. *Geochim. Cosmochim. Acta* **48**, 605–615.

BERTINE K. K. (1972) The deposition of molybdenum in anoxic waters. *Mar. Chem.* **1**, 43–53.

BOESEN C. and POSTMA D. (1988) Pyrite formation in the anoxic environments of the Baltic. *Amer. J. Sci.* **288**, 575–603.

BOSTROME K., JOENSUM O. and BROHM J. (1974) Plankton: Its chemical composition and its significance as a source of pelagic sediments. *Chem. Geol.* **14**, 255–271.

BOULEGUE J. and MICHARD G. (1978) Constantes de formation des ions polysulfures S_6^{2-}, S_5^{2-}, et S_4^{2-} en phase aqueuse. *J. Fr. Hydrologie* **9**, 27–34.

BOULEGUE J., LORD, III C. J. and CHURCH T. M. (1982) Sulfur speciation and associated trace metals (Fe, Cu) in the pore waters of Great Marsh, Delaware. *Geochim. Cosmochim. Acta* **46**, 453–464.

BRADBURY J. P., FORESTER R. M. and THOMPSON R. S. (1988) Late Quaternary paleolimnology of Walker Lake, Nevada. (abstr.) *1988 Annual Meeting, Amer. Soc. Limnol. Oceanogr.*

DISNAR J. R. (1981) Etude experimentale de la fixation de metaux par un materiau sedimentaire actuel d'origine algaire—II. Fixation 'in vitro' de UO_2^{2+}, Cu^{++}, Ni^{++}, Zn^{++}, Pb^{++}, Co^{++}, Mn^{++}, ainsi que de VO_3^-, MoO_4^{--}, et GeO_3^{--}. *Geochim. Cosmochim. Acta* **45**, 363–379.

DOMAGALSKI J. D. (1988) Trace metal and organic geochemistry of closed basin lakes. Ph.D. Dissertation, The Johns Hopkins University, Baltimore, MD.

EUGSTER H. P. (1985) Oil shales, evaporites and ore deposits. *Geochim. Cosmochim. Acta* **49**, 619–635.

EUGSTER H. P. and HARDIE L. A. (1978) Saline Lakes. In *Lakes-Chemistry, Geology, Physics* (ed. A. LERMAN), pp. 237–293, Springer-Verlag.

GIBLIN A. E. and HOWARTH R. W. (1984) Porewater evidence for a dynamic sedimentary iron cycle in salt marshes. *Limnol. Oceanogr.* **29**, 47–63.

HEM J. D. (1977) Reactions of metal ions at surfaces of hydrous iron oxide. *Geochim. Cosmochim. Acta* **41**, 527–538.

HOWARTH R. W. (1979) Pyrite: Its rapid formation in a salt marsh and its importance in ecosystem metabolism. *Science* **203**, 49–51.

HOWARTH R. W. and TEAL J. M. (1979) Sulfur reduction in a New England salt marsh. *Limnol. Oceanogr.* **24**, 999–1013.

JACOBS L. and EMERSON S. (1982) Trace metal solubility in an anoxic fjord. *Earth Planet. Sci. Lett.* **60**, 237–252.

JACOBS L., EMERSON S. and SKEI J. (1985) Partitioning and transport of metals across the O_2/H_2S interface in a permanently anoxic basin: Framvaren Fjord, Norway. *Geochim. Cosmochim. Acta* **49**, 1433–1444.

JELLISON R. and MELACK J. M. (1988) Photosynthetic activity of phytoplankton and its relation to environmental factors in hypersaline Mono Lake, California. *Hydrobiologia* **158**, 69–88.

JENNE E. A. (1977) Trace element sorption by sediments and soils—sites and processes. In *Molybdenum in the Environment* (eds. W. CHAPPELL and K. PETERSON), pp. 425–553. Marcel-Dekker, New York.

JONES B. F. and BOWSER C. J. (1978) The mineralogy and chemistry of lake sediments. In *Lakes: Chemistry, Geology, Physics* (ed. A. LERMAN), pp. 179–235, Springer-Verlag, New York.

KABACK D. S. and RUNNELLS D. D. (1980) Geochemistry of molybdenum in some stream sediments and waters. *Geochim. Cosmochim. Acta* **44**, 447–456.

KOROLOV D. F. (1958) The role of iron sulfides in the accumulation of molybdenum in sedimentary rocks of the reduced zone. *Geochem. J.* **4**, 452–463.

KREMLING K. (1983) The behavior of Zn, Cd, Cu, Ni, Co, Fe, and Mn in anoxic Baltic waters. *Mar. Chem.* **13**, 87–108.

LINDSAY S. S. and BAEDECKER M. J. (1988) Determination of aqueous sulfide in contaminated and natural water using the methylene blue method. In: *Groundwater Contamination; Field Methods* (eds. A. G. COLLINS and A. I. JOHNSON), Special Tech. Pub. 963, pp. 349–357. Amer. Soc. Test. Mat.

LORD, III C. J. and CHURCH T. M. (1983) The geochemistry of salt marshes: sedimentary iron diffusion, sulfate reduction and pyritization. *Geochim. Cosmochim. Acta* **47**, 1381–1391.

LUTHER, III G. W., GIBLIN A., HOWARTH R. W. and RYANS R. A. (1982) Pyrite and oxidized iron mineral phases formed from pyrite oxidation in salt marsh and estuarine sediments. *Geochim. Cosmochim. Acta* **46**, 2665–2669.

MARTIN J. M., NIREL P. and THOMAS A. J. (1987) Sequential Extraction Techniques: Promises and Problems. *Mar. Chem.* **22**, 313–341.

MORSE J. W., MILLERO F. J., CORNWELL J. C. and RICKARD D. (1987) The chemistry of the hydrogen sulfide and iron sulfide systems in natural waters. *Earth-Sci. Rev.* **24**, 1–42.

PYZIK A. J. and SOMMER S. E. (1981) Sedimentary iron monosulfides: kinetics and mechanisms of formation. *Geochim. Cosmochim. Acta* **45**, 687–698.

RETTIG S. L. and JONES B. F. (1986) Evaluation of a suggested sequence for the chemical extraction of soluble amorphous phases from clays. In: *Selected Papers in the Hydrologic Sciences, 1986* (ed. S. SUBITZKY), Water-Supply Paper 2290, pp. 127–137, U.S. Geol. Surv.

RICKARD D. T. (1974) Kinetics and mechanism of the sulfidation of goethite. *Amer. J. Sci.* **274**, 941–952.

RICKARD D. T. (1975) Kinetics and mechanism of pyrite formation at low temperatures. *Amer. J. Sci.* **275**, 636–652.

ROBBINS J. A. and CALLENDER E. (1975) Diagenesis of manganese in Lake Michigan sediments. *Am. J. Sci.* **275**, 512–533.

SCHOLL D. W., VON HUENE R., ST. AMAND P. and RIDLON J. B. (1967) Age and origin of topography beneath Mono Lake, a remnant Pleistocene Lake, California. *Geol. Soc. Amer. Bull.* **78**, 583–600.

SLAVIN W., CARNRICK G. R., MANNING D. C. and PRUSZKOWSKA E. (1983) Recent experiences with the stabilized temperature platform furnace and Zeeman background correction. *Atom. Spectr.* **4**(No. 3), 69–86.

SPENCER R. J. (1977) Silicate and carbonate sediment-water relationships in Walker Lake, Nevada. Master of Science Thesis, University of Nevada, Reno.

SPENCER R. J. (1982) The geochemical evolution of Great Salt Lake, Utah. Ph.D. Dissertation, The Johns Hopkins University, Baltimore, MD.

SPENCER R. J., BAEDECKER M. J., EUGSTER H. P., FORESTER R. M., GOLDHABER M. B., JONES B. F., KELTS

K., MCKENZIE J., MADSEN D. B., RETTIG S. L., RUBIN M. and BOWSER C. J. (1984) Great Salt Lake and precursors, Utah: the last 30,000 years. *Contrib. Mineral. Petrol.* **86**, 321–334.

SPENCER R. J., EUGSTER H. P., JONES B. F. and RETTIG S. L. (1985) Geochemistry of Great Salt Lake, Utah I: Hydrochemistry since 1850. *Geochim. Cosmochim. Acta* **49**, 727–737.

STOKES W. L. (1980) Geologic setting of Great Salt Lake. In *Great Salt Lake—a Scientific, Historical and Economic Overview* (ed. J. WALLACE GWYNN), Bull. 116, pp. 55–68. Utah Geol. Mineral Surv.

STUMM W. and MORGAN J. J. (1981) *Aquatic Chemistry—An Introduction Emphasizing Chemical Equilibria in Natural Waters* (2nd edition), John Wiley and Sons.

STURM P. A. (1986) Major ion analyses of Great Salt Lake brines. Utah Geol. and Mineral Surv. Brine Survey, Quarter 2 of 1986.

SWEENEY R. E. and KAPLAN I. R. (1973) Pyrite framboid formation: laboratory synthesis and marine sediments. *Econ. Geol.* **68**, 618–634.

TESSIER A., CAMPBELL P. G. C. and BISSON M. (1979) Sequential extraction procedure for the speciation of particulate trace metals. *Anal. Chem.* **51**, 844–851.

THOMPSON R. S., BENSON L. and HATTORI E. M. (1986) A revised chronology for the last Pleistocene lake cycle in the central Lahontan basin. *Quat. Res.* **25**, 1–9.

THOMPSON R. S., TOOLIN L. J. and SPENCER R. J. (1987) Radiocarbon dating of Pleistocene lake sediments in the Great Basin by accelerator mass spectrometry (AMS). (abstr.) *Geol. Soc. of Amer., Abstracts with Programs* **19**, 868.

TUREKIAN K. K. (1977) The fate of metals in the oceans. *Geochim. Cosmochim. Acta* **41**, 1139–1144.

VAN VALIN R. and MORSE J. W. (1982) An investigation of methods commonly used for the selective removal and characterization of trace metals in sediments. *Mar. Chem.* **11**, 535–564.

VINE J. D. and TOURTELOT E. B. (1970) Geochemistry of black shale deposits—A summary report. *Econ. Geol.* **65**, 253–272.

VINOGRADOVA Z. A. and KOVALSKIY V. V. (1962) A study of chemical composition of the Black Sea plankton. *Akademiya Nauk SSSR Doklady* **147**(61), 1458–1460.

VOLKOV I. I. and FOMINA L. S. (1974) Influence of organic material and processes of sulfide formation on distribution of some trace elements in deep water sediments of black Sea. In *The Black Sea—Geology, Chemistry, and Biology*, (eds. E. T. DEGENS and D. A. ROSS), Mem. 20, pp. 456–476. Amer. Assoc. Petrol. Geol.

Fluid-Mineral Interactions: A Tribute to H. P. Eugster
© The Geochemical Society, Special Publication No. 2, 1990
Editors: R. J. Spencer and I-Ming Chou

Groundwater evolution, authigenic carbonates and sulphates, of the Basque Lake No. 2 Basin, Canada

H. WAYNE NESBITT

Department of Geology, University of Western Ontario, London, Ontario N6A 5B7, Canada

Abstract—The Basque Valley includes four drainage basins each a few hundreds of metres long and 100 to 200 metres wide. The major aquifer of Basque #2 basin is a thin horizon of Mazama Ash (Crater Lake, 6000 years old). The present hydrological and geochemical regime has developed since deposition of the ash.

Waters of the Basque Lake #2 basin are concentrated more than 100 fold to produce the Mg-SO_4-rich brines. During the first 10-fold concentration Mg-calcite (8 to 29 mole percent $MgCO_3$), and protodolomite (32 to 43 mole percent $MgCO_3$) are precipitated sequentially from the progressively more concentrated waters. Magnesite and the sulphate salts, gypsum, epsomite ($MgSO_4 \cdot 7H_2O$) and bloedite ($Na_2Mg(SO_4)_2 \cdot 4H_2O$) are precipitated during the second 10-fold concentration. Precipitation of Mg-calcite dramatically increases Mg/Ca of the basin waters during concentration as does formation of protodolomite. Continued concentration, with precipitation of gypsum and magnesite yields Mg-SO_4-rich, and Ca-CO_3-depleted brines from which epsomite and bloedite precipitate.

Mg-calcite and groundwater compositions yield exchange constants similar to experimental values, demonstrating that exchange equilibrium is achieved between solution and calcite in the basin. The result supports application of exchange equilibrium concepts to carbonate authigenesis in natural settings. A thermodynamic model provides a simple explanation for the evolution of the waters of the Basque and Spotted Lake Basins, and has implications for the evolution of seawater concentrates in diagenetic environments.

Aqueous silica concentrations diminish whereas Al(aq) increases in the progressively more concentrated waters. X-ray diffraction studies of the fine grained silicate fraction demonstrate that extensive reaction has occurred between the Basque waters and the phyllosilicates of the muds.

INTRODUCTION

SMALL CLOSED DRAINAGE BASINS are excellent natural laboratories in which to study interactions between natural solutions and minerals (GARRELS and MACKENZIE, 1967; JONES, 1966; HARDIE, 1968; EUGSTER, 1969, 1970; HARDIE and EUGSTER, 1970; JONES et al., 1977; DROUBI et al., 1977; GAC et al., 1977). These studies also emphasize the deficiencies in our knowledge and understanding of natural processes. For example, very early in the study of these basins HARDIE and EUGSTER (1970) commented that the behaviour of Mg was both poorly documented and understood. They suggested a study of brines rich in Mg, with emphasis placed on the behaviour of Mg. The results are presented here. Numerous studies directed towards the behaviour of Mg have been conducted subsequently (DREVER, 1971; NESBITT, 1975; GAC et al., 1977). This study documents the effects of carbonate formation on the behaviour of Mg in the Basque No. 2 basin.

The carbonates represent a potentially large sink for Mg, particularly Mg-calcites (GOLDSMITH et al., 1955) and Ca-protodolomites (SKINNER, 1963). There are many reports that note the formation of these phases in natural settings (ALDERMAN and

SKINNER, 1957; SKINNER, 1963; VON DER BORSCH, 1965; GLOVER and PRAY, 1969; BARNES and O'NEIL, 1971; MULLER et al., 1972; MUCCI et al., 1985; ROSEN and WARREN, 1988; ROSEN et al., 1989). Only recently, however, have interactions between Mg(aq) and carbonates been understood in detail (MUCCI and MORSE, 1984, 1985). The compositional variations of the Mg-rich Basque Lake Basin waters, and properties of the associated authigenic minerals, are documented here and interpreted in the framework of these new findings. The results are combined with the recent experimental data to develop a simple geochemical model to explain both the evolution of the Basque Lake Basin ground waters and the production of associated authigenic carbonate-sulphate mineral zones. Some implications for the subsurface evaporative concentration of seawater are that Mg-calcite precipitation and decay of organics may affect early diagenetic nature and production of Mg-bearing carbonates and sulphates.

THE BASQUE NO. 2 BASIN

Introduction

The Basque Lakes contain the highest magnesium concentrations of known North American closed basins (GOUDGE, 1924). The Lakes are situated in south-central British Columbia (Fig. 1), 100 km west of Kamloops (see GOUDGE, 1924 for details). The Basque Valley contains four small closed drainage basins separated by natural clay dams. The climate is arid, averaging 25 to 30 cm of rainfall annually and the temperature fluctuates seasonally, averaging −10°C in January and 25°C to 30°C during July and August. GOUDGE (1924) provides a detailed description of the Basque Valley so that additional description is restricted to the Basque No. 2 Basin.

Description

Basque No. 2 Basin (Fig. 1B) is approximately 800 m by 200 m with a small lake at its southern end (Fig. 2A). Standing water is found in the lake after spring runoff but by late summer evaporation has produced a salt crust on top of circular brine pools. The pools are surface expressions of conical structures (Fig. 2B) with springs at the centre of the cones (GOUDGE, 1924). Once the crust has formed, communication between the brine and atmosphere is through a small 5 to 10 cm circular hole in each salt crust. Brine remains below the crust during the driest periods and is replenished by the 'conical springs' (GOUDGE, 1924). Springs are absent from the perimeter of the lake and from the basin, except for those within the brine pools. There is no sig-

nificant surface runoff to the lake during the summer months. A detailed description is given by GOUDGE (1924).

Bedrock of volcanic and metamorphic origin crops out on the eastern and western flanks of the basin (Fig. 2B). A highly porous, permeable sand apron flanks the central valley floor (Fig. 2B), and is composed mainly of rock fragments, quartz and feldspar. No detrital carbonates were observed. Sands on the flanks of the valley give way to nonporous, highly impermeable muds of the valley floor (Fig. 2B). Attempts to collect water samples from these muds have been unsuccessful. Sandy lenses are scarce and no continuous horizons have been observed to date (upper 2 meters of sediment). Quartz and feldspar are abundant and a 10 Å mica and 14 Å chlorite dominate the less than 2 μ fraction (Fig. 3B). Carbonates are absent from all muds *except those contacted by lake brines or ground waters of the basin.*

A highly porous, permeable volcanic ash horizon is present at ½ m to 1.5 m depth (Fig. 2B). Grass rootlets penetrate the muds to the volcanic ash horizon. The grasses thrive on the basin floor during the dry summer months, whereas the sparse grasses outside the basin die. The ash is a source of water. Very thin discontinuous sandy lenses were found occasionally 0.2 m to 0.5 m below the ash. Authigenic carbonate minerals are found within the volcanic ash horizon but they are much more abundant

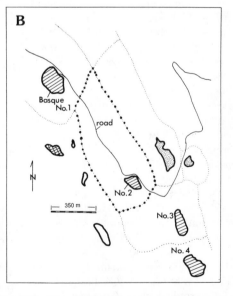

FIG. 1. Fig. 1A is a location of the Basque Lake Basin. The scale is 200 km. Fig. 1B is a plan view of the Basque Lake Basin showing the locations of the Basque Lakes (No. 1 to No. 4). The cross-hatched lake contains Na_2SO_4 brines and the stippled lakes contain Mg-carbonate salts.

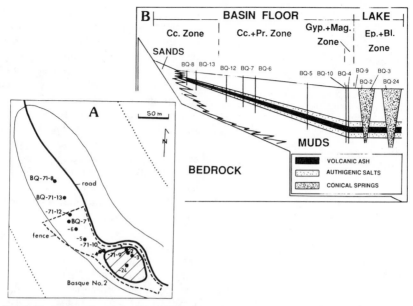

FIG. 2. Fig. 2A is a plan view of Basque Lake #2 Basin. The scale is 100 m. The hashed area is the lake proper and the dotted lines represent the margins of the basin. Dots represent sample locations. Sample labels are composed of a prefix 'BQ', a middle portion '70' or '71', (year sample was collected), and a suffix '8' (sample number). The prefix and '70' are omitted from some labels for clarity. The year, '70' and '71', is removed from labels in the text for brevity. Fig. 2B is a conceptualized cross-section showing the lake, and valley floor. Sample locations and mineral zones are included. Cc = calcite; Pr = protodolomite; Gyp = gypsum; Mag = magnesite; Ep = epsomite; Bl = bloedite.

in the muds immediately adjacent to the ash. Phyllosilicates of these muds are different from muds found well removed from the aquifer (Fig. 3A). Apparently they are altered from their original state to yield properties similar to those found in the lake muds (Fig. 3B). Immediately north of the lake at BQ-4, the ash is 10 to 15 cm thick and 1 m below ground surface. Further removed from the lake, near the head of the basin (Fig. 1B, BQ-13), the aquifer is 5 cm thick and 0.7 m below the ground surface. It thins and is found at shallower depths away from the lake. The aquifer is not exposed on the valley floor or around the perimeter of the basin. It is charged with solution in late July and where penetrated during coring, the holes filled rapidly with water from the aquifer.

Muds of the lake are different from other muds of the valley. Lake oozes are black, smell of H_2S, and contain authigenic carbonate minerals, sulphate minerals, chlorite and mica. There is no evidence that they have a different origin from the other muds, hence the difference in properties most likely result from reaction with the lake brines. Chlorite and mica, for example, display poorly defined peaks (Fig. 3B) compared with the peaks of Fig. 3A, and broad shoulders are found on the low 2θ side of

these peaks. The less than 2μ fraction is difficult to identify by X-ray diffraction methods (Fig. 3B). Ooze contacted by the brines are saturated but have very low permeability. Muds around the perimeter of the lake, between the spring high water mark and the low water mark of late summer have also been affected by contact with lake waters and brines and contain carbonate and sulphate salts.

Hydrological aspects

The volcanic ash aquifer is confined above and below by impermeable muds, does not outcrop (Fig. 2B), remains charged with water and retains a head during much of the summer period. The head is above the brine surface of the lake. Since the aquifer does not outcrop, it is not recharged directly by spring runoff. The sand apron on the flanks of the basin is sufficiently elevated and permeable to act as a reservoir. It is recharged each year by spring runoff and the volcanic ash aquifer may be hydrologically connected to the elevated unconfined sandy reservoir, as shown in Fig. 2B. The reservoir is large compared with the volume of the aquifer, thus providing sufficient capacity to replenish the aquifer into the dry summer period.

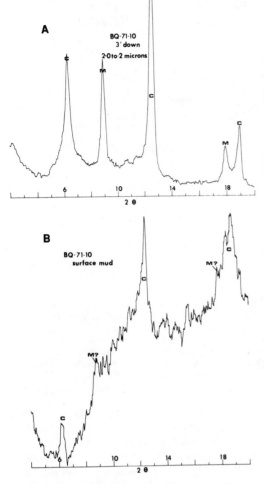

FIG. 3. X-ray diffraction patterns of fine fraction of muds at station BQ-10. Fig. 3A low 2 θ region of the fine fraction of muds from 0.9 m depth. These muds are isolated from the aquifer and lake waters. Fig. 3B, low 2 θ region of fine fraction of surface muds (ooze) contacted by spring and early summer brines. C = 14 Å chlorite; M = 10 Å mica.

The Basque Lake Basin is very small and the recharge, and probably the entire hydrological regime, is local (TOTH, 1963). Brines are present in the lake throughout the summer period and as emphasized by GOUDGE (1924), the 'conical springs' of the lake flow rapidly when exposed at depth. Apparently the springs are hydrologically connected to a reservoir but again the only likely one is the sandy reservoir. If so, the volcanic ash aquifer, and probably sandy units near the base of the sedimentary pile, provide the hydrological connection between the springs and the reservoir (Fig. 2B). The waters of the aquifer and lake are the Mg-Na-SO₄

type (Table 1), indicating a common geochemical and hydrological origin. Furthermore, the waters of the aquifer are most dilute closest to the perimeter of the basin, suggesting that they are replenished from the perimeter (i.e., connected to the sandy reservoir). Detailed hydrological studies are required to confirm the deductions.

Salinity of the waters in the aquifer increases towards the lake (Table 1, Fig. 1B). Impermeable muds overlying the aquifer inhibit evaporative concentration but evapotranspiration occurs, as is evident by the presence of luxuriant grasses on the valley floor during even the driest parts of the summer. The high salinities may result, in part, by diffusion of dissolved salts from the lake brine into the aquifer.

Unaltered glass of the ash horizon has refractive indices ranging from 1.508 to 1.512, indicating Mazama glass (Crater Lake), which has been recognized at other localities close to Basque Valley (POWERS and WILCOX, 1964). The glass is approximately 6000 years old. The thickening of the ash horizon towards the lake indicates that it has been transported somewhat after deposition. Ash deposited around the perimeter of the basin may have been transported to the basin floor, resulting in comparatively thick ash in the central, lowest portions of the basin floor and no ash on flanks.

GOUDGE (1924) collected a brine sample from Basque No. 2 Lake (Table 1, Fig. 1B, BQ-24) and its composition is almost identical to one collected in 1970 (Table 1, BQ-3). The similarity suggests that the hydrological and chemical systems have been stable over at least the last 50 years.

AUTHIGENIC SALTS OF THE BASIN

Authigenic Ca- and Mg-carbonates are found within the ash aquifer and in muds immediately adjacent to the aquifer. The distribution of carbonates in hole BQ-13 is typical. Muds at 45 cm contain little authigenic carbonates (Fig. 4A), but sediments within 7 cm of the aquifer (63–70 cm below floor) contain abundant carbonates (Fig. 4B). The ash (70–75 cm depth) contains carbonate (Fig. 4C) but less than the immediately adjacent muds. Similar relations exist below the ash horizon. The carbonates are abundant in the fine grained 2.0 to 0.2 micron fraction of the muds. The observations support an authigenic origin, where the carbonates have formed from the waters of the aquifer as a result of concentration of the aquifer waters in the subsurface.

The MgCO₃ content of calcite and protodolomite

Table 1. Basque Lake #2 Basin water compositions (mmoles/1)*

Species	BQ-7	BQ-6	BQ-5	BQ-4	BQ-3	BQ-24	SP-2
HCO$_3$	11.60	15.47	15.88	21.96	46.16	48.50	2.26
SO$_4$	146.78	142.62	185.30	1249.20	2529.63	2424.60	8.12
Cl	2.71	1.89	3.47	15.15	99.30	155.80	0.22
F	0.38	0.33	0.41	0.84	0.89	ND	0.05
Ca	8.86	9.08	9.76	43.17	21.88	ND	3.50
Mg	117.63	115.17	150.13	970.67	1711.10	1749.00	2.68
Na	51.33	51.33	66.12	452.40	1533.45	1347.80	6.09
K	2.40	2.56	3.07	21.04	127.08	163.70	0.38
Sr	0.09	0.09	0.23	0.32	0.46	ND	—
Al	3.40	3.70	4.80	35.00	59.00	ND	—
SiO$_2$	0.15	0.20	0.18	0.07	—	ND	0.03
pH**	7.25	6.98	7.04	6.86	8.10	ND	7.60
T (C)	11.50	11.50	11.00	17.00	35.00	ND	12.00
BAL (%)	−0.441	−0.119	−0.238	−1.364	−1.502	−0.871	0.32

ND = not determined; — = not detected.
* Values in M/l × 10^3 except for Al which is in mg·l.
** pH was measured in the field at the temperature indicated. Chemical analyses performed by S. Rettig, under the supervision of B. F. Jones, U.S. Geological Survey Laboratories, Reston, Va.

were determined using the data of GOLDSMITH *et al.* (1961). An error of ±3 mole percent is assigned to all MgCO$_3$ contents due to uncertainty in the data. Calcite and protodolomite peaks are separated by slightly more than 1° 2 θ and where one peak is greatly dominated by the other peak, there is the possibility of error in measuring the MgCO$_3$ content of the less abundant phase. Errors of this type, however, are insignificant. Even for the MgCO$_3$ content of calcite in BQ-9 (Fig. 5F) the corrections remain within the ±3 percent assigned error.

Calcite

X-ray diffraction patterns of muds adjacent the aquifer at BQ-8 (Fig. 2) indicates that calcite is a minor constituent compared with quartz (Fig. 5A). It is a major constituent of the muds adjacent the aquifer at BQ-12 (Fig. 5B) and at BQ-7 (Fig. 5C). Calcite is less abundant in muds adjacent the aquifer of BQ-10 (Fig. 5D), is diminished at BQ-9 (Fig. 5E) and is absent from the ooze of the lake at BQ-2 (Fig. 4F). There is systematic increase of calcite from BQ-8 to BQ-7 but it decreases in abundance in samples closer to the lake and is absent from the oozes of the lake.

The MgCO$_3$ content and relative abundance of calcite is given in Table 2. Calcite in BQ-8 contains 8 mole percent MgCO$_3$, which increases to 10 mole percent in BQ-7. The MgCO$_3$ increases still further to 29 percent in samples near the lake (BQ-10 and BQ-9).

Protodolomite

Protodolomite is absent from BQ-8 and BQ-12 but is present in moderate quantity at BQ-7 (Fig. 2, Fig. 4C). Its relative abundance increases towards the lake (Fig. 4D) compared with quartz, feldspar and calcite (Fig. 4D, 4E, and 4F) and is a major constituent of the muds fringing and within the lake (Fig. 4E and 4F). Like calcite, protodolomite is most abundant in the fine fraction of the muds. Super-

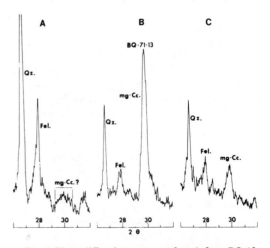

FIG. 4. X-ray diffraction patterns of muds from BQ-13. Fig. 4A is an X-ray diffraction pattern of muds 30 cm above the volcanic ash aquifer. Fig. 4B a pattern of mud 5–10 cm above the aquifer and Fig. 4C is a pattern of the volcanic ash (the aquifer). Qz = quartz; Fel = feldspar; Mg-Cc = Mg-calcite.

structure peaks have not been observed, suggesting that the protodolomite is essentially disordered and may best be assigned to the $R\bar{3}_c$ space group (calcite structure) rather than the $R\bar{3}$ space group of ordered dolomite.

Protodolomite changes composition systematically within the basin (Table 2). At BQ-7 its composition is 32 ± 3 mole percent $MgCO_3$ but the percentage increases to 43 ± 3 mole percent at BQ-10, BQ-9 and BQ-2.

Magnesite and gypsum

Magnesite and gypsum are found in the surface muds of BQ-10 (Fig. 5). The two minerals are discussed together because there is uncertainty as to which has formed first.

Magnesite has been found only in the fine fraction of the muds ($>2 \mu$ fraction) and nowhere is it abundant. It is not observed on X-ray diffraction patterns of the bulk sediment (Fig. 4 represent patterns of bulk sediment), probably because of its low abundance and is found only in fine grained separates. The magnesite lattice parameters are slightly greater than those of stoichiometric, coarse grained and well crystallized magnesite.

Gypsum is first observed at BQ-10 where it is present in minor amounts. It is, however, dispersed throughout all muds contacted by lake brine. In these oozes, gypsum crystals range from fine to coarse grained euhedral to subhedral. It is the first sulphate to form from the lake waters as they are evaporated during the early summer months.

Epsomite and bloedite

Epsomite ($MgSO_4 \cdot 7H_2O$) and bloedite ($Na_2Mg(SO_4)_2 \cdot 2H_2O$) form from the concentrated lake brines during the mid and late summer. They precipitate as individual, fine to coarse grained crystals and aggregates, forming caps and crusts over the brine.

AUTHIGENIC MINERAL ZONES OF THE BASIN

Mineral zones of the valley floor are defined by the assemblage of authigenic carbonates found in muds adjacent to the aquifer. The calcite zone (*Cc zone*) is closest to the perimeter of the basin (farthest removed from the lake). BQ-8 and BQ-13 are within this zone (Fig. 2). BQ-12 contains protodolomite as well as calcite. This assemblage represents the second zone, (*Cc-Pr zone*). The boundary separating the two zones is between BQ-13 and BQ-12 (Fig.

2). The authigenic sulphates of the surface muds within the lake are used to define additional zones. A unique mineral assemblage exists between the high-water and low-water marks of the lake. Magnesite and gypsum and protodolomite are found in the zone (Fig. 6) but gypsum is used to define the zone and is referred to subsequently as the gypsum zone (*Gyp zone*). The final zone is the area within the low-water mark of the lake. Authigenic Mg- and Na-sulphates are characteristic of this zone and it is referred to subsequently as the epsomite zone (*Ep zone*). Gypsum is ubiquitous in the zone and magnesite and protodolomite are common.

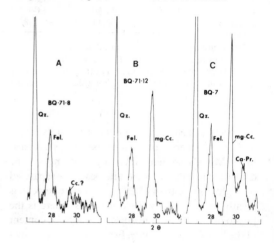

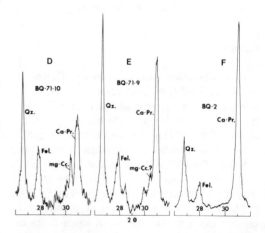

FIG. 5. X-ray diffraction patterns of muds from within the Basque Basin. The patterns illustrate the changing proportions and compositions of authigenic carbonates relative to proximity of the lake. The sites are arranged with Fig. 4A the sample farthest from the lake and Fig. 4F a sample of lake mud (see Fig. 2 for specific locations). Ca-Pr = Ca-protodolomite, other abbreviations as in Fig. 4.

Table 2. XRD data of carbonates from Basque Lake #2 Basin

Phase		BQ-8	BQ-13	BQ-7	BQ-10	BQ-9	BQ-2
Mg-Cc	2θ (104):	29.65	29.78	29.75	30.4	30.45	—
	% $MgCO_3$:	8	11	10	29	30	
	Amount:	DET.	MAJOR	MAJOR	MOD.	DET.	
	2θ (104):	—	30.55	30.5	30.9	30.9	30.9
	% $MgCO_3$:		33	32	43	43	43
Ca-Prot	Amount:		MINOR	MOD.	MAJOR	MAJOR	MAJOR

DET. = Detectable; — = Not detected; MOD. = Moderate.

COMPOSITIONS OF THE WATERS

Compositions of the waters of the basin and one inflow water (SP-2) from Spotted Lake Basin are included in Table 1 and plotted on Fig. 7. The most dilute waters from the Basque Basin were collected from the volcanic ash aquifer at BQ-6 and BQ-7, near the head of the valley (Fig. 2) but the waters become progressively more saline towards the lake. No sulphate salts have been found within the muds adjacent to the aquifer and no chloride salts have been found anywhere in the basin. Chloride ion is used as a 'tracer' to calculate the concentration factor for each water relative to BQ-6. The logarithm of the concentration factor (abscissa) and the logarithm of the concentration of the aqueous species (ordinate) are plotted on Fig. 7. Concentration-dilution trends are calculated for each species. The trends are straight, of unit slope, and emanate from BQ-6 water. If species for other waters plot on the concentration-dilution trend, then their concentrations can be derived by concentrating (or diluting) BQ-6. Species extracted from solution during concentration plot below the concentration-dilution trend and species added to solution during concentration plot above it.

General compositional trends

BQ-6 water must be concentrated almost 100-fold (concentration factor) to obtain the Cl content of the lake brines. Waters within the aquifer are concentrated 10-fold through evapotranspiration by the valley grasses and the second 10-fold concentration occurs in the lake through solar evaporation.

Anions

SO_4 (Fig. 7A, solid squares) follows the concentrative trend from BQ-6 to BQ-4 but the brines BQ-3 and BQ-24, plot below the trend, indicating that SO_4 has been extracted from the brines. Sulphate of SP-2 plots close to the concentration trend for the Basque Basin, indicating that SO_4/Cl of the very dilute input waters to the Spotted and Basque Lake Basins are similar. HCO_3 for samples of the Basque Basin plot below the concentration trend, hence it is lost from the more concentrated waters of both the aquifer and lake of the Basque Basin. Sample SP-2 plots close to the concentration trend, indicating that HCO_3/Cl of SP-2 and BQ-6 are similar. Fluoride is lost from the waters of the aquifer and lake. Although fluorite has not been found in the basin it would precipitate in small quantities and may have been present but remained undetected. F/Cl of sample SP-2 is similar to that of BQ-6.

Cations

Mg of samples BQ-6 to BQ-4 plots close to the concentration trend, thus there is little Mg lost from the aquifer waters during concentration. Carbonates, however, are found in and adjacent to the aquifer. Apparently, and although Mg is removed to form of Mg-calcite and proto-

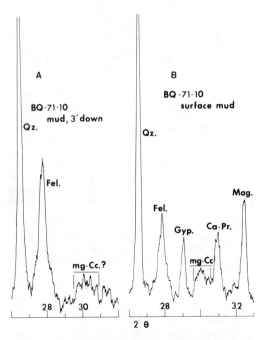

FIG. 6. X-ray diffraction patterns of muds at BQ-10. Fig. 6A is a sample of muds from 0.9 m depth (divorced from aquifer and lake waters). Fig. 6B is a pattern of surface mud. Gyp = gypsum; Mag = magnesite, other abbreviations as in Fig. 4 and Fig. 5.

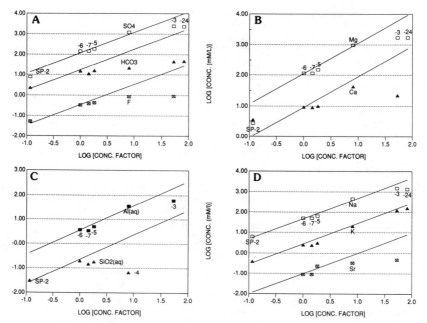

FIG. 7. Compositional trends for the Basque and Spotted Lake Basin waters. Concentrations of individual species are plotted against the concentration factor for each water (abscissa). Straight lines represent concentration-dilution curves for a water compositionally similar to BQ-6.

dolomite, the amount extracted is insignificant compared with the quantity in solution. Mg data for the lake brines (BQ-3 and BQ-24) plot well below the concentration trend, hence large amounts of Mg are removed during concentration of the brines. Epsomite and bloedite are precipitated from the brines. Mg of SP-2 plots well below the concentration curve, indicating that Mg/Cl is much lower than Mg/Cl of the dilute input waters to the Basque Basin. The calcium trend of the Basque Basin waters plots below the calculated concentration trend; Ca is removed from the waters of both aquifer and lake. SP-2 plots above the BQ-6 concentration trend indicating that Ca/Cl is greater than in the dilute Basque Basin input waters. Sodium of the aquifer waters (BQ-6 to BQ-4) and SP-2 plot on the BQ-6 concentration trend, indicating that Na/Cl are all similar. The lake brines, however, plot below the concentration trend, hence Na has been extracted from the brines during concentration. Strontium in BQ-6, BQ-7 and BQ-5 plot close to the concentration curve but BQ-4 plots below the curve and BQ-3 is still more depleted. Sr-bearing phases have not been identified but minerals such as strontianite or celestite may form; alternatively Sr may be taken into the Ca-Mg-carbonates in quantities sufficient to affect Sr contents of the basin waters.

Potassium of all waters from the Basque Basin and of SP-2 plot close to the concentration curve, thus all waters have similar K/Cl ratios. Apparently little K is removed from either the aquifer or lake waters. Aqueous aluminium of the aquifer waters plots close to the concentration trend but the Al(aq) of the brine (BQ-3) plots well below the trend. Apparently little Al is lost from solution within the aquifer but appreciable quantities are lost from the brines. Ooze contacted by brines displays much different XRD patterns from other muds of the basin (discussed subse-

quently). Al may be taken up by the oozes of the lake. Aqueous silica of the aquifer waters plots well below the calculated trend and its concentration is below detectability in the lake brines. Apparently it is extracted from the aquifer and lake waters during concentration.

Mass balances for waters of the aquifer

Cl and SO_4 are conservative in waters from the aquifers, hence either species may be used to calculate concentration factors (Table 3, bottom two rows). The Cl content of BQ-6 is low compared with SO_4 (Table 1), hence small errors in analyses or minor but real variations in Cl contents introduce relatively large errors in calculated concentration factors and mass balances. The more abundant SO_4 is used to calculate concentration factors. Gains and losses (mass balances) required to derive BQ-7, BQ-5 and BQ-4 from BQ-6 are listed in Table 3. A negative value indicates the species must be extracted from BQ-6 during concentration; a positive value indicates addition.

HCO_3 is lost during concentration of BQ-6 to BQ-7. The small negative and positive values for the other species reflect errors associated either with the analysis or with natural variations in the concentrations of the species. Mass balances between BQ-6 and BQ-7 are too small to be interpreted with confidence.

Table 3. SO$_4$-Cl derived mass balances relative to BQ-6

Species	BQ-7	BQ-6	BQ-5	BQ-4	BQ-3	BQ-24	SP-2
HCO$_3$	−4.3	0.0	−4.2	−113.5	−766.6	−1226.8	0.5
SO$_4$	-CF-	-CF-	-CF-	-CF-	−4963.6	−9332.1	−8.5
Cl	0.8	0.0	1.0	−1.4	-CF-	-CF-	-CF-
F	0.0	0.0	−0.0	−2.1	−16.4	−27.2	0.0
Ca	−0.5	0.0	−2.0	−36.4	−455.2	−748.5	2.4
Mg	−0.9	0.0	0.5	−38.1	−4339.9	−7744.9	−10.7
Na	−1.5	0.0	−0.6	2.8	−1163.4	−2883.5	0.1
K	−0.2	0.0	−0.3	−1.4	−7.4	−47.3	0.1
Sr	−0.0	0.0	0.1	−0.5	−4.3	−7.4	−0.0
Al	−0.4	0.0	−0.0	2.6	−135.4	−305.0	−0.4
SiO$_2$	−0.1	0.0	−0.1	−1.7	−10.5	−16.5	0.0
pH	7.25	6.98	7.04	6.86	8.10	ND	8.10
BAL (%)*	28.143	0.0	14.34	26.870	0.554	0.07	−0.62
CF (SO$_4$)**	1.029	1.00	1.299	8.75	17.737	17.00	0.06
CF (Cl)†	1.434	1.00	1.836	8.016	52.540	82.434	0.12

-CF- = Species used to calculate Concentration Factor of the sample.
ND = Not Determined.
* Charge balance for each water.
** Value of conc. factor based on sulphate.
† Value of conc. factor based on chloride.

BQ-5 is concentrated 1.3 times relative to BQ-6, and to derive BQ-5, from BQ-6, Ca and HCO$_3$ has to be remove in the ratio of approximately 1:2. The mass balances are consistent with the reaction:

$$Ca^{2+} + 2HCO_3^- = CaCO_3 + H_2O + CO_2. \quad (1)$$

Gains or losses of the other major constituents are insignificant.

BQ-6 must be concentrated 8.8 times to achieve SO$_4$ levels of BQ-4. Large quantities of HCO$_3$, Ca and Mg are extracted from solution (Table 3) and HCO$_3$/(Ca + Mg) of the material removed is 3:2. A more accurate mass balance can be constructed if waters BQ-6 and BQ-4 are first corrected for their charge imbalance (Table 1). Sulphate and Mg are the most abundant species by far and most likely small analytical errors in SO$_4$ and/or Mg account for much of the imbalance. The charge imbalance of BQ-4 can be corrected by increasing Mg of Table 1 by 1.8 percent, or by decreasing SO$_4$ by 1.4 percent. Correction first to SO$_4$, and next to Mg, indicates that HCO$_3$ and Ca + Mg are extracted in the ratio 1.84:1 and 1.94:1. Fluoride is lost during concentration of BQ-6 to BQ-4 and probably as fluorite. Correction to Ca for fluorite precipitation increases HCO$_3$/(Ca + Mg) to 1.87:1 and 1.98:1. The ratios are close to the that required by reaction (1), which is 2:1. The corrected data also demonstrate that more CaCO$_3$ is removed from solution than MgCO$_3$ during concentration of BQ-6 to BQ-4.

Mass balances for lake brines

Cl is used to calculate concentration factors for the lake brines. BQ-6 is concentrated 52.5 and 82.4 times to achieve Cl contents of BQ-3 and BQ-24. SO$_4$, Mg and Na are extracted from solution during concentration (Table 3) as gypsum, epsomite and bloedite (Table 4) and SO$_4$ is reduced. The amount of salts precipitated during evolution of successively more concentrated waters, normalized to unit concentration factor, is shown in Table 4. Carbonates form in large quantity both between BQ-5 and BQ-4, and from the lake brines (Table 4). The largest amounts of gypsum and epsomite (per unit of concentration) are precipitated during concentration of BQ-4 water to BQ-3. The amount of epsomite, however, is subject to error in that the amount the calculation does not include a correction for sulphate reduction. Bloedite is produced in greatest amount during the most extreme stage of concentration (BQ-3 to BQ-24). The mass balance relations correspond closely with the authigenic mineral zones.

There is loss of Al, Si and smaller amounts of K during concentration of the lake brines. These elements commonly are found in the silicate minerals and they may be removed from solution to form authigenic Al-bearing silicate minerals in the oozes. Although individual authigenic silicate minerals have not been identified from the lake muds, the XRD patterns of the clay minerals (Fig. 3) dem-

Table 4. Grams of salts produced[1] from solution

Salts	BQ-6	6-7/CF[2]	6-5/CF	5-4/CF	4-3/CF	3-24/CF
BLOEDITE[3]	0.0	0.0	0.0	0.0	109.2	337.2
CARBONATE[4]	2.2	0.0	1.6	6.3	7.5	7.6
GYPSUM[5]	0.0	0.0	0.0	0.0	36.1	18.1
EPSOMITE[6]	0.0	0.0	0.0	0.0	608.7	291.8
EPSOMITE[7]	0.0	0.0	0.0	0.0	596.7	255.3

[1] Amounts normalized to unit concentration factor.
[2] Loss/gain from BQ-6 to BQ-7 divided by CF for BQ-7 (calc. using BQ-6 as parent).
[3] Calculated assuming that Na is removed as Bloedite.
[4] ½ of HCO_3 removed forms carbonate (rxn 1).
[5] Calculated: GYPSUM = Ca(T) − 0.5∗CARBONATE.
[6] Calculated: EPSOMITE = SO_4(T) − 2∗BLOEDITE − GYPSUM.
[7] Calculated: EPSOMITE = Mg(T) − BLOEDITE − 0.5 CARBONATE. Comparison with 6 provides a check on how much of sulphate is reduced in the lake muds.

onstrate a substantial change to the properties of the bulk clays.

DISTRIBUTION OF Mg AND Ca BETWEEN SOLUTION AND CALCITE

MUCCI et al. (1985) report the compositions of Mg-calcite overgrowths on pure calcite seeds and of the associated solutions. They demonstrate that the overgrowths remain in exchange equilibrium with solution even though the solutions are slightly to greatly supersaturated with respect to the original calcite seed crystals. The findings support earlier observations (GARRELS and WOLLAST, 1978). There remains, however, well justified concern that

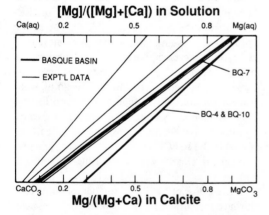

FIG. 8. Tie-line diagram illustrating the Mg-Ca distribution between solution and calcite. Upper axis represents Mg molar proportion in solution whereas the lower axis represents Mg molar proportion in calcite. Thin tie-lines are constructed from experimental data (MUCCI et al., 1985) and the thick tie-lines are constructed using field data from Basque Valley.

kinetic aspects strongly influence or even control the amount of Mg incorporated into calcite from solutions (BERNER, 1975, 1978; THORSTENSON and PLUMMER, 1977, 1978). The Basque solutions and calcite compositions can be used to test if exchange equilibrium is maintained between solution and authigenic Mg-calcite in natural settings.

The solutions used by MUCCI et al. (1985, Table 2) were subjected to a thermodynamic evaluation using the 'WATEQ' program (TRUESDELL and JONES, 1974). The activity proportion, [Mg]/([Mg] + [Ca]), was calculated (square brackets denote activities) for each water. The molar proportion of Mg in calcite, Mg/(Mg + Ca), was also calculated with all results plotted on Fig. 8. Activity or molar proportions of the solids can be used here, as discussed subsequently. Tie-lines connect the activity proportions of each experimental solution to the molar proportion in the respective Mg-calcite. The tie-lines represent exchange equilibrium between the two phases.

The same thermodynamic treatment has been performed on waters from BQ-7 and BQ-4 (Table 1) to obtain the activity proportions [Mg]/([Mg] + [Ca]), and the molar proportions of Mg in calcite associated with these waters also have been calculated and plotted on Fig. 8. Calcite of BQ-7 contains 10 ± 3 mole percent $MgCO_3$ (Table 2). Sample localities BQ-4 and BQ-10 are within 50 cm of one another. Calcite from BQ-10 contains 29 ± 3 mole percent $MgCO_3$ and the solution from BQ-4 is representative of the waters from this part of the aquifer. The tie-lines connecting BQ-7 and BQ-4 waters to their respective calcite compositions are shown on Fig. 8 (thick lines). The data of MUCCI et al. (1985) and Basque Basin data are indistinguishable. The BQ-4/BQ-10 tie-line plots at the extreme of the

MUCCI *et al.* (1985) data. Although the slope is somewhat steeper than the closest tie-line of MUCCI *et al.* (1985), the BQ-4/BQ-10 and experimental tie-lines are within analytical error (±3 mole percent for each experiment). There is strikingly good agreement between the experimental and Basque Lake Basin data. The exchange equilibrium hypothesis is used to model the evolution of the Basque Basin waters and carbonates.

Carbonate minerals produced in the Basque Basin and in the experiments of MUCCI *et al.* (1985) contain $CaCO_3$ and $MgCO_3$ as the only essential components. The solid solution is binary for both environments, thus activity coefficients are the same for the two components in Mg-calcites with the identical composition in the two settings. As a result, the conclusions drawn from this comparative study are valid whether mole proportions or activity proportions are used. Only where the actual chemical potentials of components are required (*i.e.,* for arguments based on saturation indices) are activities needed.

There is an important difference between the experimental results and results from Basque Basin; protodolomite is found in the basin. It may not have formed in the laboratory for lack of an appropriate seed on which to grow. Alternatively, supersaturation levels required to form protodolomite were not achieved in the experiments. MUCCI *et al.* (1985) provide evidence that exchange equilibrium exists between dolomite and solution. It is likely, therefore, that there is exchange equilibrium between protodolomite and solution, particularly in the Basque Basin where the protodolomite is structurally indistinguishable from very high Mg-calcite.

FORMATION OF SALT ZONES AND EVOLUTION OF SOLUTIONS

The consistent relationship between mineralogic zones and water compositions may be genetic, and a thermodynamic model is constructed to test the possibility. Development of the model is found in the appendix and the results are summarized in Fig. 9. It portrays equilibrium between solutions and the salts calcite, protodolomite, magnesite, gypsum and epsomite. The abscissa is Mg/(Mg + Ca) (solution or salts) and the ordinate is $SO_4/(SO_4 + HCO_3)$ in solution or salts. The mineral compositions are illustrated by squares and Mg-calcite and Ca-protodolomite solid solutions by bars (Fig. 9). The epsomite stability field has been expanded for clarity but geometric relations are preserved. Field boundaries delimit compositions of solutions

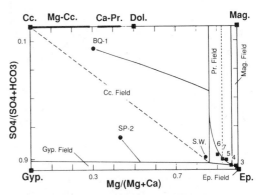

FIG. 9. Stability diagram for the system $CaCO_3$-$MgCO_3$-$CaSO_4$-$MgSO_4$. The ordinate and abscissa represent respectively, SO_4-HCO_3 and Mg-Ca molar proportions. The thick lines delimit the stability field of each salt (*e.g.* Cc field). The dotted line separates two great groups of waters (see text). The waters of Table 1 are plotted on the diagram (BQ- prefix omitted) as is seawater (S.W.) and the hypothetical water BQ-1. Thin lines emanating from BQ-1, SP-2 and S.W. illustrate their evolution during concentration. The compositions of calcite, Mg-calcite, protodolomite, magnesite, gypsum and epsomite are plotted (squares or solid bars on boundaries).

equilibrated with each of the labelled salts. The boundaries are of the eutectic (calcite-gypsum boundary) and peritectic (calcite-protodolomite) types. A solution that plots within the stability field of a salt, when concentrated, becomes saturated in that salt first, and by the equilibrium model, it is the first to precipitate. The solution evolves directly away from the composition of the precipitated salt, as required by mass balances. Similar mass balance arguments are applicable at boundaries and at triple points.

Predictions of the equilibrium model

The hypothetical water BQ-1 (Fig. 9) plots within the calcite stability field, hence during concentration of the water calcite is predicted to precipitate first. As a result, the solution composition evolves away from calcite composition (solid curve emanating from BQ-1). Mg/Ca increases in the evolved solution (Fig. 9) as a result of calcite precipitation. Equilibrium thermodynamic constraints require the composition of precipitated calcite to change towards a higher Mg content in response to changing water composition. The evolutionary path of BQ-1 therefore is curved. As concentration of BQ-1 continues, Mg/(Mg + Ca) increases in solution until the calcite-protodolomite boundary is intersected. Both calcite and protodolomite compositions plot to the left of the boundary, thus it is a peritectic, or reaction boundary. Two extreme situations arise at the boundary. One, referred to as the closed system, allows for all calcite previously precipitated to react with the evolved solution, and the other, the open system, requires all precipitated salt to be isolated from solution as it forms, and to remain isolated subsequently. Natural settings fall between these extremes.

After BQ-1 encounters the peritectic curve, continued concentration results in the conversion of calcite to protodolomite. During conversion, the solution composition evolves down the peritectic curve towards the calcite-protodolomite-gypsum peritectic triple point. The mechanism of conversion is of no consequence to the equilibrium model. If conversion is complete before the solution composition reaches the triple point, an additional degree of freedom is gained and the solution leaves the peritectic curve and migrates through the protodolomite field. The path traced by the evolving solution is determined by the composition of the protodolomite formed. If the solution encounters the triple point before all calcite is consumed, the solution remains at the peritectic point until conversion is complete. Thereafter, the solution evolves along the protodolomite-gypsum eutectic curve with the two phases co-precipitating. At the protodolomite-gypsum-magnesite peritectic point, all protodolomite previously formed reacts with solution to produce gypsum + magnesite. Once protodolomite is consumed, the solution migrates down the gypsum-magnesite eutectic curve to the gypsum-magnesite-epsomite eutectic point, where epsomite co-precipitates with gypsum and magnesite.

Interpretation of the peritectic reaction is not as simple as presented above. The peritectic curve (Fig. 9) represents reaction between a calcite and protodolomite of fixed composition. Since both carbonates display extensive solid solution and there is exchange equilibrium amongst the phases, there exist many peritectic curves (in fact a continuum), one for each combination of calcite-protodolomite compositions. A dotted peritectic curve is plotted within the protodolomite stability field (Fig. 9) as an example; it represents the peritectic boundary for the calcite and protodolomite composition of BQ-7 (Table 2). The path of a solution cannot be calculated without knowing how carbonate compositions change as the solution evolves. Exchange constants for calcite and solution are known, but there are no data for Ca-protodolomite solid solution. Data from MUCCI et al. (1985) are helpful in that exchange equilibrium between dolomite over-

growths and solution is shown to be maintained, but exchange constants are not yet available. As a result, the path of the solution while on the peritectic curve is constrained but cannot be calculated accurately. Some constraints are now discussed.

Mg-calcites of different compositions are distributed (systematically) throughout the basin, the low Mg-calcites having formed from the least evolved waters in the upper reaches of the basin, and the high Mg-calcites having formed from more evolved waters. The calcite-protodolomite peritectic boundaries applicable to each sample will be dependent on the Mg-Cc and Ca-Pr compositions. From exchange equilibrium principles, the peritectic boundary applicable to the least Mg-rich carbonates is associated with solutions of the lowest $Mg/(Mg + Ca)$ values, which are the less evolved solutions near the flanks of the valley. The peritectic curve applicable to Mg-rich carbonates is associated with solutions of high $Mg/(Mg + Ca)$ values which are the more evolved solutions found close to the lake. Peritectic curves are, therefore, spatially separated in the Basque Basin. During concentration, any solution plotting on a peritectic curve not only evolves down the curve, but shifts to new peritectic curves representative of new carbonate assemblages encountered as the solution migrates through the aquifer. As a result of migration, the solution encounters carbonate assemblages ever more enriched in Mg, hence the solution must shift to peritectic curves plotting at successively high $Mg/(Mg + Ca)$ values on Fig. 9; thus, the solution increases in $Mg/(Mg + Ca)$ even as it evolves down successive peritectic curves. The curved solution path (Fig. 9) passing through the Basque samples represents such a path. Eventually Mg-calcite is consumed, the peritectic reaction ceases and protodolomite alone forms as the solution composition evolves through the protodolomite stability field to the protodolomite-gypsum or protodolomite-magnesite boundary.

Water compositions

The BQ-1 water (hypothetical) is predicted to evolve towards high $SO_4/(SO_4 + HCO_3)$ and $Mg/(Mg + Ca)$ values. As is apparent from the data of Table 1, these predictions are correct. The exceedingly dilute waters of the aquifer associated with spring runoff have not been collected, and additional study is required in this regard. A critical test of the equilibrium model will be to collect the dilute aquifer solutions and monitor compositional changes during the spring and early summer months to see if they evolve as predicted.

Authigenic salt zones

The equilibrium model offers an explanation for the major aspects of the authigenic salt zones. The predicted authigenic salt sequence obtained during the concentration of BQ-1 are, for the open and closed systems:

Closed System		Open System
Gyp + Mag + Ep	Last to precipitate	Gyp + Mag + Ep
Gyp + Mag Ca-Pr + Gyp + Mag		Gyp + Mag Gyp
Ca-Pr + Gyp Mg-Cc + Ca-Pr		Mag Ca-Pr
Mg-Cc	First to precipitate	Mg-Cc

The sequences are similar to the authigenic mineral zones and there remains little doubt that the thermodynamic model provides a reasonable explanation for the evolution of the Basque Basin waters and salt zones. Complexities exist, however, because the zones develop in time and space. Ambiguities result where water compositions are related to these variables but an attempt is made to relate the three aspects. Dilute runoff accumulates in the aquifer during early spring. Waters are concentrated in the aquifer during the summer by evapotranspiration and the waters follow evolutionary paths similar to BQ-1 (Fig. 9). The waters of the aquifer are concentrated as they migrate towards the lake and the Mg content of progressively more concentrated waters increases due to the precipitation of calcite, hence calcites of greatest Mg content are found closest to the lake. Protodolomite forms after calcite, from still more concentrated waters, hence is formed in a zone close to the lake.

The antipathetic relationship between calcite and protodolomite contents of the muds result from the peritectic reaction among solution, calcite and protodolomite, but is complicated by the yearly hydrological cycle. During late summer, waters in the aquifer are more concentrated than spring runoff waters. Low Mg-calcite precipitated early during the summer may be contacted by solutions of the late summer. If Mg/(Mg + Ca) is sufficiently high, early formed calcite reacts with the evolved late summer solution to form protodolomite and to yield an antipathetic relationship between the calcite and protodolomite contents of the muds.

The late summer brines of the lake are restricted to local 'pools' (GOUDGE, 1924). Salts precipitated

in these pools during the spring and early summer are available to react with the evolved brines of late summer, hence closed system constraints prevail within the brine pools and peritectic reactions should occur. Salts precipitated outside of the local 'pools' (*e.g.,* mud flats) are not contacted by the late summer brines, thus constraints of the open system prevail in the mud flats, and peritectic reactions should be much less important. As a result, the assemblage gypsum + magnesite + epsomite is predicted within the pools but gypsum + protodolomite (± calcite) should be more common on the mud flats. These predictions again explain the major features of the salt zones of the lake area. Epsomite and gypsum (+ bloedite which is outside the compositional space considered here) are the dominant authigenic salts of the 'pools' whereas gypsum and protodolomite are the most abundant salts in the mud flats.

REACTIONS BETWEEN SILICATE MINERALS AND SOLUTIONS

Muds of the basin are composed of feldspars, quartz and phyllosilicates, primarily illite and chlorite (Fig. 3A). The peaks are well defined with but small shoulders developed on the low 2θ side of the 10 Å mica and 14 Å chlorite peaks. The ooze of the lake muds present a sharp contrast; the 10 Å mica peak is difficult to detect and the 14 Å ($6°$ 2θ) chlorite peak is greatly diminished and ragged. There is an exceptionally high background, in the region between 8° and 30° 2θ (Fig. 3B), perhaps resulting from the presence of X-ray amorphous Si- and Al-bearing solids. The solutions indicates reaction with silicates. SiO_2(aq) is comparatively abundant in the most dilute samples (Table 1), is less abundant in the more concentrated aquifer waters and is not detectable in the brine (Table 1, BQ-3). Apparently during approximately 100-fold concentration from BQ-6 to BQ-3, SiO_2(aq) decreases, indicating that it is removed from solution. It may be adsorbed onto mineral surfaces or it may be taken up to form authigenic silicate phases. Authigenic clay minerals may have formed in the muds adjacent the aquifer in quantities too small to be detected by the XRD studies. Alternatively, Si is incorporated into phases such as allophane (or very possibly alunite) which are either X-ray amorphous or are too fine grained to yield well defined XRD patterns.

Aqueous Al increases as Cl increases in the aquifer waters but in the brine Al is depleted relative to Cl; apparently it is removed from solution. Like

SiO_2, Al may be extracted from the brines to form fine grained clay minerals or X-ray amorphous Al-silicates. Detailed studies of the sheet silicates and fine grained fraction of the muds are required to determine, in detail, the nature of the silicate mineral-solution reactions.

DISCUSSION

The effect of calcite precipitation on the evolution of natural ground waters is well documented (GARRELS and MACKENZIE, 1967; HARDIE and EUGSTER, 1970; EUGSTER, 1970) and the influence of the formation of Mg-silicates has also been emphasized (DROUBI et al., 1977). The influence of Mg-carbonates and particularly Ca-protodolomites are less well documented. Here, the authigenic salts and solution compositions from the Basque No. 2 Basin are presented to illustrate the role that Mg-carbonate minerals have on the evolution of the Mg-Na-SO_4 brines of the Basque Basin. Even in saline, Mg- and SO_4-rich brines, Mg-carbonates have precipitated in quantity and they probably precipitate and affect the evolution of other continental groundwaters and subsurface marine waters.

The evolution of other continental groundwaters can be classified into two great groups, those which precipitate protodolomite after calcite and those that precipitate gypsum after calcite. The dashed 'join' connecting the calcite corner to the calcite-proto-dolomite-gypsum-solution triple point (Fig. 8) separates the two groups. The Basque Basin waters are an example of the group which plot above the 'join' and SP-2 water (Table 1) from Spotted Lake Basin (Ossoyos, British Columbia, Canada) is a member of the second great group; the authigenic assemblage calcite plus gypsum (no protodolomite) is predicted to form after calcite precipitation and field studies reveal this zone in the Spotted Lake Basin. The zone separates the calcite zone from the Mg-sulphate zone (lake muds) and most importantly no protodolomite was found within the calcite + gypsum zone of Spotted Lake Basin.

Waters of the Basque Basin (Table 1) span the range of salinity of seawater and its evaporative concentrates, and like seawater, the Basque waters contain high proportions of Mg and SO_4. These waters apparently have maintained exchange equilibrium with carbonates regardless of salinity, Mg or SO_4 levels of the waters; consequently the thermodynamic model may serve to accurately predict the evolutionary path of seawater as it is concentrated in subsurface environments. Seawater is plotted on Fig. 9 as S.W. It falls in the calcite field

and upon concentration in the subsurface calcite is predicted to form either by precipitation, probably overgrowths, or through biological removal. The mechanism is inconsequential to the thermodynamic model and by either mechanism S.W. evolves away from the calcite corner as the solid is extracted from solution and CO_2 is evolved (Equation 1). Decaying organic material contribute HCO_3 to subsurface seawater, adding a vertical component to its evolutionary path on Fig. 9. Where significant HCO_3 is added to seawater, the path encounters the protodolomite field before the gypsum field, hence protodolomite may form before gypsum as a result of HCO_3 uptake and calcite (or aragonite) precipitation. Precipitation of gypsum follows, because the formation of the carbonates increases SO_4/HCO_3 values of the seawater concentrate. The model implies that Mg-calcite and Ca-protodolomite may form before, as, or after gypsum precipitates. Gypsum precipitation (BUTLER, 1969) or preferential storage of Mg by algae (GEBLELEIN and HOFFMAN, 1971) are two ways to increase Mg/Ca in seawater concentrates; precipitation of either carbonate also increases Mg/Ca values of seawater. The abundance of decaying organic matter may modify the sequence of authigenic salts formed.

CONCLUSIONS

Carbonate minerals, Mg-calcite, Ca-protodolomite and magnesite precipitate from the groundwaters of the Basque Basin. The precipitation of the carbonates and particularly Ca-protodolomite is pivotal to the evolution of the brines. Its formation results in depletion of the waters in HCO_3 relative to SO_4 so that evolved groundwaters are carbonate-depleted and sulphate-enriched. The result is precipitation of the very soluble sulphate salts epsomite and bloedite from the extremely concentrated brines. The precipitation of both calcite and protodolomite also have an important effect on the cations. Calcite precipitation depletes the groundwaters in Ca relative to the other major cations. Even the formation of protodolomite depletes the waters in Ca relative to Mg because the waters have Mg/Ca values much higher than unity when formation of protodolomite commences. Large amounts of protodolomite have formed over the last 6600 years (the life of the basin).

The studies of MUCCI et al. (1985) indicate that Mg-calcite overgrowths on calcite remain in exchange equilibrium with the solution over a wide range of levels of supersaturation (with respect to calcite). The data from the Basque Basin suggests that exchange equilibrium is maintained between waters of the aquifer and Mg-calcite of varying compositions. The consequences are significant for they provide a means to model carbonate authigenesis in natural settings, including early and late diagenetic environments.

There is extensive interaction between the groundwaters and the sheet silicates of the basin muds. X-ray diffraction data demonstrate the effects to the silicates and the solution compositions confirm that Si and Al are removed from the groundwaters and/or brines. Potassium is also removed from the brines, and may be taken up by K-bearing silicates. Similarly some of the Mg lost from solutions may be taken up by the silicates. Comprehensive studies of the silicate phases are required to document and understand the nature and extent of these interactions.

Acknowledgements—I thank H. P. Eugster for encouraging me to study closed basins. The study of Mg-rich waters of closed basins was proposed by H. P. Eugster and L. A. Hardie and to both I am greatly indebted. T. Lowenstein and H. Machel read the manuscript and provided many improvements to the manuscript. The study was supported by two student grants from the Geological Society of America and by the National Sciences and Engineering Research Council of Canada.

REFERENCES

ALDERMAN A. R. and SKINNER H. C. W. (1957) Dolomite sedimentation in the south-east of South Australia. *Amer. J. Sci.* **255**, 561–567.

BARNES I. and O'NEIL J. R. (1971) Calcium-magnesium carbonate solid solutions from Holocene conglomerate cements and travertines in the Coast Range of California. *Geochim. Cosmochim. Acta* **32**, 415–432.

BERNER R. A. (1975) The role of magnesium in the crystal growth of calcite and aragonite in seawater. *Geochim. Cosmochim. Acta* **39**, 489–504.

BERNER R. A. (1978) Equilibrium, kinetics, and the precipitation of magnesian calcite from seawater. *Amer. J. Sci.* **278**, 1435–1477.

BUTLER G. P. (1969) Modern evaporite deposition and geochemistry of coexisting brines, the sabkha, Trucial Coast, Arabian Gulf. *J. Sed. Petrol.* **39**, 70–89.

CHRIST C. L. and HOSTETLER P. B. (1970) Studies in the system MgO-SiO₂-CO₂-H₂O (II): The activity product of magnesite. *Amer. J. Sci.* **268**, 439–453.

DREVER J. I. (1971) Magnesium-iron replacement in clay minerals in anoxic marine sediments. *Science* **172**, 1334–1336.

DROUBI A., FRITZ B., GAC Y. and TARDY Y. (1977) Prediction of the chemical evolution of natural waters during evaporation. *2nd Inter. Symp. Water-rock Interaction.* Strasbourg, Vol. II, 13–22.

EUGSTER H. P. (1969) Inorganic bedded cherts from the Magadi area, Kenya. *Contrib. Mineral. Petrol.* **22**, 1–31.

EUGSTER H. P. (1970) Chemistry and origin of the brines of Lake Magadi, Kenya. *Mineral. Soc. Amer. Spec. Pub.* **3**, 215–235.

GAC J. Y., DROUBI A., FRITZ B. and TARDY Y. (1977) Geochemical behaviour of silica and magnesium during the evaporation of waters in Chad. *Chem. Geol.* **19**, 215–228.

GARRELS R. M. and CHRIST C. L. (1965) *Solutions, Minerals and Equilibria.* 450 pp. Harper & Row.

GARRELS R. M. and MACKENZIE F. T. (1967) Origin of the chemical compositions of some springs and lakes. In *Advances in Chemistry Series 67.*, Washington, D.C., Amer. Chem. Soc., 222–242.

GARRELS R. M. and WOLLAST R. (1978) Discussion of Thorstenson and Plummer, 1977. *Amer. J. Sci.* **278**, 1469–1474.

GEBELEIN, C. D. and HOFFMAN, P. (1971) Algal origin of dolomite in interlaminated limestone-dolomite sedimentary rocks. In *Carbonate Cements. Studies in Geology, 19.* (ed. O. P. BRICKER), pp. 326. The Johns Hopkins University Press. Baltimore.

GLOVER E. D. and PRAY L. C. (1969) High magnesium calcite and aragonite cementation within modern subtidal sediment grains. *Bermuda Biological Station for Research, Spec. Pub. 3, Carbonate Cements*, 46–57.

GOLDSMITH J. R., GRAF D. L. and HEARD H. C. (1961) Lattice constants of the calcium magnesium carbonates. *Amer. Mineral.* **46**, 453–457.

GOUDGE M. G. (1924) Magnesium sulphate in British Columbia. *Dept. of Mines No. 642*, Mines Branch, Canada.

HARDIE L. A. (1968) The origin of the Recent non-marine evaporite deposit of Saline Valley, Inyo Co., California. *Geochim. Cosmochim. Acta* **32**, 1279–1301.

HARDIE L. A. and EUGSTER H. P. (1970) The evolution of closed basin brines. *Mineral. Soc. Amer. Spec. Pub.* **3**, 273–290.

HELGESON H. C. (1969) Thermodynamics of hydrothermal systems at elevated temperatures and pressures. *Amer. J. Sci.* **167**, 729–804.

HELGESON H. C., DELANEY J. M., NESBITT H. W. and BIRD D. K. (1978) Summary and critique of the thermodynamic properties of rock-forming minerals. *Amer. J. Sci.* **278-A**, 1–229.

JONES B. F. (1966) The hydrology and mineralogy of Deep Springs Lake. *U.S. Geol. Surv. Prof. Paper 502-A.*

JONES B. F., EUGSTER H. P. and RETTIG S. L. (1977) Hydrochemistry of Lake Magadi basin, Kenya. *Geochim. Cosmochim. Acta* **41**, 53–72.

LANGMUIR D. (1965) Stability of carbonates in the system MgO-CO₂-H₂O. *J. Geol.* **73**, 730–754.

MUCCI A. and MORSE J. W. (1984) The solubility of calcite in seawater solutions of various magnesium concentration, I₁ = 0.697 m at 25°C and one atmosphere total pressure. *Geochim. Cosmochim. Acta* **48**, 815–822.

MUCCI A. and MORSE J. W. (1985) Auger spectroscopy determination of the surface-most adsorbed layer composition on aragonite, calcite, dolomite and magnesite in synthetic seawater. *Amer. J. Sci.* **285**, 306–317.

MUCCI A., MORSE J. W. and KAMINSKY M. S. (1985) Auger spectroscopy analysis of magnesian calcite overgrowths precipitated from seawater and solutions of similar composition. *Amer. J. Sci.* **285**, 289–305.

MULLER G., IRION G. and FORSTNER U. (1972) Formation and diagenesis of inorganic Ca-Mg carbonates in the Lacustrine environment. *Naturwissenschaften* **59**, 158–164.

Table A1. Activity coefficients* of Mg, Ca, HCO₃ and SO₄

Species	BQ-7	BQ-6	BQ-5	BQ-4	BQ-3	SP-2
Mg	0.13	0.13	0.12	0.084	0.10	0.37
Ca	0.18	0.18	0.17	0.127	0.16	0.44
HCO₃	0.56	0.55	0.52	0.25	0.13	0.79
SO₄	0.13	0.13	0.11	0.037	0.018	0.44
I.S.	0.30	0.29	0.36	1.74	3.25	0.027

* WATEQ (TRUESDELL and JONES, 1974) used to calculate the activity coefficients.
I.S. = stoichiometric ionic strength.

NESBITT H. W. (1975) The study of some mineral-solution interactions. *Ph D Thesis,* The Johns Hopkins University, Baltimore. 173p.

POWERS H. A. and WILCOX R. E. (1964) Volcanic ash from Mount Mazama (Crater Lake) and from Glacier Peak. *Science* **144,** 1334–1335.

ROBINSON R. A. and STOKES R. H. (1965) *Electrolyte Solutions.* 571pp. Butterworths.

ROSEN M. R. and WARREN J. K. (1988) Dolomite occurrence in the Coorong Region, South Australia (abstr.). *Amer. Assoc. Petrol. Geol. Bull.* **79,** 242.

ROSEN R. M., MISER D. E., STARCHER M. A. and WARREN J. K. (1989) Formation of dolomite in the Coorong region, South Australia. *Geochim. Cosmochim. Acta* **53,** 661–669.

SKINNER H. C. W. (1963) Precipitation of calcian dolomites and magnesian calcites in the southeast of Australia. *Amer. J. Sci.* **261,** 449–472.

THORSTENSON D. C. and PLUMMER L. N. (1977) Equilibrium criteria for two-component solids reacting with fixed composition in an aqueous phase—Example the magnesian calcites. *Amer. J. Sci.* **277,** 1203–1223.

THORSTENSON D. C. and PLUMMER N. L. (1978) Reply to comments on Thorstenson and Plummer, 1977. *Amer. J. Sci.* **278,** 1478–1488.

TOTH J. (1962) A theoretical analysis of groundwater flow in small drainage basins. *Third Canadian Hydrology Symp.,* Calgary, 75–96.

TRUESDELL A. H. and JONES B. F. (1974) WATEQ, a computer program for calculating chemical equilibria of natural waters. *J. Res. U.S. Geol. Surv.* **2,** 233–248.

VON DER BORSCH C. C. (1965) The distribution and Preliminary geochemistry of modern carbonate sediments of the Coorong area, South Australia. *Geochim. Cosmochim. Acta* **29,** 781–799.

WOOD J. R. (1975) Thermodynamics of brine-salt equilibria, I. The systems NaCl-KCl-MgCl₂-CaCl₂-H₂O at 25°C. *Geochim. Cosmochim. Acta* **39,** 1147–1163.

APPENDIX

Experimental and thermochemical data are used to determine phase relations in the system CaCO₃-MgCO₃-CaSO₄-MgSO₄. All phase relations are determined for 1 bar total pressure and 25°C and a pH of 7.0, which is typical of the Basque Lake #2 Basin waters. The diagram is constructed for the Basque Basin waters.

Calcite dissolution can be written as:

$$CaCO_3 + H^+ = Ca^{2+} + HCO_3^-$$

and the mass action equation is:

$$10^{1.95} = [Ca^{2+}][HCO_3^-]/([H^+][Cc]). \quad (1A)$$

The mass action equation for gypsum dissolution is:

$$10^{-4.4} = [Ca^{2+}][SO_4^{2-}][H_2O]^2/[Gyp]. \quad (2A)$$

Square brackets denote activities, Cc and Gyp represent the respective CaCO₃ and CaSO₄·H₂O components in calcite and gypsum phases. Activities of Cc and Gyp are taken as 1.0. The equilibrium constants for equations (1A) and (2A) are taken from HELGESON (1969) and HARDIE (1968). For a solution equilibrated with respect to these phases:

$$10^{-6.35} = [SO_4^{2-}][H^+]/[HCO_3^-]. \quad (3A)$$

The substitution of activity coefficients and molalities for activities in Equation (3A), and taking logarithms yields:

$$-6.35 + pH = \log\{\gamma(SO_4)/\gamma(HCO_3)\}$$
$$+ \log\{m(SO_4)/m(HCO_3)\} \quad (4A)$$

where γ denotes activity coefficient and m denotes molality. For near-neutral solutions (pH = 7), as for the Basque waters, equation (4A) gives:

$$0.65 + \log\{\gamma(SO_4)/\gamma(HCO_3)\}$$
$$= \log\{m(SO_4)/m(HCO_3)\} \quad (5A)$$

Since SO₄ is doubly charged its activity coefficient is lower than that of HCO₃ in dilute solutions where complex ions are insignificant (GARRELS and CHRIST, 1965, p. 63 and p. 104). For solutions of ionic strength equal 0.1, the extended Debye-Huckel equation yields activity coefficients of 0.77 and 0.36 for HCO₃ and SO₄. These values are substituted into Equation (4A);

$$9.7 \simeq (SO_4^{2-})/(HCO_3^-) \quad (6A)$$

hence,

$$0.91 \simeq (SO_4^{2-})/(HCO_3^- + SO_4^{2-}). \quad (7A)$$

Table (A1) lists the stoichiometric activity coefficients for SO₄ and HCO₃ in the Basque Lake #2 Basin waters (calculated using WATEQ, TRUESDELL and JONES, 1974). If the activity coefficients for SO₄ and CO₃ are substituted into Equation (3A), the value of 0.91 represents a maximum for the Basque waters. The proportion of sulphate in solution therefore ranges from approximately 0.9 to 1 for the Basque Lake and Spotted Lake Basin waters and for most other natural waters. Only in solutions where the percentage of HCO₃ is appreciably more complexed than the percentage of SO₄ will the value of Equation (7A) be less than 0.9. The situation is unlikely in natural solutions, hence the calcite-gypsum-solution eutectic curve plots very close to the CaSO₄-MgSO₄ boundary of Fig. 9, regardless of the composition of the solutions.

The mass action equation for epsomite-solution equilibrium is:

$$10^{-2.13} = [Mg^{2+}][SO_4^{2-}][H_2]^7/[Ep], \qquad (8A)$$

where Ep is the $MgSO_4 \times 7H_2O$ component of epsomite. The equilibrium constant is taken from WOOD (1975) and activities of H_2O in solutions equilibrated with epsomite are taken from ROBINSON and STOKES (1965). A solution equilibrated with respect to gypsum and epsomite yields:

$$10^{2.2} = [Mg^{2+}]/[Ca^{2+}]. \qquad (9A)$$

In dilute solutions where Ca and Mg complexes are unimportant, and in the Basque waters (Table A1), activity coefficients of Ca and Mg do not differ greatly. By analogy with the derivation of Equation (7A):

$$1.0 \simeq (Mg^{2+})/(Mg^{2+} + Ca^{2+}), \qquad (10A)$$

where again the rounded brackets denote molalities.

Stabilities of the Mg-carbonates are difficult to assess. HELGESON et al. (1978) note that the experimentally determined activity products for the reaction:

$$CaMg(CO_3)_2 \rightarrow Ca^{2+} + Mg^{2+} + 2CO_3^{2-}, \qquad (11A)$$

range from $10^{-16.4}$ to $10^{-19.3}$ and analysis of hydrothermal experimental results yields a range between $10^{-16.5}$ and $10^{-18.1}$. The value applicable to a sample is dependent upon the degree of order of the sample. Protodolomite from the Basque Basin displays no ordering peaks and is considered to be completely disordered; consequently an equilibrium constant of $10^{-16.4}$ is selected to represent the Basque Lake Basin protodolomites.

For calcite-protodolomite equilibrium:

$$CaMg(CO_3)_2 + Ca^{2+} = 2CaCO_3 + Mg^{2+} \qquad (12A)$$

Addition of Equations (1A) and (11A) and the logarithms of the equilibrium constants yields the mass action equation:

$$2.51 = [Mg][Cc]/([Ca][Pr]), \qquad (13A)$$

hence,

$$0.72 = [Mg]/([Mg] + [Ca]) \qquad (14A)$$

If the activity coefficients of Mg and Ca are the same (not necessarily 1.0), then the molar proportions are:

$$0.72 = (Mg)/\{(Mg) + (Ca)\}, \qquad (15A)$$

which represents the value for the calcite-protodolomite-solution stability field boundary (Fig. 9). The boundary is applicable to all waters where there is no appreciable complexing of Mg or Ca, and where the Debye-Huckel equation is accurate (solutions of ionic strength up to approximately 0.1). In waters containing appreciable SO_4, such as seawater and the Basque Basin waters, Mg is complexed by SO_4 in preference to Ca, thus the value for Equation (15A) is greater than 0.72. The solution from BQ-7, for example, yields a value of 0.93 (and 0.907 for Equation 14A). Protodolomite is present at BQ-7, hence protodolomite saturation and precipitation has already occurred when the value 0.93 has been achieved for $(Mg)/\{(Mg) + (Ca)\}$. The calcite-protodolomite equilibrium boundary applicable to the Basque Basin waters therefore is constrained between values of 0.72 and 0.93. The equilibrium value (Equation 15A) is likely to be closer to 0.93 than 0.72 for two reasons. BQ-7 is close to the site where protodolomite is first detected (BQ-12), thus the solutions are unlikely to have evolved greatly between BQ-12 and BQ-7. Secondly, the value of 0.72 is calculated using stoichiometric compositions for calcite and protodolomite. Mg/(Mg + Ca) of 0.87 (Equation 15A) is taken as representative of equilibrium among calcite, protodolomite and solution. The stability field boundary is plotted at this value on Fig. 9. The value is between the two extremes 0.72 and 0.93, but is closer to BQ-7 water because the data from BQ-7 probably is close to the equilibrium value.

LANGMUIR (1965) notes that experimentally determined activity products for magnesite vary by almost three orders of magnitude. CHRIST and HOSTETLER (1970) indicate that the activity product of magnesite is greater than $10^{-8.2}$ but an activity quotient as high as 10^{-5} may be required to precipitate the phase. The extreme value of 10^{-5} is taken here as activity quotient required for magnesite precipitation. The equilibrium constant for protodolomite $(10^{-16.4})$ is used with the activity quotient for magnesite to calculate the magnesite-protodolomite-solution field boundary. The result is a very small stability field for magnesite (Fig. 9). It has been expanded somewhat in Fig. 9 to illustrate the geometric relations among the different fields.

The magnesite activity quotient and equilibrium constant for epsomite are used to calculate the magnesite-epsomite-solution field boundary (lower right-hand corner of Fig. 9). Equilibrium constants for the salts of Fig. 9 are used to calculate all field boundaries of Fig. 9. Assumptions and calculations are identical to those presented in the previous paragraphs.

Fluid-Mineral Interactions: A Tribute to H. P. Eugster
© The Geochemical Society, Special Publication No. 2, 1990
Editors: R. J. Spencer and I-Ming Chou

Saturation state of natural waters in Iceland relative to primary and secondary minerals in basalts

SIGURDUR R. GÍSLASON and STEFÁN ARNÓRSSON

Science Institute, University of Iceland, Dunhagi 3, 107 Reykjavík, Iceland

Abstract—Weathering and hydrothermal alteration of basaltic rocks in Iceland by natural waters include dissolution of the primary rock minerals and glass and precipitation of secondary minerals. The alteration process involves hydrogen ion metasomatism. The proton donors driving the alteration process include carbonic acid, derived from the atmosphere and the decay of organic matter (cold waters), silica (H_4SiO_4) dissolved from the rock (cold groundwaters and waters from low-temperature fields) and carbonic acid and hydrogen sulphide, which are either derived from the rock being altered or from a degassing magma (high-temperature fields). Cold groundwaters in Iceland are undersaturated with olivine, and presumably even more with basaltic glass. They are generally undersaturated with pyroxene but close to equilibrium with plagioclase. These waters seem to be heavily supersaturated with common weathering minerals like smectites, gibbsite, kaolinite and chalcedony. This may be an artifact caused by the small crystal size of the weathering products which increases their solubility. Geothermal waters are invariably undersaturated with plagioclase. They may be close to equilibrium, or somewhat undersaturated, with olivine depending on its composition. These waters tend to be pyroxene supersaturated when below about 150°C but undersaturated at higher temperatures. Heavy supersaturation with respect to the magnetite and hematite components in magnetite-ulvöspinel and ilmenite-hematite, respectively, is observed for thermal waters below about 150°C. Supersaturation decreases with increasing temperature and above about 200°C the waters seem to be characteristically magnetite undersaturated. The geothermal waters closely approach equilibrium with secondary minerals found in the altered rock such as chalcedony/quartz, calcite, K-feldspar, albite, marcasite/pyrite, pyrrhotite and sometimes with laumontite. The close approach to equilibrium is favoured by the relatively "similar" saturation state of both primary and secondary minerals. Thus, the flux of constitutents into the water by dissolution processes can easily be coped with by precipitation reactions without maintaining the solution much supersaturated with the precipitating minerals.

INTRODUCTION

THE STUDY OF hydrothermal alteration in active and fossil geothermal systems in Iceland has clearly revealed that the alteration process involves dissolution of the primary rock minerals and glass and deposition of secondary, or alteration, minerals (*e.g.,* KRISTMANNSDÓTTIR, 1978; ARNÓRSSON, 1987a). Studies of water compositions in active systems with temperatures as low as 50°C indicate that chemical equilibrium is closely approached between all aqueous major elements, except chlorine, and alteration minerals (ARNÓRSSON, 1983; ARNÓRSSON *et al.,* 1983a). Comparable conclusions have been drawn from studies of geothermal systems in many other parts of the world (*e.g.,* ELLIS, 1970; SEWARD, 1974; GIGGENBACH, 1980, 1981).

Unlike thermal waters, reaction rates involving dissolution of the primary rock constituents control the composition of cold, and warm (<35°C) groundwaters (GÍSLASON and EUGSTER, 1987a,b), although some elements like aluminium, iron and magnesium may be incorporated into weathering minerals during the earliest stages of rock dissolution. Similar conclusions have been drawn from the studies of groundwater chemistry elsewhere (REYNOLDS and JOHNSON, 1972; MILLER and DREVER, 1977; WHITE *et al.,* 1980; DEUTSCH *et*

al., 1982). Yet, there still exists an uncertainty in the kinetic process of dissolution because of discrepancies between experimentally derived rate constants and rate constants derived from field data (PACES, 1983; VEBEL, 1986).

Iceland is largely built up of basaltic volcanics (80–85%). Groundwater, cold and thermal, is among the most valuable natural resources in the country. As a result, many drillholes have been sunk into the basaltic volcanics in many areas tapping water with temperatures in the range of about 5–350°C. Large cold springs occur in places in the volcanic belts tapping extremely large groundwater bodies. These springs and, in particular, the drillholes, provide a unique opportunity to study the chemical characteristics of natural waters of widely varying temperatures that are associated with different weathering and hydrothermal mineral assemblages in a homogeneous petrological environment.

In the present contribution an attempt is made to map the characteristics of natural water compositions in Iceland with respect to dissolution and precipitation reactions. The following questions are specifically addressed:

(1) Are the primary minerals of the basaltic rock in contact with natural waters stable or unstable?

374   S. R. Gíslason and S. Arnórsson

(2) If they are unstable, how can the dissolution kinetics be described?

(3) How closely is chemical equilibrium approached between the water and specific weathering and alteration minerals found in the associated basaltic rock?

The data base on water compositions selected to answer the above questions are basically those presented by ARNÓRSSON et al. (1983a), GÍSLASON and RETTING (1986) and GÍSLASON and EUGSTER (1987b). In all we use 50 analyses of which 22 are from cold springs, 1 from a spring at 34°C, and 27 from geothermal wells with aquifer temperatures in the range 59–248°C. Sample locations are shown in Fig. 1. The wells selected have one dominant feed zone where downhole measurements have been used to obtain the aquifer temperature. Wells with multiple feeds were avoided as their discharge is necessarily a mixture of water from more than one source.

GEOLOGICAL FEATURES

Iceland is a volcanic island (103,000 km²) of Tertiary to Recent age. The exposed volcanic pile is predominantly basaltic in composition (80–85%), silicic and intermediate rocks constitute about 10%, the rest being volcaniclastics, fluvial sediments and tillites. In the Tertiary lava pile they are typically 5–10% but are much more abundant in the Quaternary rocks (SAEMUNDSSON, 1979).

As much as 50% of the lava pile erupted from major volcanic centers (WALKER, 1966). In Tertiary formations where these centers are deeply eroded intrusions of dolerite, gabbro and granophyre may constitute as much or more than 50% of the rock by volume. Above the intrusions, geothermal systems developed as indicated by the alteration of the rock, frequently into greenschist mineral assemblages (SIGURDSSON, 1966; ANNELS, 1967; FRIDLEIFSSON, 1983, 1984). Outside the volcanic centers, zeolite facies alteration is typical of the Tertiary lava pile, the extent of alteration being generally rather limited (e.g., WALKER, 1960; KRISTMANNSDÓTTIR, 1978).

Quaternary formations differ from those of the Tertiary, both with respect to structure and morphology. Pillow lavas and hyaloclastites are abundant due to eruptions under the icesheet. As a result

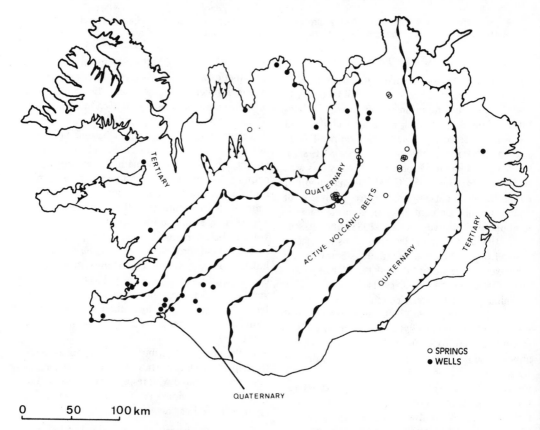

FIG. 1. Map of Iceland showing the distribution of sample sites for the cold and thermal waters considered in the present study.

Table 1. Composition of solid solution groundmass minerals and the mean activity of the pure end-member minerals in the solid solution

Mineral	Crystalline solution	Composition	Mean activity
Forsterite	olivine	Fo_{23}-Fo_{55}	0.19
Fayalite	olivine	Fa_{45}-Fa_{68}	0.33
Enstatite	orthopyroxene	En_{36}	0.3
Ferrosilite	orthopyroxene	Fs_{58}	0.58
Diopside (Ca-rich)	clinopyroxene	$En_{46}Fs_{16}Wo_{38}$-$En_{38}Fs_{29}Wo_{33}$	0.42
Diopside (Ca-poor)	clinopyroxene	$En_{53}Fs_{37}Wo_{10}$	0.11
Hedenbergite (Ca-rich)	clinopyroxene	$En_{46}Fs_{16}Wo_{38}$-$En_{38}Fs_{29}Wo_{33}$	0.22
Hedenbergite (Ca-poor)	clinopyroxene	$En_{53}Fs_{37}Wo_{10}$	0.072
Anortite	plagioclase	An_{29}	0.29
Albite	plagioclase	Ab_{71}	0.71
Magnetite	magnetite-ulvöspinel	Mt_{15}-Mt_{68}	0.35
Hematite	hematite-ilmenite	$Hem_{4.2}$-$Hem_{9.5}$	0.0054

The mole fractions are calculated from the chemical compositions of groundmass minerals in tholeiites and olivine tholeiites (sample 1 to 5, CARMICHAEL, 1967) except the one for orthopyroxene which is from a phenocryst in icelandite (sample 14, CARMICHAEL, 1967) and plagioclase which is from FISK (1978). The activities of the end member mineral compositions are calculated from the mole fractions as described in the text. The mean activity is the mean of the maximum and minimum activities.

glassy, or partially glassy volcanics are much more abundant in the Quaternary than in the Tertiary.

Post-glacial lava flows cover about 12,000 km², or more than 10% of the country (JAKOBSSON, 1972; SAEMUNDSSON, 1979). About 90% of these are basaltic, the remainder being either intermediate or silicic.

Geothermal activity is widespread in Iceland (e.g., PÁLMASON and SAEMUNDSSON, 1974; ARNÓRSSON, 1975). Geothermal areas have been classified as high- and low-temperature areas (BÖDVARSSON, 1961). The former have a magmatic heat source, occur within the zones of active volcanism and are most often associated with major volcanic centers. Drillhole data indicate temperatures in excess of 200°C in the uppermost 1000 m. Low-temperature areas are found in Quaternary and Tertiary formations and temperatures are below 150°C in the uppermost 1000 m.

The grain size and the amount of glass in basaltic rocks in Iceland are, of course, very variable. In the case of Post-glacial lavas from the Reykjanes Peninsula the average diameter for groundmass minerals at 2 m below the surface range between 5 and 200 μm (JAKOBSSON, pers. comm.). The largest phenocrysts may be as much as 1 cm across. Plagioclase is the most abundant phenocrysts and olivine is second. The amount of phenocrysts is very variable, occasionally it is as much as 50% but generally they constitute only a small portion of the basalts. In Post-glacial lava flows from the western Reykjanes Peninsula phenocrysts amount to about 4% by volume on average (JAKOBSSON et al., 1978). Phenocrysts are not, therefore, important for water-rock interactions.

In quenched rocks the composition of the glass corresponds closely to that of the bulk rock. Interstitial glass is, on the other hand, differentiated and commonly of rhyolitic composition (MAYER and SIGURDSSON, 1978). The primary minerals of basaltic rock, olivine, pyroxene, plagioclase and Fe-Ti oxides, occur both as phenocrysts and in the groundmass.

The chemistry of phenocrysts in the basaltic rocks of Iceland has been extensively studied (CARMICHAEL, 1964, 1967; JAKOBSSON et al., 1978; GRÖNVOLD and MÄKIPÄÄ, 1978; FISK, 1978; THY, 1983), but analytical data for groundmass minerals is relatively scarce. CARMICHAEL (1964, 1967) made a detailed study of groundmass minerals in rocks (basaltic to silicic) of the Tertiary Thingmúli central volcano in eastern Iceland. FISK (1978) studied some groundmass olivine and plagioclase in Post-glacial lavas from the Reykjanes Peninsula in SW-Iceland. Table 1 shows the average composition of groundmass minerals for basalts from Thingmúli and the Reykjanes Peninsula. As would be expected, individual grains show considerable zoning and are less basic than coexisting phenocrysts.

Hydrothermal alteration of basaltic rocks in active and fossil geothermal systems has been studied extensively (WALKER, 1963; SIGVALDASON, 1963; TÓMASSON and KRISTMANNSDÓTTIR, 1972; KRISTMANNSDÓTTIR, 1982; STEINTHÓRSSON and SVEINBJÖRNSDÓTTIR, 1981; EXLEY, 1982; MEHEGAN et al., 1982; RAGNARSDÓTTIR et al., 1984). It is evident from studies of eroded Tertiary and Quaternary formations that a very minor part of the volcanic pile has suffered extensive alteration. The alteration is concentrated in the blocky tops and bottoms of lava flows, along fault planes and around vesicles. The alteration mineralogy is extremely varied, with calcite, quartz, chalcedony and various zeolites being particularly abundant as amygdale minerals along with celadonite, smectite and chlorite. The alteration minerals occur, however, not only in vugs but also permeate the whole rock, occupying angular spaces between the igneous mineral grains or replacing them. A regular distribution of zeolites has been observed in the Tertiary formations (WALKER, 1960) with certain mineral assemblages occupying flat lying zones which cut across the stratigraphy. Most of the regional hydrothermal alteration appears to have occurred within active volcanic zones before the rock drifted outside these belts in response to continued crustal accretion (STEINTHÓRSSON et al., 1986).

Hydrothermal alteration in low-temperature geothermal fields is similar to the regional alteration just described. Much more intense alteration is observed in the cores of eroded central volcanoes and at depth in active high-temperature geothermal systems. The alteration typically belongs to the greenschist metamorphic facies with chlorite, albite, epidote, and quartz as the most prominent minerals but calcite, various sulphides, prehnite and actinolite are also quite common. Complete reconstitution of the rock is not uncommon when the alteration grade belongs to the greenschist facies.

Very limited information is available on the surface weathering products of the Icelandic basalts. ROALDSET (1983) studied sediment horizons interbedded with Tertiary basalts. The observed alteration mineralogy was the result of several processes, such as palagonitization, weathering and burial metamorphism/diagenesis. The alteration product included minerals such as kaolinite, halloysite, smectite, amorphous Al-Fe-oxides, and some zeolites. DOUGLAS (1987) observed that manganese-rich rock coatings are widespread in basalt lavas in Iceland. These coatings are associated with hydrothermal clay-filled fracture systems.

HYDROLOGICAL FEATURES

Precipitation in Iceland is abundant, ranges from less than 400 to more than 4000 mm a year and averages about 2000 mm and the average relative humidity is rather high, about 80% (EYTHÓRSSON and SIGTRYGGSSON, 1971). Some of the precipitation is stored temporarily in the ice-sheets, some of it evaporates directly, some of it runs off on the surface, and some infiltrates and enters groundwater bodies. In Tertiary and early Quaternary formations run-off dominates. The global permeability of the rock is low, $5*10^{-5}-10^{-7}$ cm^3/cm^2 sec (INGIMARSSON and SIGURDSSON, 1987), due to sealing by secondary mineralization. Permeable anomalies are found in these formations in association with young tectonic fractures where low-temperature geothermal systems tend to develop (ARNÓRSSON, 1987b). Late Quaternary formations have higher permeability, $10^{-2}-10^{-4}$ cm^3/cm^2 sec (INGIMARSSON and SIGURDSSON, 1987). Post-glacial lavas have no surface run-off due to extremely high permeability, 1–10^{-2} cm^3/cm^2 sec (INGIMARSSON and SIGURDSSON, 1987). Evapotranspiration may be high from such lavas due to abundant moss trapping the precipitation. Otherwise the precipitation infiltrates. In the Late Quaternary and Post-glacial rocks permeability comprises both primary and secondary fissures, scoriaceous tops and bottoms of lavas, contraction cracks in lavas and matrix permeability in sedimentary horizons.

The water in the high-temperature geothermal fields is most often local precipitation (ÁRNASON, 1977; ARNÓRSSON, 1985) but sometimes seawater or a mixture of seawater and local precipitation. The high-temperature fields are often in areas which rise topographically over the surrounding land. In the low-temperature fields the water is in most cases precipitation that has fallen in the highlands of the central part of the country (ÁRNASON, 1977). Thus, a regional groundwater flow has been envisaged from the central highlands towards the coast (ÁRNASON, 1977; EINARSSON, 1966). Opinion is divided whether the regional groundwater flow for the low-temperature systems occurs at deep or shallow levels from the recharge areas in the highlands (EINARSSON, 1966; BÖDVARSSON, 1982; ARNÓRSSON and ÓLAFSSON, 1986). Flow from the highlands at shallow levels is likely for some systems. The shallow flowing water decends to deep levels close to the boundary of the low-temperature fields forming the downflow limb of a convection cell (BJÖRNSSON, 1980; ARNÓRSSON and ÓLAFSSON, 1986).

Little is known about the residence time of water in the groundwater and geothermal systems. In extreme cases it may exceed 8000 years (ÁRNASON, 1976). The average surface area of a Post-glacial lava flow aquifer in contact with one liter of water ranges from about 600 to about 6000 cm^2 and the effective water/rock mass ratio for cold groundwater systems in NE-Iceland is of the order of 10^3 GÍSLASON and EUGSTER, 1987b). Corresponding figures for low- (30–60°C) and high-temperature (200–350°C) geothermal systems, also in NE-Iceland, are 100 and less than 10, respectively (GÍSLASON and EUGSTER, 1987b).

THE CHEMICAL COMPOSITION OF NATURAL WATERS IN ICELAND

Geothermal waters in Iceland are generally low in dissolved solids (200–2000 ppm) compared with such waters from geothermal fields in other countries. However, in some fields near the coast chloride contribution from seawater may cause the water salinity to become higher (ARNÓRSSON et al., 1989). At Reykjanes in SW-Iceland the geothermal water represents heated seawater which has modified its original composition by reactions with the basaltic rock (BJÖRNSSON et al., 1972). Cold groundwaters are lower in dissolved solids (50–100 ppm) than are thermal waters. Surface waters are comparable to cold groundwaters. However, total carbonate concentrations tend to be higher in the former, but most dissolved solids lower. Representative analyses of surface-, ground- and geothermal waters are given in Table 2.

GÍSLASON and EUGSTER (1987a,b) have elucidated the process governing the pH of cold waters in Iceland, through both, experimental work and field studies. We consider that the same process governs the pH of thermal waters. Because of the importance of pH for the dissolution/precipitation reactions considered in this contribution a brief summary will be given of the pH evolution of commonly occurring natural waters in Iceland.

The pH of pure rain water is about 5.6. Surface waters attain a higher pH (about 7) through proton uptake by the soil and rock and simultaneous dissolution of cations (GÍSLASON and EUGSTER, 1987b). Cold groundwaters develop higher pH, 9–10, and low temperature geothermal waters (less than about 100°C) even higher pH, up to 10.5. Carbonic acid derived from atmospheric CO_2 and the decay of organic matter constitutes the proton donor for surface waters. Steady state conditions are attained when proton generation by dissociation of carbonic acid into bicarbonate equals the rate of rock dissolution which dictates the rate of proton uptake. The proton donor of groundwaters isolated from the atmosphere is silica which is dissolved from the rock. Here steady state conditions are attained when proton donation rate by aqueous silica ionization equals the rate of proton uptake by the rock being dissolved (GÍSLASON and EUGSTER, 1987a,b). With increasing temperatures above about 100°C pH begins to drop. This is due to increasing mobility of both carbonic acid and hydrogen sulphide which generate protons through ionization into bicarbonate and bisulphide, respectively. The gases forming these acids may originate from the rock being dissolved or a degassing magma intrusion. It is generally agreed that shallow level intrusions constitute the heat source to high-temperature geothermal

Table 2. Chemical composition (ppm) of some natural waters in Iceland

No.	$T°C$	$pH/T°$	SiO_2	B	Na	K	Ca	Mg	Fe	Al	ΣCO_2	SO_4	ΣH_2S	Cl	F
1	—	7.00/10	3.6	—	1.48	0.09	1.02	0.18	0.025	0.07	3.3	0.2	<.01	0.47	.02
2	—	—	12.0	—	9.12	0.53	4.03	1.57	—	—	26	4.81	—	5.18	—
3	2.6	9.13/4	15.4	—	9.13	0.68	3.39	2.17	0.046	0.09	24	2.9	<.01	1.60	0.15
4	19.5	9.55/21	27.6	.016	16.2	0.63	1.88	0.18	—	0.042	7.1	3.8	<.01	14.3	0.10
5	43	9.93/22	39.0	.016	28.3	0.33	2.51	.009	—	0.12	19.2	4.31	<.01	17.4	0.07
6	75	8.46/12	68.7	.128	159	3.66	31.1	.061	—	0.043	21.0	59.1	0.02	235	0.24
7	92	9.54/22	154	.078	79.5	2.14	2.53	.023	—	0.086	22.2	54.1	1.03	34.4	0.71
8	152	7.31/152	249	.639	100	6.10	1.26	.047	0.0069	0.413	97.4	69.5	4.75	40.9	1.83
9	246	6.66/246	379	1.41	131	20.4	3.83	.072	0.016	0.085	741	41.3	383	14.1	0.36
10	240	5.31/240	425	6.87	6382	989	1066	1.29	0.156	0.056	421	32.2	5.56	13506	0.13

Sample no. 1 is a snow melt from Herdubreid, NE-Iceland that has interacted for a short time with basalt in contact with the atmosphere (Sp 12, GÍSLASON and EUGSTER, 1987). Sample no. 2 is a weighed average composition of Icelandic rivers (GÍSLASON and ARNÓRSSON, 1989). Sample no. 3 is from spring by Hrauná, NE-Iceland (Sp 15, GÍSLASON and EUGSTER, 1987). Sample no. 4 is from a well at Sveinseyri, NW-Peninsula. Sample no. 5 is from a well at Krossholt NW-Peninsula. Sample no. 6 is from well 13 at Selfoss, Southern Lowlands. Sample no. 7 is from a well at Reykjarhóll, Bökkum, North Iceland. Sample 8 is from well 1 at Reykjabóli, Southern Lowlands (ARNÓRSSON et al., 1983a), the deep water concentration and pH is shown for sample 8, 9 and 10 as calculated by the WATCH programme (ARNÓRSSON et al., 1982). Sample 9 is from well 8 at Námafjall, NE-Iceland (ARNÓRSSON et al., 1983a). Sample 10 is from well 4 in the Svartsengi geothermal field which is located on the Reykjanes peninsula (ARNÓRSSON et al., 1983a).

systems in Iceland, located within the belts of active volcanism but low-temperature systems, located in Quaternary and Tertiary formations are for the most part non-volcanic (ARNÓRSSON and ÓLAFSSON, 1986).

During the initial stages of rock dissolution by surface waters the aqueous concentrations of major components increase. This is reflected in the approximately constant aqueous ratio of the elements (GÍSLASON and EUGSTER, 1987a,b). However, some surface waters and groundwaters may show relative depletion in some elements, like aluminium, iron, magnesium and potassium, due to their uptake into weathering minerals. It is considered that all major components in geothermal waters, except chloride, closely approach chemical equilibrium with alteration minerals at temperatures as low as 50°C (ARNÓRSSON et al., 1983a).

THE CHEMICAL THERMODYNAMIC DATA BASE AND DATA TREATMENT

The WATCH programme of ARNÓRSSON et al. (1982) was used to calculate the aqueous speciation for the waters selected for the present study. The data base is that given in their Table 5 except for aqueous aluminium hydroxy complexes where the data of KUYUNKO et al. (1983) were used.

The WATCH programme calculates the ratio of ferrous to ferric ion from analyses of total iron in solution and a redox potential calculated on the assumption that equilibrium between aqueous sulphide and sulphate is attained. When sulphide is not detectable, as is the case for cold and slightly thermal waters, analysed aqueous iron is taken to be equal to ferrous iron. This is considered to be a good approximation. However, sulphide/sulphate redox equilibrium may not be closely approached, even at high temperatures, (ARNÓRSSON and GUNNLAUGSSON, 1985) causing an uncertainty in the calculated distribution of the iron-bearing aqueous species and in the calculated state of water saturation for iron bearing minerals. At low temperatures Fe^{+2} tends to be the dominant iron species but $Fe(OH)_4^-$ dominates at temperatures above 200°C.

For minerals we selected the data base of HELGESON et al. (1978) and, for the clay minerals, HELGESON (1969). We consider that the data on quartz solubility recommended by FOURNIER and POTTER (1982) are better than those of (WALTHER and HELGESON, 1977) which were used by HELGESON et al. (1978), at least in the temperature range of interest for this study (0–250°C). To use the quartz solubility data of FOURNIER and POTTER (1982) would have required correction of practically the

whole data base of HELGESON et al. (1978). This we have not done. The discrepancy so produced is quite small and does not affect our conclusions, being 0.06 Log K units at 25°C and 0.08 units at 250°C. A comparison of quartz solubility is given later using the data of FOURNIER and POTTER (1982) and WALTHER and HELGESON (1977). The calcite solubility curve used is that given by ARNÓRSSON et al. (1982). Below 100°C it is almost identical to that of PLUMMER and BUSENBERG (1982) who carried out calcite solubility experiments in the range 0–90°C.

There is considerable disagreement in the literature on the thermodynamic data base for Al-silicates. HELGESON et al. (1978) used data for gibbsite reported by HEMINGWAY and ROBIE (1977) together with the composition of mineral assemblages and interstitial water in Jamaican bauxite deposits and weathered Hawaiian basalts to obtain free energy data for kaolinite. Free energy values for all Al-silicates rest on the value for kaolinite so obtained which according to HEMINGWAY et al. (1982) may be too high, by 6.5 kJ per mole of aluminium. There is considerable and variable difference between the thermodynamic parameters for the Al-silicates given by HELGESON et al. (1978) and HEMINGWAY et al. (1982). The reason is not only disagreement on kaolinite. The data recommended by the different authors are based on different experiments. It is important to note that the conclusions of the present study on the state of mineral saturation are equally valid using either of the two data bases.

In the present study two types of dissolution reactions were considered. One involves the pure end-member composition of solid solutions of the igneous rock minerals (olivine, pyroxenes, plagioclase and iron-titanium oxides). The other involves congruent dissolution of the igneous mineral solid solution. The solid solution compositions selected for detailed discussion are those presented in Table 1 (see the discussion in the second section of this contribution). Only end-member compositions were considered for the alteration minerals. The reactions used to describe dissolution/precipitation for minerals are summarized in Table 3.

Equilibrium constants for the dissolution of pure end-member igneous and alteration minerals were calculated at one bar pressure up to 100°C and along the liquid-vapour curve at higher temperatures. The apparent standard partial molal Gibbs free energy of formation of the minerals at T and P was calculated according to the thermodynamic data and formulation given by HELGESON et al. (1978). The apparent standard partial molal Gibbs free energy of formation of the aqueous species was calculated using the revised equation of state for the standard molal properties of ions and electro-lytes (TANGER and HELGESON, 1986) using equation of state parameters from HELGESON and KIRKHAM (1974), TANGER and HELGESON (1986) and SHOCK and HELGESON (1988, 1989) with the exception of the standard partial molal Gibbs free energy of formation of Al^{+3} which we take to be −489,578 J/mole. This value is consistent with the value reported for $Al(OH)_4^-$ by SHOCK and HELGESON (1988) and is based on KITTRICK (1966) and the thermodynamic values for aqueous aluminum hydroxy complexes given by KUYUNKO et al. (1983). SHOCK and HELGESON (1988) used data on the equilibrium between Al^{+3} and $Al(OH)_4^-$ reported by COUTURIER et al. (1984) to obtain a value for the apparent standard partial molal Gibbs free energy of formation for Al^{+3}. Our value for Al^{+3} is almost the same as that reported by ROBIE et al. (1979), −489,400 J/mole.

The igneous minerals in the basalts dissolved by water represent solid solutions requiring the application of solid solution models for calculation of the activity of the end-members in these crystalline solutions. The temperature regime under consideration in the present study (0–250°C) is far below the temperature range (600–1400°C) where the experiments have been carried out that constrain empirical and theoretical solid solution models. For this reason simple ideal solid solution models have been adapted. For such models we have:

$$a_i = (X_i)^n \quad (1)$$

where a_i represents the activity of end-member i in a solid solution, X_i is its mole fraction and the exponent n refers to the number of exchange sites. The activities of forsterite and fayalite in olivine and of ferrosilite and enstatite in orthopyroxene have been obtained from Equation (1).

Clinopyroxene is taken to represent an ideal solid solution with multi-side mixing on the M_1, M_2 and the tetrahedral sites assuming that Fe and Mg are distributed between the M_1 and M_2 sites in equal proportions. It is further assumed that there is equal number of both positions in the formula unit (WOOD and FRASER, 1977). Thus, the activity of diopside ($CaMgSi_2O_6$) in a clinopyroxene solid solution is taken to be equal to the mole fraction of $MgSiO_3$ in this solution (X_{Mg}), which is assigned to the M_1 sites, times the mole fraction of $CaSiO_3$ (X_{Ca}), assigned to the M_2 sites, times the square of the mole fraction of Si (X_{Si}), assigned to the tetrahedral sites.

For the feldspars, the local charge balance model has been employed (KERRICK and DARKEN, 1975; NORDSTROM and MUNOZ, 1985). Thus, the activities of the end-member components in the solution

equal the mole fraction of that component. For example, the activity of the albite component in feldspar solid solution equals the mole fraction of $(NaSi)AlSi_2O_8$.

Mixing in the magnetite-ulvöspinel solid solution is assumed to be confined to the octahedral sites. Thus, the activity of magnetite in the solid solution is equal to the square of the mole fraction of Fe(III) occupying the octahedral sites. The hematite-ilmenite solid solution is treated in the same way; the activity of hematite in this solution equals the square of the mole fraction of Fe(III). The calculated mean activities of the various end member components in groundmass minerals in some Icelandic basalts are shown in Table 1.

Equilibrium constants for the stoichiometric (congruent) dissolution (THORSTENSON and PLUMMER, 1977) of the actual igneous minerals were obtained by assuming that they represented ideal solid solutions and choosing the standard state to be a solid solution of a fixed composition at the temperature and pressure of interest. This implies that the entalphy of mixing equals zero and since

$$\Delta S_{\text{ideal mix}} = -nR \sum_i X_i \ln X_i \quad (2)$$

we have that the free energy of mixing is equal to:

$$\Delta G_{\text{ideal mix}} = nRT \sum_i X_i \ln X_i \quad (3)$$

and for an ideal binary solution of a given composition at T and P the apparent standard state Gibbs free energy of formation, $\Delta G_{f,ss}^0(T, P)$ is

$$\Delta G_{f,ss}^0(T, P) = X_i \Delta G_{f,i}^0(T, P) + X_j \Delta G_{f,j}^0(T, P) + nRT(X_i \ln X_i + X_j \ln X_j) \quad (4)$$

$\Delta G_{f,i}^0(T, P)$ and $\Delta G_{f,j}^0(T, P)$ refer to the apparent standard state Gibbs free energy of formation of the pure end-members i and j of the solid solution at the temperature and pressure of interest; X_i and X_j are the corresponding mole fractions and $X_j = 1 - X_i$; and n is the number of exchange sites per mineral formula. R is the gas constant and T is absolute temperature. The apparent standard partial molal Gibbs free energy of formation for the pure end-members in the solid solution is obtained by the method described above and once values for X_i and X_j have been selected the stochiometry for the dissolution reactions can be written (Table 3).

METHOD OF INTERPRETATION

The degree of unstability of a mineral dissolving in aqueous solution is assessed by evaluating the Gibbs free energy for the dissolution reaction (ΔG_r):

$$\Delta G_r = \Delta G_r^0 + RT2.303 \text{ Log } Q \quad (5)$$

where ΔG_r^0 is the standard state Gibbs free energy

Table 3. Dissolution reactions for minerals

Minerals	Reactions
Forsterite	$Mg_2SiO_4 + 4H^+ = 2\,Mg^{++} + H_4SiO_4^0$
Fayalite	$Fe_2SiO_4 + 4H^+ = 2\,Fe^{++} + H_4SiO_4^0$
Olivine $(Fo_{0.43}Fa_{0.57})$	$(Mg_{.43}Fe_{.57})SiO_4 + 4\,H^+$ $= 0.86\,Mg^{++} + 1.14\,Fe^{++} + H_4SiO_4^0$
Enstatite	$MgSiO_3 + 2H^+ + H_2O = Mg^{++} + H_4SiO_4^0$
Ferrosilite	$FeSiO_3 + 2H^+ + H_2O = Fe^{++} + H_4SiO_4^0$
Orthopyroxene $(En_{0.38}\,Fs_{0.62})$	$Mg_{.38}Fe_{.62}SiO_3 + 2\,H^+ + H_2O$ $= 0.38\,Mg^{++} + 0.62\,Fe^{++} + H_4SiO_4^0$
Diopside	$MgCaSi_2O_6 + 4\,H^+ + 2\,H_2O$ $= Mg^{++} + Ca^{++} + 2\,H_4SiO_4^0$
Hedenbergite	$FeCaSi_2O_6 + 4\,H^+ + 2\,H_2O$ $= Fe^{++} + Ca^{++} + 2\,H_4SiO_4^0$
Anorthite	$CaAl_2Si_2O_8 + 8\,H_2O$ $= Ca^{++} + 2\,Al(OH)_4^- + 2\,H_4SiO_4^0$
Albite	$NaAlSi_3O_8 + 8\,H_2O$ $= Na^+ + Al(OH)_4^- + 3\,H_4SiO_4^0$
High-plagioclase $(An_{0.29}\,Ab_{0.71})$	$(CaAl)_{.29}(NaSi)_{.71}AlSi_2O_8 + 8H_2O$ $= 0.29Ca^{++} + 0.71Na^+ + 1.29Al(OH)_4^-$ $+ 2.71H_4SiO_4^0$
Magnetite	$FeFe_2O_4 + 4\,H_2O = Fe^{++} + 2\,Fe(OH)_4^-$
Hematite	$Fe_2O_3 + 5\,H_2O = 2\,H^+ + 2\,Fe(OH)_4^-$
Calcite	$CaCO_3 = Ca^{++} + CO_3^-$
Chalcedony, quartz	$SiO_2 + 2\,H_2O = H_4SiO_4^0$
Gibbsite	$Al(OH)_3 + H_2O = Al(OH)_4^- + H^+$
Kaolinite	$Al_2Si_2O_5(OH)_4 + 7\,H_2O$ $= 2\,Al(OH)_4^- + 2\,H_4SiO_4^0 + 2\,H^+$
Laumontite	$Ca(Al_2Si_4O_{12})4H_2O + 8\,H_2O$ $= Ca^{++} + 2\,Al(OH)_4^- + 4\,H_4SiO_4^0$
Microcline	$KAlSi_3O_8 + 8\,H_2O$ $= K^+ + Al(OH)_4^- + 3\,H_4SiO_4^0$
Mg-montmorill.	$Mg_{0.167}Al_{2.33}Si_{3.67}O_{10}(OH)_2 + 12\,H_2O$ $= 0.167\,Mg^{++} + 2\,H^+$ $+ 2.33\,Al(OH)_4^- + 3.67\,H_4SiO_4^0$
Ca-montmorill.	$Ca_{0.167}Al_{2.33}Si_{3.67}O_{10}(OH)_2 + 12\,H_2O$ $= 0.167\,Ca^{++} + 2\,H^+ + 2.33\,Al(OH)_4^-$ $+ 3.67\,H_4SiO_4^0$
Na-montmorill.	$Na_{0.33}Al_{2.33}Si_{3.67}O_{10}(OH)_2 + 12\,H_2O$ $= 0.33\,Na^+ + 2\,H^+ + 2.33\,Al(OH)_4^-$ $+ 3.67\,H_4SiO_4^0$
K-montmorill.	$K_{0.33}Al_{2.33}Si_{3.67}O_{10}(OH)_2 + 12\,H_2O$ $= 0.33\,K^+ + 2\,H^+ + 2.33\,Al(OH)_4^-$ $+ 3.67\,H_4SiO_4^0$
Pyrite	$FeS_2 + 4.5\,H_2O = Fe(OH)_4^- + 1.25\,H^+$ $+ 0.125\,SO_4^- + 1.87\,H_2S$
Pyrrhotite	$FeS + 0.125\,SO_4^- + 3.5\,H_2O$ $= Fe(OH)_4^- + 0.75\,H^+ + 1.125\,H_2S$

The Ca and Mn present in the orthopyroxene is disregarded for the calculations of its stoichiometric dissolution (Ca is equal to 0.036 and 0.018 mole per mineral formula). The mole fraction for the enstatite and ferrosilite in the solid solution as given in Table 2 is, therefore, raised to make their sum equal to 1. The mole fractions of olivine and plagioclase solid solutions are as given in Table 1.

for the reaction, R is the gas constant, T is temperature in Kelvin and Q the reaction quotient (activity product). Two kinds of standard states for minerals have been chosen as discussed in the previous sec-

tion. One is the pure end member composition in the solid solution at T and P. The other represents the actual mineral composition at T and P. An assumption inherent to the choice of this second type of standard state is that the respective solid solution mineral dissolves congruently.

Since $\Delta G_r^0 = -RT2.303 \, \text{Log} \, K$, where K is the equilibrium constant, it follows that

$$\Delta G_r = RT2.303(\text{Log}\,Q - \text{Log}\,K) \qquad (6)$$

At equilibrium, the Gibbs free energy of reaction is equal to zero so $K = Q$. The difference between K and Q is a measure of the degree of undersaturation/supersaturation of a mineral in a particular aqueous solution. It is also a measure of the Gibbs free energy driving the respective reaction. If a given mineral is unstable or dissolving, ΔG_r is less than zero but positive if the mineral is stable or precipitating for the reactions as written in Table 3.

Numerous experimental studies have been carried out in the past on dissolution rates for rock forming minerals and glasses far from equilibrium (*e.g.*, LAGACHE, 1976; BUSENBERG and CLEMENCY, 1976; GRANDSTAFF, 1980; SCHOTT et al., 1981; WHITE 1983; CHOU and WOLLAST, 1984; GÍSLASON and EUGSTER, 1987a). At 25°C the reaction rate becomes in most cases steady and linear after a few tens of days, even less:

$$dQ^*/dt = k^* \qquad (7)$$

where Q^* is the flux of elements entering the solution per unit area of the solid that is dissolving, t is time and k^* is the linear rate constant (mole $cm^{-2}\,sec^{-1}$).

In general dissolution rates can be described by the following equation (WALTHER and WOOD, 1986; LASAGA, 1981):

$$dQ^*/dt = k^*(1 - e^{\Delta G_r/RT}) \qquad (8)$$

WALTHER and WOOD (1986) pointed out that the dissolution rate can simply be described by the linear rate constant, k^* (equation (7)), if $\Delta G_r/RT$ is smaller (greater negative number) than -6. The value of -6 corresponds to a Log (Q/K) value of -2.6 and a ΔG_r value of 15 kJ at 25°C and 26 kJ at 250°C. If Log (Q/K) lies in the range from -2.6 to -0.17 dissolution rates are best described by equation (8), but if the difference is greater than about -0.17 (smaller negative number) the dissolution rate is satisfactorily formulated as follows:

$$dQ^*/dt = -k^* e^{\Delta G_r/RT} \qquad (9)$$

This is because the first factor within the parenthesis in equation (8) contributes little to the reaction rate when Log Q/K (and ΔG_r) take such low numbers.

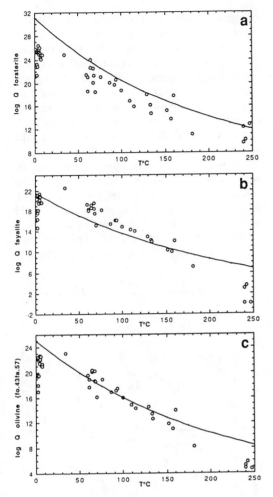

FIG. 2. The state of olivine saturation in natural waters in Iceland. Parts a and b show the state of saturation for the forsterite and fayalite components in olivine of the composition fo$_{.43}$fa$_{.57}$. Part c shows the state of saturation for stoichiometric dissolution of olivine of the composition indicated.

By assessing the value of ΔG_r or the Log (Q/K) for specific igneous and alteration minerals, and natural water compositions, the kinetic relationship best describing the rates of igneous mineral dissolution and secondary mineral deposition under natural conditions may be determined.

RESULTS AND DISCUSSION

Igneous minerals

The WATCH programme (ARNÓRSSON et al., 1982) has been used to evaluate the state of saturation of selected cold spring and geothermal well waters relative to igneous minerals of basalt (icelandite in the case of orthopyroxene). The evalua-

tion was made for end member compositions as well as solid solutions of fixed composition (Table 1). The results are shown in Figs. 2 to 6. The solid solution compositions selected are based on groundmass plagioclase data from the Reykjanes Peninsula (FISK, 1978) and the average for groundmass minerals presented by CARMICHAEL (1964, 1967) for basalts from the Thingmúli central volcano in eastern Iceland (Table 3). The activities of the end member components in the solid solutions are incorporated in the reaction quotient and were evaluated as described previously. Equations for the reaction quotient corresponding to the dissolution reactions in Table 3 are shown in Table 4. The curves in Figs. 2 to 6 depict the equilibrium con-

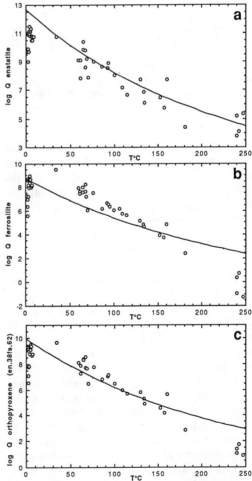

FIG. 3. The state of orthopyroxene saturation in natural waters in Iceland. Parts a and b show the state of saturation for the enstatite and ferrosilite components in pyroxene of the composition en$_{.38}$fs$_{.62}$. Part c shows the state of saturation for stoichiometric dissolution of orthopyroxene of the composition indicated.

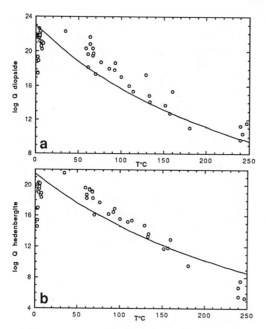

FIG. 4. The saturation state of diopside and hedenbergite components in calcium-rich clinopyroxene in natural waters in Iceland. The activities selected for diopside and hedenbergite were 0.377 and 0.658, respectively (Table 4).

stant. These curves probably represent minimum solubility of the respective minerals and end members since the minerals may be considerably disordered crystallographically and of such small crystals that their resulting surface free energy could increase their solubility.

Olivine

The saturation state (reaction quotient) for the selected natural waters with respect to the forsterite component in olivine with the composition $fo_{43}fa_{57}$ (see Table 4) is plotted against water temperature in Fig. 2a. All but four of the waters are undersaturated with respect to the forsterite component. Increasing forsterite content in the olivine will decrease the Log Q value, thus increasing the degree of undersaturation. However, the position of the data points in Fig. 2a are not very sensitive to the composition of the olivine. Changing from fo_{43} to fo_{81} would cause a decrease in Log Q corresponding to the diameter of the data points. Similarly, decreasing the forsterite content from fo_{43} to fo_{23} would shift Log Q up equivalent to the diameter of the data points.

The saturation state with respect to the fayalite component in the olivine is depicted in Fig. 2b. Most of the waters are somewhat supersaturated. The only undersaturated waters are from cold

springs with "relatively low" pH (8–9) and thermal waters with temperatures above about 180°C.

It is not known whether olivine dissolves congruently (stoichiometrically) in natural waters or if ion exchange reactions are involved although the former seems more probable since olivine solid solution (fo_{82}) dissolves congruently in experiments carried out in the temperature range 1 to 50°C (GRANDSTAFF, 1980, 1986) and the same is true for other solid solutions (*i.e.*, BUSENBERG and PLUMMER, 1989). If ion exchange was associated with olivine dissolution over the temperature range considered here (0–250°C), the forsterite component alone would dissolve in cold waters with pH above about 9 and in thermal waters up to about 150°C making the remaining olivine more fayalite-rich. In cold waters with pH of less than about 9,

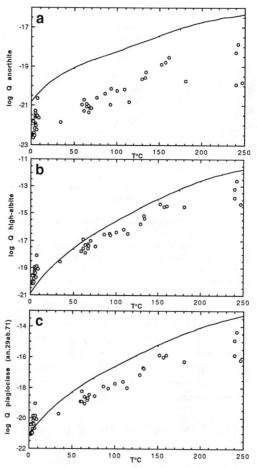

FIG. 5. The state of saturation of plagioclase in natural waters in Iceland. Parts a and b show the saturation state for the anortite and high-albite components in plagioclase of the composition $an_{.29}ab_{.71}$. Part c shows the state of saturation for stoichiometric plagioclase dissolution with the composition indicated.

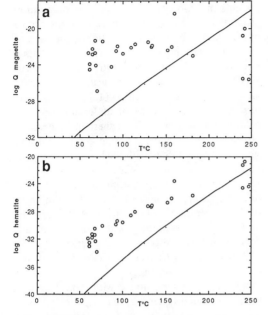

FIG. 6. The saturation state of magnetite and hematite components in ulvö-spinel and ilmenite in natural waters in Iceland.

both components would dissolve, but forsterite preferentially. The data points at 240–250°C (Fig. 2a and b) indicate that the forsterite component is somewhat unstable but the fayalite component is highly unstable. Accordingly, dissolution involving ion exchange would lead to increasing magnesium content of the olivine.

If olivine dissolves stoichiometrically, as seems likely, it can be deduced from Fig. 2c that this mineral is highly unstable in cold groundwaters and in waters above about 150°C. In the range 30–150°C most waters are relatively close to equilibrium with the olivine solid solution. What is important to note here is that the stability of olivine is very sensitive to its composition assuming that it dissolves stoichiometrically. Increasing forsterite content decreases the stability of olivine. Thus, for example, all but six of the waters plotted in Fig. 2c would be undersaturated with olivine containing fo_{70}.

The results presented in Fig. 2a to c are based on the average composition ($fo_{43}fa_{57}$) of groundmass olivine in basalts from Thingmúli (Table 1). As can be seen from Table 1, the composition of the groundmass olivine of the Thingmúli basalts varies between fo_{23} and fo_{55}. Olivine phenocrysts are much more forsterite rich, up to fo_{90} (GRÖNVOLD and MÄKIPÄÄ, 1978). Thus, it can be concluded that, if olivine dissolves stoichiometrically, phenocrysts of this mineral are unstable in contact with natural Icelandic waters at all temperatures. Groundmass

olivine may, on the other hand, be stable in the temperature range of about 30–150°C but unstable at lower and higher temperatures.

Cold waters and waters above about 180°C are

Table 4. Expressions for the logarithm of the reaction quotient (Log Q) for primary and some secondary minerals in basalts

Log Q	Expression
Log $Q_{forsterite}$	$= 2 \text{ Log } a_{Mg^{++}} + \text{Log } a_{H_4SiO_4^0} - 4 \text{ Log } a_{H^+} + 0.721$
Log $Q_{fayalite}$	$= 2 \log a_{Fe^{++}} + \text{Log } a_{H_4SiO_4^0} - 4 \text{ Log } a_{H^+} + 0.481$
Log $Q_{olivine}$	$= 0.86 \text{ Log } a_{Mg^{++}} + 1.14 \text{ Log } a_{Fe^{++}} + \text{Log } a_{H_4SiO_4^0} - 4 \text{ Log } a_{H^+}$
Log $Q_{enstatite}$	$= \text{Log } a_{Mg^{++}} + \text{Log } a_{H_4SiO_4^0} - 2 \text{ Log } a_{H^+} + 0.444$
Log $Q_{ferrosilite}$	$= \text{Log } a_{Fe^{++}} + \text{Log } a_{H_4SiO_4^0} - 2 \text{ Log } a_{H^+} + 0.237$
Log $Q_{orthopyroxene}$	$= 0.38 \text{ Log } a_{Mg^{++}} + 0.62 \text{ Log } a_{Fe^{++}} + \text{Log } a_{H_4SiO_4^0} - 2 \text{ Log } a_{H^+}$
Log $Q_{diopside}$*	$= \text{Log } a_{Mg^{++}} + \text{Log } a_{Ca^{++}} + 2 \text{ Log } a_{H_4SiO_4^0} - 4 \text{ Log } a_{H^+} + 0.377_{Ca\text{-rich}}$
Log $Q_{hedenbergite}$*	$= \text{Log } a_{Fe^{++}} + \text{Log } a_{Ca^{++}} + 2 \text{ Log } a_{H_4SiO_4^0} - 4 \text{ Log } a_{H^+} + 0.658_{Ca\text{-rich}}$
Log $Q_{anorthite}$	$= \text{Log } a_{Ca^{++}} + 2 \text{ Log } a_{Al(OH)_4^-} + 2 \text{ Log } a_{H_4SiO_4^0} + 0.538$
Log $Q_{high\text{-}albite}$	$= \text{Log } a_{Na^+} + \text{Log } a_{Al(OH)_4^-} + 3 \text{ Log } a_{H_4SiO_4^0} + 0.149$
Log $Q_{plagioclase}$	$= 0.29 \text{ Log } a_{Ca^{++}} + 0.71 \text{ Log } a_{Na^+} + 1.29 \text{ Log } a_{Al(OH)_4^-} + 2.7 \text{ Log } a_{H_4SiO_4^0}$
Log $Q_{magnetite}$	$= \text{Log } a_{Fe^{++}} + 2 \text{ Log } a_{Fe(OH)_4^-} + 0.456$
Log $Q_{hematite}$	$= 2 \text{ Log } a_{Fe(OH)_4^-} + 2 \text{ Log } a_{H^+} + 2.268$
Log $Q_{calcite}$	$= \text{Log } a_{Ca^{++}} + \text{Log } a_{CO_3^{--}}$
Log $Q_{quartz, chalcedony}$	$= \text{Log } a_{H_4SiO_4^0}$
Log $Q_{gibbsite}$	$= \text{Log } a_{Al(OH)_4^-} + \text{Log } a_{H^+}$
Log $Q_{kaolinite}$	$= 2 \text{ Log } a_{Al(OH)_4^-} + 2 \text{ Log } a_{H_4SiO_4^0} + 2 \text{ Log } a_{H^+}$
Log $Q_{laumontite}$	$= \text{Log } a_{Ca^{++}} + 2 \text{ Log } a_{Al(OH)_4^-} + 4 \text{ Log } a_{H_4SiO_4^0}$
Log $Q_{microcline}$	$= \text{Log } a_{K^+} + \text{Log } a_{Al(OH)_4^-} + 3 \text{ Log } a_{H_4SiO_4^0}$
Log $Q_{Mg\text{-}montmorill.}$	$= 0.167 \text{ Log } a_{Mg^{++}} + 2 \text{ Log } a_{H^+} + 2.33 \text{ Log } a_{Al(OH)_4^-} + 3.67 \text{ Log } a_{H_4SiO_4^0}$
Log $Q_{Ca\text{-}montmorill.}$	$= 0.167 \text{ Log } a_{Ca^{++}} + 2 \text{ Log } a_{H^+} + 2.33 \text{ Log } a_{Al(OH)_4^-} + 3.67 \text{ Log } a_{H_4SiO_4^0}$
Log $Q_{Na\text{-}montmorill.}$	$= 0.33 \text{ Log } a_{Na^+} + 2 \text{ Log } a_{H^+} + 2.33 \text{ Log } a_{Al(OH)_4^-} + 3.67 \text{ Log } a_{H_4SiO_4^0}$
Log $Q_{K\text{-}montmorill.}$	$= 0.33 \text{ Log } a_{K^+} + 2 \text{ Log } a_{H^+} + 2.33 \text{ Log } a_{Al(OH)_4^-} + 3.67 \text{ Log } a_{H_4SiO_4^0}$
Log Q_{pyrite}	$= \text{Log } a_{Fe(OH)_4^-} + 1.25 \text{ Log } a_{H^+} + 0.125 \text{ Log } a_{SO_4^{--}} + 1.87 \text{ Log } a_{H_2S}$
Log $Q_{pyrrhotite}$	$= \text{Log } a_{Fe(OH)_4^-} + 0.75 \text{ Log } a_{H^+} + 1.125 \text{ Log } a_{H_2S} - 0.125 \text{ Log } a_{SO_4^{--}}$

* This is the reaction quotient for aqueous ions and the pure end member minerals (diopside and hedenbergite) in Ca-rich groundmass pyroxene (augite) but the activity of the end member minerals, diopside and hedenbergite, in the Ca-poor groundmass pyroxene is 0.959 and 1.143, respectively.

so strongly undersaturated with olivine that its dissolution is expected to be described by the linear rate law (Equation 7). The same situation would generally hold for olivine phenocrysts at all temperatures. Groundmass olivine may be stable in some instances or slightly undersaturated so that Equation (9) appropriately describes its dissolution rate.

Orthopyroxene

Figure 3a to c shows the saturation state of the selected natural waters relative to orthopyroxene of the composition $en_{38}fs_{62}$, which is the average composition of orthopyroxene phenocrysts in icelandite from Thingmúli (CARMICHAEL, 1967) (Table 1). Orthopyroxenes are not found in the basalts of Thingmúli. Fig. 3a shows that the enstatite component is generally unstable for all cold water samples and the majority of thermal water samples. By contrast, most waters from low-temperature fields (30–150°C) are close to equilibrium or somewhat supersaturated with respect to the ferrosilite component. Cold waters with pH above about 9 are close to ferrosilite saturation. Cold waters with lower pH are undersaturated as are waters from high-temperature fields (>180°C).

If orthopyroxene dissolution is associated with ion exchange, the enstatite component would go into solution preferentially in the case of cold waters and thermal waters below about 150°C. As a result the remaining orthopyroxene would become enriched in the ferrosilite component. Waters of higher temperature would, on the other hand, tend to loose the ferrosilite component preferentially to enstatite causing the residual pyroxene to be magnesium enriched.

If orthopyroxene of the composition $en_{38}fs_{62}$ dissolves stoichiometrically, it is undersaturated in cold waters and in waters above about 180°C (Fig. 3c). Waters of intermediate temperatures are close to saturation, some even significantly supersaturated (Fig. 3c). As for olivine, the stability of orthopyroxene is quite sensitive to its composition. Increasing magnesium content leads to decreasing stability, particularly at low temperatures.

It is only in cold groundwaters with relatively low pH (<8) that orthopyroxene undersaturation is so large that its dissolution would be expected to follow the linear rate law (equation 7). The dissolution rate of orthopyroxene in other waters, if it dissolves, would be in accordance with equation (9).

Clinopyroxene

The saturation state of the cold and thermal waters relative to the diopside and hedenbergite components in Ca-rich pyroxene of the composition given by the mole fractions in Table 1, are shown in Fig. 4a and b. This composition is the average for Ca-rich groundmass clinopyroxenes in Thingmúli basalts (CARMICHAEL, 1967) and the calculated average activities of diopside and hedenbergite are 0.42 and 0.22, respectively (Table 1). Cold waters show varying degree of undersaturation depending on their pH. The highest pH waters are just saturated with the diopside component and slightly undersaturated with the hedenbergite component (Fig. 4a and b). Waters in the range 30–150°C are generally supersaturated with both components, the degree of supersaturation falling with rising temperature. Waters at 240–250°C are saturated and supersaturated with diopside but undersaturated with respect to hedenbergite.

With decreasing Ca content of clinopyroxene the waters would become progressively more supersaturated or less undersaturated with respect to both the diopside and hedenbergite components. Thus, taking the composition of Ca-poor clinopyroxene in the groundmass of Thingmúli basalts (Table 1), the reaction quotient for the diopside end member would increase by 0.58 Log units shifting all data points in Fig. 4a up by this value. Similarly, the reaction quotient for hedenbergite would increase by 0.50 Log units. Our results indicate that all thermal waters are supersaturated with respect to Ca-poor clinopyroxenes, except possibly for the waters at and above 180°C. Cold waters with pH somewhat in excess of about 9 are close to saturation but lower pH cold waters are undersaturated.

In this brief contribution an attempt has not been made to assess the stability of clinopyroxene for its stoichiometric dissolution. The pyroxene grains tend to show exsolution into calcium rich and calcium poor components and it is not certain how reliable it would be to treat such exsolved grains as ideal solid solutions.

Plagioclase

Figure 5a to c shows the saturation state of the cold and thermal waters relative to anorthite, high-albite, and plagioclase of the composition $an_{29}ab_{71}$. This composition is that of groundmass plagioclase in Reykjanes Peninsula basalts (Table 1). All the waters are undersaturated with respect to the anorthite component in plagioclase of this composition. Thermal waters above about 60°C are also undersaturated with respect to high-albite in this plagioclase, although to a lesser extent, but cold waters are supersaturated.

Plagioclase dissolution in cold ground waters involving ionic exchange would lead to increasing albite content of the residual plagioclase. The data plotted in Fig. 5c indicate that the groundmass plagioclase is unstable in all the thermal waters, if it dissolves stoichiometrically. It is, on the other hand, stable in all of the cold waters. Plagioclase phenocrysts are much more anorthite rich than the groundmass plagioclase for which data are presented in Fig. 5c. Increasing anorthite content causes increasing instability for plagioclase dissolving stoichiometrically. Plagioclase containing 70% anorthite would be unstable in contact with many of the cold waters plotted in Fig. 5c.

The degree of groundmass plagioclase undersaturation in thermal waters at 30–180°C generally corresponds to about one Log Q unit. This indicates that the dissolution rate will be in accordance with Equation (8) above. Groundmass plagioclase in high-temperature waters and anorthite-rich plagioclase (phenocryst) may dissolve according to the linear rate law (Equation 7).

Iron oxides

Data on titanium are not available for the cold and thermal waters. Some data are available on the composition of iron-titanium oxides in Icelandic basalts (Table 1). Due to lack of data on titanium in the waters, it is only possible to study the state of saturation of the waters with respect to pure iron components in the iron-titanium oxides. Another difficulty in assessing the saturation state is that it is not possible to evaluate the ratio of ferrous to ferric iron in the waters from total iron analysis unless sulphide and sulphate occur in detectable concentrations in the water. Sulphide is not detectable in any of the cold waters and when this is the case the WATCH aqueous speciation programme assumes that all the analysed iron is on the ferrous form. This is probably a good approximation, but it excludes the possibility of studying water-mineral reactions involving ferric iron in non-sulphide bearing waters.

The results for the state of saturation with respect to the iron components in ilmenite-hematite and magnetite-ulvöspinel are shown in Fig. 6a and b. Mineral compositions are those presented in Table 1. Thermal waters below about 180°C are supersaturated with respect to the magnetite component in the magnetite-ulvöspinel solid solution. At higher temperatures undersaturation is indicated. Supersaturation is particularly strong at low temperatures. Results for hematite are similar except that some waters near 240–250°C seem to be near equilibrium with the hematite component in the ilmenite.

It is of interest to point out here that quite strong negative magnetic anomalies have been observed in some of the high-temperature geothermal fields in Iceland indicating that the magnetite of the rock is destroyed, or partially destroyed, as a result of the geothermal processes (PÁLMASON, 1975).

The results depicted in Fig. 6 should be considered with some reservation. The calculation of the ferrous to ferric ratio in the thermal waters by assuming redox equilibrium between sulphide and sulphate may not be entirely valid, particularly at low temperatures, because this redox equilibrium may not be closely approached in the thermal waters. If the waters were actually less reducing than indicated by the sulphide to sulphate ratio, supersaturation with respect to the hematite component in ilmenite would be even greater than implied by the results in Fig. 6b. The same holds for the magnetite component in the magnetite-ulvöspinel solid solution.

Glass

Thermodynamic data are not available to assess the solubility of basaltic glass in aqueous solution. It is, however, evident that such glass will have higher solubility than any of the basaltic minerals as the glass has retained relatively more energy from its parent magmatic environment than the minerals. This is indeed supported by experimental data (GÍSLASON and EUGSTER, 1987a) which show that basaltic glass dissolves considerably more rapidly in water than any of the basaltic minerals. The relatively high solubility of basaltic glass could cause stabilization of the igneous minerals. For example, release of calcium, magnesium, iron and silica from the glass could reduce the degree of undersaturation with respect to olivine and may even cause the water to become supersaturated with respect to pyroxene. If glass was not present for dissolution, either due to its absence or previous dissolution, the rock dissolution pattern would probably be different. Under these circumstances the water would remain at greater olivine undersaturation and pyroxene would be less supersaturated or more undersaturated. In other words, the dissolution of the constituents of the basaltic rock could follow a specific sequence depending on the relative amount and composition of the igneous minerals and glass present, glass disappearing first, then, in succession, olivine and plagioclase, and finally pyroxene. As the chemical composition of cold water is largely affected by the dissolution process, this would lead to some changes in its chemistry depending on how far the alteration of the rock had proceeded. The composition of the thermal water could also be affected by this evolutionary process, but probably to a lesser extent because thermal water compositions are controlled by close approach to chemical equilibrium with alteration minerals (see discussion below). However, changing dissolution processes could affect the composition and even the type of alteration minerals forming, and in this way affect the chemistry of the thermal water.

Hydrogen ion metasomatism

Weathering and hydrothermal alteration of basalt involves uptake of protons and simultaneous release of other cations as the primary constituents of the rock dissolve (Table 3) and secondary minerals precipitate. The extent of H^+ metasomatism is much affected by the supply of protons to the system. In surface waters, the source of protons is carbonic acid derived from atmospheric CO_2 and the decay of organic matter. In groundwaters and thermal waters of relatively low temperature ($<150°C$) ionization of silica being dissolved from the rock ($H_4SiO_4^0 \rightarrow H_3SiO_4^- + H^+$) is the main source of protons to drive hydrogen ion metasomatism. The mobility of CO_2 and H_2S in thermal waters of high temperature is much higher than that at low temperature and it increases with increasing temperatures (ARNÓRSSON and GUNNLAUGSSON, 1985). These gases form weak acids which constitute the main proton donors in high-temperature waters. Their source could be the rock being dissolved or a degassing magma intrusion.

Natural waters in Iceland, especially those of ambient to low temperatures, tend to have relatively high pH compared to such waters in many other parts of the world. There are basically two reasons for this. One is the high reactivity of basaltic minerals and, especially, basaltic glass. The other is limited supply of organic CO_2. On high ground in Iceland vegetation is sparse and, as a result, organic soil formation is negligible. Thus, surface waters seeping into the ground pick up little organic CO_2 and relatively limited proton uptake through rock dissolution is required to raise the water pH to 9 or more after it has been isolated from the atmosphere.

In some lowland areas in Iceland where the bedrock is covered with a peat soil several meters thick, surface waters tend to have much higher CO_2 contents than surface waters in unvegetated areas. Where mixing of such water and geothermal water has taken place in the upflow of low-temperature fields, the composition of the mixed water indicates extensive reaction with the enclosing rock linked with the conversion of carbonic acid into bicarbonate and the simultaneous increase in pH and major cation concentrations, especially magnesium and potassium (MAGNÚSDÓTTIR, 1989). It is evident that surface waters rich in carbonic acid have a much higher potential to dissolve the minerals and glass of basalt than do waters, poor in carbonic acid.

The effect of increasing proton availability is to enhance the dissolution process, but to a variable extent for individual minerals. To express this in a different way, the lowering of water pH leads to a higher degree of undersaturation with respect to the various igneous minerals. pH lowering has a particularly profound effect on the degree of undersaturation for olivine, but it also decreases the stability of pyroxene as may be deduced from the respective reactions in Table 3 and the Log Q expressions in Table 4. pH in the range occurring in the Icelandic natural waters (about 6–10) has insignificant effect on the stability of plagioclase. This can be inferred from the respective reactions in Table 3 because $Al(OH)_4^-$ is the dominant aluminium-hydroxide species over this pH range. However, pH variation in this range at any particular temperature could affect the value taken by the dissolution rate constants, k^*. In the case of albite, experiments show that its dissolution rate decreases by about a factor of 5 when pH changes from 10 to 8 at 25°C (CHOU and WOLLAST, 1984). The rate of olivine dissolution decreases by a factor of about 6 for the same pH change at 25°C, in water with partial pressure of CO_2 equal to, or less than $10^{-4.5}$ atm. (WOGELIUS and WALTHER, 1989). However, if the partial pressure of CO_2 is equal to the atmospheric one ($10^{-3.5}$ bars), the rate is independent of pH in the range 6–12 and is equal to the minimum dissolution rate for olivine in the low P_{CO_2} waters (WOGELIUS and WALTHER, 1989).

Magmatic gases are known to migrate into overlying groundwater systems, whether thermal or not (ARNÓRSSON, 1986). Such gas emission is most likely linked to relatively short-lived events of magma intrusion. These gases include CO_2, SO_2 and HCl, all of which are acid gases and, therefore, upon dissolution, proton donors. It is known that gas emission of this kind has significantly lowered the pH of geothermal reservoir waters, for example to values as low as 2 in the Krafla geothermal field in northeast Iceland (ARNÓRSSON, 1986). Such pH lowering will enhance dissolution of all the igneous basalt minerals and glass and upset equilibria with alteration minerals.

Hydrothermal and weathering minerals

Quartz and chalcedony

Cold waters are somewhat supersaturated with chalcedony whereas thermal waters when less than about 180°C are very close to equilibrium with this fine crystalline variety of quartz (Fig. 7a). Waters

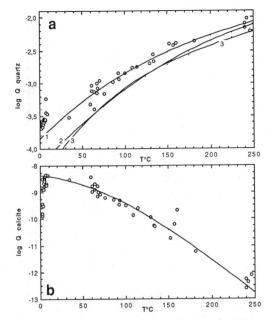

FIG. 7. The state of chalcedony/quartz (a) and calcite (b) saturation in natural waters in Iceland. Curve 1 in part a represents chalcedony solubility according to FOURNIER (1977). Curves 2 and 3 indicate quartz solubility according to FOURNIER and POTTER (1982) and WALTHER and HELGESON (1977), respectively.

at higher temperatures closely approach equilibrium with the more stable phase, quartz (ARNÓRSSON, 1975; ARNÓRSSON et al., 1983a,b) (Fig. 7a). The chalcedony solubility curve used here is that of FOURNIER (1977). It corresponds to the solubility of quartz crystals of 0.04 μm diameter at 0°C to about 0.15 μm diameter at 200°C (GÍSLASON et al., 1989). The supersaturation observed for the cold waters for quartz in this size range could result from formation of even smaller crystals, that would therefore be more soluble. Alternatively "chalcedony crystal size" could still be forming but its precipitation rate would not be sufficient to cope with the silica being dissolved from the rock until the observed degree of supersaturation is attained.

Calcite

Cold waters are generally undersaturated with calcite but all geothermal waters are very close to being saturated (Fig. 7b). As rain water seeps into the ground and reacts with the rock, calcite saturation is approached through dissolution of calcium from the rock and an increase in the carbonate ion concentration with increasing pH. Extensively reacted cold waters (pH greater than about 9) and all slightly thermal waters seem to have closely approached calcite saturation.

Smectite

Montmorillonite group clay minerals (smectites) constitute the most abundant hydrothermal alteration product in low-temperature Icelandic geothermal systems. Practically no data are available on the composition of these minerals and their crystal and thermodynamic properties may be complicated by various types of interlayering with illite, vermiculite and chlorite. Fig. 8 indicates that the cold waters invariably are supersaturated with the calcium, magnesium, sodium and potassium end member components. Supersaturation would be even greater in real montmorillonite mixtures where the activity of the end member components is less than unity. This apparent supersaturation could be due to sluggish precipitation rates causing the water to be maintained in a state of supersaturation. Alternatively the minerals forming at these low temperatures may be of such small crystal size that surface energy contributes significantly to their increased solubility.

Contrary to the cold waters, the geothermal waters are, with few exceptions, undersaturated with the pure end member montmorillonite components. For the calcium and magnesium end members the degree of undersaturation is generally equivalent to 1–1.5 Log Q units. It is slightly greater for the sodium end member, and close to 2 Log Q units for the potassium end member. This apparent but rather constant undersaturation could be due to equilibration of the geothermal waters with montmorillonite having the following end member component activities: $a_{Mg\text{-mont.}} = 10^{-1}$, $a_{Ca\text{-mont.}} = 10^{-1.5}$, $a_{Na\text{-mont.}} = 10^{-1.5}$, $a_{K\text{-mont.}} = 10^{-2}$. Because data on the chemical composition of montmorillonite are lacking the above explanation can only be regarded as tentative.

Low albite and K-feldspar

Cold waters are substantially supersaturated with respect to both low-albite and microcline (the stable K-feldspar at low temperatures). Geothermal waters are close to equilibrium although there appears to be slight supersaturation at the lowest temperatures (below about 75°C) and data at 240–250°C are erratic and inconclusive (Fig. 9a and b).

Albite is an abundant mineral in high-temperature geothermal fields and K-feldspar has also been identified, although it is much less abundant as a result of the low potassium content of the basaltic rock. Igneous plagioclase in hydrothermally altered rock has typically been replaced by albite and calcium bearing minerals such as calcite and epidote. Evidence, thus, favours that albite and K-feldspar

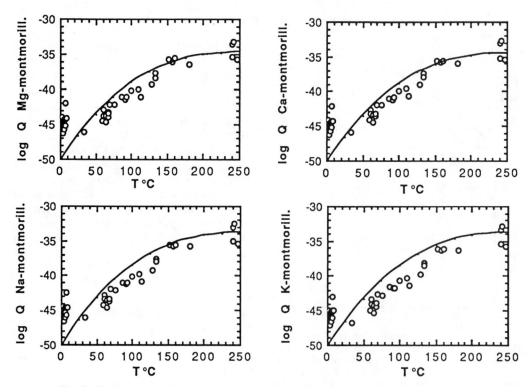

FIG. 8. The saturation state of natural waters in Iceland with respect to pure Mg-, Ca-, Na- and K-smectites.

form as alteration minerals from the Icelandic natural waters, at least, if they are considerably thermal (ARNÓRSSON et al., 1983a).

As for the smectites, the apparent albite and K-feldspar supersaturation in cold waters could be due to sluggish precipitation of these minerals at low temperatures, or it may be an artifact of their small crystal size.

Laumontite

Laumontite is one of many zeolites found in altered rocks of low-temperature geothermal fields (KRISTMANNSDÓTTIR, 1978) and in Tertiary basalts altered by burial metamorphism (WALKER, 1960). Measured temperatures in geothermal wells indicate a lower stability limit for laumontite at about 120°C (KRISTMANNSDÓTTIR, 1978). Other studies indicate that this stability limit may occur at higher temperatures, may be around 200°C (e.g., MIYASHIRO, 1971).

The data plotted in Fig. 9c indicate that geothermal waters at and above about 180°C are significantly laumontite-undersaturated. Waters in the range 100–150°C are close to saturation but as temperatures decrease below 100°C supersaturation progressively increases. Cold water are enormously

laumontite-supersaturated. Even if laumontite formed from 50–70°C waters, where supersaturation is highest among the thermal waters, its precipitation rate would be described by Equation (8) above as the degree of supersaturation is at most 2 Log Q units.

Pyrite, marcasite and pyrrhotite

Cold groundwaters do not contain detectable H_2S and for this reason their state of equilibration with respect to the iron sulphide minerals can not be assessed. Waters from the low-temperature geothermal fields are supersaturated with pyrite but close to marcasite saturation (GUNNLAUGSSON and ARNÓRSSON, 1985) (Fig. 10a). Waters above 200°C, however, seem to be close to equilibrium with pyrite. Pyrite is abundant in hydrothermally altered rocks at depth in high-temperature geothermal fields. Marcasite and pyrite have both been identified in low-temperature geothermal fields (GUNNLAUGSSON and ARNÓRSSON, 1985).

Low-temperature waters are close to equilibrium with pyrrhotite, or are slightly supersaturated especially at the lowest temperatures (Fig. 10b). Results for high-temperature waters are conflicting. The data points in Fig. 10b indicate undersatura-

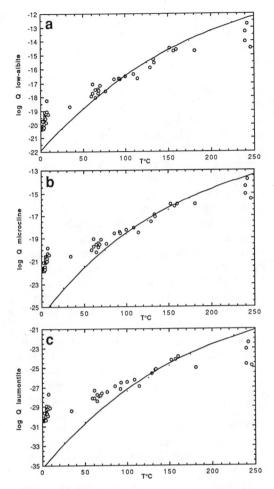

FIG. 9. The saturation state of natural waters in Iceland with respect to pure microcline (the stable K-feldspar at low temperatures), low-albite and laumontite.

tion. However, pyrrhotite is found frequently in altered rocks of high-temperature geothermal systems and data on H_2 and H_2S indicate that the partial pressure of these gases are controlled by equilibrium with a mineral buffer including both pyrite and pyrrhotite (ARNÓRSSON and GUNNLAUGSSON, 1985).

Gibbsite and kaolinite

Cold groundwaters are supersaturated with respect to both gibbsite and kaolinite (Fig. 11) indicating that these minerals, which frequently form weathering products, are stable. The supersaturation may result from sluggish precipitation or be apparent, due to the small crystal size of the minerals.

Geothermal waters above about 50°C are undersaturated with gibbsite, the degree of undersaturation increasing with increasing temperatures,

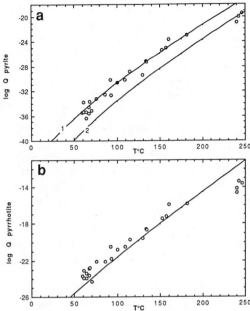

FIG. 10. The state of marcasite (1)/pyrite (2) (a) and pyrrhotite (b) saturation in thermal waters in Iceland.

indicating that gibbsite is unstable in the geothermal waters relative to aluminium silicates.

The geothermal waters are close to saturation with respect to kaolinite or somewhat undersaturated, especially at the highest temperatures. The

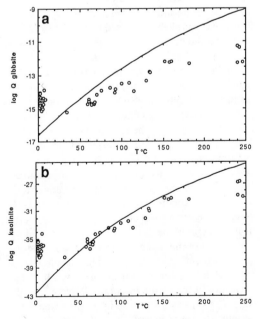

FIG. 11. The state of saturation of pure gibbsite and kaolinite in natural waters in Iceland.

slight supersaturation in waters below about 100°C may not be sufficient to drive kaolinite deposition. This mineral is generally not found at depth in geothermal systems although it has been reported and is then accounted for by acid leaching. The acid waters form most probably by dissolution of magmatic gases in the geothermal water.

SUMMARY

Cold groundwaters are generally highly undersaturated with respect to olivine, and presumably even more so with basaltic glass. The degree of undersaturation increases with increasing forsterite content of the olivine. Waters that have reacted less with the basaltic rock and have, therefore, lower pH than more reacted waters, are more unsaturated.

Cold groundwaters with relatively low pH (<9) are somewhat undersaturated with respect to pyroxene, but higher pH waters are close to equilibrium, even supersaturated. Groundmass plagioclase is close to equilibrium with the cold groundwaters irrespective of pH. Increasing anorthite content leads to progressively greater undersaturation.

Unlike cold waters, geothermal waters of intermediate temperature are close to saturation with groundmass olivine although they are undersaturated with respect to the more magnesian olivine phenocrysts. High-temperature waters are invariably olivine-undersaturated.

Geothermal waters of low-temperature fields tend to be supersaturated with respect to the end members components of both ortho- and clinopyroxenes. The supersaturation and, therefore, stabilization of the pyroxene probably is caused by rapid dissolution of olivine and basaltic glass. High-temperature waters are pyroxene-undersaturated.

Unlike the cold waters, geothermal waters are significantly undersaturated with plagioclase due to the high instability of the anorthite component.

The geothermal waters are generally supersaturated with respect to the magnetite and hematite components in magnetite-ulvöspinel solid solution and ilmenite, respectively. The degree of supersaturation increases with decreasing temperatures. However, high-temperature waters are undersaturated with magnetite, a result which conforms with observed negative magnetic anomalies over many high-temperature fields considered to be caused by destruction of the magnetite in the basaltic rock.

The dissolution kinetics of olivine in cold waters is expected to follow the linear rate law. The same may hold for pyroxene in cold, relatively low pH water (<8–9) and for olivine in thermal waters especially at high temperatures and if the olivine is forsterite rich. Otherwise dissolution rates of the primary basaltic minerals is expected to be described by the general rate law expression (LASAGA, 1981), equation (8), because departure from equilibrium is not so excessive.

Cold waters tend to be strongly supersaturated with common weathering minerals such as smectite, chalcedony, alkali feldspars, gibbsite and kaolinite. This supersaturation may be real and due to sluggish precipitation or be an artifact produced by the inferred small crystal size of the weathering products. These waters are generally calcite-undersaturated unless they have reacted sufficiently with the rock to bring pH to about 9 or more.

Geothermal waters tend to be somewhat undersaturated with both gibbsite and kaolinite but very close to equilibrium with many hydrothermal minerals found in the altered basalts such as calcite, chalcedony (<180°C) or quartz (>180°C), low-albite, microcline, laumontite, pyrrhotite and marcascite (<180°C) or pyrite (>180°C). However, it is noteworthy that waters in the range 50–100°C are somewhat supersaturated with feldspars and laumontite, the degree of supersaturation decreasing with increasing temperatures. The reason for this supersaturation is considered to be the limited precipitation rate at these low temperatures needed to balance the flux of respective chemical components into the water by primary rock mineral dissolution.

With few exceptions equilibrium between water and hydrothermal minerals is so closely approached that the precipitation rate will be described by Equation (9).

The basic chemical reaction involving weathering and hydrothermal alteration of the basaltic rock, and the accompanying changes in the water chemistry, is hydrogen ion metasomatism. The water is "titrated" through the dissolution of the igneous minerals and glass which act as bases, thus causing solution pH to rise (GÍSLASON and EUGSTER, 1987a,b). Proton donors accelerate the "titration" process. Sources of protons for surface waters constitute carbonic acid of atmospheric and or organic origin. For cold and low-temperature groundwaters ionization of silica dissolved from the rock is the main proton donor (GÍSLASON and EUGSTER, 1987a,b). In the case of high-temperature water dissolved carbonic acid and hydrogen sulphide constitute the dominant source of protons for driving the alteration process. The gases forming these acids through dissolution in water may be of magmatic origin or derived from the rock being dissolved.

There are basically two main reasons for the close approach to chemical equilibrium between thermal waters and alteration minerals, even at temperatures as low as 50°C. One is that the saturation state of the igneous minerals (Fig. 2 through 6), in the thermal waters and presumably also the basaltic glass, is not so much different from that of the hydrothermal minerals (Fig. 7 through 11). Thus, dissolution rates will not differ much from precipitation rates when the solution is close to equilibrium with the precipitating phases. The other reason is that the basaltic rock is low in chloride, *i.e.* soluble salts. Therefore, there is low flux of cations into the water from salt dissolution—cations which participate in secondary mineral deposition.

Acknowledgements—The work described in this article has been supported by the Icelandic Science Foundation and the Research Foundation of the University of Iceland. Sigurdur Jakobsson, Kristín Gestsdóttir, Anna María Ágústsdóttir of the Science Institute and Níels Óskarsson of the Nordic Volcanological Institute are sincerely acknowledged for their assistance with the data handling. Review of an earlier version of this paper by H. J. Abercrombie is sincerely appreciated.

REFERENCES

ANNELS A. E. (1967) The Geology of the Hornafjördur Region, S. E. Iceland. Unpubl. Ph.D. Thesis. Univ. London.
ÁRNASON B. (1976) Groundwater systems in Iceland traced by deuterium. *Soc. Sci. Islandica*, Rit 42.
ÁRNASON B. (1977) Hydrothermal systems in Iceland traced by deuterium. *Geothermics* 5, 125–151.
ARNÓRSSON S. (1975) Application of the silica geothermometer in low-temperature hydrothermal areas in Iceland. *Amer. J. Sci.* 275, 763–784.
ARNÓRSSON S. (1983) Chemical equilibria in Icelandic geothermal systems—Implications for chemical geothermometry investigations. *Geothermics* 12, 119–128.
ARNÓRSSON S. (1985) The use of mixing models and chemical geothermometers for estimating underground temperatures in geothermal systems. *J. Volc. Geothermal Res.* 23, 299–335.
ARNÓRSSON S. (1986) Chemistry of gases associated with geothermal activity and volcanism in Iceland: A review. *J. Geophys. Res.* 91, 12,261–12,268.
ARNÓRSSON S. (1987a) Fluid-rock interaction in geothermal systems (abstr.). *Terra Cognita* 7, nos. 2–3, 128.
ARNÓRSSON S. (1987b) On the origin of low-temperature activity in Iceland (abstr.). In *Proc. "Vatnid og landid"*. p. 95–97. National Energy Authority. Reykjavík.
ARNÓRSSON S., SIGURDSSON S. and SVAVARSSON H. (1982) The chemistry of geothermal waters in Iceland. I. Calculation of aqueous speciation from 0° to 370°C. *Geochim. Cosmochim. Acta* 46, 1513–1532.
ARNÓRSSON S., GUNNLAUGSSON E. and SVAVARSSON H. (1983a) The chemistry of geothermal waters in Iceland II. Mineral equilibria and independent variables controlling water compositions. *Geochim. Cosmochim. Acta* 47, 547–566.
ARNÓRSSON S., GUNNLAUGSSON E. and SVAVARSSON H. (1983b) The chemistry of geothermal waters in Iceland. III. Chemical geothermometry in geothermal investigations. *Geochim. Cosmochim. Acta* 47, 567–577.
ARNÓRSSON S. and GUNNLAUGSSON E. (1985) New gas geothermometers for geothermal exploration—Calibration and application. *Geochim. Cosmochim. Acta* 49, 1307–1325.

ARNÓRSSON S. and ÓLAFSSON G. (1986) A model for the Reykholtsdalur and the Upper-Árnessýsla geothermal systems with a discussion on geothermal processes in SW-Iceland. *Jökull* 36, 1–9.
ARNÓRSSON S., GÍSLASON S. R., GESTSDÓTTIR K. and ÓSKARSSON N. (1989) Chlorine and boron in natural waters in Iceland. In *Proc. Sixth Intl. Symp. Water-Rock Interaction.* Malvern 3–12 August, 37–40.
BJÖRNSSON S. (1980) Heat, groundwater and geothermal systems. *Náttúrufraedingurinn* 50, 271–293 (in Icelandic).
BJÖRNSSON S., ARNÓRSSON S. and TÓMASSON J. (1972) Economic evaluation of the Reykjanes thermal brine area, Iceland. *Bull. Amer. Assoc. Petrol. Geol.* 56, 2380–2391.
BÖDVARSSON G. (1961) Physical characteristics of natural heat resources in Iceland. *Jökull* 11, 29–38.
BÖDVARSSON G. (1982) Glaciation and geothermal processes in Iceland. *Jökull* 32, 21–28.
BUSENBERG E. and CLEMENCY C. V. (1976) The dissolution kinetics of feldspars at 25°C and 1 atm. CO_2 partial pressure. *Geochim. Cosmochim. Acta* 40, 41–49.
BUSENBERG E. and PLUMMER L. N. (1989) Thermodynamics of magnesian calcite solid-solutions at 25°C and 1 atm total pressure. *Geochim. Cosmochim. Acta* 53, 1287–1291.
CARMICHAEL I. S. E. (1964) The petrology of Thingmúli, a Tertiary volcano in Eastern Iceland. *J. Petrol.* 5, 435–460.
CARMICHAEL I. S. E. (1967) The mineralogy of Thingmúli, a Tertiary volcano in Eastern Iceland. *Amer. Mineral.* 52, 1815–1841.
CHOU L. and WOLLAST R. (1984) Study of the weathering of albite at room temperature and pressure with a fluidized bed reactor. *Geochim. Cosmochim. Acta* 48, 2205–2218.
COUTURIER Y., MICHARD G. and SARAZIN G. (1984) Constantes de formation de complexes hydroxydés de l'aluminium en solution aqueuse de 20 a 70°C. *Geochim. Cosmochim. Acta* 48, 649–659.
DEUTSCH W. A., JENNE E. A. and KRUPKA K. M. (1982) Solubility equilibria in basaltic aquifers: The Columbia Plateau, eastern Washington, U.S.A. *Chem. Geol.* 36, 15–34.
DOUGLAS G. R. (1987) Manganese-rich rock coatings from Iceland. *Earth Surface Processes and Landforms* 12, 301–310.
EINARSSON T. (1966) The origin of geothermal activity. *J. Eng. Ass. Iceland* 51, 23–32.
ELLIS A. J. (1970) Quantitative interpretation of chemical characteristics of hydrothermal systems. *Geothermics Special Issue 2*, 2, 516–528.
EYTHÓRSSON J. and SIGTRYGGSSON H. (1971) The climate and weather of Iceland. In *The Zoology of Iceland*. Vol. I, Part 3. Einar Munksgaard. Copenhagen and Reykjavík.
EXLEY R. A. (1982) Electron microprobe studies of Iceland research drilling project high-temperature hydrothermal mineral geochemistry. *J. Geophys. Res.* 87, 6547–6558.
FISK M. R. (1978) Melting relations and mineral chemistry of Iceland and Reykjanes ridge basalts. Ph.D. Dissertation, University of Rhode Island.
FOURNIER R. O. (1977) Chemical geothermometers and mixing models for geothermal systems. *Geothermics* 5, 41–50.
FOURNIER R. O. and POTTER R. W. II. (1982) An equation correlating the solubility of quartz in water from 25°C

to 900°C at pressures up to 10,000 bars. *Geochim. Cosmochim. Acta* **46**, 1969–1973.

FRIDLEIFSSON G. Ó. (1983) Mineralogical evolution of a hydrothermal system. *Geothermal Resources Council Trans.* **7**, 147–152.

FRIDLEIFSSON G. Ó. (1984) Mineralogical evolution of a hydrothermal system, II. Heat sources—fluid interactions. *Geothermal Resources Council Trans.* **8**, 119–123.

GIGGENBACH W. F. (1980) Geothermal gas equilibria. *Geochim. Cosmochim. Acta* **44**, 2021–2032.

GIGGENBACH W. F. (1981) Geothermal mineral equilibria. *Geochim. Cosmochim. Acta* **45**, 393–410.

GÍSLASON S. R. and RETTING S. (1986) Meteoric water-basalt interactions in N.E. Iceland: Methods and analytical results. U.S. Geol. Survey Open-File Rept. 87–49.

GÍSLASON S. R. and EUGSTER H. P. (1987a) Meteoric water-basalt interactions. I: A laboratory study. *Geochim. Cosmochim. Acta* **51**, 2827–2840.

GÍSLASON S. R. and EUGSTER H. P. (1987b) Meteoric water-basalt interactions. II. A field study in N.E. Iceland. *Geochim. Cosmochim. Acta* **51**, 2841–2855.

GÍSLASON S. R. and ARNÓRSSON S. (1989) The chemistry of river water in Iceland and the rate of chemical denudation. *Náttúrufraedingurinn* **58**, 183–197 (in Icelandic).

GÍSLASON S. R., VEBLEN D. R. and LIVI K. J. T. (1989) Experimental meteoric water-basalt interactions: TEM characterization and thermodynamic interpretation of alteration products. (in preparation).

GRANDSTAFF D. E. (1980) The dissolution of forsterite olivine from Hawaiian beach sand. In *Proc. Third Intl. Symp. Water-Rock Interaction.* 72–74.

GRANDSTAFF D. E. (1986) The dissolution of forsterite olivine from Hawaiian beach sand. In *Rates of Chemical Weathering of Rocks and Minerals.* (eds. S. M. COLEMAN and D. P. DETHIER). Academic Press, p. 41–59.

GRÖNVOLD K. and MÄKIPÄÄ H. (1978) Chemical composition of Krafla lavas. Nordic Volcanological Institute report 78-16. Reykjavík.

GUNNLAUGSSON E. and ARNÓRSSON S. (1982) The chemistry of iron in geothermal systems in Iceland. *J. Volc. Geothermal Res.* **14**, 281–299.

HELGESON H. C. (1969) Thermodynamics of hydrothermal systems at elevated temperatures and pressures. *Amer. J. Sci.* **267**, 729–804.

HELGESON H. C. and KIRKHAM D. H. (1974) Theoretical prediction of the thermodynamic behavior of aqueous electrolytes at high pressures and temperatures: I. Summary of the thermodynamic/electrostatic properties of the solvent. *Amer. J. Sci.* **274**, 1089–1198.

HELGESON H. C., DELANY J. M., NESBITT H. W. and BIRD D. K. (1978) Summary and critique of the thermodynamic properties of rock forming minerals. *Amer. J. Sci.* **278A**, 1–229.

HEMINGWAY B. S. and ROBIE R. A. (1977) The entropy and Gibbs free energy of formation of the aluminium ion. *Geochim. Cosmochim. Acta* **41**, 1402–1404.

HEMINGWAY B. S., HAAS J. R. JR. and ROBINSON G. R. JR. (1982) Thermodynamic properties of selected minerals in the system Al_2O_3-CaO-SiO_2-H_2O at 298.15°K and 1 bar (10^5 Pascals) pressure and at higher temperatures. *U.S. Geol. Surv. Bull.* 1544.

INGIMARSSON J. and SIGURDSSON F. (1987) Permeability of Icelandic rocks and sediments (abstr.). In *Proc. "Vatnid og landid".* p. 66. National Energy Authority. Reykjavík (in Icelandic).

JAKOBSSON S. P. (1972) Chemistry and distribution pattern of Recent basaltic rocks in Iceland. *Lithos* **5**, 365–386.

JAKOBSSON S. P., JÓNSSON J. and SHIDO F. (1978) Petrology of the western Reykjanes Peninsula, Iceland. *J. Petrol.* **19**, 669–705.

KERRICK D. M. and DARKEN L. S. (1975) Statistical thermodynamic models for ideal oxide and silicate solid solutions, with application to plagioclase. *Geochim. Cosmochim. Acta* **39**, 1431–1442.

KITTRICK J. A. (1966) The free energy of formation of gibbsite and $Al(OH)_4^-$ from solubility measurements. *Soil Sci. Amer. Proc.* **30**, 595–598.

KRISTMANNSDÓTTIR H. (1978) Alteration of basalt by hydrothermal activity at 100–300°C. In *Internat. Clay Conf.* (eds. M. M. MORTLAND and V. C. FARMER). pp. 359–369.

KRISTMANNSDÓTTIR H. (1982) Alteration in the IRDP drill hole compared with other drill holes in Iceland. *J. Geophys. Res.* **87**, 6525–6531.

KUYUNKO N. S., MALININ S. D. and KHODAKOVSKIY I. L. (1983) An experimental study of aluminium ion hydrolysis at 150, 200 and 250°C. *Geochem. Inter.* **19**, 76–86.

LAGACHE M. (1976) New data on the kinetics of dissolution of alkali feldspars at 200°C in CO_2 charged water. *Geochim. Cosmochim. Acta* **40**, 157–161.

LASAGA A. C. (1981) Rate law of chemical reactions. In *Kinetics of Geochemical Processes* (eds. A. LASAGA and R. KIRKPATRICK), Reviews in Mineralogy, Vol. 8, p. 1–68.

MAGNÚSDÓTTIR G. TH. (1989) The chemical composition of thermal waters in the vicinity of Selfoss. Unpubl. BS Thesis, University of Iceland (in Icelandic).

MAYER P. and SIGURDSSON H. (1978) Interstitial acid glass and chlorophaeite in Iceland. *Lithos* **11**, 231–241.

MEHEGAN J. M., ROBINSON P. T. and DELANEY J. R. (1982) Secondary mineralization and hydrothermal alteration in the Reydarfjördur drill core, Eastern Iceland. *J. Geophys. Res.* **87**, 6511–6524.

MILLER W. R. and DREVER J. I. (1977) Chemical weathering and related controls on surface water chemistry in the Absaroka Mountains, Wyoming. *Geochim. Cosmochim. Acta* **41**, 1693–1702.

MIYASHIRO A. (1971) *Metamorphism and Metamorphic Belts.* George Allen & Unwin Ltd., London.

NORDSTROM D. K. and MUNOZ J. L. (1985) *Chemical Thermodynamics.* Benjamin/Cummings. Menlo Park, California.

PACES T. (1983) Rate constants of dissolution derived from the measurements of mass balance in hydrological catchments. *Geochim. Cosmochim. Acta* **47**, 1855–1863.

PÁLMASON G. (1975) Geophysical methods in geothermal exploration. In *Proc. Second United Nations Symposium on the Development and Use of Geothermal Resources.* San Francisco 20–29 May, 1175–1184.

PÁLMASON G. and SAEMUNDSSON K. (1974) Iceland in relation to the Mid-Atlantic Ridge. *Ann. Rev. Earth Planet. Sci.* **2**, 25–63.

PLUMMER L. N. and BUSENBERG E. (1982) The solubilities of calcite, aragonite and vaterite in CO_2-H_2O solutions between 0 and 90°C, and an evaluation of the aqueous model for the system $CaCO_3$-CO_2-H_2O. *Geochim. Cosmochim. Acta* **46**, 1011–1040.

RAGNARSDÓTTIR K. V., WALTHER J. V. and ARNÓRSSON S. (1984) Description and interpretation of the composition of fluid and alteration mineralogy in the geothermal system at Svartsengi, Iceland. *Geochim. Cosmochim. Acta* **48**, 1535–1553.

REYNOLDS R. C. and JOHNSON N. M. (1972) Chemical weathering in the temperate glacial environment of the Northern Cascade Mountains. *Geochim. Cosmochim. Acta* **36**, 537–544.

ROALDSET E. (1983) Tertiary (Miocene-Pliocene) interbasalt sediments, NW- and W-Iceland. *Jökull* **33**, 39–56.

ROBIE R. A., HEMINGWAY B. S. and FISHER J. R. (1979) Thermodynamic properties of minerals and related substances at 298.15K and 1 bar (10^5 Pascals) pressure and at higher temperatures. *U.S. Geol. Survey Bull.* 1452.

SAEMUNDSSON K. (1979) Outline of the geology of Iceland. *Jökull* **29**, 7–28.

SCHOTT J., BERNER R. A. and SJOBERG E. L. (1981) Mechanism of pyroxene and amphibole weathering: I. Experimental studies of iron free minerals. *Geochim. Cosmochim. Acta* **45**, 2123–2135.

SEWARD T. M. S. (1974) Equilibrium and oxidation potential in geothermal waters at Broadlands, New Zealand. *Amer. J. Sci.* **274**, 190–192.

SHOCK E. L. and HELGESON H. C. (1988) Calculation of the thermodynamic transport properties of aqueous species at high pressures and temperatures: Correlation algorithims for ionic species and equation of state predictions to 5 kb and 1000°C. *Geochim. Cosmochim. Acta* **52**, 2009–2036.

SHOCK E. L. and HELGESON H. C. (1989) Corrections to Shock and Helgeson (1988) *Geochim. Cosmochim. Acta* **52**, pp. 2009–2036. *Geochim. Cosmochim. Acta* **53**, 215.

SIGURDSSON H. (1966) Geology of the Setberg area, Snæfellsnes, Western Iceland. *Soc. Sci. Islandica,* Greinar IV, No. 2.

SIGVALDASON G. E. (1963) Epidote and related minerals in two deep drill holes, Reykjavík and Hveragerdi, Iceland. U.S. Geol. Surv. Prof. Paper **450E**, E77–79.

STEINTHÓRSSON S. and SVEINBJÖRNSDÓTTIR Á. E. (1981) Opaque minerals in geothermal well No. 7, Krafla, Northern Iceland. *J. Volc. Geothermal Res.* **10**, 245–261.

STEINTHÓRSSON S., ÓSKARSSON N., ARNÓRSSON S. and GUNNLAUGSSON E. (1986) Metasomatism in Iceland: Hydrothermal alteration and remelting of oceanic crust. In *NATO ASI Chemical Transport in Metasomatic Processes* (ed. H. C. HELGESON). pp. 355–387. D. Reidel.

TANGER J. C. and HELGESON H. C. (1988) Calculation of the thermodynamic and transport properties of aqueous species. *Amer. J. Sci.* **288**, 19–98.

THORSTENSON D. C. and PLUMMER L. N. (1977) Equilibrium criteria for two-component solids reacting with fixed composition in an aqueous phase—example: The magnesian calcites. *Amer. J. Sci.* **277**, 1203–1223.

THY P. (1983) Phase relations in transitional and alkali basaltic glasses from Iceland. *Contrib. Mineral. Petrol.* **82**, 232–251.

TÓMASSON J. and KRISTMANNSDÓTTIR K. (1972) High temperature alteration minerals and thermal brines, Reykjanes, Iceland. *Contrib. Mineral. Petrol.* **36**, 123–137.

VELBEL M. A. (1986) Influence of surface area, surface characteristics, and solution composition on feldspar weathering rates. In *Chemical Processes at Mineral Surfaces* (eds. J. A. DAVIS and K. F. HAYES). Symp. Series 323. pp. 615–634. Amer. Chem. Soc. Washington D.C.

WALKER G. P. L. (1960) Zeolite zones and dike distribution in relation to the structure of the basalts in eastern Iceland. *J. Geol.* **68**, 515–528.

WALKER G. P. L. (1963) The Breiddalur central volcano, eastern Iceland. *Quart. J. Geol. Soc. London* **119**, 29–63.

WALKER G. P. L. (1966) Acid volcanic rocks in Iceland. *Bull. Volcanologique* **29**, 375–406.

WALTHER J. V. and HELGESON H. C. (1977) Calculation of the thermodynamic properties of aqueous silica and the solubility of quartz and its polymorphs at high pressures and temperatures. *Amer. J. Sci.* **277**, 1315–1351.

WALTHER J. V. and WOOD B. J. (1986) Mineral-fluid reactions. In *Fluid-rock Interactions During Metamorphism* (eds. J. V. WALTHER and B. J. WOOD), 194–212. Springer-Verlag, New York.

WHITE A. F. (1983) Surface chemistry and dissolution kinetics of glassy rocks at 25°C. *Geochim. Cosmochim. Acta* **47**, 805–815.

WHITE A., CLAASSEN H. and BENSON L. (1980) The effect of dissolution of volcanic glass on the water chemistry in a tuffaceous aquifer, Rainier Mesa, Nevada. U.S. Geol. Surv. Water-Supply Paper, 153-Q.

WOGELIUS R. A. and WALTHER J. V. (1989) Olivine dissolution: Effects of pH, carbon dioxide and organic acids (abstr.). *Geol. Soc. Amer. Abstr. Prog.* **21**, 101.

WOOD B. J. and FRASER D. G. (1977) *Elementary Thermodynamics For Geologists.* Oxford University Press, Great Britain.

Fluid-Mineral Interactions: A Tribute to H. P. Eugster
© The Geochemical Society, Special Publication No. 2, 1990
Editors: R. J. Spencer and I-Ming Chou

Origin of potash salts and brines in the Qaidam Basin, China

RONALD J. SPENCER

Department of Geology and Geophysics, The University of Calgary, Calgary, Alberta T2N 1N4, Canada

TIM K. LOWENSTEIN and ENRIQUE CASAS

Department of Geological Sciences, S.U.N.Y. at Binghamton, Binghamton, New York 13901, U.S.A.

and

ZHANG PENGXI

The Institute of Salt Lakes, Academia Sinica, Xining, Qinghai, China

Abstract—The Qaidam Basin of western China (120,000 km^2) is a closed, non-marine basin filled with over 10,000 metres of Jurassic and younger sediments. The Qarhan salt plain is one of the most active sites of evaporite deposition in the basin. The major source of water to Qarhan is Na$^+$-HCO$_3^-$ rich river water. Perennial rivers drain the Kunlun Mountains and enter the southern edge of the salt plain. The chemical composition of the river waters is similar to average world river water. Spring waters rich in Ca^{2+}-Cl$^-$, similar to many oil field brines and hydrothermal fluids, flow into the basin along deep seated faults at the northern edge of Qarhan.

Mixtures of different proportions of the river and spring inflow waters produce a variety of brine compositions in the lakes of Qarhan. These lakes precipitate a wide range of evaporite minerals, including mineral sequences typical of modern seawater evaporation as well as sequences similar to those found in many ancient evaporites. Lakes located along the southern margin of Qarhan, with the highest proportion of river inflow, contain Na$^+$-Mg^{2+}-Cl$^-$-SO$_4^{2-}$-rich brines and precipitate a mineral sequence similar to that obtained from the evaporation of modern seawater (calcite, gypsum-anhydrite, halite, polyhalite, epsomite-hexahydrite-kieserite, carnallite, bischofite). Other lakes on Qarhan do not follow the seawater evaporation sequence, because they lack MgSO$_4$ salts (polyhalite, epsomite, hexahydrite and kieserite) but precipitate potash salts (sylvite and/or carnallite). The major potash deposits at Qarhan are MgSO$_4$-free, because of a significant volume of spring inflow. These deposits are similar in mineralogy to many ancient evaporites (including the Devonian Prairie Formation). Lake and groundwater brines located along the northern edge of Qarhan contain the highest proportion of spring water inflow and precipitate salts (carnallite and tachyhydrite) which are important components of Cretaceous evaporites of the Sergipe-Alagoas basin, Brazil, Congo basin, West Africa and Khorat Plateau, Thailand.

INTRODUCTION

THE ORIGIN of ancient evaporites containing the very soluble potash salts (sylvite or carnallite) is controversial. Many of these evaporites do not contain MgSO$_4$-bearing salts such as polyhalite, kainite or kieserite (see HARDIE, 1990, for a compilation). The MgSO$_4$-bearing salts should precipitate before the potassium or magnesium chloride-bearing phases, such as sylvite, carnallite or bischofite, during the evaporation of modern seawater (HARVIE and WEARE, 1980; HARVIE et al. 1980; EUGSTER et al. 1980). These discrepancies between the mineral sequences found in many ancient evaporites and those derived from the evaporation of modern seawater have been recognised by a number of workers. BRAITSCH (1971), for example, classifies many evaporites as "MgSO$_4$ deficient"; VALYASHKO (1972) refers to "abnormal sea deposits"; and WARDLAW (1972) uses the term "unusual marine evaporite." There is a strong tendency to at-

tribute most ancient salt deposits to a seawater parentage and since the major element chemistry of seawater is thought to have remained nearly constant, at least during the Phanerozoic (HOLLAND, 1972; HOLLAND et al., 1986), these discrepancies have led to problems in the interpretation of the potash deposits.

The Qaidam Basin is one of the few areas in the world where significant quantities of potash salts are accumulating. The Qaidam Basin is, therefore, an important natural setting in which to study the origin of potash salts and the origin of the brines which precipitate the salts. It is important to note that this is a large, non-marine basin which contains extensive salt deposits, including potash salts. Although non-marine, the processes which lead to potash salt deposition in the Qaidam Basin have application to ancient potash deposits in both marine and non-marine settings.

The Qaidam Basin is located at high altitude

(~3000 meters) on the northern edge of the Qinghai-Tibet Plateau (Fig. 1a). The 120,000 km² area
is comparable in size to many ancient evaporite
basins. The Qaidam Basin contains economically
important oil and gas fields of continental origin,
and a variety of evaporitic ore minerals, including
several borates, strontianite, celestite, halite, sylvite,
carnallite and bischofite. Several saline lakes and
dry saline pans are located in the basin (Fig. 1b).
The 6,000 km² Qarhan salt plain is one of the most
active sites of evaporite deposition in the basin and
contains the only significant source of potash salt
fertilizer in China.

Two fundamentally different types of inflow waters feed the Qarhan salt plain. The majority of the
water which flows onto the salt plain is Na^+-
HCO_3^--SO_4^{2-}-rich river water. However, much of the
salt brought onto the modern salt plain comes from
Ca^{2+}-Cl^--rich spring waters. The major potash salt
on Qarhan salt plain, carnallite, forms as layered

halite-carnallite along the shores of perennial saline
lakes such as Lake Dabusun. Carnallite also forms
as an early diagenetic void-filling cement in Ho-
locene-Pleistocene layered halite-siliciclastic sequences of Qarhan salt plain. The petrographic textures of the potash minerals are similar to those in
many ancient evaporite deposits (for example the
Permian Salado Formation, New Mexico and the
Devonian Prairie Formation, Saskatchewan).

The salts found on Qarhan salt plain are, relative
to the dominant inflow water (river water), deficient
in $MgSO_4$-bearing salts, as are many ancient evaporite deposits relative to modern seawater. The lack
of $MgSO_4$ minerals in ancient marine evaporites
has been explained in several ways. Removal of sulphate by biologic reduction has been proposed
(BORCHERT and MUIR, 1964; BRAITSCH, 1971;
WARDLAW, 1972; SONNENFELD, 1984), as has syndepositional modification of seawater parent brines
by non-marine inflow. Types of inflow include me-

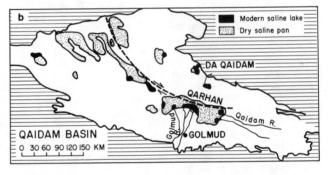
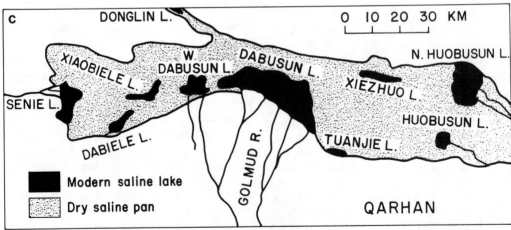

FIG. 1. a) The Qaidam Basin is located along the northern margin of the high altitude Qinghai-
Tibet Plateau. The basin is surrounded by the Kunlun, Altun and Qilian Mountains. b) The Qaidam
Basin has an area of about 120,000 km² and contains numerous saline lakes and dry saline plains at
an average elevation of 2800 m above sea level. Dashed line through the northern margin of Qarhan
and extending northwest across the Qaidam Basin is the trace of a deep seated fault zone along which
springs and karst pools are found. c) Qarhan salt plain contains many modern saline lakes. Perennial
rivers feed several of the lakes. Springs and karst pools are located along the north edge of Qarhan.

Table 1. Inflow waters to the Qarhan salt plain, analyses in molalities

Type	Density	Na^+	K^+	Ca^{2+}	Mg^{2+}	Cl^-	SO_4^{2-}	HCO_3^-
Spring	1.058	1.207	0.007	0.072	0.137	1.580	0.026	—
Spring*	1.056	1.167	0.004	0.081	0.129	1.518	0.036	—
Spring	1.051	1.131	0.004	0.040	0.131	1.464	0.007	—
Spring	1.178	2.786	0.309	0.231	0.944	5.394	0.0005	0.0014
River	1.000	0.00474	0.00015	0.00096	0.00157	0.00384	0.00122	0.00367
River*	1.000	0.00278	0.00009	0.00093	0.00119	0.00227	0.00061	0.00364
River	1.000	0.00832	0.00025	0.00045	0.00272	0.00785	0.00148	0.00410
River	1.000	0.00232	0.00010	0.00107	0.00102	0.00158	0.00072	0.00358

* Analyses used in calculating brine evolution.

teoric (D'ANS, 1933; STEWART, 1963; VALYASHKO, 1972), diagenetic, volcanogenic or hydrothermal waters (HARDIE, 1984), and waters of uncertain origin (WARDLAW, 1972). VALYASHKO (1972) has proposed that metastable or even non-equilibrium conditions prevented deposition of a "normal" marine saline mineral assemblage. Several workers (STEWART, 1963; BORCHERT and MUIR, 1964; WARDLAW, 1968; EVANS, 1970; BRAITSCH, 1971; HOLSER, 1979; HARDIE et al. 1985) suggest that the original primary evaporite mineralogy has been modified during diagenesis. HARDIE (1984) has also proposed that many of these evaporites are not of strictly marine parentage. Recent work by LOWENSTEIN and SPENCER (1990) has shown that the potash salts found in the Oligocene Rhine Graben, Permian Salado and Devonian Prairie evaporites formed from syndepositional brines which deviate from modern seawater composition. These deposits appear to be of mixed marine-non-marine or non-marine parentage.

The Qaidam Basin demonstrates that extensive non-marine potash deposits exist. The evolution of inflow waters to form these salt deposits also serves as a guide in understanding mixed-source evaporites. Below, the chemical characteristics of the inflow waters are discussed first. Theoretical evaporation paths for the two "end-member" inflows (rivers and springs) and for mixtures of these are then discussed. The compositions of brines and the minerals found on Qarhan salt plain are then compared with our theoretical predictions. The varied composition of brines of Qarhan salt plain can be produced by evaporitic concentration of mixtures of the two "end-member" inflows. Finally, possible applications of this type of mixed inflow to understanding the parent waters of ancient evaporites are discussed briefly.

INFLOW WATERS

Meteoric waters

Several perennial rivers flow onto Qarhan salt plain, including the Golmud, Wutumeiren and Qaidam Rivers. The source areas for these rivers (and the Yellow and Yangtse Rivers) are in the Kunlun Mountains, which form the southern border of the Qaidam Basin (Fig. 1). Permanent lakes are present where rivers flow onto the Qarhan salt plain. North Huobusun Lake is fed by the Qaidam River, Senie Lake by the Wutumeiren River and Dabusun Lake by the Golmud River (Fig. 1b and c). There are several ephemeral lakes, including West Dabusun Lake; which receives inflow from the Golmud River as a result of avulsion of one of the many distributory channels of the lower alluvial fan and river delta (Fig. 1c). Many ephemeral lakes are located where dry stream beds intersect the Qarhan salt plain. These lakes appear to be fed by ephemeral streams, or by meteoric groundwater flowing below the stream beds. Xiezhuo Lake, for example, has a crescent shape defined by a deltaic fan located at the terminus of a dry river bed.

The chemical compositions of the river waters which feed the lakes on Qarhan are typical of meteoric waters derived from weathering reactions. Rivers which flow onto Qarhan are similar in composition to that of average world river water (LOWENSTEIN et al., 1989). The total dissolved solid content of average world river water (LIVINGSTON, 1963) is lower than the river waters entering Qarhan (105 ppm versus >400 ppm for Qarhan rivers). The increased salinity results from evaporative concentration in this extremely arid environment and from dissolution of ephemeral salts. Evaporative concentration occurs as the rivers flow toward Qarhan, resulting in precipitation of calcium carbonate. Like average world river water, the rivers which flow onto Qarhan have molal anion abundances of HCO_3^- > Cl^- > SO_4^{2-} (Table 1). However, the relative proportion of HCO_3^- is smaller in these arid climate rivers than in average world river water and it decreases downstream toward the lakes. The molal abundance of the major cations in these rivers (Na^+ > Mg^{2+} > Ca^{2+} > K^+) also differs from average world river water which contains Ca^{2+} as the dominant cation, but otherwise is similar in composition. The lower HCO_3^- and Ca^{2+} content of the river waters entering Qarhan salt plain is probably the result of $CaCO_3$ precipitation.

The river waters flowing onto Qarhan and average world river waters are classified as Na^+-HCO_3^--SO_4^{2-} waters (Fig. 2). This is because, on an equivalence basis, Ca^{2+} is greater than HCO_3^-. Therefore, after the precipitation of the relatively insoluble calcium carbonate minerals, Ca^{2+} is depleted in the resulting brines, and both HCO_3^- and SO_4^{2-} remain. As a consequence, sodium bicarbonate and/or sodium sulfate minerals are part of the normative mineral assemblage for these waters (see BODINE and JONES, 1986 and this volume, for explanation of normative assemblages). These salts are predicted to precipitate during the later stages of evaporative concentration of these waters (see below).

Brine springs

Groundwaters of a markedly different chemical composition reach the surface along the north edge of Qarhan as a series of springs and seeps associated with a karst zone (Fig. 1b). These groundwaters were sampled about five km west of Xiezhuo Lake, where they flow from pipes which were set several meters into the salt and mud, and which project about 50 centimeters above the salt surface. This spring inflow is undersaturated with respect to halite, even though several metres of halite are present in this area. The spring waters contain Cl^- as the dominant anion (>99%) and have molal cation abundances of $Na^+ > Mg^{2+} > Ca^{2+} > K^+$ (Table 1). Brines in the surrounding karst pools fed by these groundwaters are saturated with respect to halite because of dissolution of halite and evaporative concentration.

The location of the karst springs and seeps may be controlled by the contrast between relatively porous sediments along the boundary of Qarhan (alluvial fan and dune deposits) and the less porous salts and muds found beneath the salt plain or by a deep seated fault which runs along the north edge of Qarhan (Fig. 1b). The trace of a deep seated fault along the northern edge of Qarhan coincides with the position of the karst.

The karst springs and seeps are classified as Ca^{2+}-Cl^- type waters because the equivalents of Ca^{2+} in these waters exceed the combined HCO_3^- and SO_4^{2-} equivalents. After evaporation and precipitation of the relatively insoluble calcium carbonates and moderately soluble calcium sulfates, both HCO_3^- and SO_4^{2-} are depleted in the resulting brines.

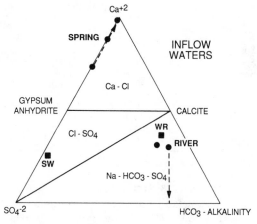

FIG. 2. Ternary Ca^{2+}-SO_4^{2-}-HCO_3^- phase diagram (in equivalents) illustrates the early stage evaporation paths for river and spring inflow and chemical divides for evolved brines. The body of the diagram is the primary crystallization field for calcite. Gypsum and anhydrite crystallize along the Ca^{2+}-SO_4^{2-} join. Two chemical divides, from calcite to SO_4^{2-} and from calcite to gypsum-anhydrite, separate the Na^+-HCO_3^--SO_4^{2-}, Cl^--SO_4^{2-} and Ca^{2+}-Cl^- brine fields. The compositions of average world river waters and seawater are shown for comparison.

However, Ca^{2+} remains, and in the late stages of evaporation, will precipitate as a chloride salt (see below). The spring waters contain Ca^{2+}-Cl^- salts as part of the normative mineral assemblage (BODINE and JONES, this volume). These waters are similar in composition to many oil field brines (GRAF *et al.*, 1966; CARPENTER *et al.*, 1974; COLLINS, 1975; CARPENTER, 1978; SPENCER, 1987) and hydrothermal brines (ROEDDER, 1972; HARDIE, 1983). No analyses of the deep formation waters which underlie Qarhan are yet available. However, deep brines from oil fields in the western portion of the Qaidam Basin are of the Ca^{2+}-Cl^- type. We believe that upward movement of similar deep seated brines beneath Qarhan is responsible for the karst springs and seeps. These waters are probably diluted to different extents by mixing with meteoric groundwaters near the surface.

To summarize, the inflow waters to Qarhan are divided into two types. Meteoric waters, especially rivers, with a Na^+-HCO_3^--SO_4^{2-} composition are the dominant source of water. The location of the lakes distributed over Qarhan is controlled by the positions of rivers which flow onto the salt plain. Groundwater inflow with a Ca^{2+}-Cl^- composition emerges along the north edge of the basin. The amount of water brought in by these groundwaters is much less than the river inflow. However, these waters are much more concentrated (by about 200

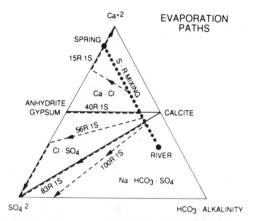

FIG. 3. The compositions of river inflow and spring inflow mixtures lie along the line S-R mixing. Evaporation paths for river waters, spring waters and several mixtures (by weight) of river inflow and spring inflow (including the mixtures for chemical divides in Fig. 2) are shown. Evaporation paths calculated using the thermochemical model of HARVIE et al. (1984).

times) than the river waters. Therefore, significant quantities of salt are delivered by the Ca^{2+}-Cl^- type groundwaters.

EVAPORATION PATHS

HARDIE and EUGSTER (1970) describe the principles of brine evolution in terms of evaporative concentration and mineral precipitation. The evolution of dilute waters into different types of brines is determined early in the evaporative concentration because of chemical divides during the precipitation of calcite and gypsum. This is illustrated in Fig. 3, which is a ternary phase diagram with Ca^{2+}, SO_4^{2-} and HCO_3^- (all in equivalents) at the corners of the diagram. The stability field for calcite occupies the body of the diagram, with gypsum and anhydrite fields along the Ca^{2+}-SO_4^{2-} join. During calcite precipitation the fluid composition moves directly away from the $CaCO_3$ (calcite) compositional point and

GOLMUD RIVER INFLOW	83 RIVER: 1 SPRING	MIXED RIVER AND SPRING INFLOW		40 RIVER: 1 SPRING	SPRING INFLOW, KARST ZONE
C C + H C + H + NaHCO₃ - Na₂CO₃ -Na₂SO₄ - Assemblages		C C + H C + H + A C + H + A + Po C + H + Po C + H + Po + Hx C + H + Po + Ki C + H + Po + Ki + Car C + H + A + Ki + Car C + H + A + Ki + Car + Bi	C C + G C + A C + A + H C + A + H + Car C + A + H + Car + Bi C + A + H + Car + Bi + Ki		C C + G C + A C + A + H C + A + H + Car C + A + H + Car + Bi C + A + H + Car + Bi + Tc
Na - HCO₃-SO₄ Brine		Na - Mg - K - Cl - SO₄ Brine			Ca - Cl Brine
TUANJIE L. SENIE L. W. DABUSUN L. DABIELE L. XIAOBIELE L.			DABUSUN L.		DONGLIN L. XIEZHUO L. HUOBUSUN L. N. HUOBUSUN L. KARST PONDS

FIG. 4. Calculated mineral precipitation sequences are shown from earliest (top) to latest minerals to precipitate (bottom) during evaporation for river inflow, spring inflow and mixtures of river inflow and spring inflow near the two chemical divides in Fig. 3. River inflow and mixtures of greater than 83 parts river inflow to 1 part spring inflow (by weight) evolve to Na^+-HCO_3^--SO_4^{2-} brines. Spring inflow and mixtures of less than 40 parts river inflow to 1 part spring inflow evolve to Ca^{2+}-Cl^- brines. Several lakes on Qarhan salt plain contain Ca^{2+}-Cl^- brines. Mixtures of between 40 and 83 parts river inflow to 1 part spring inflow evolve to Cl^--SO_4^{2-} brines. Mineral precipitation sequences differ for the Cl^--SO_4^{2-} brines (although all of these contain the same invariant mineral assemblage), depending on the particular blend of river inflow and spring inflow. The two sequences shown are for blends near the chemical divides. Minerals are calcite (C), halite (H), anhydrite (A), gypsum (G), polyhalite (Po), hexahydrite (Hx), kieserite (Ki), carnallite (Car), bischofite (Bi) and tachyhydrite (Tc).

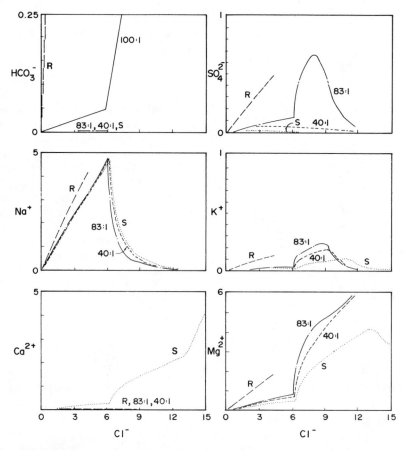

FIG. 5. Calculated evaporation paths are shown for Cl⁻ against the other major elements for river inflow (R, shortened to avoid crowding as none of the Qarhan brines have this type of composition), spring inflow (S) and mixtures of river and spring inflow along the chemical divides in Fig. 3 (83:1 and 40:1). Concentrations of all species are in molalities. The HCO_3^- content of mixtures with less than 83 parts river inflow to 1 part spring inflow is low during evaporative concentration and mineral precipitation. Evaporation of waters with a larger proportion of river waters (for example the 100:1 path) results in high HCO_3^- waters. Evaporation of inflow with less than 40 parts river inflow to 1 part spring inflow results in low SO_4^{2-} brines. Maximum SO_4^{2-} concentrations for the 83:1 path occur at the beginning of hexahydrite precipitation. Maximum Na^+ concentrations occur at halite saturation. The concentration of K^+ is higher in evolved brines with a larger component of river water inflow. Maxima for K^+ occur at the beginning of carnallite precipitation, which is later in the evaporation sequence for brines with a larger spring inflow. The concentration of Ca^{2+} is low for brines evolved from mixtures with greater than 40 parts river inflow to 1 part spring inflow. The Mg^{2+} content is highest for brines evolved from a larger proportion of river inflow. Brines concentrated beyond halite saturation contain Mg^{2+} as the dominant cation, except for Ca^{2+}-Cl⁻ brines beyond the point of bischofite saturation.

as gypsum or anhydrite precipitate, the fluid composition moves away from the $CaSO_4$ compositional point. Therefore, waters are divided into three fields (Ca^{2+}-Cl⁻, Cl⁻-SO_4^{2-} and Na^+-HCO_3^--SO_4^{2-}) by the two chemical divides in Fig. 3 (lines from $CaCO_3$ to $CaSO_4$ and SO_4^{2-}).

Mineral precipitation sequences and brine evolution are more difficult to predict from simple phase diagrams during the later stages of evaporative concentration. Evaporation paths and mineral pre-

cipitation sequences at 25°C for river and spring inflow, and mixtures of the two, are calculated using the thermochemical model of HARVIE *et al.* (1984). A number of solid phases have been withheld from the model including dolomite and other $MgCO_3$ minerals, Mg-oxides and Mg-hydroxides. These phases are predicted to form at low temperatures, but have not been observed. Additional mineral precipitation sequences are calculated at temperatures other than 25°C using the variable temperature thermochemical model of SPENCER *et al.* (1990).

Evaporative trends for river inflow

Calculated evaporation paths for the river inflow yield mineral precipitation sequences as indicated in Fig. 4. All of the river waters are calculated to precipitate calcite, followed by halite. The mineral precipitation sequence beyond the point of halite saturation varies slightly from one river analysis to another. All calculated river evaporation sequences contain sodium sulfate (mirabilite), one or more of the sodium carbonate or sodium bicarbonate minerals and magnesium sulfate minerals. The minerals precipitated vary depending on the partial pressure of CO_2 and as a function of temperature. Because the equivalents of HCO_3^- are greater than the equivalents of Ca^{2+} in these waters, they evolve into Na^+-HCO_3^--SO_4^{2-} brines (Fig. 3), and HCO_3^- increases during evaporative concentration and calcite precipitation (Fig. 5). It is important to note that no sodium carbonate or bicarbonate minerals or sodium sulfate minerals have been reported from Qarhan. Furthermore, the present-day lakes on the salt plain do not contain the concentrations of HCO_3^- expected from evaporative concentration of the river waters.

The model used for the evaporation of the river waters does not take into account the possible precipitation of Mg as a carbonate or a clay mineral. The carbonate minerals reported from Qarhan do not contain high Mg, but Mg-rich clay minerals do occur. The formation of Mg-rich clays can be described schematically by the following reaction:

$$nMg^{2+} + mH_4SiO_4 + 2nHCO_3^- \rightarrow$$

$$Mg_nSi_m \text{ (clay)} + 2nCO_2.$$

This reaction consumes HCO_3^- in the same manner as occurs during the precipitation of a carbonate mineral. The precipitation of Mg-rich clay minerals may therefore lead to a brine depleted in HCO_3^-. Removal of Mg^{2+} from solution to form clays in the lacustrine brines of the Great Salt Lake, Utah, has a pronounced influence on the brine evolution (SPENCER, 1982). Diagenetic formation of Mg-rich clays in the Great Salt Lake system has removed up to three fourths of the long term Mg input to the system (SPENCER *et al.*, 1985b), where the resulting clays are highly enriched in Mg (SPENCER *et al.*, 1984). The removal of Mg in the Great Salt Lake system is not apparent in the short term (SPENCER *et al.*, 1985a). The river waters which flow onto Qarhan may become depleted in HCO_3

through the precipitation of Mg-rich clay minerals. Removal of all HCO_3^- requires about three fourths of the Mg^{2+} entering the system to also be removed. This process leads to the evolution of high Cl^--SO_4^{2-} brines. The resulting paragenetic mineral sequence includes the minerals halite, glauberite, polyhalite, bloedite, epsomite, hexahydrite and kieserite in addition to calcium carbonate and Mg-silicate.

Evaporative trends for spring inflow

Calculated evaporation paths for the most Ca^{2+}-rich of the karst springs and seeps are given in Fig. 4. However, all spring waters modeled yielded the same mineral precipitation sequence. The order of precipitation, calcite-gypsum-halite-carnallite-bischofite-tachyhydrite is not strongly influenced by temperature. However, antarcticite, rather than tachyhydrite, is predicted to form below 18°C. Spring waters evolve to precipitate Ca^{2+}-Cl^- salts (antarcticite or tachyhydrite) in the latest stages of evaporation because the equivalents of Ca^{2+} in the inflow waters exceed the combined equivalents of HCO_3^- and SO_4^{2-} (Fig. 3). Therefore, precipitation of the calcium carbonate and calcium sulfate salts exhausts HCO_3^- and SO_4^{2-} in solution. The remaining brine contains Ca^{2+} and Cl^- is the only anion present in significant quantities (Fig. 5). Several lakes on Qarhan contain Ca^{2+}-Cl^- brines (see below).

Evaporative trends for mixtures of river and spring water

The evaporation paths for mixtures of the two "end member" inflow waters (rivers and springs) have been calculated. Mixtures of about 83 parts river water to one part spring water by weight contain equivalents of Ca^{2+} equal to equivalents of HCO_3^- and are, therefore, along the Cl^--SO_4^{2-} and Na^+-HCO_3^--SO_4^{2-} chemical divide in Fig. 3 (proportions will vary slightly depending on which river and spring analyses are used; the analyses of the Golmud River and karst spring asterisked in Table 1 are used for the calculations here and below). Mixtures with a higher proportion of river water evolve into Na^+-HCO_3^--SO_4^{2-} brines. The phases to form upon evaporation of the Na^+-HCO_3^--SO_4^{2-} waters are the same as for the river waters (Fig. 4). A higher proportion of sodium sulfate and magnesium sulfate salts are precipitated and these salts precipitate earlier in the evaporation sequence as the chemical divide is approached. However, all brines of this type contain sodium carbonate or bicarbonate phases in the final invariant assemblages.

The chemical divide between Ca^{2+}-Cl^- and Cl^--SO_4^{2-} brines is defined by waters in which the equivalents of Ca^{2+} equal the total equivalents of HCO_3^- plus SO_4^{2-} (Fig. 3). Mixtures of about 40 parts river water to one part spring water fall along this divide. Mixtures with a higher proportion of spring water evolve into Ca^{2+}-Cl^- brines. The spring waters are high in Ca^{2+} relative to HCO_3^- and SO_4^{2-}, so that mixtures with a larger proportion of spring water will contain Ca^{2+} in solution after HCO_3^- and SO_4^{2-} are depleted in the brines during evaporation and precipitation of calcium carbonate and calcium sulfate. The sequence of minerals formed during evaporation of all mixtures of inflow resulting in Ca^{2+}-Cl^- brines (greater than 1 part spring inflow to 40 parts river inflow by volume) is the same as for the spring water (Fig. 4).

Mixtures of between 40 and 83 parts river water to one part spring water evolve into Cl^--SO_4^{2-} brines (Fig. 3). All of the Cl^--SO_4^{2-} brines modeled contain the invariant assemblage of calcite, anhydrite, halite, carnallite, bischofite and kieserite (Fig. 4). The sequence of minerals precipitated from these brines varies depending on the specific mixture. The Cl^--SO_4^{2-} brines that originate from a larger proportion of spring inflow produce a mineral sequence similar to that formed by evaporation of spring inflow, except that kieserite instead of tachyhydrite (or antarcticite) is the final phase to crystallize (Fig. 4). Those Ca^{2+}-Cl^- brines that originate from larger proportions of river inflow produce a distinctly different mineral sequence on evaporation. These brines are modeled to precipitate calcite, halite, anhydrite, polyhalite (with anhydrite reacting out of the solid assemblage), hexahydrite, kieserite (replacement of hexahydrite), carnallite, anhydrite (with polyhalite reacting out of the solid assemblage) and bischofite (Fig. 4).

Other mixtures of inflow waters evolve into Cl^--SO_4^{2-} brines with additional mineral sequences (between the two sequences given in Fig. 4). Brines formed from a larger proportion of spring inflow precipitate gypsum and anhydrite at earlier stages of evaporation and the magnesium sulfate salts polyhalite and hexahydrite do not appear in these evaporation sequences. The larger the proportion of spring water inflow, the later the magnesium sulfate salts (polyhalite, hexahydrite and kieserite) precipitate. Mixtures formed from a larger proportion of river inflow result in mineral precipitation sequences which are similar to that of modern seawater. A small range of river and spring inflow mixtures have mineral precipitation sequences identical to the seawater sequence given by HARVIE *et al.* (1980), which is calcite, gypsum replaced by anhydrite, halite, polyhalite, epsomite replaced by hexahydrite and then kieserite, carnallite (with polyhalite reacting out of the solid assemblage) and bischofite.

Calculated evaporation paths of river and spring water, and mixtures of river and spring inflow defining the chemical divides in Fig. 3 are given for the major elements, HCO_3^-, SO_4^{2-}, Na^+, K^+, Ca^{2+} and Mg^{2+} against Cl^- in Fig. 5 (note the differences in scale for the various plots). Trends for all mixtures of river and spring inflow that produce Ca^{2+}-Cl^- brines lie between the spring and 40:1 curves, Cl^--SO_4^{2-} brines lie between the 40:1 and 83:1 curves and Na^+-HCO_3^--SO_4^{2-} brine trends lie between the river and 83:1 curves for all species in Fig. 5. The content of HCO_3^- is low for all Cl^--SO_4^{2-} and Ca^{2+}-Cl^- brines. The HCO_3^- content of brines with even a slight excess of river inflow (for example the 100 river inflow to one spring inflow curve for HCO_3^- in Fig. 3) becomes large during evaporation. None of the modern brines on Qarhan contain high HCO_3^-, so this type of brine is not discussed further.

The SO_4^{2-} concentration is low in all Ca^{2+}-Cl^- brines, but increases sharply for Cl^--SO_4^{2-} brines during the initial precipitation of halite and prior to the precipitation of the magnesium sulfate salts hexahydrite or kieserite. The concentration of Na^+ is larger relative to Cl^-, prior to halite precipitation, for brines with a higher proportion of river inflow. The Na^+ content of the brines decreases sharply during halite precipitation because the equivalents of Na^+ are less than the equivalents of Cl^- for both Ca^{2+}-Cl^- and Cl^--SO_4^{2-} brines. Brines derived from a larger proportion of river inflow are modeled to contain relatively higher concentrations of K^+, except during the later stages of evaporation because the potassium salts polyhalite and carnallite precipitate earlier in these brines. The Ca^{2+} concentration is low for all of the Cl^--SO_4^{2-} brines. High levels of Ca^{2+} are generated during the later stages of evaporation of the Ca^{2+}-Cl^- brines; at the invariant point the Ca^{2+} is greater than 5 molal. The Mg^{2+} molality is modeled to increase sharply after the initial precipitation of halite; Mg^{2+} is the dominant cation in most concentrated Cl^--SO_4^{2-} and Ca^{2+}-Cl^- brines.

LAKE BRINES

The modern lake brines found on Qarhan are discussed in three groups. Lakes and groundwaters located along the northern edge of the salt plain associated with the karst springs and seeps are Ca^{2+}-Cl^- brines. Lakes along the southern edge of the salt plain contain Cl^--SO_4^{2-} brines. The largest lake on Qarhan, Lake Dabusun, contains brines depleted in Ca^{2+}, HCO_3^- and SO_4^{2-}. All of the lake brines analysed in this study, as well as those reported by Zhang (1987) contain less than 0.04 molal

Table 2. Modern lake brines, Qarhan salt plain, analyses in molalities

Lake	Year	Density	Na$^+$	K$^+$	Ca^{2+}	Mg^{2+}	Cl$^-$	SO$_4^{2-}$	HCO$_3^-$
Donglin	1966	1.37	0.0822	0.0006	1.510	4.530	12.02	0.0002	—
Xiezhuo	1980	1.248	0.782	0.222	0.495	2.990	7.989	0.0023	—
N. Huobusun	1980	1.1995	4.082	0.085	0.104	0.981	6.302	0.0182	0.004
Karst pool	1987	1.202	4.588	0.014	0.277	0.457	6.059	0.0063	—
Karst pool	1987	1.200	3.904	0.048	0.318	0.885	6.171	0.0097	—
Karst pool	1987	1.207	3.709	0.041	0.390	1.021	6.251	0.004	—
Karst pool	1987	1.205	3.798	0.042	0.403	1.063	6.448	0.004	—
Karst pool	1988	1.291	0.307	0.232	0.929	3.962	9.601	—	0.0026
Karst pool	1988	1.263	0.634	0.144	0.859	3.175	8.345	0.0001	0.0028
Karst pool	1988	1.317	0.166	0.081	1.388	4.096	10.301	—	0.0039
Senie	1980	1.2144	4.628	0.209	0.0077	0.565	6.006	0.265	0.0054
Dabiele	1980	1.2419	4.062	0.247	0.0288	1.452	5.703	0.749	0.021
Dabiele	1980	1.2402	1.151	0.672	0.0123	2.990	7.667	0.079	0.0051
Xiaobiele	1980	1.2626	0.544	0.553	0.0024	3.834	8.449	0.147	0.032
Tuanjie	1980	1.3023	0.288	0.211	—	4.632	8.801	0.469	0.030
W. Dabusun	1981	1.24	2.536	0.0629	0.0105	2.915	8.262	0.090	0.0124
W. Dabusun	1982	1.07	0.633	0.0330	0.0087	0.104	0.874	0.0008	0.0003
W. Dabusun	1988	1.257	0.872	0.554	0.0097	3.387	7.727	0.115	0.038
W. Dabusun	1988	1.257	2.120	0.365	0.0192	1.910	6.158	0.087	0.018
Dabusun	1958	(1.19)	3.081	0.187	—	1.136	5.540	—	—
Dabusun	1959	1.195	3.626	0.239	—	1.510	6.884	—	—
Dabusun	1959	(1.19)	2.187	0.218	—	1.919	6.243	—	—
Dabusun	1960	1.220	3.698	0.226	—	1.301	6.525	—	—
Dabusun	1960	1.203	3.241	0.290	—	1.937	7.405	—	—
Dabusun	1964	1.220	1.152	0.577	—	2.717	7.164	—	—
Dabusun	1965	1.213	2.904	0.372	—	1.873	7.021	—	—
Dabusun	1966	1.228	1.800	0.444	—	2.465	7.174	—	—
Dabusun	1966	1.246	0.734	0.592	—	3.638	8.602	—	—
Dabusun	1980	1.3186	0.007	0.022	—	5.840	11.53	0.0813	0.031
Dabusun	1980	1.3186	0.041	0.021	0.0025	5.755	11.40	0.0806	0.016
Dabusun	1980	1.2196	2.051	0.549	0.062	2.156	6.979	0.0273	0.004
Dabusun	1987	1.239	1.219	0.524	0.023	3.131	7.700	0.176	—
Dabusun	1987	1.281	0.264	0.157	0.002	4.731	9.492	0.197	—
Dabusun	1988	1.256	0.832	0.471	0.021	3.392	7.592	0.055	0.042
Dabusun	1988	1.255	0.897	0.486	0.021	3.237	7.594	0.051	0.041
Dabusun	1988	1.256	0.883	0.462	0.022	3.257	7.471	0.060	0.041

Analyses from 1987 and 1988 this study. All others from ZHANG (1987).

HCO$_3^-$. All of the brines lie along the evaporation paths for Na$^+$ and Cl$^-$ in Fig. 5.

Ca^{2+}-Cl$^-$ brine lakes

Brines from Donglin, Xiezhuo and North Huobusun Lakes (analyses from ZHANG; 1987) and brines from the karst pools and springs (this study) along the northern edge of Qarhan (Fig. 1c) have a Ca^{2+}-Cl$^-$ composition. They are all low in SO$_4^{2-}$ and high in Ca^{2+} (Tables 1 and 2). The spring waters are dilute relative to the lake and karst pool brines. The less concentrated brines in the karst pools fall along the evaporation path of the spring waters (Fig. 6). All of the Ca^{2+}-Cl$^-$ waters modeled precipitate the same sequence of minerals on evaporation as is precipitated by the spring waters (Fig. 4). How-

ever, simple evaporation and mineral precipitation of the spring waters does not fully account for many of the more concentrated brines.

The more concentrated brines are lower in Ca^{2+} (Fig. 6) and higher in Mg^{2+} (Fig. 7) than is predicted for evaporation of the spring inflow. The deviation from the spring inflow evaporation path is exactly that expected for mixtures of spring and meteoric inflow for both elements. North Huobusun Lake receives inflow from the Qaidam River in addition to the spring inflow. Xiezhuo Lake is located at the mouth of a dry river channel, a likely conduit for meteoric groundwater. Additional meteoric water inflow is also likely along the entire North edge of Qarhan. Still, the major element composition of these brines appears to be strongly influenced by input of spring inflow.

404   R. J. Spencer *et al.*

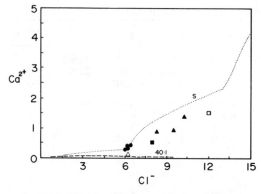

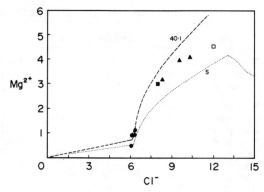

FIG. 6. Calculated evaporation paths for Ca^{2+} against Cl^- (in molalities) for the chemical divides bounding the Ca^{2+}-Cl^- brine field and the concentrations of these ions in the Ca^{2+}-Cl^- lakes and karst pools. Donglin Lake—open square, Xiezhuo Lake—filled square, North Huobusun Lake—open triangle, karst pools north of Dabusun Lake—filled triangles, karst pools near springs (west of Xiezhuo Lake)—filled circles.

FIG. 7. Calculated evaporation paths for Mg^{2+} against Cl^- (in molalities) for the chemical divides bounding the Ca^{2+}-Cl^- brine field and the concentrations of these ions in the Ca^{2+}-Cl^- lakes and karst pools. Donglin Lake—open square, Xiezhuo Lake—filled square, North Huobusun Lake—open triangle, karst pools north of Dabusun Lake—filled triangles, karst pools near springs (west of Xiezhuo Lake)—filled circles.

The K^+ content of North Huobusun and Xiezhuo Lakes, and of some of the more concentrated brines from the karst zone is somewhat higher than expected for evaporation of the spring waters or spring water-river water mixtures (Fig. 8). We interpret this to result from recycling of carnallite. The influence of carnallite dissolution is relatively large for K^+ in solution, though it amounts to less than 0.2 m. Increases of this magnitude in Mg^{2+} or Cl^- (expected for the dissolution of carnallite, $KMgCl_3 \cdot 6H_2O$) are not apparent because of the relatively high background for these ions in solution.

Cl^--SO_4^{2-} brine lakes

Xiaobiele, Dabiele, Senie and Tuanjie Lake brines all contain equivalents of SO_4^{2-} in excess of Ca^{2+}. The HCO_3^- alkalinity for each of these lakes reported by ZHANG (1987) is low as is the Ca^{2+} molality of these waters (Table 2). The SO_4^{2-} content of these lake brines varies, and Cl^- is the major anion in solution (Fig. 9). Based on their chemical compositions, these lakes are grouped together as Cl^--SO_4^{2-} brines. Their brine compositions indicate a larger proportion of river water inflow relative to spring inflow (greater than 40 parts river to 1 part spring inflow) as discussed above under evaporation paths. These lakes are located along the southern edge of Qarhan, near the Kunlun Mountains, and away from the spring inflow (Fig. 1).

The composition of Tuanjie Lake is very close to that predicted for evaporation of waters along the Cl^--SO_4^{2-} and Na^+-HCO_3^--SO_4^{2-} chemical divide (a ratio of about 83 parts river to 1 part spring

inflow). The Tuanjie brines are quite concentrated, and calculated to be at or near saturation with respect to calcite, halite, polyhalite, kieserite and carnallite. The SO_4^{2-}, Mg^{2+} and K^+ concentrations are all along the 83:1 evaporation path as shown in Figs. 9, 10 and 11. The SO_4^{2-} (Fig. 9) and Mg^{2+} (Fig. 10) concentrations in the Senie Lake brines also lie along the 83:1 evaporation path. However, K^+ is somewhat higher than the model value (Fig. 11). Senie Lake brines are less concentrated than Tuanjie Lake brines. They are at or near saturation with respect to only anhydrite and halite. The high K^+

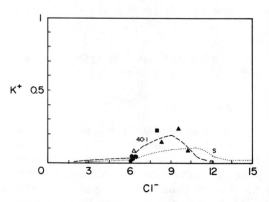

FIG. 8. Calculated evaporation paths for K^+ against Cl^- (in molalities) for the chemical divides bounding the Ca^{2+}-Cl^- brine field and the concentrations of these ions in the Ca^{2+}-Cl^- lakes and karst pools. Donglin Lake—open square, Xiezhuo Lake—filled square, North Huobusun Lake—open triangle, karst pools north of Dabusun Lake—filled triangles, karst pools near springs (west of Xiezhuo Lake)—filled circles.

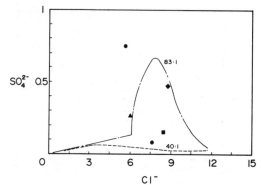

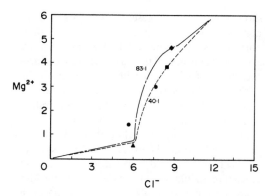

FIG. 9. Calculated evaporation paths for SO_4^{2-} against Cl^- (in molalities) for the chemical divides bounding the Cl^--SO_4^{2-} brine field and the concentrations of these ions in the Cl^--SO_4^{2-} lakes. Tuanjie Lake—diamond, Xiaobiele Lake—square, Senie Lake—triangle, Dabiele Lake—circles.

FIG. 10. Calculated evaporation paths for Mg^{2+} against Cl^- (in molalities) for the chemical divides bounding the Cl^--SO_4^{2-} brine field and the concentrations of these ions in the Cl^--SO_4^{2-} lakes. Tuanjie Lake—diamond, Xiaobiele Lake—square, Senie Lake—triangle, Dabiele Lake—circles.

content for Senie Lake may result from recycling of polyhalite or carnallite, formed when the lake was more concentrated, and later dissolved during dilution of the lake brines.

Brines from Xiaobiele and Dabiele Lakes fall within the Cl^--SO_4^{2-} field (between the 83:1 and 40:1 chemical divides) for SO_4^{2-} (Fig. 9). Both lake brines are close to saturation with respect to calcite, anhydrite, halite and carnallite. The Mg^{2+} concentrations of these lake waters are along the Cl^--SO_4^{2-} and Ca^{2+}-Cl^- chemical divide (40:1 curve on Fig. 10), slightly lower than expected from the SO_4^{2-}. Both lake brines are highly enriched in K^+ relative to the model evaporation paths (Fig. 11). The high K^+ and low Mg^{2+} may result from recycling of carnallite, and the precipitation of $MgSO_4$ salts such as epsomite, hexahydrite or kieserite. If this were the case, a proportionately larger decrease in SO_4 would be expected (SO_4^{2-} is present in lower amounts than Mg^{2+}, and therefore precipitation of $MgSO_4$ salts should be more apparent in the SO_4^{2-}). Another possibility is that these brines may have a higher than estimated proportion of river inflow. Formation of significant quantities of Mg-rich clays could alter the evaporation paths, resulting in higher K^+ and lower Mg^{2+} in the brines.

Lake Dabusun

The brines from Lake Dabusun reported by ZHANG (1987) and in this study are highly concentrated (at or beyond halite saturation) and contain low concentrations of HCO_3^-, SO_4^{2-} and Ca^{2+} (Table 2). Lake Dabusun brines appear to have evolved from source waters along the Ca^{2+}-Cl^- and Cl^--SO_4^{2-} chemical divide, indicating a ratio of about

40 parts river water to one part spring water (Fig. 3). The Golmud River flows into the lake providing Na^+-HCO_3^--SO_4^{2-} waters. Groundwaters, derived from the karst zone springs, also enter the lake along its north edge, providing Ca^{2+}-Cl^- brine input. These two inflow types mix in proportions such that the equivalents of Ca^{2+} are nearly equal to the combined equivalents of HCO_3^- and SO_4^{2-}. Therefore, nearly all of the Ca^{2+}, HCO_3^- and SO_4^{2-} precipitate as $CaCO_3$ and $CaSO_4$ salts at relatively low brine concentrations.

The K^+ content of most Lake Dabusun brines analysed is much higher than that predicted along the Ca^{2+}-Cl^- and Cl^--SO_4^{2-} chemical divide (the 40:1 curve in Fig. 12). This may be the result of recycling of K^+ from carnallite deposits along the lake

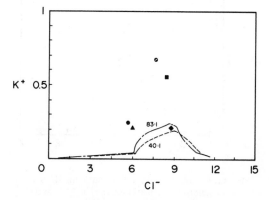

FIG. 11. Calculated evaporation paths for K^+ against Cl^- (in molalities) for the chemical divides bounding the Cl^--SO_4^{2-} brine field and the concentrations of these ions in the Cl^--SO_4^{2-} lakes. Tuanjie Lake—diamond, Xiaobiele Lake—square, Senie Lake—triangle, Dabiele Lake—circles.

shore. Carnallite was present in salt flats along the north shore of the lake in 1987, and along both the north and south shores of the lake in 1988. The extent and thickness of the carnallite beds has varied over the past few decades, with complete dissolution of the carnallite beds occurring during some years (ZHANG, 1987). However, the Mg^{2+} concentrations of Lake Dabusun brines are lower in Mg^{2+} than predicted, especially if recycling of carnallite occurs (Fig. 13). Based on the excess of K^+ in the brines relative to the predicted evaporation path, carnallite dissolution should result in an elevation of the Mg^{2+} in the brines of up to 0.4 m. The majority of the Lake Dabusun brines are near the evaporation curve for the Ca^{2+}-Cl^- and Cl^--SO_4^{2-} chemical divide, but the Mg^{2+} contents are mostly lower than predicted, not higher. The discrepancy may indicate significant removal of Mg^{2+} from solution (possibly 20% or more of the total input) to form Mg-rich clays.

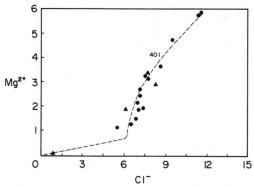

FIG. 13. Calculated evaporation paths for Mg^{2+} against Cl^- (in molalities) for brines along the Ca^{2+}-Cl^- and Cl^--SO_4^{2-} chemical divide and the concentrations of these ions in Lake Dabusun (circles) and West Dabusun Lake (triangles) brines. Brine analyses scatter along the calculated 40:1 evaporation path (as do Na^+, Ca^{2+}, HCO_3^- and SO_4^{2-}, not shown), unlike K^+ in Fig. 12.

SUMMARY OF LAKE BRINE ORIGIN

The Qaidam Basin, located at high elevation on the Qinghai-Tibet Plateau, contains the largest modern accumulations of potash salts found anywhere in the world. These deposits leave no doubt as to the possibility of forming large, non-marine potash deposits. Potash salts develop because of the mixed inflow parent waters. "Normal" meteoric waters flow into the basin from the surrounding mountain ranges. The rivers are the major source of water to the basin, and control the position of the many lakes located on the Qarhan salt plain. The chemical evolution of the river waters is altered because of the inflow of Ca^{2+}-Cl^- spring waters. The

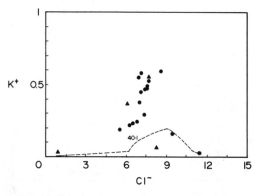

FIG. 12. Calculated evaporation paths for K^+ against Cl^- (in molalities) for brines along the Ca^{2+}-Cl^- and Cl^--SO_4^{2-} chemical divide and the concentrations of these ions in Lake Dabusun (circles) and West Dabusun Lake (triangles) brines. Brines at carnallite saturation (above about 9 m Cl^-) are along the calculated trend, less concentrated brines are generally higher in K^+ than predicted.

amount of water brought in by these springs is small, relative to the river inflow. However, the spring waters are far more concentrated than the river waters and contribute significant quantities of dissolved solutes.

Addition of small amounts of Ca^{2+}-Cl^- spring waters produce evolved brines which are depleted in HCO_3^- relative to evolved river inflow. The HCO_3^- precipitates early in the evaporation sequence as a calcium carbonate, rather than remaining in solution and precipitating as sodium carbonate or sodium bicarbonate during the later stage of evaporation. Mixed inflow brines may evolve to yield mineral paragenetic sequences very similar to the seawater sequence.

Slightly higher proportions of spring inflow produce evolved brines depleted in both HCO_3^- and SO_4^{2-} relative to evolved river waters. Progressive increase in the proportion of Ca^{2+}-Cl^- spring inflow results in the precipitation of HCO_3^- and SO_4^{2-} early in the evaporation sequence as calcium carbonate and calcium sulfate. The $MgSO_4$-bearing salts precipitate at increasingly later stages of evaporation, and finally evolved brines form no $MgSO_4$-bearing salts. These brines produce "$MgSO_4$ deficient" mineral sequences, and may precipitate CaCl-bearing salts upon extreme evaporation.

The modern lake brines and salts found on the Qarhan salt plain can be accounted for by mixing varying proportions of Na^+-HCO_3^--SO_4^{2-} river water and Ca^{2+}-Cl^- spring water. Lakes located along the north edge of Qarhan, in the area of spring inflow, have the highest proportion of spring water input, and contain Ca^{2+}-Cl^- brines. Lakes along the south edge of Qarhan, away from the spring inflow, have the highest proportion of river water input, but are influenced by spring inflow. These lakes contain Cl^--SO_4^{2-} brines. A simple evaporation-mineral precipitation model accounts for most of the observed brine compositions. Recycling of salts and the formation of Mg-rich clays may also influence lake brine compositions.

IMPLICATIONS FOR ANCIENT EVAPORITES

Many ancient evaporites, thought to have a sea-water parentage, are deficient in $MgSO_4$ salts (BORCHERT and MUIR, 1964; BRAITSCH, 1971; VALYASHKO, 1972; HARDIE, 1990). The potash deposits of the Qaidam Basin may be used to develop two possible explanations for these $MgSO_4$ deficient evaporites. First, the deposits may be non-marine, and have formed in a similar manner to the Qaidam Basin evaporites with mixed river water and spring water inflow. Second, these deposits may have had a mixed seawater and spring water parentage, with or without significant meteoric water influence.

Mixed-source evaporites may be difficult, or impossible, to identify in the geologic record. In part, this is because the water and solutes may be predominantly derived from one, or more than one source. For instance, at Qarhan the major source of water, HCO_3^- and SO_4^{2-} is meteoric, whereas spring inflow is the major source of Cl^- and a significant source of the major cations. Therefore, the stable isotopes (H, O, C, S) in these mixed source brines are dominated by the meteoric source. However, the major element composition and mineral precipitation sequence is strongly influenced by the spring water input. Great Salt Lake, Utah, has a similar mixed source inflow, with nearly all of the water from Na^+-HCO_3^--SO_4^{2-} meteoric sources and a significant input of Ca^{2+}-Cl^- spring water (SPENCER et al., 1985b). Great Salt Lake contains Cl^--SO_4^{2-} brines.

Modification of the saline mineral sequence in seawater-sourced evaporite deposits may occur from the input of similar types of spring waters as are present at Qarhan. As is true for non-marine systems, relatively small proportions of highly concentrated Ca^{2+}-Cl^- waters may significantly alter the composition of evolved brines. Higher proportions of Ca^{2+}-Cl^- waters mixed with seawater yield $MgSO_4$-bearing salts later in the evaporation sequence. Evolved brines, deficient in $MgSO_4$-bearing salts, can be generated by addition of a few percent of these waters to seawater. Because only small amounts of Ca^{2+}-Cl^- spring waters are necessary to alter the paragenesis, the water isotopes (H and O)

may retain the seawater signal. Waters rich in Ca^{2+}-Cl^-, in general, contain only small quantities of HCO_3^- and SO_4^{2-}, therefore, isotopes of C and S from mixed source brines may also give a seawater signal. Trace elements may also have a dominant seawater or spring water source in such mixed source brines. The salts found in mixed source evaporites, especially the highly soluble chloride salts, may come dominantly from the spring water source.

Mixtures of small quantities of Ca^{2+}-Cl^- spring waters to seawater can explain the mineralogy of the $MgSO_4$ deficient class of ancient evaporites. Many large ancient evaporites belong to this class. The proportion of spring to seawater in the inflow could determine the saline mineral paragenesis. Evaporites such as those found in the Early Cretaceous rift deposits of Brazil and the Congo Basin (WARDLAW, 1972) or the Late Cretaceous of the Khorat Plateau of Thailand (HITE and JAPAKESETR, 1979), which contain the CaCl-bearing salt tachy-hydrite, appear to evolve from inflow with a relatively large proportion of non-marine, Ca^{2+}-Cl^- waters. The $MgSO_4$ deficient Devonian Prairie Evaporite of Canada, along with many other deposits with a similar mineral assemblage, can be derived by a smaller influx of Ca^{2+}-Cl^- water added to a seawater source. Evidence for circulation of Ca^{2+}-Cl^- hydrothermal waters in western Canada during the Devonian is given by AULSTEAD and SPENCER (1985), SPENCER (1987) and AULSTEAD et al. (1988). Therefore, the Prairie Evaporite appears to have a mixed seawater-hydrothermal water parentage. Deposits such as the Permian Salado Formation of New Mexico, which contain $MgSO_4$ salts, but in smaller amounts than expected from the evaporation of seawater, may also have received inflow from Ca^{2+}-Cl^- waters.

Acknowledgements—We are indebted to Hans Eugster, who kindled our interest in the potash deposits of Qarhan after his visits in the early 1980's and for his assistance in setting up our cooperative research. Comments by Blair Jones and Wayne Nesbitt improved the manuscript. This work is supported by grants from the Petroleum Research Fund, administered by the American Chemical Society (PRF-18652-G2 and 21130-AC2), the National Science Foundation (U.S.A.), the Natural Science and Engineering Research Council (Canada) and Acàdemia Sinica (China).

REFERENCES

AULSTEAD K. L. and SPENCER R. J. (1985) Diagenesis of the Keg River Formation, northwestern Alberta: fluid inclusion evidence. *Bull. Can. Petrol. Geol.* **33**, 167–183.

AULSTEAD K. L., SPENCER R. J. and KROUSE H. R. (1988) Fluid inclusion and isotopic evidence on dolomitization, Devonian of western Canada. *Geochim. Cosmochim. Acta* **52**, 1027–1035.

BODINE M. W. and JONES B. F. (1986) The salt norm: a quantitative chemical-mineralogical characterization of natural waters. U.S. Geol. Surv., Water Res. Invest. Rep. 86-4086.

BODINE M. W. and JONES B. F. (1990) Normative analysis of groundwaters from the Rustler Formation associated with the waste isolation pilot plant (WIPP), southeastern New Mexico. In *Fluid-Mineral Interactions: A Tribute to H. P. Eugster* (eds. R. J. SPENCER and I-MING CHOU), Spec. Publ. No. 2, pp. 213–269. The Geochemical Society.

BORCHERT H. and MUIR R. O. (1964) *Salt Deposits: The Origin, Metamorphism and Deformation of Evaporites.* 300 pp. D. Van Nostrand Co., Princeton, New Jersey.

BRAITSCH O. (1971) *Salt Deposits, Their Origin and Composition.* 279 pp. Springer-Verlag, New York.

CARPENTER A. B. (1978) Origin and chemical evolution of brines in sedimentary basins. Oklahoma Geol. Surv. Circ. 79, 60–77.

CARPENTER A. B., TROUT M. L. and PICKETT E. E. (1974) Preliminary report on the origin and chemical evolution of lead- and zinc-rich oil field brines in central Mississippi. *Econ. Geol.* **69**, 1191–1206.

COLLINS A. G. (1975) *Geochemistry of Oilfield Brines.* 496 pp. Elsevier.

D'ANS J. (1933) *Die Losungsgleichgewichte der Systeme der Salze Ozeanischer Salzblagerungen.* 245 pp. Kali-Forschungsanstalt, Verh. Gesel. Ackerbau, Berlin.

EUGSTER H. P., HARVIE C. E. and WEARE J. H. (1980) Mineral equilibria in a six-component seawater system, Na-K-Mg-Ca-SO_4-Cl-H_2O, at 25°C. *Geochim. Cosmochim. Acta* **44**, 1335–1347.

EVANS R. (1970) Genesis of sylvite and carnallite-bearing rocks from Wallace, Nova Scotia. In *Third Symposium on Salt* (eds. J. L. RAU and L. F. DELLWIG), pp. 239–245. Northern Ohio Geol. Soc., Cleveland, Ohio.

GRAF D. L., MEENTS W. F., FRIEDMAN I. and SHIMP N. F. (1966) The origin of saline formation waters., III: Calcium Cl^- waters. Illinois State Geol. Surv. Circ. 397.

HARDIE L. A. (1983) Origin of $CaCl_2$ brines by basalt-seawater interaction: Insights provided by some simple mass balance calculations. *Contrib. Mineral. Petrol.* **82**, 205–213.

HARDIE L. A. (1984) Evaporites: marine or non-marine? *Amer. J. Sci.* **284**, 193–240.

HARDIE L. A. (1990) Potash evaporites, rifting and the role of hydrothermal brines. *Amer. J. Sci.* **290**, 43–106.

HARDIE L. A. and EUGSTER H. P. (1970) The evolution of closed-basin brines. Mineral. Soc. Amer. Spec. Paper 3, 273–290.

HARDIE L. A., LOWENSTEIN T. K. and SPENCER R. J. (1985) The problem of distinguishing between primary and secondary features in evaporites. In *Sixth International Symposium on Salt* (eds. B. C. SCHREIBER and H. L. HARNER), pp. 11–39. Salt Institute, Washington D.C.

HARVIE C. E., MOLLER N. and WEARE J. H. (1984) The prediction of mineral solubilities in natural waters: the Na-K-Mg-Ca-H-Cl-SO_4-OH-HCO_3-CO_3-H_2O system to high ionic strengths at 25°C. *Geochim. Cosmochim. Acta* **48**, 723–751.

HARVIE C. E. and WEARE J. H. (1980) The prediction of mineral solubilities in natural waters: the Na-K-Mg-Ca-Cl-SO_4-H_2O system from zero to high concentration at 25°C. *Geochim. Cosmochim. Acta* **44**, 981–997.

HARVIE C. E., WEARE J. H., HARDIE L. A. and EUGSTER H. P. (1980) Evaporation of seawater: calculated mineral sequences. *Science* **208**, 498–500.

HITE R. J. and JAPAKESETR T. (1979) Potash deposits of the Khorat Plateau, Thailand and Laos. *Econ. Geol.* **74**, 448–458.

HOLLAND H. D. (1972) The geologic history of sea water—an attempt to solve the problem. *Geochim. Cosmochim. Acta* **36**, 637–651.

HOLLAND H. D., LAZAR B. and McCAFFREY, M. (1986) Evolution of the atmosphere and oceans. *Nature* **320**, 27–33.

HOLSER W. T. (1979) Trace elements and isotopes in evaporites. In *Marine Minerals* (ed. R. G. BURNS), Short Course Notes 6, Ch. 9, pp. 295–346. Mineral. Assoc. Amer.

LOWENSTEIN T. K. and SPENCER R. J. (1990) Syndepositional origin of potash evaporites: Petrographic and fluid inclusion evidence. *Amer. J. Sci.*, **290**, 1–42.

LOWENSTEIN T. K., SPENCER R. J. and ZHANQ PENGXI (1989) Origin of ancient potash evaporites: Clues from the modern nonmarine Qaidam Basin of western China. *Science* **245**, 1090–1092.

ROEDDER E. (1972) The composition of fluid inclusions. U.S. Geol. Surv. Prof. Paper 440-JJ.

SONNENFELD P. (1984) *Brines and Evaporites.* 613 pp. Academic Press Inc., Orlando, Florida.

SPENCER R. J. (1982) Geochemical Evolution of Great Salt Lake, Utah. Ph.D. Dissertation, The Johns Hopkins University, Baltimore, Maryland. 309 pp.

SPENCER R. J. (1987) Origin of Ca-Cl brines in Devonian formations, western Canada sedimentary basin. *Appl. Geochem.* **2**, 373–384.

SPENCER R. J., BAEDECKER M. J., EUGSTER H. P., FORESTER R. M., GOLDHABER M. B., JONES B. F., KELTS K., McKENZIE J., MADSEN D. B., RETTIG S. L., RUBIN M. and BOWSER C. J. (1984) Great Salt Lake, and precursors, Utah: the last 30,000 years. *Contrib. Mineral. Petrol.* **86**, 321–334.

SPENCER R. J., EUGSTER H. P., JONES B. F. and RETTIG S. L. (1985a) Geochemistry of Great Salt Lake, Utah, I: Hydrochemistry since 1850. *Geochim. Cosmochim. Acta* **49**, 727–737.

SPENCER R. J., EUGSTER H. P. and JONES B. F. (1985b) Geochemistry of Great Salt Lake, Utah, II: Pleistocene-Holocene Evolution. *Geochim. Cosmochim. Acta* **49**, 739–747.

SPENCER R. J., MOLLER N. and WEARE J. H. (1990) The prediction of mineral solubilities in natural waters: A chemical equilibrium model for the Na-K-Ca-Mg-Cl-SO_4-H_2O system at temperatures below 25°C. *Geochim. Cosmochim. Acta* **54**, (in press).

STEWART F. H. (1963) Marine evaporites. U.S. Geol. Surv. Prof. Paper 440-Y.

VALYASHKO M. G. (1972) Playa lakes—a necessary stage in the development of a salt-bearing basin. In *Geology of Saline Deposits* (ed. G. RICHTER-BERNBURG), Earth Sciences Series 7, pp. 41–51. UNESCO, Paris.

WARDLAW N. C. (1968) Carnallite-sylvite relationships in the Middle Devonian Prairie Evaporite Formation Saskatchewan. *Geol. Soc. Amer. Bull.* **79**, 1273–1294.

WARDLAW N. C. (1972) Unusual marine evaporites with salts of calcium and magnesium chloride in Cretaceous basins of Sergipe, Brazil. *Econ. Geol.* **67**, 156–168.

Fluid-Mineral Interactions: A Tribute to H. P. Eugster
© The Geochemical Society, Special Publication No. 2, 1990
Editors: R. J. Spencer and I-Ming Chou

Control of seawater composition by mixing of river waters and mid-ocean ridge hydrothermal brines

RONALD J. SPENCER

Department of Geology and Geophysics, The University of Calgary, Calgary, Alberta T2N 1N4, Canada

and

LAWRENCE A. HARDIE

Department of Earth and Planetary Sciences, The Johns Hopkins University, Baltimore, Maryland 21218, U.S.A.

Abstract—The two major input streams to seawater are Na-HCO_3-SO_4-rich river waters and Ca-Cl-rich mid-ocean ridge hydrothermal brines. A model is presented which accounts for the major element composition of modern seawater by mixing of river water with mid-ocean ridge hydrothermal brines and the precipitation of calcium carbonate. A steady-state composition for the major ionic species in seawater, Na^+, Ca^{2+}, Mg^{2+}, Cl^-, SO_4^{2-} and HCO_3^-, can be obtained within the range of published volume fluxes for river water and mid-ocean ridge hydrothermal brines using reasonable estimates of river water and mid-ocean ridge hydrothermal brine compositions. A steady-state concentration of K^+ in modern seawater requires additional sinks for potassium.

The steady-state concentrations of the major ionic species in seawater are very sensitive to changes in the volume ratios of river water and mid-ocean ridge hydrothermal brines input to the oceans. It is likely that the ratio of these two major input streams to seawater has changed through geologic time, and that the major ion concentrations in seawater have changed. A smaller mid-ocean ridge hydrothermal brine flux results in lower concentrations of Ca^{2+} in seawater and higher concentrations of Na^+, Mg^{2+}, SO_4^{2-} and HCO_3^-. Larger mid-ocean ridge hydrothermal brine fluxes result in higher concentrations of Ca^{2+} in seawater and lower concentrations of Na^+, Mg^{2+}, SO_4^{2-} and HCO_3^-. The compositional variations in seawater that result from even modest changes of a few percent in the ratio of river water and mid-ocean ridge hydrothermal brines are significant because of the potential influence on chemical sediments produced from seawater. Smaller mid-ocean ridge hydrothermal brine volume fluxes are likely to favor aragonite precipitation rather than calcite and result in marine evaporites which contain magnesium sulphate minerals. Larger mid-ocean ridge hydrothermal brine volume fluxes favor calcite precipitation and result in marine evaporites containing potash salts (sylvite and carnallite) without magnesium sulphate minerals.

INTRODUCTION: THE PROBLEM OF THE CHEMICAL COMPOSITION OF SEAWATER

SINCE THE CLASSIC PAPER OF RUBEY (1951) on the problem of the history of seawater, a number of workers, using heterogeneous equilibrium models or mass balance methods, have attempted to account for the chemical composition of the present oceans (*e.g.,* SILLEN, 1961, 1967; GARRELS and THOMPSON, 1962; MACKENZIE and GARRELS, 1966; HELGESON and MACKENZIE, 1970; GARRELS and MACKENZIE, 1971; HOLLAND, 1972, 1984; LAFON and MACKENZIE, 1974; MAYNARD, 1976; DREVER, 1982). There is wide agreement among these workers that:

1) the basic salinity and chlorinity, and perhaps also the essential major ion composition of the oceans were, inherited from the primordial reaction between crustal igneous rocks and outgassed acid volatiles such as HCl and SO_2 at the time the oceans were formed (RUBEY, 1951; GARRELS and MACKENZIE, 1971; LAFON and MACKENZIE, 1974),

2) seawater has remained close to its present

composition for the last 1–2 billion years (GARRELS and MACKENZIE, 1971; HOLLAND, 1972). Based on these assumptions, then, the principal question that a number of workers have addressed is how the chemistry of the oceans is buffered against the massive inflow of river waters which have compositions so widely different from that of modern seawater (Table 1).

SILLEN (1961, 1967) first suggested that seawater is in heterogeneous equilibrium with the silicate and carbonate sediments on the ocean bottom. In this approach, "reverse weathering" reactions such as detrital kaolinite forming authigenic illite-smectite were called on to control the ratios of the major cations at or near those of modern seawater (see also MACKENZIE and GARRELS, 1966; HELGESON and MACKENZIE, 1970). Recently, however, it has been argued that authigenic silicates are too sparsely distributed in the ocean, and that reactions involving detrital marine clays are too slow, to buffer seawater composition against the river input (DREVER, 1974, 1982; MAYNARD, 1976; HOLLAND, 1978).

409

This, together with the discovery of extensive hydrothermal alteration of seawater at mid-ocean
ridges (see THOMPSON, 1983, for a summary), has
given rise to steady-state models involving the bal-
ancing of sources and sinks (*e.g.*, MAYNARD, 1976;
DREVER, 1982; HOLLAND, 1978). In these models
the present day river water fluxes of ions are compared to the estimated rates of ion uptake from
ocean waters by:

1) deposition of marine carbonates, cherts and
evaporites,
2) storage of seawater in the pores of marine
sediments,
3) low temperature interaction between seawater
and ocean floor basalts,
4) high temperature interaction between seawater and basalt at mid-ocean ridges,
5) ion-exchange by marine clays,
6) precipitation of authigenic silicates and sulfides.

If the chemical composition of the oceans is to remain constant, then the input fluxes must remain
in dynamic balance with the output fluxes throughout geologic time, that is, for each ion:

$$dA_i/dt = \sum f_{ir} \text{ (inputs)} - \sum f_{ir} \text{ (outputs)} = 0$$

where A_i is the concentration of ion i, t is time, and
f_{ir} is the flux of ion i from reservoir r.

It is the view of HOLLAND (1978, p. 5) that "earth
reservoirs are not at steady state" and that excursions in seawater composition through time are
likely (see also GARRELS and MACKENZIE, 1971, p.

297) given that global tectonic cycles operate over
time spans that are longer than the residence times
of all the major ions in the oceans. In addition,
climatic changes such as occurred during the major
ice ages would have been capable of influencing the
fluxes of those components with relatively short
residence times. These components include bicar-
bonate-carbonate, silica, water and perhaps calcium.
HOLLAND (1972, 1984) has attempted to determine
the limits of these chemical excursions by inverse
modelling using the mineralogy of marine evaporites deposited over the last 900 million years. He
has concluded that while "the concentration of the
major ions in seawater has not varied a great deal
during the Phanerozoic . . . the imposed limits on
excursions . . . are . . . fairly broad" (HOLLAND,
1984, p. 536). In particular, concentrations of Mg^{2+},
Ca^{2+} and SO_4^{2-} could have been two or three times
larger or smaller than the present day ocean values.
Such changes, coupled with variations in biologi-
cally-controlled or biologically-moderated components (oxygen, carbon dioxide, silica, phosphate,
nitrate, trace elements, stable isotopes, etc.), may
have considerable effects on the nature and composition of chemical and biochemical sedimentary
deposits through geologic time.

With all this in mind we have explored the consequences of a simple model that assumes that the
composition of seawater is mainly controlled now,
and has been controlled through geologic time, by
the mixing of river waters and mid-ocean ridge
(M.O.R.) hydrothermal brines. Using this model
we have attempted to predict the direction and pos-

Table 1. Major ion composition of seawater, mid-ocean ridge hydrothermal brines and average world river water
in ppm

	1	2	3	4	5	6	7	8	9
Na^+	10800	9610	10444	11725	9932	11243	11794	6.3	5.15
K^+	407	1348	1382	1009	907	1896	1450	2.3	1.3
Ca^{2+}	413	1530	1812	805	625	1162	1663	15.	13.4
Mg^{2+}	1296	16	8	0	0	0	0	4.1	3.35
Cl^-	19010	19260	20745	20530	17338	21310	22586	7.8	5.75
SO_4^{2-}	2717	31	72	0	0	0	0	11.2	8.25
HCO_3^-	137	1926*	2650*	0	0	647	396	58.4	52.

1. Seawater, RILEY and CHESTER (1971).
2. Hydrothermal brine, Borehole #8, Reykjanes, Iceland, BJORNSSON *et al.* (1972).
3. Hydrothermal brine, Borehole #2, Reykjanes, Iceland, BJORNSSON *et al.* (1972).
4. Hydrothermal brine NGS, 21° North, VON DAMM *et al.* (1985a).
5. Hydrothermal brine OBS, 21° North, VON DAMM *et al.* (1985a).
6. Hydrothermal brine #1, Guaymas Basin, VON DAMM *et al.* (1985b).
7. Hydrothermal brine #3, Guaymas Basin, VON DAMM *et al.* (1985b).
8. Average world river water, LIVINGSTON (1963).
9. Natural average world river water, MEYBECK (1979).
* Reported as total CO_2.

sible magnitude of the changes in the major ion composition of seawater that would result from likely secular changes in the relative magnitudes of the fluxes of river water and M.O.R. hydrothermal brines. In turn, we have used the predicted seawater composition variations to determine the nature of variations likely to have occurred in the primary mineralogy of ancient marine carbonates and marine evaporites.

SEAWATER AS A MIXTURE OF RIVER WATER AND M.O.R. HYDROTHERMAL BRINES: A WORKING MODEL

We present in this paper a very simple model that explores how the chemical composition of seawater is influenced by the two major input streams of fluid to the oceans, that is, river waters and M.O.R. hydrothermal brines. To explain the evolution of brine compositions in non-marine saline lakes HARDIE and EUGSTER (1970) developed the idea of "chemical divides" in which early precipitation of the relatively insoluble minerals calcite and gypsum determined the ultimate chemical signature of the brines. The success of this simple concept in accounting for the composition of lacustrine brines, where mixing of several source waters is the norm, has encouraged us to apply it to the evolution of seawater.

The concept of chemical divides can be illustrated on a simple Ca-HCO$_3$-SO$_4$ ternary phase diagram (Fig. 1). Two chemical divides (one from CaCO$_3$ to SO$_4$ and one from CaCO$_3$ to CaSO$_4$) separate three fields (Na-HCO$_3$-SO$_4$, Cl-SO$_4$ and Ca-Cl waters) on the phase diagram. The body of the phase diagram is the primary stability field for calcite; the stability fields for gypsum and anhydrite are located along the Ca-SO$_4$ join. Waters within the body of the phase diagram precipitate CaCO$_3$ solid phases and the solution compositions evolve directly away from the CaCO$_3$ composition point. In this scheme modern seawater is a Cl-SO$_4$ type water (Fig. 1). The mole-equivalent concentration of Ca^{2+} in seawater is greater than the equivalents of HCO$_3^-$, but is less than the combined mole-equivalent concentration of HCO$_3^-$ and SO$_4^{2-}$. Therefore, after the precipitation of calcite and gypsum, evolved seawater brines become enriched in both Cl$^-$ and SO$_4^{2-}$, and depleted in Ca^{2+}.

Evolved continental Cl-SO$_4$ brines, similar in composition to modern seawater, that have resulted from mixing of Na-HCO$_3$-SO$_4$ river waters and Ca-Cl spring waters, have been documented for the Great Salt Lake, Utah (SPENCER et al., 1984, 1985) and several lakes on the Qarhan salt plain, Qaidam Basin, China (LOWENSTEIN et al., 1989; SPENCER

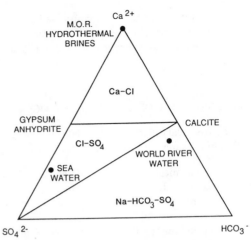

FIG. 1. Ternary Ca^{2+}-SO$_4^{2-}$-HCO$_3^-$ phase diagram, in mole-equivalents. The body of the diagram is the primary stability field for calcite; gypsum and anhydrite stability fields are along the Ca^{2+}-SO$_4^{2-}$ join. Lines from calcite to gypsum/anhydrite and from calcite to SO$_4^{2-}$ are chemical divides for the system which separate Ca-Cl, Cl-SO$_4$ and Na-HCO$_3$-SO$_4$ type waters. Average world river water has a Na-HCO$_3$-SO$_4$ composition, modern seawater is a Cl-SO$_4$ water and M.O.R. hydrothermal brines have a Ca-Cl composition.

et al., 1990). The compositions of these non-marine brines can be explained by our simple model. What is particularly relevant here is that the basic chemical signatures of the two major inflow waters to these non-marine basins are similar to the two major inflow waters to the oceans. In this sense the Great Salt Lake and Qaidam Basin mixed inflow systems are valuable analogues for the modern ocean system, and provide ample support for the basic concept behind our simple model.

In the chemical divide scheme, biological and chemical precipitation of calcium carbonate phases is the most important process in the early stages of brine evolution. Since seawater is a relatively dilute brine that falls into the calcium carbonate precipitation regime, we have restricted our modelling of seawater to simple mixing of Na-HCO$_3$-SO$_4$ river water and Ca-Cl M.O.R. hydrothermal brine combined with precipitation of calcium carbonate. We have not included more complex mechanisms such as cation exchange by river-borne clays, authigenic silicate mineral precipitation, or sulphate reduction by microbes, although their exclusion undoubtedly introduces some error (see estimations by MAYNARD, 1976, of the contributions of several processes to the chemical mass balance in the present day oceans). However, our experience in non-marine systems indicates that these reactions do not play more than minor roles in the chemical evolution of brines. We have not considered passive sinks such as evaporite precipitation and connate seawater storage that do not change the composition of seawater.

THE MIXING MODEL AND THE COMPOSITION OF THE MODERN OCEAN

The basic data needed to calculate the composition of seawater with this simple model are:

1) the chemical composition of average river water and M.O.R. hydrothermal brine,
2) the annual inflow volumes of river water and M.O.R. hydrothermal brine,
3) information on the anthropogenic input of solutes to the oceans.

River water composition

The average composition of modern rivers flowing into the oceans has been estimated by LIV-INGSTON (1963) and MEYBECK (1979) (Table 1). This average world river water has a Na-HCO$_3$-SO$_4$ composition, as illustrated in Fig. 1. The concentration, in mole-equivalents, of HCO$_3^-$ is greater than that of Ca^{2+} in this average world river water and therefore, on evaporation and precipitation of alkaline earth carbonates the resulting brines would be enriched in HCO$_3^-$ and SO$_4^{2-}$ and depleted in Ca^{2+}. With continued evaporative concentration these river water-derived brines will eventually precipitate NaHCO$_3$ and NaSO$_4$ salts and become alkaline brines (HARDIE and EUGSTER, 1970). Significant uncertainty in the river inflow composition is introduced by the uncertainties in the amounts of anthropogenic sources of solutes in modern river waters. MEYBECK (1979) estimates the anthropogenic input of sulphate to be 25% of the total river sulphate while HOLLAND (1978) suggests a higher value around 40%. For bicarbonate or carbon in modern river waters MEYBECK (1979, 1982) estimates 2 to 9% of the total is anthropogenic.

M.O.R. hydrothermal brine composition

For the M.O.R. hydrothermal brine input we have used the chemical analyses of BJORNSSON *et al.* (1972), EDMOND *et al.* (1979) and VON DAMM *et al.* (1985a,b) (Table 1). M.O.R. hydrothermal brines belong to the Ca-Cl composition type (Fig. 1). In these brines the mole-equivalent concentration of Ca^{2+} exceeds the combined mole-equivalent concentrations of HCO$_3^-$ and SO$_4^{2-}$ and therefore, on evaporation and precipitation of alkaline earth

carbonates and calcium sulphate minerals, the residual brines will become depleted in HCO$_3^-$ and SO$_4^{2-}$ and enriched in Ca^{2+} and Cl$^-$.

Volume estimates

Volume estimates of river water inflow into the oceans today range from 3.0×10^{16} to 5.0×10^{16} litres per year (l/yr) (see HOLLAND, 1978, p. 65). LIVINGSTON (1963), for example, estimates 3.25×10^{16} l/yr while MEYBECK (1979) gives 3.74×10^{16} l/yr. The equation of PROBST and TARDY (1989, p. 276) yields runoff values from 3.90×10^{16} to 4.02×10^{16} l/yr. PROBST and TARDY (1989) suggest that anthropogenic sources have led to an increase in discharge with time. For the model calculations we have selected a value of 3.75×10^{16} l/yr for the total flux of river waters to the oceans. As a guide to the flux of M.O.R. hydrothermal brine we have used the estimates based on heat flow measurements (see, for example, HOLLAND, 1978). These values range from 6.0×10^{13} to $9. \times 10^{14}$ l/yr.

A constant composition modern ocean model

A constant composition modern ocean model requires that the net flux of each species be zero. In our model the net flux for each of the major ionic species in seawater is examined for various combinations of river water and M.O.R. hydrothermal brine volumes within the limits given above. We have made minor alterations in the composition of river inflow in order to simplify the model.

The starting point for our model calculation is the average world river water composition, for which we have used the estimate of MEYBECK (1979) (Table 2, line 1). The composition of this average river water is adjusted to zero chloride by subtracting ions in proportion to the composition of modern seawater (Table 2, line 2) on the assumption that the chloride in river waters is recycled from seawater as salt aerosols dissolved in rainwater. The chloride-adjusted composition of river water is used along with the estimate for the volume of inflow to arrive at an initial value of the total flux of solutes to the ocean from rivers.

The analysis of M.O.R. hydrothermal brine from borehole #8, Reykjanes, Iceland, given by BJORNS-SON *et al.* (1972) (Table 1) is used to represent the composition of M.O.R. hydrothermal brines. The bicarbonate alkalinity is zero, by charge balance (Table 2, line 5). Seawater is assumed to be the original source water which interacted with oceanic basalts at greenschist facies temperatures to produce the hydrothermal brine (for details see THOMPSON,

Table 2. Composition of seawater inflows (meq/l)

	Na$^+$	K$^+$	Ca^{2+}	Mg^{2+}	Cl$^-$	SO$_4^{2-}$	HCO$_3^-$
				River			
1	0.224	0.036	0.669	0.275	0.162	0.172	0.852
2	0.085	0.033	0.663	0.244	0.0	0.155	0.852
3	0.083	0.032	0.644	0.237	0.0	0.126	0.852
4	0.089	−0.062	0.713	0.237	0.0	0.126	0.852
				M.O.R.			
5	418.0	37.3	76.3	1.3	532.3	0.6	0.0
6	430.2	38.4	78.5	1.3	547.9	0.7	0.0
7	469.7	10.3	20.6	106.6	547.9	56.6	2.2
8	−39.4	28.1	57.9	−105.3	0.0	−55.9	−2.2

1—natural river input (MEYBECK, 1978).
2—adjust for Cl$^-$ = 0.
3—adjust for mass balance on SO$_4^{2-}$.
4—adjust for mass balance with constant M.O.R. hydrothermal flux.
5—M.O.R. hydrothermal brine.
6—M.O.R. hydrothermal brine at seawater Cl$^-$.
7—seawater.
8—net M.O.R. flux.

1983; HARDIE, 1983). Thus, all solutes are adjusted proportionately to modern seawater chlorinity (Table 2, line 6) and the difference between the adjusted composition and the average composition of modern seawater (from RILEY and CHESTER, 1971) as given in Table 2 (line 7) is used to estimate the net addition or subtraction of each solute from the M.O.R. hydrothermal brine system (Table 2, line 8).

The net flux of Cl$^-$ is zero because we adjusted both of the inputs to be zero. The dissolved solute fluxes of all other river water components are positive. The fluxes of K$^+$ and Ca^{2+} from the M.O.R. hydrothermal brine source are positive, while those of Na$^+$, Mg^{2+}, SO$_4^{2-}$ and HCO$_3^-$ are negative. In order to maintain a constant composition ocean in our simple model, the net flux of Na$^+$, K$^+$, Mg^{2+} and SO$_4^{2-}$ must be zero and the net flux of Ca^{2+} must equal that of HCO$_3^-$.

One test of whether this simple model can account for the composition of modern seawater is to calculate the residual net flux of each solute after mixing the adjusted compositions of river water and M.O.R. hydrothermal brine in proportion to their estimated volume inflow fluxes. The mixing of river water with M.O.R. hydrothermal brine leads to supersaturation with respect to calcite (and aragonite) and thus in the model calculations, appropriate amounts of Ca^{2+} and HCO$_3^-$ must be subtracted. If after this the net solute fluxes are zero, or within the uncertainty of the input data, then the model is a viable one. Unfortunately, the range of estimates

of the volume flux of M.O.R. hydrothermal brine is large enough that an unequivocal answer does not result from this approach. The range of mixtures calculated from the volume estimates are shown on the Ca-HCO$_3$-SO$_4$ ternary phase diagram (Fig. 2).

A test of the model that avoids the uncertainties

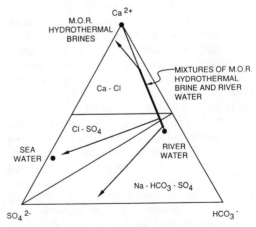

FIG. 2. Ternary phase diagram (see Fig. 1) showing the mixing line for M.O.R. hydrothermal brines and average world river water, heavy portion of the line is the range of mixtures obtained using published values of river water and M.O.R. hydrothermal volume fluxes to the oceans. Mixtures of M.O.R. hydrothermal brines and river waters are supersaturated with respect to calcite. Arrows indicate direction of brine evolution during calcium carbonate precipitation and range of compositions obtained from mixtures.

in the present day M.O.R. hydrothermal brine volume flux is to make this flux a dependent variable. In this approach, the volume flux of M.O.R. hydrothermal brine required to achieve a net zero flux for each major ion species is calculated. The calculated fluxes are compared with the published estimates of M.O.R. hydrothermal brine and with each other. This comparative approach will reveal whether or not our model is able to account for the composition of modern seawater within the range of uncertainty of the estimates of the volume flux for M.O.R. hydrothermal brine and whether or not a net zero flux for each ionic species can be obtained from the same M.O.R. hydrothermal brine volume flux.

With the initial estimates given in Table 2 (lines 2 and 8) the volume fluxes for M.O.R. hydrothermal brines required for net zero fluxes of the major ion species are given in Table 3. Zero net fluxes for Na^+, Mg^{2+} and SO_4^{2-} are obtained for very similar M.O.R. hydrothermal brine volume fluxes, indicating the ability of the mixing model to explain the proportions of these species in modern seawater. An improvement in the balance of these solutes can be obtained if more sulphate is attributed to anthropogenic sources, closer to HOLLAND'S (1978) estimate of close to 40% of the total river borne sulphate (as per Table 2, line 3). Larger M.O.R. hydrothermal brine inflow requires additional sources for these species, while a lesser inflow requires additional sinks.

The Ca^{2+} and K^+ fluxes for both the river and M.O.R. sources are positive. Excess Ca^{2+} is removed by "precipitating" an amount equal to the total mole-equivalent concentration of HCO_3^- to simulate biogenic uptake by organisms and chemical precipitation of $CaCO_3$. The M.O.R. hydrothermal brine volume flux required to balance the river-borne HCO_3^- is higher than the fluxes required for Na^{2+}, Mg^{2+} and SO_4^{2-} (Table 3), although an increase in river-borne Ca^{2+}, or decrease in HCO_3^- of less than 10% will yield volume fluxes for M.O.R. hydrothermal brines identical to those for Mg^{2+} and SO_4^{2-}. There is a net positive flux of K^+ that cannot be accounted for without introducing additional sinks such as the uptake of K^+ by terrestrial clays brought down to the sea by rivers or perhaps "reverse weathering" of oceanic basalts.

Overall, apart from the K^+ problem (see also MAYNARD, 1976), our model seems to be able to account for the present composition of seawater within the limits of the input data available in the literature. A river water composition with K^+ adjusted to a negative value to simulate exchange of Ca^{2+} on river clays for seawater K^+ as the clays

Table 3. M.O.R. hydrothermal brine volume fluxes required for zero net flux of species using river volume of 3.75×10^{16} l/yr and river water compositions from lines 2 and 3, Table 2, volumes in 10^{13} l/yr

	Line 2	Line 3
Na^+	8.09	7.90
Mg^{2+}	8.69	8.44
SO_4^{2-}	10.40	8.44
Ca^{2+}-HCO_3^-	11.60	12.98

enter the ocean which yields a net zero flux for all species using a M.O.R. hydrothermal brine flux of 8.45×10^{13} l/yr is given in Table 2, line 4. The model described allows for a constant composition ocean based on mixing of Na-HCO_3-SO_4 river waters and Ca-Cl M.O.R. hydrothermal brines, coupled with the precipitation of calcium carbonate and exchange of Ca^{2+} on river-borne clays for seawater K^+. Several reactions, such as sulphate reduction or Mg^{2+} uptake on clays, that other workers have suggested might have an influence on seawater composition have not been taken into account in our model. The acceptable balance achieved by our model suggests that these reactions do not play a major role in determining the composition of seawater, or that they have the same net effect as the processes underlying our model.

VARIATIONS IN SEAWATER COMPOSITION IN RESPONSE TO SECULAR VARIATIONS IN RIVER WATER AND M.O.R. HYDROTHERMAL BRINE FLUXES

Variations in the fluxes of river water and M.O.R. hydrothermal brine are almost certain to have occurred back through geologic time as a function of variations in the rates of sea floor spreading, mountain building, continental accretion, global climate and so on. Therefore, if simple mixing of river waters and M.O.R. hydrothermal brines controls the composition of seawater, then it follows that any changes in the river water/M.O.R. hydrothermal brine flux ratios should result in changes in seawater major ion chemistry. If these changes in seawater composition are large enough they could affect the mineralogy of marine carbonates and marine evaporites. Such changes might explain, for example, the problem of "aragonite seas" vs. "calcite seas" (SANDBERG, 1983) or the restriction of $MgSO_4$-rich evaporites to the Permian and Neogene (HARDIE, 1990).

Our simple model is used to test the sensitivity of seawater composition to variations in the ratio of river water and M.O.R. hydrothermal brine fluxes. We use the composition of river water from line 4, Table 2, and hold the volume flux of river waters constant at 3.75×10^{16} l/yr. Steady-state seawater compositions are calculated assuming that the removal of Na^+, Mg^{2+} and SO_4^{2-} are proportional to the M.O.R. hydrothermal volume flux, and that the addition of K^+ and Ca^{2+} is controlled by charge balance. For instance, Mg^{2+} is removed nearly quantitatively from seawater at the mid-ocean ridge (compare lines 6 and 7, Table 2), so that if the M.O.R. hydrothermal flux were to double, then to balance the river input a seawater Mg^{2+} concentration one-half of the modern concentration is required. Our calculations indicate that steady-state concentration values for Mg^{2+} and SO_4^{2-} are approached rapidly (on the order of a few million to a few tens of millions of years), but that much longer times (several tens of millions of years) are required to reach steady-state values for Na^+ (because only a small proportion of Na^+ is removed from a given volume of seawater at the mid-ocean ridge). The sum of the change in K^+ and Ca^{2+} concentrations are calculated in order to balance the charge removed or gained in obtaining the new Na^+, Mg^{2+} and SO_4^{2-} steady-state concentrations. Distributions of K^+ and Ca^{2+} are calculated in proportion to the flux ratios for these species determined for the modern M.O.R. hydrothermal flux. Results of our calculations of steady-state seawater compositions for various river water/M.O.R. hydrothermal brine volume flux ratios are given in Table 4.

Modest changes in the ratio of river water inflow and M.O.R. hydrothermal brine inflow result in what we consider to be very significant changes in the major ion composition of seawater, significant because of the potential effect on chemical and biochemical sediments produced from these waters. A M.O.R. hydrothermal brine volume flux of between 95 and 96% of the modern flux results in a steady-state seawater composition along the Cl-SO$_4$ and

Na-HCO$_3$-SO$_4$ chemical divide (see Figs. 1 and 2), a lesser M.O.R. hydrothermal brine volume flux leads to alkaline Na-HCO$_3$-SO$_4$ brines. An increase of 10% in the relative proportion of M.O.R. hydrothermal brine results in a steady-state seawater composition along the Cl-SO$_4$ and Ca-Cl chemical divide. Higher relative proportions of M.O.R. hydrothermal brine inflow result in Ca-Cl brines. Our calculations indicate that a 25% increase in the M.O.R. hydrothermal brine influx yields a seawater composition with Ca^{2+} rather than Mg^{2+} as the third most abundant ionic species in seawater (Table 4).

SOME IMPLICATIONS FOR CHANGES IN CHEMICAL SEDIMENTS THROUGH TIME

Calcium carbonate precipitation

Although calcite is the more stable polymorph of calcium carbonate at earth surface conditions, aragonite, rather than calcite, is the dominant form in the present ocean. The dominant polymorph precipitated from seawater may have changed through the Phanerozoic (see SANDBERG, 1983). FUCHTBAUER and HARDIE (1980) precipitated either calcite (with varying Mg) or aragonite or both from Na$^+$-Mg^{2+}-Ca^{2+}-CO$_3^{2-}$ solutions depending on the Mg^{2+}/Ca^{2+} in solution. For solutions with Mg^{2+}/Ca^{2+} less than two (mole ratio) only calcites formed, and for low ionic strength solutions with Mg^{2+}/Ca^{2+} greater than five only aragonite formed, both formed from solutions of intermediate ratios. Comparison of our model ratios with the values found by FUCHTBAUER and HARDIE (1980) to control the polymorph of calcium carbonate precipitated indicates that aragonite is favored to precipitate from seawater with a relatively low M.O.R. hydrothermal brine volume flux (high Mg^{2+}/Ca^{2+}). Calcite precipitation is favored from seawater with relatively higher M.O.R. hydrothermal brine volume fluxes (low Mg^{2+}/Ca^{2+}). It may be possible to use the presence of "aragonite seas" as indicators of low M.O.R. hydrothermal brine volume flux and "calcite seas" as indicators of high M.O.R. hydrothermal brine volume flux.

Composition of evaporites through time

Evaporation paths and equilibrium mineral precipitation sequences are calculated using the thermochemical model of HARVIE et al. (1984). Mineral precipitation sequences for the seawater compositions presented in Table 4 are given in Table 5. The mineral precipitation sequences vary significantly. Waters with a Cl-SO$_4$ composition (including modern seawater) all end their crystallization with a final

Table 4. Steady-state composition of major ions in seawater (meq/l) as a function of M.O.R. hydrothermal brine flux

Flux*	Na$^+$	K$^+$	Ca^{2+}	Mg^{2+}	Cl$^-$	SO$_4^{2-}$	HCO$_3^-$
0.95	495	2.0	0.5	112	548	59.6	3.0
0.96	490	3.7	4.8	111	548	59.0	2.7
1.00	470	10.2	20.3	106	548	56.6	2.2
1.05	448	17.6	36.2	101	548	53.9	0.6
1.10	427	24.3	51.8	97	548	51.4	0.4
1.25	376	41.2	91.4	85	548	45.3	0.3

* Ratio of M.O.R. hydrothermal brine flux relative to modern.

Table 5. Equilibrium mineral precipitation sequences for seawater compositions in Table 4

0.96	1.00	1.05	1.10	1.25
C	C	C	C	C
C+G	C+G	C+G	C+G	C+G
C+A	C+A	C+A	C+A	C+A
C+A+GL	C+A+H	C+A+H	C+A+H	C+A+H
C+GL	C+A+H+GL	C+A+H+P	C+A+H+S	C+A+H+S
C+GL+H	C+A+H	C+A+H+P+S	C+A+H+S+CR	C+A+H+S+CR
C+GL+H+A	C+A+H+P	C+A+H+P+S+CR	C+A+H+CR	C+A+H+CR
C+H+A	C+A+H+P+E	C+A+H+P+CR	C+A+H+CR+B	C+A+H+CR+AT
C+H+A+P	C+A+H+P+HX	C+A+H+P+CR+K	C+A+H+CR+B+T	C+A+H+CR+AT+T
C+H+P	C+A+H+P+K	C+A+H+CR+K		
C+H+P+E	C+A+H+P+K+CR	C+A+H+CR+K+B		
C+H+P+HX	C+A+H+K+CR			
C+H+P+K	C+A+H+K+CR+B			
C+H+P+K+KA				
C+H+K+KA+A				
C+H+K+A+CR				
C+H+K+A+CR+B				

A	anhydrite	AT	antarcticite	
CR	carnallite	E	epsomite	
H	halite	HX	hexahydrite	
P	polyhalite	S	sylvite	

B	bischofite	C	calcite
GL	glauberite	G	gypsum
KA	kainite	K	kiesertie
T	tachyhydrite		

invariant assemblage of calcite-anhydrite-halite-kieserite-carnallite-bischofite. The paths taken in order to reach this assemblage vary; glauberite appears earlier from waters with a lower M.O.R. hydrothermal brine input (lower Ca^{2+} and higher SO_4^{2-}). The mineral kainite also appears in these sequences, but not in the sequence for modern seawater.

Slight increases in the M.O.R. hydrothermal brine input to seawater result in the disappearance of glauberite as a replacement of anhydrite and the precipitation of magnesium sulphate-bearing phases (polyhalite, epsomite, hexahydrite and kieserite) at later stages in the evaporation sequence. Progressive increase in the proportion of M.O.R. hydrothermal brine to the mixture (between 1.00 and 1.05 times our estimated modern flux) results in carnallite precipitation prior to magnesium sulphate phases (epsomite, hexahydrite and kieserite), and eventually to sylvite precipitation prior to carnallite. Still higher input of M.O.R. hydrothermal brines results in Ca-Cl type waters, which do not precipitate $MgSO_4$-bearing phases (flux ratios greater than 1.10, Tables 4 and 5). All Ca-Cl brines follow similar mineral precipitation sequences during evaporation and vary significantly from that of modern seawater. The final invariant assemblages include calcite-anhydrite-halite-carnallite-tachyhydrite, with either bischofite or antarcticite.

Ancient evaporites have been used to evaluate changes in the composition of surface waters, including seawater, through time (for example see: HOLSER, 1983; HARDIE, 1984; HOLLAND, 1984; DAS et al., 1990). Evaporites have the potential to give a great deal of information about the composition of ancient surface waters, however, there are problems in using evaporites to estimate seawater composition. The most important problem is the determination of the purely marine origin of a given evaporite (HARDIE, 1984). Our models show that relatively minor changes in the input of Ca-Cl brines greatly influence the composition and mineral precipitation sequence of evaporating seawater. Mixtures of Ca-Cl hydrothermal brines with river waters can produce brine compositions and mineral precipitation sequences similar to or very different from those of modern seawater. This is demonstrated for the entirely non-marine evaporites of the Qaidam Basin, China, reported by LOWENSTEIN et al. (1989) and SPENCER et al. (1990). The same types of changes we expect for seawater as a result of changes in M.O.R. hydrothermal brine volume fluxes can be produced on a local or regional scale by mixtures of Ca-Cl brines and river waters with or without a marine component. Reading the composition of

ancient seawater from ancient evaporites is not straightforward and is unlikely to yield unique solutions. Examples of Permian evaporites are used to illustrate the problems.

HARVIE et al. (1980) compare the evaporation sequence predicted from modern seawater with that of the Permian Stassfurt Series of the German Zechstein II. They find a remarkable agreement in the mineral assemblages, precipitation sequence and even in the relative proportions of various minerals. This sequence of minerals includes glauberite as a replacement of gypsum or anhydrite, as well as polyhalite as a replacement of gypsum or anhydrite at relatively early stages in the evaporation sequence. These minerals are not found in our evaporation sequences with increased M.O.R. hydrothermal input. Further, glauberite is reported from the Permian Salado Formation, and polyhalite is reported to replace glauberite, gypsum and anhydrite (LOWENSTEIN, 1983). Magnesium sulphate salts are also present in the Salado Formation (JONES, 1972; LOWENSTEIN, 1983). The presence of these "key minerals," which indicate compositional ratios for the major elements close to that of modern seawater, appear to give strong evidence for a similar composition of Permian and modern seawater.

However, glauberite and polyhalite have not been reported from a number of other Permian evaporites, including those found stratigraphically above and below, and time equivalents of the Stassfurt Series of the German Zechstein II (see HARDIE, 1990). Many of these deposits lack these "key minerals" which form at relatively early stages during the evaporation of seawater, yet contain potassium chloride salts (sylvite or carnallite), which are found much later in evaporation sequences. The mineral assemblages for these Permian evaporites are similar to those formed during the evaporation of our calculated seawater compositions with a relatively high M.O.R. hydrothermal brine input, compositions which vary greatly from that of modern seawater.

If we assume that the Stassfurt Series of the Zechstein II and the Salado Formation evaporites resulted from the evaporation of Permian seawater, then Permian seawater must have been similar in composition to modern seawater. The M.O.R. hydrothermal brine volume flux during the Permian would have been similar, possibly slightly lower, than the modern flux. Mineral sequences found in the other Permian evaporites may have resulted from local or regional input of Ca-Cl brines which altered the evaporation sequence. However, if the Permian evaporites which lack the "key minerals" glauberite and polyhalite and contain potassium chloride salts, such as sylvite and carnallite, but do not contain $MgSO_4$ minerals, are the result of the evaporation of Permian seawater, then Permian seawater must have been compositionally very different from modern seawater. The minerals indicate a higher M.O.R. hydrothermal brine volume flux. In this case the Stassfurt series of the Zechstein II and the Salado Formation evaporites may have received a significant input of river water. The work reported by LOWENSTEIN et al. (1989) and SPENCER et al. (1990) on modern salt deposits of the Qaidam Basin, China, demonstrates that alteration of evaporation sequences by changes in the relative input of river water and Ca-Cl brines does occur.

CONCLUSIONS

The major ion composition of modern seawater can be explained by mixing of river water and M.O.R. hydrothermal brine. A steady-state composition for the major ions in seawater can be obtained by removal of river-borne Na^+, Mg^{2+} and SO_4^{2-} during seawater circulation through mid-ocean ridges, without significant additional sources or sinks. The mole-equivalent concentration of bicarbonate input by river waters to the oceans is in excess of Ca^{2+}. Additional Ca^{2+} added to the oceans at mid-ocean ridges is precipitated along with river-borne Ca^{2+} and HCO_3^- as calcium carbonate. Additional sinks for K^+ are required, as the dissolved potassium fluxes from river waters and from M.O.R. hydrothermal brines are positive.

The major element composition of seawater is very sensitive to changes in the M.O.R. hydrothermal brine volume flux. This flux is likely to have changed during geologic time. Therefore, we expect that the major element composition of seawater has also changed through time. The compositional variations in seawater that result from even modest changes of a few percent in the ratio of river water and mid-ocean ridge hydrothermal brines are significant because of the potential influence on chemical sediments produced from seawater. Smaller mid-ocean ridge hydrothermal brine fluxes result in lower concentrations of Ca^{2+} in seawater and higher concentrations of Na^+, Mg^{2+}, SO_4^{2-} and HCO_3^-. These compositional variations are likely to favor aragonite precipitation rather than calcite as the dominant marine carbonate mineral, and result in marine evaporites which contain magnesium sulphate minerals. Larger mid-ocean ridge hydrothermal brine volume fluxes result in higher concentrations of Ca^{2+} in seawater and lower concentrations of Na^+, Mg^{2+}, SO_4^{2-} and HCO_3^-. These compositional variations favor calcite as the dominant marine carbonate mineral. Marine evaporites

formed from such calcium-rich waters contain potash salts (sylvite and carnallite) without magnesium sulphate minerals.

Acknowledgments—Our ideas of brine evolution have been influenced by discussions with many individuals, we particularly thank Hans Eugster, Blair Jones and Tim Lowenstein for many discussions on the origin of brines.

REFERENCES

BJORNSSON S., ARNORSSON S. and TOMASSON J. (1972) Economic evaluation of Reykjanes thermal brine area, Iceland. Bull. Amer. Petrol. Geol. **56**, 2380–2391.

DAS N., HORITA J. and HOLLAND H. D. (1990) Chemistry of fluid inclusions in halite from the Salina Group of the Michigan Basin: Implications for Late Silurian seawater and the origin of sedimentary brines. Geochim. Cosmochim. Acta **54**, in press.

DREVER J. I. (1974) The magnesium problem. In The Sea (ed. E. D. GOLDBERG) Vol. 5, Ch. 10. Wiley-Interscience.

DREVER J. I. (1982) The Geochemistry of Natural Waters. Prentice-Hall.

EDMOND J. M., MEASURES C., MCDUFF R. E., CHAN L. H., COLLIER R., GRANT B., GORDON L. I. and CORLISS J. B. (1979) Ridge crest hydrothermal activity and the balances of the major and minor elements in the ocean: The Galapagos data. Earth Planet. Sci. Lett. **46**, 1–18.

FUCHTBAUER H. and HARDIE L. A. (1980) Comparison of experimental and natural magnesian calcites (abstr.). International Association Sedimentologists, Bochum. 167–169.

GARRELS R. M. and MACKENZIE F. T. (1971) Evolution of Sedimentary Rocks. W. W. Norton.

GARRELS R. M. and THOMPSON M. E. (1962) A chemical model for seawater at 25°C and one atmosphere total pressure. Amer. J. Sci. **260**, 57–66.

HARDIE L. A. (1983) Origin of $CaCl_2$ brines by basalt-seawater interaction: Insights provided by some simple mass balance calculations. Contrib. Mineral. Petrol. **82**, 205–213.

HARDIE L. A. (1984) Evaporites: marine or non-marine? Amer. J. Sci. **284**, 193–240.

HARDIE L. A. (1990) Potash evaporites, rifting and the role of hydrothermal brines. Amer. J. Sci. **290**, 43–106.

HARDIE L. A. and EUGSTER H. P. (1970) The evolution of closed-basin brines. Mineral. Soc. Amer. Spec. Paper **3**, 273–290.

HARVIE C. E., MOLLER N. and WEARE J. H. (1984) The prediction of mineral solubilities in natural waters: The Na-K-Mg-Ca-H-Cl-SO$_4$-OH-HCO$_3$-CO$_3$-H$_2$O system to high ionic strengths at 25°C. Geochim. Cosmochim. Acta **48**, 723–751.

HARVIE C. E., WEARE J. H., HARDIE L. A. and EUGSTER H. P. (1980) Evaporation of seawater: calculated mineral sequences. Science **208**, 498–500.

HELGESON H. C. and MACKENZIE F. T. (1970) Silicate-sea water equilibrium in the ocean system. Deep Sea Res.

HOLLAND H. D. (1972) The geologic history of sea water—an attempt to solve the problem. Geochim. Cosmochim. Acta, **36**, 637–651.

HOLLAND H. D. (1978) The Chemistry of the Atmosphere and Oceans. Wiley.

HOLLAND H. D. (1984) *The Chemical Evolution of the Atmosphere and Oceans.* Princeton University Press.

HOLSER W. T. (1984) Gradual and abrupt shifts in ocean chemistry during Phanerozoic time. In *Patterns of Change in Earth Evolution* (eds. H. D. HOLLAND and A. F. TRENDALL), Dahlem Konferenzen, Springer-Verlag.

JONES C. (1972) Permian basin potash deposits, southwestern United States. In *Geology of Saline Deposits.* Procedings Hanover Symposium 1968. UNESCO.

LAFON G. M. and MACKENZIE F. T. (1974) Early evolution of the oceans: A weathering model. In *Studies in Paleo-Oceanography* (ed. W. W. HAY), Spec. Pub. 20, pp. 205–218. Society Economic Paleontologists Mineralogists.

LIVINGSTON D. A. (1963) Chemical composition of rivers and lakes. *U.S. Geol. Surv. Prof. Paper* 440G.

LOWENSTEIN T. K. (1983) Deposition and alteration of an ancient potash evaporite: The Permian Salado Formation of New Mexico and west Texas. PhD. Dissertation, The Johns Hopkins University, Baltimore.

LOWENSTEIN T. K., SPENCER R. J. and ZHANG PENGXI (1989) Origin of ancient potash evaporites: Clues from the modern nonmarine Qaidam Basin of western China. *Science* 245, 1090–1092.

MACKENZIE F. T. and GARRELS R. M. (1966) Chemical mass balance between rivers and oceans. *Amer. J. Sci.* 264, 507–525.

MAYNARD J. B. (1976) The long-term buffering of the oceans. *Geochim. Cosmochim. Acta* 40, 1523–1532.

MEYBECK M. (1979) Concentration des eaux fluviales en elements majeurs et apports en solution aux oceans. *Rev. de Geol. Dynam. et de Geogr. Phys.* 21, 215–246.

MEYBECK M. (1982) Carbon, nitrogen, and phosphorous transport by world rivers. *Amer. J. Sci.* 282, 401–450.

PROBST J. L. and TARDY Y. (1989) The Global runoff fluctuations during the last 80 years in relation to the world temperature change. *Amer. J. Sci.* 289, 267–285.

RILEY J. P. and CHESTER R. (1971) *Introduction to Marine Chemistry.* Academic Press.

RUBEY W. W. (1951) Geologic history of seawater: An attempt to state the problem. *Bull. Geol. Soc. Amer.* 62, 1111–1147.

SANDBERG P. A. (1983) An oscillating trend in Phanerozoic nonskeletal carbonate mineralogy. *Nature* 305, 19–22.

SILLEN L. G. (1961) The physical chemistry of seawater. In *Oceanography* (ed. M. SEARS), Pub. 67, pp. 549–581. American Association Advancement of Science.

SILLEN L. G. (1967) The ocean as a chemical system. *Science* 156, 1189–1197.

SPENCER R. J., BAEDECKER M. J., EUGSTER H. P., FORESTER R. M., GOLDHABER M. B., JONES B. F., KELTS K., MACKENZIE J., MADSEN D. B., RETTIG S. L., RUBIN M. and BOWSER C. J. (1984) Great Salt Lake and Precursors, Utah: The Last 30,000 Years. *Contrib. Mineral. Petrol.* 86, 321–334.

SPENCER R. J., EUGSTER H. P. and JONES B. F. (1985) Geochemistry of Great Salt Lake, Utah. II: Pleistocene-Holocene evolution. *Geochim. Cosmochim. Acta* 49, 739–747.

SPENCER R. J., LOWENSTEIN T. K., CASAS E. and ZHANG PENGXI (1990) Origin of potash salts and brines in the Qaidam Basin, China. In *Fluid Mineral Interactions: A Tribute to H. P. Eugster* (eds. R. J. SPENCER and I-MING CHOU), pp. 395–408. The Geochemical Society, Special Publication No. 2.

THOMPSON G. (1983) Basalt-seawater interaction. In *Hydrothermal Processes at Seafloor Spreading Centers* (eds. P. A. RONA, K. BOSTROM, L. LAUBIER, K. L. SMITH), NATO Conference Series IV, Marine Sciences, pp. 225–278. Plenum Press.

VON DAMM K. L., EDMOND J. M., GRANT B., MEASURES C. I., WALDEN B. and WEISS R. F. (1985a) Chemistry of submarine hydrothermal solutions at 21°N, East Pacific Rise. *Geochim. Cosmochim. Acta* 49, 2197–2220.

VON DAMM K. L., EDMOND J. M., MEASURES C. I. and GRANT B. (1985b) Chemistry of hydrothermal solutions at Guaymas Basin, Gulf of California. *Geochim. Cosmochim. Acta* 49, 2221–2237.

Subject Index

423

426

427

Low-magnesium calcite, 23, 24, 26, 27, 29, 31, 32, 38

M

Mackinawite, 322, 333, 342, 345
Mafic, 143, 148, 150, 154
  volcanic rock, 147
Magenta Dolomite, 213, 220, 224, 262, 264, 265
  aquifer, 213, 220, 223, 226, 256, 257, 264, 265, 266, 267, 268
  salt norms, 248, 249, 250, 251
Magma, 65, 66, 70, 72, 78, 80, 81, 113
  degassing, 373, 377, 386
  felsic, 154
Magmatic, 375, 386, 390
  gas, 386, 390
  heat, 375
Magnesia, 66
Magnesite, 23, 35, 355, 360, 365, 366, 367, 368
Magnesium, 17, 23, 24, 26, 27, 28, 31, 32, 33, 34, 36, 42, 172, 175, 176, 199, 203, 363, 365, 367
  bicarbonate, 33, 38
  carbonate, 33, 34, 35, 36, 37, 38, 358, 359, 360, 362, 363, 365
  poisoning, 27
Magnesium calcite, 26, 27, 28, 31, 32, 37, 38, 355, 364, 365, 366, 367, 368, 369
  biogenic, 23, 26, 29, 36, 37
  free energy of formation, 32, 33, 34, 35, 36, 37, 38
  solid solution, 23
  solubility, 23, 28, 29
  solvus, 23
  synthetic, 24, 36
Magnetite, 3, 4, 9, 10, 11, 12, 14, 59, 105, 109, 110, 111, 112, 113, 115, 157, 158, 159, 160, 161, 162, 163, 164, 165, 166, 192, 373, 379, 390
  -fayalite-quartz buffer, 9, 10, 12, 13, 17, 18, 98, 108, 109, 112, 191, 192
  -hematite buffer, 4, 9, 10, 12, 13, 17, 18, 19, 61, 105, 179, 181, 182, 183, 186, 191
Magnetometer, 144
Manganese, 157, 158, 159, 160, 161, 162, 163, 167, 171, 173, 174, 175, 176, 203
  oxide, 315, 320, 326, 330, 331, 336, 339, 350
  oxide buffer, 8, 12, 13, 61
Mantle xenolith, 85, 86
Marcasite, 388, 390
Marine, 23, 316, 345, 351, 368, 417

clays, 410
evaporite, 395, 396, 397, 409, 410, 411, 414, 418
Mass, 25, 28, 45, 50, 362, 363, 365, 409, 411
  action, 45
  balance, 45, 50, 362, 363, 365, 409, 411
  transfer, 25, 28
Material transport, 121, 122, 124, 125, 126, 127, 128, 129, 130, 131
Mazama Ash, 355, 358
Megacryst, 65
Melt, granitoid, 59
Membrane, 8, 11
  filtration, 286
  palladium, 19
Meromictic, 340, 351
Mesostasis, 19
Metagenesis, 271, 272, 282
Metal, 315
  -ligand, 180
  oxide, 336, 338, 340, 345, 347, 348, 349, 350, 351
Metamorphic, 95, 101, 148, 205
  grade, 133
  minerals, 143
  rock, 356
  textures, 121, 122, 125, 126, 127, 128, 129, 130, 131
Metamorphism, low grade, 199
Metasomatism, 171, 373, 386
Metastable, 23, 24, 29, 31, 37, 121, 122, 123, 124, 125, 126, 127, 130, 131
Meteoric, 193, 194, 213, 215, 216, 223, 254, 257, 258, 259, 262, 263, 264, 266, 267, 286, 295, 297, 298, 299, 396, 397, 398, 403, 406
Methane, 3, 4, 5, 6, 7, 8, 13, 200, 204, 206, 271, 272, 281, 282
  -graphite buffer, 4, 5, 11, 12
Mica, 357
Mic Mac Formation, 200
Microbial, 272
Microbiological, 340
Mid-ocean ridge, 42, 410, 412, 414, 415, 418
  heat flow, 412
  hydrothermal brine, 409, 411, 412, 413, 414, 415, 417, 418
Mineral, 41, 133, 138, 139
  assemblage, 133, 138, 139
  equilibria, 41
  gangue, 41
  stability, 41
Mirabilite, 401
Miscibility, 199, 206, 209, 210
Missisauga Formation, 200
Mixing, 285, 286, 288, 294, 295, 298, 305
Moldanubian crystalline complex, 85

Mono Lake, California, 315, 316, 317, 318, 335, 336, 337, 338, 339, 340, 349, 350, 351
  brine chemistry, 331, 333
  sediment chemistry, 333, 335
Monohydrocalcite, 320, 322
Monolayer, 32
Monomictic, 316, 329
Muscovite, 121, 122, 123, 124, 125, 126, 127, 128, 129, 130, 131, 179, 188, 191, 192, 193, 203, 209

N

Natural gas, 200
Neoblasts, 87
  orthopyroxene, 88, 93, 99
  pseudomorph after kyanite, 122, 127, 128, 129, 130, 131
  pyroxene, 85, 92
Nesquehonite, 32
Nevada, 294
New Mexico, 213, 218, 294
Nickel oxide buffer, 9, 10, 11, 12, 13, 17, 18, 19, 61, 66, 158, 179, 180, 181, 182, 183, 186, 191, 192
Nitric acid, 42, 318
Nitrogen, 13, 271, 272, 273, 278, 279, 282
Non-equilibrium, 23
Non-marine, 395, 396, 397, 406, 411, 417
Normative albite, 81
Normative salt, 213, 214, 215, 398
  analysis, 213, 230, 231, 232, 233, 234, 235, 236, 237, 238, 239, 240, 241, 242, 243, 244, 245, 246, 247, 248, 249, 250, 251, 252
  assemblage, 214, 228, 260, 266, 398
  calculation, 214, 215
  diagenetic waters, 216, 217
  diagnostic chart, 216
  interpretation, 215
  marine waters, 215, 217, 253, 260
  meteoric waters, 215, 217
  minerals, 214, 398
  simple salts, 214, 217, 228
  solutes, 215
North Sea, 199, 208
Nova Scotia, Canada, 199, 200
Nova Scotia Group, 200
Nuclear waste, 213
  repository, 17
Nucleation, 31, 121, 122, 124, 125

O

Ocean, 409, 410, 411, 412, 413, 414, 415, 418
Ochoan Series, 213, 218, 220, 266

Oil, 285, 286, 396
  shale, 279, 281
Oil field brine, 395, 398
Olefins, 273, 274, 277
Olivine, 67, 76, 81, 85, 87, 90, 94,
      95, 96, 101, 103, 104,
      105, 106, 108, 110, 111,
      112, 113, 373, 375, 378,
      379, 382, 383, 384, 385,
      386, 390
  composition, 98, 100
  inclusions in, 86, 87, 88, 97, 99
Open reactor, 43
Oquirrh Basin, 298
Ore, 41, 171, 179, 180, 188, 190,
      192, 193, 315, 316, 350,
      351, 396
  deposit, 41, 171
  fluids, 188, 192, 193
  formation, 315, 316, 350, 351
  -forming, 179, 180, 188, 190
  minerals, 396
Organic, 24, 273, 274, 315, 355,
      368
  acid, 208, 320, 340, 351
  carbon, 318, 319, 320, 322, 324,
      331, 333, 335, 339, 340,
      342, 343, 344, 349, 350
  carbon dioxide, 386
  coatings, 315
  complexation, 315, 319, 331,
      333, 335, 336, 338, 339,
      344, 345, 347, 348, 349,
      350
  decay, 355, 368
  decomposition, 315, 340, 348,
      349, 350
  deoxygenation enthalpy, 273,
      276, 278, 280, 281
  deoxygenation free energy, 274,
      275, 276, 277, 278, 279,
      280, 281, 282
  deoxygenation quotient, 274,
      275, 279, 280, 281
  matter, 24, 320, 326, 327, 329,
      330, 336, 338, 339, 340,
      342, 345, 347, 348, 349,
      350, 351, 373, 377, 386
  maturity, 200
Orthoamphibole, 85, 87, 88
Orthoamphibole-like phase, 85, 91,
      92, 99
  composition, 90, 93, 94
  unit-cell, 89
Orthoclase, 59
Orthophosphate, 24, 29, 38
Orthopyroxene, 76, 85, 95, 103,
      104, 105, 106, 108, 109,
      110, 111, 112, 113, 114,
      115, 379, 384, 390
  composition, 89, 90, 91, 93, 94
  unit-cell, 89
Osmotic, 4, 5, 287
  equilibrium, 4, 5
  membrane, 287
Over pressure, 199, 200, 201, 203,
      204

Oxalic acid, 275, 279
Oxidation, 61, 181, 182, 325, 327,
      340, 350, 351
  sulfide, 215, 255
Oxide minerals, 103, 104, 107,
      108, 109, 110, 111, 113,
      114
  buffers, 3, 4, 5, 6, 7, 8, 9, 10, 11,
      12, 13, 17, 18, 19, 20,
      61, 66, 105, 108, 109,
      112, 158, 179, 180, 181,
      182, 183, 186, 191, 192,
Oxidizing, 7, 18, 19, 115, 322, 324,
      326, 327, 335, 338, 339,
      344, 345, 347, 348, 349
Oxygen, 3, 4, 60, 61, 63, 65, 66,
      67, 68, 72, 73, 271, 318,
      320, 345, 350
  buffer, 3, 4, 8, 11, 13, 14, 180,
      181, 191
  fugacity, 3, 9, 10, 13, 14, 103,
      104, 105, 106, 107, 108,
      109, 110, 111, 112, 113,
      114, 115
Oxygenated, 315, 316, 321, 329

P

Paleoclimate, 285, 288
Paleoenvironment, 285
Paleogroundwater, 285, 305
Paleohydrogeology, 285
Paleolake, 294
Paleotemperature, 291, 305
Paleotopographic, 285
Palladium, 17, 18, 19, 20
  sulfide, 19
Paradox Basin, 285, 287, 294, 295,
      297, 298, 299, 301, 303,
      305
Paraffins, 275, 277, 281, 282
Pargenetic sequence, 203
Partition, 50, 157, 158, 159, 160,
      161, 162, 163, 164, 165,
      166, 167, 171, 176
  coefficient, 23, 158, 165, 167,
      171, 172, 174, 175, 176
Partial melting, 59, 60, 61, 95
Pasco Basin, 17, 18
Pelite, 147, 149, 153
Pelites, 121, 129
Penesaline, 254
Peritectic, 365, 366, 367
Permeability, 8, 220, 222, 225,
      285, 286, 287, 291, 295,
      303, 376
Permeable, 291, 296, 298, 301
Permeation, 17
Petrographic, 201, 202, 210
Petroleum, 271, 272, 281
pH, 17, 23, 24, 25, 26, 31, 33, 35,
      37, 41, 43, 44, 45, 47,
      172, 179, 180, 181, 188,
      191, 192, 194, 208, 209,
      210, 226, 315, 316, 317,
      318, 319, 320, 327, 329,
      330, 331, 339, 340, 342,

345, 348, 349, 350, 351,
      377, 382, 384, 386, 387,
      390
  -stat, 43
Phase boundary, 88, 89
Phase diagram, 133, 134, 135
  alkalinity-calcium-sulfate, 398,
      399, 411, 413
  aluminum-iron-silica, 152
  aluminum-potassium, 123, 125,
      131
  aluminum-pressure, 61, 62
  anorthite-temperature, 68, 72,
      74, 75
  bromide-temperature, 7
  cadmium, 165, 166
  calcium-magnesium, 27, 28, 365
  carbon dioxide-temperature,
      206, 209, 210
  cassiterite, 184, 185, 186, 187,
      190, 193
  clinopyroxene, 382
  copper, 166
  corundum-diopside-olivine, 100
  Fe-Mg-Ti-Si-O, 104, 107, 108
  free energy, 36, 276
  gibbsite, 389
  glass-crystal, 69
  hydrochloric acid-hydrogen, 9
  hydrogen-temperature, 10, 12
  iron-oxygen, 109
  iron-magnesium-silica, 135
  kaolinite, 389
  magnesium carbonate, 36, 364
  manganese, 158, 161
  microcline, 389
  olivine, 381
  orthopyroxene, 381
  oxygen fugacity, 18
  oxygen-temperature, 110, 111,
      192
  pH, 49
  plagioclase, 69, 382
  pressure-silica, 112
  pressure-temperature, 77, 114,
      122, 207
  pyrite, 389
  pyroxene, 90, 91
  quartz, 387
  smectite, 388
  spinel, 92, 383
  ternary, 134
  tin, 192
  zinc, 164, 165
Phenocryst, 78, 79, 81, 375, 383,
      384, 385
Phlogopite, 171, 173, 174
Phosphate, 41
  poisoning, 36, 37
Photosynthesis, 271, 279, 281, 282
Phyllosilicate, 355, 357, 367
Physiochemical, 200
Pigeonite, 113
Piston-cylinder apparatus, 59, 60
Piston, floating, 44
Plagioclase, 59, 65, 66, 67, 68, 69,
      70, 71, 72, 73, 74, 76,

salt norms, 230, 231, 232, 233, 234, 235, 236
Rutile, 201

S

Sabkha, 285, 294, 296, 299
Salado Formation, 213, 220, 222, 223, 254, 259, 261, 267, 396, 397, 406, 417, 418
Saline, 199, 200, 286, 294, 298, 305
  fluid, 199, 200
  water, 286, 294, 298, 305
Salinity, 286, 287, 288, 294, 301, 303
Salt norms, 213, 226
  analyses, 213, 230, 231, 232, 233, 234, 235, 236, 237, 238, 239, 240, 241, 242, 243, 244, 245, 246, 247, 248, 249, 250, 251, 252
  calculation, 214, 215
  diagenetic waters, 216, 217
  graphic presentation, 216, 217, 218
  interpretation, 215
  marine waters, 215, 217, 253, 260
  meteoric waters, 215, 217
  minerals, 214, 226
  simple salts, 214, 217, 226, 228, 266
Salton Sea, 199
San Gabriel Mountains, 59
San Juan Basin, 285, 287, 299, 301, 303, 305
Sandstones, 199, 200, 201, 210
Sanidine, 3, 11, 12, 14
Santa Catalina Island, 59
Sapropelic, 279
Saturation quotient, 45, 47
Scanning electron microscope, 19, 20, 24, 29, 40, 201
  photograph, 30, 42, 202
Schist, 188, 192
Schreinemakers, 122
Scotian Shelf, 199, 200, 201
Sea urchin, 23, 24, 27, 29, 31
Seawater, 32, 37, 286, 287, 288, 294, 295, 296, 297, 298, 299, 302, 303, 355, 368, 376, 377, 406, 409, 411, 412, 415, 417
  composition, 409, 410, 411, 412, 414, 415, 417, 418
  evaporation, 395, 402
  ionic species, 409, 411, 414, 415
  modern, 395, 396, 397, 402, 409, 411, 412, 413, 414, 417, 418
  Permian, 417, 418
Sediment, 315, 320, 323, 324, 326, 327, 329, 330, 333, 335, 336, 338, 339, 340, 342, 410

Sedimentary ironstone, 143, 152
Selenite, 224
Sensor, 3, 4, 6, 7, 9, 10, 11, 13, 14
  measurement, 3, 6, 7, 9, 10, 11, 13, 14
  technique, 4
Sequential extraction, 315, 319, 324, 325, 326, 327, 329, 330, 335, 337, 344, 346, 347, 348, 350
Sergipe-Alagoas basin, 395
Shelf carbonate, 200
Siderite, 151, 203, 204
Silica, 66, 68, 72, 80, 81, 122, 129, 130
  activity, 103, 104, 105, 106, 108, 110, 111, 112, 113, 115
Silicate, 199, 207, 208
  hydrolysis, 199, 208, 209, 210, 213, 215, 254, 255, 256, 258, 263, 267
  solution model, 66
Sillimanite, 121, 122, 123, 124, 125, 126, 127, 128, 130, 131
Sillimanite segregations, 125, 126, 127, 128
Silver, 18
  azide, 13
  oxalate, 13
Simple salt, 214, 217, 226, 228, 258, 259, 260, 265
  norms, 214, 217, 226
  rose diagrams, 226, 228, 258, 259, 260, 265
Singular value decomposition, 133, 134, 137
Site-mixing model, 108
Skarn, 14, 171, 174, 175
Skeletal grain, 23
Smectite, 201, 207, 208, 209, 210, 373, 376, 387, 388, 390
Soda, 66, 68, 69, 72, 73, 78, 79, 80, 81
Sodium, 34, 35, 36, 37, 67, 203, 208
  chloride, 60, 199, 200, 203, 205, 206, 207, 210
Solid-solution, 25, 26, 27, 29, 31, 32, 34, 35, 37, 67, 70, 203
Solidus, 59, 65, 66, 67, 68, 70, 73, 74, 76, 81
SOLMINEQ, 45
Solution models, 106, 113
Speciation, tin, 179
Sphene, 59, 201
Spessartine, 147, 149
Spinel, 85, 87, 88, 94, 95, 96, 108, 109
  aluminum, 96
  composition, 90, 93, 94, 97
  exsolution, 98, 99, 100
  lamellae, 85, 87, 88, 89, 90, 91, 92, 93, 96, 97, 98
  unit-cell, 89

Spotted Lake, Canada, 355, 361, 368
Springs, 356, 358, 373, 381, 395, 396, 397, 400, 402, 403, 404, 405, 406
  analyses, 397
  evaporation, 401, 403, 404
Stabilization, 23
Stable isotope, 406
Stainless steel, 43
  autoclave, 43
  wire, 43
Stassfurt Series, 417, 418
Staurolite-grade, 143
Steady-state, 7, 8, 23, 24, 28, 31, 34, 36, 37, 409, 410, 415
Steam, 41, 199
  flood, 199
  saturation, 41
Stirring, 25, 27, 29, 31, 32, 37, 38
  autoclave, 43, 44
  effect, 24
  rate, 41, 43, 44, 45, 46, 47, 49, 50, 51, 52
Stochiometric saturation, 25, 26, 28, 31, 34, 37
Stratification, 316, 320, 329, 349, 350
Stratiform sulfide, 143
Strontianite, 396
Structural disorder, 29, 33, 36, 37
Subsolidus, 85
  exsolution, 87
  crystallographic orientation, 89, 94
Sulfate, 17, 19, 20, 34, 35, 36, 37, 41, 363
Sulfide, 17, 19, 20, 180, 193, 215
Sulfur, 279
Sulfuric acid, 215, 255, 318
Supercritical, 41, 45, 179, 180, 193, 194
  conditions, 179, 180
  fluid, 41
  solution, 41, 45, 179, 193, 194
Surface, 25, 29, 42, 43, 45, 47, 48, 51
  area, 24, 42, 43, 45, 47, 48, 51
  calcite, 47
  poisoning, 29
  reaction, 51
Sylvite, 13, 395, 396, 417, 418

T

Tachyhydrite, 395, 401, 402, 406, 417
Talc, 159, 161
Taupo, 103, 112, 114
Tectonic cycles, 410
Texture, 210
Thermocouple, 44, 60
Tholeiite, 17
T-magnetite, 103, 104, 109, 112, 113, 114, 115